Guidelines for the Technical Examination of Bronze Sculpture

Guidelines for the Technical Examination of Bronze Sculpture

Edited by David Bourgarit, Jane Bassett, Francesca G. Bewer,
Arlen Heginbotham, Andrew Lacey, and Peta Motture

J. PAUL GETTY MUSEUM, LOS ANGELES

The free online edition of this open-access publication is available at getty.edu/publications/bronze-guidelines/ and includes videos and zoomable illustrations. Also available are free PDF and EPUB downloads of the book.

First edition 2025

Any revisions or corrections made to this publication after the first edition date will be listed here and in the project repository at https://github.com/thegetty/bronze-guidelines, where a more detailed version history is available. The revisions branch of the project repository, when present, will show any changes currently under consideration but not yet published here.

Published by the J. Paul Getty Museum, Los Angeles
Getty Publications
1200 Getty Center Drive, Suite 500
Los Angeles, California 90049-1682
getty.edu/publications/

Ruth Evans Lane, *Project Editor*
Lindsey Westbrook, *Manuscript Editor*
Greg Albers, *Digital Publications Manager*
Kurt Hauser, *Cover Design*
Molly McGeehan, *Production*
Kelly Peyton, Nancy Rivera, and Leslie Rollins, *Image and Rights Acquisition*
Erin Cecele Dunigan and Dina Murokh, *Digital Assistants*

Distributed in the United States and Canada by the University of Chicago Press

Distributed outside the United States and Canada by Yale University Press, London

Library of Congress Cataloging-in-Publication Data
Names: "Bourgarit, David, editor. | Bassett, Jane (Jane L.), editor. | Bewer, Francesca G. (Francesca Gabrielle), 1960– editor. | Heginbotham, Arlen, editor. | Lacey, Andrew, 1969– editor. | Motture, Peta, editor. | J. Paul Getty Museum, issuing body."
Title: "Guidelines for the technical examination of bronze sculpture / edited by David Bourgarit, Jane Bassett, Francesca G. Bewer, Arlen Heginbotham, Andrew Lacey, and Peta Motture."
Description: "First edition. | Los Angeles : J. Paul Getty Museum, 2025. | Includes bibliographical references and index. | Summary: "This illustrated volume offers a new framework for advancing the understanding of bronze sculpture, explaining how to identify the evidence of process steps, metals used, casting defects, and surface work and alterations before moving on to address analytical techniques ranging from visual exams to imaging, material analyses, and dating"— Provided by publisher."
Identifiers: LCCN 2022055307 (print) | LCCN 2022055308 (ebook) | ISBN 9781606066904 (paperback) | ISBN 9781606066928 | ISBN 9781606067611 (adobe pdf) | ISBN 9781606066911 (epub)
Subjects: LCSH: Bronze sculpture—Technique. | Bronze Sculpture—Expertising. | Bronze founding. | Bronze—Analysis. | Sculpture materials—Analysis.
Classification: LCC NB1230 .G85 2023 (print) | LCC NB1230 (ebook) | DDC 731.4/56—dc23/eng/20230315
LC record available at https://lccn.loc.gov/2022055307
LC ebook record available at https://lccn.loc.gov/2022055308

Illustration Credits
Every effort has been made to contact the owners and photographers of illustrations reproduced here whose names do not appear in the captions or on p. 546. Anyone having further information concerning copyright holders is asked to contact Getty Publications so this information can be included in future printings.

Front cover: Bronze pour (still, video 18); X-radiograph of Venus *déhanchée* (fig. 65); *Head of a Ruler* (fig. 310)
Back cover: Modeling a metal armature for Andrew Lacey, *The Anatomy of Bronze* (detail, fig. 33); Barthélemy Prieur, *Funerary Genius* (detail, fig. 48); *Bodhisattva Avalokitesvara* (detail, fig. 151)
Inside cover: Bright field photomicrograph of an etched polished cross-section (fig. 419)

Contents

Director's Foreword

The technical study of works of art is critical to understanding how artists of all cultures and periods worked, to the materiality and processes that underlay their creativity, and ultimately to how they were able to captivate and inspire viewers, both in their own day and through the centuries to museum goers today. This is certainly true of the history of bronze sculpture, the subject of this study. Recent decades have seen great strides in the application of ever broader and more sophisticated methods of examining bronze sculptures, providing insights not only into their art history but also into the trade in raw materials, the transfers of technical know-how over space and time, and the development of new technologies and industries. This research relies increasingly on international collaboration among scholars from many disciplines and institutions. Essential to our continuing success in this pursuit will be a shared understanding of the most informative avenues of technical investigation and the development of a common vocabulary for reporting and recording the results of this research.

This richly illustrated publication offers a guide to the interdisciplinary study of bronzes through clear and concise summaries of best-practice methodologies for the technical examination (Volume I) and analytical methods (Volume II) applicable to this medium, thus providing researchers from all relevant disciplines with a common grounding in this rapidly evolving field. The multilingual vocabulary presented here will allow experts in multiple specializations to communicate the results of new research on the materiality and making of bronze sculpture in a more precise, consistent, and timely manner. The section of case studies illustrates the real-world application of these methods and how they have enhanced our understanding of individual sculptures from a variety of cultures and periods. Reflecting Getty's commitment to open content, these resources are here provided in an open-access digital format, along with a print-on-demand book option, ensuring they will reach the widest possible audience of scholars and enthusiasts.

This project is the fruit of collaboration among over fifty specialists from a diversity of fields, founded on the hope that it will inspire other researchers to further expand the appreciation and understanding of the art of bronze casting. In thanking our partners and their respective institutions, we also acknowledge with gratitude the publication's editors and their institutions: cultural heritage scientist David Bourgarit of the Centre de recherche et de restauration des musées de France, who was key to the project from its inception; curators Francesca G. Bewer and Peta Motture (retired), respectively of the Harvard Art Museums and the Victoria and Albert Museum; independent artist-founder-researcher Andrew Lacey; and the initiators of the project at the Getty Museum, our own conservators of Sculpture and Decorative Arts, Jane Bassett and Arlen Heginbotham.

Timothy Potts
Maria Hummer-Tuttle and Robert Tuttle Director
J. Paul Getty Museum

Acknowledgments

This publication is the fruit of a years-long collaboration of a broad international, interdisciplinary community. Throughout the process of the project's development and culmination, its members have drawn on the support of an extensive network of individuals and organizations. First and foremost the editorial team would like to thank the various members of the Steering Committee, which has evolved in composition over the years, for joining them in this endeavor and for their indispensable contributions to the shaping of the overall project and of these Guidelines. These include Brian Considine, Pete Dandridge, Susanne Gänsicke, Philippe Malgouyres, Carol Mattusch, Benoît Mille, Ruven Pillay, Frits Scholten, Donna Strahan, Jeremy Warren, and Ittai Weinryb.

The editorial committee would especially like to thank the sponsors and hosts of the meetings and workshops organized to convene the Copper Alloy Sculpture Techniques and history: an International iNterdisciplinary Group (CAST:ING) members on a yearly basis, starting with the launch meeting, which coincided with the Nineteenth International Bronze Congress on Ancient Bronzes at the J. Paul Getty Museum in Los Angeles (October 2015), followed by the second gathering in France (July 2016) at the Fondation et Fonderie de Coubertin. Meetings were held in May of 2017 at both the Centre de recherche et de restauration des musées de France (C2RMF), Paris, and the National Gallery of Art, Washington, D.C. The January 2018 meeting in Cambodia was organized by Brice Vincent of the École française d'Extrême-Orient, Paris, and the CAST:ING members were hosted in turn by the APSARA National Authority, Siem Reap, the National Museum of Cambodia, Phnom Penh, IT Sopheap's Art Foundry, Siem Reap, and the Royal University of Fine Arts, Phnom Penh. A special meeting focused on patina was organized in Paris (September 2018) and hosted by the Laboratoire de recherche des monuments historiques. The Getty and C2RMF were also the welcoming venues of several smaller editorial meetings. We are grateful to Art Fund for awarding Peta Motture a Jonathan Ruffer curatorial grant, which allowed her to attend the Getty and Coubertin meetings.

We owe the French Culture Ministry a great thanks for supporting the creation and maintenance of the CAST:ING project's website http://www.cast-ing.org/, which has served as a key tool almost from the inception of the project. It has captured the activities and extensive vocabulary and image databases from which selections included in the publication were made. Ruven Pillay, Dominique Dupuis-Labbé and Vincent Lefèvre have been instrumental in those efforts.

We gratefully acknowledge all the contributors to the contents and writing of this work by listing them along with the authors in their respective sections. With few exceptions the contributors are members of CAST:ING. Through their writing and/or participation in the planning or discussion meetings and other conversations over the years, this publication is based on the expertise and support of this group's members: Denise Allen, Elena Arias, Antoine Amarger, Bastian Asmus, Aurélia Azéma, Christophe Béry, Ann Boulton, Clotilde Boust, Geneviève Bresc-Bautier, Valérie Carpentier, Manon Castelle, Yi Chen, Annabelle Collinet, Pete Dandridge, Isabel Delgado, Sophie Descamps-Lequime, Jean Dubos, Nicolas Escanilla Artigua, Susanne Gänsicke, Laurence Garenne-Marot, Samnang Huot, Joachim Kreutner, Elsa Lambert, Emmanuel Lamouche, Susan LaNiece, Elisabeth Lebon, Linda Ying-Chun Lin, Jeff Maish, Carol Mattusch, Mathilde Mechling,

Benoît Mille, Lorenzo Morigi, Ruven Pillay, Barbara Plankensteiner, Chandra Reedy, David Reid, Dominique Robcis, Lise Saussus, Dylan Smith, Donna Strahan, Shelley Sturman, Annick Texier, Nicolas Thomas, Brice Vincent, Jeremy Warren, Ittai Weinryb, the late Jean-Marie Welter, and Changping Zhang.

In addition to those listed above, we wish to thank others who contributed their invaluable advice, expertise and feedback at various stages of the project, including Victoria Avery, Peter Barnet, Tony Beentjes, Andrew Butterfield, Yannick Chastang, Ilaria Ciseri, Peggy Ellis, Deena Engel, Robert Erdmann, James H. Franz, Irene Gunston, Sharon Hecker, Catherine Hooper, Susan Jenkins, Emma Jones, Robert van Langh, Marjee Levine, Siân Lewis, Jonathan Marsden, Arie Pappot, Gary Radke, Frits Scholten, Harold Schulze, Pamela H. Smith, Jeffrey Springer, the late Richard E. Stone, Iskra Karniš Vidovič, Quanyu Wang, Frank Willer, and Dimitrios Zikos.

We are greatly indebted to colleagues at our individual institutions for their most valuable support. At the C2RMF: Anne Bouquillon, Stéphanie Courtier, Yvan Coquinot, Victor Etgens, Vanessa Fournier, Mireille Klein, Alexis Komenda, Marie Lionnet de Loitière, Jean-Michel Loyer-Hascoët, Anne Maigret, Michel Menu, François Mirambet, Astrid Moitrieux, Isabelle Pallot-Frossard, Emmanuel Plé, and Pascale Richardin. At the Harvard Art Museums and Harvard University: Susanne Ebbinghaus, Katherine Eremin, Narayan Khandekar, Josef Koerner, David Odo, Sara Schechner, Martha Tedeschi and former colleagues Henri Lie, Jessica Martinez, and Tony Sigel. At Getty: Murtha Baca, Deni Bernhart, Kellie Boss, Jessica Chasen, Brian Considine, Madeline Corona, James Cuno, Anne-Lise Desmas, B. J. Farrar, Patricia Harpring, Kenneth Lapatin, Lynn Lee, Timothy Potts, Richard Rand, Erik Risser, Tina Segler, Marie Svoboda, and Julie Wolfe. At the Victoria and Albert Museum: Kira d'Alburquerque, Nicholas Barnard, Antonia Boström, Lucia Burgio, the late John Clarke, the late Diana Heath, Tristram Hunt, Gregory Irvine, Beth McKillop, Anthony Misquitta, Joanna Norman, Angus Patterson, Boris Pretzel, James Robinson, the late Martin Roth, Lois Salter, Holly Trusted, Joanne Whalley, Paul Williamson, and Michaela Zöschg.

Special acknowledgment goes to Getty Publications team members Erin Cecele Dunigan, Molly McGeehan, Chloe Millhauser, Dina Murokh, Kelly Peyton, Nancy Rivera, Leslie Rollins, and freelance copyeditor Lindsey Westbrook for their essential contributions to an admittedly challenging project and to Kara Kirk, Karen Levine, and Nola Butler for their support. Thanks are due to cover designer Kurt Hauser as well as graphic designers Hespenheide Design and Anthony Paular Design for their contributions to the overall style of the publication. We'd also like to thank the staff at Getty Digital for the technical work they did on Quire to build this publication's many features, in particular Matthew Hrudka, Hannah Balenda, Anders Pollack, and Neal Johnson, though certainly it takes a village. We owe our deepest gratitude to senior editor Ruth Evans Lane and digital publications manager Greg Albers who have worked with us tirelessly to translate our complex, multidimensional and multimodule offering into a handsome, an interactive, and a highly accessible online publication as well as a paperback. Most importantly, we gratefully acknowledge the infinite patience and support of our friends and families throughout this endeavor, a sacrifice made even greater by the complication of time zones.

CAST:ING editorial committee:

David Bourgarit
Jane Bassett
Francesca G. Bewer
Arlen Heginbotham
Andrew Lacey
Peta Motture

General Introduction

Francesca G. Bewer
David Bourgarit

Since the fourth millennium BCE, cultures around the world have used bronze and other copper alloys to produce sculptural forms—a process of translation that involves collaboration and a complex sequence of procedures that reflect the specific technologies and skills of a particular period and society. Consequently, through its technical details, any given bronze sculpture may offer a unique opportunity to add to our understanding of the people who fostered its creation. The scholarship on bronze sculpture from all regions and periods has benefited tremendously in the last decades from advanced technical studies. It too requires skills drawn from a wide range of disciplines. Yet there are few (if any) places where training in such interdisciplinary study is offered.

The Copper Alloy Sculpture Techniques and history: an International iNterdisciplinary Group (CAST:ING) sprang from a collaboration between research curator Francesca G. Bewer, objects conservator Jane Bassett, and archaeometallurgist David Bourgarit on the technological study of French bronzes that was instigated by Geneviève Bresc-Bautier, former director of the Sculpture Department at the Musée du Louvre, in preparation for the exhibition *Cast in Bronze: French Sculpture from Renaissance to Revolution* (co-organized by the Louvre, Paris; the Metropolitan Museum of Art, New York; and the J. Paul Getty Museum, Los Angeles, and presented at each over 2008–9).[1] The momentum generated by the show led them to organize a three-day symposium in 2012 at the Musée du Louvre and at the Centre de Recherche et de Restauration des Musées de France to promote further research on the development and cross-fertilization of ideas and technology related to the making of bronzes in France and by French artists abroad. It brought together a diverse group of specialists to engage in an interdisciplinary exchange.[2] Bourgarit's ensuing guest scholarship at the Getty Conservation Institute to develop a database with the rich material gathered thus far highlighted the pressing need for a more standardized approach to technical examinations and reporting, and a shared vocabulary. With the support of colleagues from the Getty (Arlen Heginbotham, Brian Considine, and Murtha Baca) and scholars from other institutions (Philippe Malgouyres and Benoît Mille) they reached out internationally to colleagues who had a stake in technical research, and officially launched the project at the Getty in 2015.

The aim of the CAST:ING project has been to bring together an interdisciplinary group of experts to create a framework that facilitates advances in the understanding of bronze sculpture through the use of shared vocabulary and protocols for the more technical aspects of art-technological studies. The international team that contributed to the project includes some fifty-two members, comprising conservators, scientists, curators, art historians, historians, archaeologists, metallurgists, and craftspeople studying bronze production of different eras and cultures. They work with or for museums and other cultural heritage institutions and universities where much of this kind of work has been carried out thus far.

From the outset, publishing a set of guidelines for best practices in methods of investigation and the reporting of technical data on bronze sculpture has been a key aim of CAST:ING. In the process of this multiyear project, which has involved numerous larger international meetings and benefited from digital technologies in ways too numerous to recount, we have created a model of an open forum dedicated to the scholarly exchange of methods, data, and ideas key to a collaborative study of the production of bronzes. In so doing, we have identified communication gaps between specialists from differing disciplines, and sought to address a fair number of them. That said, the scope of this publication cannot help but reflect both the strengths and the imbalances in the existing expertise and the research to which we have had access. We did not aspire to be universal in our coverage. Rather, the publication is an invitation for more people to further engage in, and support, this kind of work.

The purpose of the present *Guidelines* is to advise on how to investigate, interpret, and document the fabrication process of sculptural works of art made of cast copper alloys. Three main questions help structure the chapters: Why? What? and How? The "why" is discussed in the present general introduction at GI§1, together with the various reasons to investigate the fabrication techniques of bronze sculpture. It includes a section on the life cycle of a bronze, which serves as a reminder that objects bear the marks not only of their production, but often of long lives, and lists the range and disparate nature of evidence that a bronze sculpture may contain on that front. A brief overview of the historical context for the technological study of bronzes points to its roots in archaeological research, where close inquiry of the physical evidence of making and use preserved in the object was first combined with scientific analysis, interpretation of historical texts, and experimental reconstructions.

The question of "what" (is a bronze)—bronze meant here in the sense of the object—is tackled in GI§2 through the lens of the main fabrication processes used to produce bronze sculptures. While other methods can serve to fashion a sculptural form out of a copper alloy (for example forming and joining of sheet metal, 3D printing, et cetera), these *Guidelines* will focus exclusively on casting. "How" (to use the *Guidelines*) will be the topic of GI§3.

1 Why? The *Guidelines*

1.1 Current issues

The thousands of bronze sculptures that have been produced over millennia around the world and that still survive in museums, other cultural institutions, and elsewhere constitute exceptional resources with potential to add to our understanding of a variety of issues around the objects themselves, as well as the diverse cultures in which they were created, used, and collected. The questions that provide the impetus for any given study will vary, but in all cases the inquiry into the materials and techniques of the bronze objects involved and the meaningful interpretation of the physical evidence gathered require close study of related primary sources and documentation, be it historical, art historical, archaeological, ethnographic, or other. Technical treatises and instruction manuals, archival records such as accounting books, contracts and inventories, artists' notebooks, correspondence, interviews, ethnographic documentation, and records of archaeological excavations are all key sources, among others, to reconstructing the how, why, when, where, who, and other questions we seek to answer. The technological study thus relies on a combination of methodologies and skills that ensue from the collaboration of experts in diverse fields such as art history, the history of science and technology, materials science, archaeology, metallurgy, artistic practice, and artisanal foundry knowledge, among others.

While the contributions of such technological research to the art historical field is significant, it can also be perceived as a threat to the established authority in a given field.[3] But it is heartening to find that cross-disciplinary collaborations are being embraced more. To date they have highlighted:

- the value of sharing expertise and data among specialists from a variety of backgrounds studying bronze production of different eras and/or cultures;[4]
- the need for a shared multilingual vocabulary to communicate effectively across disciplines and among scholars working on different cultures and periods (from ancient to modern);
- and the need for an agreement on investigation methods and reporting protocols. This last is crucial to facilitate correct understanding, encourage nonspecialist researchers to take into consideration the available scientific data, allow accurate

comparisons by encouraging the acquisition of sound data, and facilitate the sharing of data.

As mentioned above, few (if any) places offer training in such interdisciplinary study, and there is currently no standardized vocabulary or method for approaching such investigations. Both of these are crucial in order to answer some of the broader questions about technological transfer, for instance, which requires the acquisition and sharing of comparable data. A vocabulary and protocol for technical description will benefit the technological study of bronzes of any period and place.[5] It should also encourage most nonspecialists to venture more boldly into the technical realm.

As a tool, the present *Guidelines* aim:

- to foster research and collaboration among a variety of experts with a particular interest in cast bronze sculpture—curators, conservators, conservation scientists, technicians, metallurgists, sculptors, archaeologists, dealers, and collectors as well as academics and other specialists in allied fields—by facilitating more accurate and consistent communication across fields;
- to enable a better synthesis of the rich volume of data that continues to be gathered on such works than has been possible to date;
- to sensitize a wider range of scholars and museum professionals to the importance of issues related to the materiality of these works, and facilitate the integration of technical interpretation and correct vocabulary into the scholarly (and public) discourse.

It is assumed, most importantly, that whoever considers undertaking a technological study will be sensitive to the ethical issues in play. This includes respecting the protocols for handling active sacred or ritual objects, and weighing the pros and cons (and possible irreversible alterations) that the removal of material for analysis and examination for the sake of gaining new knowledge might demand.

1.2 Why undertake technological studies?

The technological study of bronze sculpture may arise in connection with a variety of contexts, including:

- acquisition (questions of dating, provenance, authentication, correct description and interpretation, condition);
- cataloging of an existing collection (which addresses issues similar to acquisition);
- conservation (understanding the structure and condition of the object will help determine and avert causes of deterioration);
- in-depth scholarly research on an individual object or larger group of related objects (for instance to learn more about the practice of the sculptor, workshop, foundry, and/or the culture in which the object was made);
- scholarly research into the history of technology (for instance, about the technological challenges, or exchanges of know-how). Related questions often require data on a broad selection of objects to yield meaningful interpretations.

The kinds of questions that bronze sculptures may raise are many. For instance:

- What was the model? What material was it made of? Who was responsible for creating the model, the mold, and for choosing the metal? (see Case Study 5).
- Is the cast a one-off? If it is one of multiple iterations of the same form, how does this cast relate to other versions? How best to achieve useful comparisons, especially if the sculptures cannot be brought together for study?
- Can one discern patterns in the way certain artists or workshops worked? Can one use such information to discern developments in the artist's or workshop's œuvre and date works, determine workshop organization, or ascertain transfer of technical know-how within or between workshops or different centers of production?
- Is it possible to tell when a particular cast was produced, given that certain models and molds can be used by others than those who originally crafted them, even well after that person's lifetime?
- What features may relate to the post-creation—later stage—in the life of a bronze, and what can we learn from them?

Answering these questions can require a fair amount of comparable data on related objects. Monographic exhibitions have so far provided some of the main impetus for such systematic technological studies of particular workshops or artists, as they also rally the necessary international cooperation and funding for resources needed to even begin to tease out trends. Larger studies of related groups such as the ones represented in Case Study 2 and Case Study 4 signal the necessity of a broader sampling in order to recognize what some of the questions might even be, before one might attempt to draw definitive conclusions.

It is also important to bear in mind the role of the founder and other hands that may be involved in the production of any bronze object. Depending on the challenges and solutions presented by a particular project, these problem solvers may adopt or experiment with different processes before settling on the one(s) most likely to produce successful results. This is particularly true of larger casts.

Then there are questions about the condition of the artwork and its original appearance, function, and display. Most often such questions relate to altered patination, other surface coloration, or decoration and tool marks, which are covered in I.6, I.7, I.8, and I.9. But remains of molding materials or surface scratches that result from mold taking could be clues to after-casting, which falls under what we will call here the life cycle of a bronze.

1.3 Life cycle of a bronze

Most sculptures subjected to technological study have had a life story subsequent to their production, and thus bear traces of use (and possible abuse). The ability to distinguish such subsequent features and not mistake them for original marks of the makers is crucial, and constitutes the subject of much of the rest of this volume. These traces can provide rich information about the object's condition and a range of other topics, including its conservation history and evidence of provenance. Such evidence may fall into a number of categories:

Traces of original use (and/or abuse) may include:

- ritual dressing (pigments, jewelry);
- patterns of wear (see I.6§1.4);
- organic and other accretions, coatings, or gilding (see I.7, I.8, I.9);
- traces of ritual burial conditions (soil, patina and pseudomorphs, see I.8§1.1.2);
- remains of ritual contents, such as consecration objects (see Case Study 4).

Traces of reuse may include:

- alterations related to change of style, taste, politics, religion, or function;
- traces of remounting.

Many traces of reuse are briefly described in I.6§1.4.

Traces of patronage and collecting may include:

- applied, written, or inscribed markings (see I.6§1.3), including evidence of labels (for example glue patches);
- removal and/or addition of aesthetic coatings applied by later owners;
- wholesale cleaning treatments (for instance to archaeological bronzes) and other traces of conservation/restoration (see more below).

Traces of display and environment(s) that the object has been designed for or exposed to may include:

- different types of mount/base (and later remounts) (see GI§2.9 below);
- varying degrees of finishing on different parts of the object that might suggest creation for, or insertion in, a particular architectural location;
- alterations and accretions such as corrosion particular to specific environments (soot; paint drips; burial accretions such as pseudomorphs, see I.8§1.1.2).

Traces of conservation and/or restoration may include:

- traces of mechanical and chemical cleaning (see I.8§1.3, I.8§2.1);
- waxing or lacquering; over-gilding (see I.7§2.7);
- protection from destruction by concealing with overpaint;
- inscriptions or messages left by an earlier restorer.

Traces of replication include (fig. 1):

- evidence that the cast was the model for an after-cast or *surmoulage* (see I.1§5.2, I.6§1.4.2);
- evidence that the cast is itself an after-cast.

Traces of scientific sampling of materials may be discernible:

- in the metal (drilling, sawing, abrading, see II.5§3.5, II.5§6.2);
- in the core (drilling, breaking, prying, see II.7§3);

- on the patina from the surface (not easily visible, see II.6§3.1).

Most evidence of scientific sampling is briefly described in I.6§1.4.2.

Damage and disfigurement may be due to any number of reasons, including:

- iconoclasm and other crude dismemberment (for instance reuse or resale of metal, see I.6§1.4);
- expansion of core due to rusting iron armature (see I.3§1.2.4);
- graffiti;
- later addition of signatures (perhaps for the purpose of deception);
- ritual destruction/burial;
- accident;
- poor construction and/or failure of repairs over time (for example loss of patches, see I.4§1.1.1);
- cannibalizing of sculptural components to create a novel composition.

1.4 A short historiographical note on the technological study of bronze sculpture

Interdisciplinary research into how bronzes were made draws on a marriage of text-based historical research and object-focused investigations of physical evidence, as well as experimental reconstructions. Depending on the approach taken and who was involved, the results see the light in a range of specialized publications in the diverse fields of conservation, archaeology, science, history of science and technology, foundry work, art history, and museum work, which exist in many languages.

Precious little has survived of the vast knowledge and skill, and of the transactions, that went into the making of bronze sculptures over the millennia. But some snippets were captured in writing. The nineteenth century saw a great push toward more systematic archival research on the part of historians. The transcription and publication of large collections of documents has led to the discovery of information pertaining to bronze production centers and the commissioning and execution of bronze sculptures. Such sources form essential evidence in the reconstruction of chronologies, fabrication processes, and distribution and attribution of labor, especially with regard to large sculptural projects (see for instance Case Study 5).[6]

Founders and scientists (in particular chemists and metallurgists) with an interest in historical metallurgical processes have also made seminal contributions to the history of science and technology, which include, among other things, translations of some of the rare technical treatises that feature descriptions of medieval and later European bronze production. The information in these texts about the preparation of raw materials as well as the steps of casting processes have been invaluable in the interpretation of both technical features on bronze sculptures and analytical data on alloy and mold compositions.[7]

Archaeological excavations, collections, and methodology have played a key role in the study of bronze technologies, as it is in this context that object-focused studies of bronzes first developed. The challenge of how to deal with archaeological bronzes that either were already disfigured by corrosion or began to break out into active "bronze disease" following excavation was one of the impetuses for the establishment of early scientific labs in museums. It kept scientists busy for decades, and their evolving understanding of the mechanism has been reflected in the attitude toward treatment and the aesthetic choices that account for the appearance of such objects to this day (fig. 157).[8]

Increasingly methodical, scientific excavations unearthed a whole range of copper alloy artifacts and shed light as well on the more ephemeral evidence of their production and use, such as mold and furnace remains and ritual contexts.[9] These include discoveries of foundry sites in different parts of the world, for instance from the fifth century BCE onward in Houma, China, and the Athenian Agora.[10]

The taste for collecting ancient materials is centuries old and has been accompanied by a market for copies and forgeries for almost as long, many of which coexisted peacefully with the genuine early works. But the shift to a more scientific classification went hand in hand with the development of forensic methods to characterize the materials and facture of objects, which served to describe and identify objects more precisely. The discoveries of fabulous archaeological works both on land and at sea, and the publicity that surrounded them, also fed broader public interest in how the marriage of science and the humanities may unravel the mysteries of these objects and help to conserve them. The two bronze warriors from Riace are one such find.[11]

Similarly, a greater awareness and understanding of the damaging effects of pollution on well-known public monuments led to conservation projects that provided further opportunities for technical investigations. The research data and results of these scientific and historical investigations—and the often-dramatic treatments—were not infrequently featured in exhibitions and related publications.[12]

During the early twentieth century there was an increase in the application of science to the study and treatment of works of art, and the international exchange of data and ideas through conferences and dedicated journals.[13] After World War II, the trend continued with the establishment of the first professional conservation organizations as well as the creation of new scientific museum labs. Over the decades these applied ever more sophisticated scientific tools and methods (for instance thermoluminescence dating, atomic absorption spectroscopy, lead isotope analysis, X-ray fluorescence analysis) to the identification and characterization of materials, tracing of sources, and dating. And the new archaeological science (also known as archaeometry) took off with a rise in, among other things, technical research on bronzes.[14]

By the 1970s, projects covered bronzes from around the world, from the Himalayas and Southeast Asia to West Africa, from ancient Egypt to Renaissance Europe, from ancient China to nineteenth-century France.[15] Some of these were seminal for bringing together experts in archaeology, art history, and the history of technology, conservation, and science, and were the first to be closely associated with, and even integrated into, sculpture exhibitions.[16] Other research projects consisted of systematic and thorough technological studies of collections.[17] For a good decade the Boston area was an important nexus of interdisciplinary collaborations and exchanges, including work on ancient bronzes, nineteenth-century European sculpture, and the Application of Science in Examination of Works of Art seminar series.[18]

With the ninth meeting of the International Congress on Ancient Bronzes in 1986, the topic of technological studies assumed a permanent place in the hitherto solely archaeological and art historical gathering.[19] The 1980s also marked the establishment of the Beginning of the Use of Metals and Alloys (BUMA) conferences in Asia, which promote exchanges of scholarship between Western and non-Western researchers. The 1990s witnessed several experimental archaeology workshops and seminars on various aspects of the production of large bronzes, which contributed to validating hands-on experience in scholarly work and also modeled rich collaborations among experts from different disciplines.[20] It also sealed a turn in the Renaissance field toward the integration of technological studies, which thereafter became almost de rigueur in scholarly bronze publications.[21]

During the past two decades, archaeologists, art historians, and museum curators have increasingly collaborated with scientists and conservators as well as craftspeople, and the technology of metalworking has become central to the study of bronze sculptures, often by international teams.[22] Research on bronzes from classical antiquity has flourished and been published as monographs and presented at meetings such as the International Congresses on Ancient Bronzes.[23] Medieval European bronzes are also gaining more attention (see Case Study 2).[24] Interdisciplinary, systematic studies have been carried out on a wide array of small bronzes from the Himalayas.[25] The technological study of Chinese, South Asian, and Southeast Asian bronzes is very active and the findings are shared, among other places, at international symposia and conferences, including the Forbes Symposia at the Freer Gallery of Art and BUMA.[26] Research on the production of bronze sculpture from West Africa is also continuing apace.[27] And archaeologists and archaeometallurgists are directing their attention toward a greater variety of periods, and thus also excavations of ancient metalworking sites.[28]

Among the more recent phenomena are studies of entire collections and larger corpuses of related works (see examples of this in Case Study 2 and Case Study 4) as well as research focused on the work of a particular artist, in which technical findings are woven into the art historical narrative.[29] Conservation projects also continue to provide unique opportunities for close investigation. Reconstruction, replication, and reenactment of historical processes and objects are gaining traction as methods of research and pedagogy.[30] Newer technical means of analysis, including 3D scanning, are being called upon to help address the complex and challenging questions around originality and authenticity (for instance with regard to the posthumous production of editions of certain nineteenth- and early twentieth-century European sculptures).[31] And other digital tools, from imaging to databases, are instrumental in the sharing and interpretation of large bodies of data and communication across international teams.[32] It is within, and from, this broader context that the CAST:ING project took shape.

2 What? The basics of making a bronze sculpture

Casting a bronze consists of creating a negative impression of a desired form in a refractory mold material, generally by using a model, and filling that with a molten copper alloy. Control over the mold is crucial for a successful casting. An overview of main materials and techniques used to make a bronze sculpture is given below. The processes comprise several steps. Table 1 and figure 3 give some clues on the relative time these different steps may require.

2.1 Casting: Primarily the impression of a model in a refractory mold

2.1.1 The refractory mold and core material

The refractory mold is the temporary, fire-resistant mass that holds the impression of the model to be reproduced. In lost-wax casting it is often referred to as the investment. Mastery of the refractory mold is essential for the successful outcome of a cast, including the quality of the as-cast surface of the metal (compare figs. 4, 5).

In designing it (see also videos 1, 2), it is important to ensure that it:

- is heat resistant;
- can capture fine detail;
- can withstand the molten metal coursing through it;
- has sufficient porosity integrated into the mold design and/or in its makeup to assist in the escape of air and gases to ensure that these do not stay trapped in the metal;
- can be broken down. Ideally it will be just strong and cohesive enough to withstand the casting process, but easy to remove from the cast (note: this is not the case with a ceramic shell).

In addition to the outer mold, a hollow bronze will require an internal mold form known as the core, made of similar refractory materials (see GI§2.1.4 below).

The refractory materials used for the casting of most bronze sculptures can be categorized into four main groups:

- clay-based material rich in sand and often organic materials that can be shaped either directly by hand or indirectly in a mold. Clay-based cores and refractory molds may be the result of a complex manufacturing sequence comprising up to five stages, namely the selection, possible modification, and mix of raw materials, and the shaping and baking of the core. Clay-based materials may be called loams;
- sand-based refractory mold materials bound by clay or small amounts of resin (fig. 6);
- plaster-based slurries made of gypsum with added sand and crushed ceramic powder or "grog";
- colloidal silica slurry layered with larger refractory particles that form a "ceramic shell" through sintering of the colloidal particles (fig. 7).[33] See also Case Study 7.

Monumental bronzes may be built over cores that have an internal fired-brick structure.[34] Bricks can also be found in life-size and medium-scale casts (fig. 8).[35] The cores themselves may be hollow. This is the rule in bell casting, and has also been witnessed on bronze statuary, notably on ancient bronzes in the South Arabian peninsula.[36]

How the core is held in place in relation to the outer mold in order to ensure the desired gap between them during casting is described in GI§2.6 below on armatures and core supports.

2.1.2 Models and molds

Over the centuries in different cultures around the world, expertise on how best to cast a bronze has evolved, based partly on the availability of materials, the level of control over pyrotechnic technologies, the range of forms to be represented, and the metal and other materials that will be used. The context of production is also key. For instance, a large workshop or foundry producing great numbers of multiples will most likely use a different process than a sculptor-founder working alone making a one-off cast (see Case Study 7). Every step of the casting process may leave diagnostic traces, which the technical investigator must tease out (see I.1).

The production of a bronze may involve a sequence of models, often in different materials. For instance, in lost-wax casting the original model may have been made of clay, which would then have been molded to produce wax models designed for casting, which are now generally referred to as casting models or inter-models (see video 3). In sand casting that function is performed by the pattern or chef-modèle (fig. 9). But specific terms have been developed in different cultures, periods, and areas; see the Vocabulary.

Similarly, a sequence of molds may have played a part in the translation from an original model to a bronze sculpture. Indeed, while the metal is ultimately cast into a refractory mold, in some instances other intermediary matrices may have served to create the casting model or pattern (see below).

2.1.3 Direct and indirect casting

Art historians use the term "direct cast" for a bronze in which the original model is sacrificed during the process of casting (for instance when the original model is made in wax, which is then burned out). "Indirect cast" is generally associated with the lost-wax process and refers to the use of an intermediary and generally reusable mold. The reusable mold is usually not a refractory one; it is created specifically to make replicas that stand in for the original model. However, the term "indirect" could (and should) in principle refer to any bronze that can be replicated without the loss of the original model (see GI§2.2 below for details). Sand casts are definitely indirect casts. These are very general terms. There are myriad combinations of the above, not to mention instances in which it makes sense to cast various sections of a sculpture by different methods and then assemble them (see Case Study 7).

2.1.4 Solid and hollow casts

A cast may be solid or formed as a metal shell in which the internal hollow shape is defined by an internal heat-resistant mold, referred to as the core (fig. 10, video 4). Casts of a certain size or volume tend to be created hollow to reduce the amount of metal. This not only saves expensive metal, but also minimizes the risk of shrinkage and potential cracking, as thicker volumes of metal solidify in less controllable ways. The thicker the metal wall, the larger the difference of temperature between the wall's middle and surfaces, so as the surface of the wall solidifies, large amounts of metal may still be liquid inside (see I.3§1.3.1, I.3§1.3.2). Solid casting is therefore generally (but not always) done with smaller casts or in areas of a cast—such as extremities—where a core would be too fine and therefore too delicate to withstand the violence of the casting process. The decision as to whether a cast was to be solid or hollow may also depend on a number of nontechnical parameters.[37]

2.1.5 Creating an impression of the desired shape in the refractory mold

There are three basic ways to create an impression of the desired shape in the refractory mold.

The first consists of encasing a model designed to be burned or melted out in a malleable, heat-resistant mold, as in:

- lost-wax casting, which encompasses other organic modeling agents such as latex or carved banana stems;[38]
- casting directly from life, or life-casting (fig. 11);
- lost foam casting (a modern process adopted from industry).

The second consists of compacting a heat-resistant mold material over a hard, reusable model that will need to be removed before casting to make room for the metal, as in:

- sand casting;
- clay piece molding.

The third consists of carving the negative form directly into a refractory mold material, as is used for:

- stone mold casts;
- cuttlefish casts;
- loess piece mold casts (for example to add certain decorative elements in Chinese bronze vessels);
- sand casts (also to create the casting channels).

Most of these casting processes are explained in greater detail below. The terms used to describe these processes and how they are referred to in several other languages are captured in table 2.

2.2 Lost-wax casting overview

2.2.1 Direct lost-wax casting

Although it was not a sculpture, the first known direct lost-wax (or indeed any kind of) cast dates from the fifth or fourth millennium BCE (fig. 12). The earliest known bronze statuette was also a direct lost-wax cast and dates from the beginning of the third millennium BCE.[39] The direct technique has been used worldwide throughout history and is still in use today, though some of the materials have changed (table 3).[40]

The process basically consists of translating a unique model that is shaped in wax or another modeling medium that can be burned out of the mold into metal by direct replacement. A sprueing system is attached, and the sprued model is embedded in or coated with a refractory mold material, which is heated to melt out the wax, thereby creating the hollow cavity to be filled with bronze (fig. 13). The direct process allows the artist the freedom to create forms of great intricacy without having to worry unduly about undercuts (videos 5, 6).

Other combustible materials, such as plant and animal materials, including latex, can also be used. Small objects can be made from solid wax models; larger ones are generally formed hollow by building up the wax model over a refractory clay core. Sometimes the core may be added after the wax is formed in what is referred to as the "direct wax-slab process," see GI§2.5.2 below. This layer of wax is built up in any number of ways: by laying sheets of wax over it, by modeling, or by dipping the core into molten wax.

The main advantage of the direct process is the creative flexibility it allows the artist or artisan. It does entail the loss of the original model, which may not be an issue in mass productions of devotional figurines, but becomes more of an issue with for instance a large, one-off sculptural commission. A disadvantage is, as already noted, the loss of the original model in the process.

2.2.2 Indirect lost-wax casting

Indirect lost-wax casting appeared comparatively early in the history of bronze sculpture,[41] and several variants are still in use today (table 3). In this process, reusable molds are taken from an original sculpture model in order to replicate the desired form in bronze. The original model can therefore be made of any number of materials. Undercuts in the model are dealt with either by using a flexible mold (for instance one made of rubber) or by creating a (rigid) piece mold[42] designed in small, removable sections to fit around the model like a three-dimensional jigsaw puzzle, and usually held together by an outer "mother mold" (fig. 15). These reusable molds enable the maker to create one or more wax replicas or inter-models for use in the casting process (the "casting model" in modern parlance).

The main advantages of indirect lost-wax casting include the preservation and potential reuse of the original model, and the ability to cast multiple, more or less identical replicas of the original model. These can, of course, also be manipulated intentionally to create quite distinct works.[43]

2.3 Variations on indirect lost-wax casting: slush molding, indirect wax slabs, lasagna, cut-back core

The wax replica that is made for casting in the indirect process (the inter-model) can be formed in a variety of ways, which fall under two main categories, where:

- the core is created after the hollow wax model has been produced;
- the core is created before the wax model.

Within these, four processes have been most commonly documented in historical bronzes.

2.3.1 Wax first, then core

Slush molding

Slush molding (fig. 16, videos 7, 2) may be among the oldest processes used in indirect lost-wax casting, and is still widely used today (table 3). Slush molding consists of pouring liquefied wax into a mold and swirling it around to ensure that the wax builds up relatively evenly. Small molds can be filled to the rim. The wax will first solidify at points of contact with the mold. The remaining molten wax is poured out, and the process repeated until the thickness of wax matches the desired wall thickness of the bronze. With larger molds, a fine layer of liquefied wax may be painted onto the surface first, to make sure that all the details are captured, before pouring the liquefied wax into the mold. Larger molds may be only partially filled and then agitated and rolled around to ensure that the wax slushes sufficiently to coat the entire inner surface.

The main advantages of slush molding are its rapidity and simplicity. Because the mold is generally manipulated by hand, size is a limiting factor, requiring larger bronzes to be molded in manageable sections.

Indirect wax slab process

Another indirect process consists of lining the mold with preformed wax slabs of the desired thickness (figs. 17, 18). As far the authors know, there is no standardized nomenclature for this latter process. The name we propose, "indirect wax slab process," refers directly to the slab pottery technique. The earliest examples of bronzes for which the indirect wax slab process may have been used are Egyptian statues dated to the beginning of the first millennium BCE.[44] The process appears to have been widely used during Greek and Roman antiquity,[45] and probably long after as well.[46] Some modern-day founders also use it in one form or another (table 3).

This process might be used with mold sections of larger bronzes. A variant involves building up a wax layer of the desired thickness by pressing small masses of softened wax into the mold (figs. 19, 50).[47] Alternatively, molten wax may be painted into the mold first, to ensure good coverage of finer details, and then built up layer by layer in a variety of ways—painted, backed with wax slabs, or reinforced with softened wax or slush molding. For all the aforementioned processes other than slush molding, it is necessary for the mold to be open before the application of wax on the internal surface of each of its sections.

The main advantage of the slab process is the ability to precisely control the thickness and evenness of the sculpture's metal walls (fig. 20).

Wax painting

The process entails painting liquefied wax into one or more sections of an open mold. While it could be used on its own there is little evidence of that; most often it might serve to supplement other indirect processes such as slush molding as a way to ensure good coverage of the mold and to capture fine details (figs. 21, 22).

2.3.2 First core, then wax

The so-called lasagna process (fig. 23) was first described by sculptor, goldsmith, and writer Benvenuto Cellini (Italian, 1500–1571),[48] whose use of the term "lasagna" has been generally adopted. Its likely use has been identified on sixteenth-century Italian and French bronzes,[49] and seems to draw from the practices of cannon casting (see GI§2.4.2 below). It is still employed in some modern foundries (table 3). Instead of wax slabs (see previous section), slabs of clay or another dough-like substance of even thickness—the so-called lasagna, which Cellini describes as being the width of a knife's back—are laid into the inner surface of a piece mold of a larger sculpture (fig. 24). The core is built up to fit the hollow space in the lasagna-lined mold using material that is either clay-based or in the form of a plaster-grog slurry. Thereafter, the lasagna is removed from the mold. The piece mold is reassembled around the core, and wax is poured into the space previously defined by the lasagna layer. As with the other processes, a fine layer of wax may be painted into the mold first. See GI§2.4.2 below for a variant of the lasagna process without using wax. The lasagna can be applied to the piece mold much more rapidly than wax slabs (whereas wax slabs are usually applied cautiously onto the model surface to capture all the details, the lasagna clay slabs are applied more "approximately," at least in modern foundries).[50] This allows for more control over the condition and surface of the core, and thus also over the outcome of the cast.

The cut-back core process (fig. 25) has been identified on eighteenth-century French bronzes, but might have been used elsewhere and earlier.[51] It is still in use today (table 3). Here the core is produced by using the piece mold to cast a replica of the model in refractory material. This is then pared down to the desired thickness, secured back into the piece mold, and liquefied wax is filled into the space between the core and the mold. As with the lasagna method, this process allows control of the core, limiting shrinkage of the wax and thus affording a more predictable and even metal thickness. This is a highly skilled operation.

2.4 Refractory piece molding processes

This consists of packing a refractory material (sand and clay) over a hard, reusable model in order to form the desired matrix of rigid pieces, like a three-dimensional jigsaw puzzle designed to circumvent undercuts. Several processes of this kind have been used for sculptural casting. The one most familiar today is sand casting. Piece mold casting, such as that used to make ancient Chinese vessels, is a much older form.

2.4.1 Sand casting

Various forms of sand casting have been used to create a wide range of items since the beginnings of metallurgy.[52] From the Renaissance period until the nineteenth century, the process involved a range of materials and binders not strictly classified as "sand" and was limited to small-scale sculptural works.[53] Sand casting as we now know it was first used for bronze sculpture in Europe at the end of the eighteenth or the beginning of the nineteenth century,[54] and is still in use today for this purpose, and prolifically for industrial casting, especially of iron.

In this process, special casting sand that is naturally bound with clay (or synthetically bonded materials such as resin)[55] is rammed around a chef-modèle positioned in stacked metal frames (or flasks) called the cope (top) and the drag (bottom) (fig. 9). Smaller piece mold segments are often fashioned within these larger sections and designed for ease of construction and disassembly, as it will be necessary to remove the pattern and reassemble the mold precisely for casting. Should a core be needed, it is formed using a sand version of the cut-back core process (see GI§2.3.2 above and Case Study 6).[56]

The predominantly sandy composition of the mold minimizes shrinkage of both mold and core, which assures greater control of the thickness of the cast. Access to the inner surface of the mold before casting also allows for greater quality control and last-minute alterations or repairs.[57] Furthermore, that surface is often coated with a layer of soot or graphite, which helps produce a reducing environment at the point of contact with the hot metal. The lower cost of sand relative to wax is also seen as an advantage,[58] and natural sands (meaning, those that contain a natural clay binder) are reusable. Piece molding requires great skill, especially when working with models with complex textures and undercuts (see for instance Case Study 6). It should be noted that high-quality sand casts such as those made in France in the nineteenth century can be essentially seam-free. This is accomplished through precise ramming of the sand—not too loose, not too dense—so that the sand swells just slightly upon baking to seal the gap.[59]

2.4.2 Piece mold casting

Piece mold casting is mainly associated with early Chinese vessels, including some with very figurative forms (see Case Study 3),[60] and with bells and artillery in early modern Europe.[61] Known examples of bronze sculpture made by this method are rare.[62]

This process, which is also known as section-mold casting (but recall, as noted above, that a number of other processes use piece, or section, molds), consists of creating a clay-based, refractory piece mold around a model that has also been fashioned in refractory clay. The rigid mold must, as with other piece molds, be designed to circumvent undercuts. Once the mold has been made, the model is pared down to form the core (fig. 26).

Some literature points to the use in nineteenth-century Germany of a version of piece mold casting combined with the lasagna process that omits the wax.[63]

The piece mold process allows for a very even thickness and a good dimensional stability of the resulting cast. Shrinkage is very limited (see II.4§1.1). Details can be both molded and modeled/carved. As with sand casting, the design and production of the complex piece mold required for intricate forms requires great skill.

2.5 Other casting methods/variations for bronze sculpture

2.5.1 Direct life-casting

Casts from life are found in Roman antiquity and were also widespread during the Renaissance and the nineteenth century.[64] Modern artists are still using the technique. Casting "from life" refers to the use of natural forms (flora and fauna, fig. 27), but also human-made materials (for instance textiles), as models. These are embedded in a refractory mold and burned out to create a void that reproduces the model in fine detail (fig. 11). Such a process eliminates the need to model the object afresh, and allows for the reproduction of fine details present in the original. As with all direct processes, the original model in this case is destroyed in the process.

Note: It may be difficult to distinguish a direct life-cast from an indirect cast from life, in which a reusable mold is taken from a found object or creature. Life and death masks, for instance, are indirect life-casts. Also, contrary to belief, a direct cast from life can be hollow.[65]

2.5.2 Direct wax-slab process

The ancient Greeks are known to have used the direct wax slab process to create bronze multiples, such as the Griffin Protomes.[66] Theophilus, the twelfth-century artisan monk and compiler of a technical compendium on diverse arts, describes a similar process used in his time for making incense burners in Germany.[67] See Case Study 2. It is also used in modern-day Nepal and Thailand for sculptural work (table 3).[68]

The process involves bending wax slabs freehand in order to shape the model, with the core inserted later (figs. 28, 29). This is not to be confused with the direct lost-wax process described above in GI§2.2.1, where for hollow casts the wax is built over the core.[69] It allows for quick modeling with a very good control of thickness. Its main drawback is the loss of the original model.

2.5.3 Carving into a mold

Another approach that has been used mostly for simpler sculptural forms involves carving a mold directly into a hard, refractory material such as stone, hardened refractory clay, casting sand, or cuttlebone (fig. 30). This is the purest direct process, as the tool marks create the negative space. As a result, the thickness of the cast is more likely to be uneven.

2.6 Armatures, core supports, and core vents

Many models used for casting may be built up on an armature for structural support. This can apply not only to models with a refractory core intended for hollow casts, but also to solid ones.[70] In both lost-wax and sand casting, an armature will further help support the core during the pour. As the name implies, core supports help to both strengthen and support the core. The inclusion and placement of armatures and other supporting elements (and in some cases, core vents as found in sand casting) are crucial to the success of a hollow cast in particular. Shaped most often out of metal, their remains might be preserved in the cast.

2.6.1 Armatures

The model may need structural support to prevent it from drooping, collapsing, or breaking during its making and manipulation, as well as during the pouring of the metal. Such an interconnected structural framework is usually made of ferrous metals (notably iron, and in recent times steel), though parts could be made of other materials such as wood.[71]

These different elements may (but not always)

- form a complex structure like a skeleton (figs. 31, 32, 33) bound together by various means (for instance welding, twisting, or entwining with finer wires, fig. 34);
- extend down into the base;
- protrude beyond the core and model into the refractory mold, therefore serving the dual function of core support.

Not all sculptures necessitate armatures. And any kind of armature can act as core support if it extends from the core into the refractory mold. The armature in a cast bronze may be preserved to serve as a structural/mounting function post-casting, depending on the shape of the sculpture and location of the armature. Iron left inside an outdoor bronze may cause damage due to expansion of the iron as it rusts (fig. 149).

2.6.2 Core supports

As the name implies, core supports serve to reinforce or strengthen the core at different stages of the casting process. Two definitions appear in the literature. The term may apply to inserts or spacers that hold the core in place during the pour. These come in a variety of configurations:

- Core pins are positioned perpendicularly to the wall of wax or bronze. They protrude both into the outer mold and into the core to hold it firmly in place. These could be wires, rods, or nails (figs. 35, 36, 37).
- Chaplets serve as spacers and tend to be more or less the width of the gap between the core and the outer model. They are nestled at intervals in that space, balancing the core in relation to the outer mold. Chaplets are often fused into the body of the cast during the pouring (figs. 38, 39, 40).
- Mold extensions form bridges between the core and outer mold and are made of refractory material (fig. 41).

The term may also apply to internal wires or rods that help strengthen the core during assembly of the casting model and/or during the pour. Often made of single wires or rods that do not connect into the base, they may have one or more of the following characteristics:

- They may be inserted into more fragile core segments, for instance the arms or legs (fig. 2).
- They may be inserted into the core perpendicular to wax-to-wax joins to help hold them together (common in indirect lost-wax casts; fig. 42).
- They may be used to support joins between core sections, as may occur in sand casting (fig. 9).

2.6.3 Core vents

Core vents (also called lanterns, figs. 6, 43) provide an exhaust route for gas buildup inside the core. They run through much of the core and extend into the outer mold, thus also serving as core supports and sometimes as armatures all in one. Core vents were commonly used in sand casts from the late nineteenth century onward (fig. 9, Case Study 6).

2.7 Sprues

A sprue is any channel that feeds metal to the mold, in contrast to a vent, which lets air out. The sprue system is the entire network of channels designed to distribute the metal through the refractory mold efficiently while ensuring that air and vapors can escape so as to avoid casting flaws (figs. 21, 44, 45, 46, video 8). In sand casting these channels are carved into the mold once the impression has been formed in the sand mold parts. In lost-wax casting they are usually formed with wax rods of various sizes, which are affixed strategically to the model by melting the ends onto the surface with a hot knife before the casting model is invested with the outer mold material.[72]

Sprues (generally referred to as "gates" in sand casting) are usually located on higher points of the cast surface (as opposed to depressions) and, if possible, in areas of less detail (inside, back, edges) to avoid having to re-create complex surfaces of the model after casting. More substantial castings will need a greater number of feeds/ gates to allow the alloy to reach all areas of the model at the optimal temperature, minimize turbulence, and reduce risk of shrinkage (video 9).[73]

There are two main approaches to feeding the metal into the mold. The cast is described as "bottom fed" when the "feed" directs the metal straight to the bottom of the mold from whence it pushes its way up through the cavities in the mold (fig. 13). In foundry terms it might also be considered "indirectly cast" (not to be confused with the indirect casting method described above in GI§2.2.2). It is considered "top fed" if the feed is attached directly to the top of the figure and the metal allowed to follow whatever paths it might find down through the mold (video 4).[74] The former reduces the chances of erosion of the mold, or "scouring," which can result from forceful impact of the metal on its fragile inner surface.

The orientation of the cast—upright, upside down, horizontal, or at an angle[75]—can vary depending on the shape and size of the form as well as on cultural context and practice of the founder.[76]

2.8 Alloys

The preparation of the metal that will be poured into the mold consists of many preliminary steps, including alloying, in order to achieve the desired properties (color, castability, workability, corrosion resistance). This final mixing of elements, including the use of scrap metal, may happen in advance of a pour, or at the time of casting of a given sculpture. The overview of the range of alloys used in "bronze" sculpture and some of their salient qualities/ properties is provided in I.2.

2.9 Finishing

While castings can come out of the refractory mold with near-perfect surfaces, most need some repair or reworking of the metal (see I.4§1, videos 10, 11), and quite a few need more than that. Many of the kinds of imperfections found in casts are described in I.3.

Most cast bronze sculptures will at the very least need to have their sprues cut off and core pins extracted. Removal of the coarser traces of the casting process falls under the rubric of fettling, often followed by refining of the metal surface via chasing (see I.6§1.2). Surface treatments that do not fall under the rubric of tooling of the cast metal may include various forms of metal plating, gilding, the application of patina, and/or inlays and overlays. And there is a long tradition of combining bronze elements with other materials to create composite objects (fig. 47).

Not all sculptures are cast in one piece, and the rich variety of assembly technologies and related processes are fully described in I.5. Although not covered in the present *Guidelines*, bases and mounts may represent a highly useful source of information to document fabrication techniques. This is the case with ancient large bronzes—notably where only the bases survive—and also for statuettes and groups from the Renaissance and Baroque periods.[77]

3 How? User guide

The *Guidelines* are accessible in a variety of ways.

3.1 Evidence of the fabrication process (volume I)

Volume I is designed to help the reader identify and investigate the physical evidence present on a bronze sculpture. It is divided into nine chapters that correspond to the different aspects of the creation of a bronze, more or less in chronological sequence of the fabrication steps. It goes into more detail on the variety of casting processes described in this general introduction; discusses the properties of the different copper alloys encountered in artistic castings; and characterizes defects, repairs, assembly techniques, tool marks, as well as plating and other surface decoration techniques. Each chapter follows a similar "What? / Why? / How?" format. The first section lays out how distinctive features are produced, and thus sheds light on what to look out for and also possible misinterpretations. That is followed by FAQs that address why such evidence might be interesting and/or useful to investigate. A checklist at the end outlines the primary methods of investigation and analytical techniques that are relevant to the investigation of this topic, and points to the more in-depth discussions of these techniques in volume II. The definitions of terms included in the Vocabulary are hyperlinked upon their first mention in each chapter.

3.2 Methods of investigation (volume II)

Volume II provides a summary description of each of the multifarious techniques of examination and scientific analysis that have been usefully applied to the study of bronze sculpture. The volume is intended for both specialists in technical examination and analysis (e.g., conservators, conservation scientists, archaeologists) and nonspecialists (curators, historians). Specialists will find practical guidance on how each of the techniques may be specifically adapted and optimized for the study of bronze sculpture, along with extensive references to additional technical resources and published examples of their application. Nonspecialists will find succinct descriptions of the techniques and the scientific principles on which they are based. The descriptions are followed by discussion of the capabilities and limitations of each particular technique, along with a description of any sampling requirements that may be required. This information is presented to help those in positions of responsibility make informed decisions regarding the selection of appropriate examination and analysis techniques in the context of a technical study.

In the first chapter, the reader will find practical advice and resources on how to plan and document a technological study (with a reminder of the importance of cross-disciplinary collaboration at every step). This is followed by chapters on image-based examination and documentation (photography and radiography) and measurement. Analytical techniques that further aid in the identification and characterization of materials are described in separate chapters dedicated to the specific materials type (base metal, surface layers, and core). In the penultimate chapter, methods for dating a bronze sculpture are described, and the final chapter discusses the design and implementation of experimental simulations.

3.3 Vocabulary

The rich terminology associated with sculpture, and bronzes and foundry work in particular, is very specialized. It can also vary to a surprising extent across different groups of specialists such as professional foundry practitioners and historians of bronze sculpture, or even across experts focused on different chronological or geographical areas within a field. One of the main goals of the interdisciplinary CAST:ING project was to shed light on some of these ambiguities, wrangle definitions, and where possible suggest a common preferred term. The selection of agreed-upon terms offered here should be useful to a range of researchers—as an aid for interpreting written sources or features on a cast, and for recording and communicating those discoveries and observations. A shared vocabulary ultimately also facilitates a wider sharing of data. Definitions for each term are provided in English and French. Translations are provided in the three other modern languages that reflect the expertise of the CAST:ING members: Italian, German, and Chinese. Early French and Italian translations are also proposed here and there. Earlier terminology is also provided for Italian and French.

It was beyond the scope of this project to produce such a thesaurus for the more than three hundred technical terms collected by its members. We agreed on a selection of fifty-eight based on their frequency of use in the study of bronze sculpture, and on the possible definitional ambiguities. For each entry, we have aimed to include a translation in the selected languages, along with references to two or more authoritative bibliographical sources, as these terms are also to be incorporated in the structured online vocabulary of the Getty Art and Architecture Thesaurus (AAT), which is used broadly by art historians. Where necessary, multiple translations are provided with notes to explain differences between them. Entries are illustrated with relevant images drawn from the Visual Atlas of Features. And hyperlinks to the select terms are provided throughout the *Guidelines*.

3.4 Case Studies

The case studies offer readers a representative example of investigations into the technology of bronze sculpture. The selection draws on the varied experience of the CAST:ING project members. Some cases focus on one object, others on a group of related works. They cover a variety of contexts in which such studies are undertaken, issues that may be addressed, and the kinds of research tools and historical sources—whether written or oral—that are essential to formulating an informed interpretation of the physical evidence. Their structure varies, but each of them ends with a synopsis of the findings and further questions, and tools applied in the study. The case studies complement the descriptions of processes, technical evidence, and scientific techniques of investigation discussed in volumes I and II, and hyperlinks to particular videos are included where relevant.

3.5 Visual Atlas of Features

The dictum "an image is worth a thousand words" is apt when trying to describe technical details, and this is particularly true for bronze sculpture. Therefore, the *Guidelines* rely heavily on illustrations. This section gathers all of the illustrations used in the publication, including radiographs, 3D models, sketches, charts, and diagrams. A fair number of these images have visual annotations to aid in their interpretation. In the online version these appear in layers that can be activated (turned on and off) by the reader. In the print-on-demand version, these are merged onto the image. In the digital publication it is possible to search for and browse through the photos and diagrams via keywords. In the print-on-demand version, the images all print together separately from the text in the order of appearance in the *Guidelines*.

The visual atlas also gathers short videos illustrating a range of fabrication processes described in the text. Links to them appear in the relevant sections of the *Guidelines*. Their inclusion is made possible by the digital format and affords readers a better sense of the gestures, timing, choreography, and sound involved in these processes—all of which are not readily conveyed in words.

3.6 Tables

This section includes fifteen synthetic tables created to provide an overview of the technical data.

3.7 Bibliography

One overall bibliography serves as reference for all sections of the *Guidelines* and is unique in its diversity, as it draws on the expertise of an interdisciplinary cast of contributing members. Yet while many of the key publications on bronze casting and related disciplines have been included, the bibliography does not pretend to cover every publication about this expansive field. The bulk of the publications cited are in English and French, reflecting the makeup of the primary organizers of the publication, but every effort has been made to incorporate important works in other languages.

4 Overall disclaimer

The information and images presented here have been selected to illustrate specific technical features without necessarily suggesting that they are characteristic of a given artist's production or period. This resource is *not* primarily intended to provide a guide to the attribution or dating of specific bronzes.

The interpretation of a feature as evidence of a particular process, technique, or material is often complex, carrying frequent risks of misidentification. And while we have done our best to illustrate each feature with a representative photograph, these images cannot replace the experience of close examination of an object under different angles, lighting, and magnification conditions. Where applicable, we strongly invite the reader to seek out the publications referred to in the image captions for more in-depth discussions of the examples.

Making meaningful sense of the physical evidence preserved in an object requires weighing these technical observations with a wide range of contextual information from related/comparative objects as well as historical, archaeological, or other relevant sources. Ultimately, these *Guidelines* cannot replace the hard-won experience of experts.

Notes

1. The term "technological" is used here rather than "technical" to characterize the type of studies carried out on bronzes, as a reference to the so-called technological approach in anthropology (Lemonnier 1992). Although the term is rarely used by scholars studying the techniques of bronze fabrication, the objectives they pursue—or at least the questions and conclusions that the results might lead them to—generally pertain to this technological approach (see GI§1.2). This includes the reconstruction of the *chaîne opératoire* (operational sequence of different steps involved in the production of an artifact) and a better understanding of the social, cultural, and economic environment surrounding the fabrication and subsequent life of the bronze. For a recent essay on the historiography of the use of the term "technique" in art historical contexts see Dupré and Hendrikson 2023.
2. Bourgarit et al. 2014.
3. Formigli 2010.
4. Rehren 2014.
5. This very issue of the lack of shared protocols was raised at a workshop sponsored by the Dutch Organization for Scientific Research (NWO) in collaboration with the Metropolitan Museum of Art, New York, "A New Model for Scientific Research on Cultural Heritage: Joint US-NL Workshop on Integrated Collaborative Research on Technical Art History, Conservation, and Scientific Research," New York, April 2014. Private communication from Francesca G. Bewer.
6. For example Bresc-Bautier 1989 and Bresc-Bautier 2003 on archival work related to sixteenth-century Parisian founders; Somigli 1958 on the *Perseus* by Benvenuto Cellini; and Baxandall 1965 and Baxandall 1966 for a brilliant reconstruction of the production of the Fugger altarpiece in Augsburg, Germany (1581–84), based on the account book for the commission. More recently, Avery 2011 reconstructed the complex network of Renaissance Venetian founders and their work.
7. For example Theophilus [ca. 1122] 1979; Agricola [1556] 1950; Biringuccio [1540] 1990.
8. Beale 1996.
9. The specialized study of historical metal production from extraction to refinement and trade is more commonly referred to as archaeometallurgy.
10. Houma in China's southwest Shanxi province was first excavated in 1957: Shanxi Provincial Institute of Archaeology 1993; Institute of Archaeology of Shanxi Province 1996. On the Greek site see Mattusch 1975.
11. On the Riace bronzes see Formigli 1984. For recent publications of the Riace Bronzes and for new developments in studies of classical bronzes, see Daehner and Lapatin 2015. For another opinion on the production of

these statues, not widely adopted, see Konstam and Hoffmann 2002; Konstam and Hoffmann 2004.

12. For example Bearzi 1950; Leoni 1979.

13. The 1930 conference "Study of Scientific Methods Applied to the Examination and Conservation of Works of Art," organized by the International Museums Office, was the first of such conferences. The organization's journal *Mouseion* was one vehicle of communication for such work. And *Technical Studies in the Field of the Fine Arts*, published by the Fogg Museum (1932–42), was the first international journal to be dedicated to this interdisciplinary area of research. The International Institute for the Conservation of Museum Objects was established in 1950. See Bewer 2010.

14. For instance, in the 1960s and 1970s, the Oxford University Research Laboratory for Archaeology and History of Art led in the development of thermoluminescence (TL) dating of archaeological materials. And in the 1970s and 1980s, the Rathgen-Forschungslabor of the State Museums of Berlin and the British Museum's Department of Scientific Research performed large-scale systematic analyses of alloy composition for works from different cultures and also entire museum collections. X-ray fluorescence (XRF) was applied to (among other things) the study of alloys used in Renaissance medals, and TL dating made a splash almost immediately, with controversy over the dating of medieval statuettes, alerting people to the possible pitfalls of interpreting analytical data in the absence of the full life history of an artwork (Glinsman and Hayek 1993; Craddock 2009, 143–44).

15. For example Werner 1972; Beale 1975; Ashley-Smith 1977; Larsson 1979; Stone 1981; von Schroeder 1981; Riederer 1982; Craddock 1985; Craddock and Picton 1986; Bagley 1987; Reedy and Meyers 1987.

16. Mitten and Doeringer 1967; Wasserman 1969; Doeringer, Mitten, and Steinberg 1970; Wasserman 1975; Born 1985.

17. Gettens 1969; Milam, Reedy, and Sussman 1988.

18. The Application of Science in Examination of Works of Art seminars were held at the Museum of Fine Arts, Boston, in 1958, 1965, and 1970, and the proceedings were published respectively as Museum of Fine Arts (Boston) Research Laboratory 1970; Museum of Fine Arts (Boston) Research Laboratory 1967; Young 1973.

19. The initial meeting of the series took place in Nijmegen, the Netherlands, in 1970, and the most recent (XX International Congress on Ancient Bronzes) took place in Tübingen, Germany, in 2018.

20. Among these was a series of experimental archaeology workshops on ancient bronzes organized by Edilberto Formigli (Formigli 1993; Formigli 1999a).

21. Examples include Bewer 2001; Dillon 2002; Bewer, Stone, and Sturman 2007; Bassett 2008; Stone 2011; Van Langh 2012; Smith 2013; Bourgarit et al. 2014.

22. The 2012 conference in Paris on French bronzes is one example among others (Bourgarit et al. 2014).

23. Janietz 2000; Peltz 2011; Zimmer et al. 2011; Mattusch 2014; Descamps-Lequime and Mille 2017; Willer, Schwab, and Mirschenz 2017.

24. See Thomas and Dandridge 2018 for a recent survey of technological studies on medieval bronzes.

25. Reedy and Meyers 1987.

26. Technical papers on bronzes were presented at three of the five Forbes Symposia (Bourgarit et al. 2003; Reedy and Meyers 2007; Bewer 2012). See also Vincent, Bourgarit, and Jett 2012; Mechling et al. 2018; Strahan 2019.

27. For an early technical examination of an African bronze see Fagg and Underwood 1949. More recent work includes Junge 2007; Craddock 2009; Pernicka and Berswordt-Wallrabe 2008; Peek 2021.

28. Polkinghorne et al. 2014; Meyer, Thomas, and Wyss 2014.

29. The Metropolitan Museum's catalogue of Italian bronzes (Allen et al. 2022) offers an exemplary integration of technical and historical information. For the work on French bronzes see Bewer, Bourgarit, and Bassett 2009; Bourgarit et al. 2014. The production of specific sculptors is a subject of increasing investigation, and the publications are too many to mention exhaustively here. See for example Adriaen de Vries (Netherlandish, 1556–1626) (Bewer 1999; Bassett 2008), Edgar Degas (French, 1834–1917) (Lindsay, Barbour, and Sturman 2010), Gaston Lachaise (French, 1882–1935) (Day et al. 2010), Michelangelo (Italian, 1475–1564) (Avery 2018), Medardo Rosso (Italian, 1858–1928) (Cooper and Hecker 2003; Hecker 2020).

30. For recent examples of this see Smith and Beentjes 2010; Bilak et al. 2016; Lacey 2018; Lacey and Lewis 2020. For a broader discussion of this methodology see Dupré et al. 2020.

31. Boulton 2006; Boulton 2007; Lindsay, Barbour, and Sturman 2010; Beentjes 2019; Hecker 2020.

32. For example, the Louvre and C2RMF have collaborated on a technical database on large bronzes from classical antiquity that includes data on over three hundred statues (Descamps-Lequime and Mille 2017). See also the aforementioned studies on Southeast Asian bronzes.

33. Developed in the 1950s for industry (Campbell 2011, 915), ceramic shell has been increasingly applied to fine art casting since the 1970s.

34. See for example the monumental equestrian statues by François Girardon (French, 1628–1715), *Louis XIV on Horseback* (1685–87), cast by Balthasar Keller (Helvetian, 1638–1702) in 1692 in Paris, and Jean-Louis Lemoyne (French, 1665–1755) and Jean-Baptiste II Lemoyne (French, 1704–1778), *Louis XV on Horseback* (1731–35), cast by Pierre Varin (French, active 1736–53) in 1758 in Paris (Desmas 2014).

35. As seen notably in numerous sixteenth- and seventeenth-century life-size French bronzes (Castelle 2016) and medium-scale functional bronzes, for instance bases of sculptural Venetian firedogs and andirons (e.g. early seventeenth-century casts). See Motture 2003b, 296, no. 4.

36. Mille 2012.

37. For instance, in antiquity the production of hollow casts may be related to the quality required by the commission and may therefore be encountered in very small figures (see Mille 2019a). Conversely, relatively large statuettes may be cast solid, as seen notably in South India or in Indonesia (Mechling et al. 2018).

38. For more on the use of latex (a West African practice) see Herbert 1984, 85, and on carved banana stems (specific to the Kingdom of Congo, now Congo Republic) see Herbert 1984, 87.

39. The first known metallic sculpture is a bull, found at Nausharo, Pakistan (inv. NS.90.09.00.12) (Mille 2017).

40. As with alloys, a variety of wax compositions have been used by founders and sculptors for a variety of reasons (cost, availability, properties, et cetera). See notably Lebon 2020 and the appendix for French nineteenth-century recipes (Rome and Young 2003, 139–40) and Case Study 7 for examples of current practices.

41. The earliest examples are Egyptian statues dated to the beginning of the first millennium BCE (Mille 2017). See also GI§2.3.1.

42. Rigid material may be plaster or clay. Flexible materials such as gelatin and then silicon have been extensively used from the nineteenth century onward, although flexible molds are mentioned in connection with small reliefs in the Italian Renaissance (see Motture 2019, 61, with refs, 242nn119–20).

43. For an example of this see Diemer 1996.

44. Mille 2017. It was not possible to characterize more precisely which process(es) were used (wax slabs or painting).

45. Mille 2017; see also Case Study 1.

46. It has been suggested that the *putti* from the Neptune fountain in Piazza della Signoria, Florence, by Giovanni Bologna (called Giambologna, Flemish, active Florence, 1529–1608, fountain completed in 1565) were made this way (Morigi 1990). A similar hypothesis has been made for a Venus, possibly by Hubert Le Sueur (French, ca. 1580–1658), Louvre (inv. MR3278, see Castelle 2016).

47. Verbal communication, Christophe Bery, director, Coubertin Foundry, July 2016.

48. Cellini [1568] 1967, 114–26.

49. Bewer, Bourgarit, and Bassett 2008; Castelle, Bourgarit, and Bewer 2018.

50. Jean Dubos, former director of Fonderie de Coubertin, personal communication to authors, June 2019.

51. Bassett and Bewer 2014.

52. Such a process and material have been used to cast copper-based objects since protohistoric times (Ottaway and Seibel 1998).

53. Casting with "green sand" or powders was used to make medals, plaquettes, small lamps, and so on, but these are still often classed and studied as sculpture by Renaissance art historians. See for example Motture 2019, 55–57, for various recipes and references.

54. Lebon 2012.

55. In the Indian state of Himachal Pradesh they use molasses (Reedy 1987), and one expects other traditional workshops in various geographic regions may add other substances to obtain a better impression of the model.

56. This is by far the most common technique. However, in large commercial art foundries that use resin set sand, it is common to use thin foam sheets to form the metal thickness. The foam sheets are placed or stuck on to the internal surface of the sand mold and back filled with more sand to form the core. Once set with gas, the foam sheets are removed and the mold reassembled as per normal. If the mold is complicated, the foam is burned out as the bronze is poured in.

    This process is clearly illustrated in this two-part video, https://www.youtube.com/watch?v=ZaPeYeEIuv0 and https://www.youtube.com/watch?v=DSw0lHTf7oc.

57. "It occurred to Mr. Eugene Aucaigne to ask the sculptor to make his last touches in the sand mold instead of on the model, in other words to finish with intaglio carving rather than by going over the round, and the result justified his reasoning completely." *New York Times*, March 25, 1904, cited in Boulton 2018.

58. Lebon 2012.

59. Suverkrop 1912.

60. Chase 1991; Strahan 2010.

61. For bells see Neri and Giannichedda 2018 and Thomas and Bourgarit 2014, 56. For cannons see Biringuccio [1540] 1990, 220–21. Loam piece molds such as those used for bells and artillery inspired Leonardo da Vinci's (Italian, 1452–1519) plans for the *Sforza Horse*, although it was never executed (Bernardoni 2009). For connections between cannon casting and sculpture see Day and Allen 2014; Castelle, Bourgarit, and Bewer 2018.

62. It has been speculated that this technique was used for the fabrication of at least several of the bronze statues of the tomb of Maximilian I, Holy Roman Emperor, in Innsbruck, which started in the early sixteenth century (Knitel 1987).

63. Wallack 1840, cited by Lebon 2012. In the 1950s, the Gorham foundry, headquartered in Providence, Rhode Island, set up a variant of piece mold casting to produce bronze statuettes in series using what they advertised as "plaster molding." Here, rather than filling the void left by the lasagna, it is filled directly with the molten metal, much like the process used for casting cannons in the Renaissance.

64. Papet 2001; Smith and Beentjes 2010.

65. See for example the molding of a turtle in Making and Knowing Project et al. 2020, folio 148v.

66. Mattusch 1990; Mattusch 2014.

67. Theophilus [ca. 1122] 1979, 132–38.

68. Craddock 2015; Craddock 2017; Furger 2017, 26–36.

69. It is used for production of series of statuettes in contemporary Nepal (Craddock 2015).

70. Jett 2010.

71. See for example Washizuka, Tomii, and Friello 1997, 60; Strahan 1997, 28, fig. 10.

72. Other materials, such as reeds dipped in wax, have also been used. Benvenuto Cellini described how he adapted clay water piping to use as sprues, of which traces remain on the bust of Bindo Altoviti (1549; Isabella Stewart Gardner Museum, Boston, inv. s26e21)—an astute idea, as it would prevent "scouring," or erosion, of mold along the channels during the pour. Cellini 1956, 344; Bewer and MacNamara 2012, 73.

73. For more on sprueing systems see Rama 1988; Rome and Young 2003.

74. See for example how Pierre-Jean Mariette (1694–1774) criticized Germain Boffrand's (1667–1754) choice for the casting of a monumental equestrian statue of Louis XIV by not having used a bottom-to-top gate system (Desmas 2014, note 39).

75. For a brief synthesis on this see Mechling et al. 2018.

76. This is one main area of research in modern industry (Campbell 2011, 939).

77. Based on traces of attachments and inscriptions on 387 bases of Greek statues from the end of the seventh century to the first century BCE (in Delphi, Delos, Athens), Rachel Nouet (Nouet forthcoming) was able to infer not only attachment techniques, but also materials (bronze versus marble), sizes, and artisans. Warm thanks to the author for sharing this information. For Renaissance bases and socles see for example Bewer 1995, 704; Bewer 1996b, 86–88; and for further references see Motture 2019, 250n168.

Volume I: Evidence of Process Steps

Edited by David Bourgarit and Peta Motture

Volume I is designed to help the reader identify and investigate the physical evidence present on a bronze sculpture. It is divided into nine chapters that correspond to the different aspects of the creation of a bronze, more or less in chronological sequence of the fabrication steps. It goes into more detail on the variety of casting processes described in this general introduction; discusses the properties of the different copper alloys encountered in artistic castings; and characterizes defects, repairs, assembly techniques, tool marks, as well as plating and other surface decoration techniques. Each chapter follows a similar "What? / Why? / How?" format. The first section lays out how distinctive features are produced, and thus sheds light on what to look out for and also possible misinterpretations. That is followed by FAQs that address why such evidence might be interesting and/or useful to investigate. A checklist at the end outlines the primary methods of investigation and analytical techniques that are relevant to the investigation of this topic, and points to the more in-depth discussions of these techniques in volume II. The definitions of terms included in the Vocabulary are hyperlinked upon their first mention in each chapter.

I.1

Casting Processes

Francesca G. Bewer
David J. Reid
Peta Motture
David Bourgarit

Additional Contributors: Denise Allen, Valérie Carpentier, Annabelle Collinet, Pete Dandridge, Nicolau Escanilla Artigas, Susanne Gänsicke, Laurence Garenne-Marot, Arlen Heginbotham, Andrew Lacey, Elsa Lambert, Susan La Niece, Benoît Mille, Chandra L. Reedy, Annick Texier, Jeremy Warren

This chapter serves as a guide to assist in distinguishing the disparate features that indicate the process by which a model was translated into a bronze—the sequence of steps that led up to the pouring of the bronze, or casting process. The chapter focuses on the two most common approaches: lost-wax and sand casting. Life-casting and piece mold casting other than sand casting will also be discussed. For a detailed description of the different processes, see GI§2. Within these broad categories there are a variety of alternative procedures whose traces are explained. These often leave characteristic features discernible via visual examination of the work's external and/or internal surfaces.

However, many of the features that relate to the construction and structure of hollow bronzes, such as diagnostic variations in the thickness of the bronze, or evidence of the various metal inserts such as core supports and armatures, are only discoverable through radiography. In many cases, access to the inside of a bronze may be impeded either by the shape and nature of the cast or by the remains of core material. The description of each process will, therefore, fall into three sections: evidence on the outer surface, evidence on the inner surface, and radiographic evidence. A synthesis of all physical evidence related to the different processes is presented in table 4.

The way in which the metal was designed to flow through the refractory mold is called here the "casting disposition." Though most traces of this step, such as sprues, are removed during fettling, some do survive, and the chapter provides guidance on how to identify them.

Please be warned that the following provides only an indication of which features related to specific casting processes might be encountered on a bronze. It does not offer firm evidence, and it is essential to consult experts in the field for advice where required. For more, please read the overall disclaimer for the *Guidelines* (GI§4).

1 Characteristic features associated with lost-wax casting and direct life-casting

This section covers disparate lost-wax casting processes as well as a few other methods that also use a model that is sacrificed in the course of creating the void in the refractory mold into which the metal will be cast.

1.1 General evidence of lost-wax casting

How is it possible to identify whether a bronze was made using lost-wax casting? Evidence may come from both the working characteristics of the wax itself (used for the model or inter-model) and the particularities of the processes. Intricate shapes, the presence of complex undercuts, or naturalistic textures in a cast may point to the use of lost wax, although this is not always the case. The very nature of wax means that it can be softened or melted by heat and shaped in numerous ways, for example modeling, carving, or casting. Therefore, bronzes that were produced from models made with wax often preserve traces of such manipulation on both the exterior and interior.

Warnings:

- A wax sculpture may not necessarily be translated into bronze using the lost-wax process.[1]
- Intricate shapes and undercuts may also be encountered in sand casting.
- The transference of natural textures (such as fur or scales) could equally have occurred as a result of life-casting.[2] Conversely, not all lost-wax casts display such features.

1.1.1 Evidence on the outer surface

When softened, wax is easily manipulated into a variety of shapes without tools: rolled into balls or coils, pinched into pointed tips, flattened and smeared, or just poured. Such shapes and certain marks—like the parallel lines left by a fingerprint as it pushes wax across the surface—may be clearly captured in soft wax, and later translated into bronze (figs. 48, 49, 50).

Furthermore, any number of tools may be used to shape, incise, stamp, or texture a wax model (see I.6§1.1). In most cases it is hard to identify the material from which the modeling tool was made, but generally:

- Wooden modeling tools are often used to work with softer waxes.
- Metal tools such as punches, pointed modeling tools, engraving tools, and files are more successfully used on hardened wax when crisper details are required (figs. 51, 52).
- Either type of tool can be used to create signatures and edition marks. And founders' marks are traditionally stamped in the wax model before casting (see I.6§1.3).

1.1.2 Evidence on the inner surface (of hollow casts)

Several types of features are characteristic of lost-wax casting, and can be diagnostic of the exact process used (as will be discussed in more detail in I.1§1.3 and I.1§1.4 below; see GI§2.2 for a description of the different lost-wax processes):

- Drip marks and brushstrokes may have been produced during the creation of the wax layer by an indirect lost-wax process such as slush molding (see I.1§1.2, I.1§1.3, and I.1§1.4 below).
- A smooth internal surface often occurs with lost-wax casting and particularly slush molding (fig. 53, I.1§1.2, I.1§1.3, and I.1§1.4). Some sand-cast animalier sculptures have been confused with wax (see I.1.n.21).
- Uneven, thickened lips of metal around the circumference of a limb, sometimes accompanied by a distinct difference in wall thickness at that juncture, are often the result of wax-to-wax joints. These were created by fusing separately formed wax parts at the abutting edges using a hot tool. They are usually invisible from the outer surface, since the wax model is generally refinished to unify the form and conceal such features. The joints may be observable on the inner surface (fig. 54, Case Study 4).
- Soft-edged lips of metal around core pins or core pin holes reproduce the displaced wax formed when the core pin is pushed into a hollow wax model (fig. 55).
- Small spherical globules are often the result of air bubbles trapped on the surface of the model when an investment and core formed as a slurry are used, and are eventually filled by the liquefied metal during casting (figs. 56, 57). Similarly, organic materials from the core may be translated into the bronze (fig. 58). These are characteristic of lost-wax casting using clay-based core.
- Sharp, uneven fins of metal—or the jagged remains of those that have been coarsely cut down—tend to be flashing, which occurs when the liquefied metal enters

cracks that have formed in the mold or core. These are more commonly found in lost-wax casts with clay- or plaster-based cores (figs. 59, 60) as they are more prone to shrinkage due to moisture content. Flashes in sand casts are relatively rarer, and most often form along the seam lines of ill-fitting piece mold sections either in the core (fig. 61) or, in extreme cases where the mold maker is less experienced, on the outer surface (fig. 62).[3]

Risks of misidentification/misinterpretation

- Wax-to-wax joints may be confused with certain types of metal joints, as in both cases there is often an uneven thickening of metal along the joint line (see I.5).

1.1.3 Radiographic evidence

Radiography may be the only way of detecting the aforementioned features if the internal walls of the bronze are not visible. This is particularly the case for the small spherical globules testifying to the use of a plaster core or ceramic shell. And because of their location on the interior, wax-to-wax joints are generally hard to see and often detectable only by these means. They usually generate a thickening and/or thinning of the metal around the circumference of a body part that has been attached, for example below the shoulders, at the top of the thighs, or at the neck. They may appear on the radiograph as:

- a discontinuity in the metal wall thickness (fig. 63);
- different densities due to the resulting difference in thickness of the parts that are joined (fig. 63);
- a more or less straight line (fig. 64), or a circular or elliptical line (fig. 31).

Risks of misidentification/misinterpretation

- Wax-to-wax joints may be confused with the juncture between two separately formed cores (fig. 65 point 8). The core may be built or filled in several steps, thus possibly generating joints visible on radiographs.
- The absence of most or all of these features does not preclude the possibility of a bronze sculpture being a lost-wax cast or a life-cast.

1.2 Evidence of direct versus indirect lost-wax casting

There is a broad distinction between a direct cast, in which the original (wax) model is destroyed in the process of being translated into bronze (that is, the wax is burned out), and an indirect cast, in which a wax replica of the original model—a casting model or inter-model—is created for the purpose of casting (see GI§2.2). That said, it is not uncommon to find combinations of both.

1.2.1 Evidence on the outer surface

In direct casting, the overall shape is modeled from the outer surface (fig. 13) (in the direct wax-slab process, the model may also be worked from inside). This may leave specific marks, particularly when the modeling is done without tools. Among the characteristic features that point to modeling in wax that has been done freehand (without tools), the most easily recognized are:

- rolled threads and rods of different sizes, which are an almost universal visual design form (figs. 12, 20, 66, 67, 68, 69),
- amorphous blobs of softened wax squished or pinched between two fingers (figs. 70, 71).

But note that these are not *definitive* evidence of direct lost wax; see risks of misinterpretation enumerated below.

Conversely, in the indirect lost-wax casting process, the wax model that is intended to be cast is made from a mold. As a consequence, a firm identification of any feature relating to the mold used to create wax replicas allows us to reject the possibility that it is a direct cast. This includes seam lines:

- raised ridges, which may correspond to traces of the piece mold possibly used to create a wax inter-model as part of the indirect process (figs. 72, 73, 74);
- raised ridges, which may correspond to traces of the piece mold possibly used to create a plaster model that was then used in an indirect process (fig. 75).

Risks of misidentification/misinterpretation

- The fact that the original model is fashioned in wax, and thus that the resulting bronze bears all aforementioned features, does not necessarily imply a direct lost-wax casting—or lost-wax casting at all. The waxy features may relate to an earlier model one or more steps removed from the actual casting model. This would be the case, for instance, when a wax model is replicated in plaster, perhaps amended with more wax, and then cast into a bronze chef-modèle, which is in turn used to create a sand cast.[4]

- Wax models may be created by a combination of both indirect and direct processes. The outer surface of the wax inter-model, for instance, may be reworked during finishing by adding or altering details (fig. 68). In cases where there are multiple casts of the same model, this often accounts for variations in the different casts (fig. 76).[5]
- If the original model is made of clay, similar evidence of modeling may be found on the final bronze as on a lost-wax cast. Both clay and modeling wax are soft materials that may be handled in a similar way, and thus an indirect lost-wax cast may be misidentified as directly cast. Sand casting may also be used to translate a clay model into bronze.
- Seam lines from the piece mold used to make the wax may easily be confused with those produced by other processes using piece molds (for example, sand casting and piece mold casting, see I.1§2 below).
- The remnants of flashing or feathering can look very similar to seam or mold lines once filed down (compare fig. 72 and fig. 77).

1.2.2 Evidence on the inner surface (or verso of a relief)

It is not always possible to identify evidence that is characteristic of direct casting on the inner surface of hollow casts. The internal contours of direct casts can vary widely, and can be smooth (see Case Study 2; the direct slab process also yields very smooth internal contours) or roughly shaped. As the wax layer is formed over the core it will capture its shape and textures, including for instance:

- tool and finger marks (fig. 78);
- reinforcing wires that were used to bind the core (fig. 96).

An exception is the direct wax-slab process, or possibly the use of a "false core" (where the sculptor models over a core that is subsequently removed and replaced with another for casting).[6]

Risks of misidentification/misinterpretation

Reinforcing wires to bind the core have been observed for the lasagna technique, which is considered an "indirect" casting process (see I.1§1.4 below)

1.2.3 Radiographic evidence

The greater or lesser evenness in wall thickness of a bronze reflects the level of conformity between the shape of the core and the final shape of the sculpture. Some sculptors (assuming that they work on their own) will produce a core with a high level of detail and use some form of markers, such as chaplets, to help them control the overall thickness of the wax layer. Using preformed wax slabs of the same thickness will also ensure a relatively even layer of wax (fig. 28). But complicated features such as hair or smaller decorative elements, or any changes in the design that were not foreseen in the core, may result in thicker or thinner areas in the wax layer (and consequently in the bronze). Such larger discrepancies are often associated with a direct technique, especially with one-off casts (as opposed to multiples).

An armature reflecting the overall shape of the sculpture is more likely to be found in a direct cast, especially larger ones (figs. 13, 79), as it constitutes the structural "skeleton" of the sculpture over which the model is designed. But armatures might also be found in an indirect cast, inserted into the hollow wax inter-model before it is filled with core material.

A synthesis of all physical evidence related to the direct lost-wax process is presented in table 4.

Risks of misidentification/misinterpretation

- An indirect process such as the lasagna technique may lead to uneven thicknesses as well (see I.1§1.4 below).
- With slush molding there is likewise no assurance of an overall even thickness, as molten wax may pool and solidify more readily in some areas, such as the extremities of figures (fig. 80).

1.3 Characteristic features of slush molding, wax painting, and indirect wax-slab processes

These types of indirect lost-wax processes share the creation of a core *after* the wax model is made (figs. 16, 18, see GI§2.3 for a detailed description).

1.3.1 Evidence on the outer surface

The reusable mold used to form the wax inter-model may leave distinctive traces on the outer surface of the bronze. Gelatin molds, rubber molds, and plaster piece molds may all produce seam lines on the wax. Usually these features are removed during the refinishing of the wax inter-model, but not always fully. These may, however, not be considered singular to these processes (see I.1§1.2 above). See also I.1§1.1 above for the evidence common to all lost-wax processes. Securely identified examples of the indirect wax-slab process are rare. The way in which the metal curves in on the back of a relief could be the result of the bending of existing wax slabs around the core (figs. 81, 82).

1.3.2 Evidence on the internal surface

The following features can point to any of the three processes:

- traces of working of the wax such as tool marks (figs. 83, 84), like those left by the hot spatulas that were used to melt together separate wax sections at the joints;
- evidence of localized masses of soft wax pressed into areas that needed thickening, and related fingerprints (fig. 50);
- evidence of core pins, whether inserted from inside[7] or outside, or by waxy lips, molten drips, and/or metal excrescences when core pins were pushed into the hollow wax before the core was filled in (fig. 85);
- internal sprues (fig. 86).

The following features are characteristic of slush molding:

- a smooth, flowing surface on the interior of the cast that is conformal with the outer contour (that is, the inner surface echoes that of the outer), while rounding out any sharp transitions (figs. 53, 87);
- tide lines and/or drips left by the wax as it was slushed around and poured out of the mold (fig. 88).

Brushstrokes are usually evidence of the molten wax being painted into the mold (figs. 21, 22, 55, 89).

Risks of misidentification/misinterpretation

- Drips may stem from heated pins (see above), from wax slabs being cut by a hot tool (fig. 60), or (rarely) from thickening the edges of a hollow direct cast if it is possible to gain access to the interior, such as with the use of a false core (see I.1§1.2 above).
- It is also possible for brush marks to be formed on the core and subsequently picked up by the wax.
- Conformal casts may also be found in direct processes.

1.3.3 Radiographic evidence

The thinness of the wall and the close correspondence of the inner and outer contours are the most characteristic features of expert slush molding and of indirect wax slabs. Radiography is the ideal tool to reveal this. It can even help distinguish between the two processes because:

- Slush molding leads to the rounding out of transitions (fig. 80).
- Wax slabs of even thickness may ensure the most even thicknesses on a bronze (figs. 90, 91).
- The overlap of wax slabs, typical of the indirect wax-slab process, might be revealed by radiography as a greater density due to thickening in the overlapping area (fig. 92).
- Tide lines and drips produced during slush molding may be revealed as well (fig. 88)

Risks of misidentification/misinterpretation

- Even and thin metal walls may be achieved by the direct wax-slab process.
- Drips may stem from heated pins, from wax slabs being cut by a hot tool (fig. 60), or (rarely) from thickening the edges of a hollow direct cast if it is possible to gain access to the interior, such as with the use of a false core (see I.1§1.2 above).
- Rounding out of transitions in the core is also found in direct lost-wax casts and bronzes formed with the lasagna technique.

Table 4 synthesizes all features that might be identified on a bronze related to these processes.

1.4 Characteristic features of the cut-back core and lasagna processes

Unlike with slush molding and the use of indirect wax slabs, the core in both of these processes is produced before the wax model. In fact, the core serves to define the inner boundaries of the molten wax that is poured into the reusable outer mold, where it will fill the space created either by paring down the core or by the removal of the lasagna (figs. 23, 25).

Please be advised that these guidelines do not provide an exhaustive list or description of the variety of indirect processes and combination of processes that may have been used, but rather offer a firm basis of more common, documented examples. Table 4 synthesizes all features that might be identified on a bronze related to these processes.

1.4.1 Evidence on the external surface

There is no specific evidence. See I.1§1.1 above for the evidence common to all lost-wax processes.

1.4.2 Evidence on the internal surface

The following evidence is common to both processes: the internal surface of the wax model will capture the shape and tool marks, or fingerprints left on the core.

The following evidence is specific to the *cut-back-core process*:

- The pared-down core often displays more angular transitions around the contours of the shapes (fig. 93).
- Any tool marks left from the paring-down process may be captured in the metal surface (fig. 94, see I.6§1.1.3).

The following evidence is specific to the *lasagna process*: the core is not pared down, but built up to fit the mold, thus possibly generating different features:

- The transitions around the contours of the shapes are very simplified, if not absent, instead of being angular (figs. 24, 95).
- Specific marks due to the working of the lasagna from inside may be captured, such as the joints of clay sheets (figs. 24, 95).

Risks of misidentification/misinterpretation

- The method of producing a cut-back core closely resembles that used in sand casting and can therefore lead to possible confusion between the two.
- The less-defined contours of the lasagna layer resemble those generated by slush molding or possibly of direct lost-wax casting (see Case Study 5).
- If wires were used to bind the core they may have become embedded in thc inner surface of the bronze (fig. 96). Such wires are evidence that the core was formed directly, before the wax, and reinforced in this way. Although our sole known written mention of the procedure is in Benvenuto Cellini's (Italian, 1500–1571) description of the lasagna technique (see Case Study 5), this procedure is not necessarily exclusive to the lasagna process.

1.4.3 Radiographic evidence

Both the lasagna and cut-back-core processes share some common radiographic features:

- Inner and outer surfaces of the bronze are not always very conformal (that is, they do not closely follow the contours of the exterior), and the thickness of the walls, especially in areas of fine raised detail, tends to be uneven (figs. 80, 97).
- Wires or tethers wrapped around the core for reinforcement before the wax is added and subsequently embedded in the inner surface of the metal will produce patterns of a slightly different density in a radiograph (fig. 34).

The differences between the two processes as evident through observation of the inner surface (see I.1§1.4.2 above) may also be seen on radiographs.

Risks of misidentification/misinterpretation

- Non-conformality of the interior and exterior surfaces, and thick and uneven metal walls, are similar to those associated with direct casting (see Case Study 5).
- An even thickening of the metal along a straight section may be evidence both of overlapping wax slabs and of a wax-to-wax joint.
- As stated above (I.1§1.4.2), wrapping the core with wires is not necessarily exclusive to either the lasagna or the cut-back-core processes.

1.5 Characteristic features of the life-casting process

This process shares common features with lost-wax casting, as it entails the use of organic materials (for instance animal and vegetable forms or textiles) that can be fully or partially burned out of the mold (fig. 11, see GI§2.5.1).

1.5.1 Evidence on the external surface

On the external surface, this process:

- produces a 1:1 scale of replication (although note the potential for shrinkage);
- reproduces the texture of the original in every detail—something that would be a challenge through the use of modeling (figs. 27, 98, 99, 100);[8]

- may result in seam lines. Natural objects that do not burn out so readily like crab or crayfish are often piece molded directly in the refractory material, which allows removal of the hard, carbonized remains. Therefore, seam lines may appear on the bronze (fig. 27).[9]

Risks of misidentification/misinterpretation

Animals and plant forms can also be reproduced in bronze by an indirect process, which can look very similar to these life-casts. In this case, reusable molds are taken from such models in order to produce wax inter-models that are often hollow and can be altered further at the wax stage.[10] It can be difficult to distinguish between realistic casts that were molded directly versus indirectly on an animal or plant. For example, both may have seam lines.[11]

1.5.2 Radiographic evidence

A hollow cast may reveal drip marks or the use of core pins, both of which will suggest that a wax inter-model was used and therefore point to a cast from life.

Risks of misidentification/misinterpretation

A direct life-cast can be hollow, albeit rarely (see GI§2.5.1).

2 Characteristic features associated with sand casting and piece mold casting

Both of these processes consist of making the refractory mold in discrete pieces and removing the model.

2.1 Characteristic features of sand casting

In sand casting, a special sand is pressed onto the pattern within a sturdy frame structure to create removable sections (fig. 9, see GI§2.4.1). Simple shapes (for example medals and small, low reliefs such as Renaissance plaquettes) may only require a two-piece mold. Complex models will necessitate more complex molds.

2.1.1 Evidence on the outer surface

- Relative lack of undercuts may point to a simplification of the model (or filling of the undercuts in the model) to create a piece mold.
- A distinctive, even, granular roughness on unfinished areas on the surface (for example in hard-to-reach places) may represent the texture of a coarse, sand-molded surface before chasing (fig. 101). Most first layers of sand on the outer layer are quite fine; such textures are more common for the core.
- The location of divisions between the separately cast parts (the casting plan) will be different on a sand cast than on bronzes cast using other processes—something that only a founder can really explain (figs. 102, 103).
- A thin, continuous, raised linear feature may correspond to a mold seam line formed at the joint lines of piece molded sections (see Case Study 3).
- An offset, stepped, linear feature that either follows an obvious piece mold joint line, or disrupts or cuts across the modeling of an area of a cast, could be evidence of a misaligned piece mold.

Risks of misidentification/misinterpretation

- None of the above features are specific to sand casting. In particular, complex undercuts can be achieved in sand casting (figs. 102, 103), and simplification of undercuts may also appear in indirectly cast bronzes.
- Surface roughness may be due instead to fine pitting caused by corrosion, to acid cleaning, or to sandblasting typically used to remove investment in the ceramic shell process.
- As-cast surface roughness due to sand molds and clay molds (for example, the lost-wax process, figs. 104, 105) may be hard to distinguish.
- A relative lack of undercuts may reflect the artist's composition rather than the process.
- As mentioned above for indirect lost-wax casting, the remnants of flashing or feathering can look very similar to seam lines once filed down (see I.1§1.2 above).
- Seam lines could equally testify to the use of piece mold casting (see I.1§2.2 below). More generally, the possible variety of molds used during the whole making process—mold used for creating the original model, wax mold, refractory mold—may generate a variety of seam lines that are easily confused.
- Identifying whether or not the casting plan used for a specific bronze is indicative of the sand-casting process requires specialized skills, typically those of a founder.[12]

2.1.2 Evidence on the inner surface

The only feature that may be specific to sand casting is the grainy texture on the inner surface of a hollow cast or back side of a relief (figs. 101, 106, 107). The following features could also be observed inside or at the back of a sand cast, but are not specific to sand casting:

- evidence of particular ways of planning the assemblage, often only visible from the inside if the finishing work used to conceal such joints was artfully done (fig. 108);
- raised inscriptions produced by carving or impressing into the core (fig. 109);
- geometric and/or cut-back shapes of the inner surface resulting from the characteristic paring down of the core (figs. 53, 107);
- less core flashing compared to lost-wax casting (see I.1§1.1 above);
- seam lines or flashing corresponding to joints between separate core sections (figs. 43, 61).

Risks of misidentification/misinterpretation

- The evidence of the core having been pared down may also pertain to the cut-back core technique.
- The limited amount of core flashing is not necessarily indicative of the process.
- Although sand casting is the most common reason for a coarse, grainy texture, a coarse clay or plaster-based core can produce a similar appearance in the bronze.

2.1.3 Radiographic evidence

Much of the aforementioned evidence will be visible on radiographs (for example metal joints, shape of core, core flashing).

2.2 Piece mold casting

Here, the mold is formed over the model in separate sections made directly of refractory investment material (for instance kaolin or loam). The piece molds thus obtained may then be further reworked from the inside to create or enhance the decoration, before being reassembled over the pared-down core (fig. 26, Case Study 3).

2.2.1 Evidence on the outer surface

- Seam lines develop along mold joints (see figs. 110, 111, 112, 113, 114, 115, 116, 117 in Case Study 3).
- The form of the surface details can reflect features carved directly into the mold.

2.2.2 Evidence on the inner surface

This may include:

- marks indicating that the core has been shaved down (fig. 118);
- raised inscriptions on the inside that represent carving or impressing into the core.

2.2.3 Radiographic evidence

This may include:

- walls of uneven thickness due to a more cursory shaving down of the core;
- seam lines;
- signs of chaplets or mold extensions (also known as core extensions).

Risks of misidentification/misinterpretation

All of the aforementioned evidence may be encountered in other processes. The combination of a large number of these clues, however, may be indicative of this technique.

3 Characterization of features related to the armatures and core supports

Armatures, core pins, chaplets, and other core supports are generally made of metal, and this section aims to provide guidance for identifying their characteristics. Their functions are fully described in GI§2.6, together with the reasons for investigating them. The features that provide a focus for documenting these various features are their shape, size, and placement, as well as the material from which they are made.[13]

3.1 Armatures

- Armature rods may project from the base or bottom of the sculpture and can be used to mount the work.

- The exposed end of an iron armature rod that has been cut down flush to the surface may remain visible in the bronze surface and can be recognized by the buildup of rust, unless covered with patches (fig. 119) or hidden by an opaque coating.
- Historically, many armatures were made of iron rods, in which case it may be possible to locate them using a magnet.
- The recurrence of rods with similar dimensions and profiles among related sculptures may signal possible patterns in production.

Risks of misidentification/misinterpretation

- The absence of an armature does not mean that there was none; armatures were often removed or cut off flush with the surface. Also, some armatures may have been made of organic materials such as wood and burned out or removed sometime after the sculpture was cast into metal (either before assembling separately cast parts, during fettling, or at some later stage). See GI§2.6.
- Several radiographic views may be necessary to determine the shape and location of an armature, but may not be sufficient to discern other characteristics such as the profile of the rods or wires or the exact materials from which they were made (fig. 34).
- Radiography may not always fully reveal armatures, since iron (and in rare cases wood) is less X-ray opaque than copper. Rusted iron is even less visible, and when rusted into laminated form may look like a sheaf of thinner rods. Access to the internal surface would allow for a visual assessment of such components.
- Although there is a tendency to look for patterns and consistency in the materials used within a workshop, there are likely to be variations. Explaining these remains a matter of speculation. For instance, it is plausible that a founder made use of whatever rods and wires were available in the workshop.[14] Also, the complexity of the armature is inevitably linked to the form of the sculpture (compare for example fig. 31 and fig. 34).
- Relatively high amounts of iron in a copper alloy may be the cause of magnetic attraction as well, and so may confound identification of the presence of an iron armature using a magnet.[15]

3.2 Core supports (metal and other)

Below are clues that help characterize the categories of metal features that support the core, the so-called core supports (core pins, chaplets), as well as mold extensions, all of which are defined in the general introduction and Vocabulary. Armatures may also serve as core supports.

Evidence of core pins includes:

- magnetic attraction, which if ferrous and not fully rusted will suggest the presence of an iron core support or core pin on the interior;
- ends of extant wires, rods, or nails visible on the inside—a nail head on the inside means there was access to the interior to insert the nail, something occasionally found on larger sculptures (fig. 36);
- linear features in radiographs that stand out from the forms of the cast bronze (figs. 37, 120);
- paired holes in a radiograph, which might indicate the use of transverse core pins that pierce through the model from one side to the other (fig. 64);
- round or square holes that pierce through the external surface and that have been left unpatched (or may be filled with organic filler such as wax or resin) (fig. 121);
- a logical pattern of plugs or patches of similar shape and size distributed over the entire bronze that would cover core pin holes (figs. 35, 64, see I.4§2.2);
- wire acting as an internal core support that may be embedded in the core, and is usually only detectable by radiography (fig. 32).[16]

Chaplets, or "spacers," are larger than core pins and are often made of the same copper alloy as the cast (figs. 38, 39, 122). They do not always bond with the cast metal and for that reason the outlines may be visible. Access to the internal surface or radiography is often key to identifying them.[17]

Mold and core extensions usually leave large regular openings in the metal wall, and are often repaired (figs. 28, 123).

Risks of misidentification/misinterpretation

- It may not be easy to distinguish patched or plugged core-pin holes from other repairs (see I.4§2.2).
- Core pins and chaplets are often well hidden by chasing and patination.

- Core pins, chaplets, and internal core supports may be difficult to distinguish from armatures when they break through the surface of a bronze, as they are often made of materials of similar density, and so can be confused in radiographs.
- Magnetic attraction may be due to an iron-rich copper alloy rather than indicating the presence of an iron armature (see I.1§3.1 above).

4 Sprueing and casting orientation

4.1 Sprueing system

It is virtually impossible to determine how the metal was fed through the mold (for instance in the case of a lost-wax cast), despite close study of a finished cast. In most finished bronzes, sprues tend to be cut off and the surface chased, making reconstruction of the sprueing disposition very difficult. On rare occasions the entire network, including the casting cup, is preserved. This can range from designing the sculpture itself in such a way as to take into account the distribution of the metal (fig. 71) to preserving the full network of sprues for demonstration purposes (fig. 46). In most cases only occasional traces survive. These can take various forms:

- The occasional sprue may be preserved in its entirety, either on the outside of the bronze (fig. 124) or incorporated as an integral feature of the sculpture. They may also survive intact on the inside of a bronze (fig. 86).
- More often, remnants of sprues survive only as small protuberances when they were cut off and only crudely leveled with the surrounding surface. Such remnants are found more readily along the bottom edges or inner or rear surfaces where removal was not as carefully executed.
- More rarely, shrinkage, porosity, or differentiated corrosion due to microstructure differences may mark where the sprues or gates were located. Cooling in those areas is slower because there is a greater volume of metal, thus impacting the microstructure (see I.2§3).
- Core vents, also known as lanterns (figs. 6, 43), may be visible by radiography (see GI§2.6.3, Case Study 6).

Risks of misidentification/misinterpretation

- Some sculptural features and/or decorative elements may be confused with remnants of a sprue.
- Differential corrosion may stem from a variety of reasons other than the aforementioned microstructural differences in sprue areas (fig. 125).

4.2 Evidence of the orientation of a cast

It is a challenge to determine whether a sculpture was cast upside down, vertically, or horizontally, and also whether the metal was poured directly into the sculpture or fed in in a more controlled fashion by directing the metal to the bottom of the mold first (that is, via an "indirect feed").

The following features may help to determine some of this, and are all considered casting defects:

- Porosity tends to form in the regions of the cast that are higher during casting, since bubbles rise. Any concentration of porosity may, therefore, be a good indication of the direction of the casting disposition, as mainly seen using visual and/or radiographic examination (figs. 126, 127).
- Similarly, wash from the refractory mold—mainly from sand molds—will rise to the surface of the cope. The resulting floating debris would cause easily recognized casting defects.[18]
- Potential discrepancies in the lead content in the case of leaded alloys should see the heavier lead sinking to the bottom of the cast. In reality, it is nearly impossible to ascertain this through analysis.[19]

Risks of misidentification/misinterpretation

Surface cavities may originate from a variety of reasons, not just floating debris.

5 Why investigate casting processes? and other FAQs

5.1 Can I tell whether a bronze is cast by the lost wax or another method, and can I precisely figure out which variation has been used?

This is a big question. It is possible, but depends on many factors. See all the indications above, and bear in mind

- the complexity of the process, which may be the result of a combination of several variations (for example indirect and direct; see fig. 68);[20]
- that different processes may produce similar features (see all "risks of misidentification/misinterpretation" enumerated above, and Case Study 5).[21]

5.2 Can I determine more clearly how different casts relate (or not) to the same model?

Casts of the same model may be any number of things: parts of an edition, replicas, after-casts, et cetera (see GI§1.3). But a few clues can point to the relationship between two casts:

- A reproduction of the topography of a repair on the surface of the object that served as model (but absence of the actual feature), or misinterpretation of a feature from the object that served as a model, would point to a bronze being an after-cast.[22]
- Comparing measurements may reveal a discrepancy in size, and if one of the casts is found to be overall consistently smaller, this will suggest that it is an after-cast (see II.4).
- Traces of the mold-making process such as scratches, remains of the molding material, or more rarely drawings or marks may remain on the original statue, indicating that an after-cast has been taken (figs. 128, 129, see I.6§1.4.2).

While this evidence can point to the fact that other versions are likely to exist, it will not necessarily identify the bronze after which it was cast. However, replication of distinctive marks can point to a specific bronze original, such as lacunae filled during the after-casting process (for example suspension holes in medals).[23]

5.3 Can I determine what the original model was made of?

This requires both traditional art historical research and close observation. Given the variety of intermediate models between the original and the finished bronze, unless the model is known, it is often impossible to be sure. However:

- With direct lost-wax models, the answer is clearer (wax).
- Life-casts preserve the intricate texture of the model.
- Tool marks may be diagnostic (figs. 101, 130, see I.6§1.1.1).

5.4 Can different artists, founders, or workshops be distinguished based on specific casting processes?

To make such an assessment, art historical (or archaeological) and technical data must be available in connection with a significant enough group of securely attributed, provenanced (this means known excavation context in the case of archaeological materials), and/or dated comparable bronzes. The specificity of a workshop's practices will be reflected in a combination of parameters in addition to the casting process (metal and core composition, chasing, patina, et cetera).

In cases where enough data is available, the answers can vary. For instance:

- Certain idiosyncrasies seen in the interior and exterior surfaces of bronzes (as in subject matter and style) have allowed scholars to identify the work of particular antique foundries.[24] The technical evidence from some of Barthélemy Prieur's (French, 1536–1611) large bronzes tends to indicate that these were produced in the same workshop (see Case Study 5).[25]
- Conversely, the greater standardization of processes that is not specific to one particular workshop, such as that found in late sixteenth- to seventeenth-century Venice, may render a precise attribution more difficult.[26]

Remember as well that for various reasons, artists and/or founders experiment. This seems to have been the case, for instance, with sculptors who were closely involved in the entire process, such as Jean-Antoine Houdon (French, 1741–1828).[27] See also Case Study 7.

5.5 What can be learned from understanding the original casting orientation?

This can help us better understand some of the defects and challenges faced by the founder. Certain orientations may be characteristic of a culture, region, or period. For instance, we have little evidence of horizontal casting so far in the West, but it is still current in South and Southeast Asia (see Case Study 4).[28]

6 Checklist: How do we investigate the casting technique?

The most important and readily available way of gathering information is close visual examination. This primary and necessary step in the investigation may be assisted by the following (see table 5 for a synthesis of available techniques):

- Good lighting shone at different angles over the surfaces can reveal the topography of tool marks and other features (see II.2§1.2 for guidance).
- Magnification through portable loupes or binocular microscopy can help characterize details in modeling, defects, and flaws related to the forming of the model and of the mold.
- Small magnets can help locate iron armatures or core pins that have not been removed but may not be readily visible on the surface (but be warned that some copper alloys may be magnetic; see I.1§3.1 above).
- A borescope (or endoscope) can afford visual access to internal recesses of hollow casts that are otherwise impossible to reach so as to reveal information about core pins, armature, and the forming of the core that mirrors that of the inner surface of the bronze (see II.2§2.2).
- Radiography, as seen throughout this chapter, is a source of invaluable information, especially when the interior is not readily accessible (see II.3).
- 3D scanning may be useful for comparison of measurements with other pieces connected to the same model, as it might help to establish the genealogical relationship between them (see II.2§5, II.4§2.2.2).
- A number of other techniques, including ultrasonic testing and thermography, may be of use to track surface features related to several aspects of the casting process, for instance holes left by armatures and/or core pins (see II.2§3.2 and II.4§2.3.3).

Notes

1. Pingeot 2002; Lebon 2019.
2. These include animal skin captured during indirect casting (see Smith and Beentjes 2010) or waxed cloth as seen on Donatello's *Judith and Holofernes* (see Stone 2001).
3. Andrew Lacey, personal communication; see also Heginbotham 2014.
4. This is notably the case for a number of Barye bronzes (Wasserman 1975; Pingeot 2002; Lebon 2019).
5. See also the sculptor Antico's (Italian, ca. 1455–1528) 1519 reworking of casts for Isabella d'Este, made at the Gonzaga court in Mantua, from the same model and/or molds (Motture 2019, 164–67, with earlier refs).
6. See for example the sculptor-founder Andrew Lacey's (British, b. 1969) suggestion in relation to Andrea Riccio's (Italian, 1470–1532) *Shouting Horseman* made in ca. 1510–15 (V&A, A.88-1910), cited in Bresc-Bautier and Scherf 2009, 79n50. Also Bewer, Stone, and Sturman 2007 regarding the suggested process that Lorenzo Ghiberti (Italian, 1378–1455) used to create the large relief panels for the *Gates of Paradise* of the Bapistery in Florence (1425–52).
7. See for example Donatello's (Italian, ca. 1385/86–1466) *Judith and Holofernes*, bronze with traces of gilding, ca. 1455–60, at the Palazzo Vecchio, Florence. See Stone 2001.
8. See also Donatello's *Judith and Holofernes* (see note 7), where the textile has been added to the wax model (Stone 2001).
9. Andrew Lacey, personal communication, June 2019, based on his experience with Pamela H. Smith on the Making and Knowing Project et al. 2020.
10. Examples include the numerous animal-shaped inkwells often linked to north Italian Renaissance production. See also Thomas Eakins's (American, 1844–1916) collection of anatomical casts from dissection experiments now in the Philadelphia Museum of Art.
11. See note 9, above.
12. Andrew Lacey, personal communication with the authors, May 2019.
13. See II.4§1.4 and II.5, although the latter chapter deals only with copper-alloy analysis. For iron analysis, refer to Dillmann and L'Héritier 2007 as well as II.8§2.2.2 for dating of iron armatures.
14. This is probably what happened for the four *Virtues* on the funerary monument of Henry II and Catherine de Médicis, Basilica of Saint-Denis, Saint-Denis, monument erected in 1567. In three of the Virtues, the main vertical armature has a round profile; on the fourth, the profile is hexagonal (Castelle 2016).
15. 0.5wt% of iron in a copper or copper alloy is enough to stimulate a rare earth magnet, as notably witnessed on the West Mebon Vishnu, made in the Khmer Kingdom during the Angkorian period (CAST:ING 2018).
16. These are referred to as leashes in Stone 2006. For examples in X-rays of Venetian bronzes, see Motture 2003b, 295, fig. 21; Motture 2019, 46.
17. For example, the specific chaplets evidenced in Barthélemy Prieur's (French, 1536–1611) cast discussed in Case Study 5 could only be revealed by endoscopy. For radiographs of chaplets (called spacers in this context) and core extensions in Chinese bronze sculpture see Strahan 2010.
18. Some Indian bronzes were cast face-down to ensure a perfect front; backs are messy and unfinished (see Schorsch, Becker, and Caro 2019). This is still the case, for instance, in sand casting, with the important surface facing down into the drag (David Reid, personal communication, 2019).
19. There is little evidence to support this as an invariable phenomenon. For more, see for example Motture 2019, 22 and 239n80.
20. For example, direct lost-wax elements have even been found incorporated in traditional section-mold casts from the Eastern Zhou Spring and Autumn Period (770–476 BCE) in China, at a transitional moment when lost-wax casting was slowly beginning to be adopted (Strahan 2019).
21. It can be difficult to distinguish between lost-wax and sand casts. For example, a cast by Antoine-Louis Barye (French, 1796–1875) (see note 9) in the Victoria and Albert Museum, London (S.EX. 65-1882), appears waxy, with a potential wax-to-wax joint underneath. It appears instead to be sand cast, judging from the flashing that points to the use of mold pieces (notes made during a Barye Study Day held at the V&A, November 14–15,

2002; it was also noted that this type of cast was unusual for Barbedienne). The distinction between piece molding and lost-wax casting in ancient China is still a matter of debate among bronze specialists (Notis and Wang 2017).

22. For example, in later versions of Barthélemy Prieur's *Gentleman* (or *Young Man*) *Holding a Pair of Gloves* (probably late nineteenth or early twentieth century), the gloves have been misunderstood and reproduced as an ill-defined block (as seen in the example in the Walters Art Gallery, Baltimore); see Seelig-Teuwen, Bourgarit, and Bewer 2014; Motture 2019, 206 and 252n78.

23. Numerous examples exist, but see for example a portrait medal of Federico Zuccaro, the original made in 1578 by Pastorino de' Pastorini (Italian, 1508–1592), in the collection of the Victoria and Albert Museum, London (V&A A.44-1978), which is most likely an enhanced after-cast: http://collections.vam.ac.uk/item/O312298/federico-zuccaro-medal-de-pastorini-pastorino/.

24. For instance Mattusch 2009.

25. See also the work that has been done on Renaissance bronze workshops, for instance Antico (Stone 1981; Smith and Sturman 2011); Giambologna (Flemish, active Florence, 1529–1608) and workshop (Bewer 1996b); Severo da Ravenna (Italian, active 1496–before 1538) (Stone 2006; Smith 2008b; Smith 2013); Andrea Riccio (Italian, 1470–1532) (Stone 2008; Sturman et al. 2009); and Adriaen de Vries (Netherlandish, 1556–1626) (Bewer 2001; Bassett 2008).

26. Motture 2003b.

27. Houdon's production in the final quarter of the eighteenth century in Paris is discussed in Bassett and Scherf 2014.

28. Mechling et al. 2018. Leonardo da Vinci's (Italian, 1452–1519) proposals for the monumental Sforza horse included a suggestion for casting horizontally; see Bernardoni 2009, esp. 119–24, incl. fig. 81. Note that in modern art foundries, an edition may be cast after months or years of experimentation by the founder in search of the ideal sprue system, and the casting orientation may change over this time (Andrew Lacey, personal communication, May 2019). See also Case Study 7§5.

I.2

Metals

David Bourgarit

Additional Contributors: Aurélia Azéma, Ann Boulton, Joachim Kreutner, Andrew Lacey, Kenneth Lapatin, Elisabeth Lebon, Carol Mattusch, Benoît Mille, David J. Reid, David Scott, Donna Strahan, Jean-Marie Welter

This chapter overviews the composition and properties of the copper alloys used to make cast sculptures and sculptural objects, and presents the reasons why it might be helpful to identify them as part of the technological investigation of a bronze sculpture. The text aims to explain and unravel common approaches to the varied nomenclature applied to such alloys, and introduces two main parameters by which metals can be characterized: their chemical elemental composition and their microstructure. These two parameters control an alloy's chemical properties (for example its corrosion resistance) and physical properties (for example its color and strength). This chapter will also outline the reasons why and how elemental analysis and metallography, among other analytical techniques, can contribute to a technological study of bronze. The more complex discussions will be most relevant to those who are directly involved in the technical investigation of bronzes. Recommendations for determining metal composition are set out in II.5.

1 What is bronze?

The term "bronze" has various meanings. It may designate a copper alloy that has tin as the primary added element, or any other copper-based alloy. In fact, in common parlance, "bronze" is used to describe most copper alloy sculptures regardless of the actual elemental composition of the alloy. Either inherently or by design, bronze sculptures generally include a number of other elements, but we recommend reserving the term specifically for copper-tin alloys, as will be explained.

1.1 Why create an alloy?

Adding different metallic elements to copper makes it possible to control the resulting alloy's properties (for instance castability, strength, resistance to corrosion, workability) as desired by the sculptor or founder (see table 6, I.2§2 below, and Case Study 7§9). Cultural influences, the cost of raw materials, the desired color, and/or the need for the cast to take gilding may also play a role in the choice of alloy. Historically, except for brass, copper alloys were produced by combining the different metals in a crucible or furnace.[1]

1.2 Elemental composition of metals

Elemental composition should be distinguished from both structural composition (see II.6§2.1) and isotopic composition (see II.5§5). Most metals occur in nature as ores in a mineral form, and are extracted from the earth. Even when refined, the extracted metal is rarely 100 percent pure; other elements will be present in small amounts as impurities, described as the "impurity pattern." The elements are set out in the periodic table, where each is given a symbol (table 7, fig. 131)—for example, Cu for copper, Sn for tin—so that scientists can quickly discern its properties. Analysis of any historic or contemporary bronze sculpture will likely reveal a variety of elements even within a simple alloy. In addition to the impurities or trace elements present within each metal, other elements may have been deliberately added for practical or aesthetic reasons.[2] There is also the common practice—even today—of adding metal that has been previously used, for instance the sprues and casting cups cut off of previous casts.

1.3 Alloy nomenclature

1.3.1 Different users, different nomenclatures

"Bronze" has become a generic term to describe sculpture that may in reality have been made from a large variety of copper alloys, from non-alloyed copper to high-zinc brasses as well as ternary and quaternary alloys (those made up of primarily three or four elements, respectively). The usage and understanding of alloy nomenclature varies considerably across historical sources, as used by craftspeople and modern industry and/or by those studying bronze sculpture.[3] The complexity and lack of uniformity of terminology in modern industry alone exemplifies the situation.[4] Technical studies of large groups of related copper-based items tend to show that alloys and their uses evolve greatly over time and place, and thus our nomenclature for them varies accordingly.[5]

The discrepancies are partly related to varying viewpoints on the percentage at which an element present in the metal (lead, tin, zinc, arsenic, et cetera) should be considered an alloying element (intentionally added or kept in the metal) or an impurity (unintentional). The still-loose understanding of the effect of the different elements on the properties of alloys is partly responsible for the lack of clarity.[6]

1.3.2 Pragmatism should be the rule

The present *Guidelines* do not aim to impose any norm and cannot address all issues, but the following are some parameters:

- "Bronze" is used extensively in this volume, and continues to be the preferred term for many artistic objects or sculptures cast from copper alloys regardless of their actual composition, even if they are known to be technically different (for example brass) or more nuanced (for example leaded bronze, quaternary alloy).
- When writing a report or publication it is important to detail the results of analysis (if any) and identify which alloy definition (if any) will be used, either referring to existing formulas or defining new ones.[7]
- The alloy nomenclature compiled by scholars for the present *Guidelines* is intended as a useful starting point (see table 8).
- The terms "major," "minor," and "trace" in reference to elements do not convey any judgment regarding whether an element is present intentionally or not, unlike the terms "alloying elements" and "impurities." As such, these terms might be a good compromise. Symbolic thresholds are often put forward in both industry and cultural heritage fields to define these three terms:
    - 1wt% (1 percent of the total weight) is considered a major element, however, a copper containing 1wt% lead is not necessarily leaded copper (see table 8).
    - Under 1% means the lead (or any other element) is considered a minor element.
    - Under 0.01% (100 ppm), the lead is considered a trace element.[8]

1.4 Chemical symbols and formulas

Scientists generally use the chemical formula to describe the known alloy content (for example, after analysis). Using a chemical formula to describe an alloy could avoid misunderstandings, although even then, various nomenclatures coexist, and nonscientists may not be familiar with chemical symbols (table 7). For a generic designation, Cu-Sn can be used for copper-tin alloys, Cu-Zn for brasses, Cu-Sn-Zn for ternary alloys, and so on. For semi-quantitative compositions, the ECS (European Committee for Standardization) and ISO (International

Standardization Organization) norms are often used, for example CuZn7Sn3Pb2 when the number refers to the rounded %wt. Note that only whole numbers are reported, which means that any chemical element in an amount lower than 0.5wt% is not represented.

2 What are the main physical and chemical properties of copper alloys?

Leaving aside the impact that cultural influences and availability of raw materials may have on the choice of alloy, key considerations will be its melting temperature, castability, hardness, weldability, and suitability for patination and/or metal plating. Other properties that may be factored in are linked to the function and life cycle of the statue, such as its color, luster, strength, and symbolic properties.[9] These properties may vary greatly across different copper alloys.[10] While the influence of the composition on copper alloy properties is far from being fully characterized, especially for historical alloys, it is worth attempting to describe the role of the alloying elements in a qualitative rather than quantitative manner in the following sections and in table 6.[11]

2.1 Melting temperature

The melting temperature of copper alloys varies from 1084°C (pure copper) to approximately 1000°C for a copper bearing 10wt% tin (fig. 132) or 20wt% zinc, and down to 955°C for a highly leaded copper (36% lead). Melting temperatures as low as 900°C can be reached, for example, with 20wt% tin or 40wt% zinc. These temperatures hold for "normal" pressure conditions, namely 1 bar. Low melting temperatures are a distinct advantage for casting, as long as the other desired properties are not compromised.

2.2 Castability

The castability of a metal or alloy is best defined as its ability to fill in and pick up every detail of a mold. The castability will be affected by two main parameters, namely the metal properties and the mold characteristics, and of course the pouring temperature. Numerous metal properties, mainly controlled by the alloy composition, impact the castability. As an example, because unalloyed copper easily oxidizes, it is difficult to cast.[12] Conversely, lead is known to greatly enhance castability.[13]

2.3 Color

While all metals are opaque, and more or less shiny when polished, only gold and copper are not white or whitish. The color palette of copper and its alloys depends primarily on the alloy composition (see table 6).[14] The addition of zinc to copper has a particularly pronounced impact on its color, lending it a gold-like appearance (fig. 133). The addition of silicon or aluminum yields similar colors. The visual differences between tin bronzes with different tin contents are not necessarily marked (fig. 134), unless high quantities (more than 15wt% of tin) are added, such as those used for mirrors and bells.

The color palette of copper and its alloys, and their visual appearance more generally, depends also on the surface texture. For this reason, and because metals are glossy, determining how to characterize or describe the color of copper alloys is quite complex, not to mention the various alterations of the surface, including the patina (for more on color measurement, see II.2§4).

2.4 Hardness: cold working and/or machinability, wear resistance

Cold working includes fettling, chiseling, chasing, and polishing (see I.6§1.2). Cutting and compressing are the main actions involved. The principal related material property is hardness, and what modern metal handbooks report as "machinability" (the ease with which a material can be formed with a cutting tool).[15] The hardness of a metal depends greatly on its composition (see table 9). As will be seen below, microstructure also greatly impacts hardness, but by definition cast bronzes are mainly in an as-cast state, and thus do not generally show large variations in microstructure. Soft metals such as unalloyed copper are particularly difficult to cut; they tend to smear under a cutting tool or during polishing. For hard metals, such as 10% tin bronzes, a small addition of lead improves machinability.[16] Several standardized tests are available to measure hardness; sampling is usually necessary.[17]

2.5 Suitability for patination, gilding, and metal plating, and/or corrosion resistance

The chemical composition of an alloy controls its reactivity to chemical attack—in other words, what might be called its suitability for patination (see I.8§1.1.1). This property is also directly linked to corrosion resistance.[18] An alloy's suitability for plating (for example gilding and tinning) depends on both the technique used and the metal. For example, tinning based on tin sweat can obviously only be performed on tin-bearing alloys (see I.7§1.3.2). Also, leaded alloys are thought not to be adapted to fire gilding although this is not a universal rule (see I.7§1.1.2).

2.6 Mechanical ductility and strength

Most common metals, and notably copper alloys, offer one remarkable property when compared to stone: they are ductile, as opposed to brittle. Metals can be heavily deformed before breaking. In addition, a number of copper alloys, including bronzes, have relatively high tensile strengths: much effort is needed to deform a bronze sword by stretching or bending (and even more so a statue).[19] The direct consequence of these two properties—ductility and strength—is that bronze sculptures can support heavy loads, allowing artistic compositions of considerable freedom (figs. 135, 136). While a support is often integral to the composition of a stone or marble sculpture—such as a tree trunk attached to the legs of a standing figure, or a large strut beneath a rearing horse—the bronze equivalent will not require such a support.[20]

A finished bronze sculpture is rarely supported by its internal armature, as its own metallic structure is sufficient. Hollow statues are even more resistant because of their tube-like shape; a tube is much harder to bend than a solid rod. A number of large sculptures, including Greek and Roman bronzes, demonstrate this, as the armatures (if they once existed) were usually removed before the separately cast sections were assembled (figs. 137, 138).

Other mechanical properties include elasticity, fracture toughness, and more, but these are of little interest in relation to most bronze sculptures.[21]

2.7 Weldability

The ability of copper alloys to be assembled, particularly by welding, is an important consideration, as founders select metals for the cast and the weld. This must have been as true for metalsmiths using flow fusion welding during antiquity (see I.5§1.1.1, Case Study 1) as it is for welders in modern-day foundries.

2.8 Symbolic properties

In certain cultures, metals, and notably copper alloys, have been (and still may be) imbued with specific spiritual powers. These may be attributed to the metal and ascribed to various intrinsic properties, such as color or smell.[22] Rarer metals and other materials may also be added to the melt as part of a symbolic or spiritual ritual.[23]

2.9 Is there an "ideal" copper alloy for bronze sculpture?

The variety of alloys encountered in bronze sculpture throughout history tends to show that there is no ideal alloy. There are a variety of reasons for this, starting with the potential competition between properties—including hardness and castability—depending on the requirements of the finished work. The destination of the sculpture and the way it is worked (amount of chiseling, nature of the patina, gilding, et cetera) are key, not to mention tradition and habits (see Case Study 7).

However, very satisfying compromises may be found. Bronze with approximately 10% tin provides a good balance between castability, corrosion resistance, mechanical strength, and hardness. This alloy became a standard as early as the Late Bronze Age for tools and weapons,[24] and for large bronzes during antiquity.[25] That said, copper alloys containing zinc are also widespread, notably in modern times.[26]

3 What is microstructure? What does it reveal about the making of a bronze? And how might it influence some of the alloy properties?

3.1 How can we distinguish whether a sculpture has been left as cast or has been cold worked, and why is this significant? Dendritic versus recrystallized microstructure

The atoms that make up a metal or an alloy align themselves in geometric patterns, as opposed to most glass types, whose atoms are distributed randomly. These patterns are cubes for copper and copper alloys (fig. 139), with various displays of the atoms depending on the composition of the alloy: these are the phases. These patterns in turn form specific structures called microstructures. Once in a while, one might encounter a fine textile- or branch-like pattern (fig. 140) or a geometrical one (fig. 141) within the surface of a bronze. This is the microstructure of the metal, and is not to be confused with the mineralized remains of organic materials (pseudomorphs) that may form upon contact with a bronze surface during prolonged burial.

As when studying the crystalline structure of a rock, a cross section with specific preparation is necessary to investigate the microstructure of a metal sample through metallography (see I.2§3.3 below; also II.5§6). This structure can give clues as to whether the sculpture has been reworked or not, and potentially whether there are cast-on repairs. Two very different metallurgical states—and consequently microstructures—are found in cast bronze sculpture. The first corresponds to the as-cast state and exhibits a dendritic microstructure (figs. 142, 143, 144, 145). The second reflects disturbance of the dendritic structure by cold working, and sometimes by subsequent heating (metallurgists use the term "annealing"), in the areas closest to the worked surface. This may result in a recrystallized granular microstructure (figs. 144, 146, 147, 148). The microstructure affects physical and chemical properties dramatically, particularly corrosion resistance and mechanical properties. For this reason, for instance, hammered-in repair patches may stand out from the surrounding cast bronze setting because their crystallized microstructure oxidizes or otherwise patinates in a different way (fig. 125). The hardness of an as-cast bronze may be greatly increased by distorting the dendritic as-cast microstructure through hammering (see I.6§3.3).

3.2 Primary versus secondary casting: welding, soldering, cast-on repairs

In addition to cold working, the metallic microstructure may record another type of technical process, namely the addition of a metal melted *onto* the primary cast. Such cast-on additions may occur as repairs (see I.4§1.2) and/or assemblies such as brazing, welding, or soldering. However, the microstructure of the primary cast will only be impacted if the incoming metal is hot enough. This only occurs with welding.

3.3 Are metallographic investigations necessary?

Sampling for the purpose of metallographic study should only be considered in order to address very specific questions. Some assembly techniques and repairs may be advantageously investigated through metallography. Metallographic investigations have so far proven the only way to positively confirm the use or not of flow fusion welding and to characterize the degree of mastery involved (see I.5§2.2). Similarly, in some instances only metallography can distinguish between cast-on and mechanical repairs (see I.4§2.1).

Metallography also provides excellent evidence of the presence or absence of cold and/or hot working.[27] Theoretically, it might also provide evidence of conditions during casting.[28] Metallographic sampling is not readily done on relatively pristine works, especially since the areas that would be most interesting to sample tend to be on the outer surface, in clearly visible areas. Less invasive methods are available, but there are limitations to their ability to answer certain questions (see notably thermography in II.2§3.2).

4 Why investigate metal composition and properties? and other FAQs

4.1 Why analyze the metal composition?

A number of questions may be addressed by investigating the metal composition. Alloy determination may be required for broad documentation—defining whether a sculpture is bronze or brass, for instance. Some specific alloys may be associated with specific periods, production centers, or workshops, and/or may refer to specific constraints (technical, economic, political, cultural).[29] Impurity patterns in the metal may help to group statues or fragments of statues (in conjunction with other data),[30] tackle the provenance of the raw metal,[31] and/or detect particular applications and processes.[32] See table 10 and II.5§1 for a discussion about limitations.

4.2 Can metal composition help to authenticate a bronze sculpture?

Both alloy composition and impurity patterns may be used for authentication,[33] although the ubiquity of compositions in both space and time often renders the task complex.[34] A combination of technical markers and archaeological and/or historical information is often necessary.

4.3 Can metal recycling be detected analytically?

There are no specific characteristics of metal composition useful for definitively distinguishing fresh metal from recycled. One way to detect the probable occurrence of metal recycling is to investigate the evolution of metal composition within a large group of well-contextualized bronzes. This has been tested for archaeological bronze artifacts, since some chemical elements such as arsenic are volatile, and their content may slowly decrease with repeated recycling (remelting).[35] But the authors do not know of any such studies on bronze sculpture. Archives such as commission contracts sometimes specify that scrap metal should be used, or, conversely, that it should not be used (although such documents cannot be blindly trusted as reliable).[36]

4.4 Is it useful to know the physical and chemical properties of the metal in a sculpture?

Leaving aside conservation issues, there are a good number of reasons to want to understand the properties of the metal, for all the aforementioned points (I.2§2 above) and more. Knowing a given property may add to the discussion of the intentionality of the presence and/or the amount of a given element, and the objective targeted by the commissioner, the artist, and/or the founder. For example, one may be able to discuss the presence of high amounts of lead in a given sculpture: if intentional, was it to enhance castability, to lower the cost of the metal, or because of tradition or beliefs? This may then open up large avenues for research (transfer of know-how and knowledge, trade in materials, et cetera).

Notes

1. In Europe, brass was once made by so-called cementation, a relatively complex process where metallic copper was mixed with zinc ore. The use of metallic zinc to create brass was not mastered until the mid-nineteenth century (see Bourgarit and Thomas 2015).
2. Young and Pernicka 1999.
3. See Motture 2019, 18–21, for a brief summary of medieval and Renaissance terminology and issues around the different interpretation of early sources, including the interchangeability of the use of *aes* (bronze) and *aurichalcum* (brass, but literally "golden copper"). See also Thomas 2009, 498–510, and Welter 2018. The nomenclature of copper alloys is an old issue among scholars (see Rickard 1932).
4. For example, in modern industry, the term "red brass" designates two very different alloys depending on its use. For wrought alloys, red brass is a binary alloy, typically CuZn12-15, sometimes called tombak. For cast alloys it is a quaternary alloy, typically CuSn5Zn5Pb5.
5. For example, two different sets of nomenclatures have been proposed for medieval common metalware found in London (Bayley 1991) and in Paris (Bourgarit and Thomas 2012). Note that in both cases, compositions with up to 2–3wt% of zinc, tin, and lead are still considered unalloyed copper.
6. For cultural heritage copper alloys, some answers are given in Young and Pernicka 1999. See also Welter 2007.
7. Nomenclatures published to date for historic coppers are mainly related to medieval and modern European contexts (Bayley 1991; Glinsman and Hayek 1993; Motture 2001; Bourgarit and Thomas 2012). Modern standards can be used as well (ASM, AFNOR, et cetera), although they may not be as appropriate for historic alloys. For example, for the ASM, 2.5% lead would be a so-called extra-high leaded alloy, whereas in ancient Greek and Roman sculptures such alloys would be characterized as unleaded (high-leaded bronzes bear up to 30% lead in Roman statues). Similarly, a 10% zinc alloy is called a commercial bronze. Although terms such as "leaded bronze" or "brass" are used, they should always be followed by a summary of the alloy by percentage if available.
8. An interesting definition of "minor element" has been proposed (Welter 2007, 95): "Elements which by amount range between the intentionally added major elements to adjust the properties of copper and the unavoidable trace elements which are basically considered as a nuisance." However, these guidelines are not standard, and not always adhered to in the art historical literature, for example.
9. There are a plethora of references on this subject; see for example Stewart 2014; Motture 2019.
10. Cottrell 1967, 39, neatly explains that all these properties specific to metals (when compared to other materials such as stone or ceramic) are due to the "pervasive glue" stemming from the specific metallic bond between atoms. See also Smith 1981.
11. Metal handbooks provide the most comprehensive synthesis for industrial alloys; see for example Davis 1998.
12. However, unalloyed copper sculptures of various sizes and from diverse cultures were successfully cast, as for example an anthropomorphic solid figure dated from the early second millennium BCE, India, H. 45cm (Musée des Arts Asiatiques de Nice, inv. 2002.2.1, see C2RMF Internal Report 2002), some of the Tibetan statuettes from the seventeenth and eighteenth centuries in the British Museum collections (Craddock 1981), and the sixteenth-century Wolsey Angels in the Victoria and Albert Museum (V&A A.1 to A.4-2015; h. between 101 and 108 cm each; see Motture 2019, 197–98, with additional references) (as yet unpublished analysis undertaken as part of the collaborative V&A Wolsey Angels Research Project, forthcoming).
13. A comprehensive description of which metal and mold properties control castability is beyond the scope of the present guidelines; please refer to the specialized literature (Lesoult 1986; Beeley 2001, 17–25; Campbell 2003, 75–95). A very clear synthesis has been proposed by Mille 2017, 378–404, including the role of lead in castability of protohistoric and historic copper alloys. This latter topic has been long debated among archaeometallurgists. A recent experimental development (Mille 2017) has demonstrated, for the first time, the influence of high lead content (above 10wt%) on the castability of copper alloys under specific conditions (mold made of low thermal diffusivity material such as clay or plaster, preheated mold).
14. For ancient copper, see Mödlinger et al. 2017. For the most recent measurement attempts and an updated bibliography, see Radivojević et al. 2018.
15. According to US standards, machinability is scaled against the most machinable alloy, namely the "free-cutting brass C3600" (35.5% Zn, 3% Pb) for copper-based alloys; see Tyler and Black 1992, 760. Hardness is reported either quantitatively (Tyler and Black 1992, 779) or qualitatively according to the metallurgical state, annealed, et cetera (hard, 1/2 hard, 1/4 hard, etc.); see Tyler and Black 1992, 817.
16. This was notably clear on the Greek Vix crater (Châtillon-sur-Seine, France, sixth century BCE, one of the largest bronze vessels known in antiquity). It was shown that around 1wt% lead had been deliberately added to the 10% tin bronze to facilitate engraving of the cast elements of the frieze and handles (Mille and Bourgarit 2003).
17. Revankar 2000; François 2004.
18. It is beyond the scope of this essay to discuss when, where, and how founders paid attention to the alloy composition with respect to patination.
19. Most weapons and tools were made of bronze during the Middle Bronze Age and Late Bronze Age.
20. Actually this is a complex matter, and the presence (or not) of struts is now understood as not simply a question of bronze versus marble, notably for Greek and Roman sculpture. Bronzes can have struts too, although it is not clear why (see for example the statue of a young Dionysos from the Chicago Art Institute, published in Mattusch 1996, no. 23, 224–31). And marbles can have them or not (see Hollinshead 2002; Anguisola 2018), as

kindly indicated by Carol Mattush, June 2019. And it is not clear whether all marbles really need the struts that appear; in some instances they may be due to workshop practice (Anguisola 2018).

21. For more on mechanical properties of cast metals, see Campbell 2015, chapter 9.

22. See the "red gold" in Africa (Herbert 1984); the evil properties of copper in the pre-Hispanic Caribbean (Martinón-Torres et al. 2007); the cosmological and gender connotations of copper in Colombia due to color and smell (Falchetti 2003); and the curative properties of the Khmer *samrit* (Vincent 2012, 297–301). See also Stewart 2014 for ancient Greece; Motture 2019, esp. 15–17, for Renaissance; and Droth et al. 2005 generally.

23. Gold and jewelry are added in the melt for Buddhist statues today. The metal of church bells is still blessed before casting in various countries, including France (see http://www.youtube.com/watch?v=5gg3THv4vfI, thanks to Alice Chéron, Ecole du Louvre). Also still common today is the practice of adding remelted metal that has been previously used, as recommended for example by Pliny the Elder for its "seasoned brilliance . . . tamed by perpetual use" (Pliny the Elder [77 CE] 1857, 34.20, p. 199) and by Pomponius Gauricus (Italian, 1482–1530): Gauricus [1504] 1886, 222–23; Gauricus [1504] 1969, 218–19; Gaurico [1504] 1999, 228–29.

24. Pernot 2000.

25. Mille 2012; Mille 2017; Descamps-Lequime and Mille 2017. Zinc is detrimental to welding, and since welding was systematically carried out on ancient large bronzes, the works never contain more than 1% zinc.

26. Although related to decorative arts gilt bronzes rather than statuary, an innovative study carried out by French chemist Jean-Pierre-Joseph d'Arcet (1818) to determine which brass would meet most technical requirements is of interest in this context. Eight alloys were tested by craftspeople involved in the production, namely founders, chasers, turners, and gilders. Only one was rated very good, which happens to be very similar to the alloy found in eighteenth- and nineteenth-century French gilt bronze (Heginbotham 2014).

27. The authors do not know of such investigations on bronze sculpture but are aware of some on other types of prehistoric and historic bronzes (see notably Chase 1994; Pernot 2000; Scott 2014).

28. The investigation of the microstructure (such as grain size, the interdendrite spacing, and the composition of the various phases, see I.2§3 in the present chapter) may provide a lot of information about the casting conditions (pouring temperature, mold materials, and so on). Yet given the number of parameters potentially controlling the microstructure, the task of distinguishing which parameters are responsible for the observed microstructure is very complex. Given the invasive aspect of metallography, it is understandable that no such study has been carried out on bronze sculpture.

29. See Case Study 5 and a number of other studies on ancient large bronzes (Mille 2012; Mille 2017; Descamps-Lequime and Mille 2017); Khmer bronzes (Vincent, Bourgarit, and Jett 2012); and Venetian sixteenth- to seventeenth-century bronzes (Motture 2003b). For composition of Renaissance bronzes see for example Motture 2019, 22–25.

30. Numerous examples are available for a variety of periods: Bouquillon et al. 2006; Azéma et al. 2012; Vincent, Bourgarit, and Jett 2012; CAST:ING 2018. See also Case Study 1 and Case Study 5. Note that in all cases, elemental composition alone was not sufficient to ascertain the grouping; other analysis (core) and/or approaches (style, archaeology) proved necessary.

31. Bourgarit and Mille 2014.

32. For example, the presence of phosphorous in the welding metal on the monumental Roman bronze foot from Clermont-Ferrand (fig. 68) has revealed the use of flux (Darblade-Audoin and Tavoso 2008).

33. For example, an alloy containing zinc can hardly be dated to an Angkorian production or found in a large antique bronze. And silicon bronzes and additions of phosphorous did not appear until the mid-twentieth century.

34. The same bronze composition may be found in a Javanese ninth-century statuette and a French sixteenth-century large bronze (see Case Study 3, Case Study 5).

35. For example Bray et al. 2015.

36. See for example Welter 2014.

I.3

Casting Defects

Jean-Marie Welter
David Bourgarit
Francesca G. Bewer
Andrew Lacey

Additional Contributors: Benoît Mille, Peta Motture

This chapter overviews some of the most significant types of flaws found in bronze sculptures, together with a brief explanation of their possible causes. Few, if any, casts come out of the mold looking perfect. Some imperfections are inherent to the casting process, and damage may occur at any point during the lifetime of the finished sculpture (table 11). This chapter's primary focus is on flaws that occur during casting. We will provide advice on how to differentiate the various casting defects and other imperfections, and how to interpret their causes, which amounts to trying to reverse engineer what went awry in a cast.

In most periods and cultures, evidence of the casting process and flaws were generally removed and/or concealed as part of the fettling, chasing, and finishing of the sculpture. Surviving casting defects are thus, for the most part, not immediately visible. They may lie hidden below a surface coating or a well-executed repair, or have always been invisible to most eyes because they occurred within the metal wall or on the sculpture's inner surface. For this reason, bear in mind the connection between defects versus repairs, cold work, and patination, which are considered elsewhere in the *Guidelines*.

1 What kinds of casting defects can be found on a bronze sculpture?

1.1 Broad description of casting defects

The term here is reserved for unexpected flaws that result from some dysfunction that occurs during the metal casting process. These include defects resulting from faulty sprueing (judicious gating-system design is one key parameter, among others), flawed mold or core construction, or a malfunction during the actual pour (see GI§2.1, GI§2.6, GI§2.7). Many factors can have a bearing on the genesis of casting defects. These include the mold makers' and founders' general level of expertise and care (compare fig. 4 and fig. 5), the equipment on which they rely, and the materials they use. These variables all impact the various steps of the casting process that ultimately govern the flow of the metal through the mold. A good knowledge of materials science, including castability, and of the practices related to the production of a given sculpture are invaluable in assessing which step of the process may have led to the detected defects (see also I.3§2.1 below).

Defects can occur on any area of a bronze—on the external or internal surfaces of the cast, as well as through or within the metal wall. Depending on their location and size, defects may affect our aesthetic perception of the sculpture, and are therefore often removed, repaired, or covered over.[1] The most significant flaws include holes or cavities, cracks, excrescences and depressions on internal or external surfaces, and distortions of the form. Variations in the alloy composition might also be perceived as casting defects, as they may strongly influence the color of the alloy and its resistance to mechanical and corrosive attacks during the lifetime of the object.

Repairs may also leave traces that could be confused with inherent defects. Furthermore, alterations or damages to a bronze that have occurred over time can look like defects and potentially be misconstrued as such. Clarifying the cause of a defect is important.

1.2 Defects inherent in the casting process and later damages

Not all blemishes and imperfections are considered casting defects. Before delving into what constitutes such flaws, let's identify the categories of imperfections that are *not* included—for instance the features to be expected upon breaking a fresh cast out of its refractory mold, or blemishes or damage that occur from mishandling the cast surface. These fall into four categories (table 11).

1.2.1 Unavoidable imperfections inherent to the casting process

These include:

- shrinkage of the cast in comparison to the model due to several parameters, including the thermally induced contraction of the metal during cooling (see II.4§1.1.1 for more on shrinkage, its impact on dimensions, and the issues faced by founders to control dimensional variations within their cast);
- the armature rods or core pins breaking through, or holes marking their location;
- the remains of sprues, usually appearing as protrusions;
- low, raised ridges known as seam lines due to the meeting of two or more sections of piece mold.[2]

1.2.2 Avoidable defects occurring before the casting

These include flaws resulting from:

- mishandling of the model (for instance sagging of a wax model placed in an overheated environment, or wear on the surface);
- distortion occurring during the construction of the mold and/or core system (for instance caused by inhomogeneous temperature distribution during the baking of the investment or uneven compression of the sand mold).

1.2.3 Avoidable defects occurring during the finishing of the cast

These include:

- incompetent or incomplete finishing that reveals and/or leaves unwanted evidence of the casting process (see I.3§1.2.1 above);
- poor application of patina;
- issues related to fire gilding.

1.2.4 Damage that may occur during the lifetime of the sculpture

Examples include:

- mechanical damage resulting from intentional or accidental dropping or striking of the object, which may cause it to be scratched, bent, dented, punctured, or broken;
- removal of precious materials such as gilded surface or inlays;
- surface degradation induced by aggressive environments leading to layers of corrosion products, or air- and waterborne deposits and localized defects such as pitting. Note that corrosion may affect not only the metal wall, but also any iron armatures, which can expand while rusting, potentially pressing into or breaking through the bronze from within (fig. 149);
- surface disfigurement due to harsh cleaning treatments and/or vandalism, whether mechanical, chemical, or electrolytic.

For more relevant information on the life cycle of the bronze, see also GI§1.3.

1.3 Five categories of casting defect

The characteristics of a specific defect include its morphology; its frequency of occurrence; whether it is unique, repeated, and/or clustered; and its location on the object. The morphology of the defects provides a simple guiding principle for their identification and their grouping into meaningful categories. Table 12 shows the most popular catalog adopted by the metallurgical industry. It takes into account seven families of structural defects as well as one for chemical anomalies.[3] For bronze sculpture, we consider the following five main categories (fig. 150).

1.3.1 Cavities, porosity

A cavity may be located:

- through the entire wall (figs. 73, 151);
- opened on just one surface of the bronze (figs. 152, 153);
- within the wall (fig. 152).[4]

The forms that such defects take relate to the way in which they were generated and how the metal solidified. Round and smooth cavities are mostly (small) bubbles created after the release of gaseous elements from the solidifying melt. Cavities with a granular and complex surface are often caused by material that has fallen into the mold and become encapsulated in the metal. This could comprise a loose piece of investment or core, or it could be slag, dross, or charcoal (when used) coming with the melt (fig. 154).

Larger cavities, or lacunae, are often caused by trapped air pockets or the mold failing to fill completely, resulting in the melt becoming too viscous while cooling down and ceasing to flow (fig. 73). Poor feeding of the metal may also lead to the formation of cavities with angular shapes as a consequence of internal shrinkage (fig. 155, video 9). Internal shrinkage should be distinguished from external shrinkage, which leads to the reduction of dimensions.

Cavities generated by gas or internal shrinkage are collectively known as porosity.[5]

An extreme case of porosity can result in an incomplete casting. Malformation of an entire section of a bronze is not uncommon (fig. 151). This is described as an "incomplete piece" in modern industrial parlance.[6] However, it may be caused by a single defect or a combination of defects from different sources.

Risks of misidentification/misinterpretation

- As noted above, cavities resulting from the removal of armature rods and core pins may potentially be misidentified as casting flaws. Such cavities are often revealed when patches, plugs, or cast-on repairs have been lost.
- Missing inlays may also be mistaken for defects (see I.9§2.1).
- Sampling by drilling creates holes that unless documented might be mistaken for porosity, though the perfectly round hole is usually recognizable (fig. 156).
- Pitting from corrosion that has been removed or from electrolytic cleaning can be mistaken for a casting defect (fig. 157).
- On radiographs, metal porosity may be confused with gaps in the core. The latter are usually coarser and have more angular shapes (fig. 65 point 1).

1.3.2 Cracks and fractures

Cracks appear as long, linear or angular gaps in the metal. They can start on any surface and pass partly or all the way through the metal wall. Fractures can even be found within the wall and are often associated with inclusions and cavities (fig. 150). Cracks and fractures that have traveled through the entire wall of the bronze can lead to breakage (fig. 158), even the loss of a portion of the sculpture such as an arm, hand, or foot.

The metal may tear for several reasons. The main one is excessive tensile stress, which arises when the cooling metal cannot contract freely in the mold, notably when the temperature in the metal is heterogeneous. Another cause of cracks may be the stress induced by an overly rigid incompressible core.

Cold shuts are another type of fracture. They occur during the filling of the mold when two flows of molten metal meet but are not hot enough to fuse together (fig. 159).

Risks of misidentification/misinterpretation

As a rule, damages are repaired and any broken part reattached by one of the many available assembly techniques (see I.5), or re-created and cast on. It can, therefore, be difficult to differentiate between a breakage that originated during the casting process, such as a thin section broken off during careless fettling or chasing, and one that occurred through later damage.

1.3.3 Surface excrescences

Aside from cavities, the most apparent defects are excess metal in various forms of excrescences on the different surfaces of the raw cast. These include flashing, which results during the pour when liquid metal flows into cracks that formed in the refractory mold and/or in the core. Such flashing can occur on the external surface as mold flashing (fig. 77), or the internal surface as core flashing (figs. 59, 60, 90). Both types appear in many forms, from thin, raised ridges to feathery extensions that emerge from the surface of the cast.

Furthermore, larger areas of both surfaces can be textured in an undesirable manner (fig. 160). Such textures appearing on the interior of the bronze wall are usually, but not always, associated with specific problems caused by the core. But do not confuse these with features specific to the fabrics of the core, such as bubbles in plaster core (figs. 56, 57) or evidence of organic fibers (fig. 58), which are defects inherent to the process.

Textures forming on the outer surface of a bronze can be caused by numerous and diverse issues. The surface may for example resemble the skin of an orange, hence the term "orange peel" effect (fig. 161). This occurs notably when dross or oxide films flowing with the melt during pouring push against the inner surface of the mold (fig. 154). When the investment material is insufficiently compacted, it may create a rough and granular surface that is translated into bronze. The mold's surface can also roughen due to overheating, creating a texture somewhere between the granular appearance of the latter and the "orange peel."

Risks of misidentification/misinterpretation

- It is often difficult to judge whether a textured surface is intentional or due to a faulty cast.
- Sandy or grainy textures covering large areas of the cast, or even its entire surface, can mislead the investigator into thinking that the cast was made using a different method than was actually the case, for example a lost-wax casting appearing as a sand casting, or vice versa.[7]
- Residues of excrescences on the outer surface can look very similar to seam lines. Note: seam lines should not be referred to as "flashes" because they are not considered defects; they are integral to the casting process.

1.3.4 Distortions due to mold and/or core shift

Distortions of the sculpture may occur for a variety of reasons, including refractory mold shift (fig. 162) or heterogeneities in the refractory mold that developed during the fabrication process or during casting (localized and rapid heating, metal static pressure, et cetera).

When improperly anchored in the mold, the core can be displaced by pressure differences of the melt inflowing through different channels, thereby increasing the wall thickness on one side and decreasing it on the opposite side (fig. 65 points 6, 7). Note that the core shift may arise from mishandling before casting.

Risks of misidentification/misinterpretation

- Variations in wall thickness may be intentional, or may be due to unintentional variations in the wax thickness, as for example in slush molding (due to lack of experience and/or awkwardness of the mold).
- Distortions may also arise in a damaged wax model or one that has sagged, as sometimes occurs in very warm temperatures.

1.3.5 Heterogeneous alloy composition

A variety of factors may prevent the homogeneous distribution of all the elements in the alloy throughout the cast. It may be that the ingredients in the melt have failed to mix completely.[8] Another cause is the unintentional segregation of various elements—mainly zinc, tin, and lead—on the surface, such that one or more of the elements segregates out and becomes more abundant toward the surface of the cast (fig. 163).[9]

Consequences may include:

- an uneven color, since the metallic color of a bronze directly corresponds to the composition of the alloy (for a more detailed description and warnings, see I.2§2.3);
- difficulties in using some coloring techniques such as fire gilding (the presence of excessive lead on the surface makes fire gilding difficult; see I.7§1.1.2);
- different reactions to environmental conditions over time, leading to subtle changes in the formation of corrosion products;
- misleading results from metal analysis, especially when surface techniques such as X-ray fluorescence (XRF) are used (see II.6§2.2).

Risks of misidentification/misinterpretation

- Repairs, even when made with the same alloy, are a primary source of uneven coloration of a bronze sculpture because their alloy composition and structure are usually different from that of the body metal. The differential coloring is mostly caused over time by corrosion.

- More generally, wear and/or corrosion are a major source of uneven coloration.

2 Why investigate casting defects? and other FAQs

2.1 Can the investigation of casting defects help reconstruct the casting process?

Some casting defects are typically associated with specific materials used in cores and refractory molds, and thus may point to specific techniques and/or setups as mentioned in I.1:

- Sinuous core flashes are more characteristic of clay-based cores used in lost-wax casting, while linear ones are more common to sand casts (compare fig. 59 with fig. 61).
- Some defects may indicate or help to identify the casting orientation, as detailed in I.1§4.2 (gas porosity, wash, lead content).

2.2 Can the investigation of casting defects help distinguish between founders' workshops?

Defects, due to the very nature of their formation, are highly descriptive of the issues encountered during the casting process. Knowledge of how defects occur can reveal something of the artist's or founder's working parameters and know-how, including material usage and aesthetic considerations.[10] However, matters may prove exceedingly complex. A defect (for example a crack) can originate for a variety of reasons (feeding system, refractory mold, temperature of metal, et cetera). Also, as noted, defects may be due to inherent peculiarities of the process as well as to accidental circumstances. Therefore, defects discovered in two virtually identical statues cast within a short time period—even in the same workshop—may vary considerably.[11]

Similarly, different qualities of cast may occur at different points on the founder's learning curve, especially when they are produced over time. In general, defects are not useful to determine the sequence of castings or the relationship between versions of the same model. An exception may be the after-cast of a bronze that has many repairs or flaws, where traces may help reveal features of the (unknown) precursor (fig. 164).

2.3 Can casting defects help us better appreciate the artistic and technological "value" of a bronze?

The perception and acceptance of faults is subjective (fig. 165). To what level were defects acceptable? The answer depends, among other considerations, on cultural and economic factors. Repairing might have been less expensive than recasting, even for heavily flawed bronzes (fig. 166). And the size of the bronze will also determine what kinds of flaws are deemed important. Structural flaws are of greater concern in a monumental cast than in a small figurine, while surface blemishes are more disturbing in the latter than on the back of a large sculpture destined for display above eye level.

It is always understood that a bronze will have to be reworked upon casting to remove or repair flaws, though this does not always occur.[12] Several heavily defective sculptures have been maintained as such throughout time, although the reasons for doing so are not always known (fig. 151). It was the rare sculptor who considered defects as a part of their creation and thus opted to preserve them rather than repair or remove them.[13] And even if the artisans (and patrons) were aiming for a perfect cast, we cannot assume that a sculpture free from defects is superior in quality, or more technologically advanced.

2.4 Can defects occurring during the casting process and those arising later on during the lifetime of a bronze be distinguished?

This can be difficult. What may look like later damage—notably cracks—could have been created during the casting process, or vice versa (see above for potential misinterpretations). Yet some apparently similar features, such as open cavities, may be sufficiently specific in shape and/or distribution on the sculpture to allow for clear distinction. For instance, holes occurring during casting are often diverse in size and unevenly distributed throughout, as opposed to core-pin holes, holes made as part of an attachment mechanism, or holes due to electrolytic cleaning. In any case, consider the sculpture as a whole, with all its flaws, before deciding whether a specific flaw is a casting defect, one that occurred in another phase of the production, or later damage.

3 Checklist: How do we investigate defects?

- Visual inspection supported by optical magnifying instruments can prove very efficient for studying defects that traverse the metal wall as well as those located on the exterior or interior surfaces.
- For defects on the interior walls, endoscopy is the preferred technique.
- Image-analyzing algorithms enable the dimensions of the defects to be quantified.
- To look into the metal wall (or when defects occurring elsewhere are hidden, for example by coatings) a number of significant nondestructive testing techniques are increasingly used, including X-radiography, eddy currents, ultrasonic testing, and thermography.

For a synthesis of available investigation techniques, see tables 13, 10, 5, and associated chapters of volume II.

Notes

1. Some authors separate them into categories as negative and positive defects (Rome and Young 2003). For an inventory of casting defects and their origin in the modern industrial foundry see Mascré, Thomas, and Hénon 1952; Hénon, Mascré, and Blanc 1971; Reuter and Schneider 1971; Campbell 2004; Rajkolhe and Khan 2014; Siddalingswami and Dulange 2015; Campbell 2015.
2. An exception is French sand molding, which, when expertly done, results in a seam-free cast. This is accomplished through precise ramming of the sand—not too loose, not too dense, so that the sand swells just slightly on baking to seal the gap. See Case Study 6.
3. A list of observed defects occurring in industrial cast products was first produced in France in 1952, resulting in the publication of a defect atlas (Mascré, Thomas, and Hénon 1952). This initiative led to international cooperation aiming to systemize the defect classification. The guiding line was their appearance, which led to the seven basic categories presented in table 12. Atlases giving the code and the description of commonly occurring defects were issued in the 1970s in various countries.
4. See Mascré, Thomas, and Hénon 1952; Ammen 1980; Campbell 2015. Hénon proposes grouping cavities into three categories according to their shape and distribution: isolated round cavities, groups of small round cavities, and cavities with a rough surface.
5. Ammen 1980; Campbell 2015.
6. Mascré, Thomas, and Hénon 1952.
7. Notes made during a Barye Study Day held at the Victoria and Albert Museum, November 14–15, 2002; see I.1.n.19.
8. This is particularly true with reverbatory furnaces. The relatively low thickness of the liquid bath heated from above prevents thermal convection and thus any mixing. For a general description of reverbatory furnaces, see Mödlinger, Asmus, and Ghiara 2024.
9. This occurs when the founder has not provided a sufficient head of liquid metal to ensure a proper metallostatic pressure in the mold. The still-hot, contracting solid metal can detach itself from the mold surface and reheat up to the melting point. Zinc may evaporate in the hotter parts, circulate in the metal-mold gap, and redeposit on colder surfaces. Tin and lead are usually pushed out from the inner subsurface layer to the top of the external surface. This segregation phenomenon is known as tin or lead sweat, and gives the surface a silvery-gray appearance. Tin sweat and also arsenic sweat may be intentional; see I.7§2.8.
10. Andrew Lacey: "In the course of making my own bronzes I often allow or even encourage certain defects into the casting. Grainy or pitted surfaces in small discrete areas can work well especially in conjunction with the highly polychromatic patinas. Also, fine flashing lines formed by the mold cracking under pressure from the bronze can create structures reminiscent of bodily wounds or scars; this helps me suggest a fragility or tenderness in the form that is difficult to create by modeling alone." Personal communication, April 2019.
11. The technological study has shown that Donatello's (Italian, ca. 1386–1466) workshop possibly produced both *Spiritelli* in the Musée Jacquemart André, Paris (inv. MJAP-S 1773-1 and MJAP-S 1773-2, heights 60 and 65 cm, respectively), based on similar models, but there are notable differences in the defects observable on the two casts, due mainly to significant variance in metal wall thickness (Castelle et al. 2019).
12. In Guido Mazzoni's (Italian, ca. 1450–1518) early sixteenth-century estimate for his unrealized project for the tomb of King Henry VII the repairs were costed in advance; see Motture 2019, 197, with additional references.
13. A number of artists have left evidence of the casting process on the final cast, including casting defects, for example Adriaen de Vries (Netherlandish, 1556–1626) (Bewer 2001), Medardo Rosso (Italian, 1858–1928) (Cooper and Hecker 2003; Hecker 2021), and Auguste Rodin (French, 1840–1917) (Hecker 2017).

I.4

Repairs

Jane Bassett
Lorenzo Morigi

Additional Contributors: Denise Allen, Aurélia Azéma, Ann Boulton, Manon Castelle, Laurence Garenne-Marot, Arlen Heginbotham, Joachim Kreutner, Andrew Lacey, Elisabeth Lebon, Jeffrey Maish, Peta Motture, Jeremy Warren, Jean-Marie Welter

This section focuses primarily on metal repairs, and considers what kinds of repairs may be present on a bronze, why it may be interesting to investigate them, and how best to do so. Repairs undertaken in the foundry address problems in the metal that have occurred during casting (see I.3§1). Nonmetal repairs, such as those made of plaster, wax, and resin, may also be found on sculptures, but fall outside the remit of this chapter.

1 What kinds of metal repairs might be encountered? And where, when, why, and how were they made?

Repairs may vary considerably in nature, shape, and scale, from shallow flaws that occur only on the outer surface of the bronze wall, to those that pierce the wall (for instance filling of holes left by core pins and armature rods), to the larger replacements of mis-cast sections. The range of commonly encountered repairs is presented in figure 167. Although repairs may be needed to restore the structural integrity of a cast, they are most often made for aesthetic reasons—to repair unsightly flaws on the bronze's surface. Whereas surface repairs can be carefully hidden by chasing (figs. 168, 169, 170), artists and founders rarely take steps to hide evidence of repairs on the interior, which is why access to these areas for examination can be very informative. That said, some types of set-in patches and plugs only extend partway through the wall of the bronze and are not visible from the hollow interior (figs. 152, 171).

Repairs may be undertaken concurrently with and are related to—other finishing techniques, such as the joining of separately cast elements (see I.5). Repairing flaws is a time-consuming but unavoidable and necessary step in the casting process.[1]

While organic materials such as resins or waxes may be used at times to make fills, the vast majority of repairs found on copper-alloy sculptures are made of metal, and these are primarily fashioned out of copper-based alloys. There has been no broad survey of repairs that would enable us to identify the regional occurrence or chronological development of repairs. Among the few exceptions are the evident lack of machine-threaded plugs in Europe before the middle of the sixteenth century (see I.4§1.1.2 below) and the advent of modern welding processes in the twentieth. In light of the paucity of data, this chapter will not attempt to outline when and in what geographic area specific types of repairs might be expected.

All metal repairs require the same general steps:

- the edges around the flaw are prepared;
- the repair is made;
- the surface is finished to render the repair invisible.

Depending on the size and location of defects and the skill of the person making the repair (often the founder), two main categories of metal repairs are encountered on historic bronzes: mechanical repairs and metallurgical repairs. (We use the term "metallurgical repair" in this volume for simplicity to designate all repairs using molten metal, as an echo of the term "metallurgical joint"). Many bronzes contain both types of repair.

1.1 Mechanical repairs

Mechanical repairs (often referred to as set-in repairs) are carried out by fitting a metallic-alloy fill into a flaw. The fill may be made from discarded (cut-off) sprues or other excess metal from the original cast, from sheet metal, or from separately cast sections. The alloy of a set-in repair does not necessarily match that of the surrounding cast. Due to the amount of hammering, filing, and punching required to secure and chase a set-in repair, the founder may choose a softer and more malleable copper alloy than that used for the surrounding surface, or even pure copper. For this reason, the color of set-in repairs can differ from that of the surrounding as-cast surface, and will be accentuated over time by oxidation or corrosion. The differences of metallurgical state between the repair (hammered) and the surrounding metal (cast), and/or the degree of porosity, may also impact the corrosion behavior, and consequently the surface appearance (fig. 125).

For the purpose of these guidelines, mechanically attached repairs that are set into the surface of a sculpture are called patches. Repairs that are inserted into holes that extend through the full thickness of the metal wall are called plugs.

1.1.1 Patches

Patches can be found in a wide range of sizes and shapes and are often rectangular or polygonal (fig. 172). A common process for setting in patches is described in figure 173 and a variation for small flaws in figure 174.

Patches may be applied to repair a range of flaws, including superficial surface porosity or holes that pass clear through the metal wall (fig. 35). They often fall out and are lost, leaving the prepared geometric hole visible on the surface (fig. 175). When flaws in a sculpture are located in areas where the surface has complex modeling, patches can be cast separately and set into place. In such instances, the patches can be directly modeled in wax in the flawed area or can be formed using the original molds (fig. 176).

1.1.2 Plugs

Flaws that extend through the entire thickness of the bronze, including holes left when core pins or armature rods are removed, are often repaired with circular plugs inserted after the defective spot has been drilled out. As observed on European bronzes from the second half of the sixteenth century, plugs are often (but not always) threaded.[2] Once screwed into place, the visible end of the rod is chiseled down close to the surface, then hammered and chased to bring the repair flush with the surrounding wall (fig. 177).

Methods for forming threads on a plug and in its corresponding hole have changed over time. In the early modern period, the first use of threaded metal fasteners seems to have appeared in the realm of arms and armor manufacture as well as clock making, around the middle of the sixteenth century. Although no comprehensive study has been done, it can be assumed that their use in sculpture would have followed. Although thread cutting lathes may have existed quite early, forming threads in copper alloys in this period was probably primarily achieved by "swaging," or displacing the metal, rather than by cutting. Female threads would have been formed by forcing a handmade, hardened steel threaded rod, or "tap," into a slightly undersize hole in the bronze; as the tap was twisted into the hole, the steel threads displaced, but did not cut, the surrounding metal. For forming male threads on a plug, a "screw plate" could be made by using the original tap to form female threads in a plate of annealed steel. After hardening the steel, a slightly oversize rod of bronze could be twisted through the screw plate to form a threaded rod, again by displacement of the metal, not by cutting.[3]

In the second half of the eighteenth century and into the nineteenth, mechanized machine lathes, improved screw plates, and fluted taps that cut threads started to become more and more widespread, a development that presumably would have increased the use of threaded plug repairs after this date.[4] Beginning in the mid-eighteenth century, systems of standardized dimensions for threading began to appear in Europe.[5] Today, precision tap and die sets are used to cut threads in both repair rods and corresponding drilled-out flaws.[6]

Threaded plugs inserted at the edge of larger geometric patches also may be used to help secure the large patches to the surrounding metal (fig. 178).

It is important to remember that various combinations of techniques may be found on a single sculpture.

Risks of misidentification/misinterpretation

- Set-in repairs are often geometric in shape; cast-on repairs are often asymmetric. However, metallographic investigation of several rectangular repairs on Roman statues has revealed examples of their having been cast on (fig. 145)[7] and later examples also have been observed (fig. 123).[8]
- As patches and plugs are effectively inlays, they may be visually similar (see I.9§2.3).
- Also, a repair can be confused with plugs used to assemble separately cast parts (such as pinned sleeve joins; see I.5§1.2.1). Separately cast sections can almost always be identified by examining the surface with strong raking light to reveal the thin joint around them, but occasionally these are only revealed through radiography.

1.2 Metallurgical repairs

Metallurgical repairs are made by pouring or melting metal directly onto flaws. Molten metal repairs can be designed to fit precisely even the most complex losses.

1.2.1 Cast-on repairs

These are the most common metallurgical repairs, created using a process in which a copper alloy is melted and poured directly into a flawed area. Cast-on repairs may be used to replace a range of losses, from relatively small surface flaws to large sections of a composition (figs. 179, 180). The process of casting on using a localized lost-wax process is described in figure 181; see also figure 182. Clay or other modeling materials could be used to model the repair instead of wax.

Cast-on repairs may have their own flaws, as evidenced by patches located within a cast-on repair.

In order to secure larger cast-on repairs in place, dovetails or holes of any shape were sometimes cut around the borders of the defect into which the molten metal flowed, locking the repair in place as the metal cooled (figs. 183, 184).

Cast-on repairs can be made of copper alloys that are similar to the bulk alloy, or of alloys with a lower melting temperature and containing more tin, zinc, and/or lead. Over time, cast-on repairs—as is also the case with mechanical repairs—may become more visible as the dissimilar alloys oxidize or corrode differently (fig. 184), although this is not always the case.

Although best known as a technique to join separately cast parts of a sculpture, examples of flow welding to repair Greek and Roman bronzes are also recorded (fig. 185).[9] A chain of oval-shaped color variations on the surface of a bronze is typical of the technique. Flow fusion welding does not appear to have been used in medieval or Renaissance foundries for bronze sculpture, although welding is attested for repairing Renaissance cracked bells (fig. 186).[10]

In light of the high degree of artistry required to achieve successful cast-on repairs, it is often assumed that they are made in the original foundry.

1.2.2 Soldering, brazing, and welding

Soldering, brazing, and welding are techniques of using molten metal to repair flaws—to apply patches and/or fill losses directly. The difference between them is determined by the alloy of the repair metal and the temperature used for the repair. As a general rule for the purpose of this publication, solders are white in color; brazing and welding metals are copper-alloy based and therefore yellow in color. In all three methods, the heated, molten metal is drawn by capillary attraction directly into the flaw or into a gap between the flaw and the repair. For more on the different techniques, see I.5§1.1.

Lead-based soft solder can be used to fill flaws. With its low melting point and soft surface, such repairs are relatively easy to pour and to chase, yet are gray in color and therefore may be hidden by opaque coatings. More commonly, hard and soft solder repairs can be found as a method for securing copper alloy patches and plugs (fig. 187).

As with soldering, brazing is undertaken at a low enough temperature that the surrounding metal is not melted. When the sculpture is heated, small filings of the brazing metal scattered near the edges of the repair melt and flow into the join, drawn by capillary attraction. Because the brazing metal is an alloy similar to that of the surfaces to be joined, a brazed repair can be difficult to see. However, over time, corrosive flux residues that may remain on the surface will cause raised corrosion (figs. 106, 188).

Welding is the most common method of repair currently used in many parts of the world.[11] Undertaken at temperatures high enough to cause melting of the metal surrounding the repair, once chased and polished, a weld repair can be difficult to discern on the surface of a sculpture. Surface porosity along the weld or differences in corrosion over time due to alloy variations between the cast and welding metals may make it easier to see (fig. 189).

Risks of misidentification/misinterpretation

- It may be difficult to determine whether a section of a bronze that was cast on (for example the arm of a figure) was added as an intentional second step in the casting process, or became necessary owing to a flaw in the original casting (see I.5§2.3).
- When well cleaned and chased, it may also be very difficult to distinguish between the different copper-based metallurgical repairs (cast-in repairs, brazing, and welding). When sampling is possible and appropriate, metallography will confirm the degree of heating and whether or not the surrounding metal has previously been melted (see I.4§2.1 below).
- Although it might be assumed that lead repairs would be reserved for lower-quality castings, this is not always the case.
- Equally, it may be difficult to determine whether a lead fill was applied in the workshop or as part of a later intervention (fig. 190). Lead repairs may be found in bronzes that also have other repairs made in a copper alloy. The presence of more skillful copper-alloy repairs should not automatically suggest that the lead ones must be later additions.
- Corrosion along a joint may be due to remains of flux applied during soldering, brazing, or even some types of welding.

1.3 Nonmetallic fills

It is not uncommon to encounter even quite large fills made of nonmetallic materials such as painted plaster, wax, or resin. Nonmetal fills found on the same sculpture with metal fills may be evidence of a later intervention (one would not expect both metal and nonmetal fills to be carried out simultaneously). Nonmetal fills may have been applied at any time in the sculpture's life cycle and, with the exception of modern restoration materials, may be difficult to date. Depending on the skill of the person applying them, these fills can vary from well hidden to ghastly. They are usually easy to see with good lighting and/or ultraviolet radiation (UV) (fig. 191) and by touching (their warmth contrasting with the cold feeling of metal repairs).

1.4 Metal fills anticipated before the casting

In modern foundries voids such as core pin holes that can be predicted before the bronze is cast may be filled using the following method: a ball of wax is added to the model adjacent to where rods or wires extend through the wax casting model, and when the metal is cast, this excess wax adjacent to the rod becomes a lump of bronze that can be hammered into the void, filling the hole with the exact alloy used for the surrounding surface (fig. 192). It is unknown how far back in history the technique goes, but once mastered, it is relatively quick and straightforward.[12] It would be relatively difficult to identify on the surface of a bronze or in radiographs.

2 Why investigate repairs? and other FAQs

Below are some frequently asked questions related to the identification of repairs as well as broader questions that the investigation of repairs may help to address.[13]

2.1 Is it possible to distinguish a cast-on repair from a mechanical repair or a weld?

Because the foundry often carefully hides surface repairs on the exterior, examination of the interior (figs. 177, 193) and/or cross sections or the use of radiography (figs. 194, 195, 196, 197) can potentially reveal the method of repair.

2.2 Is it possible to distinguish a patch used to hide a casting flaw from a plug used to fill a core pin hole?

- Holes left when core pins are removed are geometric (often circular or square section) and extend through the bronze wall to the interior; casting flaws may or may not be geometric. It is not unusual to find core pins remaining inside a bronze adjacent to the repair.

- The presence of patches or plugs of similar diameter in a pattern over the surface suggest that they fill core pin or armature holes (see fig. 35, I.1§3).

2.3 Can the presence of repairs give information about the process that was used to cast the bronze?

Examination of the repairs alone will not conclusively identify the specific method of casting that was used. However:

- As stated above, the presence of repairs can help to indicate the placement of core pins and armature rods.
- Repairs can also indicate casting defects. Large numbers of repaired porosity holes in one area of a bronze may indicate that that part was oriented upright during the pour (see I.1§4.2).

2.4 What can repairs tell us about the quality of the cast, the skills of the foundry, and the commission more generally?

Due to the skills and time necessary to carry out repairs, their quantity and quality may lend insight into the abilities of those responsible for them. Two potentially contradictory situations are possible: extensive repairs are evidence of extensive casting flaws, yet the presence of well-made repairs is evidence of a skilled finisher.[14]

Without written documentation, making assumptions regarding the intention of a commission based on visual observations is a matter of conjecture (see also I.3§2.3). When faced with a flawed cast, the artist or foundry would consider various options: remelt the cast and begin again; repair some or all of the flaws; or leave the flaws unrepaired. The reasons why one option might be chosen over another are various and not necessarily evident, so drawing conclusions is risky.

However, all cold work, including repairs, is extremely time consuming, and extensive repairs therefore reflect considerable labor.[15] Indeed, unrepaired casting flaws are often found in places where they are not easily visible. The aesthetics of a sculpture are certainly affected by decisions regarding whether or not to repair. Whatever the culture, casts were often substantially repaired before finishing (figs. 166, 198), yet some artists, for example the sculptor and expert founder Jean-Antoine Houdon (French, 1741–1828), at times left extensive surface flaws unrepaired (fig. 153).

2.5 Can original workshop repairs be distinguished from later repairs?

Comparison of a repair with the surrounding surface may help identify repairs made in the original workshop. For example, given that foundry repairs are often made before the final finishing of the surface, tool marks such as polishing or texturing may run continuously from the bulk surface over repairs that were applied at the foundry (figs. 168, 169).

Evidence of later repairs may include differences in color and/or texture of the patina, which may be thinner, scratched, or discolored by heat applied during the repair process. In contrast, uniformity of the patina may suggest an old (and possibly original) repair (fig. 170), but keep in mind that patina uniformity may reflect repatination of the surface after repairs were undertaken such that a much-later repatination will be of no help in dating any repairs it covers.

If it is possible to access the inside of the bronze, later repairs may have a different appearance from original repairs. If the bronze has been repatinated and the interior is not accessible, identifying later repairs may be very difficult without radiography (fig. 199).

Repair alloys on the same bronze may vary considerably, so attempts to determine later repairs by looking at the alloy alone may be misleading. Although it was common to remelt cut-off sprues for use in cast-on repairs, such repair alloys are, nonetheless, generally different in one element or possibly many. This could be due to loss of alloying elements during the previous casting process (for example, heating an alloy with zinc causes some loss due to evaporation), or because another element could have been added to the repair melt to lower the melting point or alter the cold-working properties.

The degree of corrosion on a repair might not reflect its age. As an example, the patch repairs on the monumental figure of the twelfth-century Khmer West Mebon Vishnu were applied at the time of manufacture, yet their alloy and metallurgical state left them far more resistant to corrosion than the original cast (fig. 125).

2.6 More generally, can repairs help to date a bronze sculpture?

Under certain circumstances, and with caution, the type of repair chosen by the founder or sculptor may be indicative of practices from a particular geographical area or period.[16] Identifying the method by which the threads were cut into a repair rod can help date the repair (see I.4§1.1.2 above). Occasionally, the end of a plug will extend far enough into the interior cavity of a hollow-cast bronze to allow close viewing of the threads (fig. 177). But be warned: threaded plugs that can be observed either visually inside the bronze or in radiographs are often cut very short, making it difficult to determine how the threads were formed. Recall that handmade threaded screws used to secure metal parts have been found in the sixteenth century, whereas machine-cut threads were introduced in the second half of the eighteenth century. Evidence of the use of welding suggests a date after the nineteenth century.

Note of caution: though we assume that practitioners develop certain preferences and patterns in their practices, it is always possible that conditions requiring improvisation may arise (fig. 200). Second, repairs may form part of later restorations and, as with all comparisons, the relative lack of detailed technical studies of well-provenanced bronzes limits our ability to draw firm conclusions.

3 Checklist: How do we investigate repairs?

For details of examination and analytical techniques, please refer to tables 5, 10, 13.

3.1 Visual examination

Examination of the surface under strong lighting (including raking light) often reveals the presence of repairs. As seen above in I.4§1, quite commonly the patches acquire a different-colored patination due to differences in corrosion (figs. 125, 176, 184, 185). Even with very carefully chased bronzes, the outer edges of patches and cast-on repairs are often visible due to shifting of the repair or darkening of the fine gap between the bulk material and the repair due to accumulation of soil or darkened coatings.

Repairs on bronzes with opaque coatings may be quite difficult to detect by visual examination of the surface alone. When access to the interior of a sculpture is limited to small gaps or holes, an endoscope can help detect and identify repairs. For thinner-walled sculptures with access to their hollow interiors, shining a bright light inside may reveal gaps along the edges of repairs where the borders are not perfectly closed.

3.2 Other techniques without sampling

3.2.1 Radiography and computed tomography

Radiography can be highly diagnostic in determining the number, type, and location of repairs. It may also indicate the type of flaw that is being repaired. CT (computed tomography) scanning permits a more nuanced understanding of a repair, as it allows sectioning in any direction through the repaired area (see II.3§2).

3.2.2 Other nondestructive testing

Nondestructive techniques, including ultrasonic testing (see II.4§2.3.3), eddy currents (see II.5§2.3.2), and thermography (see II.2§3.2), may be useful for detecting hidden repairs, either instead of radiography (when that is not possible) or to complement it.[17]

3.2.3 Surface elemental analysis: XRF/PIXE/PIGME

Surface elemental analysis—for instance X-ray fluorescence spectroscopy (XRF) and particle-induced X-ray emission spectroscopy (PIXE)—including 2D chemical mapping, allows analysis of numerous locations on the surface, including repairs and solder, brazing, and welding metal (see II.5).

3.2.4 UV

Examination of the surface under ultraviolet (UV) illumination can be helpful in locating organic fills, coatings, or painted retouching, including faked patina. Note that thickly applied organic coatings will block such fills from UV illumination. See II.2§3.1 for more on which materials fluoresce under UV light.

3.2.5 Workshop tip: knocking

As when investigating the structure of a wall, try knocking the surface with a wooden stick or a pencil to detect a patch. You'll hear a different sound if a patch is present and has a different thickness or is not perfectly connected to the metal. This technique is most effective on large outdoor monuments and sculptures. Correlating differences in sound to specific structures in the cast is much easier when the interior is visible. This empirical method can be employed to decide where to investigate further, but it requires considerable experience.

3.3 Techniques requiring sampling

3.3.1 Metallography

In order to perform a metallographic study of a repair, the sample removed must include material from the repair zone. Such a study has the potential to give information about the nature of the repair, such as (see also I.2§3):

- the method by which the repair was formed (hammered or cast, see fig. 145);
- the degree of cold working;
- the method by which the repair was locked into place.

Metallography is generally considered too damaging to undertake on works of art unless there are detached repairs that can be easily sampled or loose parts that can be easily removed. Imaging techniques such as radiography and CT scans are alternative methods to ascertain how the repair was formed and locked into place.

3.3.2 Analysis of organic fills

A number of techniques may be used to determine the identity of organic fills. These techniques require micro samples (Raman and infrared spectroscopies, chromatography, et cetera).[18]

3.3.3 Carbon-14 dating

It may be possible to date organic fills such as wax or animal-glue-based pastes using carbon-14 (see II.8§2).

Notes

1. Benvenuto Cellini (Italian, 1500–1571) took four or five years to fettle and repair his *Perseus*, completed in Florence in 1554 (Pope-Hennessy 1985, 179–81). An extreme case is the monumental equestrian statue by Pierre Hubert L'Archevêque (Swedish, 1721–1778) and Johan Tobias Sergel (Swedish, 1745–1814) of King Gustav II Adolf in Stockholm, inaugurated in 1796, for which no fewer than seventeen years were needed by the chaser and two assistants to complete fettling and finishing (Desmas 2014, 239). See also table 1.
2. Although no definitive study has been carried out, the earliest published examples the authors know about are found in the work of Giambologna (Flemish, active Florence, 1529–1608), for example *Kneeling Woman Drying Herself*, ca. 1560, Museo Nazionale del Bargello (Sturman 2001).
3. Stone 2006.
4. Brooks 1991; Rybczynski 2005.
5. For tables describing many standard thread types, see Camm 1942, 127–86.
6. Rome and Young 2003, 266–68.
7. Mille 2012.
8. Smith and Sepponen 2019, fig 11j, 47, 54.
9. Mattusch 1996, cat. no. 24, *Statue of a Youth*, fig. 24j, 232–36.
10. Biringuccio [1540] 1990, 275–77; Motture 2019, 49 and note 61.
11. Rome and Young 2003, 268–71.
12. Andrew Lacey, personal communication, 2020.
13. Determining the nature of repairs is important for decision making during subsequent conservation. For example, a Chinese bronze horse came apart during conservation when R. J. Gettens subjected it to an electrolytic bath. Unbeknownst to him, it had been previously restored using plaster (Bewer 2012). Similarly, repatination of a modern bronze using heat would not be wise if it has resinous repairs.
14. Without documentation, these observations must be considered assumptions. As an example, bronzes by Hubert Le Sueur (French, 1580–1658) in Westminster Abbey, London, are heavily flawed although Le Sueur was acknowledged as a skilled founder (Evelyn 1995, 91). Repairs may also be carried out by a workshop or foundry that did not cast the bronze.
15. See note 1 above.
16. For example, the investigation of repairs on a Roman bronze representing a *Child with bulla* (second century BCE) in the Louvre, Paris (inv. Br 17), combined with other investigations, including metal analysis, led to the identification of three major restoration phases: two before 1809, a third before 1820; see Descamps-Lequime, Mille, and Robcis 2008. See also the recent experimental work made around other Roman bronzes found in France (Adamski, Pernot, and Bouet 2020).
17. Orazi et al. 2016; Mercuri et al. 2017; Mercuri et al. 2018. Different profiles and consequently techniques of mechanical repairs were detected by thermography on the medieval *Capitoline Wolf* in Rome; see Mercuri et al. 2017.
18. See Artioli 2010.

I.5

Assembly

Donna Strahan
Benoît Mille

Additional Contributors: Aurélia Azéma, Ann Boulton, Joachim Kreutner, Andrew Lacey, Susan La Niece, Mathilde Mechling, Dominique Robcis

Not all bronze sculptures are cast in one piece. Some are intentionally cast in sections, and these parts are considered "primary castings" (as opposed to subsequent cast-on additions such as decorative elements, repairs, and reworkings, here called "secondary castings"). Casting in parts may reflect a particular founder's know-how as well as the limitations of local technology of the period and/or place. It may be done because it is easier, more efficient, or necessary to cast a complex or large work in smaller sections (see notably Case Study 7§4). Smaller elements allow for easier sprueing and limit difficult geometries that may impede metal flow or trap air (see GI§2.7), and they do not require the manipulation of large quantities of molten metal.

Casting in parts may also pertain to specific workshop or workflow organization.[1] In some instances, parts that did not cast correctly need to be remade. The number of separately cast sections, their position, and the sequence in which they were put together is sometimes synthesized visually by researchers in the form of a casting plan.

Regardless of why a bronze was cast in separate sections, the parts can be joined in a variety of ways. This chapter provides examples of the different types of joining methods, how to recognize them, and the visual and analytical techniques that assist in their characterization.

1 What types of assembly features might be encountered?

Joining methods used to assemble bronzes fall into two basic categories: metallurgical and mechanical, both of which encompass a variety of different methods (fig. 201). That said, a single joint may itself combine metallurgical and mechanical components.

1.1 Metallurgical joining: fusion welding, flow fusion welding, interlock casting, brazing, soldering

All metallurgical joining processes in bronze sculpture have in common the use of liquefied metal.

1.1.1 Welding: Fusion welding and flow fusion welding

"Welding" in modern metallurgy refers strictly to joining metal parts by creating atomic bonds. This can be done in two ways: with relatively low temperatures by forging (as is done by blacksmiths with iron, but is not used for bronze sculpture), or by partially melting the metal parts, a process referred to as fusion welding. Most forms of fusion welding used for bronze sculpture include the addition of a welding metal of a copper alloy more or less similar to that of the primary cast. High temperatures are required to achieve the melting temperature—typically above 1000°C for copper alloys. As the metals begin to melt, the parts can be joined by very careful temperature control before they become fully molten or start to lose their shape. The degree of metallurgical fusion and the strength of the joint depend on both the heat achieved during the process and the two alloys involved, namely the primary casting and the welding metal.

The methods used to melt the metal that is to be joined have changed over time. To join the separately cast parts of large bronzes during classical antiquity, a quantity of bronze was melted separately and poured along the joint until it melted the edges of the adjacent primary cast parts while filling the void between them. It acts both as a filler metal and as the main source of heat (see fig. 202, video 12, Case Study 1). This process, known as flow fusion welding, is specific to ancient large bronzes and was used systematically in their production. None of these statues was cast in a single pour.[2] So far, the only known use of this process in subsequent periods is in connection with repairs during the Renaissance in Europe.[3]
Careful control of heat is essential in fusion welding because the melting temperature of the repair metal is close to that of the cast surface. Beginning in the late nineteenth century, electric arc welding made it possible to create a pinpointed concentration of heat. In the early twentieth century, the blowtorch led to the widespread use of fusion welding.[4] Common types of modern welding equipment include TIG (tungsten inert gas) and MIG (metal inert gas) (see Case Study 7).

If the alloy of the joint (secondary casting) varies from that of the primary casting, differential corrosion may develop, thereby making the joint visible—perhaps even stand out—on the surface of a bronze (figs. 59, 203). Ancient Greek and Roman flow fusion welds often show up as a chain of oval shapes of varied color (corresponding to basins for increasing the contact area and for accumulating heat), possibly ringed by metal with a higher concentration of porosity, that are easily recognized by the naked eye (figs. 204, 205).[5]

In both the TIG and MIG processes, the molten welding metal builds up at the joint. Unlike brazing or soldering, in which the majority of the added metal wicks into the joint, fusion welding results in a considerable amount of excess metal on both the internal and the external surfaces. For aesthetic reasons, the metal is usually ground or filed away from the outside, but often remains visible on the inside of the sculpture (fig. 59, see I.5§2.1 below).

Risks of misidentification/misinterpretation

Because both welding and brazing metals are copper based (see next section), it may be difficult to distinguish between the two by visual observation alone, thus requiring specific analytical techniques, including metallography and nondestructive testing techniques (see I.5§2.2 below).

1.1.2 Soldering and brazing

In both soldering and brazing, a metal with a lower melting temperature than the primary cast is added to the joint, aided by a flux that prevents oxidation, which could interfere with the bond. The metal of the pieces to be joined by soldering and brazing do not reach temperatures high enough to melt them (unlike with welding), and the resulting joints are therefore usually weaker (see also I.4§1.2.2). Solder or brazing metal may be added to a variety of joint types (fig. 201 D, E, F, G).

Both brazing and soldering may appear as straight lines of a slightly different color than the assembled pieces (figs. 188, 203). As a rule of thumb, soldering metals are white (silver or lead with added tin, bismuth, or other components) and typically operated at lower temperature (below 450°C), whereas brazing metal is yellow (copper alloys) and operated above 450°C, but such color distinctions may be difficult to discern visually due to patination and/or corrosion, and brazed joints may be hidden by chasing. Any traces of flux left on the surface will eventually cause corrosion along the joint line (fig. 188).

Brazing and soldering have been used in a variety of periods and areas, including different parts of Asia (fig. 206) and modern Europe.[6] As the brazing or soldering metal is of a different alloy than that of the surrounding parts, it may be revealed over time through corrosion.

Risks of misidentification/misinterpretation

Corrosion, patina, and coatings may make it difficult to see a joint, which would be a different color. Later repairs may also complicate the identification of original soldering and brazing (see I.5§2.3 below and I.4§1.2.2). Radiography often remains the best way to identify the techniques used, and may even give an indication of the materials (figs. 187, 206).

1.1.3 Interlock casting or lock-on casting

In interlock casting, the temperature does not get high enough to produce a metallurgical joint. The poured-in metal may fill the gap between the joined pieces (fig. 201 F) and may also fill holes along the edges of the sections to be joined, ultimately acting as a kind of "staple" (fig. 201 G). Any metal can be used, including molten brazing or soldering metal. In a similar process, lap joints may also be secured by pouring in molten metal (fig. 201 I, J, see I.5§1.2.1 below). Interlock casting is found on many a bronze from Renaissance Italy (figs. 100, 207), as fusion welding was not practiced at the time, and also in a variety of other contexts.

Interlock casting appears as a combination of features, including a more or less continuous and thick joint area, which may be filled with brazing or soldering metal or surrounded by patch-like forms.

Risks of misidentification/misinterpretation

A rhythmic pattern of patch-like features in a concentrated area may pertain to any number of things (for instance core pin hole or armature fills) rather than joints (fig. 119).

1.1.4 Casting on parts

Casting on parts is a particular type of interlock casting that refers to the joining of two parts by casting one element onto another, preexisting cast section. An entire section, such as an arm, may be cast onto a precast unit (primary casting). This can be used on large bronze sculptural objects to add parts (figs. 208, 209) as well as on smaller objects. Usually, this does not result in a fusion weld. Casting onto existing tangs or projections on the primary casting, as seen notably in ancient Chinese and Cambodian bronzes, helps strengthen the assembly.[7] And casting on a part may also constitute a repair (see I.4§1.2.1).

Because the cast-on section will tend to shrink during cooling, a gap between the sections will often remain visible unless the joint area is reworked by hammering. Radiography is often necessary in order to identify a cast-on piece, whether it is an original assembly or a repair.

Risks of misidentification/misinterpretation

It may be difficult to distinguish cast-on parts from additions made in the wax model (see I.5§2.1 below).

1.2 Mechanical joining

Mechanical joining can be achieved in a number of ways. One of the benefits (and potential weaknesses) is that some mechanical joining processes also allow for disassembly of the object.[8] This was, for instance in early modern and modern Europe, key for chef-modèles that were cast in parts and assembled with removable rivets so that the sections could be separated for molding as needed (fig. 210).

1.2.1 Lap joints

Greater strength can be achieved by overlapping the two edges of a joint. Once overlapped, any number of methods may be used to secure the joint, such as solder, brazing metal, or inserted pins or rivets (fig. 201 I, J, K).

A sleeve joint is a specific type of lap joint in which a recessed sleeve is integrally cast onto one section (the male part) that is designed to slide inside the section to which it is affixed (the female part) (fig. 201 K). The attachment may be secured by one or more pins or rivets. This joining technique is common in a variety of cultures (fig. 65 point 4, figs. 127, 211, 212); it is often referred to as comprising mortise and tenon joints in the study of Southeast Asian bronzes and Roman joints by scholars of European bronze sculpture. Instead of (or in addition to) pins or rivets, molten metal may be poured into the joint to secure the parts.

Without strong lighting, a perfectly designed sleeve joint may be nearly invisible on the exterior of the bronze when the outer contours of both sections meet in a narrow line and the pins are filed flush. The best designs often conceal even these very fine joint lines beneath clothing, drapery, or other design elements, making them very difficult to detect without radiography (figs. 212, 213) or access to the interior.

Risks of misidentification/misinterpretation

Repair plugs that are in the area of a joint may be misidentified as pins used to secure the latter. Joints in which the two edges are flush with one another (for instance sleeve joints, interlock casting, and end-to-end joints) may be difficult to distinguish from one another (fig. 201).

1.2.2 End-to-end joints (butt joints)

As the name suggests, end-to-end joints butt up to each other without overlapping (fig. 201 D, H). They may be strengthened with some form of mechanical interlocking (such as rivets, fig. 201 H) and/or reinforced across the back or on the inner hidden surface with a layer of melted metal. Without such additional reinforcement, end-to-end joints are intrinsically weak, and with the exception of butt welding, they are most of the time mechanically reinforced.

Risks of misidentification/misinterpretation

From the outer surface, end-to-end joints look very similar to interlock casting joints.

1.2.3 Tangs and other extensions

A tang is an extension cast integrally with the figure and used to attach parts, mainly but not exclusively to mount a figure to its base (figs. 214, 215). A tang may be solid (for instance an integrally cast sprue extending from the foot of a figure) or hollow. The attachment process may simply consist of slotting the tang into a prepared hole in the base or feature to be added. In the case of a base, it may be left loose to allow it to be separated easily, or hammered down and splayed to lock it in place (fig. 216). Or one or more tapered pins, rivets, or screws may be used to hold the two pieces together (fig. 217).

We recommend that the term "tang" be used for sleeve joints that secure a separately cast element to a base (figs. 217, 218). Another variation common with sand casting in France is the use of threaded rods cast integrally with the figure and secured with nuts (fig. 219, see also Case Study 6).[9] An extra metal ridge may be cast around the area planned for the joint so that this metal can be hammered over the seam to disguise it after joining. The use of tangs and other types of extensions is common in many cultures.

1.2.4 Miscellaneous

- A simpler way to attach a figure to a base is with screws inserted through holes drilled and tapped in the separately cast base, and then into tabs cast on the interior of the feet through which holes have also been drilled and tapped (fig. 220). See I.4§2.6 for more about nuts and bolts and how they may help to date a bronze.
- Armatures that are preserved after casting may be used to attach sculptures to separate bases (fig. 221). And in the case of bronzes cast in parts, or damaged archaeological bronzes that need support, a structural armature may be fabricated to fit inside the bronze post-casting to hold the parts in place and anchor them to a base (fig. 222).
- Metal tabs attached to one piece can be bent over the adjoining piece, usually on the interior. Examples on cast bronze sculpture are unknown to date. This technique is more usual on sculptures made of hammered sheet metal (fig. 223).
- Dovetail joints, more commonly found in woodwork, have also been used to join separately cast elements (fig. 224). These too may be designed to slot easily together and come apart.
- Cold-hammer joining, also known as pressure joining, consists of hammering two pieces of metal together with enough pressure to deform the metal until a joint is attained. It is difficult to achieve on copper alloys and lacks strength, as it does not produce a metallurgical joint. A rare example of this technique is found on solid-cast Chola dynasty (855–1280 CE) Indian bronzes (fig. 216).
- A consecrated sculptural image may contain sacred relics that were deposited inside after casting. To close the aperture, sheet metal is generally mechanically joined and held in place using metal burrs chiseled around the opening (figs. 225, 226).
- Organic adhesives such as waxes, bitumen, or resins may also be used to join components. Plaster, cement, and other such materials have also been used to hold sculptures together.

2 Why investigate assemblies? and other FAQs

As with other technical features, the location, pattern, and type of assembly can be informative regarding choices that were made at the time of the sculpture's creation (the original structure potentially represented by a casting map), technologies available, and skill of the founder. They can also reveal information about flaws and damage—the structural integrity and condition of an object. Discovery of anachronistic technologies can lead to a reevaluation of the dating of a bronze or to the presence of restoration work.

2.1 Is it possible to distinguish between a metal-to-metal joint and a wax-to-wax joint?

In some circumstances, both types of joints may look alike, particularly on radiography, where they may both exhibit circular or elliptical metal wall thickening. Here are some clues to make the distinction:

- The exterior of wax joints can be smoothed relatively easily and is generally not visible on the outside of the subsequent cast. In contrast, separately cast sections that are then joined in the metal may have a gap that needs to be hidden with cold work. Strong lighting may reveal such metal-to-metal joints. Special attention is required for large bronzes from classical antiquity, where fusion welding was often carried out so perfectly as to be completely undetectable.[10]
- Unlike wax joints, metal joints imply separately cast parts, and thus possible differences in the alloy composition of the different parts. Also, should they be present, brazing and soldering generate varying alloy compositions at the joint.
- Wax joints present either a lack of metal or a small amount of extra metal on the inside of the joint. Fusion welding (including flow fusion welding) may generate extensive additional metal along the joint.[11]

2.2 Is it possible to distinguish between different metallurgical joints (brazing, soldering, and flow fusion welding)?

Brazing and soldering usually appear on the surface as straight lines (fig. 206), whereas flow fusion welds often look like chains of ovals (figs. 204, 227). However, these features may be hidden, and flow fusion welds have been known to appear as straight lines. Radiography and/or surface elemental analysis may prove necessary to distinguish between flow fusion welding and brazing on the one hand (copper alloy), and soldering on the other (denser metals such as lead or tin alloys). Whereas brazing necessarily uses a slightly lower-melting-point copper alloy than that of the sculpture (primary metal), a weld will use a circa identical melting-point alloy. However, surface elemental analysis may not be sufficient to distinguish fusion welding from brazing, since results may be highly affected by patina or corrosion, thus necessitating a very accurate and well-located sampling of sound metal, or a metallographic analysis (figs. 146, 228, 229). In many cases the location of the area of interest may prove difficult to access in order to take a sample.

2.3 Is it possible to distinguish between features associated with the original assembly in the foundry and joints linked to later repairs?

- Elements such as tenons or sleeves cast with the bulk of the sculpture to accommodate the securing of a limb suggest that the figure was designed to be cast in parts.
- Repairs undertaken at a later date are sometimes made of a significantly different alloy than the original. This usually shows up in the manner used to attach the repairs and through differences in color.
- If significant time has elapsed, the difference in or lack of patina on the new repairs may help to identify them.
- If the object dates to before the mid-nineteenth century, repairs attached using more modern methods such as machine-made screws and bolts will be identifiable as new.
- It is often possible to distinguish modern solders by their composition, which might include elements that have only been available in recent times, but this requires analysis.

2.4 What can we learn by investigating the way in which a sculpture is assembled? Can it indicate something about the workshop or the period?

The very fact of a sculpture having been cast in sections may be specific to a period and therefore provide an initial indication of the date of production.[12] The sequence and method by which a sculpture was assembled may be characteristic of a foundry, workshop, or period. It may reflect a founder's ability to cast large pieces or indicate preferences for certain types of complex molds, idiosyncratic sprueing systems, and mastery of efficient joining processes. To date not enough studies have been carried out on related bronzes with documentation using casting plans to characterize a workshop or period. Adopting the practice of making a casting plan to document the separately cast elements of a sculpture could be helpful in comparing the joining methods used.

2.5 What can the assembly technique tell us about the workshop or the period?

A number of assembly techniques such as sleeve joints are ubiquitous throughout human production of bronzes. However, some may be specific to a foundry or characteristic of a culture. For instance, flow fusion welding to date is only known to have been used broadly in the fabrication of bronzes during classical antiquity.[13] Then, also, the aforementioned Chola dynasty bronzes made during the ninth to thirteenth centuries in south India show unique joining techniques (fig. 216). And in

Renaissance Europe, the use of handmade screws were designed to add elements to Severo da Ravenna's (Italian, active 1496–before 1538) bronzes.[14] More studies are needed to draw firm conclusions for other contexts (for example modern Europe or Asia).

3 Checklist: How do we investigate and differentiate assembly features?

Detecting what assembly methods have been used can prove difficult because the evidence may have been obscured during the chasing and finishing or hidden by corrosion (the latter tends to reveal to the naked eye more than it hides, but prevents any accurate surface analysis). Pay particular attention to areas that might have been added as appendages, such as arms, hands, legs, heads, objects held in hands (including attributes), bases, elaborate drapery folds or coiffures, halos, or jewelry, as well as seam lines, variations in metal color, rivets, or plugs. For details in examination and analytical techniques please refer to tables 5, 10, 13.

3.1 Visual examination, including raking light, microscopy

Direct visual examination is usually the first step when looking for joints. These may be indicated by a variation in surface coloration, uneven surfaces, or thin gaps between sections, whatever the process. The aid of a binocular microscope and the use of raking light may be useful. As indicated above, the presence of screws, bolts, nails, staples, plugs, or pins may indicate a joint area.

3.2 Nondestructive testing, including radiography, ultrasonic testing, eddy currents, thermography

Radiography (whatever the source: X-ray, gamma, neutrons) may help identify the presence and type of joint. Any additional metal that might have flowed around the joint on the interior is easily detected by radiography (figs. 187, 206), as are most mechanical joints (figs. 211, 212). An industrial CT scan may prove necessary to resolve specific joining queries.[15]

Ultrasonic testing may detect discontinuities in the metal surface that indicate the presence of joints. This type of testing is also sensitive to variations of metal thickness and density and is thus able to track metallurgical joints (figs. 230, 231).

By using eddy current measurements across a suspected joint, it might be possible to detect a change in variation in the electrical conductivity and magnetic permeability of an object that is not otherwise visible, thereby identifying a joint (but hardly the precise process of joining).

Thermography may also reveal a thermal anomaly at a joint and/or changes in thermal behavior between two separately cast parts.

3.3 Analytical techniques without sampling

Surface elemental analysis such as X-ray fluorescence spectroscopy (XRF) or particle-induced X-ray emission spectroscopy (PIXE) can detect differences in the alloy composition when scanned across a suspected joint (see fig. 188 and II.6§2.2 for details and limitations). It can help distinguish a metallurgical joint (variations in alloy composition) from a mechanical joint (no added metal, and therefore no variations in alloy). It can also reveal the presence of solder (a non-copper alloy).

3.4 Analytical techniques requiring sampling

Sampling is generally only carried out when nondestructive methods have failed to produce clear results (see II.5§4, II.6§3). In order to identify slight variations in metal composition (for example from separately cast parts, later additions, or repairs) it may be necessary to drill metal samples for analysis using high-sensitivity techniques. Metallographic cross sections may be necessary to distinguish between flow fusion welding and brazing by revealing the different microstructures (figs. 228, 229) Although an unusual step to take, cross sections may also reveal cast-on joints (fig. 209).

Notes

1. Separate casting may be part of serial production systems, as seen in a number of cultures, including the Khmer Angkorian period (Bourgarit et al. 2003) and nineteenth-century bronze editions in France (Lebon 2003, 45). It can also be used to allow iconographic flexibility, as seen for example with Roman imperial statuettes (Hill 1982, kindly advised by S. Descamps) and in sculptures by Hubert Le Sueur (French, ca. 1580–1658) and contemporaries, whose small equestrian sculptures had interchangeable heads (see for example Leithe-Jasper 1986, 246–53, cat. nos. 66a and b [Manfred Leithe-Jasper]; Evelyn 2000, 144–45).
2. Haynes 1992, 95; Descamps-Lequime and Mille 2017.
3. For instance, the Sienese metallurgist Vannoccio Biringuccio (Italian, 1480–before 1539) describes such a process for repairing cracks in bells (Biringuccio [1540] 1990, book VI, ch. 15, 275–77).
4. Carlisle 2005, 365.

5. Formigli 1984; Descamps-Lequime and Mille 2017.

6. Massimiliano Soldani Benzi (Italian, 1656–1740) describes in a letter to his London agent that in making bronzes, solder is never used to join separate pieces together, but only molten bronze of the same alloy so that the color will be identical. Unpublished document, referred to in a 1996 conference paper by Dr. Charles Avery in Berlin. We are grateful to Dr. Avery for allowing us to include his discovery prior to publication.

7. Gettens 1967.

8. One example is Leone Leoni's (Italian, 1509–1590) fabulous sculpture of Charles V and the Fury (1551–55) in the Museo del Prado, Madrid, whose armor can be taken off, revealing the nude king.

9. See Suverkrop 1912, 928.

10. See for example the *Piombino Apollo* (Greek, 120–100 BC, H: 115.5 cm, Musée du Louvre, inv. Br2). The weld joint between the legs and the body is so perfect that despite thorough examination (X-radiography, endoscopy, phased-array ultrasonic testing), it was only revealed at one small area, namely a platform joint between the two legs. Based on this observation, either each leg was cast separately, or one or the other leg was cast separately and joined to the body (Mille and Descamps-Lequime 2017, figs. 42.5, 42.7).

11. For example, traditional X-radiography showed the presence of an increased flow of metal on the interior of the neck area of a thirteenth-century Khmer sculpture depicting Seated Brahma, now in the collection of the Walters Art Museum, Baltimore (inv. 54.2734, H. 29.5 cm). This could be interpreted as the thickening that occurs in a wax casting model when a separately cast head is attached to the body using a wax-to-wax joint. However, examination of sequential images in a CT scan showing the neck in profile and face-on revealed a thickness of metal across the interior of the neck and extensive metal flow along the armature extending into the body, implying a cast-on technique was used to cast the head onto an already-cast body (fig. 232). This was then verified via an X-ray fluorescence spectroscopy (XRF) scan across the joint (Lauffenburger, Strahan, and Gates 2016, 33).

12. For example, all ancient large bronzes studied to date have been shown to be cast in sections and assembled (Mille 2017). Early Italian Renaissance large bronzes were also cast in sections, but a number of Renaissance bronzes from Italy and France were cast in one pour, without any assembly; see for example Bourgarit et al. 2014. For a summary of attitudes toward the single pour see Motture 2019, 227–28, with references. See Case Study 7§4 for the way Andrew Lacey, a contemporary artist and founder, envisions his casting plans.

13. Descamps-Lequime and Mille 2017. In addition, a typology has been set up for the oval-shape variant (welding in basins) of flow fusion welding (Azéma 2013, 140; Azéma et al. 2017). The distinguishing features include the way the sections to be welded are prepared, and whether the welding is superficial or goes through the entire thickness of the metal wall. It is not yet clear if these variants have a chronological significance. Hopefully, this typological approach will be extended to objects from other periods and/or geographical areas, for example Asian bronzes.

14. Stone 2006; Smith 2008b; Smith 2013; Warren 2014, 140–41.

15. Formigli 1984; Descamps-Lequime and Mille 2017.

I.6

Tool Marks

Pete Dandridge
David Bourgarit

Additional Contributors: Antoine Amarger, Aurélia Azéma, Francesca G. Bewer, Nicolau Escanilla Artigas, Laurence Garenne-Marot, Joachim Kreutner, Andrew Lacey, Elisabeth Lebon, Jeffrey Maish, Mathilde Mechling, Benoît Mille, Lorenzo Morigi, Peta Motture, Barbara Plankensteiner, Dominique Robcis, Brice Vincent

This section discusses and provides examples of the different types of tool marks that might be encountered on either the interior or exterior of a bronze sculpture. It outlines the working processes that can create tool marks, and focuses on how these marks may be identified and associated with a specific action. A flowchart summarizes the different possible tool marks (fig. 233). Features derived from the casting and fabrication process are discussed in I.1.

1 What types of tool marks and how they occur

Since the external surface of a sculpture is more accessible than its internal surface, a large majority of the tool marks mentioned below will be found on the exterior. However, a number of marks may also appear on the inside. If a mark is specific to an action associated with either the interior or exterior surface, its location will be indicated. Tool marks can derive from four different stages in the fabrication of a bronze: those made in the model; those made in the mold; those made directly on the metal surface after casting; and those stemming from later intervention(s). Although technically similar to many other tool marks, inscriptions are treated as a separate category of features (fig. 233).

1.1 Tool marks made prior to casting

1.1.1 Tool marks made in the original model and/or the inter-model

In creating a model or a wax inter-model for an indirectly cast sculpture, a variety of generic tools and/or tools specific to a particular medium might be utilized (tool marks made in the original model will predate those in the inter-model). The associated marks would be transferred to the external surface of the finished sculpture, whether the model was made in wax (fig. 51), wood (fig. 101), clay, plaster, stone, or another material (fig. 130).[1] The fidelity with which these marks are transferred to the cast-metal surface will reflect the physical properties and composition of its refractory mold and/or alloy, including the latter's castability, as well as the success of that particular casting, the subsequent working of the surfaces, and possible later interventions.

Most marks appear on the external surface of the finished work. However, those that appear on the internal surface usually derive from the tools used to work the interior surface of a hollow wax model (figs. 83, 234) or, occasionally, the surface of the core for a sand casting.

1.1.2 Tool marks made in the mold

In indirect casting, the artist or founder has the opportunity to work the interior surface of the mold either to introduce new details or to reinforce features not sufficiently well reproduced from the original model. How those marks then transfer to the casting will depend upon the technique utilized. If a wax inter-model is created, the tool marks will be transferred to its surface, with the artist then being able to assess the fidelity of the tooling with the option of further reworking. In sand casting, any alteration within the mold would be transferred directly to the casting.[2] All of these marks appear only on the external surface and can be very difficult to differentiate from one another.

1.1.3 Tool marks made in the casting core

Marks made on the surface of the casting core will be transferred to the internal surface of the bronze.[3] Some of the most common marks result from shaping the core (figs. 78, 118). Obvious examples are the parallel lines cut into a bell's core during its shaping on a lathe, with individual tools or with a strickle cut into the shape of the required form.

1.2 Marks of manufacture in the metal

After casting, the sculpture is generally cold worked with a variety of tools used to disguise manufacturing features, to sharpen and/or modify details transferred from the model (some of which may not have been sufficiently defined in the cast), and/or to decorate the surface.

1.2.1 To disguise

When the casting is removed in its raw state from its mold or investment, it will retain its sprues or gating system now translated into metal, and possibly core pins, as well as flaws such as flashes or casting defects (figs. 104, 235, 236). If there are separately cast elements joined together, the joints also may need to be disguised. In modern art foundries, the multistep process of removing or disguising coarse exterior features is divided into two separate steps: fettling and chasing. Fettling is usually carried out directly after casting to remove the bulk of those features that have become superfluous and in the way—that is, sprues, pins, flashing, and so on (videos 11, 13).[4] The person carrying this out may be less trained than the chaser and might use saws, hammers, chisels, and/or coarse files.[5] Power tools such as grinders, power files, rotary tools, drills, and air- or electric-powered pencils with various stones and burrs have been used for fettling since the second half of the twentieth century (fig. 237).[6] Once fettled, the chasing may begin.

1.2.2 To remove the core

The core is frequently removed after the casting using a variety of tools, potentially leaving scratches or chisel marks on the internal surface of the statue (fig. 60). It is often difficult to identify exactly what implements were used. Dipping the bronze into an acid bath has also been a way to break down plaster-based investment, but that does not tend to leave traces.

1.2.3 To sharpen, enhance, or decorate

The amount of work required to bring a casting to its finished state depends upon the quality of its as-cast surface. While it is possible that no significant chasing will be necessary,[7] more usually the founder, chaser, or artist will finish all of the broadly modeled surfaces and introduce or reinforce details, textures, and patterns (figs. 105, 238, 239, 240, 241, 242, video 10).[8] All evidence of such working would be restricted to the external surface.

The tools used to finish a cast surface can be divided into two broad groups: those that remove metal and those that compress metal. The former category includes files, chisels, scribes, gravers, burins, scorpers, scrapers, and abrasives, while the latter encompasses hammers, planishers, punches, grounders, and burnishers.[9] The number of variants within each group and tool type is potentially vast, since chasers often make the precise tools they need for the job at hand (figs. 243, 244, 245).

Expediency might dictate flexibility in the use of a tool. For instance, a graver or a chisel could be used to remove metal and then as a punch to create a decorative pattern or a line, respectively. Peening, which involves the use of a small ball-peen hammer with a rounded head, can be used both to disguise porosity or other surface flaws as well as to enliven the bronze surface with myriad compressed facets that catch and reflect light (figs. 238, 246).

Broad, undecorated metal surfaces with imperfections such as flashing or porosity might first be cut back with files (figs. 247, 248), chisels (fig. 249), and scrapers (fig. 250). Sometimes these tools produce incidental textures, such as chatter marks (fig. 251), which should not be confused with file marks. Subsequently, the metal might be compressed by hammering, punching, and/or burnishing (figs. 238, 239, 252). A final polish could then be achieved by soft metal brushes or a series of graded abrasives, the scratches from which might be readily observable to the naked eye or under a microscope (fig. 253). A variety of natural abrasive substances might be used, potentially including a range of stones with varying degrees of hardness or coarseness, shave grass, and powdered stones or clays (table 14). More recent are a variety of sandpapers and air abrasives.

Reliefs, designs, or details such as strands of hair, eyelids, and fingernails or toenails may be reinforced, underscored, highlighted, or created by chasing, that is by punching, engraving, and/or chiseling (figs. 169, 254, 255). An added design element might first be drawn or incised on the surface with a pointed tool such as a scribe.

Further decorative surface textures and finishes might be struck into the metal by any variety of punches (figs. 239, 240, 242, 256).

1.3 Additional marks, including inscriptions

Marks may have been added during the fabrication process to serve as aids in assembly (see I.5)[10] or registration; as votive inscriptions (fig. 257); or to identify the model or title of the piece (fig. 258), the owner (for instance for an inventory number; fig. 259), the commissioner, the artist (fig. 260), the process (fig. 261), the edition number (fig. 261), the date, and/or the founder or foundry (figs. 14, 262, 263, 264).[11]

Inscriptions may be added at any stage of a model's production. These might be drawn in a wax or clay model, allowing for the easy creation of fluid, curved shapes that might displace the wax into slightly raised lips along their edges. Marks could also be made by stamping or scoring the core or mold prior to casting, with the understanding that these will result in raised markings (figs. 264, 109). Marks of various kinds can also be punched, chiseled, or engraved in the cast piece. This requires more force and a different set of skills, especially to shape smooth, curved lines. They all produce letters or shapes with very crisply defined edges, and in the case of chiseled letters, for instance, may still preserve the fine stepped (chatter) marks representing the piecemeal progression of the tool's edge as it is propelled along by each hammer blow (figs. 259, 262, 263, 265). As with decorative marks, it is not always easy to establish exactly how an inscription was made, since a signature drawn in wax, for instance, might be reworked in the metal.[12]

1.4 Later intervention

1.4.1 Reuses and other alterations

Cleaning, polishing, and wear can introduce scratch patterns and/or erode or etch the surface either in localized areas or across the entire surface (figs. 266, 267).[13] A sculpture might have been reworked at a later date due to its being damaged or reinterpreted. The "reuse" of statues may be accompanied by a number of specific tool marks, including saw or chisel marks left as a result of a statue's removal from its initial position or base, scratches from a file used to eradicate an inscription, or marks from modern tools used to deface an object (fig. 268)[14] or rework or reintegrate a blemished area. Such alterations might also involve dismemberment or fragmentation of the statue; a connection to or combination with new elements (for instance in remounting) or marriage with pieces from other works; and/or the removal of marks of ownership, such as when changing hands. Ritual sculptures might undergo multiple alterations, including the application of repeated surface coatings over time, thereby obscuring original tooling, or deconsecration, generating new tool marks such as those associated with the removal of an attribute (fig. 269), or when they might be otherwise altered to become an "artwork" in a collection.

1.4.2 Other interventions

Marks on the surface might equally relate to deliberate actions associated with technical examination. For instance, to identify an alloy or the generic type of the metal (for instance unalloyed copper or brass), a patina might have been scratched to reveal the metal below, or a sample drilled (fig. 156) or incised with a graver or scalpel. These marks may be found on internal and external surfaces.

Taking a mold from an existing bronze to produce an after-cast may result in scratches in the patina or coating, or even into the metal of the bronze model along the seam lines of that new mold (fig. 128). Such damage can occur in the process of finessing the walls needed to define the boundaries of the mold pieces. Other traces of the after-casting process such as sketched outlines (fig. 129) or the remains of molding materials may also provide evidence that such an operation occurred. This may either confirm the relationship between a group of objects or lead to a search for one or more matching after-casts.

A sculpture's patina may have been removed deliberately in order to repatinate the work, potentially destroying or affecting existing tool marks as well as creating new ones. Archaeological bronzes have often been mechanically cleaned of surface mineralization products to find or to re-create the "original" surface, and these treatments may leave marks as well. Although not directly generated by "tools," electrochemical and acid treatments can etch the surface, leaving distinctive marks, typically tiny holes in the surface or a visibly enhanced grain structure.

Finally, other markings may provide invaluable clues to an object's life cycle and provenance. These may include painted, stenciled, or printed inventory numbers, lot numbers from auction sales, customs stamps inked directly on the metal or on labels adhered to the underside or back of a bronze, and exhibition labels.

Risks of misidentification/misinterpretation

- Air abrasives using grit, sand, glass beads, and so on are often used to remove a patina. The resulting surface can seem softened by this action, resulting in "cold worked" tool marks in the metal appearing very similar to as-cast tool marks made in the wax (fig. 270).
- Large or monumental sculptures are sometimes cleaned using very coarse abrasives resulting in a heavily abraded surface, which can have the appearance of sand casting.

2 Why investigate tool marks? and other FAQs

Investigating tool marks on a bronze sculpture may address a number of issues, some of which are listed below.

2.1 Can tooling on the model be distinguished from cold working?

Several models may precede casting—typically clay and wax models. These models may generate as-cast tool marks in the bronze (fig. 130). There are several ways to distinguish between as-cast marks and cold working:

- Usually, the tool marks made on the model prior to casting have a more fluid feel than those made in the metal due to the relative ease with which the materials can be worked (especially on a clay or wax model), with the edges being softened during casting as the more or less viscous liquid metal flows and solidifies (figs. 51, 52, 271, 272, Case Study 4). Under the microscope, it may be possible to observe that the mark retains its as-cast surface (fig. 273).
- Conversely, during cold working, significantly greater force is required to remove or deform the metal, with the resulting mark having a sharper, more defined profile that is likely to have eliminated all or most traces of the as-cast surface (fig. 254).
- Marks made prior to casting on the model(s) and/or on the mold might be deepened or reinforced in the metal, potentially leaving traces of both as-cast and chased detailing still visible on the surface.
- If decorative details on a part of a cast appear inaccessible to a tool, it suggests the tooling was made on the model and that the model was made in different parts that were decorated before being assembled prior to casting. Alternatively, the sculpture may have been cast in parts that were chased in the metal after casting and then assembled.
- When tool marks overlay work carried out after casting, such as repairs, it is obvious that the decoration has been made or at least enhanced by cold working (figs. 168, 169).

Risks of misidentification/misinterpretation

- Identifying the origins of a tool mark can be complicated by the fact that, over time, wear and cleaning can erode the surface significantly, thereby smoothing or softening its details (figs. 266, 274, 275, 276).
- If multiple castings are similar in every other respect but have different detailing, one may be tempted to suppose that the additional detailing was made either in the model or in the mold, but variable detailing can also be added during chasing.

2.2 Can as-cast surfaces be distinguished from cold-worked surfaces?

- In many instances, an as-cast surface will exhibit porosity, bumps, flashes, and other surface blemishes (fig. 104).[15] The most efficient way to detect an as-cast surface is to find the associated surface imperfections that are often retained in areas of the casting that would have been difficult to access for subsequent tooling (fig. 277).

- Transitional areas from as-cast surfaces to chased surfaces or between adjacent areas with intentionally different finishes or textures can retain tool marks associated with earlier steps in the finishing process (fig. 278).

Risks of misidentification/misinterpretation

If there is visual access to the interior of the casting, it may be tempting to use those raw surfaces as an indication of the original appearance of the exterior surface. However, the composition of the core and the investment may have been different, potentially imparting different textures to the internal and external surfaces of the metal.

2.3 Can cold working marks be differentiated?

- The cross section of a tool's working face can help characterize the kind of mark it might leave (figs. 243, 245). Gravers, chisels, and scrapers have sharp cutting edges (figs. 276, 279). Other tools used in chasing have rounded faces or blunt edges.
- Depending upon the amount of metal to be removed, a V-shaped burin or graver might be held in the hand and a delicate line incised into the surface of the metal in one continuous cut, leaving a tapering entry and exit mark (figs. 276, 279).
- Increasing the depth of the cut might require rocking the graver back and forth, creating a faceted quality to the line that can serve a decorative function as well.
- More depth yet might entail using a hammer, with each successive strike eliciting a jump in the graver and a consequent step in the cut channel. As a line changes direction or turns to create an arc, the frequency of the strikes increases, leaving a greater density of corresponding steps (fig. 280). Irrespective of the depth of the incised line, the tool creates a curl of metal at its cutting face. If the cut truncates abruptly, that curl will break off, leaving a slight nub at the end of the cut.
- All punch work is performed with the aid of a hammer. Generally, the tool is held perpendicular to the surface, with each strike compressing the metal (video 10). If the tool is used to define a linear pattern, each individual strike in the series will leave a mark variable in its depth and imparting a pattern similar to that left by a graver, but much subtler (fig. 169).

Risks of misidentification/misinterpretation

- The final shape/aspect of the tool mark may depend on the hardness of the metal surface (figs. 281, 282, 283). Also, tool marks become obscured and/or altered over time.
- Peening and punch marks are similar in shape—compare for instance figure 238 and figure 242 with figure 246. Although the latter are usually smaller, large punches exist as well. There is no definitive way to distinguish between them.

2.4 Can original cold working be distinguished from later interventions, including reuse and wear?

See I.6§1.4 above.

2.5 Can different hands be distinguished?

Experienced scholars may recognize chasing associated with a specific period or style (fig. 284). Identifying the hand of a particular artist or workshop based on tool marks is generally more complex. The fact that most sculptors and founders make and alter their own tools (see Case Study 7) is useful when comparing decorative punch marks, for instance.[16] But this avenue of research is complicated for a number of reasons:

- the generic nature of both the tools and the techniques of surface working (with the possible exception of identifiable punches; see figs. 242, 285, 286);[17]
- the fact that multiple people may have worked on the same object;[18]
- the possibility that tools, models, molds, et cetera might have been passed down or on to others.[19]

Such a quest is only possible with the systematic cataloguing of all working processes. Surface metrology techniques may provide an appropriate methodology for documenting the repertoire of tools (see I.6§3.1 below), and particularly those marks that might serve as a fingerprint.

2.6 Is it possible to assess technical aptitude?

Prehistorians, and particularly those studying fabrication techniques of lithics, have developed methods to quantify technical ability based on the statistical analysis of tool traces.[20] The objective is to evaluate the level of tool control and techniques in order to infer a number of socio-anthropological aspects linked to the activity: degree of specialization, modes of transfer of know-how, and so on. While there are numerous technical studies of bronze sculpture cataloguing a specific work's creation, an artist's working processes, and the evolution of specific technologies, to date there have been no attempts to apply such a statistical model to bronze sculpture. It may well be that the number of steps involved, the multiplicity of skills required, the potential involvement of multiple artisans, and the complexity of an artist's intention are too varied for objective analyses.

2.7 Can electric tooling be distinguished from manual tooling?

Marks from rotary electric tools will usually produce parallel curved lines if they are not removed by finer grits or filing (fig. 237).

3 Checklist: How do we investigate tool marks?

See table 5 for a summary of available techniques. See tables 10, 13, and associated chapters in volume II for details of the techniques.

3.1 Visual examination and surface metrology

The dimensions of tool marks are mostly small, often requiring classical stereomicroscopy and digital microscopy for their examination and characterization (figs. 273, 279, see also II.4§2.2). If the surface of the sculpture has been painted or patinated, possibly multiple times, the original tool marks will be masked or obscured. The surfaces of archaeological sculptures may be obscured by layers of passive and/or active corrosion products, which can either hide tool marks or diminish their legibility. Details can also be softened by handling. Under certain circumstances, it may be appropriate to selectively clean a discreet area to determine the nature of the retained tool marks.

Specific light sources, including raking light and reflectance transformation imaging (fig. 241), can enhance the reading of the surface.

In addition to digital microscopy, an increasing number of approaches and techniques are being developed that enable measurement of precise profiles of tool marks. Such details can be derived either by noncontact methods such as interferometric microscopy and photogrammetry, or by contact methods such as profilometers. Some of the methods are limited to either horizontally oriented or planar surfaces.

Silicone rubber impressions have been utilized to record surface metrology in numerous instances.[21] The technique may have the potential to record tool marks on sculpture. Analogous techniques used in other disciplines that could be applied to sculpture include technical studies of goldsmiths' work, bones, and lithics as well as the forensic sciences generally.[22]

3.2 Nondestructive testing

Radiography, whatever the source (X-rays, gammas, neutrons), might help to detect tool marks not otherwise visible. Other nondestructive techniques (thermography, ultrasonic testing) used for surface inspection may be useful as well, although the authors do not know of any application to bronze sculpture to date.

3.3 Identification/measurement of mechanical stress on the metal

Cold working induces stress on the metal, which alters its microstructure and thus also its hardness and other properties. The presence of stress can be identified and measured by a variety of methods, including hardness testing, metallography, and/or neutron or X-ray diffraction. Such measurements have been fruitfully applied to the study of hammer-hardened protohistoric bronze axes, and these experiments may serve as a potential model for sculptural bronzes.[23] Be warned that later interventions and wear may also induce stress on a bronze surface, thus complicating the reliability and interpretation of such measurements.

Notes

1. See for example Theophilus [ca. 1122] 1979, 135, which describes the incising of the wax model. See also all the possible preparation of sockets for inlays described in I.9§1.3.
2. In traditional bell casting, the decoration is added to the wax model of the bell, or to the clay "false bell," and is thereby translated into the inner surface of the cope or investment. However, it is common practice in England (and occasionally elsewhere) to create the decoration by punching the motifs into the interior surface of the loam cope (or outer mold) in reverse, so as to produce decoration on the exterior of the bell. See Motture 2001, 28, 57n14.
3. For an example inside a bell see Motture 2001, 159, cat. 49, fig. a.
4. For examples from the Italian Renaissance naming those who worked in different roles on a specific bronze, including fettling, see Heikamp and

Paolozzi Strozzi 2014, 380–85, cat. 34 and cat. 35 (Volker Krahn) and Heikamp and Paolozzi Strozzi 2014, 300–301, cat. 15 (Dimitrios Zikos).

5. This is at least the case in a number of modern art foundries, personal communication between Andrew Lacey and David Reid and the authors, 2018.

6. Personal communication between Jean Dubos, former director of the Fonderie de Coubertin, and the authors, 2018.

7. For example Giorgio Vasari (Vasari [1550] 1960, 166, sec. 69) wonders at the quality of figures that do not need any cold work after casting.

8. For a medieval European description of the finishing of a cast censer, see for example Theophilus [ca. 1122] 1979, 138, also 86–98, 102, 149. See also Dandridge 2006 for medieval aquamaniles. A variety of authors have described the chasing of European Renaissance bronzes. See for example Edilberto Formigli's essay on Ghiberti's *Gates of Paradise* in Florence (1425–52) (Formigli 2007). Modern European treatises on chasing are numerous; see for example Garnier 1903; De Bois 1999.

9. Untracht 1968, 84–85, 111–13. For a more detailed version, see Untracht 1982. See also McGrath 2005.

10. See for example assembly marks on Andrea Riccio's (Italian, 1470–1532) *Paschal Candelabrum* (1507–16) in the Basilica of Saint Anthony in Padua (Sturman et al. 2009).

11. It is not the scope of this section (nor of I.1) to identify when the signature of an artist or founder might have been introduced into a mold or what relevance that might have in copyright issues, the dating of a cast relative to the life or death of the artist, et cetera. For a useful series of nineteenth-century signatures see Beale 1975, 50–53, and also Berman 1974 for signatures and foundry marks. Founder's marks can change over time, vary in style, and/or include numbers.

12. Some pieces signed by an artist in the model, mold, or cast might provide a means of differentiating the marks on other, lesser-known examples. Similarly, there are sculptures where the working method is known that can be used for comparison. For instance, the sculptor Chaim Jacob Lipschitz, known as Jacques Lipchitz (Lithuanian French American, 1891–1973), used to mark the wax model with his thumb just before casting in order to distinguish each edition and to prove he personally oversaw the casting (Wilkinson 2000).

13. To the authors' knowledge, existing studies focus on how wear can indicate the function of utilitarian objects, but none have focused on sculpture (Dolfini and Crellin 2016).

14. See for example the *damnatio memoriae*: the names of prominent figures in ancient Egypt and Rome were often erased from their bronze portraits when the individuals represented fell from favor.

15. For a study focused on distinguishing cold worked from as-cast surfaces see Van Langh 2012, 65–74.

16. See Formigli 2007.

17. The dimensions of clearly similar punch marks on the neck of the Medusa head of Benvenuto Cellini's (Italian, 1500–1571) *Perseus with the Head of Medusa*, now in the Loggia de' Lanzi, Piazza della Signoria, Florence, and the forehead of his bust of Bindo Altoviti (see fig. 242) have been shown to match with the use of a caliper. The punch was probably a square rod filed in a crosshatch pattern of two-by-two grooves. The nine-points pattern was used in two slightly different ways: keeping the tool parallel and running along an ideal line on the Perseus to create a defined texture, and more freely on Bindo's forehead to obtain a more confused texture imitative of skin. Personal communication between Lorenzo Morigi and the authors, 2018.

18. In 2002, 2004, and 2006, CAST:ING member Barbara Plankensteiner witnessed a division of tasks in contemporary brass casting workshops in Benin City, Nigeria, and Foumban, Cameroon, where several individuals might be involved in the process. While one single individual creates the wax model—and thus may be identified through marks made in wax—several others might work on the casting, potentially erasing part or all of the model designer's marks in the wax. Such distribution of workload and/or collaboration is a historically common practice in the production of bronze (notably large-scale) sculpture. For instances in the Italian Renaissance, see, for example, Avery 2011, ch. IX; Motture 2019, ch. 4.

19. Among the numerous examples in the Renaissance is Vasari's claim that Donatello (Italian, ca. 1386–1466) passed his "equipment, designs and models" for reliefs in the Basilica of Saint Anthony in Padua to his follower Bartolomeo Bellano (Italian, 1437–1496) (Vasari [1550] 1996, 1:433). Giambologna's (Flemish, active Florence, 1529–1608) workshop and models were passed on to his successors (Zikos 2012, 389), with a 1687 inventory citing Giambologna and Pietro Tacca's (Italian, 1577–1640) models inherited by Giovanni Battista Foggini (Italian, 1652–1725) (Lankheit 1962, 269, doc. 258).

20. See Rivero 2016 for a study of 280 engraved pieces of Cantabrian and Pyrenean Middle Magdalenian portable art. Marks considered accidents or errors in the tracing were counted negatively, and those that reflected control of a tool were counted positively.

21. See Larsen 1987; Gwinnett and Gorelick 1998; Li et al. 2011; Li et al. 2012.

22. Burd and Greene 1948; Baldwin et al. 2013.

23. Artioli et al. 2003.

I.7

Gilding and Plating

Susan La Niece
Dominique Robcis

Additional Contributors: Aurélia Azéma, Ann Boulton, Manon Castelle, Jens Daehner, Joachim Kreutner, Emmanuel Lamouche, Annick Texier, Brice Vincent

This chapter summarizes the different types of gilding and plating, how to identify them, and what technical analysis can assist.

1 What are gilding and plating?

Different metals can be applied to the surface of bronze sculptures to produce polychromatic effects, to imitate precious metal, or as protection against corrosion. (Overlays and inlays are other ways of applying metal to the surface; see I.9.) Many metals in addition to gold can be used, including platinum, silver, copper, and aluminum. This is broadly called "plating," and more specifically "gilding" when the applied metal is gold. Thus, the presence (or absence) of plating, the identification of metal used, and the method of attachment or adhesion are key elements in the technical study of any sculpture.

1.1 What are the most common gilding techniques on bronze sculpture?

1.1.1 Leaf gilding

The application of very thin sheets, known as leaf, is a common technique requiring very little precious metal.[1] Gold leaf is typically 0.1–1.2 µm thick,[2] normally 7–11 cm square, although smaller pieces are encountered (fig. 287).[3] Found as early as the second millennium BCE in Egypt, gold leaf was widely used in the Greek and Roman world (fig. 288), then declined in the West during the medieval period due to the development of mercury gilding, then gained popularity again in the Renaissance.[4] In Asia it is still used for religious sculptures, and devotees apply gold leaf to Buddha figures as an act of worship (fig. 289).[5] Leaf may be applied directly to clean metal or attached with adhesive.[6]

Variants

Three prevalent adhesive techniques are used for applying metal leaf to bronze sculpture: water gilding, oil gilding, and lacquer gilding. Gold leaf can also be applied using mercury, as described below (I.7§1.1.2). Little is known about the adhesives used on ancient sculptures because they degrade easily. In Asia, the use of lacquer is well known.[7] In the West the choice of adhesive is more complex, with a distinction generally made between water gilding and oil gilding (terminology originating from traditional wood gilding). Water gilding normally has an animal-glue adhesive over a gesso ground that is sometimes covered with a clay layer to allow the gold to be burnished for a more compact and reflective surface. This process was used in ancient Mesopotamia and Egypt, but on bronze it is uncommon in later periods.[8] Oil gilding is a process in which gold leaf is adhered to the metal surface using an oil-based adhesive; it is more suited to outdoor conditions than water gilding. Many metals in addition to gold can be applied by the oil technique, including platinum, silver, copper, and aluminum.

Identification

- Leaf gilding is very thin and therefore delicate.
- It is usually golden yellow because of the high gold content (about 90–98%).[9]
- The prime indicator that leaf has been used is the overlapping of the straight edges, visible as squares on the surface (figs. 287, 290). With wear, the overlapping edges are sometimes all that remains because of the greater thickness of material.
- Leaf has a tendency to flake off as the adhesive ages.
- Traces of a gesso coating may indicate that gilding was once present.
- Traditionally, the siccative (drying agent) added in the oil gilding process was a lead compound such as litharge (lead oxide). In the twentieth century, lead was gradually replaced by cobalt, manganese, and iron compounds, which can be detected by elemental analysis.
- For the gilding and regilding of outdoor sculptures, yellow epoxy paint is often applied as a preparation layer to prevent corrosion of the bronze, but also to hide wear. Its presence is an indicator of modern gilding.

Risks of misidentification/misinterpretation

- Dutch leaf, around 0.3 μm thick and made of brass, is sometimes used in imitation of gold leaf. It may be detected because it darkens (tarnishes), and elemental analysis such as X-ray fluorescence spectroscopy (XRF) can be used to further distinguish it.
- Gold-colored paint, often used in restoration, can be identified under magnification by its granular appearance (see I.8§1.2.1). This may be made with powdered gold or brass in a binder. Modern "gold" paints may contain nonmetallic luster pigments.

1.1.2 Mercury gilding (also known as fire gilding or amalgam gilding)

Gold (and less commonly silver or tin) can be amalgamated with mercury to form a metallic paste to apply to the casting. The gilded metal is heated to about 250–400°C to vaporize the free mercury, leaving a matte layer of gold.[10] The resulting porous surface is then usually burnished to form a compact layer. If the substrate is carefully prepared prior to gilding, burnishing can produce a highly reflective and durable surface, with the gold effectively fused with the surface of the bronze. A further process known as *mise en couleur* was used at least as early as the Renaissance period to alter the reflective qualities and color of the gilding by applying various chemicals. Historical texts describe the process, which involved heating chemicals such as sulfur, saltpeter, and other metal salts.[11]

Mercury gilding has been widely used on sculpture (fig. 291), appearing in China as early as the Warring States period (475–221 BCE) (fig. 292).[12] In Europe, it can be found a little later, in the Late Iron Age and Roman periods, flourishing in the Middle Ages, the Renaissance,[13] and into the nineteenth century.[14] Safety legislation brought its use almost completely to an end in Europe, though not in all parts of the world,[15] by the close of the nineteenth century, due to the toxicity of the mercury vapors. Because of the semiliquid application, it is more suited to gilding complex shapes than foil or leaf. It has also been used to gild selected parts of an object (known as parcel gilding) (fig. 293).

Variants

An alternative method is to apply mercury directly onto the surface of the object, lay on gold leaf, and then heat.[16]

Identification

- The detection of mercury is the key identifier: mercury compounds remain in the gilding even after the free mercury is driven off.[17]
- Amalgam gilding is applied as a semiliquid, so may show splashes or runs (figs. 293, 294), and adherence to the bronze substrate is good, though the gilding may show wear at high points.
- The thickness of the gilding layer can vary from 1 to 10 μm.[18]

- A metallographic cross section of mercury gilding will show good bonding to the bronze substrate (fig. 295).
- The interior of a sculpture is never fully gilded, though drips may be present.
- If the lead content of a sculpture exceeds 1% it has been thought unlikely that the surface was mercury gilded because lead globules on the surface spoil the gilding,[19] but this does not seem to be a universal rule.[20]

Risks of misidentification/misinterpretation

- Mercury may be present on a sculpture when there is no mercury gilding, for example if the red pigment cinnabar (mercury sulfide) is present. This was observed on some Tibetan bronzes that were gilded using other techniques.[21]
- Minute traces of mercury may be present as residues of some metal polishes,[22] or from quicking for electroplating (see I.7§1.1.3 below), although these traces are usually too small to be detected by XRF.

1.1.3 Electroplating (also known as electrogilding or voltaic gilding)

Electroplating was developed in the first half of the nineteenth century.[23] It relies on an external electrical current to dissolve the plating metal (the anode) and move the ions through an electrically conductive electrolyte solution to the object to be plated (the cathode). It has good adhesion and is able to consistently produce a very thin, continuous plating layer, consequently becoming the most commercially used technique for plating from the mid-nineteenth century until today.[24]

The metals that are commonly applied by electroplating include gold, silver, nickel, chromium, and zinc. The plating can be deposited as pure metal, as successive layers of different metals, or as alloys. Any surface, not only metal, can be electroplated, including plastic, plaster, and resin, if prepared with a conductive coating. Chemical and/or electrolytic treatments have also been used in electroplating to obtain a wider color palette.

Identification

- Unless the piece is marked as electroplated, it can be a difficult technique to identify.
- If concealed areas such as the base of a sculpture are gilded, it is likely that the sculpture was electroplated because the effort required to mask these areas during plating would be greater than the value of the gold saved.
- Visible indicators of electroplating include loss of the thin plating due to wear, and also blistering of the plating layer due to the substrate being imperfectly cleaned or the electroplating bath not being adequately maintained (fig. 296).
- If nickel is detected analytically through the gilding, it may be a base layer for electroplating.[25]
- Electroplating is very often less than 1 μm thick for modern sculptures, but it may exceed 20 μm on nineteenth-century sculptures.
- If possible, examine a cross section of the plating and substrate layers at high magnification; the plating layer will be even in texture and thickness.

Risks of misidentification/misinterpretation

- If the plating is very thick, it may be difficult to distinguish gold plating from solid gold by surface analysis.
- Confusion with amalgam plating may occasionally occur because of the use in the nineteenth and twentieth centuries of mercury salts to prepare the surface, a process known as quicking,[26] which leaves small traces of mercury.
- The electrogilding of iron is preceded by plating with copper. In such cases the under-plating may be detected.

1.2 What other gilding techniques might be found on a sculpture?

1.2.1 Mechanical plating

Mechanical attachment of metal foil is one of the earliest methods (fig. 297).[27] Foil is generally defined as a thin sheet of metal formed by hammering or rolling, and unlike metal leaf, it is thick enough to support its own weight when held at the edge. A sheet of metal foil can be wrapped around an object and held in place by folding and crimping the joints. Alternatively, it can be fixed by rivets or by punching the edges of the sheet into grooves that are hammered shut or filled with solder (fig. 298). Inserting gold foil into grooves is a method mentioned by Pliny for a sculpture of the young Alexander.[28] Rubbing thin gold or silver sheet into fine grooves or punch marks is a method used on ironwork and copper alloys.[29]

Identification

- Foil gilding tends to blur the detail of a casting (fig. 299) and can be detected by the presence of wrinkles or folded overlaps in the plating layer, or the presence of grooves in the casting (fig. 300).
- On corroded or worn metal surfaces, X-radiography may be necessary to reveal evidence of keying (roughening of the surface) to hold the gilding or silvering; traces of precious metal in the keying will show up because of its greater opacity to X-rays than the baser metal of the substrate.

1.2.2 Diffusion gilding

Diffusion gilding is the attachment of gold foil by heating and burnishing to promote interdiffusion between the gold and the metal substrate, thus creating a metal-to-metal bond without solder. It is difficult to achieve on copper alloys, as a layer of copper oxides forms easily during the heating process, and is a barrier to the plating's bonding. The technique is more common when gilding silver, both in antiquity prior to the introduction of fire gilding[30] and in modern times. It has rarely been identified on copper-alloy statuary, but it has been reported on Roman period bronzes.[31]

Identification

- Diffusion gilding can be identified by metallographic examination of a polished cross section. It cannot be identified by surface observation.
- In section it is recognizable by the bonding of the gilding layer. Effects induced by heat, in particular fine fingers of gold penetrating the surface of the bronze and grains crossing at the interface, will be visible.[32] Very fine porosity may also be apparent at the interface.
- No mercury will be present, and the thickness of the bonded gold-foil layer will be 3–10 μm.

Risks of misidentification/misinterpretation

- Because of the moderate use of heat, diffusion gilding can be similar to fire gilding in appearance, but analysis will show no mercury.
- It may also be confused with leaf gilding applied without heat, as outlines of the foils may be seen on the surface, but the penetration of gold into the bronze will not be present unless heat was used.

1.2.3 Depletion gilding

Unlike other gilding processes, depletion gilding is a subtractive process rather than an additive one and can therefore only be used on a copper alloy that contains a significant amount of gold. The process is based on the resistance of gold to oxidation: acids and/or heat are used to remove the copper from the surface of the copper alloy, leaving behind an ultra-thin surface layer of almost pure gold. It is a standard goldsmithing technique[33] for improving the golden appearance of baser gold alloys and is known from the third millennium BCE onward.[34] But it is rare on sculpture, except on small figures, for example from the pre-Columbian cultures of South and Central America.[35] Modern goldsmiths use mineral acids such as sulfuric or nitric, and it is thought that prior to the discovery of these strong acids, organic acids such as oxalic were used. The resulting matte surface is burnished or hammered to create a continuous layer of gold.[36]

Identification

- This technique is best detected by examining a polished cross section through the surface plating and measuring the gradient of gold composition from the core metal to the surface by, for example, energy-dispersive spectroscopy (EDX) in a scanning electron microscope (SEM).
- Burial in a corrosive environment can cause loss of baser metals from the surface of a gold alloy, but in that case the surface will be pitted and porous, not burnished to a smooth, continuous layer of gold.

1.2.4 Electrochemical displacement/replacement plating

Electrochemical displacement/replacement plating uses no external electrical current source. Instead it takes advantage of the natural galvanic potential difference between metals, causing atoms of the metal with the higher galvanic potential, for example gold, to attach themselves to a metal object of lower galvanic potential, for example copper. This method can be used to apply gold, silver, or tin onto copper and has been documented in pre-Columbian metalwork from Peru and Ecuador.[37] It is a method commonly used now to restore worn plating, often using proprietary pastes applied by rubbing.[38]

Identification

- This form of plating is not suited to large sculpture, but can be useful for restoring areas of worn plating on objects of any size.

- Because the reaction stops when the surface is completely coated, the plating is extremely thin (typically 0.5–2 μm) and fragile.
- No mercury is present.
- It can be distinguished from fusion plating (see I.7§1.2.5 below) since the plating layer is devoid of any metallographic structures such as dendrites, which are typical of plating applied molten.
- Electrochemical displacement plating is similar to electroplating, but usually results in a thinner layer.

1.2.5 Fusion plating

Fusion plating is a metal coating applied in molten form, which generally limits its use to small objects or components. The metal best suited to this form of plating is tin (with a melting temperature of 232°C; see I.7§1.3.2 below). Gold and silver are more difficult to apply as molten and are relatively rare, although small Iron Age and Roman items are silvered by this method,[39] and some small pre-Columbian cast-copper objects exhibit fusion silvering and gilding.[40] Fusion plating of both silver and gold requires the precious metals to be alloyed with copper to lower the melting temperature. In pre-Columbian examples these plating layers were found to have also been treated to remove copper from the gilding or silvering alloys—that is, depletion gilding/silvering took place after the fusion plating was applied.[41]

Identification

- As it is applied molten, it has the typical microstructure of a cast alloy (see I.2§3.1).
- Drips or runs may be visible on the surface, and the plating is likely to be of uneven thickness, but with no mercury present.
- Loss of copper from the surface of the plating, caused by either corrosion or deliberate removal to improve the color, may also be evident.

1.3 What are the features of silvering and tinning?

1.3.1 Silvering

Many of the methods used for gilding have been used to apply silver to sculpture.[42] The commonest is electroplating: the silver can be deposited with a satin or bright metal finish. Mercury silvering is much less common than mercury gilding, especially for statuary, but it was sometimes used in conjunction with mercury gilding in Han dynasty China (206 BCE–220 CE) (fig. 292, see I.7§1.1.2 above).[43]

Examples of mechanical application of silver-foil plating survive on small statues from such periods as the Late Bronze Age and Roman.[44] Foil attached with a continuous layer of tin-lead alloy, a technique of silvering (but not gilding) known as close plating, can be used on small sculpture. Silver coatings applied in molten form are also only suited to small figures such as those found in pre-Columbian South America.[45] Silvering with leaf, French plating (multiple layers of leaf applied using heat and pressure), and electrochemical plating are techniques not well documented on sculpture, but may have been used for repairs to plating.

Identification

- Silvering is much less durable than gilding, and where it survives it may not appear silver in color, so first the metal should be identified by elemental analysis, for example XRF.
- Techniques of application may be distinguished by the same methods as for the different gilding techniques outlined above.
- Detection of compounds containing antimony, bismuth, selenium, and sulfur indicate the use of electroplating.
- The base metal for electroplating with silver is frequently a white metal alloy of nickel, copper, and zinc. This alloy is known by various trade names and may be marked, depending for example on place of origin: EPNS (electroplated nickel silver, UK), Alpaca or Alpacca (South America), Alfenide (used by Christofle for some silver-plated pieces). If plated, the sculpture should not carry the marks "925" or "sterling."
- Evidence of mechanical plating may be seen in the form of grooves even where plating is lost (see I.7§1.2.1 above).
- Examination of a polished cross section through the substrate and the plating layer can be helpful.

Risks of misidentification/misinterpretation

- White metal coatings other than silver have been widely used, the commonest being tin (see I.7§1.3.2 below).
- Arsenic-rich surfaces are also silvery in color. A bronze bull statuette from Horoztepe, Anatolia, dating to circa 2100 BCE, has a pale coating of arsenic alloy.[46] On

small castings of arsenic-copper alloy the silvery appearance is the result of segregation of the arsenic compound from the surface of the casting during cooling.[47]

1.3.2 Tinning

Tinning dates back to at least the eighth century BCE.[48] It is applied to copper, bronze, brass, iron, and steel, either by wiping molten tin onto the heated base metal or by hot dipping into molten tin.[49] Tinning can coat whole surfaces with silvery-colored plating or, when applied selectively, can be used to make decorative patterns. It is most commonly found on small items.

Identification

- Tinned surfaces are frequently mistaken as being silvered from their visual appearance.
- Elemental analysis can confirm this, but it can be difficult to distinguish tinning from the bright silvery surfaces of cast high-tin bronzes without informed microscopic study of the surface and cross section.[50]
- Silvery surfaces are sometimes found on cast bronzes caused by a casting phenomenon known as tin sweat or inverse segregation (fig. 163).[51] This may be an accident of casting but could be deliberately encouraged to create a silvery surface on bronze.

2 Why investigate gilding and plating? and other FAQs

2.1 What can I learn from the identification of plating on a sculpture?

Correct identification can further the understanding of the technological and cultural significance of the object and its makers. Knowing the type(s) and technique(s) of plating may also help to determine the date it was applied and the subsequent history of interventions. It also informs decision making in cultural heritage conservation, for example when assessing the need for a special microclimate to protect a sculpture.

2.2 How easily can I identify whether a bronze has been plated?

Gilding and plating are often clearly visible, but it may be difficult to see when the plating is worn or concealed under corrosion. Furthermore, bronze and brass are typically golden in color, and areas of clean or varnished metal on a corroded statue can appear to be gilded when no gilding is present. Nevertheless, a combination of microscopic examination and elemental analysis can confirm the presence of plating. It should also be remembered that the plating now visible may not be original. Regilding over the lifetime of a sculpture is common (figs. 301, 302), and several techniques may be encountered on a single piece.

2.3 Are sculptures gilded using pure gold or alloyed gold?

Although gilding can be of pure gold,[52] gold alloys with several percent of copper and/or silver have often been used, making the golden color redder or whiter (fig. 303). In addition, imitations are common from at least as early as the medieval period.[53] Yellow metal is not necessarily gold, nor is white metal plating necessarily silver, so elemental analysis is advisable.

2.4 What does the presence of mercury prove?

Mercury is used for fire gilding, but it may also indicate the presence of cinnabar, the red mercury sulfide pigment. Quicking from electroplating and some metal polishes may leave traces of mercury, although the residual level of mercury in these is usually too low for XRF to detect.

2.5 Does the absence of mercury prove that mercury gilding was not carried out?

A large proportion of the mercury will have evaporated during the gilding process but some will remain as mercury compounds, so it is unusual for it to be completely undetectable by XRF analysis.

2.6 Can the thickness and composition of plating layers be linked to specific workshops?

The thickness and composition of the layers can indicate the plating technique used, but they are unlikely to be helpful in identifying a specific workshop.

2.7 Can traces of lost or hidden metal plating be detected, and can original and later intervention be distinguished?

Potentially, yes. Visual examination of the whole surface can reveal evidence, and radiography can detect plating hidden by corrosion, at least when high-density metals such as gold or silver are applied and with sufficient thickness so that they appear whiter than the surrounding bronze on the radiograph. To understand the full history, it may be necessary to take small samples from areas that are believed to represent the sequence of gilding layers for examination in cross section. The layer closest to the cast surface will be the earliest, but if all are applied by the same technique, they may be contemporary to build up thicker plating. The presence of different techniques is a sign of later interventions. Recent leaf gilding may be indicated if cobalt, manganese, or iron compounds are detected by elemental analysis, as in the twentieth century they replaced the lead-based siccative used in early oil-gilding processes. Gold-colored paints and electrochemical plating pastes are frequently used in modern restoration of damaged areas.

2.8 Is white metal plating usually made with silver?

No, it is very often tinning, especially on small sculptures from the Roman period. Arsenic-rich surfaces are also silvery in appearance.[54]

3 Checklist: How do we investigate gilding and plating?

The flowchart in figure 304 will guide you in the visual inspection of plating (see also fig. 305). For more details on how to identify plating processes, refer to the *Identification* sections in this chapter, which describe all potential processes. For details on analytical techniques see tables 5, 10, and II.6.

XRF surface analysis can determine the composition and thickness of a plating layer, though the best estimate of the thickness is obtained by ion beam analysis (Rutherford backscattering spectrometry). For further investigation, sampling may be required. SEM-EDX analysis of cross sections are often performed, but with high-purity gold the thickness can be overestimated due to smearing caused by mechanical polishing of the soft gold. Ionic polishing helps to avoid this problem. For more, see II.6§1.2.

Keep in mind that plating is often renewed and restored during the lifetime of a sculpture (fig. 289). Traces of lost or hidden metal plating are potentially detectable, and original and later interventions are sometimes distinguishable (see I.7§2.7).

Notes

1. Oddy 1993. For example, only 12 kg of gold was necessary to regild the Dome of Les Invalides, Paris.
2. Nicholson 1979.
3. For a brief introduction to contemporary practices in leaf gilding see: https://www.philamuseum.org/learn/conservation/case-study-gilding-diana.
4. Darque-Ceretti, Felder, and Aucouturier 2011.
5. Strahan and Maines 2000.
6. Darque-Ceretti, Felder, and Aucouturier 2011.
7. Strahan and Maines 2000.
8. There is evidence from analysis that animal glue was used for the very fugitive gilded detail on Donatello's (Italian, ca. 1386–1466) Chellini roundel (ca. 1450) in the V&A (A.1-1976) (Museum departmental records, cited in Motture 2019, 68). See also Oddy, Pearce, and Green 1988.
9. All compositions in this publication are in weight %.
10. Theophilus [ca. 1122] 1979; Anheuser 1997; Anheuser 1999.
11. d'Arcet 1818; Cellini [1568] 1967.
12. Lins and Oddy 1975.
13. Cellini [1568] 1967, 175; Motture 2019, 68.
14. Goodison 1974.
15. Oddy, Bimson, and La Niece 1981; Furger 2017.
16. Oddy, Bimson, and La Niece 1981; Furger 2017. Cellini did not recommend this method as too much mercury could dull the gold (Motture 2019, 68n152).
17. Margreiter et al. 2022
18. Anheuser 1997.
19. Because lead does not dissolve in copper and copper alloys but forms separate inclusions, it risks weeping out when the surface is heated for the amalgam process (lead melts at a low temperature). See Oddy, Bimson, and La Niece 1981; Oddy 2000; Vincent, Bourgarit, and Jett 2012.
20. d'Arcet 1818; Jett 1993; Strahan and Maines 2000; Cowell, La Niece, and Rawson 2003.
21. Oddy, Bimson, and La Niece 1981.
22. La Niece 1990.
23. Hunt 1973; Raub 1993; Lins 2000.
24. Oddy 1995; Grant and Patterson 2018.
25. Selwyn 1950.
26. Lins and Malenka 2000.
27. Oddy 2000.
28. Pliny the Elder [77 CE] 1857, 34.19n94.
29. Oddy 1993. The *Queen Napirasu* (1340–1300 BCE) in the Louvre (inv. Sb 2731) also shows large grooves that probably have been used for gilding. Examples are known also for Khmer bronzes (Vincent 2012, 288–90).

30. Oddy, La Niece, and Meeks 1981.

31. See Bott and Willer 2014; Willer, Schwab, and Mirschenz 2017.

32. Bott and Willer 2014; Willer, Schwab, and Mirschenz 2016b; Willer, Schwab, and Mirschenz 2017.

33. Grimwade 1999.

34. La Niece 1995.

35. Lechtman 1984; Scott 1983; Scott 2000.

36. Grimwade 1999.

37. Lechtman 1979; Scott 2000.

38. Stalker and Parker 1960.

39. Northover and Salter 1990; La Niece 1993.

40. Scott 1986.

41. Scott 1983.

42. La Niece 1990.

43. La Niece 1993.

44. La Niece 1990.

45. Scott 1986.

46. Museum of Fine Arts Boston, 58.14; Smith 1970.

47. Craddock 1981; La Niece 1989.

48. Meeks 1993.

49. Welter 2019.

50. Meeks 1986.

51. Meeks 1986.

52. Here pure gold means unalloyed gold.

53. Smith and Hawthorne 1974.

54. Smith 1970; Craddock 1981.

I.8

Patina

Ann Boulton
Susanne Gänsicke
Shelley Sturman

Additional Contributors: Denise Allen, Aurélia Azéma, Jane Bassett, Sophie Descamps-Lequime, Laurence Garenne-Marot, Joachim Kreutner, Susan La Niece, Lorenzo Morigi, Luc Robbiola, Dominique Robcis, Annick Texier, Jean-Marie Welter

This section discusses the meaning of the term "patina"; the natural and artificial formation of patinas, including chemical patinas; basic techniques of patination, including various forms of coatings (except for gilding and metal plating; for those see I.7); and how to distinguish between different types of patinas.

1 What is patina?

The widely used but elusive term "patina" comprises at least three different meanings. One prevalent view is that every object—whether a bronze or marble sculpture, a piece of furniture, or a painting—acquires a patina over time, often perceived as a pleasing surface alteration that may add aesthetic value and, perhaps more significantly, forms part of its history.

Some use the term "patina" more strictly to refer to the chemical transformation of a metal surface to a mineral layer that usually has a different color from, and reduces the bright metallic reflectance of, the polished original cast surface (fig. 306).[1] The resulting surface layer is often referred to as chemical patina. (Some people restrict the term "chemical patina" to intentional patina, but here it is used for any kind of patina resulting from a chemical transformation of the bronze surface, including corrosion.)

In addition to chemically induced patinas, the term can also encompass organic coatings applied to the surface of a sculpture to enhance and/or complement the form by changing its color, texture, saturation, and/or reflectance. A coating may be wax, varnish, lacquer, gum, oil, grease, or paint.[2] The related term "patination" may be used as a synonym, though some investigators use it more narrowly to describe only an intentionally applied coating or an induced chemical alteration layer, or both, rather than for patinas formed naturally. The diagram in figure 307 synthesizes both chemical actions and organic coatings in patina formation.[3] For the purposes of this chapter, the term "patina" is used broadly to mean any kind of surface alteration on bronze other than gilding and metal plating[4] or inlay (see I.7, I.9).

1.1 Chemical patinas

1.1.1 How do chemical patinas form?

Chemical patinas are produced by a chemical reaction, namely the transformation of the bronze surface, initially a metal, into nonmetallic compounds such as oxides. Chemical patinas can form naturally over time or be intentionally created by human intervention. Copper alloys develop surface alterations or corrosion layers almost immediately after cooling from the melt. Note that the very term "layer" may be misleading, as this process of mineralization can develop in all three dimensions rather than as a uniform stratification that follows the surface of the sculpture. Indeed, it may evolve along the complex microstructure of the metal and even take the shape of organic materials that are in close contact with the surface (as for example pseudomorphs, see I.8§1.1.2 below).

In addition, layers consisting of various copper corrosion products, of colors ranging from green and blue to red and black (such as copper chlorides, carbonates, oxides, or sulfates),[5] can form naturally over time due to burial in soil (figs. 308, 309, 310), submersion in water in aerobic or anaerobic conditions (fig. 311), or exposure to the atmosphere either indoors or out (fig. 312).[6] In this case, patinas are referred to as "natural." Alternatively, it is also possible for a layer of copper corrosion products to be formed by an artist, founder, restorer, et cetera, using chemicals with or without heat to create a controlled corrosion of the surface that changes the color (video 14). In this case, the patina is considered "artificial."[7] It is perfectly possible, indeed common, for both artificial and natural patinas to occur together on the same sculpture.

Certain preparatory steps are involved in producing a stable, artificial chemical patina. Following casting, fettling, and chasing, the bronze may be cleaned in an acidic bath to strip away oxides or other residue from the casting process (fig. 313). This is current practice, but the use of acid for tarnish removal goes back to antiquity, and might also have been a method of cleaning the metal prior to patination. Patina creation begins when chemical solutions are applied either hot or cold to the surface and allowed to react with the metal (video 14).[8] The final desired appearance is achieved using an often quite complex sequence of chemical applications selected for their specific color reaction with that particular bronze alloy.[9] Indeed, the composition of the substrate metal will influence the color of the chemically induced patina (fig. 314). Particular combinations of chemicals and alloys have been established over millennia to create specific patina colors. A striking example of this are the so-called black bronzes from antiquity.[10] Note also the well-documented phenomenon occurring in nineteenth-century Paris where patina creation became extremely elaborate and complex and drew on various patina technologies and fashions from diverse time periods and cultures around the world.[11]

1.1.2 What do chemical patinas look like?

The chemical alteration of a bronze surface may take many forms: a layer of corrosion may be smooth or rough (fig. 315), continuous or irregular (figs. 316, 317), powdery or well adhered (fig. 308). All have in common that they obfuscate the shiny surface of the polished metal, though the mineralized layer may be rubbed through on certain high points of a patinated form, either intentionally or through repeated handling or abrasion. On ancient and historic bronzes, a chemical patina can enhance value and beauty when the surface alteration is a continuous and smooth layer and preserves the original detail (fig. 309). Conversely, a thick, uneven corrosion buildup, sometimes amplified by additional surface accretions as on a bronze sculpture recovered from an underwater environment, can obscure, if not distort, the surface. Often such disfiguring features will be deemed as warranting partial removal (fig. 311).[12] There is great risk, however, in removing such corrosion products, as the topography of the original surface may itself have been disturbed, and evidence of an original patina, paint, or other surface decoration may be irretrievably lost in the process.[13] And it can also lead to the removal of invaluable, contextual evidence. In fact, under certain burial conditions, the copper corrosion products may seep into, and eventually take the shape of, animal or vegetable materials such as textiles that touch the surface, thereby forming pseudomorphs (fig. 318).[14]

When it comes to intentionally created chemical patinas, a wide range of colors and effects may be produced.[15] The patination process is highly specialized, and to create a range of color and depth that complements the sculpture is an art form in itself.

1.1.3 Why apply an artificial chemical patina?

There are various reasons for the intentional application or creation of patinas on bronze sculpture. Primarily, these relate to the final appearance of the surface as desired by the artist or the commissioner (see Case Study 7), as dictated by cultural or religious conventions,[16] or as part of a conservation or restoration treatment (figs. 306, 319). More specifically:

- A freshly cast, unpatinated bronze surface (fig. 313) is highly susceptible to its environment and will immediately begin to react with the oxygen and chemicals in the air to form a protective mineralized layer or tarnish. This is usually considered undesirable, and to circumvent this natural corrosion, artists or founders impose their chosen surface by altering the color of the as-cast surface or by reducing the bright metallic sheen of polished metal, notably to enhance appreciation of the form (fig. 320).
- If the artwork is intended for display outdoors, a patina can serve to protect the surface from damage by exposure to the elements. This may be a chemical patina that forms a passive protective layer by slowing down further oxidation.[17]
- Diverse patinas may be applied selectively to differentiate the color of different parts of a sculpture (fig. 321).
- Patinas may be applied for other aesthetic considerations, such as to make an object appear aged.[18] For example, green patinas were intentionally created as early as the thirteenth century CE on Song dynasty bronzes[19] to imitate much older bronzes from previous historical periods. Similarly, there may be curatorial consensus during conservation treatment of a damaged bronze surface that provides an acceptable rationale for re-creating an aged surface to conform to accepted ideas of how the piece should appear (figs. 306, 322).[20] However, some curators and/or institutions do not agree with repatination, regardless of whether it is in imitation of existing remnants or in an attempt to re-create what was assumed to be the original surface treatment. In the nineteenth century some ancient bronzes were chemically treated and patinated black to reinstate what was believed to have been their original color.[21] Archaeological bronzes were more likely to have been heavily restored in past centuries, while contemporary conservators generally favor a minimalist approach.
- Artificially aged surfaces are sometimes induced for willful deception, as in the creation of a forgery or to conceal a restoration.[22]

1.2 Applied coatings

1.2.1 What are applied coatings?

Patinas also come in the form of organic coatings and may be applied by an artist, founder, restorer, maintenance worker, or others (fig. 323). These may also be referred to as "artificial patinas." Applied coatings composed of natural substances such as resin, oil, wax, or, more recently, synthetic resins create distinct surface layers that are usually chosen to protect rather than react chemically with the metal surface. Such coatings may be chosen to produce a range of diverse effects:

- Most paints or lacquers loaded with pigment produce an opaque coating (fig. 324).
- Lacquers, resins, heated oils, waxes tinted with dye or translucent pigments let the underlying surface show through (fig. 325).[23]
- Paint or lacquer mixed with metallic powders, mica, or "shell gold" lend a more sparkling, though not shiny, "metallic look" to a cast (figs. 326, 327).[24]

Surface coatings may be applied directly onto the polished metal, but they are often applied over chemically patinated surfaces as part of the original patination color/gloss/translucency scheme, or for added protection (see I.8§1.2.2).

1.2.2 Why apply coatings?

As with chemical patinas, coatings may be applied for both aesthetic and protective reasons:

- Outdoor sculptures now are often exposed to highly corrosive environments and can become severely disfigured and damaged (fig. 317), although prior to the Industrial Revolution outdoor bronzes left to their own devices could develop patinas described as "attractive" (fig. 328).[25] Clear surface coatings may be applied over bright metal or intentionally induced chemical patinas to prevent corrosion caused by weather and air pollution.
- Coatings may protect indoor bronzes as well, notably from damage caused by frequent handling.[26]
- Coatings can increase color saturation and surface gloss to change the aesthetic appearance of the object.[27]
- Regular maintenance and application of protective coatings are important to preserve the bronze and the patina, a practice that has been in use since at least ancient Roman times.[28] In spite of such maintenance, bronze surfaces may still change in appearance and color over time.

- Tinted surface coatings may also be applied over chemically patinated surfaces to hide casting flaws (fig. 329), to integrate uneven coloration of repairs, and/or later as a conservation or restoration treatment to visually integrate uneven weathering or wear.
- Devotional bronzes may have been coated during their lifetime with ritual dressings of various kinds, including libations and pigments, and may preserve traces on their surfaces.[29]

1.3 Complex variations of patina

It is important to recognize that any given patina, whether chemical or coating or mixed, may be a complex multilayered sandwich whose layers could have developed naturally or could have been applied at different times by zdifferent people for different reasons.[30] A patina may have changed considerably over time due to aging and/or various interactions, including handling, rubbing, or any other repeated contact. For instance it is becoming increasingly clear that some bronzes with natural chemical patinas formed during burial were also intentionally patinated at some point prior to burial. On occasion, a natural corrosion layer on an excavated bronze may have been deliberately removed and replaced artificially using chemicals and heat.[31]

Some chemically induced surface changes occur unintentionally due to, for example, the reactions of the different components of the alloy with:

- remnants of core or investment material that were not completely removed after casting and, due to their hygroscopic nature, introduced moisture and encouraged chemical reactions (figs. 330, 331);
- residual chemicals from cleaning or other treatment (fig. 332);[32]
- salts, oils, and acids on human skin that can lead to etching or corrosion (figs. 141, 333);
- air pollution (often sulfur compounds) indoors or outside (figs. 316, 334);[33]
- the burial environment.

Chemically induced compounds and applied layers of organic coatings of various kinds (for instance resins or waxes) can be present on the same object, and could have been applied at the same time by the same person. For instance, one nineteenth-century French patination manual describes the smoking of bronze surfaces treated in special ovens with either chemical solutions, lacquers, or both to create a variegated greenish-to-black patina in imitation of antique sculptures.[34]

Ritual practices may also contribute complexity to the patina (fig. 335). Conversely, repeated touching of raised areas of a sculpture can remove patina in discrete areas, exposing bare metal, which will then likely oxidize to a dull color (figs. 266, 267).

2 Why investigate patina? and other FAQs

An investigation of patina may answer a number of questions regarding attribution and more generally the history of a bronze sculpture. Bear in mind, though, that the history of a patina may be complex (see above), and the necessary technical investigation may be equally complex (see I.8§3 below, and II.6). Even with the best technical analysis available, it may not be possible to characterize the patina.

2.1 Can one determine whether the original patina has been preserved intact?

Without a record of the conditions that the bronze has been exposed to during its lifetime, it is hard to reverse engineer its original appearance without documentation from the time that clearly describes what it looked like. As mentioned above, a number of factors may affect the appearance and chemical makeup of the original patina, and consequently complicate the assessment of the appearance of the original surface.[35] Organic coatings can darken and opacify over time, and if reactive can cause unintended corrosion of the bronze surface. Weather, pollution, cleaning, handling, or burial for extended periods may result in extreme alteration of features, including total removal of the original patina (figs. 267, 336).[36]

It has been normal practice since at least the Renaissance to repatinate—that is, to completely remove any remaining patina and begin again—often using materials and methods quite similar to those used originally. On occasion this might have been done for commercial reasons, with shortsighted focus on "improving" patina to enhance sales providing a powerful incentive for repatination. Therefore, the condition of the patina does not necessarily mean that it dates from either the initial fabrication of the sculpture or a more recent moment.[37]

The identification of more recently developed chemical patinas,[38] pigments, or resins in colored coatings on a historical bronze may be a sign that the work has been repatinated at a later date, restored, or in extreme cases faked. Traces of the original patina may be preserved in recessed areas that have been protected from touch and/or environmental change. If sampling by cross section down to the substrate metal is warranted, a multilayered patina may be found that retains original layers at the bottom of the section unless previous wholesale repatination destroyed them.

Sculptures intended to be displayed outdoors should be documented prior to installation and monitored carefully for patina changes, as regular protective coating maintenance will slow but not prevent alterations. If available, photographic documentation from the foundry or the dedication of a monument, for example, can assist in ascertaining the original appearance.[39] Regular maintenance is key for optimal preservation of outdoor bronzes.

2.2 Can a later patina be distinguished from an original patina?

As mentioned above, this is complex. For modern sculpture, archival records may help determine an artist's or maker's original intent or concept and offer informative comparison to the existing patina. Otherwise, if a patina is not appropriate for a known region or time period, the authenticity of the object or its surface may be called into question.[40] However, the relative lack of detailed technical studies of well-provenanced bronzes limits our ability to draw firm conclusions (see I.8§2.6 below).

2.3 Can a natural indoor patina be distinguished from an outdoor one?

Here again, a variety of phenomena may interfere with common-sense perception. While natural patinas generally form much faster outdoors, indoor fountain sculpture and bronzes exposed to chemically treated water or indoor air pollution may also exhibit accelerated natural patina or corrosion formation (figs. 316, 336). Chemical color changes from indoor air pollution are well documented.[41] Natural resin coatings can be expected to break down over time with exposure to heat, intense light, and/or elevated humidity. The presence of an intact organic coating may suggest that the sculpture has aged indoors.

2.4 Can a corrosion patina created naturally be distinguished from one created artificially?

Though the corrosion products produced—either naturally through exposure or burial, or artificially by human intervention—may be chemically similar, their formation and structure may differ.[42] Artist- or foundry-made patinas are generally thinner and more homogeneous compared to naturally formed patinas, which can be quite lumpy and thick and mixed with nonmetallic inclusions (for instance soil) from the burial environment. However, this is not always the case. Notably, delicate and pleasing natural patinas may develop during burial (fig. 309).[43] A clear understanding of the chemical and physical nature of the corrosion layers is often necessary to help identify the age of the patina.[44]

2.5 Can a chemical patina be distinguished from an applied coating?

Generally speaking, an applied coating will be thicker than a chemical patina and sit on top of the surface. As it is not chemically bound to the metal surface, it may flake off with a distinct edge to the loss. Distinguishing between the two by eye may be difficult, especially if loss is due to wear. Organic solvents may help in this case because chemical patinas composed of metallic compounds are usually insoluble. But certain alkaline compounds, such as monoethanolamine, can change or strip a chemical patination from a bronze, as can strong acids. This is further complicated when in the presence of cross-linked organic coatings, as these may be too insoluble in organic solvents.

2.6 Can the patina help with attribution?

Ensure, first, that the patina under investigation is original (see above). Specific types of artificial patinas were developed in different parts of the world over many centuries of practice, for example, in ancient Egypt, Greece, and Rome.[45] Those dating from the Italian Renaissance and later periods are also sometimes documented and/or well known.[46] At times their presence can indicate place of origin, or help in characterizing the work of individual artists and workshops.[47] However, caution is again required, as with all comparisons, since the relative lack of detailed technical studies of well-provenanced bronzes limits our ability to draw firm conclusions.

2.7 Can we determine who was responsible for choosing an artificial patina?

This will usually depend on documentation such as correspondence between an artist and foundry, known foundry practices, or styles of patina.[48] Conservation records may be a useful source of information if the object has undergone treatment for patina damage or instability. Identification of synthetic resins or pigments also provides clues. The technical study of a large number of bronzes produced for an artist by one foundry may reveal standard foundry practices or the preferences of the artist, as with Henri Matisse (French, 1869–1954) and Jean Arp (French, 1886–1966).[49] Unfortunately such studies are rare.

2.8 Why are some outdoor sculptures green or brown, or show black stripes or other disfiguring patterns?

This is due to aggressive corrosion of the bronze by air pollution, especially sulfur compounds mixed with soot (fig. 334). Repeated washing by rainwater preferentially removes some of these compounds in more exposed areas of the bronze, leaving black crusts in more protected areas.

2.9 What is bronze disease?

Many forms of copper corrosion are disfiguring, and some are severely damaging. One specific type that has become infamous is a copper chloride corrosion termed "bronze disease".[50] Under favorable conditions of elevated relative humidity, it can rapidly progress and lead to loss of solid metal, causing not only disfigurement, but in the worst case the disintegration of an artifact. The term alone terrorized early collectors and curators—back in the early days of scientists tackling this issue, it was believed that the outbreaks were a kind of bacteria-like canker that could spread from one object to another—and the "disease" was frequently called corrosion "cankers."[51]

"Bronze disease" is caused by chloride salts, which are commonly introduced into corrosion layers during burial by surrounding soil or water, or if the object is near an ocean, although other causes are possible, including chloride-containing chemicals used during patination or chemicals in the environment (fig. 337).[52] Within the corrosion stratigraphy, a copper chloride typically forms on the surface of the metal. Ambient moisture causes this compound to disintegrate and in the process create small traces of hydrochloric acid, which dissolves further metallic copper from the substrate. Characteristic small mounds of light-green corrosion are indicative of this active corrosion, which can be confirmed by a chemical spot test.

Early treatments aimed at removing all corrosion layers through aggressive chemical or electrochemical treatment turned bronzes in many museum collections into thin-walled, porous, "stripped" metalware. Today, chloride corrosion on cupreous artifacts is controlled through desiccated microclimates or with chemical corrosion inhibitors and applied coatings (fig. 315). Depending on the location of bronze objects in less-controlled settings, such measures may be difficult to achieve and regular corrosion checks and retreatments are often necessary.

3 Checklist: How do we investigate patinas?

Most techniques are similar to those used for the characterization of gilding and metal plating, and are therefore presented extensively in II.6, including those selected below. A list of examinations and examination protocols for the characterization of a patina layer is presented in figure 338. Flowcharts set out in figure 305 indicate what might be achieved depending on the possibility of sampling and the nature of the sample.[53] For details of examination and analytical techniques see tables 5, 10, 13.

3.1 Visual examination (magnification, ultraviolet light)

Visual examination using a variety of lighting sources and angles (raking, specular) should always be the first means of studying a patina (see II.1§2). Normal museum exhibition/gallery illumination levels are generally not sufficient to observe the subtle variations in patina color from golden to browns, reds, pale greens, deep greens and blues, to almost black. Raking light and ultraviolet illumination may help clarify the presence and/or nature of a patina, especially if there is an applied organic coating. Use microscopes and other magnification lenses with bright, adjustable light sources when possible.

Observations that assist in gathering data on the existence and nature of the patina involve:

- describing and mapping the color and texture variations (or lack thereof) on the surface, including examination of multiple locations on the object, such as in recesses where dirt or dust might accumulate, but which may also be protected from the wear that typically occurs on high spots (fig. 323). It is common practice to describe in broad terms the variable colors of the patina on a sculpture as part of the cataloging and/or treatment process. For example, an object may be pale green in one area compared to dark green with mottled black spots in another (fig. 334). However, there is no standard nomenclature for assigning exact color notations to the different elements of a patina with chromatic variations over a bronze. Color charts may be helpful.[54] Precise color measurement by scientific instruments may be challenged by the interplay of various limitations such as colors, textures, and the size of the area of analysis (see II.2§4, fig. 339);

- noting whether the patina consists of a well-adhered layer, or if it is flaking or powdery and friable; lumpy or smooth; opaque, transparent, translucent, or variable;
- assessing whether the patination was potentially altered or added to, for example if different layers of color are visible, or if there are areas missing from layers that are evident on another part of the sculpture;
- searching for drip or brush marks that may indicate an applied coating or paint;
- comparing patinated areas to those where it is lacking, such as under the base or in areas of damage;
- examining (with magnification) the stratigraphy of the corrosion layers in an area of damage.

The accumulation of water on parts of an outdoor sculpture can result in areas exhibiting a different color and/or texture from areas that stay mostly dry. Areas that collect organic debris may have yet another appearance.

3.2 Analysis with and without sampling

As discussed in II.6, analysis without taking samples rarely provides sufficient data, although this will depend on the questions being posed.[55] If sampling is possible and acceptable, a polished metallographic cross section through the entire patina layer, including the underlying metal, provides the most information. This can include an indication of the chronology of the patina formation in the location sampled. Numerous other useful analytical techniques may be applied to both cross sections and powder samples. For more information, including guidance for sampling, see II.6.

Notes

1. For discussion of the metallurgical significance of patina see Scott 2002 and Aucouturier et al. 2003. For example, the copper (Cu) surface may be transformed into red cuprite (Cu_2O) by a chemical reaction with oxygen (O_2): $Cu+1/2\ O_2$ --> Cu_2O.
2. For examples of patina in ancient Egypt see Shearman 2010. For ancient Greek bronzes see Formigli 2013b and Descamps-Lequime 2015. For Italian Renaissance see notably Stone 2010, 107–24; Smith 2008a, 18; Motture 2019, 68–70, with further references, esp. 242n155. For current practices in Nepal see Furger 2017, 116–31.
3. For a broad overview refer to Scott 2002. For more detailed historical focus see Weil 1996.
4. However, Michel 1922, 186, includes a patina recipe for bronze entitled "Imitation of Florentine Bronze," for which the first step is to electroplate a thin layer of red copper to the surface of the bronze, over which is applied a paste of sanguine (iron oxide) and graphite, which is in turn covered with drying oil or varnish, followed by further application steps.
5. For a comprehensive list of all compounds that may be found in patinas see Scott 2002.
6. Schweizer 1994; Robbiola, Blengino, and Fiaud 1998; Robbiola and Portier 2006.
7. For a comprehensive and synthetic view on patina formation mechanisms see Scott 2002; Aucouturier et al. 2003; Aucouturier 2007.
8. Fucito 2013, 138–40.
9. See Hughes and Rowe 1989; Kipper 1995; Runfola 2014; Motture 2019, 70 (for summary of black and other Renaissance patinas with references and sources).
10. See Craddock and Giumlia-Mair 1993; Chase 1994; La Niece et al. 2002; Benzonelli, Freestone, and Martinón-Torres 2017; Aucouturier, Mathis, and Robcis 2017.
11. Hughes 1993, 10.
12. Scott 2002, 10, makes a distinction between patina, "a smooth, continuous layer that preserves detail and shape," and corrosion, "mineral deposits that do not form a continuous and smooth layer."
13. Shearman 2010, 48–49.
14. With careful preparation and examination, the fossilized vegetal or animal fiber may be precisely characterized (Moulherat et al. 2002).
15. A richly illustrated book on patinas is Hughes and Rowe 1989. See also Adil and DePhillips 1991.
16. See Falaschi 2017 for a fascinating account of blue-colored Greek bronzes; Furger 2017 for practices in Nepal; and Oddy, Bimson, and La Niece 1981 for other observations on Himalayan bronzes.
17. The bronze sculpture of Horace Greeley cast in New York by the sculptor J. Q. A. Ward (American, 1830–1910), cast by the Henry-Bonnard foundry in 1890, was not artificially patinated before being placed outdoors in New York City. Presumably it was meant to develop a natural patina over time by exposure to weather; see "Casting in Bronze" 1891, 866. For an apparently similar approach by the sculptor Adriaen de Vries (Netherlandish, 1556–1626), see Bassett 2008, 270–71.
18. Craddock 2009, 365–68; Fucito 2013, 137–38; also Risser and Saunders 2017.
19. Craddock 2009, 356.
20. Newman 2011, 36.
21. Craddock 1990, 259–61.
22. Weil 1996, 403, avers that by the Renaissance, corrosion products had acquired value for their "antiqueness" so as to testify to the age of antique bronze. This new value spurred forgeries.

23. Hiorns 1907; Michel 1922.

24. Lauffenburger 2006, 76, found metal powders incorporated into varnish patinas on several sculptures by Antoine-Louis Barye (French, 1795–1875) cast by the founder Jean-Honoré Gonon (French, 1780–1850) in the 1830s.

25. Weil 1996, 397, describes these attractive patinas as thin, compact, translucent, generally red-brown, and more or less tinged with green, depending upon moisture levels.

26. Giambologna's (Flemish, active Florence, 1529–1608) apparent use of a transparent varnish seems to suggest the desire for sheen and protection when handling, while retaining the somewhat golden color of the bronze. See for example Stone 2010, 118–20, suggesting that the use of heat (or stoving) made the varnish more durable for handling.

27. Stone 2011, 178, believes that the vast majority of Italian Renaissance patinas were organic oil-resin varnishes and notes that although there were various chemical patination recipes in the Renaissance, "it has proven surprisingly difficult to find any undoubted examples of this method ever actually being employed. A conspicuous exception is the work of the sculptor Antico."

28. Descamps-Lequime 2015; see also Formigli 2013b.

29. Hindu and Buddhist rituals in Asia include applications of sacred substances such as libations and pigments, and even dress and jewelry to bronze sculptures of divinities. See Gänsicke 2016; Gänsicke and Sharma 2021.

30. See above, and for example Massimiliano Soldani-Benzi (Italian, 1656–1740), who describes in a letter to his London agent how to refresh the patina on bronzes using clear walnut or linseed oil and red hematite (lapis rosso); unpublished document, referred to in a 1996 conference paper by Dr. Charles Avery in Berlin. We are grateful to Dr. Avery for allowing us to include his discovery prior to publication.

31. Risser and Saunders 2017.

32. Schrenk 1994, 60, discusses supposed "protective" coatings, like neat's-foot oil, often applied to Benin bronzes by collectors that have instead caused extensive disfiguring metallic fatty-acid corrosion products to form on the surface. Michel 1922, 160, describes a five-year-long experiment in Germany beginning in 1864 whereby daily washing and monthly application of grease to a bronze bust displayed outdoors resulted in the formation of a beautiful patina compared to ugly ones on the busts used as controls.

33. See also Marabelli 1994, 14, who found the surface of the equestrian monument of Marcus Aurelius, outdoors since late Imperial Roman times, to have extensive sulfation, including brochantite, antlerite, and chalcanthite.

34. Debonliez and Malepeyre 1979, 8.

35. For a thorough definition of the original surface of a bronze see Bertholon 2004.

36. For a brief case study of patina formation see Scott 2002, 328–29.

37. For a sample of bronzes in the Wallace Collection, London, where it seems evident that they have been repatinated, see Warren 2016, vol. II, esp. pp. 503–31, nos. 113–14; p. 474 (no. 108), p. 550 (cat. 117) on nineteenth-century patination; p. 666 (no. 137).

38. For modern patination recipes, see Hughes and Rowe 1989.

39. The sculpture of Thomas Hart Benton by Harriet Goodhue Hosmer (American, 1830–1908) was described at its unveiling as having a bright golden color (Weil 1996, 407).

40. Schrenk 1994.

41. Grzywacz 2006, 11–13.

42. See Robbiola and Hurtel 1997, a study carried out on the Roman busts of Livia (fig. 309) and Augustus that demonstrated that the "noble" patina originated from a natural corrosion process: a presence of compounds from the soil in the deeper layers and tin enrichment of the upper dark-green layer due to decuprification.

43. Gettens 1970.

44. Relatively few studies have been carried out on the subject; see Graedel, Nassau, and Franey 1987.

45. For the introduction of artificial patinas in the Hellenistic period see Eggert 1994; Heilmeyer 1994; Descamps-Lequime 2015, 151–65.

46. Pier Jacopo Alari Bonacolsi (known as Antico, Italian, ca. 1455–1528) is renowned for using a particular black patina; see figure 293 and Stone 2011. Artists such as Giambologna, his followers, and later sculptors are known for the reddish Florentine/Tuscan patina (fig. 325); see for example Stone, White, and Indictor 1990; Radcliffe and Penny 2004, 162, 168, 174, 179, 190, 198, 202; Stone 2010; esp. 116; Pitthard et al. 2011.

47. See Vauxcelles 1905, 195, for a lengthy description of the process for the "wet crow" patina applied to a cast of Rodin's (French, 1840–1917) *Thinker*, made by the Hébrard foundry in 1905. An example of foundry practice is the French Valsuani foundry's famous "noir Valsuani" patina used in the early twentieth century for bronzes by various sculptors; Lebon 2003, 260. See also Le Normand-Romain 2007 and Barbour and Glinsman 2015.

48. For example, Rick Stewart's research on the bronzes of sculptor Charles Russell (American, 1864–1926) at the Amon Carter Museum in Fort Worth, Texas, revealed that in 1941 the collector who later sold them to Amon Carter had most of the collection repatinated at the Roman Bronze Works in New York. A green cold patina was applied over the original brown. Stewart 1994, 141.

49. Boulton 2006, 86; Hamilton 2018.

50. For a history of the treatment of this issue see for instance Beale 1996.

51. Drayman-Weisser 1994, 143.

52. For a case study of corrosion see Scott 2002, 334–38.

53. For a comprehensive and critical review of most available methods see Aucouturier 2007; Stone 2010.

54. Munsell color charts are currently being employed by Dorothy Cheng at the Smithsonian American Art Museum in Washington, DC, in her technical study of bronzes by the artist Paul Manship (American, 1885–1966).

55. The patina demonstration carried out during the October 2018 CAST:ING meeting (see video 14) and the subsequent analysis by Aurélia Azéma of LRMH provides a good example of how difficult it may be to characterize even a relatively simple patina. The patina consisted of subsequent hot application of three aqueous solutions, namely a copper nitrate, an iron nitrate, and a potassium polysulfide. Half of the lion was covered with wax at the end. Surface X-ray fluorescence spectroscopy (XRF) analysis was only able to detect the presence of iron, potassium, sulfur, and titanium. X-ray diffraction (XRD) analysis on powder carefully sampled under a microscope using a blade revealed a copper sulfate, a copper nitrate, and lead oxide. No wax could be detected. In this example it was impossible to reconstruct the exact patina process.

I.9

Inlays and Overlays

Jeffrey Maish
Annabelle Collinet

Additional Contributors: Aurélia Azéma, Valérie Carpentier, Andrew Lacey, Elsa Lambert, Carol Mattusch, Mathilde Mechling, Peta Motture, Dominique Robcis, Brice Vincent, Jean-Marie Welter

This chapter describes the types of inlays and overlays that might be encountered on a bronze sculpture, identifies working processes, and describes how inlays and overlays may be identified and associated with a specific technique. It applies to metal, stone, ivory, glass, and bone additions, but does not go into detail about how these additional elements are initially formed. Gilding and metal plating, which may sometimes be considered as overlay, are addressed in I.7. Other surface ornaments and dressings that enhance color, including attachments such as jewelry, garments, hair, or feathers, are not considered here. While inlays and overlays do not always survive, evidence of their prior existence may remain, and the descriptions below should, in some cases, help to identify them and assist in interpreting what the sculpture may have looked like earlier in its life.

1 What are the different inlay and overlay techniques and materials?

Inlays and overlays enhance the appearance of sculpture mostly by adding discrete elements of a different color to what is basically a monochromatic surface in which texture and patina generally serve to create coloristic variations. Inlays and overlays may add realism by enhancing details of human or animal forms, or may be applied more extensively to create overall decorative surface effects (fig. 340). The term "damascene" is sometimes used to describe inlays.

The applied materials may be clearly visible or presented more subtly, especially if small in scale or hidden by patina or corrosion. Overlays and inlays are also by their nature surface effects and susceptible to loss. Overlays are easily abraded, and mechanically attached inlays can be completely lost, leaving empty cavities.

1.1 Inlay or overlay?

The terms "inlay" and "overlay" are often used indiscriminately, although they differ in one main regard: inlays are inserted into the bronze surface, whereas overlays are set onto the surface. Another distinction has been proposed: inlays are mostly attached mechanically without any fasteners, sometimes cast in, whereas overlays are joined to the surface using a variety of techniques, including riveting, soldering, and adhesives.[1] This chapter subscribes to the first categorization—

inserted into versus set onto the surface—to distinguish between the two.

Inlays are inserted directly into the bronze surface, generally after casting, following the design of the cut, cast, or chased recess (fig. 341). Inlays may create fine linear designs or larger fields of color by using a variety of materials, including metal, glazed fritware, glass, enamel, ivory, or stone. The surfaces of inlays themselves may be chased to form more precise designs so that a simple figure becomes more textured and detailed (fig. 342). Inlays have been used in copper-alloy works over a broad time period and in different parts of the world (see examples below).

Overlays comprise materials such as stone, enamel, or metal laid over the metal surface (see I.9§1.2 below). They are less frequent on bronzes than inlays. Good examples of overlays may be found in a variety of cultures (figs. 47, 223).

1.2 Inlay and overlay materials

Inlay and overlay materials may be inorganic—for instance colored stones, vitreous materials, contrasting metals—or organic materials such as shell, ivory, or plant-based items.

1.2.1 Inorganic materials: metal and metal-based

Metals for inlay are generally selected based on color, ductility, and melting point. They include gold (fig. 343), silver (figs. 293, 343, 344), copper (figs. 344, 345, 346),[2] and iron, as well as alloyed (figs. 346, 347)[3] and patinated metals.[4] A copper-alloy inlay with a composition similar to the surface metal and with no color difference may suggest that it was originally patinated. Gray-colored inlays also occur, for instance in Buddhist statuary from Kashmir (eighth century CE).[5] Lead-based compounds have been observed on Khmer bronzes (fig. 348). On ancient bronzes and Renaissance emulators, including Antico (Italian, ca. 1455–1528), tin may be confused with silver.[6] Greater polychromy can be achieved by using multiple metals as inlays and overlays (figs. 349, 350),[7] but also with gilding and silvering (see I.7).

Powders of both metal and metal oxides may be used in a manner similar to powdered glass (see I.9§1.4.2 below). Niello is a well-known example, although rarely encountered on bronze sculpture.[8]

1.2.2 Inorganic materials: mineral-based

Inlays and overlays of mineral-based material include stones and semiprecious stones (figs. 297, 351, 352, 353), gems (fig. 354), glass (figs. 309, 351, 355), and mixed material such as frit (glass and ceramic mix) and pigmented glaze (fig. 356). A variety of stones can be employed to add detail to a sculpture, as on reliquaries and jewelry. They may be roughly shaped or cut into fine lamina for inlay, and may be opaque or translucent. Generally found on jewelry, they allow light reflection from a patterned underlayer of gold (fig. 340 G). Stones may also have symbolic or religious significance. Stone and/or pigment can be combined with metal to create multicolored effects, as seen notably on ancient Egyptian and medieval Indian statuary, and also on Chinese ware (fig. 357).

1.2.3 Organic materials

The use (or surviving evidence) of organic materials is more common in art objects from the medieval period onward, as such objects have not been subjected to aggressive bio-deterioration processes. In general, organic inlay materials include tortoiseshell, horn, claws, seashell (fig. 358), mother of pearl, coral, amber, pearl, wood, charcoal, and dyed resins. Secondary or ancillary compounds may include tinted waxes, resins, bitumen, or plaster. Organic resins may be tinted with pigments or dyes and applied in a manner similar to enamel. Organic materials such as ivory, bone, and teeth are sometimes present. Materials such as fibers and hairs may be used as attachments (figs. 359, 360).

Risks of misidentification/misinterpretation

- Glass inlays and overlays are easily confused with enamel. Close examination with a microscope may be sufficient to make the distinction.
- Pigments may be applied thickly (resembling overlay) or as thin surface glazes (figs. 360, 361). In some instances, pigment may be used in imitation of inlaid stone (fig. 362).

1.3 Surface preparation

Steps are required to effect the bond between the cast bronze surface and the added material.

1.3.1 Preformed cavity or perforation

Many Egyptian, Greek, and Roman bronzes were produced with empty eye sockets for later insertion of naturalistic eye inlays (figs. 309, 311). Roman eye production was a specialized fabrication process and occupation (*ocularis*). Each eye includes multiple parts, such as copper supports and eyelashes, white stone or ivory sclera, and stone or enamel irises and pupils (fig. 351). During the Italian Renaissance these inlays were imitated (fig. 293).[9] These apertures have no rear wall, as the inlays are held in place within folded copper sheets, often with eyelashes (fig. 340 I, fig. 351). The entire assembly may be lost during burial, revealing the empty eye socket.

1.3.2 Elevated decorative borders

Decorative inlay or overlay may be enclosed by elevated decorative borders. Small chambers are created by soldering metal wires or strips to the surface (referred to as cloisonné).

1.3.3 Engraved or cast surface cavities

Different methods are used to shape cavities for inlays. These may be preestablished in the model, or they may be cut into the metal post-casting (champlevé). Additional attachment holes may be created at various stages. The cavities may have smooth surfaces, or be roughened to increase the surface area and thereby assist with adhesion. Furthermore, as with repair patches, the perimeter of the recesses prepared for metal inlays may be slightly undercut, as the act of hammering these in will splay the metal edges and thereby help lock the added metal into place (fig. 340 A–C, F).

1.3.4 Cast around

In some instances, preformed inlay elements might be inserted into the wax model or mold (fig. 340 D), and the bronze cast around them. Since the inlay is formed first, the bronze does not need surface preparation (figs. 346, 363).

1.4 Attachment methods

For all but the last type of inlay listed above, once the area that is to receive the inlay or overlay has been prepared, the material of choice is attached to the bronze. This may require one or more methods of attachment.

1.4.1 Mechanical methods

Metal inlays are generally formed by hammering the material into the prepared recess in the cast sculpture. This requires the inlay metal to be ductile enough to flatten and spread during the operation—for example, gold, silver, or copper (fig. 344). In addition to roughening the surface with a chisel, for instance (fig. 340 H), it may be desirable to bevel the edges of the recessed design. As the metal is hammered in, it spreads and locks into the undercuts around the perimeter (fig. 340 A). Conversely, the surface metal may be channeled and hammered around the inlay's perimeter to secure it (fig. 340 E).

As noted previously, chambers cut into the substrate (fig. 340 C) or formed within elements soldered to the surface may be used to hold stone cabochons or enamels (fig. 340 J, K, L; see also I.9§1.3.2 above).

In some instances, an opening may be formed in the bronze substrate either before or after casting to allow for the insertion of larger inlays from the reverse side (fig. 340 I).

Riveting requires perforating the bronze (fig. 340 B). Hammering of the rivet either spreads and locks the rivet head or expands the entire feature, locking the inlay or overlay to the surface. Rivets and pin ends may be designed to function as decorative elements, or finished to be less visible on the surface, for instance by countersinking them.

Pinning is similar to riveting, but pins are usually smaller in diameter and more numerous. Compared to rivets, they may also be less visible. Riveting relies on the flattening or splaying of either end to secure an attachment, while pinning relies on simple pressure or friction within the substrate (as with a nail in a wall).

1.4.2 Methods employing heat

Casting-in consists of pouring molten metal into a recess in the surface to fill a prepared cavity. The final surface may require further leveling by filing, polishing, or burnishing (fig. 340 C).

More elaborate metal inlays with several parts may be more easily fabricated separately and incorporated into the actual casting of the bronze. This is the case, for instance, with lips and nipples of ancient Greek statues that were formed in copper and set into the wax model before casting (fig. 363). Such an inlay needs to be made of metal with a higher melting point than the surrounding matrix to avoid being destroyed in the process. The use of such a process is characterized in part by a lack of clear borders, as the surrounding bronze is likely to run over its edges (fig. 340 D). Precast inlays may be held in a mold with tab-like extensions, for example. These would be concealed by the cast bronze and visible only by radiography (fig. 363). A decorative technique using liquefied metal may show characteristic evidence of flow or trapped bubbles.

Preformed inlay elements may also be joined onto the sculpture using a metal with a lower melting point than the surface metal (for example soldering with a lead-tin alloy, fig. 364). Solders may be made of wire, or a powdered solder may be applied between two elements and heated locally. When the melting point of the solder is reached, it flows and creates the join. The fact that solder bond is relatively weak and may deteriorate over time accounts for the frequent loss of inlays and overlays applied in this fashion. Remaining traces of the solder can be mistaken for evidence of tinning.

Enameling is a specialty in itself and is used on a variety of metals, including silver, gold, copper, and sometimes bronze. In the vitreous enamel technique, powdered glass is fused to the metal substrate by heating at high temperature to create opaque and translucent colors. Molten glass enamels flow on heating and are used for both inlay and overlay decoration. To form detailed inlay patterns, glass may be used within engraved or cast-in depressions (fig. 309). Small partitions of metal are soldered onto the surface in order to enclose the enamels that are added later (figs. 223, 340 K, L). Enameling may also be applied directly to the prepared surface (fig. 47).[10] A related material, frit, combines glass with ceramic and is used in a similar manner; elements may be molded or preformed prior to insertion.[11]

1.4.3 Adhesive methods

Whereas metal inlays and overlays are generally attached mechanically, nonmetallic materials (stone, glass, ivory, et cetera) may require an alternative attachment method such as an enclosed chamber used in conjunction with an inorganic bedding medium such as lime or gypsum plaster, or organic binder such as a resin (rosin, shellac, or gum arabic).[12] Various binders were available in ancient Greece and Rome,[13] and adhesives were presumably used for the attachment of glass and lapis inlays on Late Period Egyptian statuettes.[14] These materials are subject to deterioration over time, ultimately causing inlays to detach, leaving only the associated recesses in the metal.

2 Why investigate inlays and overlays? and other FAQs

Characterizing the materials and techniques of inlays and overlays contributes to any study of the technological know-how of a bronze-producing craftsperson, workshop, or culture. Needless to say, it is also useful in the dating of a work and more nuanced description of its condition, especially if one finds evidence of later interventions. The distinction between overlays and inlays may, admittedly, be somewhat problematic given that the preparation and setting is often inaccessible for examination without resorting to a destructive sampling process. To complicate matters, both processes may be combined in the same area, or the inlay itself may rise above the level of the surrounding substrate.

2.1 Can I detect traces of lost or hidden inlay and/or overlay?

Given that inlays and overlays are particularly susceptible to wear, damage, and loss, a gaping cavity may account for the former presence of an inlay or overlay. Empty voids, especially those that are deep and with sharp borders (fig. 364), or the presence of preparatory traces on the surface, such as chasing marks or specific reliefs (fig. 348), may indicate lost inlays. An ancient statue with voids for eyes is likely to have lost these. Sheet or leaf metal applied with an adhesive is also more likely to become detached as the glue deteriorates, or to be worn down in exposed areas and preserved in protected recesses.

Traces of metal, solders, and adhesives, for example, may be a good indication of lost inlays or overlays (fig. 364). Later protective coatings, patination, or corrosion may obscure ancient inlays and overlays (fig. 311). Surface elemental and/or structural analysis, for example using particle-induced X-ray emission spectroscopy (PIXE), portable X-ray diffraction (XRD), or Raman spectrometry, may help to detect inlays or overlays and even establish chemical compositions and the original color schemes.[15] Radiographs may also be useful in detecting material composition and density differences, seams, and other features (tabs, extensions) associated with inlays and overlays (see I.9§3.2.1 below).

2.2 Can I distinguish between original inlay and/or overlay and later intervention?

Careful comparison of all existing inlays and overlays and tooling traces throughout a sculpture can often help distinguish original versus replacement parts. Substitution of materials or restorations can generally be identified by inconsistencies in material composition (and the presence of anachronisms) or method of attachment.[16] Even if lost, the former presence of an inlay may be inferred from jarring voids or recesses, or by comparison to related objects on which the inlay is preserved. Analytical methods may also be necessary to identify and compare materials.

2.3 Is there any chance of misinterpreting the nature of an inlay or an overlay?

Materials used in small areas or deteriorated material may be hard to characterize or distinguish. Visually, tin may be confused with silver, and slightly oxidized silver may appear golden. A dark or black glass may resemble black marble.

Finely surfaced marble or alabaster may be hard to distinguish from ivory or bone. (This may require close microscopic examination or nondestructive surface elemental analysis using X-ray fluorescence spectroscopy, or XRF.) For example, under high magnification, marble will have a granular appearance and XRF may reveal a high calcium content. Separation of bone from ivory may rely on morphological study, although newer protein identification methods have been used successfully to distinguish animal species (see I.9§3.2.2 below).

Inlay methods may also be used for repair patching rather than as decorative methods. Similarly, repairs might be mistaken for decorative features (see I.4§1.1).

2.4 Can the characterization of inlays and overlays help distinguish between different craftspeople, workshops, or production centers?

Trace analysis of materials used in inlays and overlays may help characterize materials more precisely and help group related objects, as can a comparison of bulk metal and core analysis (see I.2§4, II.7§1.2). The trace elements may point to sources of materials, processing methods, or production centers and reveal whether there is consistency or variety in their use. It is therefore important to balance analytical results with the identification of technical features, such as attachment techniques, that may be more characteristic of a particular workshop. Characterization of processes and tool marks can also indicate methods used in a given period or region. The study and careful documentation of tool markings can help establish production chronology and date production sequences.[17]

3 Checklist: How do we investigate inlays and overlays?

Inlays and overlays can be identified and defined using different basic methods of visual examination. Initial identification of features can help establish what further testing and analysis may be beneficial. As noted, identification of missing and hidden inlays or overlays is not always obvious, and different tools may be needed to detect them. For details of examination and analytical techniques please refer to tables 13, 10, 5.

3.1 Basic identification techniques

3.1.1 Visual examination

Visual examination is the logical first step in identifying seams, textural variations, and color differences. In addition to the naked eye, low-power magnification with a handheld magnifier or microscope is useful. Raking and variable-angle illumination using a simple handheld lamp or even a smartphone light can help with texture assessment and evaluation of the condition or state of deterioration. Ultraviolet light may help in detecting and identifying particular organic materials (see II.2§3.1) or adhesives that fluoresce under this lighting and can be signs of recent restoration.

3.1.2 Chemical and physical analysis

Basic methods include the use of a magnet to detect iron. Surface elemental analysis using handheld XRF can help to detect inlays.

3.2 Advanced techniques

3.2.1 Imaging

Radiography is useful in detecting and characterizing inlay, including the material used to make it. For example, gold is much denser than silver and copper, and thus a gold inlay is easy to distinguish from a silver or copper inlay using radiography as long as the inlay thicknesses are comparable (see II.3§1.4). Computed tomography may aid in imaging the overall shape of the inlay and isolating the material, based on radio-opacity. Reflectance transformation imaging (RTI) and infrared thermography (IRT) may be used to detect inlays.[18]

3.2.2 Chemical and physical analysis

Without sampling

XRF is useful as long as inlays and other features are more than a few millimeters in size. Better surface resolution can be achieved through PIXE analysis. Small statuettes less than a few centimeters deep may be introduced in a laser ablation cell and analyzed by laser ablation inductively coupled plasma mass spectroscopy (LA-ICP-MS). In addition to elemental analysis, other analytical approaches include XRD, ultraviolet (UV), surface-enhanced Raman (SERS), Rutherford backscattering spectrometry (RBS), and nuclear reaction analysis (NRA) (fig. 305 A). Materials applied with an adhesive may be detached, allowing analysis of traces of the adhesive. Color or hyperspectral analysis may also be very useful, although alteration may impact the original appearance (see II.2§4).

With sampling

Elemental and structural analysis may require a drilled sample (fig. 305 B, C). As for repairs, cast inlay can only be distinguished from one mechanically attached through cross-sectional analysis (see I.4§2.1). Similarly, an inlay or overlay may be too altered for proper surface analysis: sampling for cross section may be required. Adhesives can also be identified in cross section, and elemental and spectral analysis can characterize the adhesive composition.[19] This may help to distinguish an ancient technology or material (for instance a bituminous or resinous compound) from a modern repair (made with a synthetic glue).

Notes

1. For a precise definition see Untracht 1982, 315.
2. Copper inlay was also reportedly used in the historical restorations of Herculaneum bronzes in the eighteenth and nineteenth centuries (Lahusen and Formigli 2006). Inlay wire is also described in material from the *Mahdia* shipwreck, Tunisia, carrying Greek works of art for Romans (Cüppers 1994).
3. Alloys can be employed to achieve a variety of more subtle color effects. Alloys include pewter (85–99% tin with copper, bismuth, and antimony); copper alloyed with tin; lead (at sufficiently high concentrations that it can affect the color tonality of bronze, see fig. 346); zinc; gold (for example, copper alloyed with 25% gold to form Roman "pyropus," see Pliny the Elder [77 CE] 1857, 34.20); and electrum, a gold-silver alloy. Gold could also be alloyed to become more malleable for thin inlays. One of the most liberal uses of colored inlay is exemplified in the Alexandrian bronze table Mensa Isiaca in Turin, Italy (fig. 365). Alloy inlays are reported notably on Egyptian objects (La Niece et al. 2002; see also fig. 350). A bronze bull of the Alaca Höyük type (ca. 2000 BCE) from Eastern Anatolia, now in the collection of the Museum of Fine Arts, Boston, is inlaid with copper-gold stripes and is partially silvered (inv. 58.14, https://collections.mfa.org/objects/267288). See Smith 1970.
4. The authors do not know of examples of patinated copper inlays on sculpture, but do on bronze implements (Robcis, Bourgarit, and Mille 2000; Descamps-Lequime 2005).
5. See Pal 2007, 62.
6. A layer of cassiterite (tin oxide), inferred as degraded tin sheet, has been identified on the eyes of an ancient Greek statue of Dionysos as a Youth (Mattusch 1996, cat. no. 23, 224 and 231n6). Similarly, the Charioteer happens to be inlayed with tin (Descamps-Lequime, Mille, and Psalti 2019). See also Snodgrass 2000. For Antico, see Smith and Sturman 2011.
7. For ancient Egypt see La Niece et al. 2002, 97–102. For Greek and Roman antiquity, see Descamps-Lequime 2015 and 2019.
8. Niello is composed of one or more metal sulfides (silver, copper, or copper and lead), with the composition varying depending on period and place. Its composition determines its method of application. Niello can be an inlay or an overlay of black material with a gray or blue reflection/appearance, and can be confused with black enamel. It is generally used on gold and silver, but can also be found on copper alloys. It can be inlaid in chased or overlaid on smooth surfaces. See La Niece 1993.
9. Smith and Sturman 2011.
10. For example certain works by Charles Cordier (French, 1827–1905) in the collection of the Musée d'Orsay, Paris; see https://www.musee-orsay.fr/fr/evenements/expositions/aux-musees/presentation-detaillee/article/charles-cordier-1827-1905-sculpteur-lautre-et-lailleurs-4210.html. See also Papet 2018.
11. See also the Egyptian bronze figurine of Osiris (ca. 3rd Intermediate Period through the 26th Dynasty, Research Collection of the School of Religion, University of Southern California, inv. USC 5407) inlaid with blue glass, analyzed in Scott and Swartz Dodd 2002. Inlays of preformed blue and green enamel are reported in Byzantine doors (Kleinbauer 1976).
12. Untracht 1968, 441.
13. Formigli 2013a, 8: figures 23 and 24 describe the use of an organic binder in the manufacture of eye inlays in the Greek statue known as Riace A (460–450 BCE, H. 198 cm., National Archaeological Museum of Reggio Calabria). Although not specific to inlay, Pliny the Elder [77 CE] 1857, 33.64, mentions adhesives such as albumen, fruit juices, fish glue, milk, and blood (Giumlia-Mair 2002, 33).
14. La Niece et al. 2002, 101.
15. For example see Robcis, Bourgarit, and Mille 2000; Delange, Meyohas, and Aucouturier 2005; Alessandri et al. 2013; Descamps-Lequime, Mille, and Psalti 2019 ; Collinet 2021; and I.9§3.2.2.
16. For example, see the restoration of the head of the seated Hermes from Herculaneum (Mattusch 2006) and the eighteenth-century repairs to Roman bronzes from the Villa dei Papiri at Herculaneum (Mattusch 2005; Mattusch 2013).
17. See the study on medieval Iranian inlaid metalware Collinet 2021. To the authors' knowledge no such extensive study has been carried out on bronze sculpture.
18. Mercuri et al. 2018.
19. Possible additional methods include peptide mass fingerprinting (PMF, see Kirby et al. 2011) and antibody-based (ELISA, see Heginbotham, Millay, and Quick 2006).

Volume II: Analytical Techniques

Edited by Arlen Heginbotham and Elisabeth Lebon

Volume II provides a summary description of each of the multifarious techniques of examination and scientific analysis that have been usefully applied to the study of bronze sculpture. The volume is intended for both specialists in technical examination and analysis (e.g., conservators, conservation scientists, archaeologists) and nonspecialists (curators, historians). Specialists will find practical guidance on how each of the techniques may be specifically adapted and optimized for the study of bronze sculpture, along with extensive references to additional technical resources and published examples of their application. Nonspecialists will find succinct descriptions of the techniques and the scientific principles on which they are based. The descriptions are followed by discussion of the capabilities and limitations of each particular technique, along with a description of any sampling requirements that may be required. This information is presented to help those in positions of responsibility make informed decisions regarding the selection of appropriate examination and analysis techniques in the context of a technical study.

In the first chapter, the reader will find practical advice and resources on how to plan and document a technological study (with a reminder of the importance of cross-disciplinary collaboration at every step). This is followed by chapters on image-based examination and documentation (photography and radiography) and measurement. Analytical techniques that further aid in the identification and characterization of materials are described in separate chapters dedicated to the specific materials type (base metal, surface layers, and core). In the penultimate chapter, methods for dating a bronze sculpture are described, and the final chapter discusses the design and implementation of experimental simulations.

II.1

Planning and Documenting Your Technical Exam

Arlen Heginbotham
David Bourgarit

This chapter aims to guide the reader in preparing for a technical examination and its subsequent reporting and publication. Regardless of the intended outcomes and products of a technical study, it is useful to build an effective interdisciplinary team, prepare the appropriate tools and equipment, and think carefully in advance about how the results will be collected and managed.

1 Initial planning

Before initiating a technological study of a bronze sculpture, we recommend establishing a multidisciplinary examination team. Such a group would ideally include at a minimum a specialist from the humanities (for instance a curator, art historian, or archaeologist), a materials or conservation scientist, and a conservator whose expertise encompasses some of both in addition to a good knowledge of artistic processes. Together they can clarify the questions surrounding the sculpture that the study should aim to answer and also identify which examination and analytical methods are likely to provide informative results. Additional questions may arise in the course of the examination.

Also essential is to identify the background research needs and literature to be reviewed in advance or in parallel to the technological investigation, and clarify desired outcomes and outputs, particularly with respect to the reporting of results (for instance in monographic studies or journal publications, or for integration in shared databases). Budget considerations and a desired schedule for completion are likewise important up-front elements of planning.

Given the broad range of questions to address and clues to look for on a bronze sculpture, it can be extremely useful to begin with an examination checklist that enumerates subjects to be investigated. A checklist, or report template, can facilitate the systematic recording of observations that may ultimately help answer questions about the sculpture under study. Therefore, when designing an examination checklist or report template, it is critical to consider the specific questions and desired research outcomes that underlie the study. Several useful examples of examination checklists/reports developed in the past by bronze sculpture researchers are presented in "Appendix: Document Library" (online at https://www.getty.edu/publications/bronze-guidelines/appendix/). It is noteworthy that there is considerable variation in the length and specificity of the documents presented here; the particular circumstances of any given research program will determine which type of checklist may be most useful. Note that electronic notebooks are increasingly used by conservation scientists and archaeologists in the field.

As those who engage in such research can attest, it is crucial that the various team members involved in the project be kept abreast of—if not actually personally involved in—the various steps of the technical examination and analysis to ensure that as new discoveries and observations are made, each participant may contribute their interpretations and identify potential new lines of inquiry.

The amount of time required to conduct a technical examination of bronze sculpture can vary dramatically. The simple examination of a small statuette without the application of any scientific analysis might be completed in less than a week, including reporting, while an in-depth technical study of a series of large bronzes from different international collections might take years to organize and execute. Some indication of the time required to pursue technical studies may be gleaned from several of the case studies presented in this work (see the "Synopsis of Technical Parameters" sections at the end of Case Studies 2, 3, 4, and 5).

2 Lighting the examination space

Normally, the physical study of a bronze sculpture begins with a close visual examination. Lighting conditions in galleries, in storage rooms, and even in dedicated laboratories are often insufficient to observe surface details properly (an activity that is often key for understanding the work) or subtle variations in patina. Both the intensity and the spectral distribution of examination lighting are important.

2.1 Light intensity

For an optimum examination environment that allows full visualization of color and detail, a minimum illuminance of 1,500 lux is recommended.[1] This is substantially higher than the 500 lux that is typical for office light levels.

2.2 Type of light

The color perceived by the eye depends fundamentally on the spectral distribution (or "color") of the incident light. Light with a smooth, continuous distribution, similar to sunlight, is optimal for human color perception. The color rendering index (CRI, a measure of how close the spectral content of a light source is to a perfect black-body radiator) is now commonly used to measure the quality of a light source for optimal color perception. In industry, a CRI higher than 95 is considered mandatory for precise color work (for color measurement, see II.2§4).

Three main types of electric lighting are now commonly available: tungsten, LED, and fluorescent. Tungsten halogen bulbs or standard incandescent bulbs offer the best CRI (100) and are excellent for examination purposes, though they can produce significant radiant heat. LED light sources are highly variable in CRI, though so-called high CRI and ultra-high CRI LEDs are now available with ratings from 95 to 99. These can provide suitable light while minimizing heat. Few fluorescent or compact fluorescent bulbs can provide CRIs of 95 or above, making them generally less desirable for examination purposes.

3 Examination tool kit

Some tools and equipment are nearly indispensable in the initial visual examination of a bronze sculpture. Following are recommendations from experienced examiners:

- **Overall lighting**. A bright and evenly lit room is a tremendous asset (see above).
- **Notebook** and/or **audio recording device**.
- **Gloves**. Gloves protect the sculpture from fingerprints and soiling. Nitrile gloves are widely preferred, though latex and cotton are commonly used.[2]
- **Flashlights** or other directional light sources. Intense, focused light is important for seeing into and through dark patinas as well as for creating localized raking light.
- **Measuring tools**. A cloth measuring tape, as used by tailors, is useful for measurements directly on objects, and a traditional, stiff measuring tape is also essential. A variety of calipers can be helpful, including Vernier, dial, or digital calipers for fine measurements, as well as different sizes of outside and inside calipers for measuring wall thicknesses, void dimensions, and so on (fig. 366). For overall measurements, a drafting or framing square, combined with square sheets of cardboard or foam core, and a bubble level are useful to make accurate measurements. See II.4.
- **Camera** with macro capability and tripod. See II.2§1.

- **Color reference card**, a **scale bar**, and a **pointer** are necessary for good photo documentation (see II.2§1.1).
- **Magnifying eyewear**. A wide range of specialized head-mounted magnification systems are available. Reading glasses with magnification of +7 are a low-cost alternative.
- **Tools for macro and microscopic examination**. Higher magnification than is available as eyewear is often desirable. Tools can range from an affordable hand lens (loupe) to very expensive articulated stereomicroscopes or high-end digital microscopes. Relatively low-cost USB microscopes or cell phone adapters may be an attractive option. The ability to record images of good quality is advantageous for documentation.
- **Ultraviolet light** (and protective eyewear). For initial examination, a handheld ultraviolet lamp is usually satisfactory (see II.2§3.1, I.4§1.3, I.8§3.1, fig. 191) Relatively low-cost LED flashlights are now available, with even distribution and high intensity; the most useful emit at 365 nm. The ability to darken the examination room is important.
- **Mirrors**. A selection of small inspection mirrors of different sizes can be useful for looking under or inside sculptures. Inspection mirrors often have an adjustable swivel mount at the end of a telescoping handle.
- **Magnets**. A strong magnet is useful for detecting core pins and other ferrous components (see I.1§3). Many practitioners place a small rare-earth magnet into a folded piece of tape. This provides a "handle" for the magnet and prevents direct contact with the object's surface. Even slight attraction of the magnet can be detected by observing flexing of the tape handle.[3]
- **Simplified line drawings or photographs** from different angles are useful for graphic documentation of features such as joints, repairs, or damage. Printed images can be marked up with colored pencils or marker, and digital files can be annotated digitally.
- **Tools and containers** for sample collection. These will vary depending on the type of sample to be collected. See II.5§3.5, II.6§3.1, II.7§3, II.8§1.4, II.8§2.4.

4 Guidelines for reporting technical results

4.1 How to report raw data

Although intended for a wide audience, including nonspecialists, the results obtained from a technological study should be recorded and reported in a scientifically reliable and useful manner for future researchers and for integration in shared databases. Therefore, it is strongly suggested that technical reports follow several basic rules observed in academic papers in the natural sciences.

4.1.1 Report on the operating conditions

Operating conditions shall be recorded and reported as precisely as possible, including all reference materials used for analysis, if relevant. Based on the report, another team of researchers should be able to carry out the same investigations using the same approaches to confirm the reliability of the results.

4.1.2 Raw results should be made available

Raw data should be made publicly available as much as possible, particularly measurements and analytical results. For quantitative data, tables are preferred to text files to allow for further processing (statistics, databases, and so on; see below). Attention should be paid to reporting significant figures based on the precision of the method.[4]

4.1.3 Distinguish objective physical evidence from interpretive data

As was made clear in volume I, the interpretation of physical evidence in the technical examination of bronze sculpture is rarely simple or straightforward. Different experts may draw different conclusions from the same X-radiograph or alloy analysis. For this reason, it can be beneficial to separate the reporting of results and observations from discussion and interpretation (as is standard practice in scientific publications).

4.1.4 Define your technical vocabulary

As is evident from a careful reading of the Vocabulary section of this publication, the terminology used to describe technical aspects of bronze sculpture can have different meanings to researchers from different disciplines or areas of specialization. Technical terms used in reports and publications should therefore be clearly defined within the document, and preferably refer to published authorities.

4.2 A picture is worth a thousand words

In natural sciences there is a rule of thumb that a report should be understandable by only looking at the illustrations, images, and diagrams, which means that illustrations and captions should be both clear and complete. The following four types of graphics are commonly provided in technical reports on bronze sculpture. For each, some general guidance is provided:

4.2.1 Image scale

For documentary photography of all types, a scale bar or indicator is of great importance. This is particularly true for macro photography and photomicrography.

4.2.2 Annotated figures

Annotated photographs, radiographs, line drawings, and other images, using arrows and/or shapes to focus attention on specific features, are highly recommended (see figures 159 and 188 for examples of simple annotations, and figures 65 and 367 for more complex annotations).

Make digital annotations using software that allows for superposition of an unlimited number of vector-based layers. In this manner, many different features may be recorded in a single, scalable file (for instance core flashes in green, mechanical joints in red, armatures in blue). Other image types may be included as layers, such as UV photographs (fig. 191) or radiographs. Each layer can be made visible or not to create different versions of the annotated image. In such documents a legend is crucial; as much as is possible, keep the legend consistent from one study to another.

4.2.3 Graphs synthesizing analytical results and/or measurements

Whereas a table is useful to check details or for the investigator to recheck their own calculations, graphs are well adapted to synthetize complex data sets and to highlight particular trends. Clear captions are mandatory (see Case Study 4, figs. 512, 518; and Case Study 5, figs. 525, 528).

4.3 Samples, images, and data management

Management of sample material, images, and data is key. This covers two aspects, namely conservation and record keeping, for which serious efforts are being made in the cultural heritage field.[5] The ideal way to document samples, technical images, and data for easy retrieval and further processing is to integrate them in structured databases.[6] The most important aspects to bear in mind before creating a database are:

- Define the scope of the database. What will the data be used for? For simple cataloguing for the operator and institution, or for exchanges with other teams and/or databases? In the latter case, what will be the purpose of the exchange? Do you really need a database?
- Ensure data reliability.
- Ensure data "interoperability" (readable and processable by other teams).[7]
- Structure the data as much as possible.

A simple spreadsheet may already constitute a database as soon as all these requirements are met. A number of databases related to cultural heritage may be used as examples.[8] The Hephaistos database is to date the only example of a large technical database dedicated exclusively to bronze sculpture (fig. 368).[9]

4.4 Statistics

One use for a large set of data may be statistical analysis. If the body of data is large enough, statistics may help characterize and quantify observed trends.[10] This may start with simple correlations (for instance between date and composition) and progress to more sophisticated approaches (cluster analysis, principal component analysis, machine learning, or artificial intelligence). For bronze sculpture, most statistical analysis carried out so far has been applied to chemical composition.[11] So-called unsupervised methods can be used to discover previously unrecognized patterns in data sets.[12] Before engaging in statistical analysis, confirm that the data to be analyzed are truly reliable and comparable. If statistical analysis becomes significant in the context of a technical examination, and especially if it is to be published, it is highly recommended to engage the assistance of a qualified statistician.

4.5 Publication

The dissemination of technical investigation of bronzes suffers currently from two handicaps. First, while the technical investigation of bronzes builds upon multidisciplinary teams and the results are normally directed toward a multidisciplinary audience, peer-reviewed publications that reach a broad, multidisciplinary audience are rare. Fortunately, in recent years, technical essays within exhibition catalogues have been used successfully to disseminate the results of technological studies.

Second, many studies remain as unpublished internal reports or as non-peer-reviewed conference publications due to restricted time allocated for research and publication. Some attempts have been made to facilitate access to this so-called gray literature,[13] and we hope these initiatives will multiply. That said, every effort should be made to place the results of technical studies in peer-reviewed publications. The transition to peer-reviewed publication will be facilitated by adhering to the aforementioned basic rules while preparing internal reports.

Notes

1. Ezrati 2015.
2. When selecting nitrile gloves, see Stephens et al. 2022.
3. Warning: 0.5 wt% of iron in a bronze is enough to attract magnets. Note that in the late eighteenth century, spelter brasses were preferred to cementation brasses for compasses because they are lower in iron and thus less magnetic (Watson 1786).
4. For a concise and clear discussion of significant figures and rounding see Bevington and Robinson 2003, 4–5.
5. For image management see Warda 2011, chapter 5. For sample management see the ongoing Heritage Samples Archives Initiative: https://www.iccrom.org/projects/heritage-samples-archives-initiative.
6. Specific labels are given to collections of data according to their structure and complexity: thesauruses, ontologies, and so on. For simplicity's sake, the term "database" is used here.
7. A key aim of the present *Guidelines* is to help researchers produce interoperable data. It is beyond our scope to guide on all the steps necessary to build a sound database (structure standards; interfaces for management, retrieval, and use). For more on databases see the introduction for nonspecialists in Baca 2008.
8. See for example the FLAME database at https://flame.arch.ox.ac.uk/ (Pollard et al. 2018); http://grossbronzenamlimes.de/database/begruessung; http://rembrandtdatabase.org/; https://www.getty.edu/projects/appear-project/; https://arachne.dainst.org/.
9. It has notably allowed comparison of Greek and south Arabian casting techniques of large bronzes (Mille 2012) and inference regarding new ideas about transfers of know-how (Mille 2017).
10. Baxter 2003.
11. See for example Glinsman and Hayek 1993; Bourgarit et al. 2003; Heginbotham, Erdmann, and Hayek 2018. For an overview of the subject see Baxter 2001.
12. Baxter 2006.
13. See for example CHARISMA, an initiative involving twenty-two European institutions: https://cordis.europa.eu/project/id/228330.

II.2

Photography and Other Imaging Techniques for the Visualization of a Sculpture

Clotilde Boust
David Bourgarit

Additional Contributors: Jane Bassett, Arlen Heginbotham, Alexis Komenda, Anne Maigret, Ruven Pillay, Yosi Pozeilov

This chapter aims to provide specialists and nonspecialists with insight into a range of photography and 3D modeling techniques for the study of bronze sculpture. Nonspecialists—those who find themselves in situations where professional photographers cannot be present—will find general guidance regarding production of high-quality images and what results can be expected. Advantages and drawbacks of available methods are discussed, including relative capability, costs, and time considerations when possible and relevant. For a synthesis, see table 13.

This chapter provides, by necessity, only abbreviated and general guidelines. For detailed guidance on best practice for photographic documentation of cultural heritage, including image and metadata management, see Warda 2011, Pozeilov 2015, and other references cited below.

1 Visible-light photography: most common techniques

1.1 Lighting a bronze sculpture

Many bronze sculptures have reflective or glossy surfaces (fig. 293), and particularly in these cases, a uniform and diffuse light will help minimize specular reflection. Diffuse lighting means that light is directed at the subject from large-area sources, usually accomplished by use of reflection umbrellas or diffuser screens in front of the light sources or by using flat-panel LED arrays. Normal flash photography, which involves a single small light source near the camera, is usually undesirable. Normally, at least two light sources are recommended. For documentary photography, shadows should not be too "hard" and there should be more even light than for catalogue or commercial images, where dramatic lighting is often preferred.[1]

Use lights that are designed for photography and have a high color rendering index (CRI) and always include a standard color and scale chart. For documentary photography, use only one type of illumination source at a time; if light sources of different color temperatures are used (such as window light in combination with photo lamps), the color rendition may be significantly different on different parts of the subject. In most circumstances, it is recommended to use a tripod with the camera set on "aperture priority." Selection of a high f-stop will significantly improve sharpness and depth of field, but will require longer exposures, which in turn necessitates a tripod. In situations where a high f-stop is not adequate to capture the required depth of field, a series of images may be captured, with the focal point moving stepwise from the nearest to farthest point on the sculpture. So-called focus stacking software has become increasingly available in common image processing programs, allowing the image set to be blended into a single image where all areas appear in focus. A tripod is essential for focus stacking, and all individual images should be archived along with the stacked image.

Particularly with dark sculpture, the full dynamic range (the difference in luminance between the darkest and lightest areas) may not be captured in one exposure. In such cases multiple, "bracketed" images of different exposures should be captured. High dynamic range (HDR) algorithms can composite different exposures into a single image, but always retain the original files.

1.2 Raking-light photography

For the documentation of fine surface topography such as tool marks and evidence of wear, raking-light photography may be useful. Here, instead of diffuse lighting, a single, strongly directional light source is positioned at a very low angle to the surface, typically around 10 degrees. This arrangement causes minor irregularities in surface topography to cast shadows, making them easier to perceive. Altering the direction of the light source can reveal different features on any given surface. Raking light can be useful for standard photography, macro photography, and even photomicrography (see II.2§2.1 below).

1.3 Macro photography

In many circumstances, close-up or macro photography will be necessary to resolve details of technical interest. Specialized macro lenses, which are designed to minimize distortion, are preferred for this purpose. The closer-up the image is, the smaller the depth of field will be, and thus the greater importance of using a high f-stop. In situations where it is desirable to hold the camera by hand, dedicated macro flash units, such as lens-mounted ring flash units, can provide sufficient diffuse or directional light for high speed and high f-stop macro imaging.

1.4 Image quality and resolution

The quality of images produced will depend on both the resolution of the camera's sensor (fig. 369) and the quality of the lenses used. For this reason, a professional-quality digital camera is recommended. As of the year 2023, a typical mirrorless camera may be equipped with a 8200 × 5500 pixel (45 megapixel) sensor, which will capture a one-meter-high statue with a vertical resolution of 82 pixels/cm, corresponding to a pixel size of 0.12 mm on the statue. Professional medium-format cameras with 100+ megapixel sensors can be of great benefit, but are cost prohibitive for many labs.

1.5 Time required to properly photograph a bronze

The time required to photograph a bronze sculpture in a professional studio can vary dramatically depending on the nature of the object and the number of details required. The simple documentation of six low-gloss statuettes via four or five general views of each might take two and a half days: half a day for the setup, one day for the photographs, and one day for post-processing. On the other hand, the full documentation of a single large, dark, and glossy bronze with numerous details may take more than six days: half a day for the setup, two days for capturing images (which will require several lenses), and four days for post-processing.[2]

2 Visible-light photography: less common techniques

2.1 Photomicrography

The imaging of surface features substantially smaller than one millimeter (polishing marks, details of tool marks, small inclusions, porosities, excrescences) may require higher magnification than macro photography, namely photomicrography. Traditional stereomicroscopes installed on articulated arms are very useful for this purpose, and these can be fitted with cameras for image capture.

High-quality digital microscopes (without eyepieces) offer additional capabilities and are increasingly widely used for technical examination of bronze sculpture. Most importantly, these microscopes usually incorporate automated focus stacking procedures that overcome the shallow depth of field inherent to photomicrography. Because the focus adjustments are precisely controlled by stepper motors, the microscope's software can reconstruct a scaled 3D model of the field of view. Such topographic reconstructions can be useful for documenting, characterizing, and comparing fine features such as tool marks (figs. 273, 285, 286, see Case Study 4). With this type of 3D model, the sources of errors for precise calculations are numerous and measurements should therefore be considered with caution. Digital microscopes often have polarizers incorporated that can greatly reduce specular reflection, and some can be adapted for use with ultraviolet illumination.

If extremely fine, high-precision topographic measurements are required, a method called microtopography may be useful. Various techniques are available, with resolutions as small as 0.1 μm. This technique is expensive, slow, and can generally be used only on samples smaller than 20 cm across.[3]

There are also a wide range of low-cost handheld digital microscopes available with a variety of magnification ranges that can be tethered directly to a laptop computer. Some models are equipped with ultraviolet and/or polarized light sources. While the image quality is generally not as good as more expensive laboratory- and research-grade equipment, they may be adequate for routine examination.

2.2 Endoscopy

Endoscopes (alternately called borescopes) are common tools for those studying the techniques of bronze sculpture. They are rigid or flexible tubes with a lens at one end and some optical or digital means to transmit an image through the lens to an eyepiece, camera, or digital sensor. In most cases, a light source is incorporated into the endoscope's tip. They can be used to examine the interior of a hollow sculpture by inserting the boroscope into the base (or any aperture) (figs. 38, 370). Image quality tends to be low, and images generated by a boroscope tend to be difficult to interpret because there is no external frame of reference, so careful attention to image annotation and captioning is critical. Recording video with audio narrative can also be an effective method of documentation.

The cost of endoscopes can vary tremendously. Models produced for industry or the medical field that have articulated heads (whose direction can be controlled from the exterior) and software that enables making measurements may be priced in the tens of thousands of US dollars, while simple USB models are available for less than one hundred US dollars.

2.3 Reflectance transformation imaging

Reflectance transformation imaging (RTI) can be of great utility in documenting and visualizing fine surface topography as well as color and gloss. In this method, both the object and the camera are fixed; only the point light source (typically a flash unit) moves. A large number of photographs are taken of the object (usually between twenty-four and one hundred) as the light source is systematically moved around the object (fig. 371). A reflective sphere is placed into the frame of each image and the position of the specular reflection of the light source on the surface of the sphere is used to calculate the direction of the light source. The software for compositing and viewing RTI images is open-source and free, making the technique readily accessible (fig. 241). Several mathematical image enhancement features are available that allow interactive relighting and enhancement of color and surface shape attributes.[4]

3 Imaging with nonvisible light

3.1 Ultraviolet fluorescence photography

Organic materials such as varnish, pigments, glue, resins, and a variety of restoration materials may fluoresce visible light when exposed to (invisible) ultraviolet light (UV). Photography of UV-induced visible fluorescence (often misleadingly called UV photography for short) may thus prove useful to visualize such materials on the surface of a sculpture (fig. 191). The method may be used in conjunction with macro photography and photomicrography.

Several types of UV light sources are available, including LEDs, fluorescent tubes, and arc lamps. Generally, lamps emitting at 365 nm with minimal visible light emission are preferred. Standard digital cameras can be used for UV fluorescence photography, but require special filters in front of the lens to block UV and infrared (IR) transmission to the camera's sensor. For details see Pozeilov 2015.

Risks of misidentification/misinterpretation

Although some materials emit important and specific color under UV, a variety of heterogeneous materials may emit similar fluorescence. Moreover, fluorescence is also affected by long-term light exposure and thermal aging. As a consequence, fluorescence does not generally allow for precise material characterization. Once fluorescent materials are localized by UV examination and photography, complementary analysis may be required to identify the materials present (see II.5).

Additionally, organic coatings may be present even if little or no fluorescence is detected. Over time, copper ions may migrate into organic coatings, and this can cause dramatic quenching of fluorescence. In addition, some organic materials possibly present on bronze sculpture, including many synthetic resins, do not fluoresce.

3.2 Thermography and infrared (IR) photography

Repair patches, inlays, and other discontinuous areas of a bronze are often thermally isolated from their surrounding regions. When the surface is heated, for example by the sun or a tungsten lamp, the heat will be kept longer there, dissipating more slowly than on the surrounding large surfaces. IR thermography (IRT) typically uses a camera sensitive to long-wave IR light (about 9000–14,000 nm) to image heat transfer and buildup on an object's surface. The thermograms thus obtained have been used on bronze sculpture to investigate and map defects, mechanical and cast-on repairs, as well as inlays.[5] Thermography in the mid-IR (about 3000–5000 nm) has also been applied to cultural heritage.[6] Imaging in the near-IR (about 700–10.1 µm) or short-wave IR (1100–2500 nm), commonly used on many other types of artwork, is generally of limited utility in the examination of bronze sculpture.

4 Color measurement

Color measurement has been used on occasion for research related to bronze sculpture, either to evaluate conservation treatment methods,[7] to characterize the color of different alloys,[8] or to characterize the color of patinas.[9] Most of these applications have been made in the context of experimental simulation (see II.9§1.2), where sample size, uniformity, and geometry can be controlled.[10]

A colorimeter or spectrophotometer may be used to measure color. These instruments illuminate a sample area under controlled conditions and measure the light reflected by the object. Colorimeters use filters to measure the amount of red, green, and blue light reflected from the sample, while spectrophotometers generate a detailed spectrum of the reflected light. In both cases, the resulting color measurements are normally made using the CIELAB color space, defined by three variables, L*, a*, and b*, as defined by the Commission Internationale de l'Éclairage (CIE) (fig. 372).[11] Color measurement instruments commonly used for museum objects are usually handheld and cost anywhere from several hundred to several thousand US dollars.

Color measurement of bronze patinas can also be accomplished using visual matching to standard color swatches such as Munsell soil color charts as described in Devogelaere 2017.[12]

Risks of misidentification/misinterpretation

The science behind color generation, color perception, and color measurement is complex. This is particularly true in the context of bronze sculpture because the color of bare metal is generated through a completely different physical mechanism than the color of patinas and oxidation layers.[13] Bronze sculpture is thus much impacted by goniochromism—that is, the change of color with the angle of the observer.[14] Where patinas (either organic, inorganic, or mixed) are not entirely opaque, meaningful and reproducible color measurement by any method may be difficult to achieve. Recently, new methods have been proposed that characterize surface appearance (particularly of patinas and gilding layers) using more complex procedures that take into account color, gloss, and translucency at several viewing angles.[15]

Color measurement is theoretically possible with digital photography following a color calibration protocol. However, in practice, color measurement of bronze sculpture with photography is extremely challenging. In addition to the difficulties mentioned above, accurate color measurement requires that the light falling on the measured area be the same color and intensity as the light falling on the calibration reference card and have the same directionality. This poses a practical problem with three-dimensional sculpture, since any change in the angle of the surface, as well as local inconsistencies in intensity or color temperature of the lighting, can affect the measured color.

5 Photogrammetry and 3D scanning

Photogrammetry and 3D scanning can produce three-dimensional renderings of objects useful for documentation and dissemination of information on the appearance of a sculpture. There are currently three main techniques available: photogrammetry, structured light scanning, and laser scanning. These can be combined in various ways to produce mixed models. All of these technologies are developing rapidly, and costs for both hardware and software are falling. In addition to recording appearance, these methods can also be useful for physical measurement; this latter aspect of their use is addressed in II.4§2.2.2 and II.4§2.3.4. All of these methods require specialized knowledge and training, though photogrammetry is probably the easiest to learn and has the additional advantage of not requiring specialized hardware.

5.1 Photogrammetry

In photogrammetry, many photographs are taken of the subject from different angles and a 3D model is derived from a computational analysis of the individual images using dedicated software (fig. 373). Most photogrammetric work done in cultural heritage has been carried out with professional-grade digital cameras, though it is also possible to build very high resolution models using medium-format cameras that record images of 100 megapixels and higher. Applications for smartphones are also now also available at lower resolution.

Depending on the quality and resolution of the source images, photogrammetry can provide a quite high-fidelity model of a bronze sculpture with good color accuracy, particularly if the photography is done in a studio with controlled lighting. These images can be useful for both public and scholarly audiences, providing the opportunity for high-quality web-based interactive viewing. Photogrammetry can also be carried out using UV and IR photography.[16] Photogrammetry is being used more and more for bronze sculpture, notably to design models (fig. 374) and mounts (fig. 375), or to test reassembly of fragmentary statues. Large-scale bronzes can be captured using scanners mounted to drones.[17]

5.2 Structured light scanners

Structured light scanners shine a regular pattern of light (usually a series of parallel lines) onto an object and then record an image of the pattern with a camera that is offset from the projection source. Software then analyzes the apparent distortion of the light patterns by the three-dimensional surface and generates a computer model of the form. Most structured light scanners also record standard photographic images of the subject using onboard flash lighting, which alternates at high speed with structured light pulses.

The lenses and sensors used in structured light scanners are generally of lower quality than a professional digital camera, and lighting and color management are often a challenge.[18] As a result, structured light scans often yield a final model that is less realistic and detailed in appearance than a photogrammetric model (figs. 376, 377), though the dimensional accuracy may be as good or superior. It may thus be used for precise measurements (fig. 378, see II.4§2.2.2).

5.3 Laser scanners

Most laser scanners used in cultural heritage employ a rapidly rotating mirror to deflect a visible or invisible laser beam in a pattern that sweeps across the subject. The scanner integrates a range finder that accurately measures the distance to each of the millions of points that the beam traces, building up a high-resolution 3D model of the form. Often multiple scans from different vantage points are combined to produce the final model (fig. 379). Most laser scanners integrate cameras to provide "texture," or photographic detail, that can be overlaid onto the 3D model (figs. 380, 381). However, as with structured light scanning, the relatively low quality of the cameras used can result in models that are less realistic and detailed in appearance than a good photogrammetric scan.

Notes

1. Yosi A. Pozeilov recommends that the main light point toward the object at an angle of 50–60 degrees. The second lamp should be at 35–45 degrees. For this and other pointers, see Pozeilov 2015, 88.
2. These data were kindly provided by two photographers at the C2RMF, Anne Maigret and Alexis Komenda.
3. Mélard et al. 2016; Page et al. 2016.
4. Dellepiane et al. 2006. Free software and viewers can be downloaded from the Cultural Heritage Imaging website: http://culturalheritageimaging.org/Technologies/RTI/.
5. Orazi et al. 2016; Mercuri et al. 2017; Mercuri et al. 2018. Different profiles, and consequently techniques of mechanical repairs, are detectable on the *Capitoline Wolf*.
6. See Gavrilov, Maev, and Almond 2014.
7. Heginbotham, Beltran et al. 2014; Letardi et al. 2016.
8. Devogelaere 2017; Mödlinger et al. 2017; Radivojević et al. 2018.
9. Benzonelli, Freestone, and Martinón-Torres 2017; Formigli 2013a.
10. Letardi et al. 2016 carried out direct measurements on statues. Devogelaere 2017 focused on ancient bronzes but working on experimental coupons. See also Formigli 2013a.

11. For an accessible synthesis of color science in the context of cultural heritage see Berns 2016. For more in-depth scientific insights see Hunt and Pointer 2011 and Johnston-Feller 2001. And for a deep dive into color models and their application see Fairchild 2013.

12. Devogelaere 2017; Munsell Color 2000.

13. Nassau 1987.

14. The term was first proposed in McCamy 1996.

15. Arteaga et al. 2021; Arteaga et al. 2022.

16. Lanteri, Agresti, and Pelosi 2019.

17. For more, see http://culturalheritageimaging.org/Technologies/Photogrammetry/.

18. Felicísimo and Polo 2022.

II.3

Radiography and Tomography for the Visualization of the Metal Wall and the Interior

Elsa Lambert
David Bourgarit

Additional Contributors: Aurélia Azéma, Arlen Heginbotham

Radiography is widely used among specialists studying bronze sculpture because it can reveal a great deal of information regarding production methods and previous restorations. Figure 65 illustrates some of the many technical features of a bronze sculpture that can be observed using radiography. In principle, bronzes can be radiographed regardless of their size or the nature of the alloy, but the amount that can be learned depends heavily on the experience of the operator of the radiographic instrumentation, and also on the experience of the researcher interpreting the results.

This chapter aims to guide both specialists and nonspecialists in the use of radiography and its 3D corollary, tomography, for the technical investigation of the metal wall and interior of a bronze sculpture. Note that only radiography and tomography are dealt with in this chapter. Other techniques that may be applied to the study of the metal wall are dealt with elsewhere in the *Guidelines* (see II.4§2.3.3 for ultrasonic testing, II.2§3.2 for thermography, and II.5§2.3.2 for eddy currents).

The *interpretation* of radiographs is not the subject of this chapter, but some guidance in this respect may be found in relevant chapters of volume I. Here, the basic scientific principles of several radiographic techniques will be described, including X, gamma, and neutron radiography as well as X-ray and neutron tomography. Specific recommendations for working with bronze sculpture will be given with respect to both common examination methods as well as cutting-edge techniques. The advantages and drawbacks of available methods will be discussed, including costs and time requirements (see table 13 for a summary). Given the large range of techniques available and their complexity, the descriptions of each technique are necessarily brief. For a synthesis, see table 13. For more detailed information, references to specialized literature are provided.[1]

1 X-radiography

1.1 Scientific principles

Radiography is considered a nondestructive testing technique; normal application of radiographic techniques as practiced in the field of cultural heritage should pose no risk to objects.[2] X-rays are, however, a form of ionizing radiation that can damage living tissues, and therefore the use of X-radiation is regulated in all countries, requiring appropriate safety training and controls. In addition, for bronze sculpture, an important consideration prior to radiography is that X-radiation exposure will interfere with subsequent trapped charge dating techniques—thermoluminescence (TL) and optically stimulated luminescence (OSL)—resulting in age estimates that are older than the actual age (see II.8§1).

1.1.1 Attenuation

A radiographic image is obtained by transmitting a beam of X-ray photons through an object. The transmitted X-rays are recorded by a detector placed at the rear of the object (fig. 382). Depending on the thickness of the object and the material from which it is made, the direct transmission of X-rays through the object is blocked, or attenuated, to a greater or lesser degree. Attenuation can be caused either by absorption or by scattering of the photons. Where X-rays pass through the object easily, the detector is strongly exposed, causing the image to appear dark. Where X-rays are more heavily attenuated, the image will appear light (only rarely is this convention inverted).

The actual attenuation of X-rays is dependent on two main parameters of the object being imaged. The type of atoms that compose the object (copper, lead, tin, iron, et cetera) greatly influences the attenuation. In general, the higher the average atomic number of the matter, the more attenuating it will be, so an object made of lead will be dramatically more attenuating that an object of the same dimensions made of aluminum.

The thickness of the object to be imaged also directly affects the attenuation of X-rays. Given two objects made of the same material, one twice as thick as the other, the thicker object will attenuate an X-ray beam twice as much as the thinner (fig. 383). Naturally, a porous object will attenuate X-rays less than a similar solid object, depending directly on the degree of porosity.

1.1.2 X-ray energy

The attenuation of X-rays by an object is also dependent on the energy (or wavelength) of the X-rays used for imaging (fig. 384). Higher-energy X-rays are generally more penetrating than lower-energy X-rays. An X-ray tube may be set to produce X-rays at a range of energies. The tube is set to a certain energy level (expressed in kilovolts, or kV), which determines the maximum energy of X-rays produced (expressed in kiloelectron volts, or keV).

The nominal voltage setting may be somewhat misleading, since X-ray tubes produce a continuous spectrum of X-rays, with energies ranging from close to zero up to the nominal voltage setting. The dominant or average energy of the resulting X-ray beam will normally be between 25% and 50% of the nominal energy, with very little radiation actually being produced at the maximum level.

1.1.3 X-ray sources

Most radiography for cultural heritage is conducted with X-ray tubes. Common industrial X-ray tubes have maximum nominal voltages of 100–450 kV (rarely up to 750kV). Conventional industrial X-ray tubes suitable for studying bronze sculpture are heavy and require large auxiliary transformers as well as circulating liquid cooling systems, making them difficult or impossible to transport for use in situ. Increasingly, portable X-ray tubes that are air cooled and can be plugged into an ordinary electrical wall socket are becoming available.

Also becoming available are pulsed X-ray systems that are compact and battery powered, and generate multiple short pulses of X-radiation (measured in nanoseconds), with maximum voltages approaching 400 kV. The short pulse length significantly reduces the health and safety hazard associated with their use, and their small size makes them suitable for in situ imaging. Unfortunately, the low total photon output of these generators means that they can only be used effectively with digital radiography (DR) detectors, and not with film or computed radiography (CR) systems (see below).

1.1.4 Gamma and synchrotron radiography

In instances where conventional X-radiation is too limited in energy to penetrate high-density and/or thick artifacts (figs. 385, 386), gamma radiation may be used for imaging. Gamma rays are, like X-rays, photons of high-energy electromagnetic radiation, but are by definition produced by the decay of a radioactive source. For industrial gamma radiography, the two most common source materials are iridium-192 and cobalt-60. Radioactive sources have fixed, nearly monochromatic energy output. Iridium-192 has its primary gamma emission at around 380 keV, while cobalt-60 is at around 1250 keV. Iridium-192 allows bronze statues with thickness of up to 5 cm to be examined,[3] while Cobalt-60 will reach 10 cm. Contrast is rarely as good as for conventional X-ray radiography (fig. 194). For very thick objects, an alternative to gamma radiography is to use high-energy X-rays (6000–30,000 keV) produced by linear accelerators or at larger particle accelerator facilities such as synchrotrons.

1.1.5 Electron emission radiography

When exposed to high-energy X-rays (300–400 keV), secondary electrons are emitted by some of the atoms present in the upper surface layers of an artifact. The heavier the atoms present (that is, the higher their atomic number), the more intense the electron emission. These electrons may be collected on a film held in close contact with the surface; this is called electron emission radiography (figs. 387, 388). X-ray film is much more sensitive to electron exposure than to high-energy X-rays, so the dominant image produced on the film will reflect the distribution of heavy elements on the surface.

As opposed to transmission X-radiography, only the upper 50–100 μm of surface will contribute to an electron emission radiograph, and areas of higher density will appear darker rather than lighter. Electron emission may be used to great effect in imaging inlays of different metals (particularly silver or gold) in flat or very low relief objects such as bronze plaques (fig. 389).

Intimate contact between the film and the surface is required in order to limit the absorption of the emitted electrons by air. Since the film is pressed onto the surface, it cannot be used on fragile surfaces. Unfortunately, the preferred (single-sided) industrial films are not produced anymore. CR plates are much less sensitive than film, and DR detectors are unsuitable for electron emission work.

1.1.6 Detectors

Currently, three main types of X-radiographic detectors are in common use: silver halide film, CR plates or sheets, and DR detectors.

Film offers a wide range of sensitivity and very high resolution, and it is very easy to cut and bend in order to adapt to the different shapes and sizes of the objects. Exposure times are typically on the order of several minutes, though exposures measured in hours are not unheard of for large sculptures. Processing of film radiographs is relatively time and labor intensive, taking about ten minutes to develop a single sheet. Silver halide film is becoming somewhat more difficult to obtain, though a wide range of silver halide films are still produced for industrial radiography. The chemicals used to process film are becoming increasingly subject to regulation as pollutants.

CR uses photostimulable storage phosphor plates to record images. These plates may be housed in rigid cassettes or used as loose, flexible sheets similar to traditional film. First the plates are exposed, and the phosphor layer captures the energy of the incident X-rays. The plate is then inserted into a scanner that scans a focused laser across the surface, stimulating the release of the stored energy as light. The intensity of the luminescence at each point on the plate is recorded by an optical sensor, which generates a digital image. Scanning can take between one and ten minutes, depending on the resolution settings and the size of the plate. Plates can be erased and reused hundreds of times if well cared for. The range of sensitivity of CR systems is at least as good as film, while the resolution is not quite as good, with pixel dimensions of around 25–50μm. The plates are available in many sizes, including long strips around 35 cm wide and 150 cm (or more) long. They can, if necessary, be cut and bent like film. Exposure times are typically a quarter or half as long as film. Modern CR systems usually incorporate digital filters for edge enhancement and other image optimization tools. These tools can dramatically improve the visualization of subtle features in radiographs.

DR systems use fixed arrays of X-ray-sensitive pixels that send image data directly to a computer through a wired or wireless interface, where it can be captured, viewed, and saved in real time. DR detectors have a wide range of sensitivity, but overall sensitivity is much higher than with film or CR, resulting in much shorter exposures, often only one or several seconds. Averaging of multiple exposures is typically recommended to reduce noise. The short time needed to generate radiographs makes it easier to produce many different views of an object and facilitates the acquisition of short movies or image sequences that aid in the interpretation of complex three-dimensional forms (figs. 121, 390, 498, 564). With sufficient rotational precision, image sequences suitable for full computed tomography (CT) can also be generated (see II.3§2 below).

DR systems yield images with lower resolution than film or CR, with pixel dimensions on the order of 100–200 μm. The detector plates are manufactured in fixed sizes (typically from 20 × 20 cm up to a maximum of about 40 × 40 cm) and are housed in rigid cassettes that cannot be cut or bent. Most modern DR systems also include the same kind of digital image optimization filters that are found in CR systems.

Some DR systems utilize line-scan sensors that require an x, y displacement system to allow the complete scanning of the object. These detectors are often used for X-ray tomography. Their main advantage is a much lower cost than array detectors. Some systems also enable the scanning of much larger surfaces (100 × 100 cm versus around 40 × 40 cm) (see table 15).

1.1.7 Controlling image quality: exposure time, current, and distance

Correct image exposure is normally obtained by adjusting the exposure time (usually expressed in seconds or s) and the tube current (expressed in milliamps or mA).[4] The current determines the rate of X-ray production by an X-ray tube, or its brightness. The product of the current and the exposure time (mAs) determines the total quantity of photons emitted and is thus commonly used as a designation of overall exposure. The distance between the X-ray source and the detector has a strong influence on the time and current necessary for a proper exposure. Total exposure is inversely proportional to the square of the distance, so doubling the distance between source and detector requires that the exposure time (or current) be increased fourfold.

1.1.8 Contrast

Contrast in a radiographic image can be controlled by adjusting the voltage setting of the X-ray tube. The higher the voltage, the lower the image contrast will be. As a general rule of thumb, select a voltage that is just above the minimum energy required to effectively penetrate the most attenuating region of the object.

1.1.9 Sharpness: focal spot and distance

To guarantee maximum image sharpness, three main parameters may be adjusted: the distance from the detector to the object, the size of the X-ray tube's focal spot (the area from which the X-rays are generated), and the distance from the focal spot to the object. The detector should be placed as close as possible to the area of interest of the examined object (fig. 391 3, 4), keeping the detector surface perpendicular to the beam (fig. 391 5, 6).

Very often the morphology of a three-dimensional subject does not allow close contact with all parts of it. Sharpness and depth of field can be increased by using a small focal spot. Depending on the model, the X-ray tubes may have different sizes of focal spots; the largest one allows the use of the entire power range of the tube, while with a smaller focal spot only the lower power range is accessible. The smaller the focal spot, the sharper the image (fig. 391 7, 8). Compensate for a consequent reduction in intensity by increasing exposure time.

To further increase sharpness, increase the distance between the X-ray source and the object (fig. 391 1, 2). This has the same effect on sharpness as shrinking the focal spot, and has the added effect of reducing geometric distortion. The effective exposure will be reduced according to the inverse square principle, so exposure time will increase accordingly.

1.1.10 Scattering

When making a radiograph, part of the X-ray beam is absorbed by the object and another part is transmitted to the detector to form the image. But at higher energies, a significant part of the beam is also scattered in all directions by the object itself and the room's walls, floor, and ceiling. This generates fogging of the detector and a reduction in contrast (fig. 392). There are several strategies to reduce the effects of scatter. Lead shutters or a collimator may be placed in front of the X-ray source so that only the necessary area of the object is exposed by the beam, and X-rays that would fall outside the detector area are blocked. Lead sheeting several millimeters thick may also be placed immediately behind the detector and/or around the object to absorb the backscattered X-rays coming from the surroundings (walls, ceiling, floor).

1.1.11 Filters and screens

The primary X-ray beam may be filtered at the source to selectively attenuate lower-energy X-rays, and thus increase the average energy of the output spectrum. This is called beam hardening and is usually accomplished using aluminum or copper filters from one to several millimeters thick. Such filtration reduces overall X-ray output (requiring longer exposures) but can increase the effective penetration of the beam and also reduce scattering to a certain degree by attenuating non-penetrating X-rays that still contribute to scatter.

At high-energy exposures, silver halide film may benefit from being inserted directly between two thin lead screens (around 50 μm thick). High-energy X-rays cause the lead to emit electrons, which exposes the film very efficiently, thus reinforcing or intensifying the image. These screens are frequently used with film and CR detectors for bronze objects (figs. 37, 393), but are not useful for DR detectors.

1.2 Application to bronze sculpture

1.2.1 Configuration

Each sculpture is unique because of its shape, material, and facture. Thus, the geometric configuration of the X-ray source, the sculpture, and the detector must be optimized for each object, and unlike the case with medical or industrial applications, no preestablished protocol can or should be applied (compare for instance the operating conditions for three different bronzes as reported in figs. 32, 37, 394). The image obtained in an X-radiograph is the projection of a three-dimensional volume on a two-dimensional surface. The features present on many different planes are therefore merged, and determining the depth of any given feature can be very difficult.

To help locate features in three-dimensional space, it is therefore extremely helpful to make multiple images from different angles. A front view and a side view are the minimum for a radiographic study, while numerous views from different angles may be of great benefit. Complementary information will appear in the various images, leading to a clearer interpretation of the object. To aid in interpretation, it is very helpful to also have on hand high-resolution photographs taken from the same angle as the radiographs.

The speed of acquisition associated with newer DR systems allows the goal of multiple views overlaid with visible images to be more readily attainable. When combined with a digital turntable designed for product photography, it is even possible to produce radiograph sequences that capture a sculpture rotating through 360°. The angular interval between frames can be customized (5° is often satisfactory) so that an image set can be generated in around 20 minutes and can then be processed and viewed in a web browser. After radiography, the X-ray source may be removed and a digital camera may be placed in the same position allowing visible images and radiographs to be overlayed (fig. 498). If a simple easel is used (fig. 563), the detector can be repositioned, keeping the object and x-ray source fixed, so that multiple series of radiographs can be composited to yield image sets of objects much larger than the detector panel (fig. 564). This type of 360° X-ray image package provides substantially more information to the viewer than individual radiographs and yet is more accessible and simpler to produce than a full CT scan.

As a general rule, the larger the sculpture, the greater the optimal distance between the X-ray source and the object in order to minimize geometric distortion and enhance sharpness. For example, a distance of two meters is enough for a typical sculpture smaller than one meter. For a taller and broader sculpture, the X-ray tube might be positioned three meters from the object. If a complete radiograph of a medium-to-large sculpture is desired, limitations on detector size often make it necessary to make several successive exposures and stitch them together to obtain a composite radiograph. Depending on the file size of each radiograph, stitched images can generate files of several hundred megabytes, requiring computers with adequate processing power.

When images are going to be stitched, the X-ray source and sculpture should stay in the same relative position for all exposures. The film or detector should then be moved for each exposure in an overlapping grid within a single plane perpendicular to the X-ray beam, and as close to the sculpture as possible.

1.2.2 Optimizing image quality

In theory, one should use the lowest voltage possible as long as the X-ray can penetrate the object being studied. If the material is thin or if the metal is corroded, 250 kV or below may be sufficient (fig. 394). In practice, for most bronze sculptures, the variation in thickness will be significant and the contrast of the resulting radiographs will tend to be very high. As a result, voltages exceeding 250 kV are frequently required, often with significant beam filtration in order to reduce contrast and yield proper exposure in all areas of the image (figs. 32, 395). Steps to limit scatter should also be taken.

A trade-off when using very high voltages with beam filtering is that local contrast within relatively homogenous regions of the radiograph will be very low, making subtle details difficult to resolve. Advanced image filters used in CR and DR can compensate for this difficulty to a considerable degree, but an alternative is to capture multiple images at lower voltage (higher contrast), exposing each to optimize exposure in different areas of the image. By using lower voltages on selected areas, small variations in thickness may be revealed. For objects with relatively thin walls, traces of hammering can be visualized by reducing the voltage (to less than 100 kV if possible), avoiding filtration, and working with long exposures and maximum tube current (fig. 396).

1.3 Cost and availability

The use of X-rays requires special infrastructure to contain the ionizing radiation. The walls of dedicated X-ray rooms are usually made of thick concrete sheathed with lead.

Regulations for facilities and operator training vary depending on the relevant national and international agencies. X-radiography can also be carried out on-site with appropriate precautions. Here, too, detailed safety regulations will apply. For on-site work, the use of nanosecond-pulsed X-ray generators in conjunction with DR detectors offers significant advantages, since the total X-ray dose generated per image is reduced by many orders of magnitude.

A new installation of fully compliant X-ray facilities capable of imaging bronze sculpture will likely cost several hundreds of thousands of US dollars. If dedicated facilities are not available, private industrial radiography companies are often able to provide on-site mobile radiography services using tube radiation, pulsed generators, or isotope sources. The transportation, use, and control of isotope sources are all governed by specific rules defined by the International Atomic Energy Agency (IAEA).

Prior to using gamma radiography in the field, detailed compliance documents must be prepared in collaboration with the local authorities. The use of high-energy accelerator-based radiography is, of course, logistically and bureaucratically complex, involving transportation and security considerations as well as the preparation of a detailed research proposal to the relevant facility.

1.4 Possibilities for misinterpretation

The fact that three distinct parameters contribute simultaneously to the attenuation of X-rays leads to possible misinterpretation of image density. For example, a 10 mm thick aluminum foil and a 3 mm thick bronze plate may yield similar gray levels. Conversely, two sheets of equal thickness, one of aluminum and the other of bronze, will appear with dramatically different gray levels on the picture (fig. 388). At the same time, two distinct materials may be difficult to distinguish if they have a similar atomic number. For example, pure copper and brass with 30% zinc will appear nearly the same if they have similar thicknesses (see also copper and steel in fig. 388).

The interpretation and identification of specific technological features in radiographs is also fraught with difficulties. Volume I contains numerous examples of clearly defined features, and the technical literature offers some additional guidance,[5] but much still depends on the experience of the specialist in recognizing subtleties in the images and unraveling their meaning.

2 X-ray computed tomography (CT)

2.1 Scientific principles

X-ray tomography, also known as computed tomography (CT), is familiar to most from its medical application as so-called CAT scans. This technique generates a complete 3D model of an object, assigning a density value to each three-dimensional pixel, called a voxel. An X-ray tomograph is assembled by making a series of two-dimensional X-ray images as the object is rotated through 360 degrees (figs. 121, 390, 397). Each recorded image is called a projection. Traditionally, the number of projections needed to perform a workable tomography is calculated based on the size of the detector and the size of the object, with many hundreds of projections commonly required. The minimum number of projections is equal to the number of pixels occupied by the width of the object on the detector; the ideal number of projections equals the number of horizontal pixels in the image $\times \pi / 2$.[6] Recently, a great deal of work has been done developing algorithms that can create reconstructions based on many fewer projections.[7]

The set of projections is analyzed using sophisticated and computationally intensive computer algorithms in order to produce the final model. The results may be viewed as two-dimensional slices through the object at any desired angle, or voxels of similar density may be chosen for viewing, which allows a virtual three-dimensional view of selected components or features on the interior or exterior of the object. Tools for simple measurement (distance, angle, diameter; fig. 398) and for image processing with digital filters (noise reduction, sharpness, segmentation of different materials; fig. 399) are available. With dedicated CT interpretation software, image analysis is possible (measurement of wall thicknesses, volume of porosity and inclusions, detection of defects, et cetera). The accuracy of these different tools depends on the resolution of the initial set of projections.

2.2 Application to bronze sculpture

2.2.1 Configuration

When making a tomographic reconstruction of a bronze sculpture, the subject should be positioned so as to present a minimum variation in thickness or density to the X-ray beam. It is often advantageous to angle the X-ray beam slightly upward or downward to avoid passing it directly through horizontal elements parallel to their long dimension. For example, outstretched arms or the tops of integrally cast bases can be very difficult to penetrate if the X-ray beam must pass through their length.

In tomography, the subject cannot come in contact with the sensor at any point in its rotation. This means that the bronze is often placed farther from the detector than in conventional radiography. Therefore, it is best to maintain a maximum distance from X-ray source to object as well as a small focal spot in order to maximize sharpness and minimize geometric distortions.

For conventional tomography, the bronze must appear entirely within the field of view of the detector on each projection to guarantee a satisfactory reconstruction. The size of the sensor (usually less than 50 × 50 cm for plane detectors) is thus a significant limiting factor for large works. It is possible to stitch together multiple radiographs into a series of larger projections, though this requires great precision and repeatability in the placement of the detector for each view.

2.2.2 Optimizing image quality

In radiography, the parameters (voltage, intensity, exposure time) are adjusted for each shooting angle according to the thickness of the material to be X-rayed. In tomography, however, the shooting parameters cannot be changed during the scan. It is thus necessary to adjust the parameters to accommodate both the greatest and the least thickness encountered. To ensure penetration of thick parts and simultaneously avoid overexposure of thinner parts, very high voltage is often used and the beam is heavily filtered to further reduce contrast. Under these conditions, measures to reduce scatter are very important.

2.3 Cost and availability

Available CT equipment falls into two major categories: medical and industrial. Medical CT is generally limited in voltage to 160 kV or below, so industrial tomography equipment is typically used for most bronze sculpture. The size of the sculpture, the resolution needed, the thickness of the metallic wall, and the type of copper alloy are all considerations in determining the specific equipment required.[8] Many of the same companies that provide on-site industrial radiography can also provide industrial CT services using tube or isotope sources.

Micro- and nanotomography, which use micro-focus X-ray sources to produce tomographic reconstructions of very high spatial resolution, are also available. These techniques are used mainly for the investigation of fine microstructures of small items (less than several centimeters in diameter).

It is also possible to configure an in-house CT system, particularly if a DR system is already present.[9] Additional requirements would include a precision-controlled turntable with adequate weight capacity, a powerful computer, and tomography software.

The computer should be fast and have large capacity in order to view and analyze the tomographic reconstructions. The more pixels in the detector used, the larger the volume of data to be processed. The processing computer should have a minimum of 256 GB of RAM and be equipped with a powerful GPU (graphics processing unit) and CPU (central processing unit). An SSD (solid state drive) can also be advantageous to reduce the reconstruction time and also during the volume analysis process. A storage drive with capacity of several terabytes is needed for data backup. A number of commercial vendors supply software that can be used for tomographic reconstruction, and free open-source packages are also available.

In special instances, high-intensity and high-energy radiation generated at particle accelerator facilities may be used for creating tomographic reconstructions (fig. 358).

2.4 Possibilities for misinterpretation

The mathematical and physical complexity of tomographic reconstruction leads to many possibilities for the generation of "artifacts" in CT scans.[10] Ring artifacts, beam hardening artifacts, and noise artifacts may appear to be features in the reconstruction, but are purely accidental products of the reconstruction algorithms. Working closely with an experienced CT interpreter is of great importance.

3 Neutron radiography and tomography

3.1 Scientific principles

Neutron radiography uses a beam of so-called thermal neutrons rather than photons to create radiographic images.[11] A beam of thermal neutrons is typically produced at a nuclear reactor by slowing down, or moderating, the "fast" neutrons generated by nuclear fission, though radioisotopes may also be used as a neutron source. Whereas X-ray and gamma photons interact with electrons, thermal neutrons are attenuated by either scattering or absorption by the nuclei of the atoms in the beam path.

Neutron radiography produces images that are fundamentally different than those of photon-based (X-ray and gamma) methods (compare fig. 400 with fig. 401). While X-ray attenuation is highly correlated with atomic number (larger and heavier elements such as lead block the transmission of X-rays more effectively than lighter elements such as aluminum), the attenuation of neutrons by matter is not dependent on atomic mass. Elements as diverse as hydrogen, cadmium, and mercury are very high attenuators, while sulfur, tin, and lead are nearly transparent (fig. 402).[12] The extremely high attenuation of thermal neutrons by hydrogen means that organic materials or water are clearly visible in neutron radiographs while metal is relatively transparent. Neutron imaging is therefore complementary to traditional radiography.

Detectors for neutron radiography are usually made of two components: a conversion screen and a photon detector like that used for X-radiography. The conversion, or scintillation, screen (often made of a thin layer of gadolinium) converts the kinetic energy of the neutrons into visible or other radiation, which is then recorded on X-ray film or by a digital camera, CR plate, or DR detector.

Tomographic reconstructions from neutron radiographs are produced in the same manner as for X-ray tomography, with similar requirements for software and computational capability.

During neutron radiography, it is also possible to conduct neutron diffraction studies, which can characterize the structure of materials through their entire depth. This technique is described in II.5.

3.2 Application to bronze sculpture

Neutron radiography has two primary advantages for imaging bronze sculpture. First, very thick metal walls are easily imaged due to the low attenuation of neutrons by most metal elements. Second, neutron radiography can effectively image materials with lower density than the surrounding envelope. This is particularly true for organic materials inside of a bronze sculpture, as seen in the study of votive offerings in Tibetan Buddhist metal images.[13] Metal objects such as silver, copper, and gold embedded in a denser material such as lead or tin may also be detected in this way (figs. 403, 404, see Case Study 4). Core material may also be more effectively imaged with neutrons than with X-rays.[14]

3.3 Cost and availability

The main drawback of neutron radiography is that so far, large-scale research facilities are required. Increasingly, certain facilities have promoted the use of neutron imaging and analysis for cultural heritage.[15] In such cases there is often no fee per se for the imaging, but a lengthy application process may be necessary to secure the allocation of time and instrumentation at a reactor facility. Costs can, of course, be significant for transportation and security, and the administrative burden can be substantial.

Another serious consideration is that during imaging, atoms that absorb neutrons into their nuclei form radioactive isotopes that subsequently emit gamma radiation. These isotopes must be allowed to decay to safe levels before handling. Depending on the specific elements present and the length of exposure (tomography requiring significantly longer exposure than radiography), this period of quarantine may be several hours or days, or even longer, imposing additional cost and security considerations. Copper alloys are not usually subject to extended quarantine unless they contain substantial amounts of gold or arsenic. For example, no quarantine was needed for the Javanese statuettes discussed in Case Study 4.

3.4 Possibilities for misinterpretation

As with X-radiography, exposure to thermal neutrons will interfere with subsequent trapped-charge dating techniques (TL and OSL), resulting in age estimates that are greater than the actual age (see II.8§1).

Notes

1. For X-ray techniques, see in particular Lang and Middleton 2005. For neutron techniques see Chankow 2012; Van Langh 2012.
2. However, long-duration exposure to focused, high-intensity X-radiation has been shown to cause minor damage to organic materials such as wood (Kozachuk et al. 2016).
3. For example, the French sixteenth-century greater-than-life-size bronzes from Primatice at Fontainebleau, and Germain Pilon and Ponce Jacquio at the Basilica of Saint-Denis (Henri II and Catherine de Medici's funerary monument), have been radiographed using an iridium source (Castelle 2016).
4. If rigorous optimization of image quality is desired, standardized image quality indicators (IQI), as are used in industrial foundries, may be used as a guide. See Ruault 1991, 2:199–209; Halmshaw 1995, 146–60.
5. Mattusch 1996; Lang and Middleton 2005; Bassett 2008.
6. Buratti et al. 2016.
7. Sidky, Kao, and Pan 2006; Bian et al. 2010.

8. Plessis, le Roux, and Guelpa 2016.

9. Bossema et al. 2024.

10. Boas and Fleischmann 2012.

11. For a thorough overview of neutron radiography see Chankow 2012. For general principles on neutron scattering and neutron reactors see Bordenave and Mirebeau 2018.

12. See Halmshaw 1995, 284–94.

13. Mechling et al. 2018; Henss and Lehmann 2016. For an overall view of the applications of cultural heritage see Mannes et al. 2015.

14. Van Langh et al. 2009.

15. For instance Paul Scherrer Institute in Villigen, Switzerland, and Budapest Neutron Centre. The Orphée reactor in Saclay, south of Paris, closed in 2019. The trend now is to produce neutron beams using high-voltage accelerators instead of nuclear plants.

II.4

Measurements of Dimension

David Bourgarit
Francesca G. Bewer

Additional Contributors: Aurélia Azéma, Jane Bassett, Manon Castelle, Laurence Garenne-Marot, Andrew Lacey, Emmanuel Lamouche, Carol Mattusch, Sreyneath Meas, Lorenzo Morigi, Dominique Robcis, Donna Strahan, Brice Vincent, Jean-Marie Welter

This chapter overviews the different ways in which measurements such as weight, volume, and size are useful in the study of bronzes, and aims to guide the reader in how to make a variety of measurements correctly.[1] How to best report measurements (accuracy, et cetera) is mainly addressed in II.1§4. Micro-scale measurements of tool marks on a bronze surface, such as those obtained with digital microscopes, are dealt with in II.2§2.1. How to measure the thickness of surface layers, including patinas and metal plating, is covered in II.6§1.

1 What can we learn from measurements? And what kinds of measurements are useful?

1.1 Overall dimensions

Overall dimensions (height, width, depth) are important, and standard, components of a sculpture's description. Such basic measurements, and those of individual parts (arms, head, and so on) may yield important clues to the generational relationships between works related to a similar model.

1.1.1 Shrinkage and internal measurement

Because metal and wax both shrink slightly as they cool from molten to solid states, precise measurement of dimension may assist in identifying contemporaneous versions from the same model, or later after-casts or copies (fig. 1). In theory, the former should be nearly identical in size, while the latter should be smaller.

For this type of analysis, it is important to focus on so-called internal measurements, as coined by Ann Allison in her study of bronzes by the Renaissance sculptor Antico (Italian, ca. 1455–1528).[2] "Internal" refers to measurements taken within areas that do not contain wax-to-wax or metal-to-metal joints, and so should not (in theory) be subject to significant distortion due to assembly of the model or casting. Allison used internal measurements such as width across the chest, length from chin to crown of head, eye to eye, and so on to assist in establishing a "family tree" of different versions cast after the same model (and potentially from the same molds).

Even when internal measurements are used, caution should be exercised. Shrinkage is a complex physical phenomenon that is difficult to model or predict with accuracy, particularly for elaborate hollow forms. Clearly, a variety of processes other than shrinkage can affect final dimensions, including steps taken during model or inter-model production as well as cold work undertaken after casting.[3] For instance, it is important to consider that wax models can be bent or distorted (intentionally or unintentionally) prior to investment, resulting in castings that are deformed in comparison to the artist's original model. In contrast, sand casting (normally using a rigid pattern) tends to produce a more accurate and undistorted casting.

When evaluating published literature, it is important to distinguish between volumetric shrinkage and linear shrinkage. In the technical literature, volumetric shrinkage usually refers to the change in volume of a solid mass upon cooling; this may have a complex and unpredictable relationship with the change in volume of a hollow sculpture or wax model. Linear shrinkage normally refers to the change in exterior dimension of a solid mass of material on cooling; again, the relationship with actual shrinkage of a hollow sculpture or wax model may not be straightforward. For waxes and copper alloys, linear shrinkage should theoretically be about one-third the magnitude of volumetric shrinkage.

Commonly cited estimates of shrinkage suggest that wax, poured as liquid into hard molds, will shrink around 2.5–3% in linear dimension.[4] During metal casting, copper alloys shrink somewhat less and can be expected to shrink linearly on cooling by around 1–1.5% (fig. 41).[5] In general, sand- and gypsum-plaster-based investment molds will not shrink appreciably after forming (plaster actually expands slightly on hardening); ceramic shell molds may shrink around 0.5% when fired;[6] and clay molds (rarely encountered) can shrink dramatically, 4–5% or more in linear dimension.

1.1.2 Variations in overall shape

Measurement can also help determine how versions of the same model were assembled by highlighting differences in the relative orientation of different sections (fig. 76). This can provide insight into original fabrication techniques as well as into the relationships between different versions.[7] Measurements may also be useful in puzzling out the correct assembly of fragmentary pieces of sculptures found in archaeological contexts.[8]

1.2 Weight and surface area

The weight of the metal, the overall volume, and the surface area of a figure may provide useful information about the life of the sculpture and the context of its fabrication. The total metal weight may give insights into both the cost of the raw materials and the type of furnace required,[9] and consequently insights into the economic and/or social status of the workshop. (Although outside the scope of the present *Guidelines*, weight could be helpful in determining logistics related to a work's transportation, and/or the kind of mount and base it had or would need.) It should be considered, of course, that material other than the cast copper alloy may contribute to the weight of the sculpture, such as residual core material, armature, repairs, et cetera. The weight of the statue is increasingly measured by modern founders in order to track illegal copies. Diana Widmaier Picasso systematically weighs the bronzes for the Pablo Picasso (Spanish, 1881–1973) catalogue raisonné.[10] Measurements of surface area are less common, but could be used, for example, in calculating the amount of gold needed for the gilding of a statue, to provide historical context, or in planning the restoration of a monument.

1.3 Wall thickness

In hollow castings, the thickness of the metal wall and its homogeneity can be crucial for understanding how a bronze was cast. The conformality of the inner and outer surfaces can provide important clues about the fabrication process (for example direct or indirect lost-wax process, see I.1). Localized, well-defined areas of greater or lesser thickness may correspond to repairs. Thickening in select areas may represent intentional adjustments to control a sculpture's center of gravity and thereby ensure its stability.[11]

1.4 Dimensions of technical features

The form and dimensions of technical features such as core pins, armatures, repair patches, tool marks (fig. 242), and so on may help to characterize the nature of the production and tell us something about workshop practices and the skills and habits of the craftspeople. For instance, a workshop might routinely use a particular kind of iron nail for core pins.[12] Threaded plugs may also be of great technical interest. Prior to widespread industrialization, thread-making tools were made individually and were not standardized, and so the specific dimensions of threaded fasteners may be characteristic of early workshops.[13] Numerous standardized thread gauges became common over the course of the nineteenth century and can also be helpful for dating or characterizing workshop production (see I.4§1.1.2).[14]

1.5 Risks of misinterpretation

Though we assume that practitioners develop certain preferences and patterns in their practices and use of materials, it is always possible that situations or conditions may arise that require improvisation and the use of whatever is available. For example, of the four bronze Virtues of Henri II and Catherine de Medici's funerary monument in the Basilica of Saint-Denis, France (figs. 405, 406), the main armatures of three statues were made with a hexagonal section, but the fourth had a round section rod. All other technical features clearly point to the same foundry.[15] This holds true for all the fabrication steps, including repairing.

2 How to measure

The following measurement procedures are generally used for sculpture (not specifically bronzes). As a rule, international units (metric) should be used to record dimensions. Imperial units may, of course, be used as well if desired.

2.1 Measuring external dimensions

Sculptures vary greatly in form, and perhaps surprisingly, there are no standard rules on how to go about measuring them. This can, and often does, result in discrepancies between measurements taken by different operators.[16] A consistent methodology is therefore advisable, and the following basic guidelines are designed to assist.

2.1.1 Height

The maximum height of a sculpture resting on a flat surface such as a pedestal or the floor is easily determined by positioning a level (also known as a spirit level or bubble level) on its uppermost tip and establishing the distance between the level and the base by means of a ruler or rigid tape measure. In the case of a figure with an upraised arm or other feature that is higher than the head, measurement should be taken from there. Alternatively, two straightedges may be used together with a tape measure (fig. 407). More than one person may be required to take the measurement depending on the scale of the object.

If a sculpture is mounted on a separately made base (as opposed to an integrally cast one), it is standard practice to provide two height measurements: one with the base and one without. The same measurement process should be used as described above. If pins, sprues, or other elements of the casting extend below the visible sculpture (into the base), the nature of the height measurement should be clearly described in text to avoid confusion.

2.1.2 Width and depth

When measuring a sculpture in the round, it is not always clear what constitutes the "front." In cases where the sculpture has an integrally cast rectangular base, one of its straight edges may help to determine this. The width and depth may be thought of as the internal dimensions of a box (whose sides are parallel to the base) into which the sculpture will fit. The width (sometimes referred to as length) is designated as the side-to-side measurement. The depth is the longest front-to-back measurement (fig. 408). Tools such as a pair of framing squares and a tape measure can be helpful (fig. 409). If the base is round or irregularly shaped, or if there is no base, it will be important to document which side of the sculpture is being considered the front for purposes of the examination. The "internal box measurement" process applies in these cases. Where the integral base extends beyond the sculpture, a decision needs to be made whether this element is to be considered part of the sculpture and thus included in the width and depth measurements. It may be desirable to report dimensions for base and sculpture separately. Keep diagrams of where measurements were taken as part of the sculpture's documentation.

Measuring a fragmentary bronze presents challenges, especially if it is not clear what the piece's original position was intended to be. Its incomplete state may also get in the way of adequately comparing its measurements with those of a related cast (fig. 410). Annotated visuals are particularly important in such cases to help describe where the measurements were taken.

2.2 "Internal" measurements and measurements for comparison

2.2.1 Classical methods

So-called internal measurements (see II.4§1.1.1 above) can be made manually using calipers, a measuring tape, or a string (fig. 366). Digital calipers now often have direct recording capacities via connection to a computer.

Imprints of specific features using plaster or silicone casts may prove useful for comparison of measurements as well, and are quite easy to carry out (fig. 411).

2.2.2 3D computer modeling

3D computer modeling offers a more sophisticated means of making measurements (fig. 379), and in particular of comparing measurements across similar casts (fig. 378). A variety of software exists that can assist in identifying innumerable points of comparison in an intricate web.[17] While requiring a certain degree of technical and scientific skill, especially for quantitative calculations, 3D is becoming increasingly user-friendly and affordable.

Three main methods are currently in use for 3D modeling: photogrammetry, structured light scanning, and laser scanning. For basic descriptions of these methods, see II.2§5. There are no hard rules to determine which method is best for creating suitable 3D models for bronze sculpture. The measurement precision of each method can vary tremendously depending on the specific equipment used, the experience of the user, and the nature of the subject. Photogrammetry, for instance, depends heavily on the camera resolution and the software, but even with a 6–7 megapixel camera, Anestis Koutsoudis et al. showed that skilled practitioners can produce models accurate to better than ±1% for measurements 25–120 cm.[18] Professional-grade structured light and laser scanners vary in precision according to manufacturer and model and usually have a specific size range for which they are optimized to scan. Within the target range, precision of both methods can be very high, with accuracies better than the ±0.01% claimed by manufacturers.

In the end, the best method of 3D modeling for any given examination will depend on the desired end use of the measurements (how much precision is necessary) as well as the available equipment and expertise. In practice, most studies on objects in the cultural heritage realm have used either photogrammetry or structured light scanners, while laser scanning has been dominant for sites and architectural applications.

2.2.3 Cost

The cost of the digital camera used in photogrammetry can, of course, vary tremendously depending on resolution, lens quality, and features. Software packages to extract 3D data are evolving and improving rapidly, and several universities are working on 3D reconstruction algorithms.[19] The cost of photogrammetry software depends on one's needs and output expectations. Free software is very easy to use but offers relatively low quality in regards to point accuracy and overall geometric and visual quality. Professional software can cost up to US$4,000 and allows for the greatest possible quality available but requires high-end computing power and technical expertise to achieve quality results.

Consumer-grade structured light scanners are available for less than US$1,000, but professional-grade scanners more appropriate for technical study usually cost between US$15,000 and $50,000. Operating software (which may require an annual license fee) is normally provided with the scanner, but other software may be desired for additional model manipulation or comparison purposes. This can range from free and open-source packages such as Meshlab or Blender to proprietary packages costing more than US$10,000 and/or requiring annual license fees of many hundreds of dollars.

In the late 2010s, the price of laser scanners certified for metrology ranged from about US$15,000 to $150,000 depending on the resolution and the precision. In addition, proprietary software associated with the scanners can cost from US$10,000 to $50,000, though some free or low-cost data processing software is available.

2.2.4 Duration

The speed of data acquisition and processing has increased dramatically in recent years. Using any of the three methods discussed above, an experienced operator can normally accomplish standard-resolution scanning/imaging of a one-meter-high bronze sculpture and generate a finished 3D model in half a day.

2.3 Measuring metal wall thickness

There are four primary approaches to measuring the wall thickness of a hollow bronze sculpture: direct physical measurement on the bronze; radiography or tomography; ultrasonic testing; and 3D scan. Each method has advantages and limitations (table 13).

2.3.1 Direct measurement

In instances where there is direct access to both inner and outer walls of the cast, outside calipers (sometimes called external calipers) may be used to make direct measurement of wall thickness. Electronic outside caliper gauges can be useful, as they give an electronic readout of the dimension in real time and do not require that the caliper be removed to record the measurement. Depending on size, calipers do not allow measurement very far away from the access zone, thus limiting the regions that can be measured. In addition, features such as casting defects, repairs, and corrosion may alter the original thickness in the area where the measurement is carried out. In the end, the representativeness of the measurement may be difficult to assess.

2.3.2 Radiography and tomography

Accurate measurement by radiography can be challenging for two key reasons. First, divergence of the X-ray beam inevitably results in some magnification of the image on the sensor, requiring careful calibration of the image using geometric calculation or reference scales placed in the image. Second, the walls of a sculpture are often curved, which can lead to indistinct edges in X-radiographs.

To complicate matters further, the exterior edge of a roughly cylindrical metal form will present the thinnest cross section in a radiograph. This means that the surface contour of a sculpture may be easily overexposed, thus disappearing in the image and giving the impression of a thinner wall than actually exists (fig. 63). Measurement by X-ray or gamma-ray tomography will generally be more reliable (figs. 379, 398), though precision of the measurement will depend greatly on the equipment used (see II.3).

2.3.3 Ultrasonic testing (UT)

Ultrasonic testing (UT) has proved an efficient method of measuring the thickness of metal. Although used widely in industry, it has only rarely been applied to bronze sculpture (figs. 230, 231).[20] UT has the advantage that no dangerous radiation is produced, facilitating on-site and in-lab analysis. UT can also measure very thick metal that might be difficult or impossible to measure with standard X-ray equipment. Unfortunately, UT requires the application of coupling gels to the surface of the object to be inspected, and for any loose paint or corrosion to be removed. In addition, UT signals from metal walls of irregular thickness or with rough surfaces are difficult to interpret, requiring careful attention from experienced technicians.

2.3.4 3D scanning

To the extent that the interior of a sculpture is accessible for 3D scanning or photogrammetry, it may be possible to merge interior and exterior surface models and create an accurate model of the wall thickness for at least a portion of a hollow sculpture.

2.3.5 Reporting and interpreting wall thickness measurements

Once wall thickness measurements are made, reporting and interpreting results may not be straightforward. Wall thickness will almost certainly vary considerably from area to area. What, then, is the wall thickness to report? Of course, all measurements that have been made should be recorded individually, but for the purposes of characterizing a bronze and comparing it with other castings, providing some summary value(s) is necessary. Unfortunately, but unavoidably, this is bound to be a somewhat subjective process.

First of all, it may be useful to distinguish the different elements of a casting (separately prepared sections of the wax model, for instance, or separately cast sections) and treat them individually. If many measurements are possible for any given element, calculating the median value may be a useful way of characterizing the thickness. If not, then it may be up to an investigator's judgment to choose a "typical" thickness measurement to report. Some researchers suggest that investigators attempt to determine a "targeted thickness" representing the founder's envisioned thickness,[21] though this has not yet been widely accepted.

In addition to reporting quantitative results, some qualitative descriptors of a sculpture's thickness may be useful. "Homogeneous," "regular," and "conformal" are all terms that convey important information about wall thickness, though their definitions are imprecise. The thickness of the wall may be said to be homogeneous if it is more or less the same throughout the sculpture. The thickness may be termed regular if it does not show any sharp discontinuities or changes (as observed at wax-to-wax joints). The metal wall is said to be conformal if the inner and outer contours match closely such that the thickness remains constant as the form meanders.

2.4 Measuring weight, volume, and surface area

The overall weight of a sculpture can be measured rather straightforwardly using floor scales, pallet jacks with scales, or hoists. In some cases, however, it may prove more complex to deduce the weight of the bronze sculpture alone from the total weight of the object. Bases that cannot be removed pose an obvious problem. If core material remains in the interior, this may contribute dramatically to the overall weight. The weight of the sculpture may also be affected by repairs, armatures, or mounting materials. If the volume of extraneous materials can be determined with some precision, it is possible in some cases to estimate weight by using a standard average density value.

Volume and surface area measurements are most efficiently and accurately made using 3D scanning or photogrammetry. Most 3D modeling software will readily generate these measurements from a model.

3 How to compare measurements

When making rigorous comparisons between different versions, or copies, of the same model, best practice is to take the measurements using the same tools and techniques for each example. While in theory, direct measurements should be comparable to measurements made from (for instance) 3D scans, it is possible that using different measurement techniques can lead to minor systematic biases that could distort the results. At least, make sure a scale has been used and properly characterized.

Remember as well that there are many reasons why an individual pair of measurements may differ between sculptures (see II.4§1.1 above). If point-to-point measurements are being made (either directly or using 3D models), many pairs of measurements should be compared and the differences recorded as a percentage. The average percent difference may be a useful metric of comparison, but consider it with respect to the standard deviation of the individual values to ensure that it represents a real difference and is not just the result of random variation.

Where many versions or copies of a model exist, variability may be assessed and groupings may be made by making multiple dimensional measurements of each example and then analyzing the results according to the coefficient of variation (CV) or geometric morphometric analysis (GMM) (fig. 412).[22] These methods are commonly used in archaeology to compare large groups of objects and to quantify standardization.

A relatively new but potentially very powerful means of visualizing and quantifying overall differences in form and scale across different versions of an individual model is through 3D modeling software. Accurate models of two versions can be made, imported into the same virtual space, and aligned automatically to optimize registration. Different software packages offer somewhat different tools and procedures, but, particularly if the models are rendered as partially transparent, differences in the alignment of certain sections can be made clearly evident (fig. 378). In scanning sculptures that have been assembled from different pieces, one may have to switch initial alignment points to obtain good registration for the various areas.

Another useful visualization tool is to display cross section contours on corresponding planes of different versions. Many 3D modeling software packages also offer the capability to quickly calculate volume measurements and surface area measurements. These tools are useful to quantify shrinkage between versions, particularly if individual sections of the sculpture, excluding wax or metal joint lines, are compared independently.

Notes

1. For basic definitions on measurement see ISO 2007.
2. Allison 1993; see also Perrault 2006.
3. See Motture 2019, 229.
4. See for example Rome and Young 2003, 289. Some modern pattern waxes are engineered for minimal shrinkage.
5. See for example Beale 1975, 54.
6. For a highly technical investigation into the relationship between the dimensions of the pattern and the final casting in ceramic shell casting see Cannell and Sabau 2007.
7. Beale 1975; Boulton 2007.
8. Combined with metal composition, dimensions proved decisive in relating two life-size Roman bronze statues of riders to two horses (Augst, Switzerland, first century CE), see Mille 2019b.
9. Welter 2014.
10. Elisabeth Lebon, personal communication.
11. Beentjes 2019, 221.
12. The dimensional similarity of iron core pins proved decisive in establishing that a bronze finger held by the Louvre Museum in Paris (fig. 172) was the missing finger from the monumental hand of Constantine at the Capitoline Museum in Rome (Azéma, Descamps-Lequime, and Mille 2018). See Case Study 5 for how specific chaplets may help in characterizing a sculptor.
13. For guidance on measuring threaded forms see Camm 1942, 92–106.
14. For tables describing many standard types of thread gauges see Camm 1942, 136–45.
15. Castelle 2016.
16. Carol Mattusch found the measurement of ancient bronzes so problematic that she chose to designate dimensions in general terms such as "life size" and "over life size" rather than with precise measurements (Mattusch 2005, 126–27).
17. Sequenzia, Fatuzzo, and Oliveri 2021.
18. Koutsoudis et al. 2014.
19. Find useful information about photogrammetry at http://culturalheritageimaging.org/Technologies/Photogrammetry/. For more on algorithms see Verma 2019.
20. For the Marcus Aurelius bronze at the Capitoline Museum in Rome, see Marabelli 1994. For the Apollon Piombino at the Louvre in Paris, see Mille and Descamps-Lequime 2017, also Scott 2002, 394.

21. One nice and large-scale application of this concept resides in the Hephaistos database on large antique bronzes (Descamps-Lequime and Mille 2017).

22. Birch and Martinón-Torres 2019; Okumura and Araujo 2019.

II.5

Metal Analysis

Arlen Heginbotham
David Bourgarit

This chapter aims to guide both nonspecialists and specialists in understanding and/or using techniques for analyzing the metals from which bronze sculptures are made. Metal analysis encompasses elemental analysis, isotope analysis, and metallography (the study of metal microstructure). The analysis of surface layers (patina, gilding, et cetera) is addressed in II.6.

The basic information here will help nonspecialists navigate the wide range of analytical techniques now available and determine which are best suited to answer which questions (for a summary, see also tables 5, 10). A comprehensive list of existing techniques and their advantages and drawbacks may be of interest for specialists as well. Several related topics are discussed in I.2, including a fundamental description of what metal is; what types of copper alloys have been used for sculpture; what is meant by "major," "minor," and "trace" elements; and what is meant by "microstructure."

1 Elemental analysis: overview

1.1 Principles

Elemental analysis determines the proportion of different chemical elements present in a particular sculpture's casting alloy. The elemental analysis may be qualitative or quantitative. The former simply lists the elements detected, with only a rough idea of their relative amounts; the latter gives a measurement of the amount of each element present as a percentage of the total, usually reported as a weight percent (wt%). Quantitative analysis should also provide an estimate of the uncertainty associated with the measurement (the error).[1]

Some researchers also use the term "semi-quantitative," often in circumstances where the error of the measurement is relatively large, or unknown. In such circumstances, the numerical measurements are unlikely to be reproducible by other analysts and are thus not useful for comparison with the results of other laboratories.

1.2 Accuracy, precision, and sensitivity

Any quantitative analytical method can be characterized in terms of accuracy, precision, and sensitivity. In scientific terms, an accurate method is one where the average of many measurements is likely to be very close to the true value. This is sometimes called trueness. An analytical method is considered precise if the results of many repeated measurements are likely to be very nearly the same. Given these definitions, an analytical method can be accurate, precise, neither, or both. A common way of visualizing these four possibilities is shown in figure 413. The term "sensitivity" can be used in different ways in reference to chemical analysis, and there have been vigorous debates on the subject. Here, the term is used in its common sense, as "a device's ability to detect a small amount or slight change in the measured quantity."[2]

1.3 Application to bronze sculpture

A basic alloy-type determination should be standard practice for the technical study of bronze sculpture for purposes of documentation and typology. This may be achieved with qualitative analysis. Quantitative analysis provides additional important information that may help to refine the possible date of production or confirm that a sculpture's alloy is consistent with securely identified examples from a certain time period, region, workshop, or even artist (see I.2§4.1).

In the end, the information that may be derived from elemental analysis of any given sculpture depends to a large degree on the quantity and quality of reference data available from comparable, well-documented examples. Conducting new and rigorous analysis of securely attributed works is of great importance to expand the body of knowledge usable for comparative purposes. Some strides have been made in the development of large, shared databases of copper alloy compositional data,[3] which facilitate sophisticated statistical analysis that will certainly lead to new discoveries with regard to workshop, regional, and temporal characterization of bronze sculpture.

Different methods of elemental analysis may be invasive, noninvasive, or minimally invasive. The terms "destructive" and "nondestructive" are also commonly used, although in analytical sciences, "destructive" refers specifically to the sample, not to the item: the analysis is destructive if the sample is destroyed after analysis, nondestructive if not. Each type of analysis has its advantages and disadvantages.

2 Non- or micro-invasive methods of elemental analysis

2.1 X-ray fluorescence analysis (XRF)

X-ray fluorescence analysis (XRF) is a relatively accessible analytical method employed to characterize elemental composition. It is nondestructive and offers rapid, multi-element analysis with good sensitivity, often down to a few hundredths of a percent in copper alloy samples (equivalent to a few hundreds of parts per million [ppm]). A beam of X-rays is directed at the sample material to stimulate the emission of fluorescent X-rays with specific energies characteristic of the elements present.

The emitted X-rays are detected and converted into a spectrum, either with an energy dispersive (ED-XRF) or a wavelength dispersive (WD-XRF) detector. ED-XRF is far more common than WD-XRF and is used in all of the portable and handheld XRF instruments that are now widely available in museum and university laboratories. ED-XRF offers the possibility of giving a quick qualitative assessment of the type of alloy present. Newer silicon drift detectors (SDDs) take measurements in a minute or less, making it possible to take many measurements in the course of an examination.

To optimize results for historic copper alloys, voltage on the spectrometer should be set relatively high (around 40–50 kV). In most cases, moderate to heavy filtration of the beam is advantageous, using aluminum, titanium, and/or copper filters. In the case of modern sculptures, if light elements such as silicon, phosphorous, or aluminum are to be analyzed, filters should be removed, the voltage turned down to about 15kV, and a helium flush or vacuum should be used if available. These operating parameters also allow for the analysis of sulfur in both historic and modern bronzes, which may be very informative.

Due to the complex interactions between X-rays and the sample material, quantitative alloy analysis with ED-XRF requires complex calibration procedures. Many ED-XRF instrument manufacturers provide software and calibrations for metals that will generate quantitative results. Unfortunately, the use of manufacturers' proprietary calibrations comes with some difficulties: the results from different manufacturers (and even different models from the same manufacturer) are often not in agreement; it can be difficult or impossible for the user to select the list of elements to be quantified; error estimates are often unrealistically small; and the spectrum processing and calibration methods used are often not transparent to the user.[4] These and other factors make it difficult to share quantitative results confidently between researchers, and manufacturers' results should generally be considered semi-quantitative. Some protocols designed to enhance reproducibility and facilitate sharing have been developed that may be useful to researchers wishing to share results and build collaborative databases.[5]

Elemental mapping may be done using XRF. Dedicated XRF scanners are available that are capable of scanning flat (or nearly flat) areas of a square meter or more at high resolution. For three-dimensional objects, smaller areas are scanned (typically a few square centimeters) with many individual analyses of the elemental composition at definite points, together defining an elemental map (fig. 414).[6] The objective may be to detect and characterize inlays and overlays; brazing, welding, or soldering; and/or repairs. Mapping requires specialized equipment and software.

XRF is essentially a surface analysis technique, with an effective penetration in copper alloys of around 30–80 µm, or less than a tenth of a millimeter. This means that the technique is very sensitive to surface alteration or contamination. For qualitative use on historic materials, this is usually not a significant problem. But for quantitative analysis, surfaces to be measured should be clean, flat, and free from patina or corrosion.[7] It is possible to encounter both surface enhancement and surface depletion of selected elements that may result in misleading results, particularly with archaeological materials. Working with an XRF analyst who has significant experience with copper alloys can help produce reliable and useful results for comparative purposes.

In 2023, portable ED-XRF instruments cost between US$25,000 and $50,000. Commercial laboratories may be able to provide analysis for US$50 or less per sample.

WD-XRF offers significantly higher spectral resolution and sensitivity than ED-XRF, but is less common and much more expensive to own and operate. Samples for WD-XRF analysis are typically extracted from the sculpture, embedded, cut, and polished prior to analysis. Sample size requirements may vary, but are often from 1 × 1 mm to 5 × 5 mm. These instruments are found primarily in commercial and research laboratories focusing on earth sciences. In theory, WD-XRF instruments offer the possibility of more precise quantitative analysis of alloy composition than ED-XRF instruments, although many earth sciences labs may not routinely analyze copper alloys, and therefore may not have calibrations prepared in advance. Costs per sample are normally less than US$100.

2.2 Particle-induced X-ray emission (PIXE)

Particle-induced X-ray emission (PIXE) commonly utilizes a highly collimated beam of protons to excite X-ray emission from a sample. Other charged particles may be used as well, but protons are by far the most common. It is possible but not necessary to place the sample in a vacuum chamber or flush it with helium to improve detection of light elements. X-ray spectra are collected, usually by an ED detector, for processing into a quantitative elemental composition. PIXE offers better sensitivity than XRF[8] and can detect very small quantities of many materials typically present in casting alloys, often as little as a few tenths of ppm. Moreover, PIXE offers better spatial resolution (a few tenths of microns) thus enabling the analysis of very small features such as inlays (fig. 350). In the case of point analysis, no specific preparation is required, though as with XRF, PIXE analysis is a surface technique, so the sample site should be clean, flat, and free of patina or corrosion. The depth of penetration of PIXE using protons is somewhat less than for XRF. Unfortunately, few facilities provide access to ion beam analysis techniques.

As for XRF, chemical mapping can be carried out with PIXE.[9]

2.3 Other techniques

2.3.1 Laser-induced breakdown spectroscopy (LIBS)

Laser-induced breakdown spectroscopy (LIBS) uses a focused, high-intensity, pulsed laser to vaporize tiny spots of sample material. The plume of plasma generated at the sample site is immediately analyzed by optical (or atomic) emission spectroscopy (see II.5§3.1 below). Usually, many laser "shots" are recommended, and the results are averaged. LIBS instruments are fast and easy to use, but for quantitative analysis they face significant hurdles in producing consistent and reproducible results.[10] Handheld LIBS instruments are becoming more widely available, and at lower prices.

2.3.2 Eddy current testing (ECT)

Eddy current testing (ECT) is an analytical technique that uses a coiled wire probe to produce an alternating magnetic field, which then induces an alternating current in a conductive sample material (usually metal). The alternating current flow in the sample generates, in turn, a magnetic field that interacts back with the magnetic field and current flow in the probe. Monitoring electrical impedance in the coil probe yields information about the nature and structure of the sample material.

ECT can distinguish between different major alloy types in solid metal, but its main use is in detecting flaws, repairs, and inlays. For this purpose, ECT relies on the fact that current is interrupted by discontinuities in the metal—that is, where a cast-in repair, patch, or inlay has been added, or at the site of a cold shut. By mapping sites of discontinuity, it is possible to draw a quite reliable contour of the feature on the surface of the sculpture. A probe with a sharp point can help to improve precision. The technique may allow a qualitative appraisal of the nature of repairs, and help to form groups with similar behavior for subsequent in-depth investigation using other techniques.[11] ECT is often used when radiography is not an option (fig. 415).

2.3.3 Neutron diffraction

Neutron diffraction is a rarely used technique that can characterize the composition of the bulk metal in a bronze sculpture.[12] This may be accomplished in conjunction with neutron tomography (the principles of neutron radiography and tomography are described in II.3§3). Only certain elements can be detected and quantified using neutron diffraction, and the error of measurement is relatively high compared to other techniques, but analysis with neutrons has the advantage that it can measure the composition of the metal throughout the thickness of the metal wall, and it is possible to map the composition at different points onto a tomographic reconstruction that reveals spatial variations.

Neutron diffraction has also been shown to be able to discriminate between bronze surfaces that have been chased and as-cast surfaces.[13] This is done by detecting residual stresses (elastic strain) in the surface metal that are not present 1–2 mm below the surface.

Neutron diffraction is a highly specialized technique requiring large-scale research facilities, typically nuclear reactors. In such cases there is often no fee per se for the imaging, but there will be a lengthy application process to secure the allocation of a facility's time and instrumentation. Costs can, of course, be significant for transportation and security, and the administrative burden can be substantial.

Another serious consideration is that atoms that absorb neutrons into their nuclei during imaging form radioactive isotopes that subsequently emit gamma radiation. These isotopes must be allowed to decay to safe levels before handling. Depending on the specific elements present and the length and intensity of exposure (tomography requires significantly longer exposure than radiography), this period of quarantine may be several hours, days, or even longer, imposing additional cost and security considerations. The presence of gold, for instance, can dramatically increase the quarantine period.

3 Invasive techniques of elemental analysis

3.1 Atomic spectroscopy techniques: atomic (or optical) emission spectroscopy (AES or OES) and atomic absorption spectroscopy (AAS)

Atomic spectroscopy refers to a group of analytical techniques where sample material is introduced into a flame, furnace, electric arc, or plasma and breaks down into individual atoms and ions. In this excited state, the atoms emit and absorb specific frequencies of light (from near-infrared to ultraviolet, including the visible light range) that are characteristic of the specific elements present. The elemental composition of the sample is thus determined by analyzing the spectral features of the emitted light (atomic or optical emission spectroscopy, or AES/OES) or of light transmitted through the excited sample material (atomic absorption spectroscopy, or AAS).

Atomic spectroscopy was at the very origin of the rise of archaeometallurgy as a discipline.[14] In particular, flame AAS and inductively coupled plasma AES (ICP-AES) have provided the large majority of published analysis on bronze sculpture so far.[15] ICP refers to a highly controlled and reproducible method for the atomization/ionization of sample material using plasma made from an argon stream.

AAS and AES are sensitive to trace quantities of elemental components, from several hundreds of ppm to the ppm level in a copper matrix depending on the element. This allows precise quantification of a large number of trace elements, which may be useful, notably when addressing provenance questions. Preparation of the sample for analysis involves complex chemical digestion procedures to dissolve all components of the sample. The complexity of atomic spectroscopy techniques means that analysis is best performed by private or university laboratories specializing in this analysis, and with staff experienced in cultural heritage applications. Small changes to the analysis protocol can make significant differences in the quality of the outcome.

Analysis by atomic spectroscopy techniques typically requires a sample of about 10–50 mg. At the time of this publication, costs per sample range from around US$100 to $350 for academic clients.

3.2 Mass spectrometric techniques: inductively coupled plasma mass spectrometry (ICP-MS) and laser ablation ICP-MS (LA-ICP-MS)

In recent years, ICP mass spectrometry (ICP-MS) has largely superseded ICP-OES as the method of choice for the analysis of trace elements in bronze alloys. As with ICP-AES, the sample is atomized and ionized using an inductively coupled plasma. But instead of characterizing the light emitted during excitation, ICP-MS relies on direct detection of the ions produced in the plasma stream. These are sorted and counted by the spectrometer according to their mass-to-charge ratios. ICP-MS are complex instruments and are made in several different configurations, but all measure the abundance of each ion type with a high degree of sensitivity, leading to limits of detection typically 10–1000 times lower than ICP-AES (ppm to ppb), depending on the element and the instrument configuration.[16] ICP-MS can also detect different isotopes of the same element, allowing for isotopic analyses (see below).

ICP-MS systems are significantly more expensive to purchase and operate than ICP-OES systems. As for atomic spectroscopy techniques, ICP-MS analysis is best conducted in specialized laboratories dedicated to the technique and with staff experienced in the analysis of cultural heritage materials.

ICP-MS has had a disadvantage in comparison to ICP-AES, which is a difficulty in analyzing, with one single sample, a wide range of element concentrations (from sub-ppm to percent levels). When optimized for trace elements, major elements (tin, lead, and zinc) have often been measured with relative precision equal to or lower than by ED-XRF, and copper has rarely been quantified at all. Significant improvements in instrument design since the 2010s have largely mitigated this problem.

ICP-MS is sometimes used in conjunction with laser ablation (LA), a technique that uses a focused, high-intensity, pulsed laser to vaporize tiny spots of sample material that are then drawn into the ICP-MS instrument. LA-ICP-MS is generally considered a micro-destructive technique because the ablation spots are extremely small, but it is generally applied to fragments or samples that are small enough to be placed onto an instrument's analysis stage, often only several inches across. Laser spot sizes often fall in the range of 10–300 μm, so multiple sites should be analyzed within the sampling area to ensure that micro-inhomogeneities in the metal structure do not yield misleading results.

Various so-called matrix effects can complicate quantitative analyses produced via LA-ICP-MS. While the technique has been used extensively for the quantitative analysis of glass, it has only been applied relatively recently to quantitative analysis of copper alloys and is not yet widely considered an established technique. As with LIBS, its ability to repetitively ablate the material may give access to sound metal below the intentional patina and/or corrosion layers.

3.3 Neutron activation analysis (NAA)

Neutron activation analysis (NAA) is based on the principle that if a sample material is bombarded with relatively low-energy "thermal" neutrons (usually produced at a nuclear reactor), the atoms in the sample will capture neutrons into their nuclei, forming radioactive isotopes. As these isotopes decay over a period of days or weeks, they emit gamma rays characteristic of each element present in the sample. The gamma rays may be detected and identified, leading to a quantitative estimate of the sample's total elemental composition. The high penetrating power of neutrons and gamma rays means that NAA provides information from the interior as well as the exterior of the sample material. Neutron activation is a highly sensitive method for chemical analysis. Data for a wide range of elements can be obtained with sensitivity at the low- and even sub-ppm level. No specific preparation of the sample is needed.

NAA may be used as a noninvasive technique if the entire object to be studied is analyzed, though often samples are taken in order to reduce cost and transportation issues for valuable sculptures. A drilled sample also allows for removal of surface layers and corrosion product prior to analysis. Reliable results require about 50–100 mg of sample material. Sensitivity for some elements is less than for ICP-MS, and the analysis requires more time to complete than other techniques (since the rate of decay is variable, gamma-ray collection often happens over a period of days).

The restricted availability and cost of this nuclear method make it increasingly difficult to access, and so it has been largely supplanted by ICP-MS.

3.4 Scanning electron microscopy (SEM)

In scanning electron microscopy (SEM), a small sample is placed into a vacuum or near-vacuum and bombarded with a highly focused beam of electrons that is continuously rastered, or scanned, across the surface. Multiple detectors are placed inside the chamber that sense different types of emissions from the surface of the sample as the beam is scanning.

For elemental analysis, SEM may be coupled with X-ray detectors for energy-dispersive spectroscopy (SEM-EDS or SEM-EDX) or wavelength-dispersive spectroscopy (SEM-WDS or microprobe). These detectors yield X-ray spectra that are useful for elemental analysis.[17] Areas can be scanned to obtain bulk composition measurements, or spot analysis can be made of individual metallic phases or inclusions. Both techniques are capable of quantitative elemental analysis, although WDS offers substantially higher precision and sensitivity of measurement than EDS.

Samples for quantitative elemental SEM analysis of copper alloys should be at least 1 × 1 mm, but preferably 5 × 5 mm or more to take into account natural inhomogeneities in copper alloys, particularly inhomogeneity of lead (which tends to form globules in copper) and elements associated with it (such as arsenic, silver, and bismuth). Samples are normally embedded in resin, cut, and polished to provide a clean, flat surface for analysis.

Today, many major museum and academic laboratories are equipped with SEM-EDS. These instruments may be valuable for metallographic analysis (see below), but for high-precision, high-sensitivity quantitative elemental analysis, other (unfortunately less accessible) methods such as ICP-AES, ICP-MS, PIXE, SEM-WDS, or NAA (see II.5§3.3) may be preferable.

3.5 How to sample

Sampling for elemental alloy analysis (atomic spectroscopies and NAA techniques) is most often done by drilling (fig. 416). Naturally, this is usually carried out in hidden areas of the base or in the interior of the sculpture (fig. 156). A target sample weight of 50 mg of copper alloy represents approximately 5–6 cubic millimeters of metal. If a 1.5 mm–diameter drill bit (steel or tungsten carbide preferred is used, this corresponds to a drilling depth of about 3 mm. If greater depth of drilling is possible, the size of the drill may be reduced correspondingly; for instance, a 1 mm drill bit will provide enough sample if drilled to a depth of about 7 mm, while a 0.6 mm drill bit would require a drilling depth of nearly 20 mm. In practice, this last situation would almost certainly mean drilling multiple holes. It should also be noted that smaller-diameter drills are more likely to break while drilling and, particularly if a porosity is encountered, recovery of the broken end may be difficult or impossible.

When sampling bronze sculpture, drilling is often done on an edge, such that the drill penetrates parallel to the metal wall, avoiding surface effects (see II.5§2.1). For thin-walled sculptures, the thickness of the metal wall may constrain the diameter of drill that can be used; the drill should not be larger than half the wall thickness. Normally the first 0.5 mm of sample material is discarded (drilling depth adjusted accordingly) in order to remove any possibility of surface enhancement or depletion effects. Carefully clean or replace the drill bit between uses.

Preparation of sampled material for analysis is a particularly important step that is usually carried out by the laboratory in charge of the analysis; errors may result if it is not well carried out. Once collected, drillings should be carefully examined under a microscope to remove unwanted materials (corrosion, fibers from clothes, fragments of the drill, et cetera). A magnet may prove useful to remove steel fragments from the drill. For atomic spectroscopies and ICP-MS where the sample will be digested in acid, precise weighing is necessary, and digestion solutions specific to historic copper alloys should be used.[18]

Sampling for SEM analysis usually follows protocols used for metallography (see II.5§6.2).

4 Choosing a method of elemental analysis

When selecting a technique for elemental analysis, consider your aim. If only major element composition is required, techniques such as XRF may be sufficient. If provenancing and/or fine attribution are required, trace element analysis with the best available sensitivity may be preferable, requiring techniques such as ICP-AES or ICP-MS. Chemical mapping can currently be carried out using XRF, PIXE, LIBS, or LA-ICP-MS, depending on the spatial resolution required (from µm to cm).

Questions of sampling and cost/technique availability are often decisive. Many noninvasive processes, such as XRF, PIXE, and LA, are surface techniques and thus prone to producing results that may not be representative of the bulk metal composition. Corrosion may deeply alter the quantification of alloying elements (for example, a huge tin enrichment due to the selective dissolution of copper). On the other hand, noninvasive techniques tend to be rapid, and the cost of undertaking repeated measurements is low, allowing measurement of more sample sites in different surface areas.

In contrast to noninvasive techniques, invasive techniques that require drilling will represent the bulk metal well, even if surfaces are enhanced, depleted, or contaminated. Their disadvantage is that the damage done by drilling, and the relatively high cost per sample, means that very few analyses are usually carried out, and these are usually constrained to the lower regions of the sculpture. Particularly if the sculpture is composed of several separately cast sections, this means the results may not reflect the overall composition.

In summary, noninvasive techniques are vulnerable to depth inhomogeneity, while invasive techniques are vulnerable to spatial inhomogeneity but provide better sensitivities (which also means more elements are analyzed). In this sense, the two types are complementary. A particularly complementary and practical pairing is XRF alongside ICP-AES or ICP-MS. Initial study with XRF on many points of a sculpture can characterize gross overall (in)homogeneity and identify differences in composition between separately cast sections and/or areas of repair. The XRF results may also inform the choice of sample site(s) for drilling. Naturally, radiography and visual examination of the interior and exterior should also inform the choice of sampling sites. ICP analysis will provide additional information on trace elements and may help identify any layer inhomogeneity issues that could affect the XRF results.

XRF analysis may also confirm the validity of ICP results. In one instance, serious procedural errors in two ICP-MS results were detected by one of the authors when XRF results were found to be dramatically different that those reported by ICP. The errors were subsequently identified and confirmed by the commercial ICP laboratory, but would never have been noticed if not for the complementary XRF study.

It is often stated that analyses carried out by different teams cannot be compared, but this is simply not true if the methods are regularly controlled by the use of relevant certified reference materials, and the methodologies and results are published in relevant peer-reviewed journals. This has unfortunately not been standard practice for long, but the trend is clearly toward a greater emphasis on comparability, reproducibility, and transparency in analytical methods and reporting.[19]

5 Isotope analysis

5.1 Scientific principles

Atoms are made of a nucleus around which a "cloud" of electrons spins. The nucleus is made of two main types of particles, protons and neutrons, and the number of protons defines the element. Copper, for example, has 29 protons, tin has 50, and lead has 82. The number of neutrons in the nucleus of any given element can vary within a certain range, and this number of neutrons determines the isotope. Isotopes are designated according to the combined number of protons and neutrons in the core (the atomic weight). So, for example, a lead atom (Pb) with 82 protons and 122 neutrons is designated ^{204}Pb. Some isotopes are stable and do not change over time. Others are not stable (these are radioactive) and will decay at a predictable rate into other elements by loss of nuclear particles and energy. Carbon-14, or ^{14}C, is a well-known example of an unstable isotope that is used for radiocarbon dating purposes (see II.8§2).

Isotope analysis for bronze sculpture focuses on the quantification of different stable isotopes.[20] The total amount of any given isotope in a copper alloy may be extremely small. For instance, in unalloyed copper, an important lead isotope, ^{204}Pb (see II.5§5.2), may be present in quantities around 0.0001 wt% (1 ppm). A sensitive technique is thus needed, typically mass spectrometry. In addition, the difference in isotope ratios between two copper deposits is often very small, thus requiring mass measurements with very high precision (around 0.1%) that only certain types of advanced mass spectrometers can achieve. Typically, so-called multicollectors (MC) are used and the technique is referred to as MC-ICP-MS.

5.2 Application to bronze sculpture

Lead, a well-studied isotope series, has four stable isotopes as well as scores of unstable isotopes. The stable isotopes, however, make up nearly all the lead on Earth and are ^{204}Pb, ^{206}Pb, ^{207}Pb, ^{208}Pb. Depending on the age, geological location, and nature of the ore from which it is derived, lead metal contains different proportions of these different isotopes. Small amounts of lead are systematically present in copper as a natural impurity, so lead isotope studies on copper artifacts are often designed to study the geographic origin of the copper. Conventionally, the ratios $^{206}Pb/^{204}Pb$, $^{207}Pb/^{204}Pb$, and $^{208}Pb/^{204}Pb$ are used for describing results.

Few isotope analyses have thus far been carried out on bronze sculpture. The main applications have been lead isotope studies on ancient copper alloy artifacts—mainly protohistoric collections of weapons, tools, and/or everyday items.[21] Tin isotopes have recently become a focus of greater study in the context of bronze artifacts, while zinc and copper isotopes have potential applications, but have not been widely used.[22] In the long run, the study of several isotope series in combination will probably be more informative than an examination of lead alone.

Using MC-ICP-MS analysis, very small sample sizes are needed, often less than 10 mg. Samples may be obtained by drilling as described in II.5§3.5.

5.3 Cost and availability

Isotopic analysis for copper alloys is mostly carried out in earth science laboratories. In 2023, the cost of one analysis ranged from US$100 to $300.

5.4 Risks of misinterpretation

The provenancing of copper based on lead stable isotope analysis alone is far from straightforward.[23] Isotope analysis is probably best used as a complementary approach together with elemental analysis, archaeology, and/or history.

Another major issue in applying isotope analysis to bronze sculpture is that lead is commonly added to the copper by founders or even by the metal supplier, be it through direct addition or recycling. In this case, the lead isotope signature will reflect not the provenance of the copper but rather that of the lead. In bronze sculpture, it may not always be evident whether low levels of lead are naturally occurring or added. Rather than focusing on geographic provenancing of metal, isotope analysis may be better suited to authentication and attribution, by characterizing expected isotope ratios for different regions and time periods. This will only be possible if large numbers of well-provenanced sculptures are studied in the future.

6 Metallography

6.1 Scientific principles

Metallography is the study of metal microstructure using microscopic imaging techniques.[24] Microstructure, its interpretation, and its importance for the study of bronze sculpture is discussed in I.2§3. The reader may consult that section to learn about the types of information to be gleaned from metallographic study and evaluate whether such study is worth undertaking.

In the field of cultural heritage, metallographic analysis is often confused with elemental metal analysis, but the two are quite different. In addition to the study of the metallurgical state of the metal, metallography might determine the different phases and their distributions. When two or more metals are mixed together to form an alloy—for example, copper and tin—different phases may form depending on the relative proportions of the different ingredients and the cooling temperature. Phase diagrams are useful to infer what phases may form and the nature and variety of the properties of the alloys, including melting temperatures and freezing ranges (fig. 132).[25]

A metallography study involves two main steps: sampling and preparation, followed by observation and analysis.

6.2 Sampling and preparation

In situ characterization of microstructure without sampling is rarely relevant, for several reasons. First, the microstructure of the bronze's surface is rarely what requires observation. A cross section of the metal wall is often much more informative, as it may show joints, repairs, et cetera. Also, the original surface may be severely altered by wear, use, later interventions, and/or corrosion. Second, a proper polishing is difficult to achieve in situ. (Several attempts have been carried out to develop pre-polishing and diamond or alumina polishing using rotary abrasive heads.) Third, in situ observation is very limiting in terms of magnification and image quality. Fourth, the surface to be prepared and etched must be large enough to encompass the heterogeneity of the microstructure (at least 1 × 1 cm).

Metallography on an extracted sample is the standard method. For hollow bronzes, a transverse section is usually required, extending over the whole thickness of the wall. The section can be as narrow as possible (approx. 1 mm), and its length depends on the questions to be addressed (5–20 mm, for instance a weld joint), such that enough surface is covered to enable a fully representative study. The choice of sample site is obviously crucial. Bronze is hard, and metal walls on a sculpture are often several millimeters thick. A diamond saw mounted on a drilling machine is generally the best tool (figs. 417, 418). Great care must be taken when sampling to preserve the structure of surface layers (see II.6§3.1).

The preparation of the cross section may be as follows. The cross section is embedded in resin (often epoxies or acrylates), pre-polished using silicon carbide papers of different grades, then polished on cloths to a mirror surface, either with alumina or diamond paste. Many microstructural features may be observed optically, directly on a polished surface. SEM examination is also usually conducted directly on the polished sample.

In order to reveal particular features, a solution made of various acids and/or bases may be applied to the surface of the cross section. This is called etching. Etching solutions react chemically with some specific parts of the microstructure such as grain boundaries, and thus reveal these features by either removing them or coloring them. The optimal etching solution depends on both the type of copper alloy and the features under investigation. Etching is discussed at length in Scott 1991.[26] Before etching, make sure to observe and record all the microstructural features that are best observed without etching, or else repolishing will be necessary.

6.3 Observation and analysis

The polished cross sections should first be examined using an optical microscope both to get an overview of the microstructure and/or surface layers (fig. 146) and to estimate their homogeneity or heterogeneity. There are advantages to the specificities of this technique. Two viewing modes are possible. In bright field conditions (specular light), metallic and nonmetallic inclusions can be visible, as can different metals, such as gilding layers. This is usually done before etching. Bright field mode also enables the clear visualization of relief generated by etching as well as the colors of the microstructure generated by etching (figs. 142, 145, 419, 420). Dark field (diffuse reflectance) is rarely used for microstructural observation, but rather for the investigation of surface layers, patina, gilding, and coatings, similar to stratigraphic cross sections used in the study of paintings.

Scanning electronic microscopy (see II.5§3.4) may be required for higher magnification and for elemental analysis of inclusions, metallic phases, and/or surface layers (figs. 421, 422, 423, 424). Some SEM instruments may allow for electron back-scattered diffraction (EBSD) structural analysis as well.[27]

Metallography is a highly invasive procedure and (as explained in I.2§3.3) should only be considered if it will answer very specific questions. As with most other analytical techniques mentioned in these *Guidelines*, a minimum of skills and techniques (for sampling, preparation, observation, and interpretation) is mandatory to ensure that the most valuable information is gathered.

Notes

1. Bevington and Robinson 2003, 1–14; ISO 2008; Pilon et al. 2017.
2. Ekins and Edwards 1997.
3. Pollard et al. 2018.
4. Heginbotham et al. 2011.
5. Heginbotham, Bassett et al. 2014; Heginbotham and Solé 2017; Heginbotham et al. 2019.
6. For more on the study pictured in figure 414 see Mille 2019a, 182–85; Laval, Calligaro, and Mille 2019.
7. For more detailed discussion of parameters affecting XRF of bronze sculpture see Smith 2012.
8. Calligaro et al. 2011; Calligaro et al. 2000; Dran, Calligaro, and Salomon 2000.
9. Done notably on medieval Iranian bronzes to analyze the metal inlays (Collinet 2021) and on a Roman statuette (Pacheco and Mille 2019).
10. Alberghina et al. 2011; Gaudiuso et al. 2010.
11. Mille et al. 2012.
12. Van Langh 2012; Peetermans et al. 2012.
13. Van Langh et al. 2011.
14. Pernicka 2014; Killick 2015.
15. See notably all the work carried out at the Rathgen-Forschungslabor der Staatlichen Museen zu Berlin (Riederer 2002) and at the C2RMF (Mille and Bourgarit 2000; Bourgarit and Mille 2003).
16. Pollard et al. 2007.
17. Mugnaini et al. 2014; Lombardi 2009; Vincent 2014.
18. Notably to make sure all the silver, possibly present in relatively high amounts, is correctly dissolved. See Bourgarit and Mille 2003.
19. Heginbotham and Solé 2017; Heginbotham et al. 2019.
20. For a comprehensive overview of isotope analysis applications in cultural heritage see Nord and Billström 2018.
21. See Artioli 2010.
22. Stephens et al. 2021.
23. Artioli et al. 2020; Pollard 2009; Gale 2009.
24. Metallographic studies recently celebrated their 150th birthday (Philibert 2014). See also the interesting lecture by J.-M. Lago on the history of optical microscopy at the conference "150 ans de la métallographie," March 21, 2014, Paris, http://docplayer.fr/4477079-J-m-lago-150-ans-de-metallographie-a-toutes-les-echelles-2.html.
25. The most common diagrams for ancient copper alloys are reported and commented upon in Scott 1991.
26. Scott 1991, 69–74.
27. Baudin 2011; Northover and Northover 2012.

II.6

Surface Layer Analysis

David Bourgarit

Surface layers on a bronze sculpture may include gilding and other metal plating, patina, and various coatings such as varnishes. The present chapter aims to guide the reader's investigation of these layers, from thickness measurement to chemical analysis. Color measurement, which also contributes to the characterization of the surface layers, is discussed in II.2§4. Surface measurement and investigation of tool marks is dealt with in II.4.

Surfaces of bronze sculptures often comprise several complex layers resulting from complex series of events, from the time of the fabrication of the sculpture to the present day (see I.7, I.8). Unfortunately, confronted by such complexity, the researcher is, here more than elsewhere in these *Guidelines*, limited with respect to sampling. Still, because of its indispensability, sampling constitutes the backbone of this chapter, which sets out the techniques, their possibilities, and their limitations. A summary synthesis/workflow of the analytical techniques discussed in this chapter is provided in figure 305 and in tables 5, 10.

1 Thickness of surface layers

1.1 Noninvasive techniques

Due to the aforementioned tendency of surface stratigraphy to be complex—composed of several thin, heterogeneous layers—measurements of thickness without sampling may yield false results, or at best partial information. Rarely can one measure the surface layer thickness directly on the sculpture, without sampling.

The thickness of simple organic coatings such as varnishes and resin can be estimated using eddy currents.[1] In theory, eddy currents allow measurement of the thickness of any kind of surface layer, but they are very sensitive to the metal alloy underneath the coating as well as porosity, repairs, et cetera, which may lead to misleading measurements. X-ray fluorescence spectroscopy (XRF) and even handheld XRF may give quite accurate measurements of metal plating layer thicknesses, provided these layers are not too porous.[2]

The best estimate of thickness is obtained by ion beam analysis, particularly Rutherford backscattering spectrometry (RBS), which can deal with simple metal plating layers, for instance patinas made of one or two compounds (such as cuprite on black bronzes) homogeneously distributed. These can also possibly be investigated by nuclear reaction analysis (NRA).[3] The latter is among the more complex and less accessible techniques. It has proven useful in simple situations, but it has numerous limitations: facilities are rare (a particle accelerator with the beam exposed in air is necessary for a sculpture more than about 10 cm tall), as are experts able to interpret spectra; complex layers cannot be unraveled using such nuclear techniques; and the layers may not be thicker than 10–50 μm.

1.2 Invasive techniques

In most cases, sampling and subsequent metallographic preparation of cross sections is necessary. Although the preparation is the same as the one described in II.5§6.2 for bulk metal microstructure investigation, particular attention is required to avoid damaging the surface layer. Surface layers such as gilding or fragile patinas are easily altered during cutting, resin embedding, and polishing. Resin impregnation prior to sawing is often advantageous, and careful low-speed, low-pressure sawing is recommended. If the sample contains organic coatings, embedding resin may interfere with future organic analysis of the sample by methods such as micro Fourier-transform infrared spectroscopy (FTIR). One danger during embedding is that the shrinkage of embedding resin may induce separation of surface layers from the substrate. Ion beam polishing, if it is available, is highly recommended for very fragile layers.

Once the cross section is ready, optical microscopy is generally sufficient to detect and measure the thickness of the different surface layers. Scanning electron microscopy (SEM) may be necessary for micrometric and submicrometric layers, such as those obtained by electroplating, but with high-purity gold the thickness can be overestimated due to smearing caused by mechanical polishing of the soft metal. Ionic polishing helps to avoid this problem.

2 Composition of surface layers: noninvasive techniques

2.1 Elemental versus structural analysis

A surface layer may be fully characterized by its elemental composition (for a definition of "elemental composition" see I.2§1.2) in only one situation: when the layer is made of metal (gilding, silvering, tinning, et cetera; see I.7). Other types of surface layers, such as patina (including chemical patina) and coatings (varnishes and so on, see I.8), are made of compounds/molecules that elemental analysis alone often cannot fully characterize.[4] Structural analysis is necessary to complete the characterization and determine the molecular or crystalline forms in which the elements are arranged. For example, specific patina components such as cuprite, malachite, or cassiterite (to name but a few) cannot be identified using elemental analysis alone. Similarly, many natural earth pigments share a common elemental composition, but vary considerably in color and structure (for instance hematite, goethite, and limonite). Organic varnish components likewise cannot be identified merely by elemental analysis.

2.2 Elemental analysis

The most common surface elemental analysis technique for bronze sculpture is XRF. Particle-induced X-ray emission spectroscopy (PIXE) is also increasingly used (see II.5§2.2). Yet they all share the same drawback for surface analysis: their penetration is often much deeper than the layer thickness (typically several tens of microns, whereas many surface layers of interest may be less than one micron thick). Consequently, the analytical results are a mix of the studied surface layer and the copper alloy substrate.

Moreover, surfaces are often multilayered. Thus, as for thickness measurements, sampling is often required to overcome the complex layering of different materials. But all the problems are not solved: the analytical techniques used on cross sections require a low-size beam, smaller than the thickness of the layer thickness—their "surface resolution" needs to be appropriate—so most XRF devices do not allow for such analysis. That said, if areas of bare, unpatinated bronze are present adjacent to patina, it may be possible to identify patina elements by comparing the measured composition of both areas.[5] Also, some attempts have recently been made using very-high-frequency eddy currents (>>1 MHz).[6]

2.3 Non- and micro-invasive depth profiling

One might dream of a technique able to determine, without sampling, the elemental composition of each individual layer present on the surface of a bronze—in other words, capable of drawing a composition depth profile. Laser-induced breakdown spectroscopy (LIBS), laser ablation inductively coupled plasma mass spectrometry (LA-ICP-MS), RBS, and NRA allow this to be achieved in theory, but each method has its drawbacks:

- Surface layers on bronzes may be too thick for these methods (more than several tens of microns).

- Light elements such as hydrogen, carbon, and oxygen are not systematically detected, although they are frequently present in the surface layers.
- Very small areas (less than 1 × 1 mm) are usually investigated because of time and damage, thus leading to potentially nonrepresentative analysis.
- In most laboratories LA-ICP-MS can only be used on extracted samples (in most favorable cases, only small statuettes not exceeding 20 cm in their highest dimension may be investigated).[7]
- These methods are rare and costly (although handheld LIBS instruments are becoming less expensive, with prices comparable to handheld XRF instruments, about US $40,000 to $50,000).

2.4 Structural analysis without sampling: X-ray diffraction (XRD), Fourier-transform infrared spectroscopy (FTIR), Raman spectroscopy

Although rare, some analytical techniques allow analysis of molecules directly on the bronze surface, without sampling. Techniques include X-ray diffraction (XRD), microscopes equipped with Fourier-transform infrared spectroscopy (FTIR), and Raman spectroscopy. Specially adapted versions of the instruments can be used directly on objects rather than on samples in the lab, but these often yield results that are not as precise or reproducible as those made with laboratory instruments. As for elemental analysis, due to the thinness and the complexity of often intermixed layers, structural analysis without sampling rarely proves fully efficient.

3 Composition of surface layers: invasive techniques

3.1 Guidance for sampling

Samples can be in powder form, prepared as a cross section, or in solution. The main advantages and drawbacks of these sample types are:

- Powder samples are much easier to acquire and prepare, but a major disadvantage is that multiple layers may be intermingled in the sample. Careful sampling under a stereomicroscope is essential.
- Cross sections allow observation and analysis of the layering, but it can be extremely difficult to remove a small, discrete sample of a chemically bonded patina. Preparation of cross sections requires special attention (see II.6§1.2 above).
- Dissolved samples may be collected for analysis using a cotton swab or by applying and then retrieving a droplet of solvent using a micropipette. The best choice of solvent(s) may be difficult to determine, as the solubility of oil, resin, and polymer coatings will vary considerably and will be affected by aging.

Whatever the sample form, great care should be taken in identifying the most representative area for sampling.

3.2 Observation of cross sections

Careful observation of the cross section is highly recommended to assess the structure of the surface layers and their homogeneity. This may prove particularly useful to distinguish artificial from natural patina (see I.8§2.4). The same techniques as those used to investigate the metal microstructure may be used for imaging (see II.5§6). Improved imaging of samples using multi- or hyperspectral imaging, ultrasound tomography, and optical coherence tomography (OCT) are possible. In some situations, very-high-magnification electron microscopes, namely transmission electron microscopes, may prove necessary.[8]

3.3 Elemental analysis

The same techniques as those described for metal analysis (II.5) may be used, depending on the type of sample and the performances required (range of elements, sensitivity, accuracy). A variety of techniques may be used, for instance powder and solution. On cross sections, spatially resolved methods should be chosen, such as micro-XRF, SEM energy-dispersive spectroscopy (SEM-EDS), SEM with wavelength-dispersive spectroscopy (SEM-WDS), LA-ICP-MS, or PIXE.

3.4 Structural analysis

When using a powder sample, micro-chemical/solvent testing may prove very rapid and efficient, although infrequently used.[9] For powder and possibly solution samples, some of the more common techniques for characterizing the structural composition of bronze surface layers—in this context exclusively patina—include XRD, FTIR, gas chromatography with mass spectrometry (GC/MS and pyrolysis-GC/MS), and Raman spectroscopy.

On cross sections, spatially resolved techniques are recommended. This includes XRD, FTIR, and Raman microscopies (including surface-enhanced Raman, or SERS), and reflectance FTIR. In circumstances in which a synchrotron beam is available, cross sections or thin sections maybe analyzed using micro-FTIR, time-of-flight secondary ion mass spectrometry (TOF-SIMS), false color imagery, and/or photoluminescence.

Most published studies concerning structural analysis of surface layers on bronze sculpture to date focus on corrosion products. The characterization of organic coatings is significantly less common,[10] while the characterization of intentional chemical patina recipes is perhaps least common.[11] However, all these methods have limitations, starting with their incapacity to detect all compounds in a comprehensive manner. The fact that a given compound is not detected does not mean that it is not present in the surface layer.[12]

Notes

1. See Heginbotham, Beltran et al. 2014.
2. Cesareo et al. 2009; Lopes et al. 2016; Karydas 2007.
3. Ioannidou et al. 2000; Mathis et al. 2007.
4. For example, the elemental analysis of a patina may measure the amount of copper (Cu), tin (Sn), and oxygen (O). But this elemental composition may correspond to several "structural" compositions. The three types of atoms Cu, Sn, and O may be present in two distinct compounds, namely cuprite (Cu_2O) and cassiterite (SnO_2), or in other forms such as a mixed oxide (CuSnO).
5. Chromium was detected in the patina on a sculpture by Henri Matisse (French, 1869–1954) (Boulton 2007, 86). Iron was identified as a patina component on a work by Edgar Degas (French, 1834–1917) (Lindsay, Barbour, and Sturman 2010). Platinum was detected on a work attributed to Louis-Simon Boizot (French, 1743–1809) (Bewer and Scott 1995).
6. The thickness of both copper sheet (1.2–1.5 mm) and patina (approx. 40 µm) on a Renaissance statue has been investigated by eddy currents. Such a device cost around US $6,000 in 2019. The technique was initially developed to measure the thickness of nonconductive coatings in the automotive industry (JM Welter, personal communication). As for ultrasonic testing, eddy currents are very sensitive to any perturbation in the surface layers. A skilled operator is necessary.
7. Dussubieux, Golitko, and Gratuze 2016.
8. Studies of black bronzes are exemplary (Craddock and Giumlia-Mair 1993; Craddock 2009; Delange, Meyohas, and Aucouturier 2005). Elemental analysis or structural analysis alone does not enable an understanding of the patination process; only a combination of both approaches, including sampling, has allowed researchers to unravel the mystery. XRD performed directly on the surface has revealed a simple patina composition, namely cuprite Cu_2O. Surface elemental analysis (PIXE) as well as bulk metal analysis via inductively coupled plasma atomic emission spectrometry (ICP-AES) identified a specific copper alloy, sometimes referred to as Corinthian bronze, with several percent of gold. Transmission electronic microscopy on samples (thin sections) provided the clue to the problem (Notis 1988; Benzonelli, Freestone, and Martinón-Torres 2017; Mathis et al. 2005): nanoparticles of gold dispersed within the cuprite layer proved to be responsible for the specific *shakudo* color.
9. Odegaard, Carroll, and Zimmt 2005.
10. See the studies carried out on Italian Renaissance patina using gas chromatography mass spectrometry (GC-MS) (Stone 2010; Pitthard et al. 2011; Aucouturier, Mathis, and Robcis 2017), and the studies of black bronze patina (Giumlia-Mair 1995; Benzonelli, Freestone, and Martinón-Torres 2017).
11. UV-visible photoluminescence generated by a synchrotron beam has led to a useful characterization of various cuprite compounds that might be applied to patina studies (Thoury et al. 2016).
12. Stone 2010; Pitthard et al. 2011.

II.7

Core Analysis

Manon Castelle
Chandra L. Reedy

This chapter aims to elucidate why it is meaningful to study core materials, and to describe the chemical and petrographic analytical techniques that are available for this purpose. The techniques for studying cores also apply to refractory molds, which are made of similar materials but are rarely preserved. The overview is intended for nonspecialists, followed by a more detailed treatment of the relevant analytical methods. Note that this chapter does not cover thermoluminescence or radiocarbon dating of core material. For these subjects, please refer to II.8.

1 What is core?

Hollow bronze statues, by definition, were filled with a refractory material, or core, at the time they were cast. Generally, the core was removed after casting, but in some cases, part (or even all) of the original core may remain, and this material may be analyzed both chemically and petrographically. In the absence of any remnants of original core, there is still potential to find significant clues as to the nature of the core, either by visual examination of the inner surface or by radiography (see I.1§1.1.3, I.1§2.1).

The most common core materials are either clay based (mostly clay, with sand and organic temper; see fig. 10), sand based (mostly sand, bound with clay and possibly organic temper added; see fig. 6), or plaster based (normally gypsum, with or without clay, sand, or organic temper; see fig. 57). Clay-containing bricks may be encountered as well (fig. 8). In clay-based and plaster cores, the clay component may contain raw clay or pre-fired and ground clay (grog), perhaps in the form of ground ceramic or brick. Sand (silica) may be naturally present in clay based cores, or may be added separately; the extent of silica addition may be so great that refactory material commonly thought of as clay based may be more properly considered sand based.[1] For a more detailed discussion of core materials and fabrication, see GI§2.1.1. Casting core studies have often focused on clay-based materials, whereas plaster-based and sand-based cores have been less well studied.[2]

1.2 Why investigate cores?

The materials used and the method of forming the core are critical parts of the fabrication process, and reflect important technical choices on the part of the founder. Careful study of residual core can help characterize these materials and methods, which can in turn answer important questions about a bronze sculpture. The analysis of casting cores has been included in the technical study of bronze sculpture for more than thirty years.[3] Particularly in the last twenty years, a large number of publications have clearly demonstrated that core analysis is a powerful tool for understanding and characterizing the production methods of bronze sculpture.[4]

1.2.1 Geographic provenance

Analysis of core material may yield important clues regarding where a bronze sculpture was cast. Specific mineralogical or chemical signatures identified in the raw materials can be compared to databases of geological reference materials, which may establish a likely region of origin. Since foundries have historically tended to choose locally available materials, this may help to geolocate the casting site.

From a methodological point of view, analytical techniques developed for the study of ancient ceramics have been widely applied to core characterization and have influenced the overall approaches taken to the study of cores.[5] As a consequence, questions of geographic provenance have been a major focus of many casting core studies (see Case Study 5),[6] with chemical compositions used particularly to determine the geographic source of the raw materials. Special attention has focused on the relative amounts of different rare earth elements (REEs),[7] though recent studies indicate that existing reference databases should be carefully evaluated before use.[8] Shape, texture, and mineralogy of non-clay minerals have also been used for geographic provenancing, albeit less frequently.

1.2.2 Attribution

Core analysis may also shed light on temporal, geographic, or artistic attributions for bronze sculpture. If characteristic practices and raw materials can be identified and associated with particular workshops or artists, it may be possible to support or refute specific attributional hypotheses.

Recently, there has been increased interest in the characterization of workshops and/or related technical practices through characterization of core material (see Case Study 5).[9] The characteristics of raw materials (clay, temper, et cetera) may be used to identify workshop practices and recipes employed by specific foundries, in certain periods, to overcome specific technical challenges of bronze casting. Questions may then be raised about the relationships between artists and foundries, and thus, casting core studies can provide another avenue to investigate issues of authenticity.

2 How to investigate cores

The analytical techniques used to study cores fall into two main categories: chemical and petrographic. Chemical analysis generally focuses on the elemental composition of the materials, including their bulk elemental composition and the concentration of certain minor or trace elements that are useful for geographic provenancing. Petrographic analysis focuses on the nature, morphology, and proportion of the different minerals and other additives that are in the core, be they natural or added by the founders (for instance added sand in clay). As much as possible, we recommend combining chemical data and petrographic investigation results to achieve a comprehensive characterization and documentation of the core material.

2.1 Chemical analysis

Chemical analysis follows similar protocols as those described for metals (see II.5), although metal analysis can sometimes be performed in situ. For core analysis it is necessary to obtain samples. Because core material is typically not visible from the exterior, permission for core sampling may be more easily obtained than for metal sampling.

2.1.1 Inductively coupled plasma with atomic (or optical) emission spectroscopy (ICP-AES or ICP-OES) or mass spectroscopy (ICP-MS)

Inductively coupled plasma (ICP) refers to a type of high-intensity energy source (torch) that efficiently breaks sample material into individual atoms and then ionizes the atoms. Once atomized and ionized, the total elemental composition can be determined, usually either by atomic emission spectroscopy (ICP-AES) or by mass spectroscopy (ICP-MS).[10] AES determines elemental composition by analyzing spectral features of the light produced when the sample material is exposed to the ICP torch. Mass spectrometers are complex instruments and made in several different configurations, but all collect the sample ions and directly measure the mass and abundance of each ion type with a high degree of precision.

ICP analytical techniques are very sensitive to trace quantities of elemental components, often able to detect even a few ppm depending on the detection method. This allows precise quantification of a large number of trace elements, and in particular of rare earth elements,[11] which can be extremely useful when addressing provenance questions. Both ICP-AES and ICP-MS involve long, complex, and mandatory chemical preparation of the sample prior to analysis. The analyst must select and follow a dissolution procedure (acid digestion) that is compatible with the sample material and then inject the solution into the plasma torch. MS instruments are generally more sensitive and precise than AES instruments, though this varies somewhat, element by element. MS systems are also significantly more expensive to purchase and difficult to operate.

An alternative to dissolution methods is the use of laser ablation (LA). LA-ICP-MS or LA-ICP-AES uses a focused, pulsed laser to ablate sample material into a plume that is drawn into the ICP instrument directly. Laser spot diameters often fall in the range of 10–300 μm, which can complicate analysis when working with microscopically heterogeneous materials. On the other hand, the precision of the technique has been used to advantage in the analysis of individual mineral grains to trace geographic sources of ceramic material.[12] A disadvantage of LA-ICP is that various matrix effects can complicate quantitative analysis.[13]

The complexity of all ICP variants means that analysis is best performed by private or university laboratories specializing in this technique, and with staff experienced in cultural heritage applications. At the time of publication, costs per sample range from around US$100 to $350 for academic clients.

2.1.2 Neutron activation analysis (NAA)

Neutron activation analysis (NAA) is based on the principle that if a sample material is bombarded with relatively low-energy "thermal" neutrons (usually produced at a nuclear reactor), the atoms in the sample will capture neutrons into their nuclei, forming radioactive isotopes. As these isotopes decay over a period of days or weeks, they emit gamma rays characteristic of each element present in the sample. The gamma rays may be detected and identified, leading to a quantitative estimate of the sample's total elemental composition. The high penetrating power of neutrons and gamma rays means that NAA provides information from the interior as well as the exterior of the sample material.

Neutron activation is a highly sensitive method for chemical analysis. Data for a wide range of elements, including rare earth elements, can be obtained with sensitivity at the low- and even sub-ppm level. No specific preparation of the sample is needed. However, the restricted availability and cost of this nuclear method makes it more and more difficult to perform. Reliable results require about 50–100 mg of sample material, more than is sometimes possible to obtain from a bronze core. Detection limits for some elements are higher than for ICP-MS, and the analysis requires more time to complete than with other techniques. For these reasons, neutron activation analyses of casting core materials are scarce in the literature. Nonetheless, it remains a very effective tool for geographic provenancing through chemical fingerprinting, and many of the relevant compositional databases have been built using this technique.[14]

2.1.3 Scanning electron microscopy (SEM)

In scanning electron microscopy (SEM), a sample is placed into a vacuum or near-vacuum and bombarded with a highly focused beam of electrons that is continuously rastered, or scanned, across the surface. Multiple detectors are placed inside the chamber that instantaneously detect different types of emissions from the surface of the sample as the beam is scanning.

For the visualization of core and metal samples, a backscattered electron (BSE) detector is commonly used, which generates an image in which heavy elements (of high atomic number, such as lead) appear bright, while light elements (of low atomic number, such as silicon) appear dark. The precise focus and control of the electron beam allows for the extremely high magnification that SEM is known for.

For elemental analysis, SEM may be coupled with X-ray detectors for energy-dispersive spectroscopy (SEM-EDS or SEM-EDX) or wavelength-dispersive spectroscopy (SEM-WDS or microprobe). These detectors yield X-ray spectra that are useful for elemental analysis.[15] Bulk composition can be obtained as well as spot analysis of individual grains or features. Only elemental information is provided, not information about the mineral structure of the components. Both techniques are capable of quantitative elemental analysis, although WDS offers substantially higher precision of measurement than EDS. This puts SEM-EDS at a disadvantage for the purposes of geographic provenancing. The accuracy of both methods may be adversely affected by substantial porosity in the sample.

Today, many major museum and academic laboratories are equipped with SEM-EDS. These instruments may be very valuable for petrographic analysis (see II.7§2.2 below), but for high-precision quantitative elemental analysis, other (unfortunately less accessible) methods such as ICP, NAA, particle-induced X-ray emission spectroscopy (PIXE), or SEM-WDS may be preferred.

2.1.4 X-ray fluorescence analysis (XRF)

X-ray fluorescence analysis (XRF) analysis is a relatively accessible analytical method employed to characterize elemental composition. It may be used to analyze casting core materials[16] or bulk metal composition (see II.5§2.1). XRF directs a beam of X-rays at the sample material to stimulate the emission of additional X-rays with specific energies characteristic of the elements present in the sample. As with SEM, the emitted X-rays are detected either with an EDS detector (ED-XRF) or a WDS detector (WD-XRF).

ED-XRF is far more common and is used in all of the portable and handheld XRF instruments that are now widely available in museum and university laboratories. ED-XRF can provide a quick qualitative assessment of the type of core material present. In order to optimize results for light elements such as silicon, aluminum, and calcium, the voltage on the spectrometer should be kept relatively low (around 15 kV), filters should be removed, and helium flush or vacuum should be used if available. Comparison to reference materials (prepared by the research team) may be helpful for general characterization of the core type.

Due to the complex interactions between X-rays and the sample material, quantitative analysis with ED-XRF requires complex calibration procedures and sample preparation that are not well developed for core analysis. In any case, with ED-XRF, sensitivity to REEs will generally not be adequate for geographic provenancing, with the smallest detectable amounts usually in the low hundreds of ppm. Portable ED-XRF instruments now cost between about US$25,000 and $50,000. Commercial laboratories may be able to provide analysis for US$50 or less per sample.

WD-XRF offers significantly higher spectral resolution and sensitivity than ED-XRF, but is less common and much more expensive to own and operate. These instruments reside primarily in commercial and research laboratories focusing on earth sciences. WD-XRF instruments offer the possibility of more rigorous quantitative analysis of core material with sensitivity sufficient to be useful for geographic provenancing, though careful sample preparation is required. The cost per sample is normally less than US$100.

2.1.5 Particle- (or proton-) induced X-ray emission (PIXE)

PIXE commonly utilizes a highly collimated beam of protons to excite X-ray emission from a sample (other charged particles may be used as well, but protons are by far the most frequently used). It is possible but not necessary for the sample to be placed in a vacuum chamber or flushed with helium to improve detection of light elements. X-ray spectra are collected, usually by an EDS detector, for processing into a quantitative elemental composition. PIXE offers a good balance between analytical performance level and amount of time required for analysis.[17] The technique can detect very small quantities of many materials typically present in casting cores, often as little as a few tens of ppm. In the case of point analysis, no specific preparation is required other than placing the sample between two thin polymer films to ease the positioning.

In addition, recent developments in PIXE elemental mapping now permit obtaining a spatial distribution of elements of interest within the sample.[18] In this case, the sample should be embedded in resin and polished. Unfortunately, few facilities provide access to ion beam analysis techniques.

2.1.6 X-ray diffraction (XRD)

X-ray diffraction (XRD) takes advantage of the fact that crystalline materials diffract X-rays in predictable and characteristic ways. A collimated beam of X-rays is passed through the sample material, and a detector registers the angles and intensities of the diffracted radiation. Comparing with reference databases allows a determination of the specific crystalline phases present in the sample material.

With respect to core analysis, XRD allows the identification and quantification of important crystalline phases (such as quartz, feldspars, clays, different hydration states of gypsum, et cetera) in clay-, sand-, and plaster-based cores.[19] Note that the accuracy of the phase percentages is generally limited to about ±2 wt% for conventional XRD.[20] Synchrotron XRD can offer a more precise alternative by using energy ranges allowing for increased accuracy in detection and quantification.

Both conventional and synchrotron XRD can be used to analyze both macro (approx. 100 µm) and micro (less than 50 µm) samples. For macro-scale samples, depending on the geometry of the diffractometer, the sample may be a fine, homogeneous powder mounted on a sample holder or between two Mylar sheets, or a whole fragment of material. In case of micro scale setup, the sample may be placed into a glass capillary or constituted of single crystals or groups of crystallites. In every case, identifying and quantifying mineral phase can help address questions of provenance by comparison with reference material. In case of plaster-based material, the identification of anhydrite (type and proportion) can yield valuable clues about the process of preparing the core.

2.2 Petrographic analysis

Petrographic analysis is ideally done on prepared thin-section samples of intact casting core fragments.[21] It is intended to provide information on the different proportions of materials present (clay, plaster, sand, organic temper), the size and shape (morphology) of the particles, the amount and distribution of porosity, and the distribution of different mineral phases). These features are essential for precise characterization of the core.

Some qualitative petrographic studies have been undertaken, but quantitative approaches using point counting or image analysis have provided more significant data and are generally considered indispensable. Quantitative analysis allows efficient identification of clay sources according to the grain size distribution of the natural sand (quartz) components of different clays. Quantitative measures can also help determine whether some portion of the sand temper was intentionally added to clay-rich material because both the size and shape of the quartz particles will likely fall into two distinct categories, as is often the case in ancient ceramics.[22]

2.2.1 Optical microscopy (OM)

Basic petrographic analysis involves performing optical microscopy (OM) observations on transparent thin sections, usually about 30 µm thick, and preferably at least 1 × 1 cm. Commercial petrographic services will prepare appropriate thin sections from bulk samples for between US $30 and $75. Samples are observed primarily in transmitted light—either plane (linearly) polarized light (fig. 425) or crossed polarized light (figs. 426, 427, 428, see also Case Study 5), with magnifications generally in the 40–400X range, depending on the size of the mineral grains present. Sample material is always placed on a rotating microscope stage so that it can be observed at different angles with respect to the polarization angle of the transmitted light. An experienced petrographer can identify many specific minerals and other inclusions based on observation alone. To identify metallic minerals (which are not transparent), reflected light is also often added.

An optical microscope equipped with polarizing lenses is easy to use, relatively low cost, and available in many laboratories, but the expertise required for competent petrographic examination is substantial, and most researchers of bronze sculpture send their samples to specialists for analysis.

2.2.2 Cathodoluminescence (CL) microscopy

Cathodoluminescence (CL) microscopy is based on the principle that when certain minerals are bombarded by electrons in a vacuum, they emit light of different colors.[23] Among these materials, the most commonly encountered in cores are feldspars, which yield a blue luminescence (fig. 429). CL microscopy is often performed in scanning electron microscopes that have been equipped with a CL detector, but optical microscopes can also be adapted with a CL stage.

2.2.3 Scanning electron microscopy (SEM)

SEM can be used for high-magnification observation of petrographic structure. The morphology of organic temper components in loose or crushed sample material may be studied in detail using a secondary electron (SE) detector. This can allow the identification of charcoal, straw, fibers, et cetera. At very high magnification, the morphology of clay grains may be used to identify the types of clay present.

Switching to the BSE detector (see section 2.1.3 above) and working with a polished thin section can yield high-resolution images that show the size and distribution of mineral grains based on their density. These may be used for image analysis (see section 2.2.4 below). Using an EDS detector (see section 2.1.3 above) enables the elemental characterization of mineral grains. Most SEM software is now capable of producing elemental maps that show the distribution of different elements in the sample, as well as phase maps that identify regions with similar elemental compositions.

2.2.4 Image analysis

Quantitative data can be obtained from OM, CL, and SEM images by performing image analysis (fig. 430, see Case Study 5). We recommend taking into consideration a sufficient surface area of the sample (at least a few square millimeters) in order to obtain representative data. At least three different areas of the sample should be analyzed. An alternative is to analyze the entire thin section using a high-resolution scan.[24] A dedicated image analysis program may be provided with the microscope, or stand-alone comprehensive image analysis packages can be purchased. Free and open-access software is also available, including ImageJ, which offers good solutions for the automated determination of many petrographic parameters.[25]

Some important parameters to examine include the relative areas of the different components (including porosity); distribution of the area ratio occupied by each grain with regard to the total area occupied by sandy temper (area-to-area sum);[26] the grain size distribution of the different components; and the shape of the grains (including measures of sphericity and angularity).

While much of the image analysis *process* can now be automated, an experienced petrographer can make crucial judgments about its *implementation*. For instance, fragments of crushed rock are difficult to delineate, as they are composed of multiple smaller grains, so these should be removed from the automation and processed manually.

3 Sampling and sample preparation

Sampling and sample preparation are of great importance for preserving all the material evidence of the fabrication process. Whereas chemical analysis can be performed on powdery material, it is preferable to work with a coherent block of core, as it enables full characterization of the petrographic structure (fig. 431), chemical heterogeneity, and porosity. Petrographic characterization is ideally performed on a substantial thin section, made from a block of one cubic centimeter or larger if possible. Smaller samples of even a few square millimeters may still provide valuable information. The location of each sample should be clearly documented for future reference.

Cores may be layered structures, so it is important to document the orientation of the sample material as it is removed, and if possible have the thin section prepared as a cross section (with the outer, metal-side surface on one edge and the innermost portion of the sample at the opposite side). In addition, the portion of the core that was in direct contact with the molten metal at the time of casting will have been exposed to much higher temperatures than material slightly removed from the metal surface. This elevated temperature may induce phase transitions in the core material, so it is important to be aware of the orientation of the sample.

The sampling strategy depends on the degree and locations of access to the core. If there is easy access through an open underside or through casting flaws, for example, burins or chisels may be used to remove core material, which may be hard and compacted (fig. 432). If there is no obvious core material within easy view, exploration with a borescope may reveal pockets of hidden core. If access is difficult, a steel rod may be bent and sharpened into a customized sampling tool specifically shaped to reach the core location. Sampling may be monitored using the borescope, and dislodged blocks may be retrieved with a flexible claw pickup tool. Images collected during the process will ensure that the orientation of the block is documented.

Sometimes only loose, powdery sample material is accessible. For example when a very porous, friable core has partially disintegrated in an archaeological bronze over time, it may be impossible to extract a block sample. Chemical analysis may still proceed with powdery sample material, and so may other types of analysis, such as particle size distribution and shape analysis, provided enough representative sample material is available.

In rare cases, bronzes that allow no access to their cores have been drilled and core extracted through a small hole. The metal drillings have been used for elemental analysis and the core drillings for chemical analysis. Of course, in such cases the knowledge that might be gained must be carefully weighed against the consequent damage,[27] and this approach will often be considered inappropriate.

Extracting core material for analysis from a bronze does not affect the long-term stability of the object, as the core only plays a role during the casting process.

4 Risks of misidentification/ misinterpretation

Perhaps the greatest risk of misinterpretation lies in the possibility that the material sampled is not original core. In archaeological bronzes there is ample opportunity for soil and sediment to either imitate or contaminate core material. Also, it is not uncommon for cast-in repairs to be made, adding additional core material after the original flawed casting was complete and the original core removed. This may be to repair original casting flaws (executed at the time of the original manufacture) or to repair damages later in the life of the object. While core intended for repairs is, of course, interesting in its own right, it may not be the same as the original core, and should not be confused with it.

If original core is clearly present, care should be taken to extract a sample that appears representative of the bulk of the core. If the core appears to be heterogeneous, then multiple samples may be warranted.

It should also be noted that the filling of hollow bronze sculptures with plaster, long after their original production, is not unheard of. The motives for such treatment are not always apparent, but clearly it happens and may cause much confusion. Common sense is usually the best approach to determine whether the plaster is original or not.[28]

Casting core studies still suffer from nonstandardized procedures. They are often incomplete and may not be comparable with other aspects of technical examination of bronze statues due to lack of expert data interpretation.

Notes

1. Liu et al. 2013.
2. Stone 1981. To date, unpublished studies of plaster cores include several samples taken from eighteenth-century French bronzes at the C2RMF. Forty-four plaster cores have been examined at the J. Paul Getty Museum, only some of which have been published (Fogelman and Fusco 2002; Bennett and Sargentson 2008; Schmidtling 2008; Bassett and Bewer 2014). Sand casting core material from four foundries as well as eighteen sand casting cores have been examined at the J. Paul Getty Museum, only two of which have been published (Schmidtling 2008).
3. Early published analyses include Stone 1981; Milam, Reedy, and Sussman 1988; Reedy 1991b; and Formigli 1993.
4. Reedy 1997; Woodward 1997; Lombardi and Vidale 1998; Bewer 2001; Lombardi 2002; Weisman and Reedy 2002; Schmidtling 2008; Lombardi 2009; Carò 2010; Mille et al. 2010; Vincent 2014.
5. Waksman 2014; Léon 2014.
6. Holmes and Harbotte 1991; Reedy 1991b; Lombardi and Vidale 1998; Mugnaini et al. 2014.
7. For an in-depth review of analytical methods applied to rare earth element analysis see Zawisza et al. 2011.
8. Hancock et al. 2019.
9. Mugnaini et al. 2014; Castelle, Coquinot, and Bourgarit 2016. However, as early as the 1990s Edilberto Formigli already questioned the fabrication process of the core and its integration within the sculpture's overall fabrication process (Formigli 1993).
10. Lombardi and Vidale 1998; Gratuze, Blet-Lemarquand, and Barrandon 2001; Goemaere et al. 2014; Mugnaini et al. 2014.
11. Mugnaini et al. 2014.
12. Gehres and Querré 2018.
13. Sylvester 2008.
14. Holmes and Harbotte 1991; Reedy 1997; Lombardi and Vidale 1998.
15. Lombardi 2009; Mugnaini et al. 2014; Vincent 2014.
16. Smith et al. 2011.
17. Calligaro et al. 2000; Dran, Calligaro, and Salomon 2000; Calligaro et al. 2011; Castelle, Coquinot, and Bourgarit 2016.
18. Pichon et al. 2014.
19. Stone 1981; Holmes and Harbotte 1991; Lombardi and Vidale 1998; Schmidtling 2008; Lombardi 2009; Mugnaini et al. 2014.
20. Bish and Post 1989.
21. Lombardi and Vidale 1998; Newman 2006; Schmidtling 2008; Lombardi 2009; Mugnaini et al. 2014; Vincent 2014.
22. Riederer 2004; Velde 2005; Reedy 2006; Livingood and Cordell 2009; Dal Sasso et al. 2014.
23. Chapoulie, Robert, and Casenave 2016.
24. Reedy et al. 2014.
25. Reedy 2006.
26. Velde 2005.
27. Milam, Reedy, and Sussman 1988; Weisman and Reedy 2002; Reedy 2006; Reedy and Meyers 2007.
28. *Neptune* (1583) by Barthélemy Prieur (French, 1536–1611) in the Musée de Melun, France (inv. 802, Millet 17), was filled with plaster once removed from its fountain (Seelig-Teuwen, Bourgarit, and Bewer 2014). In Donatello's (Italian, ca. 1386–1466) *Spiritelli* in the Musée Jacquemart-André, Paris (inv. MJAP-s1773), nineteenth-century newspaper fragments were found in the plaster core, attesting that the plaster was not original but resulting from a later investment (Castelle et al. 2019).

II.8

Dating Methods

Arlen Heginbotham
Stéphanie Leroy
Antoine Zink

This chapter briefly presents the different methods currently available for dating a bronze sculpture. These methods generate an estimated age based on physical changes in the material that occur due to aging. Date estimates can also be made by comparing technical characteristics of a bronze sculpture to those of securely attributed examples, or through knowledge of the date when certain technological developments occurred; these *relative* dating methods are discussed throughout volume I.

Two dating techniques are suitable for the study of bronze sculpture: trapped-charge dating, including thermoluminescence (TL) and optically stimulated luminescence (OSL), and radiocarbon dating. Of these methods, trapped-charge methods have seen the most use, although even they are infrequently applied. In both cases, it is not the copper alloy itself that is dated, but rather the core material and/or the iron armatures, almost always from the sculpture's interior. These methods are thus restricted, with rare exceptions, to hollow sculpture that retains original core and/or iron armature. The main characteristics of both methods are presented in figure 433 and table 10.

Before embarking on trapped-charge or radiocarbon dating, confirm that the planned analysis is likely to yield a result with sufficient precision to answer the questions at hand. Neither method is highly precise, so other methods of estimating date (based on fabrication techniques, materials, historical sources, et cetera) may be equally advantageous, depending on the circumstances.

1 Trapped-charge dating: thermoluminescence (TL) and optically stimulated luminescence (OSL)

1.1 Basic scientific principles

Ambient radioactivity exists everywhere in the environment. Some minerals found in casting cores, notably quartz and feldspars, have the ability to store energy from ambient radioactivity in the form of free electrons that become trapped in nano-scale irregularities in the mineral's crystal structure. The longer one of these minerals is exposed to radioactivity, the higher the stored energy or "accumulated dose." When heated above 400°C (as happens to core material in contact with molten bronze at temperatures in excess of 900°C), a charge-trapping mineral releases its stored energy, and its accumulated dose is "zeroed out." Therefore, the more time that has elapsed since a bronze was cast, the higher the accumulated dose of trapped energy in its core minerals.

Analytical techniques allow the accumulated dose to be measured and the date of last high temperature exposure to be roughly calculated. When stimulated by heat or light, energy-trapping minerals can release their trapped energy in the form of light (luminescence); therefore, a measurement of the amount of luminescence produced by a sample upon stimulation serves as a measure of the amount of trapped energy in the minerals. This, in turn, can be correlated to the amount of time that has passed since casting—or, more precisely, since the minerals were last zeroed out by heat or light.

Two measuring techniques are available. Either the sample is heated to stimulate luminescence—called thermoluminescence (TL)—or the sample is illuminated with single-wavelength laser light—called optically stimulated luminescence (OSL). When possible, it is advisable to measure both TL and OSL in order to reduce uncertainty. OSL offers a better reproducibility than TL but does not allow for discrimination between the unstable components of the signal (risk of under-evaluation). While all samples have a TL signal, the OSL is more rarely present.[1]

TL and OSL dating are, from an experimental point of view, extremely complex, with numerous potential opportunities for error. Samples must be meticulously prepared to isolate luminescent minerals. Measurement of the luminescence is fairly straightforward, but converting a luminescence value to an estimated date is not. First of all, an estimate must be made of the annual dose of radiation that the sample has received during its lifetime. Much of the dose will have been delivered from minute quantities of radionucleotides (such as thorium 232, uranium 238, radium 226, and potassium 40) that are naturally present in the sample material itself. The dose rate from these sources can be measured directly from the sample, but the nature and intensity of other sources of radiation that may have impacted the sample throughout its history may be difficult to determine with certainty.

Once the estimated annual dose is determined, a prediction of the amount of energy that will be trapped per annum in a given sample depends on the energy trapping efficiency of the specific minerals present. This must be measured empirically, by repeatedly irradiating and measuring the prepared sample material using known dosages of radiation. Finally, the estimated age may be determined by calculating the estimated dose received by the sample and dividing by the estimated annual dose.

1.2 Application to bronze sculpture

As mentioned, trapped-charge dating of bronze sculpture requires the presence of residual core material. The first dating of a bronze sculpture by luminescence was performed in the early 1970s.[2] The dating method depends critically on the assumption that, during the fabrication process, the quartz- and/or feldspar-containing core was sufficiently heated to reset the signal (zero out the accumulated dose). Thus, only core material in close contact with the bronze surface (where the temperature has exceeded 400°C) can be dated.

Several special circumstances must be considered in trapped-charge dating for bronze sculpture:

- Gypsum plaster cores often do not contain quartz or feldspar in sufficient quantity to be useful for standard TL or OSL dating. Research has been conducted on trapped-charge dating of geologic gypsum deposits,[3] but this has not been applied to bronze core material.
- Soil contamination is commonly present in the interior of archaeologically recovered sculpture and may be mistaken for core material, or mixed with it. The soil can often be dated, but in this case, the dated event corresponds to the last exposure of the minerals to light, and therefore to the burial of the sculpture, not its fabrication. (For specialists: experimental solutions may exist, such as single-crystal OSL.)
- During the life of a bronze, exposure to fire or another heating event may set the accumulated energy to zero. The heating event is then dated instead of the sculpture's fabrication.[4]
- The metal wall of a bronze acts as a shield against external environmental radioactivity, which may reduce the annual radiation dose expected for the sample. This phenomenon is not encountered with ceramic materials and other sample types commonly dated by this method. For specialists: the gamma dose attenuation may be estimated by calculation[5] or, if a dosimeter can be introduced inside the sculpture, by comparative measurements on the interior and exterior.
- Since the method is based on the amount of ionizing radiation received, any exposure to an artificial source of radiation, for instance X-ray radiography, will cause the object to appear older than it is. Laboratory measurements have shown that an overestimation of the age of core material by as much as one thousand years is possible following radiography.[6] It is generally

not recommended to attempt dating an artwork that has previously been X-rayed.

1.3 Precision and accuracy

Broadly speaking, trapped-charge dating is an accurate method but not a precise one (fig. 413). For instance, the uncertainty of a TL or OSL date may be on the order of +/–140 years for a thousand-year-old object with 95% confidence. It is possible to reduce this uncertainty by increasing the number of measured samples, particularly where all the parameters are well characterized and measured. Recently, earth scientists have favored single-crystal OSL whereby each mineral grain from a sample is analyzed individually, allowing for a more nuanced interpretation of results.[7]

Artworks from museum collections are rarely sufficiently documented to enable precise estimation of the annual dose. In particular, the contribution due to the radioactivity from the environment is unknown; only the radioactivity from the object itself is measurable. As a consequence, for such artworks the uncertainty is often two to three times higher (a 95% confidence interval of +/–300 years for a thousand-year-old object).

1.4 Guidelines for sampling

To avoid disturbing the trapped-charge signal, samples for TL and OSL should be taken under photographic safelights.[8] After judging the accessibility of the core for the sample, the surface of the core should be removed to a depth of 1–2 mm before taking the sample for analysis, preferably using a small tungsten carbide drill bit. If the residual core is less than 2 mm thick, dating may not be possible. The amount of powder required to carry out an analysis is about 100–200 mg, depending on the mineral composition of the core (an average-size black peppercorn weighs approximately 60 mg). Sampling can be difficult, particularly because the vibrations associated with drilling can unintentionally dislodge pieces of core. Vibration can be minimized by adjusting the speed of the drill. It is very important that the sample material be collected cleanly, without contamination. Working in enclosed spaces under dim safelight can make the entire process challenging.

1.5 Cost and availability

Several academic and private laboratories offer luminescence dating. Most laboratories are devoted to geological sediments; only a small number work with cultural heritage artifacts, and of these, most focus on archaeological materials. As of this writing, typical costs per analysis range from about US$300 to $1,000. The lowest prices generally correspond to simple "authentication tests," which follow less robust calibration procedures and thus yield a larger uncertainty in the result. In a typical authentication test, only the trapped energy dose is measured. The inherent radioactivity of the sample material is not measured (although it determines the internal annual dose received by the sample). Instead, the expected annual radiation dose (and thus the age) is roughly estimated using typical mean values.

1.6. Risks of misinterpretation

Be aware that prior radiography can impact the results of trapped-charge dating, shifting the estimated date earlier than would be expected. This is primarily a concern if the sculpture has been radiographed repeatedly as part of a prior technical examination. Exposure to radiation for airport security screening or added cosmic radiation due to air travel will have a minor or negligible effect.

Samples taken for trapped-charge dating should be representative of the date of fabrication of the sculpture. For instance, insufficiently heated core (sampled from too far away from the metal wall) may yield ages much older than the fabrication of the sculpture; conversely, a core heated accidentally long after the fabrication of the statue (for instance by fire) will lead to a younger age. For suggestions on evaluating the condition and originality of core material, see II.7§3 and II.7§4.

2 Radiocarbon dating

2.1 Basic scientific principles

For more than fifty years, radiocarbon has been the most common dating method in archaeology. Covering a large period of time, from the beginning of the upper Paleolithic (about fifty thousand years ago) to the middle of the seventeenth century (fig. 433), it can be applied to a wide array of materials, provided they contain carbon. The dating principle is based on the radioactive decay of the carbon isotope with atomic weight 14 (carbon-14, or ^{14}C), which is formed in the upper atmosphere by cosmic radiation interacting with atmospheric nitrogen (fig. 434). As carbon-14 is continuously formed and continuously decays at a nearly constant rate, a stable, though very small, amount of carbon-14 is present at equilibrium in the atmosphere in the form of carbon dioxide (CO_2). Atmospheric carbon dioxide molecules are, in turn, incorporated into living organisms—plants and animals—throughout their lifetimes. When an organism dies, the assimilation of atmospheric carbon is interrupted, and thus the concentration of carbon-14 atoms in the organism's tissue starts to decrease by radioactive decay; it loses half its radiogenic carbon every 5,730 years. By measuring the concentration of residual carbon-14 atoms in an organic biological sample, one can calculate an estimate of time elapsed since the tissue was formed. This is called the radiocarbon determination.

Complicating matters, minor natural fluctuations of carbon-14 production and distribution around the world and through time lead to discrepancies between radiocarbon determinations and true ages. As a result, radiocarbon determinations (reported in terms of years "before present," or BP) must further be calibrated according to international calibration curves to obtain accurate calendar ages (fig. 435). These calibration curves are revised and refined on a regular basis.[9]

The advent of widespread coal burning led to a progressive reduction in the concentration of carbon-14 in the atmosphere due to the release of large amounts of fossil carbon, which is depleted in carbon-14. For this reason, carbon-14 dating becomes less and less precise after the early seventeenth century. The testing of thermonuclear weapons, beginning in the early 1950s and peaking in the early 1960s, produced large amounts of carbon-14 in the atmosphere, raising levels above the natural equilibrium state. This "bomb spike" is clearly detectable by radiocarbon dating.

For many years after its introduction in the late 1940s, radiocarbon dating was neglected in the study of art, mostly because of the large sample size required. In the last twenty-five years, the development of accelerator mass spectrometry (AMS) has allowed a drastic reduction in the amount of sample required. Whereas 5–10 g of charcoal or vegetal remains were necessary for conventional radiocarbon dating, today AMS requires only 2–5 mg to obtain a reliable date.

2.2 Application to bronze sculpture

2.2.1 Dating of organic material in the core

As far as bronze sculpture is concerned, additives to the core such as plant remains, fragments of wood, fiber, and charcoal may be present in casting core material. These materials, if they can be isolated in sufficient quantity, are suitable for radiocarbon dating, providing their age is contemporaneous with the fabrication date of the statue.

On occasion, charcoal powder may have been used as a release agent on the surface of the casting core, and remnants of such material on the interior surface of the bronze sculpture have been used successfully for radiocarbon dating. The bulk of the core material (clay, sand, and/or gypsum plaster) is not suitable for radiocarbon dating.

2.2.2 Dating of iron components

Traditionally produced iron contains small amounts of carbon derived from the fuel used during the smelting process. This carbon is trapped in the microstructure of the metal in the form of cementite, or iron carbide (Fe_3C). If the fuel source was charcoal (from recently cut wood), then the entrapped carbon can be usefully analyzed to estimate the iron's date of production. If the fuel source was coke (a purified form of mineral coal, entirely depleted of carbon-14), then no useful date can be derived. Although coke or coal is known to have been used in metallurgy as long ago as ancient Roman times and in China at least as early as the eleventh century, the use of coke for iron smelting only became widespread beginning in the mid-eighteenth century in Britain.[10] This means that iron armatures in sculptures produced prior to this time may be suitable for radiocarbon dating.

Radiocarbon dating of iron has been explored and tested since the 1960s.[11] Over the years, a number of potential sources of contamination and interference have been identified that can make results unreliable.[12] Since 2015, collaborative research between several units of the French National Centre for Scientific Research (CNRS) and the French Alternative Energies and Atomic Energy Commission (CEA) has resulted in an improved methodology that has gained wider acceptance for the dating of iron objects.[13]

2.3 Precision and accuracy

For cores and armatures, the final carbon-14 date is always given with an error correlated to both the experimental processes of sample preparation and measurement, and uncertainties associated with the subsequent calibration. Consequently, a radiocarbon age is better considered as an interval of time rather than a single date. The calibration process generates more or less large intervals of calendar ages—from eighty years to several centuries—depending on the period of time considered and the precision of the measurement (fig. 435).

2.4 Guidelines for sampling

2.4.1 Core sampling

Organic materials can be collected from the core matrix under a stereomicroscope using a scalpel and tweezers, then stored in aluminum foil pouches. Samples are then chemically cleaned at the radiocarbon laboratory with acid and alkali solutions to eliminate potential contaminants.[14]

2.4.2 Iron armature

Compared to organic materials, the carbon contents in bloomery iron are very low (<0.8wt% carbon). This considerably increases the sample size necessary to yield sufficient amounts of carbon for reliable dating. Furthermore, the carbon is often heterogeneously distributed within the metallic matrix, so a preliminary metallographic analysis of the sample is typically made to identify the areas containing the highest concentration of carbon.

Currently, standard practice is to extract at least one cubic centimeter of uncorroded iron for analysis. Sampling is usually executed using a thin diamond cutting wheel (0.6–1 mm thick; fig. 436) This sample should be examined metallographically and then resampled to yield 125–1000 mg (according to the carbon content in the metal) of iron drillings suitable for analysis.

2.5 Cost and availability

Carbon-14 dating of organic remains can be processed by any of several AMS radiocarbon laboratories for an average cost of approximately US $450 (discounts are sometimes available for nonprofit and academic institutions). For iron armatures, dating using the new methodology has thus far only been carried out as a collaboration between two laboratories, namely the Laboratoire Archéomatériaux et Prévision de l'Altération (LAPA-IRAMAT) and the Laboratoire de Mesure du Carbone 14 (LMC14-LSCE), following a comprehensive protocol that includes investigation of fabrication techniques and iron provenancing.

2.6 Risks of misinterpretation

When radiocarbon dating organic components of core material, it is very important to exclude any material that might have been introduced at a later date. In one cautionary example, fruit seeds deposited in the interior of a bronze sculpture by an animal led to erroneous radiocarbon dating.[15] For suggestions on evaluating the condition and originality of core material, see II.7§4.

Be aware that if the sample material dates to the period between the early seventeenth century and the early 1950s, the precision of radiocarbon dating is likely to be very low.

Radiocarbon dating of iron armatures is not necessarily sufficient for an accurate dating of a bronze sculpture. Iron may be wholly or partially recycled, in which case radiocarbon dating would yield a date earlier than the fabrication of the sculpture. The LAPA and LMC14 laboratories recommend a comprehensive archeometallurgical study of the armature to address this possibility.

Notes

1. For more see Aitken 1985; Aitken 1998.
2. Fleming 1971; Zimmerman, Yuhas, and Meyers 1974.
3. Mahan and Kay 2012.
4. The luminescence dating of the Hawtar'athat statue (see Mille et al. 2010) gave a much younger age (115 BCE–545 CE) than that deduced from epigraphic analysis (late seventh century BCE to first half of the sixth century BCE). It turns out that the metallographic analysis showed traces of fire. The statue of Hawtar'athat probably had a very long period of use, which unfortunately ended in a violent fire between 115 BCE and 545 CE. Radiocarbon dating of charcoal fragments trapped in the core (video 15) provided an age compatible with the epigraphic analysis.
5. Martin, Incerti, and Mercier 2015.
6. Castaing et al. 2004.
7. Jacobs et al. 2011.
8. A one-day exposure to daylight in northern Europe during winter is enough to set the accumulated dose to zero.
9. Reimer et al. 2020.
10. Taylor 1987, 85–87.
11. Merwe and Stuiver 1968.
12. Hüls, Petri, and Föll 2019; Craddock, Wayman, and Jull 2002. For a summary of the history of radiocarbon dating of iron with additional references see Taylor 1987, 85–87.
13. The collaborating institutions include the Laboratoire Archéomatériaux et Prévision de l'Altération (LAPA), Institut de Recherche sur les Archéomatériaux (IRAMAT), Laboratoire de Mesure du Carbone 14 (LMC14), and Laboratoire des Sciences du Climat et de l'Environnement (LSCE). For details of the methodology see Leroy, Hendrickson et al. 2015; Leroy, L'Héritier et al. 2015; Leroy et al. 2018; Delqué-Količ et al. 2017.
14. Dumoulin et al. 2017.
15. Michelucci 2006.

II.9

Experimental Simulation

David Bourgarit

Additional Contributors: Francesca G. Bewer, Andrew Lacey, Benoît Mille

This chapter aims to guide the reader in experimental simulation, a potentially powerful research tool for the better understanding of bronze sculpture fabrication techniques.

1 What is experimental simulation?

Experimental simulation consists of replicating and testing the materials, tools, and/or procedures of a historical art technology,[1] in this case the fabrication of bronze sculpture. In the cultural heritage field, experimental simulation has been carried out for more than seventy years, mainly in the field of archaeology.[2] The simulation experiments themselves may be undertaken following several different models, which may be usefully divided into three main types: historical, laboratory, and virtual.

1.1 Historical simulation

Historical simulations strive to reconstruct a partial or full sequence of a historical (or prehistoric) process using tools and materials that approximate those used originally. The design of the reconstructions may be based on historical texts, physical evidence from historic production sites, physical evidence from fabricated objects, or a combination of these. This approach can deepen understanding of the physical characteristics that result from different production processes. It may also be useful for simulating the original appearance of objects that have deteriorated over time. In addition, historical simulations can elucidate the social and economic context within which objects were created. The experimenters may wish to answer questions related to productivity, efficiency, costs, and the skill level required for successful production. Historical simulations may be multifaceted, long-term undertakings or small-scale experiments designed to answer a specific question.[3]

Historical simulations focused broadly on the fabrication of bronze sculpture have been carried out in European classical[4] and Renaissance[5] contexts. A long-term study of the economic factors affecting medieval European brass making was published in 2018.[6] More specific studies have focused on fusion welding,[7] sixteenth-century French casting recipes,[8] and the patina formulations of the Renaissance sculptor Antico (Italian, ca. 1455–1528).[9] A simple but effective experiment to determine whether tool marks observed on a bronze sculpture were made in the wax model or directly on the metal was conducted by Lorenzo Morigi (fig. 271).[10]

For researchers interested in historical simulation, publications beyond the scope of bronze sculpture may provide guidance and inspiration. Several archaeometallurgical publications address closely related subject matter.[11] Two proceedings focused on paintings by the Art Technological Source Research study group discuss various aspects of using historic texts to inform historic reconstructions.[12]

1.2 Laboratory simulation

Laboratory simulations may have more narrowly focused objectives than historical simulations. The experimental setup is usually smaller in scale and may be quite dissimilar to the historical technical environment. Laboratory simulations tend to answer detailed scientific questions that are only answerable via repeated experiments under carefully controlled conditions. Such simulations may increase understanding of how basic parameters such as temperature, alloy composition, chemical interactions, or variations in procedure influence the final state of a bronze sculpture.

Examples of recent laboratory simulations include fusion welding experiments[13] that have yielded evidence of the quite narrow operating window (alloy, temperature, timing) in which fusion welding can be carried out (video 12). These results have highlighted the exceptional skill of classical European founders in controlling complex heat transfers, particularly for large bronzes where welded joins may extend over several meters (see I.5, Case Study 1). The castability of various copper alloys has been tested through a series of experimental simulations that have, for the first time, properly demonstrated the role of lead in promoting the successful making of thin-walled casts (figs. 437, 438).[14] Detailed investigations into specific patina formulations in classical bronzes have shed additional light on the sophistication of ancient craftspeople,[15] and the thermodynamics of brass production by cementation have been rigorously investigated.[16]

To broaden the perspective to all archaeometallurgy, most laboratory experiments on historical copper metallurgical processes concern smelting.[17] One may add to the list all the simulations that have synthesized ancient copper alloys in order to characterize their mechanical properties[18] and color.[19] Experiments carried out on copper and copper alloys to study the behavior of major and minor elements during smelting and melting are not directly connected to any process reconstruction.[20]

1.3 Virtual simulation

In virtual simulation (some prefer the terms "computational simulation" or "numerical simulation") the materials and the phenomena are all investigated by computer modeling. This approach has only rarely been applied to historical technologies, but two studies demonstrate nicely the potential utility of virtual simulation with respect to bronze sculpture. Markus Ratka, Peter Sahm, and Wolfgang Bunk used computer modeling to infer the operating conditions (temperature, duration) and possible source of casting flaws in ancient bronze casting.[21] Welding has also been investigated by virtual simulation.[22] Beyond the specific topic of bronze sculpture, related simulation by computer modeling has mostly focused on ventilation issues for protohistoric copper and iron smelting.[23]

2 Designing an experimental simulation

2.1 Background research

Once a prospective topic for experimental simulation has been identified, a comprehensive survey of background sources is essential to inform the design of the experiment. Sources may, of course, include historical, scientific, and craft-oriented materials, or textual or audiovisual references. It is very important to consider that materials cited in historical documents may have, historically, contained significant impurities that are not present in today's supplies.[24] Publications of the Art Technological Source Research working group may provide helpful guidance in finding and interpreting historical sources.[25] It is equally important to consider existing physical evidence from relevant artifacts.[26]

2.2 Defining the question

Any experimental setup obviously depends on the objectives of the trial. Defining a clear question and setting clear goals will help determine the appropriate methodology and subsequent reporting.[27] Clearly, the more narrowly focused a specific experiment is, the likelier it will yield useful and clear results. Good experimental design is difficult to achieve in the context of bronze sculpture because so many process parameters feed into the work's final appearance. The most successful experiment will investigate these parameters, as much as possible, one at a time in order to map the effect of each contribution.

One example of this is provided by Andrew Lacey's experiments on the Rothschild bronzes.[28] After noticing that a hasty application of investment material to a wax model can have detrimental effects, the author duplicated his first casting, keeping all parameters but one unchanged; in the second iteration he applied the investment with much more care than the first time. The results speak for themselves: compare figure 4 versus figure 5. Through multiple focused experiments, the parameters of a larger process can be mapped and understood. Lacey, for example, also undertook a separate, specific simulation to find out whether the original model of a Rothschild statuette was made of wax or of clay, reconstructing only the modeling step (not the casting) based on two trials.

2.3 Special consideration for historical simulations

Laboratory simulations offer more controlled conditions than historical simulations in the field. They normally allow for faster experimental setup, better repeatability, and more accurate recording. For example, the temperature is much more homogenous in a laboratory electrical furnace than in a meter-deep charcoal-powered medieval-like brick furnace. Laboratory simulation is therefore strongly recommended when appropriate. If historical simulations are planned, it is often advisable to carry out several smaller-scale preparatory simulations in advance to help refine the materials and methods prior to full-scale implementation. This can rapidly clarify working hypotheses, and may reduce the necessary scope of costly and time-consuming field experiments.[29]

If historical simulation of an entire process is the goal, it is best to reproduce, as much as possible, the entire technical environment known to have prevailed at the time and in the cultural context under study, and at the proper scale. One usually refers to this type of undertaking as a field experiment. These require considerable investment in time and materials and are often conducted only once, and thus it is particularly important to meticulously record and document all materials, tools, and activities.[30]

2.4 How precise should the recording be?

Experimenters may be attracted by high-precision devices, such as pyrometers that record the temperature of a furnace to within 0.1°C or lasers that measure the diameter of a furnace with a precision of 0.1 mm. Particularly in the context of historical simulation, such precision is often not necessary. Historical production of bronze sculpture did not (and still does not) require a high degree of quantifiable precision for most operations. Craftspeople have traditionally relied on years of experience to make procedural judgments on the basis of observations such as the color of a flame or the sound of the hammer on the metal. Moreover, in an ancient or historical context, the operating conditions could not always be rigorously controlled.[31]

That said, when laboratory simulation aims to tackle fundamental chemical models, higher precision may be required. This was notably the case for the laboratory brass experiments that led to proposing a new thermodynamic model for the integration of zinc into copper.[32] To sum up, the precision of measurements should be consistent with the aims of the experiment, the process under study, and the experimental setup.

At least as important as precision is reproducibility. To ensure that experiments are not affected by unwanted variables, such as a faulty measurement device or some unobserved failure of equipment, experiments should if possible be repeated at least once.

2.5 Skills of the experimental team

The skills of the experimenters in the processes under study are obviously decisive for a successful experimental simulation. The researchers should also possess deep historical knowledge regarding the artifacts at hand and, if possible, textual accounts of the processes under investigation. This is particularly important to avoid biases and preconceptions based on the researchers' training or exposure to modern methods. Scientific skills are equally important to ensure rigorous and meticulous setup, recording, and reporting.[33]

Because of the varied skill sets necessary for success, many simulations should ideally be carried out as collaborations between individuals trained in craft, history, archaeology, and natural sciences. Whatever the initial skills of the experimenters, preliminary "soft" tests can be carried out to ensure that the experimenters are familiar with the materials and processes before commencing a formal simulation.[34]

2.6 Documentation and preservation of the experimental artifacts

If properly carried out, an experimental simulation can produce samples of high scientific value. These samples may constitute collections of objects that may be useful for future investigations beyond the immediate experimental program. For this reason, all sample materials should be catalogued and preserved so that they are linked in a permanent way to the relevant experimental documentation.

3 Risks of misinterpretation

The results of experimental simulation can be difficult to interpret. For instance, a simulation that yields a result similar to a historical artifact does not mean that the artifact was produced in this way; it merely points to one possible means of achieving a given outcome. Likewise, failure to yield a specific result through simulation does not necessarily prove that the process was not used successfully in the past. The experiment may have failed because the experimental setup was not appropriate, or the skill of the experimenter was not sufficient, or the process was not fully understood. What might be called a "successful unsuccessful experiment" is often of great value in helping to define the outer parameters or boundaries of a process—that is, where parameters start to break down (for example, casting in a range of temperatures whereby the upper and lower temperatures fail for different reasons).

A thorough evaluation of the results of an experimental simulation may take considerable time and effort. Success in this regard will be aided by a simple and straightforward experimental design, detailed documentation of all procedures, and careful preservation of all sample materials.

Notes

1. The foreword of Mark Clarke, Joyce H. Townsend, and Ad Stijnman's *Art of the Past: Sources and Reconstruction* defines art technology as "knowledge concerning the production methods of works of art or craft, i.e. machines, materials, studios, techniques, tools etc." (Clarke et al. 2005).
2. As a result of archaeology's dominance in this area, the term "experimental archaeology" is commonly used with regard to the experimental study or reconstruction of technological production methods. In the context of bronze sculpture, the term "experimental simulation" seems more appropriate. "Experiment" is sometimes used alone, instead of "experimental simulation." If not sufficiently contextualized, this term may be misleading. Bear in mind that the simple observation of a phenomenon or an object is already an experiment, and belongs within the realm of experimental science (as opposed to theoretical science).
3. A related pursuit is historical reenactment for didactic purposes. Such endeavors are not necessarily designed to answer specific technical questions and in this sense are not "experimental," although they can be quite informative, as in the case of this reconstruction of an aquamanile by Ubaldo Vitali: https://www.youtube.com/watch?v=tbQSAVFf-OE&feature=youtu.be.
4. Formigli 1993; Formigli and Hackländer 1999.
5. Lacey 2018.
6. Thomas and Bourgarit 2018.
7. Zwicker 1993; Formigli 1999c.
8. For instance the Making and Knowing Project et al. 2020, directed by Pamela Smith: https://www.makingandknowing.org.
9. Stone 2011.
10. Morigi 2018.
11. Dungworth and Doonan 2013; Heeb and Ottaway 2014. For a recent survey on copper extractive metallurgy experimental simulations see Bourgarit 2019.
12. Clarke et al. 2005; Kroustallis et al. 2008. See in particular Stijnman 2005.
13. Azéma et al. 2011; Azéma et al. 2013.
14. Pernot 2000; Andrieu et al. 2000; Mille 2017; see also I.2§2.2.
15. Mathis et al. 2007.
16. Newbury, Notis, and Newbury 2005; Bourgarit and Bauchau 2010.
17. Burger, Bourgarit, Wattiaux et al. 2010; Burger, Bourgarit, Frotté et al. 2010.
18. Lechtman 1996.
19. Benzonelli, Freestone, and Martinón-Torres 2017.
20. Tylecote, Ghaznavi, and Boydell 1977.
21. Bunk, Sahm, and Ratka 1999.
22. Zimmer et al. 2011.
23. Kölschbach et al. 2000.
24. Sometimes the composition of historically available raw materials can be determined. See for example Hickel 1963.
25. Stijnman 2005.
26. Stijnman 2012.
27. For more, see the concise and very explicit guidance provided by Alan Outram in Outram 2005.
28. Lacey 2018.
29. During the first three months of the Laitons Mosans project, aimed at understanding medieval brass making in Western Europe, about three hundred experimental simulations were carried out at the C2RMF laboratory to decipher the operating conditions (temperature, ratio of copper to zinc, et cetera). See Bourgarit and Bauchau 2010. This allowed the reduction of the number of resource-intensive field tests to eight (fig. 439) (Bourgarit and Thomas 2011).
30. In the 1980s John Merkel tested numerous copper smelting processes using a large brick furnace installed in the basement of the Institute of Archaeology, University College London (Merkel 1990).
31. For example, within a one-meter-diameter medieval-like furnace used for field experiments on brass making, the temperature varied from several hundreds of degrees around 1000°C from the bottom to the top, in the stationary regime (Thomas and Bourgarit 2018).
32. This demonstrated that the widely cited threshold of 30wt% zinc in ancient brasses could not be explained by the intrinsic limitation of the cementation process (Bourgarit and Bauchau 2010).
33. Outram 2005.
34. Kucera 2004 quoted by Heeb and Ottaway 2014 speaks of "soft" experiments for preparatory tests and "hard" experiments that can be scientifically exploited.

Case Studies

Edited by Francesca G. Bewer, Pete Dandridge, and Ruven Pillay

The case studies offer readers a representative example of investigations into the technology of bronze sculpture. The selection draws on the varied experience of the CAST:ING project members. Some cases focus on one object, others on a group of related works. They cover a variety of contexts in which such studies are undertaken, issues that may be addressed, and the kinds of research tools and historical sources—whether written or oral—that are essential to formulating an informed interpretation of the physical evidence. Their structure varies, but each of them ends with a synopsis of the findings and further questions, and tools applied in the study. The case studies complement the descriptions of processes, technical evidence, and scientific techniques of investigation discussed in volumes I and II, and hyperlinks to particular videos are included where relevant.

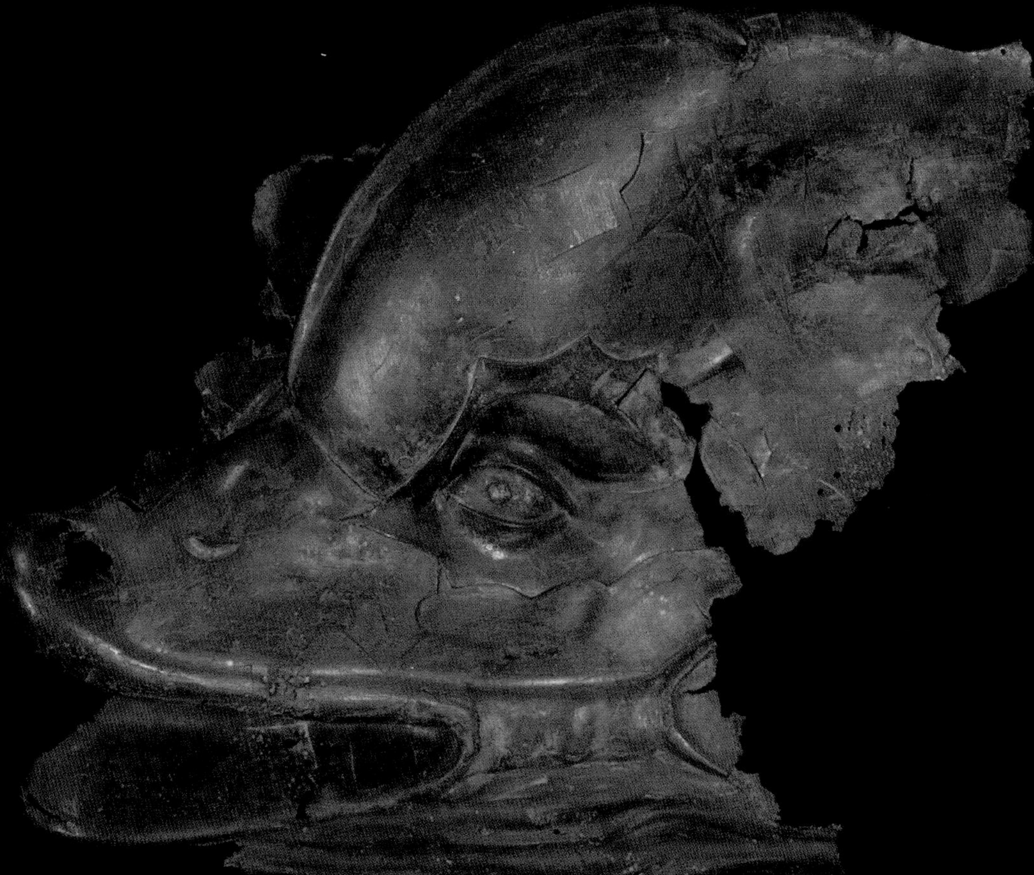

Figure 440 Front view of the first head. *Great Dolphins of Vienne*, Roman period, Vienne, France, 2nd century CE, max L. 260 cm (Musée des Beaux-Arts et d'Archéologie de Vienne, France, inv. R.1998.2.43).

Case Study 1

Reconsideration of a Monumental Roman Relief: *Great Dolphins of Vienne*, France

Benoît Mille

Introduction

(figs. 440, 441, 442, 443, 444, 445, 446, 449, 450, 451, 452, 453, 454, 455)

Fragments of a unique, large-scale, gilded Roman relief unearthed in the early nineteenth century in Vienne, France, were the subject of an extensive conservation campaign that included a multidisciplinary technological investigation. The results not only shed new light on the skillful methods of production that went into the object, but also led to a reinterpretation of the group's composition and resulted in the fragments' rearrangement for display.

Context

The discovery of the dolphins in Vienne, France

(fig. 440)

In 1839, during excavations for a new quay on the riverbanks of the Rhône in Vienne, engineers using a pile driver unearthed what were described as a group of "imposing metallic fragments." These were a rare discovery: the remains of a monumental gilded bronze relief representing swimming dolphins consisting of two heads, a body in two fragments, and three tails. This sculptural group, now known as the Great Dolphins of Vienne, is a rare surviving example of such reliefs from the Roman world.

Occasions for technical studies

(figs. 447, 448)

The bronze fragments had been studied in the 1960s. Based on the surviving tails, the proposed reconstruction of the relief's original composition was a five-meter-long frieze of three dolphins swimming in a row. The visual examination already made clear that joints were present, and that the relief surface was extensively repaired. In 2006, the Vienne museum decided to renovate the presentation of the dolphins. Given the fragmentary nature and the complicated reading of the find, the museum requested a thorough technological probing as part of the conservation process, which addressed a series of questions about the sculptural group.

The main questions

(figs. 440, 441, 442, 443, 444, 445, 446)

- How was it made?
- How do the surviving fragments fit together?

- Why was it so heavily repaired?
- What may have been the original iconography and composition of the full relief?

Visual examination

Some important new findings

(figs. 456, 459)

By puzzling through the physical evidence on the fragments, the scientist-conservator team discovered a previously undetected connection between the two dolphin heads. Closer study of the fragments proved that the dolphins were originally side by side, hitched together, thereby fundamentally modifying our understanding of the relief's composition.

At the same time, they noticed a neatly cut-out area at the top of the dolphin in the foreground: this is possibly where a rider may have been posed.

It also became clear that the deformation and fragmentation in one area of the relief was largely the result of the violent and high-speed impacts of a 13 × 13 cm implement—very probably the pile driver used in 1839.

Reconsidering the original composition

(fig. 459)

The new understanding of the arrangement resolves some iconographic inconsistencies. The dolphin in the foreground turns its head toward the viewer while the one in the background is presented in profile, reinforcing the perspective. In the same way, the wave rising behind the head of the foremost dolphin and descending over its body and tail should be read as passing between the two dolphins. The cutout at the top of the foreground dolphin could correspond to the inset of a rider such as Eros. The existence of three dolphin tail ends still raises questions. Were more than two dolphins hitched together? The newly determined perspective, the structure of the waves, and the presence of the upper edge preserved in most places make this an unlikely scenario. Instead, the tail could belong to one of a second pair of dolphins, or—given how widely it is splayed—to an unhitched, frolicking dolphin diving into the waves.

How was the wax model made?

(figs. 84, 86, 450)

Examination of the rear surfaces of the fragments revealed evidence that the sculptor worked in wax and had access to what would have been formed as a wax shell in a mold—in other words, an indirect lost-wax casting process. One finds, for instance, a series of marks made by dragging a tool through a malleable material that were ultimately translated into bronze.

The presence of numerous sprues on the reverse also proves that the wax model was worked from the back surface, and therefore not built up directly on a core.

The wax heads were made of separate parts that were joined in the wax. This can be seen in the lips, for instance, which appear to have been made up of several discrete sections. The wax joints would further confirm the use of an indirect process. The overall even thickness of the bronze walls, which was confirmed by radiography (see "Summary of main technological features observed at the back of the dolphins" below) points to the probable use of a wax slab process, as that would offer greater control than other processes such as slush molding (see I.3§2.3.1).

How were the segments of the relief assembled?

(figs. 460, 462, 463, 466, video 12)

Once all the separate parts were cast, the vast puzzle of primary castings had to be assembled. To achieve this, the Roman founders responsible for the relief used flow-fusion welding, a technique learned from ancient Greece, where it had been practiced since the fifth century BCE.

Recent scientific studies and characterizations of ancient welds, with experimental simulations in the laboratory, have shed more light on the long-forgotten process (video 12; I.5).

With its three-plus linear meters of weld joints, the Vienne dolphin relief bears testimony of the extensive application of this complicated method of assembly. And while the remains of the relief represent only a small part of the original sculpture, they provide a window into the exceptional scale and ambition of this work.

Some welds measure more than the staggering 70 cm long one seen here. The fact that these joints were achieved in a single pour bears witness to the great mastery of the founder. The welds were very wide and the flow of metal must have been contained, probably by channeling the welding metal with a refractory mold that covered the whole weld path. Despite the channeling, some leakage did occur (see also fig. 467).

Bulk alloy analyses by inductively coupled plasma with atomic emission spectroscopy (ICP-AES, see II.5§3.1) on drillings indicated a difference in the alloy used for the different sections and welds. The alloy of all of the primary castings is in the range of 3–4% tin, 5% lead, and the welds are 1–2% tin, 5% lead.

Structure of the assemblage using X-radiography

(fig. 367)

X-radiography confirmed that the metal walls of the separately cast sections are relatively thin and even across the board (measured at 4–6 mm), including the mouth area, even though it is in high relief. It also helped to clearly map out the extent of the assemblage: the welds show up as irregularly shaped, lighter swaths that correspond to the thicker—and therefore denser—metal joints.

The vast amount of repair work that the relief necessitated, especially in the central area, is visible as well. The edges of the many polygonal patches that were inset into the outer surface to repair the innumerable small flaws are recognizable, as are the copper rivets, which will be further discussed below.

Proposing a casting plan

(fig. 367)

Roman founders did not have the technological know-how to cast complex pieces such as the dolphin relief in a single large pour, so they cast larger sculptures in smaller sections (primary castings) and then joined the metal parts by flow fusion welding (see I.5§1.1). Careful examination of the fragments and the corresponding radiographs, which helped locate the joints, led to the proposal of a casting plan for the eleven extant pieces of the large relief. It provides some idea of how the original model was divided up to facilitate casting.

How was the cast repaired?

(figs. 367, 464, 465)

All large ancient bronzes, even exceptional commissions, will inevitably have some casting defects, as is reflected by the large number of repairs. The Vienne dolphins are no exception: if the extant portions are anything to go by, the relief was heavily flawed. Repairs cover 25 percent of the exterior surface. But once they were covered by gilding, they would have been nearly invisible.

Close examination of the surfaces, aided by cleaning during conservation, showed that the external bronze skin is riddled with endemic porosity (fine holes that formed in the metal during casting) and repairs. These are even more clearly visible in the radiographs (see fig. 367).

It was possible to tally no less than 265 patches made of bronze. Many of the largest ones (a total of 178) were fastened to the primary cast from the back with copper rivets.

Summary of main technological features observed at the back of the dolphins

(fig. 467)

The diagram synthesizes the evidence preserved from the various stages of production of the relief beginning with the creation of the wax model. The discrete, indirectly formed sections were joined in the wax (light blue lines). Sprues and wall fasteners (blue) were modeled in wax and joined to the back.

The separately cast bronze pieces (fine red lines indicate the edges) still preserve some core pins (black dots) and flashing (lilac).

The separate bronze pieces were joined by massive flow fusion welds (red to pink).

And the many flaws were fixed with cast-on repairs (brown) and innumerable patches (green), many of which were reinforced with copper rivets (dark green).

Conservation treatment and surface polychromy

(figs. 290, 380, 381, 468)

As conservators prepared the careful mechanical cleaning of the surface, examination of all of the fragments also served to assess their structural and physico-chemical condition and any alterations of the metal. Focus was on the removal of the bronze corrosion products that masked the entire surface. It was decided early on to clean the reverse as well in order to enhance the readability of technical features and learn as much as possible about the object's manufacture. There was no question of attempting to address deformations or fill losses.

During the cleaning, which was mostly done mechanically with an ultrasonic scalpel, it became clear that below the outer corrosion crust, the bronze surfaces of the dolphins still preserved much of an original gilding. The geometric pattern of 12 cm wide squares with denser gold at their edges was clear evidence that gold leaf had been used and applied, as is traditionally done in many cultures, overlapping neatly at the edges. The main challenge in laying bare the gilding was to avoid damaging this extremely fine layer, given its fragility and the corroded state of the underlying bronze.

Also revealed was that the decorative layer was applied selectively: the waves bore no trace of gold. Instead, their surfaces, which were only slightly corroded, were found to have been covered with a very dark (brown to black) homogeneous layer. Whether it was deliberately patinated black to create a dramatic contrast with the gilded dolphins remains to be determined.

Summary of findings

(fig. 379)

The results of the technological study led to a reassessment of the earlier interpretation of these fragments' placement. The revised composition produces an enhanced new sense of perspective, spatial depth, and movement. A new iconographic interpretation was proposed for these rare remains of an important architectural ornament: instead of a five-meter-long frieze of dolphins, it now appears to have been a hitch of two dolphins, probably pulling a marine chariot, and most likely with a rider, Eros, atop one of them. The original gilded surfaces that were uncovered during conservation treatment lend more visual power and sophistication to the relief.

The technological examination also showed that the preserved fragments were the product of one workshop, as evidenced by:

- the unified production of the lost-wax model with similarly thin and even wall thicknesses;
- similarity in the kind and quantity of casting defects (usual in the Roman period) and their repair; same alloy;
- same variant of flow fusion-welding over more than three meters of linear joints.

Synopsis of technical parameters

The Musée des Beaux Arts et d'Archéologie de Vienne coordinated the entire operation. The Centre de recherche et de restauration des musées de France (C2RMF) took charge of developing, setting up, and carrying out all of the examinations and analyses. The restoration project was conducted by the Center for Restoration and Municipal Archaeological Studies (CREAM) in Vienne. The objectives and issues of the restoration project were determined and discussed collegially by the three partners. The study consisted of:

- daylight photography: D. Vigears (C2RMF), 5 days, using a high-resolution Hasselblad numerical camera;
- X-radiography: Thierry Borel (C2RMF), 5 days, using an Isovolt 420 kV X-ray tube;
- bulk metal analyses by atomic emission spectrometry (ICP-AES): Benoît Mille (C2RMF), 10 days, using a Perkin-Elmer Optima 3000 SC atomic emission spectrometer—a protocol developed specifically for elemental analysis of cultural heritage copper-based artifacts (Bourgarit and Mille 2003);
- 3D scanning and reconstruction: Loïc Espinasse and Pascal Moral (Archeovision), 10 days, using lasergrammetry and 3D modeling;
- metallographic samples: Benoît Mille (C2RMF), 10 days;
- surveys and technical drawings: Benoît Mille (C2RMF), 15 days.

Further questions

(fig. 459)

- What did the full relief look like and represent?
- Were the rivets a distinctive way of securing patches, or commonly used?
- Was the dark patina on the waves deliberate?
- Was the low tin content (1–4%) of the alloy used for the primary casting commonly used for large Roman bronze sculptures? Or was it particular to reliefs?
- Why was the work so heavily repaired?

Further resources

https://c2rmf.fr/actualite/les-grands-dauphins-de-vienne

Azéma 2013

Azéma and Mille 2013b

Azéma et al. 2011

Boucher 1964

Mille 2017

Mille and Robcis 2012

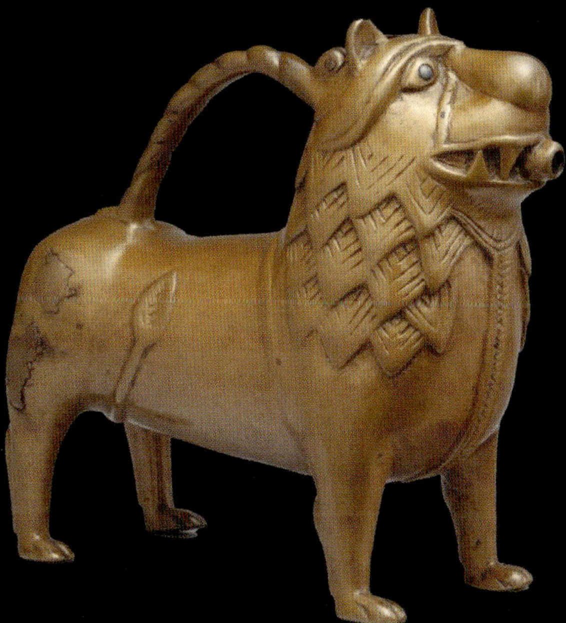

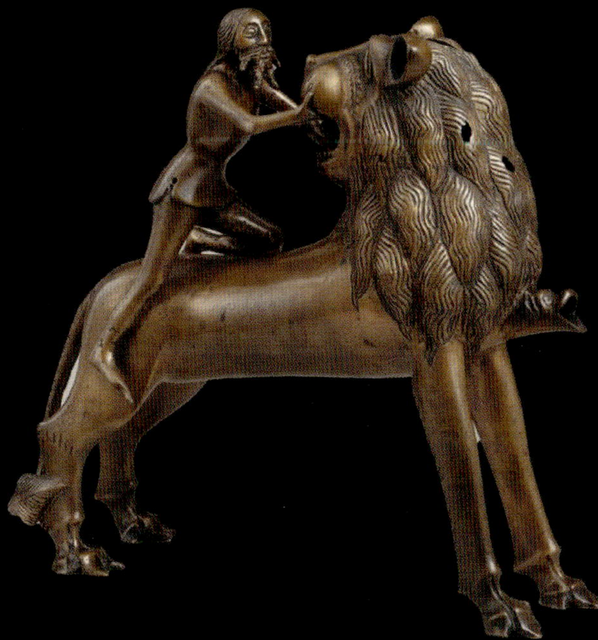

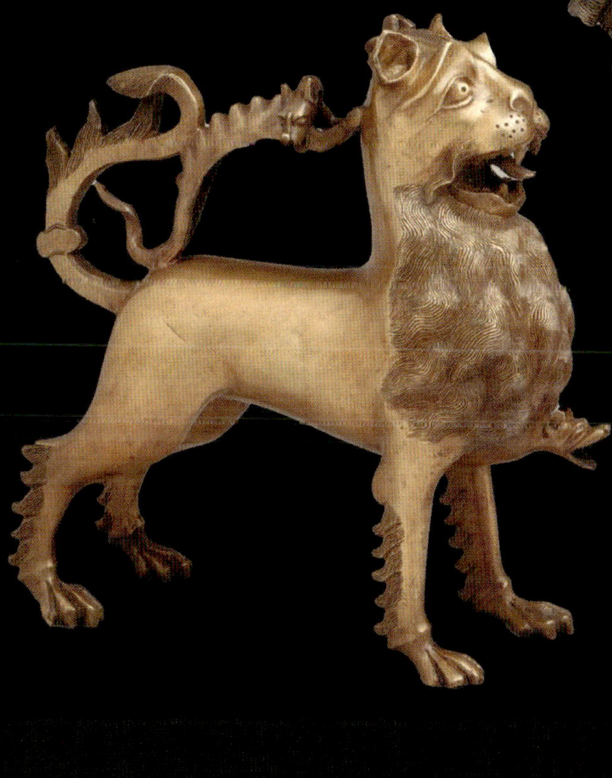

Figures 469–71 Clockwise from top left: Fig. 469, *Aquamanile in the Form of a Lion*, probably northern Germany, 12th century, H. 19.5 cm (Metropolitan Museum of Art, gift of Irwin Untermyer, 1964, inv. 64.101.1491); Fig. 470, *Aquamanile in the Form of Samson and the Lion*, northern Europe (?), possibly Germany, ca. 1380–1400, H. 34.1 cm (Metropolitan Museum of Art, Robert Lehman Collection, 1975, inv. 1975.1.1412); Fig. 471, *Aquamanile in the Form of a Lion*, Nuremberg, Germany, ca. 1400, H. 31.9 cm (Metropolitan Museum of Art, The Cloisters Collection, 1994, inv. 1994.244).

Case Study 2

Exquisite Objects, Prodigious Technique: Aquamanilia, Vessels of the European Middle Ages

Pete Dandridge

Introduction

(figs. 469, 470, 471, 486)

Aquamanilia are zoomorphic or anthropomorphic vessels for the washing of hands at the table. Cast from copper alloys, they emerged as a distinct artistic form in the twelfth century in Lower Saxony and continued to be fabricated in various areas of Germany for the next three centuries.

They were filled via lidded openings, generally at the tops of the animal's or rider's head, with the spouts either integrally cast as part of the figure's mouth or separately fabricated as a spigot and mechanically set into its chest.

They represent an important stage in the evolution of European casting technology, being the first hollow, fully three-dimensional sculptural forms produced since antiquity. More than three hundred aquamanilia are known to have survived through the centuries, some inspiring nineteenth-century replicas, and all are unique.

Context for the technical study

(figs. 469, 470, 471, 472)

The collection of aquamanilia and associated medieval German metalwork in New York's Metropolitan Museum of Art became the focus of an exhibition that integrated art historical analyses with technical examinations of the vessels in the Met's collections and related cast objects in other collections around the world. Several dozen works formed the focal point of this study, though information from many others also fed into the research.

Invaluable insights were also gleaned from the experimental replication of one of the Met's lion aquamanilia that followed closely the processes prescribed by the medieval monk Theophilus in his twelfth-century treatise *On Divers Arts: The Foremost Medieval Treatise on Painting, Glassmaking, and Metalwork*. Written around the time when aquamanilia first appeared, the treatise describes, among other things, the casting of a complex sculptural censer—a hollow, copper-alloy object.

The mane questions (pardon the pun)

(fig. 470)

- How were these fantastical objects fabricated?
- How does their facture compare to the processes outlined in Theophilus's twelfth-century treatise? Might those descriptions help in the interpretation of the technical evidence preserved in the vessels? Do they apply to the fabrication of aquamanilia as well?
- Ultimately, with data from a large enough sampling pool, would it be possible to start tracing the evolution of this casting technology as the centers of production shifted over time?
- Might it be possible to establish idiosyncratic patterns in techniques or materials that would help to identify and characterize the products of particular workshops or regional production centers?

This case study focuses predominantly on the first two questions.

Lost-wax casting

(fig. 473)

All technical evidence points to medieval aquamanilia being cast by the lost-wax process. This basically entails fashioning a model of the desired form in wax with which to create a hollow impression in a fire-resistant mold material that will serve as a matrix for the liquefied metal (see GI§2.2.1 and fig. 13).

The uniqueness of each known example suggests that aquamanilia were directly modeled over earthen cores. The internal surfaces are all relatively smooth, and there are no waxy drip marks or other physical indicators that they were formed by an indirect process.

Core (replication)

(fig. 474, video 4)

The internal volume of any hollow bronze is defined by a core made of fire-resistant material. Theophilus describes shaping a well-kneaded and sifted mixture of sandy clay and dried dung into more or less the desired form, integrating many of its physical features. The sand would temper the clay, reducing shrinkage and cracking during firing. The organic material would provide "green" strength or cohesiveness to the mixture and create a more porous body that could help absorb any vapors that might evolve during the pour.

The core would be finished to closely mirror the desired shape of the vessel. Relatively simple forms such as this one didn't require an internal armature to support the weight of the clay, while evidence of such supports was found in larger and more complex structures. After casting, the core material and armature were generally removed to clear the hollow of the vessel for use.

Core (examination)

(fig. 475)

Traces of the earthen mixture preserved in harder-to-access areas of several of the aquamanilia were sampled and prepared as thin sections for petrographic analysis (see II.7§2.2) and scanning electron microscopy (SEM; see II.7§2.2.3) imaging to characterize their material makeup. Several examples included the identifiable forms of organic inclusions. The mineral content pointed to a metamorphic rock source, but overall these analyses were (as yet) inconclusive regarding the identification of the clays' geological origin, and therefore the geographical location of the aquamanilia's production.

Wax layer (replication)

(fig. 476, video 5)

The model for the censer Theophilus describes was created by first overlaying the cured core with pieces of sheet wax rolled out to a specific depth, cut into "small pieces according to the spaces which you have cut in the clay of the censer." These were warmed and applied with the aid of a hot iron, then trimmed and smoothed out, taking care that the thickness of the wax layer was "not thicker or thinner in one place than another." Such evenness helps to ensure that the metal cools and shrinks at an even rate, avoiding distortions. The fine modeling of details can then begin. This same process was applied to the replication of the Lion aquamanile.

Wax model

(figs. 469, 477, 478, 479, 480, 481, 482)

The physical evidence points to the aquamanile being directly modeled in wax as described by the monk. The many fine details have the quality of a malleable material, and many were ultimately added in the wax. In some instances, elements such as the sculptural handles would have been formed separately and then joined to the body of the vessel.

Radiographs were invaluable in assessing the overall form of the aquamanile's core and how the wax layer (translated into metal) was built up. In such images, the denser, more X-ray-opaque materials show up whiter. The consistent opacity observed in the central part of both lions' bodies points to their modeling in a single layer of sheet wax. The slimmer parts of the legs were modeled in solid wax.

In the X-ray of the lion-shaped aquamanile from North Germany, the details of the mane are denser, showing that they were built up from an additional layer of wax.

In contrast, the undulating shapes of the mane in the flame-tailed lion aquamanile from Nuremberg were already formed in the core and overlaid by an even layer of wax as prescribed. The darker areas in this second lion's mane may have resulted from pressing the wax sheet onto the more complex core shapes, or from revisions in the final modeling that might have thinned the wax and resultant metal cast.

The more complex and larger aquamanilia required a more sophisticated construction, which could entail joining separately modeled figures (and cores) and/or the additional support of localized armatures. The retained core of the falconer, for instance, appears from the radiographs to be continuous with that of the horse. The slightly irregular darker area in the falconer's abdomen may be the location of an armature made of less dense organic material.

Conversely, the X-ray image of the small Samson wrestling on the back of the lion shows that the core of the human figurine extended only to his knees and beyond his shoulders into his head. Samson was modeled in wax separately before being joined to the back of the wax lion. A fine wire armature can be detected coming down from his head into the left calf.

Theophilus does not refer to armatures as the openwork form of the censer, and the bell-like shape of its two halves provided sufficient surface area for contact between the core and the investment to not require the use of armatures, but fine wires and other materials were certainly used in more complex objects. Again, these were often removed from the body of the vessel after casting, but were left in other, less accessible areas.

Core pins, sprueing, and investing (replication)

(figs. 483, 484, video 8)

Once the modeling was complete, metal core pins were partially inserted through the wax and into the core to secure the registration of the aquamanile's core within its outer mold or investment (see video 5). Next, the model would be fitted with a network of sprues through which to melt out the wax, feed the liquefied metal into the mold, and allow the gas and air to escape. When possible, connections helping with that flow were integrated into the design of the sculptural form itself, as can be seen in the falconer.

Then, as Theophilus describes, multiple layers of earthen mold material (similar to that of the core) were gradually built up over the sprued wax model. This clay investment mold containing the sprued model would be heated to melt out the wax, which, being an expensive material, would, according to the treatise, be recovered and presumably recycled. Heating the mold would also serve to get rid of all moisture because, the monk warns, "if the inside cores experience dampness, they immediately swell and all your work will break apart." The mold is then heated until it is red hot and buried in a pit, casting cup up, ready to receive the liquefied metal, which is poured from a crucible in which it has, meanwhile, been alloyed.

Core pins and core plugs

(figs. 180, 485, 486, 487, 488)

Examination of the aquamanilia revealed that these were made of copper or iron, and some out of bent copper sheet. The use of one type did not preclude the other.

Should the core pins fuse to the metal during casting, they were cut down to the level of the outer surface and smoothed over with some surrounding metal. But the internal portion is often still preserved on the inside, and visible in radiographs or in accessible areas.

If it could be loosened, however, the core pin was removed and the resulting hole filled with a plug, which was often a tapered, hammered-in copper pin.

Patterns may be found in the treatment of core pins and may prove characteristic of a particular workshop. In the Nuremberg aquamanilia studied here, when core pins were removed, the ensuing holes rounded with files and plugged with shallow discs of metal similar to the cast (probably cut from a sprue), and their joints secured with soldering.

Metal

(fig. 489)

Theophilus describes coloring copper by the cementation process using "a stone of yellowish color, sometimes reddish." This stone was, in fact, the zinc ore calamine and the process that of alloying "latten," a brassy metal containing zinc as the primary alloying agent and tin and lead in levels above 1%. (Zinc as a metal had not yet been isolated or identified in his time.) This is the golden-colored metal predominantly used for aquamanilia. Workshops in northern Germany and Lower Saxony, and later in Nuremburg, were close to the source of these raw materials.

Each ore source has distinctive trace minerals that can help identify where the metal comes from. To what extent the tin and lead contents were intentional cannot be readily determined, as both were also contained in copper ores. And given that metals were often recycled, with each remelting, the zinc concentration might be reduced due to the evaporation of zinc vapor. With so many variables to consider and a relatively small statistical group to work from, the results thus far of the metal analysis are far from conclusive.

That said, over the span of production of aquamanilia, the techniques used to alloy the calamine with copper became more sophisticated, which allowed for an enhanced absorption of zinc. Such an increase can be seen in the alloy analysis of aquamanilia produced in Nuremberg at the beginning of the fifteenth century compared to earlier casts.

Fettling

(figs. 104, 277)

Once the metal had cooled in the mold, the investment was broken away to reveal the casting. Fettling was the initial step in finishing the raw casting with the retained sprueing system (runners and risers) and core pins cut back with chisels and files. Once these features were removed, the artist had greater access to the as-cast surface, which in medieval western Europe was generally coarse and required significant finishing. One can get a better sense of this in areas inaccessible to tooling, like the undersides of handles, or animals' indented paws or hooves.

Cast-in repairs

(fig. 490)

Imperfections and casting flaws seem to have been an anticipated outcome for medieval metalworkers. Theophilus suggests using a localized lost-wax casting process for making repairs. "If there is any flaw, due to neglect or accident, . . . thin out that place by filing around it. Apply wax and likewise add clay on top and when it is dry, heat it, and so pour over it until the liquid bronze flows into that place and what you have poured on top adheres. If, when you examine it, it is not strong enough, solder with burnt wine-tone and the filings of mixed silver and copper as we have described above."

Many of the aquamanilia show evidence of such cast-in repairs. The metal-to-metal joint lines are generally faint on the outer surface, but the denser masses of metal that flowed into the hollowed core area are clearly defined in the X-rays.

Patches and plugs

(figs. 478, 488, 492)

Hammered-in copper-alloy plugs or patches were used on smaller flaws with no evidence of solder reinforcement. Such reinforcement is, however, apparent in aquamanilia cast in Nuremberg where plugs made of small discs of an alloy similar to the body metal (probably from a sprue) were inset into carefully prepared holes left by the core pins and soldered into position.

Chasing

(figs. 493, 494, 495, 496, 497)

Chasing required a variety of tools, including claw chisels, punches, scrapers, and burnishers to reduce marks from fettling and to close up surface porosity and small blemishes. Decorative details would be introduced or reinforced with engraving tools and punches, and a final finish might be achieved with a series of graded polishing agents.

Tool marks retained on the surfaces of the aquamanile suggest their transformation closely followed Theophilus's description that "you first file over all the grounds with various files—square, triangular, and round, then engrave with the engraving tools and scrape with scrapers. Finally, . . . clean . . . it all over with sand and with sticks slightly rushed at the top."

Summary of findings

(figs. 44, 469, 470, 471)

The physical evidence encountered during the technical study of the representative grouping of aquamanilia and a few related objects in this study resonates with Theophilus's description of the casting of a censer. By adopting as closely as possible the practices and materials described in the treatise, the experimental reproduction of one of the Met's lion aquamanilia allowed us to better understand the technical parameters of the originals, even the re-creation of an unintended casting flaw.

Techniques and materials used in casting aquamanilia remained fairly consistent over the course of their creation, but as their forms became more complex, adaptations were made. And some regional variations in methodology were noted. For instance, in addition to the idiosyncratic core plugs, the aquamanilia cast in

Nuremberg sported separately cast spigots, and the brass alloy used was consistently high in zinc.

Synopsis of technical parameters

(figs. 469, 470, 471)

The research was undertaken in collaboration with the Bard Graduate Center for Studies in the Decorative Arts, Design, and Culture and culminated in a 2006 exhibition in New York at the Bard Graduate Center accompanied by a catalogue: *Lions, Dragons, and Other Beasts: Aquamanilia of the Middle Ages, Vessels for Church and Table*. The various aspects of the technical examination were conducted as part of the day-to-day responsibilities of the Objects Conservation Department at the Metropolitan Museum. While no specific log was kept of time spent, the research, associated teaching, travel, and writing represented the better part of two years' work. Unless otherwise stated, the examination and analysis was undertaken at the Met:

- endoscopic and microscopic examination of interior and exterior surfaces: Pete Dandridge
- visible-light photography of exterior and interior surfaces: Pete Dandridge
- surface analyses by qualitative X-ray fluorescence (XRF) analysis of disparate elements incorporated either at the time of manufacture or as part of a later restoration: Pete Dandridge
- X-radiography: Pete Dandridge
- scanning electron microscopy (SEM) and polarized light microscopy of core and investment material: Richard Newman (Scientific Research Lab, Museum of Fine Arts, Boston)
- X-ray diffraction (XRD) analyses of discrete samples of the polishing compounds and pigments: Pete Dandridge
- inductively coupled plasma mass spectrometry (ICP-MS) analysis of metal samples: Josef Riederer (Rathgen-Forschungslabor, Berlin)
- replication casting: Ubaldo Vitali
- video documentation and editing: Han Vu (Bard Graduate Center for Studies in the Decorative Arts, Design, and Culture)

Further questions

- Many aquamanile and other cast medieval objects examined in European collections retain significantly more core material than the corresponding material in US collections. What might we learn from the analysis of a broader selection of core samples? Might further analysis corroborate some of the groupings made on stylistic grounds? Are there detailed enough geological maps of Germany to allow for the identification of geologic sources for the materials comprising their clays? And what additional information might be gleaned from the shape and size of the component grains?
- What technical relationships might we find between aquamanilia, other small-scale cast objects, and monumental works of art from the same period?

Further resources

Barnet and Dandridge 2006 and the DVD insert: Ubaldo Vitali and Pete Dandridge, *Medieval Alchemy and the Making of a Lion Aquamanile*, documenting the replication of a lion aquamanile

Bourgarit and Thomas 2015

Cellini [1568] 1967

Dandridge 2018

Magnus 1967

Mende 2013

Theophilus [ca. 1122] 1979

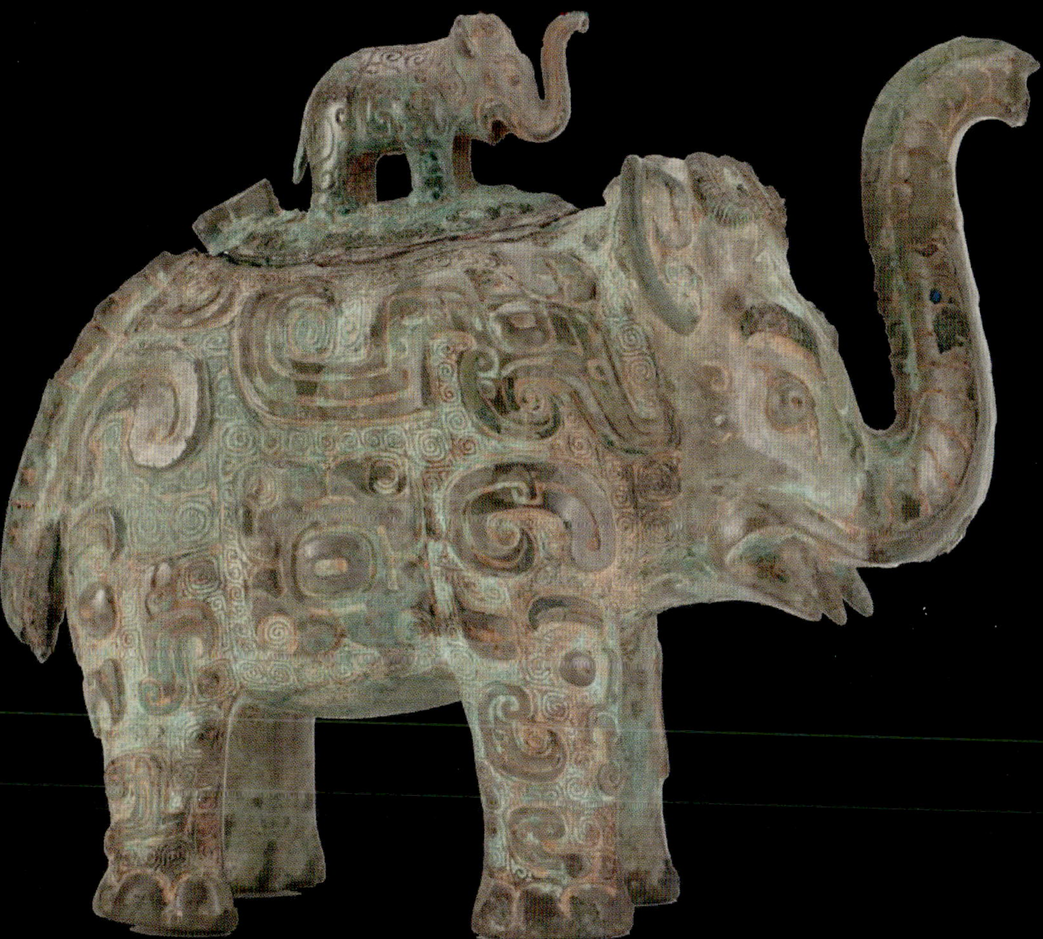

Figure 500 Side view. *Huo Ritual Vessel*, 11th century BCE, Middle Yangtze Valley, China, H. 17.2 cm (Freer Gallery of Art, Smithsonian Institution, Washington, DC, purchase—Charles Lang Freer Endowment, inv. F1936.6a–b / Department of Conservation and Scientific Research).

Case Study 3

Determining the Casting Process of a 3,000-Year-Old Bronze Elephant from China

Donna Strahan

Introduction

(fig. 500)

The question of when lost-wax casting was first used in central China has been of long-standing debate. When this rare sculptural ritual vessel, or *huo*, was first studied in the 1960s, it was thought to have been cast by such a method. But more recent excavations of foundry remains contemporaneous to this vessel at Anyang, and the refinement of examination and analytical tools used for the technological study of bronzes, provide new insights into the casting methods of such vessels in central China during the eleventh century BCE in the Yangtze River Valley. More recent technical study of the elephant-shaped vessel revealed clear evidence of quite a different method of production, namely the piece mold casting technique, the most prevalent method used to create such bronze vessels at that time.

Context

The Freer Gallery of Art in Washington, DC, acquired this vessel of unknown archaeological provenance in 1936. To date, its function is unknown. Stylistic evaluation and dating of a sample of the remaining clay core material suggest it was fabricated in China's Middle Yangtze Valley in the first half of the eleventh century BCE. The sculptural elephant form is very rare for Chinese bronze vessels of its time.

The main questions

(fig. 503)

- What characteristics point to piece mold rather than lost-wax casting?
- Can one tell whether the decorations were created on the model, the mold sections, or both?
- What accounts for the object's condition?

Piece mold casting

(fig. 26)

In general, the design of an object cast using the piece mold method is constrained by the fact that its model is made in loess, a fine-grained refractory soil found throughout northern China that gives very sharp detail when carved. The mold was made in pieces or sections (hence the terms "piece mold" or "section-mold" casting) designed like a 3D puzzle to deal with undercuts and allow removal without damage to either the model or the mold pieces. The elephant vessel and lid would have been cast separately using the piece mold casting process consisting of the following main steps (note that the diagram in fig. 26 uses a simpler model):

A model representing the final form is created, most probably including some of the decorative details (1). Multiple mold sections or pieces are built up around the model (2). The mold pieces are removed from the model (3) and used to make the core by filling the reassembled mold with loess and carving back the resulting shape (4–5). In the case of the elephant vessel, the core was sectioned to separate the legs from the container core. The decorative inner surface of the mold is further enhanced by carving (6). If needed, small rectangular metal spacers (chaplets), or loess core extensions placed at intervals, serve to stabilize the core in relation to the outer mold. The mold sections are reassembled around these, and vents and a pouring cup are cut into the piece mold. Molten bronze is poured into the reassembled mold, filling the gap between core and piece mold (7). The cast bronze is broken free of the mold and imperfections addressed (8–9).

Evidence for piece molding on the vessel

(figs. 110, 111, 112, 113, 116, 501)

Metal seam lines invariably developed along mold joints on the exterior surface of the cast object. Some were ingeniously incorporated into the design as raised, decorative flanges. Others were removed during the finishing process, but often traces can still be detected on the surface. One such seam line can be recognized running along the top of the elephant's trunk and continuing as a raised linear feature between the elephant's forehead and the rim of the vessel opening.

Visual and microscopic examination allowed us to identify the traces of mold joints across the surfaces of both vessel and lid, and thereby attempt to reverse engineer the design of the piece mold segments, which add up to at least a dozen for the main body of the vessel alone.

It was determined that the hollow of the vessel's body was defined by a core inside the belly that reached up to the oval rim of the opening and into the trunk (A). The cores of the legs were separate (B).

On the outside, two main mold sections bisected the elephant vertically, running up the tail, along the middle of the back to the back of the ears, and perpendicularly across the top of the head and out over the ears (C).

A large mold section ran along the belly, including the insides of the legs and the lower lip (D).

Two vertical face molds ran down the front of the ears and across the top of the trunk and include the exterior of the tusks (E).

A third head mold section included the rest of the mouth and tusks (F).

A further mold piece formed the front of the trunk (G).

Evidence for piece molding on the lid

(figs. 114, 115, 117)

For the lid, two main mold sections were used for either side of the body, including the ear, trunk, tail, and ridge down the tail (H).

A core filled the space inside the legs and under the tail and trunk (K).

Another core (I) filled the underside of the lid using core extensions to secure its attachment to the bottom mold section (J). The holes in the cast rim locate the extensions' positions.

Evidence of core material

(figs. 111, 112)

Currently, there is soil from burial in the interior of the elephant vessel's belly and in the trunk, but it was originally emptied of its loess core to allow it to function as a vessel. The legs retain their core material, as is often the case with elements like handles and feet in vessels that do not need to be emptied out for functional reasons. A core sample was drilled from the elephant's rear right leg to date the vessel using thermoluminescence analysis (see II.8§1). The dating test results placed it in the first half of the eleventh century BCE.

Radiographic evidence

(fig. 502)

X-radiography shows that the walls of the vessel are relatively thin and even. The more X-ray-opaque or whiter features, like the decorative elements, indicate areas of thicker metal. Compared to the belly and head, the slightly greater density of the legs and trunk is caused by the extant core material. Radiography also confirmed that the tops of the legs are closed off by the metal wall of the vessel's body.

Surprisingly, neither radiography nor visual examination revealed any evidence of metal spacers in the vessel, suggesting that it was loess core extensions that anchored the core in position in relation to the mold. One such extension would have connected the core and mold through the opening on the back of the elephant. Another may have been through the trunk. The cores of the legs would have extended out the openings at the bottom to attach to outer mold sections.

Decorative elements

(fig. 500)

The elaborate sunken and relief decorations across the surface of the *huo* were executed avoiding undercuts to facilitate piece molding. The fine loess soil used in China to make the models and molds was able to capture very sharp, clearly detailed decorative designs. Further refinements would often be added by carving directly into the mold's surface, as has been found on excavated mold remains from Anyang. In the absence of surviving mold sections from this vessel, it remains unclear whether the *huo*'s finer thunder pattern (*leiwen*) in low relief or the dragons and hooks in high relief were created in the model, the mold, or both. No refinements were made to the design after casting. The *leiwen* patterns, at least, show no signs of chasing in the metal.

Condition

Originally, the vessel's surface would have been a golden bronze color, perhaps with a pigmented inlay material in the recesses, as has been found on other pieces excavated in better condition. Its current appearance in the larger raised areas—mottled, pale gray, green, reddish brown, and very dark and shiny gray—is mostly the result of thousands of years of chemical action during burial in the ground. Despite this patina, all the details of the *huo*'s original design are maintained with little or no disturbance of the surface. What accounts for this, given that corrosion often distorts, expands, and/or destroys the surface, giving it a rough, bulging, craggy appearance?

X-ray fluorescence analysis (XRF; see II.5§2.1) identified the corrosion as having a high tin content, presumably a reflection of the alloy composition and burial environment. Tin oxide gives it a smooth appearance and is a much smaller molecule than the green copper carbonates. High-tin bronzes have relatively good resistance to corrosion in burial conditions, but at the same time are very brittle. That may explain the multiple chips on the higher ridges across the surface. Closer examination showed that the vessel had also been restored post-excavation: many edge losses had been filled and in-painted to make the surface more aesthetically appealing.

Summary of findings

(fig. 503)

Close examination of this elephant-shaped vessel provides irrefutable evidence that it was cast by the piece mold method rather than by the lost-wax process, and allowed for a precise reconstruction of the mold sections. The casting of this rare form shows great technical skill. The preservation of core material made it possible to date the sculptural vessel more accurately. The good preservation of the surface detail in spite of the *huo*'s lengthy burial is no doubt thanks to the nature of the alloy. There is not much evidence to suggest that the vessel was heavily reworked during restoration.

Synopsis of technical parameters

All of the studies referenced were undertaken by conservators using equipment at the Department of Conservation and Scientific Research of the Freer Gallery of Art and Arthur M. Sackler Gallery of Art in the Smithsonian Institution in Washington, DC.

An individual object examination is part of the day-to-day work of the department and no additional costs were incurred. The durations for each step were approximately as follows: photography and X-radiography, five hours; spectrographic analysis: eight hours; sampling for thermoluminescence testing: three hours; microscopic examination: three hours; interpretation of results: four hours.

- X-radiography: W. T. Chase, D. K. Strahan
- TL sampling: W. T. Chase
- TL testing: Research Laboratory for Archaeology at Oxford University
- photography: W. Y. Ng, D. K. Strahan

- microscopic examination: R. J. Gettens, N. Barnard, W. T. Chase, D. K. Strahan

Further questions

- How much was the surface reworked after casting?
- Could industrial computed tomography (CT) scanning provide clearer images of the intersections of the tops of the elephant's legs with the bottom of its body cavity and help clarify how these were constructed?
- If there were residues of ritual offerings on the interior of the belly, could analyses determine what the vessel contained?
- Can technical comparisons to other ancient bronze elephants recently excavated in China from provenanced sites provide further information about the Freer elephant?

Further resources

Gettens 1969

Pope et al. 1969, 228–31

Bagley 1987, 15–140

Yue and Yinxu Xiaomintun Archaeological Team 2008

Stoltman et al. 2018

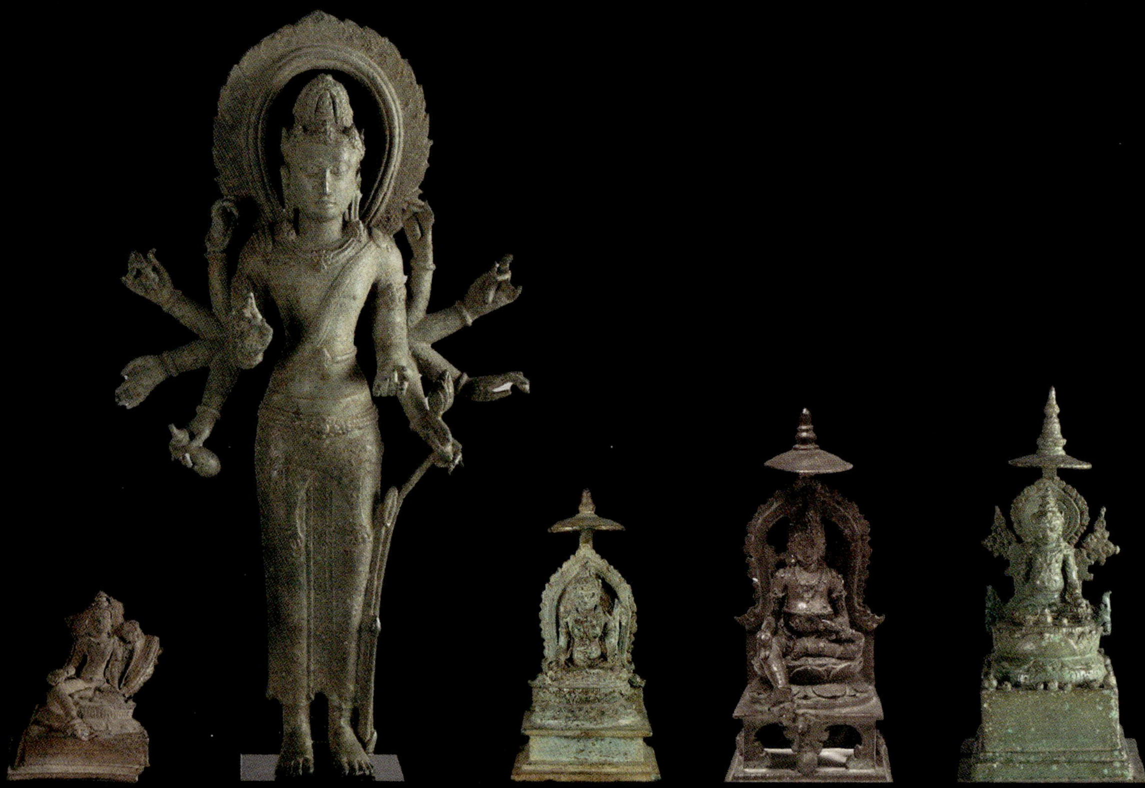

Figures 504–8 From left to right: Fig. 504, *Statuette of Bodhisattva*, style associated with Bangladesh but found in Java, 8th–early 9th century, H. 8.5 cm (Musée national des arts asiatiques – Guimet, Paris, inv. MA 507); Fig. 505, *Ten-Armed Avalokiteśvara*, Java, first half of the 9th century, H. 34 cm (Musée national des arts asiatiques – Guimet, Paris, inv. MG 3816); Fig. 506, *Statuette of Vasudhārā*, style associated with Bangladesh but found in Java, second half of the 9th century, H. 13 cm (Musée national des arts asiatiques – Guimet, Paris, inv. MG 2255); Fig. 507, *Kubera/Jambhala*, Java, ca. late 9th–early 10th century, H. 17 cm (Musée national des arts asiatiques – Guimet, Paris, inv. MG 3619); Fig. 508, *Kubera/Jambhala*, Java, first half of the 10th century, H. 18 cm (Musée national des arts asiatiques – Guimet, Paris, inv. MG 3625).

Case Study 4

The Technical Investigation of Eighth-, Ninth-, and Tenth-Century Statuettes from Indonesia

Mathilde Mechling
David Bourgarit

Introduction

(figs. 504, 505, 506, 507, 508)

The technical study of a group of thirty-nine statuettes in the collection of the Musée national des arts asiatiques – Guimet, Paris, was the first comprehensive examination undertaken of Hindu and Buddhist bronze statues from Indonesia. The wide array of visual and analytic methods utilized has provided a preliminary characterization of Indonesian bronze casting traditions. In addition, the study probed the possible exchange in technical know-how between the Indonesian islands and other regions of South and Southeast Asia, previously suggested on the grounds of stylistic connections. The research was part of a PhD project on Indonesian bronze sculpture combining traditional art historical research with technical studies.

The main questions

- Is there consistency in technical features such as the facture and alloy composition of Indonesian bronze statuary? If so, how distinctive are these features?
- Is it possible to refine the stylistic and chronological grouping of these objects by taking technical features into consideration?
- Given the stylistic connections found within bronze statuary across multiple regions of Asia, might there also be connections in their material and casting features?

Casting technique

Lost-wax casts with numerous wax assemblies

(figs. 67, 54)

Naked-eye observation of the statuettes provides evidence that they were made by the lost-wax casting technique.

This is evident in the general nature of the modeling, and in particular in the decorative elements on the outer surface, whose undercuts suggest direct modeling.

And close examination of the hollow interior surfaces provides incontrovertible evidence that the figures were affixed to their pedestals before casting and that the bases were constructed from separately assembled wax sections: the uneven ridges at the level of the lotus blossoms and base are wax-to-wax joints.

Solid-cast figures

(figs. 126, 510)

The X-radiographs confirm that the figurines, including the largest ones (30 cm tall) are mostly solid, with open pedestals.

Cold work

(figs. 276, 511)

Naked-eye and microscopic observation of the outer surfaces of many of the statuettes found crisply defined decorative details and textures that are characteristic of cold working of the metal. (Exactly how much detail was already present in the wax model is hard to say.)

Engraving, as suggested by reflectance transformation imaging (RTI)

(fig. 241)

Reflectance transformation imaging (RTI; see II.2§2.3) allowed for close documentation and study of the incised patterns. Under certain lighting conditions, the interrupted, stepped pattern along the interior walls of each incision was brought into relief. These are characteristic of engraving, formed by successive strikes on the graver as it is worked through the metal. The tapering ends of the incised lines are also consistent with engraving.

Metal punching confirmed by digital microscopy

(figs. 285, 286)

Close examination of punch marks (0.8 mm diameter) by digital microscopy (see II.2§2.1) revealed that the smooth, compressed surface inflected by the punch and the dimple of displaced metal around the edge resulted from cold working. In addition, on one statue, a 3D reconstruction of the dots and consequently of the active part of the punch showed a systematic defect in the sphere, clearly testifying to the use of the same tool for all the motifs investigated.

Metal engraving over preparatory drawing on wax confirmed by digital microscopy

(figs. 273, 512)

Digital microscopy was used to map the topography of various tool marks. The indented lines on several statuettes showed the V profiles and slightly raised edges (or burrs) along the tops characteristic of engraved rather than punch marks. In some instances, the presence of two lines very close to each other may testify to a preparatory drawing on the wax model.

Concealed offerings revealed

Cavities filled with dense material

(fig. 513)

X-radiography has revealed in a number of statuettes a cavity extending from the top of the base up into the legs and the abdomen of the deity. This cavity is filled with a denser material than the metal of the statuette, making that area more radio-opaque (and consequently whiter in the film) and effectively masking all structural details. X-ray fluorescence analysis (XRF; see II.5§2.1) of the embedding alloy on several statuettes revealed them to be tin-based with smaller amounts of lead (90–95wt% tin, 5–10wt% lead). Such low-melting-temperature alloys are very radio opaque (i.e., difficult for X-rays to penetrate).

Consecration deposits visible from underside

(figs. 54, 514)

In one of the statuettes, damage to the pedestal resulted in the partial loss of the metal fill. One can catch a glimpse of what remains embedded in the remaining metal matrix: several folded metal foils, the rim of a silver coin, a stone bead, and several more unidentified objects. These constitute a deposit to consecrate the statue before it could be used for worship. Such addition of a consecration deposit embedded in a metal matrix would have been a common feature within the group of statuettes studied, though today few of these objects preserve their interior contents. Might there be a way to learn more about these hidden contents without further disrupting them?

Consecration deposits revealed by neutron radiography and tomography

(figs. 401, 404, 403, 516)

Access to a neutron reactor allowed for the neutron radiography and tomography (see II.3§3) of several Javanese sculptures that still preserved their consecration materials embedded in metal matrices. This works because metals such as copper, tin, and lead may be X-ray opaque, but they are relatively transparent to neutrons.

The resulting images clarified the shape and location of the interior contents. Among other things, they enabled researchers to identify the typically Javanese *piloncito*-type cubic coins. Neutron radiography does not, however, allow for the precise characterization of materials that these consecration deposits were made of.

Alloy composition

Bulk metal

(fig. 517)

Inductively coupled plasma atomic emission spectrometry (ICP-AES; see II.5§3.1) analysis of the bulk metal composition of the bronzes revealed that most of the thirty-nine statues analyzed are made of unleaded bronze alloys—copper-tin alloys with less than 2–3wt% lead (Pb). Zinc (Zn) is present only in traces. With some variations, median values for each time period are higher than 8wt%. Nine statues have relatively high (12–20wt%) tin (Sn) content.

Impurities in the metal

(fig. 518)

ICP-AES allowed us to determine that the main impurities found in these alloys are silver (Ag), arsenic (As), iron (Fe), nickel (Ni), sulfur (S), and antimony (Sb).

Summary of findings

Consistent features of Indonesian production

(figs. 504, 505, 506, 507, 508)

In all of the Indonesian statues investigated, even the largest ones, the figures themselves are mostly solid. All were made via lost-wax casting, with extensive modeling and assembling work on the wax. All were originally likely consecrated with sacred objects, and therefore have a cavity in the supporting pedestal or cushion.

The systematic use of a liquefied metal, namely a tin-based alloy, to seal the consecration deposits constitutes another consistent feature of Indonesian production. Copper-based sheets hammered over the base are often encountered for Buddhist statues in other regions, notably in Tibet (Reedy 1991a) and China (figs. 225, 226).

Most statues are made of unleaded tin-bronze, with relatively high levels of tin. This is not very surprising given the vast and accessible resources of the Southeast Asian tin belt, running from central Burma down the Thai-Malay Peninsula to Bangka Island east of Sumatra, in the Indonesian archipelago.

Toward a chronological and stylistic serialization supported by metal composition

(figs. 518, 519)

Despite the common technical features apparent in the whole study group, the lowest levels of tin were found exclusively in the statues attributed to the first half of the ninth century. Tin content increased markedly in the statues attributed to the early tenth century, despite several exceptions to the rule. Some periods are marked by very specific and homogeneous alloy composition and impurity patterns. This is particularly true for two groups, namely the group attributed to the second half of the ninth century and the group attributed to the early tenth century.

Alloy connections between Java and mainland Southeast Asia

(figs. 509, 521, 522)

The high tin content may pertain to technical connections with contemporaneous casting traditions in mainland Southeast Asia, namely the Mon culture of Dvāravatī in central and northeast Thailand (notably the Prasat Hin Khao Plai Bat II bronzes in Buriram province). High-tin bronze alloys are also found in the Khmer bronze-casting tradition for objects from various periods (the first to the tenth century). Conversely, coeval bronze statuary from the Indian subcontinent shows much lower tin content, except some examples from Sri Lanka (Polonnaruwa) and northeast India (Nalanda).

Characterization of consecration deposits thanks to neutron tomography

(fig. 401)

Neutron tomography has proven a successful nondestructive technique to learn more about the consecration deposits sealed inside the statuettes. Such deposits are important archaeological artifacts and indicators of the ritualistic rules overlaying the creation and use of these religious images. Few of the known contemporaneous statues from India or other parts of Southeast Asia retain their consecration deposits, but there is ample evidence that this practice was widespread in Indonesia. Bronzes such as these also form a crucial link with the later and better-documented Tibetan tradition of consecrating Buddhist images.

Synopsis of technical parameters

This case study is based on a technical study carried out by Mathilde Mechling in collaboration with David Bourgarit and Brice Vincent as part of her PhD project on Indonesian bronze sculpture. The examinations and analyses were integrally carried out for free at the C2RMF as a service provided to French museums, in this case the Musée national des arts asiatiques - Guimet. Neutron tomography was also carried out for free at Reactor Orphée - Laboratoire Léon Brillouin, CEA-CNRS, Paris, as a service for public research. Main operating conditions are reported in Mechling et al. 2018. For more on the analytical techniques, please refer to volume II.

The approximate time required to carry out the whole study, excluding publication, was more than sixty days. The study involved:

- daylight and UV photography: Anne Maigret (C2RMF)
- digital microscopy: Mathilde Mechling (PhD candidate), David Bourgarit and Dominique Robcis (C2RMF)
- X-radiography: Elsa Lambert (C2RMF)
- neutron tomography: Frédéric Ott (CNRS-CEA)
- bulk metal analysis by ICP-AES: Nathalie Gandolfo and David Bourgarit (C2RMF)
- XRF: David Bourgarit (C2RMF)

Further questions

- Would comparative data from either similar objects or copper and tin ore deposits make it possible to determine the provenance of the metal?
- Is there physical evidence pointing conclusively to the use of either the direct or indirect process to form the wax model?
- Can the fabrication of the hosting cavity be further elucidated?
- Can it be determined whether the statuettes were cast horizontally or vertically? Since horizontal casting is known to have been utilized in South India, Sri Lanka, and Tibet, is it a feature that might help to determine intercultural exchanges?
- Can a comprehensive quantification of tool marks (depth and angle of the grooves) by digital microscopy yield interesting clues regarding know-how in certain workshops?
- Can the identification of items in consecration deposits serve as a means of tracing the regional exchange of materials, and if so, can it be done nondestructively, avoiding the deconsecration of the religious image?

Further resources

Bordenave and Mirebeau 2018

Fontein, Soekmono, and Suleiman 1971

Fontein, Soekmono, and Sedyawati 1990

Le Bonheur 1971

Lohuizen-de Leeuw 1984

Lunsingh Scheurleer and Klokke 1988

Mechling et al. 2018

Reedy 1991a

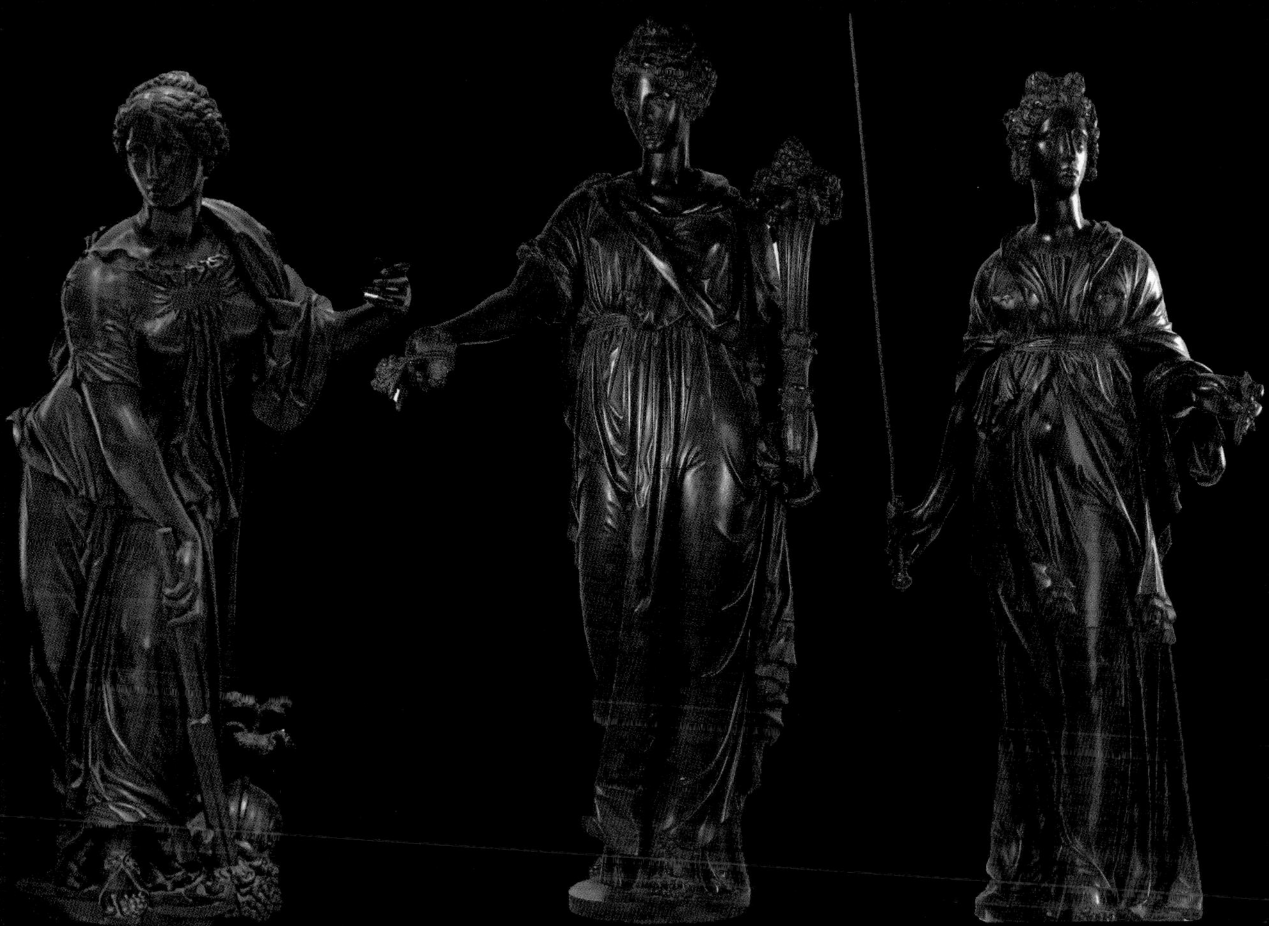

Figures 524, 34, and 97 From left to right: Fig. 524, *Peace*, cast in 1571 by Nicolas Péron, H. 128 cm (Musée du Louvre, inv. MR1683); Fig. 34, *Abundance*, cast in 1571 by Nicolas Péron, H. 128 cm (Musée du Louvre, inv. MR1681); Fig. 97, *Justice*, 1571, H. 128 cm (Musée du Louvre, inv. MR1682).

Case Study 5

One Commission, Two Sculptors, How Many Founders? A Technical Study of Three Renaissance Statues from the Monument of the Heart of Duke Anne de Montmorency

Manon Castelle
David Bourgarit

Introduction

(fig. 523)

Commissioned in 1571, the Monument of the Heart of Duke Anne de Montmorency (French, 1493–1567), Constable of France, includes three bronze statues. These were made by two sculptors, Barthélémy Prieur (French, 1536–1611) and Martin Lefort (French, active 1562–82). In the archives, only one founder, Nicolas Péron, is associated with the casting of two of the statues. A technical investigation of the statues, including metal and core analysis and visual and radiographic examinations, was undertaken to discern whether Péron might have cast all three and to shed light on the relationship between the sculptors and the founder(s), which was crucial in this period of Renaissance bronze statuary.

The commission

In 1571, Madeleine of Savoy, Anne de Montmorency's wife, commissioned the architect Jean Bullant (French, 1515–1578) to produce a funerary monument for the heart of her husband. The monument, now in the Louvre, is composed of a sculpted stone column that would originally have been topped by an urn containing the heart. Bronze allegorical personifications of Abundance, Peace, and Justice were situated around the base of the column as witnesses to the character of the deceased. Bullant charged two sculptors to produce the statues. Prieur was responsible for Abundance and Peace, and Lefort made Justice. Péron was commissioned to cast the statues by Prieur. There is no record of who cast Lefort's.

The main questions

(figs. 34, 97, 524)

- Considering the number of people who collaborated in the production of such sculptures (and there were no doubt more unnamed hands involved as well), can the physical evidence help clarify who was ultimately in charge of the production?
- Does the technical evidence suggest that the same founder or different ones worked with the two sculptors to cast the three sculptures?
- Can comparison of the casting processes used for each of the statues help answer these questions?
- What can the characterization and comparison of the materials used tell us with regard to these questions?

Visual evidence of lost-wax casting

(figs. 8, 51, 70)

Clear traces of lost-wax casting can be found in the modeling of details such as the vine leaf and cornucopia held by Abundance. They reproduce the malleable quality of the original model's material.

It is also clear from an examination of the inside of the sculptures that all three are hollow casts with relatively thick, uneven metal walls, still partially filled with the baked, sandy clay cores that were built around a system of iron armature rods, as seen here in the interior of Peace.

Internal evidence of unique chaplets

(fig. 38)

The location of core supports is nearly invisible on the radiographs. Endoscopic examination of the interior metal walls of Justice and Peace where the core had been removed revealed a rather unique feature in both: the raised outline of what looks like a rectangular bronze patch with an integral faceted tapering point extending out from its center. No such chaplets were found in any other large contemporary Parisian bronzes. That these features are idiosyncratic to these two casts is a strong indication of their production in the same foundry. Whether the Abundance has them can't be verified as the statue is still largely filled with core.

A direct lost-wax cast?

(figs. 13, 34, 97, 524)

Radiography of the statues showed the relatively thick metal walls and the shape of the preserved vestiges of the original armatures. Few wax-to-wax joints are visible, and the internal surfaces do not conform to external contours. The latter features further support the hypothesis that the sculptures were directly cast.

Armature system

(figs. 34, 97, 524)

The armature systems were mapped out based on the radiographs for each sculpture, to the extent possible. The gauge of iron rods seems similar in all three. Both of Prieur's figures show that the core of the protruding arms was bound with wires. The Lefort figure does not have these, probably because the core was almost entirely removed.

Conflicting archival evidence

(fig. 23)

It was a surprise to find out from the surviving contract between Prieur and Peron that the founder was to produce the wax model from a plaster piece mold that the sculptor had made of his original model. All of the evidence forced us to reassess our interpretation and points to some version of the lasagna technique described by Benvenuto Cellini, the Florentine goldsmith and sculptor who was active at the French court around this time (Italian, 1500–1571; see GI§2.3.2).

A characteristic core composition?

(figs. 23, 274, 405, 406, 427, 428, 525, 526, 527)

Some core could be sampled from underneath each statue. A new methodology based on the precise measurement of the size of the quartz inclusions showed that the cores of the three allegories were not only the same, but very specific with respect to sixteenth- and seventeenth-century Parisian bronzes analyzed to date.

Indeed, whereas in most sixteenth- and seventeenth-century Parisian bronzes, two sizes of inclusions coexist (very small inclusions of around 5 µm in diameter and much larger ones, up to 150 µm in diameter), in the allegories only large inclusions are present (100–200 µm).

The alloys

(fig. 528)

Inductively coupled plasma atomic emission spectroscopy analysis (ICP-AES; see II.5§3.1) of a sample drilled from each statue revealed that Prieur's Abundance and Peace are made of tin-bronze, whereas Lefort's Justice is made of brass. Moreover, the very different impurity patterns suggest disparate copper ore sources. If Péron cast all three sculptures, he would have been responsible for preparing and pouring the metal. How then to explain the variation in the alloy types?

Summary of findings

A unique founder?

(figs. 034, 097, 524)

The three figures were cast by a similar method using idiosyncratic chaplets and a similar core composition, which strongly suggests that they were cast by the same founder. The variation in metal composition would suggest, however, that a different founder may have been involved. All of Prieur's large bronzes that were cast in France are made of tin-bronze. It is possible that he imposed his choice of alloy on the founder. Conversely, Lefort was of less renown, with no documented previous experience in casting, and so he might have left it up to Péron to work with his preferred alloy, which we assume was brass (this of course is only a hypothesis; we unfortunately have no record of other productions by Péron).

In the end, the sculptors' different backgrounds and status led us to speculate that the choice of bronze alloy adopted for Prieur's two figures may have been linked to the artist's exposure to practices he encountered in Italy, and that the founder Péron was willing and able to successfully execute the casts for the monument in both alloys.

Synopsis of technical parameters

The study was carried out at the C2RMF as a public service to French museums, and the procedures undertaken by staff unless indicated. Its total duration was more than one hundred person-days, as part of Manon Castelle's PhD research. The study consisted of:

- visual examination: Manon Castelle, David Bourgarit, and Francesca G. Bewer (Harvard Art Museums). This was one of the most time-consuming steps in the study.
- daylight photography: Dominique Bagault and Anne Maigret. Of the circa 100 photos taken per statue, about half were selected.
- X-radiography: Thierry Borel, Jean Marsac, and Elsa Lambert. The particularly large thickness of the metal wall and the core filling created a very dense barrier that pushed the X-ray facilities of the C2RMF to their limit.
- endoscopy: Manon Castelle.
- bulk metal analyses by atomic emission spectrometry (ICP-AES): Nathalie Gandolfo, Jessica Legendre, Benoit Mille, and David Bourgarit.
- core analysis: Manon Castelle and Yvan Coquinot. In the framework of Castelle's PhD work, this study offered the opportunity to develop a new protocol detailed in II.7§2.2.4.

Further questions

- What was the original patina of the statues? Was it the same for all three?
- What was Péron's training and production prior to the casting of the allegories? Such information might provide insight into his alloy preference and approach to casting sculpture.

Further resources

Bewer, Bourgarit, and Bassett 2009

Bresc-Bautier and Scherf 2008

Castelle 2016

Castelle, Bourgarit, and Bewer 2018

Castelle, Coquinot, and Bourgarit 2016

Castelle et al. 2021

Grodecki 1986, 134, 135

Seelig-Teuwen, Bourgarit, and Bewer 2014

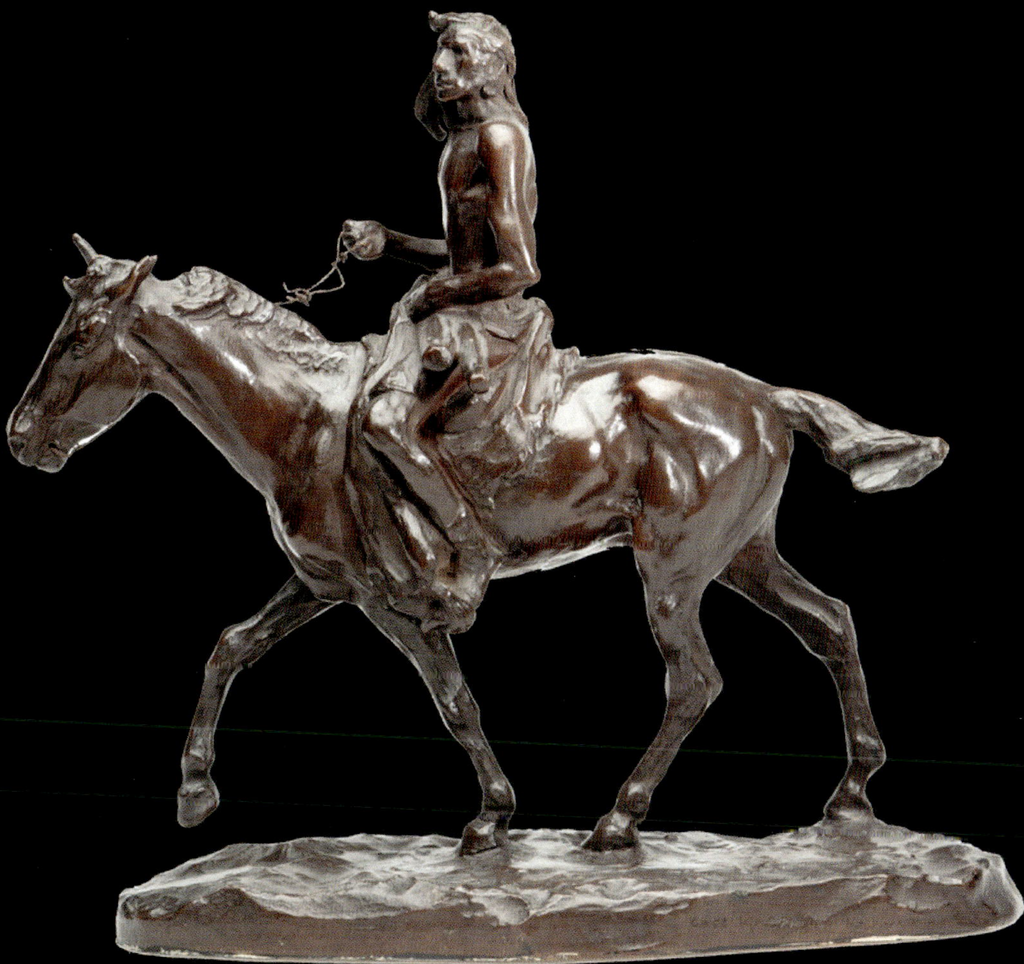

Figure 529 Charles Marion Russell (American, 1864–1926), *Medicine Whip*, modeled 1911, sand cast 1912–16, H. 14.8 cm (Gilcrease Museum, Tulsa, 0837.14).

Case Study 6

Medicine Whip (Indian on Horseback): The Making of an American Sand-Cast Bronze through Contemporary (and Modern-Day) Eyes

Ann Boulton

Introduction

(fig. 529)

In preparation for the 2014 exhibition *Frontier to Foundry* at the Gilcrease Museum in Tulsa, Oklahoma, research revealed that *Medicine Whip*, a sculpture in the museum's collection, had been the subject of an unusually well-illustrated technical article in a contemporary industrial trade magazine in 1912. Such documentation is quite rare and offers a unique opportunity both for a detailed comparison of the actual bronze with the written description and for consideration of the technology transfer of a nineteenth-century French technique to the early twentieth-century United States.

Context

The magazine article

(fig. 530)

Sometime in 1912, E. A. Suverkrop, a reporter for *American Machinist*, toured the art bronze foundry of Auguste Griffoul & Bros. Co. in Newark, New Jersey. The article that ensued, titled "The Molding of Bronze Statuary," is an extraordinary document that provides a step-by-step, photo-illustrated, detailed explanation of the French sand-molding process that had been brought to an extremely high degree of perfection for the casting of sculpture in nineteenth-century Paris. Indeed, a few years prior to moving to the United States in 1899, Griffoul had operated a foundry in Paris, where he cast *The Thinker* for Auguste Rodin (French, 1840–1917). By 1905 he had established the one in Newark, which is reported to have been the last foundry in the United States using French sand. By the date of this article, the lost-wax casting process was overtaking sand casting as the preferred method for elite sculptors, and founders of art bronze by sand casting were on the defensive and eager to advertise their careful surface-replication methods.

The sculpture

(fig. 529)

The sculpture in the *American Machinist* article is identified only as "Indian on Horseback" and the base of the Gilcrease cast is marked "Cast by Griffoul Newark, NJ" and by the artist's signature. The extensive photographic documentation makes it possible to identify the sculpture as *Medicine Whip* by the artist Charles Russell (American, 1864–1926), who modeled it in 1911. It has been estimated that fewer than ten casts were made of *Medicine Whip*, three of them by Griffoul, although the article makes clear that a bronze model or chef-modèle used in the process depicted in the article had been created in anticipation of a larger production, as was common practice in France.

Main questions motivating the technical study

(fig. 531)

- Was the equestrian sculpture indeed cast as indicated in the article?
- Is this the actual cast illustrated in *American Machinist*?
- How does this process compare with evidence of other French sand casts and of later lost-wax casts as that process supported artistic sand casting?

Diagram of sand casting (based on the Suverkrop article)

(fig. 9)

Sand molding is an exacting, complex, highly skilled endeavor. It relies on sand from specific deposits in France—fine grained and naturally mixed with clay that becomes cohesive when tightly packed. Because a densely packed sand mold is rigid, it has to be made in as many pieces as necessary to account for undercuts in the model so that the pieces might be disassembled and reassembled without breaking in preparation for casting. It is destroyed to remove the cast; the sand mold has to be made anew for each cast. The mold is created within multiple, stacked, sand-filled iron frames called flasks. The process that Suverkrop witnessed might be summarized as follows.

1. The process requires a specially designed solid model for sand casting that is hard and can withstand the repeated handling and ramming of sand. This is often made of metal, and is called a chef-modèle.
2. The first flask, or "cope," is filled loosely with sand, into which the chef-modèle—or "pattern"—is buried halfway. This pattern and the leveled surrounding sand are then coated with a fine powdery separating agent that defines the "parting line." Smaller mold pieces are created one by one around the pattern, accounting for undercuts, making sure that they will separate and slot together again precisely. "Keys" or recesses scooped into the sand along the way are molded by the subsequent section and ensure their perfect registration. The entire exposed part of the pattern is molded thus, by ramming the sand with mallets onto each carefully defined area of the model, thereby imprinting it in all its detail in the sand.
3. The many smaller mold pieces are eventually backfilled with more sand that is tightly rammed within the upper metal frame, or "drag," which slots onto the cope and is affixed to it.
4. When the molding of that initial exposed side is completed, the cope and drag are flipped. The cope, now on top, is removed to release the sand and clear access to the back side of the model. A similar piece molding process now takes place until the cope is filled with rammed sand. (In the case of this sculpture, a third flask was needed.)
5. Once all the rammed sand mold pieces have been created, the flasks are parted and the mold pieces carefully disassembled to remove the pattern.
6. In order to produce a hollow cast, it is necessary to create an internal mold or core, which will essentially define the thickness of the bronze walls. To make a core, the mold pieces are reassembled in the cope and the drag respectively. Sand is carefully rammed into the hollow impression left by the pattern. Wires or internal core supports—some in the form of loosely rolled tubes or "core vents"—are laid across the new sand, extending out into the surrounding mold pieces. The cope and drag are joined, uniting the two halves of the sand replica, then disassembled again to remove the sculpture.

7. The sand replica is shaved down evenly overall to form the core.

8. The mold pieces and core are baked, coated with a fine layer of soot to ensure a smooth flow of the metal, and reassembled in the cope and drag. The gap formed by the paring down of the core determines the thickness of the sculpture's bronze walls. The embedded internal core supports projecting from the core will serve to suspend it in place in the mold during casting. Channels are cut strategically into the mold to ensure that the metal flows through and fills it efficiently.

9. The cope and drag are reassembled and locked together. The metal alloy, which has in the meantime been liquefied, is poured into the mold.

10. Once the metal has solidified and been allowed to cool, the sand mold is broken away to reveal the bronze sculpture, which will be one with the network of sprues formed by the metal-filled channels. The sprues are cut off. The surface is cleaned and repaired as needed. Holes in the bronze created by the core supports are plugged.

11. If a bronze is cast in numerous sections, or separately from a base, it will be joined at this stage. The final steps consist in little, if any, chasing and decorative coloration through the application of a chemical patina or some form of coating. (In this sculpture, the reins were made of metal wire and added at the end.)

Visual examination

Evidence of the original modeling

(figs. 532, 533, 534, 535)

The bronze cast reproduces the loose modeling of the artist's original design, which, judging from the quality of the details and what we know of the surviving models of other Charles Russell sculptures, would probably also have been fashioned in wax around a wire armature embedded in a plaster or wooden base. The crisp-edged, rounded facets and soft, indented outlines on the mane, for instance, suggest that the artist used a fine, round-tipped modeling tool. And the faint parallel striations on the horse's belly behind the rider's left heel may be traces of a fingerprint drawn across the surface or of a serrated tool. A partial fingerprint translated from the original model of the base is discernible above the artist's signature, which was drawn with a pointed tool in a soft material—probably wax. Both the finer inscription that reads "copy v. 1911" and the foundry information marked on the base are in the same hand, which is different from that of the artist. They too appear drawn in a malleable material, rather than engraved or stamped into metal, as was more often the case with casts of that period. All of these marks must have been already present in the original model before the metal pattern was made. The subtle rows of rounded, shallow indentations between the artist's initials and the sketch of the horned skull may have been made in the wax, or could represent punch marks done in the metal during the finishing process.

Evidence of the casting process

(figs. 536, 537, 539, 540, 541)

More than fifty pieces were required, with a total construction time of fifteen hours, to create the sand mold for one cast of *Medicine Whip,* a small sculpture less than ten inches tall, and yet few if any traces of the mold seams are apparent. When properly prepared, a French-style sand mold should produce a cast that closely replicates the artist's original model and needs little further repair prior to patination. The sand evenly vents gases to prevent air holes from forming in the metal; additionally, the swelling of perfectly rammed sand pieces prevents the formation of a network of raised metal seam lines on the surface of the sculpture.

Discrete areas display quite visible tool marks made after casting. Dense clusters of punch marks mar the modeling at the top of the Indian's head and back, as well as the rear of the horse top of the tail and rump. These marks correspond to the exact location of the large sprue attachments illustrated in *American Machinist*.

X-radiography

(figs. 195, 542)

Radiography helped to explain the punch marks on the head of the Indian and the forehead and rear of the horse. They mark the locations of bronze plugs that were hammered into the holes left by the internal core vent/supports that were cut down at the surface of the cast and pushed into the sculpture. The radiograph further confirms much of what the *American Machinist* photographs show: the Indian and horse were molded and cast as one; the legs of the horse did not contain any core, indicating that they would become solid metal.

It also shows the square nuts attached to the screws that the *American Machinist* photographs depict placed into the sand mold at the horse's extremities and protruding from the horse's hooves were to serve as attachments to a base. (Visual examination of the interior of the base of the bronze cast is precluded by a sheet metal cover soldered to the bottom of the base, probably by a collector.)

Radiography also reveals that the core vents or "lanterns," made of a rolled-up sheet of metal bound with fine metal wire to prevent it from unfurling, are not one, as shown in the photographs (fig. 542), but two: one traversing the length of the horse and the other coming down from the head of the rider into his mount. The radiograph clearly shows how the plugs align with the vents on the animal's forehead (rather than the chest, as seen in *American Machinist*) and rump, as well as at the top of the rider's head. Although not visible, one may also figure at the other end of the vent, in the belly of the horse. This is clear evidence that the Gilcrease bronze is not the one illustrated in Suverkrop's article.

Summary and discussion of findings

(fig. 529)

Comparison of the Gilcrease's *Medicine Whip* with the detailed photographic documentation of French sand casting in *American Machinist* indicates that while the museum's version must clearly be part of the series cast by Griffoul, it is not the actual bronze portrayed in the magazine.

The article also makes painfully clear how little trace is left on the bronze itself of the skilled mold making that went into its production, and how speculative our technical interpretations often are. Suverkrop's fairly detailed record of the process he witnessed in the Griffoul foundry offers a precious window into know-how that would otherwise be irretrievably lost.

Suverkrop's vocabulary is that of industrial sand casting of machine parts, betraying his own expertise and that of his readership. His substitution of the industrial term "false core" for the more familiar term "piece mold" element, for example, blurs the distinction between the core and the exterior mold. The dissonance underlines the gulf between arts-bronze casting and machine-parts casting—one that still exists today. In that respect, even with this detailed documentation, we lack a precise description of the process chez Griffoul.

Synopsis of technical parameters

The sculpture was examined in normal and raking light with low-power magnification.

Photography of relevant features was done in raking and normal light. Gamma radiography was performed at the Gilcrease by Tulsa Gamma Ray with a portable source due to lack of radiography capability at the museum.

Further questions

(fig. 529, 543)

- What accounts for the different layout of the core vents in this cast versus the one illustrated in the article? Could it have been experimentation? The result of different "hands" working on the molds in the Griffoul foundry? Or a hiatus in production?
- Might the Gilcrease cast actually be the bronze chef-modèle depicted in the article around which the mold was created? Could comparison of the Gilcrease bronze to another Griffoul cast clarify this?

Further resources

Boulton 2006

Boulton 2007

Boulton 2018

Lebon 2003

Stewart 1994, 192

Suverkrop 1912

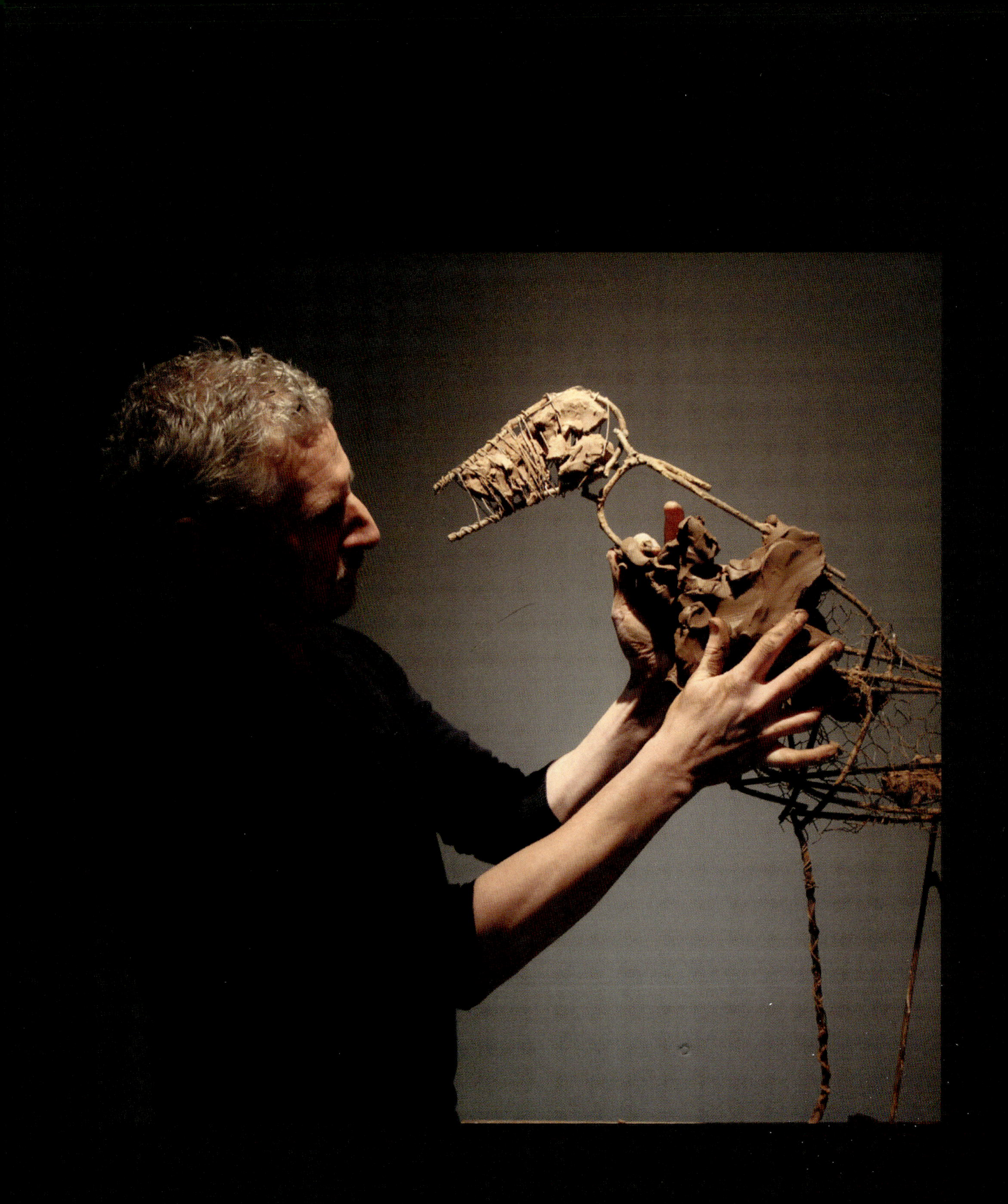

Figure 33 (detail) Detail of fig. 33, Andrew Lacey (British, b. 1969) beginning the modeling process on a metal armature for *The Anatomy of Bronze*, cast by the artist in 2019, Devon, UK, H. 45 cm (artist's collection).

Case Study 7

Interview | A Contemporary Artist-Founder on a Horse: Choices of Materials and Processes

Andrew Lacey
David Bourgarit

Introduction

This case study takes the form of an interview with contemporary British sculptor-founder Andrew Lacey. Archaeometallurgist David Bourgarit, who has been intimately involved in the technological study of bronze sculptures from different periods and cultures over many years, engages Lacey on his process in creating *The Anatomy of Bronze* from its initial modeling to the application of the patina.

The objective is to explore topics that specialists and nonspecialists may wish they could ask the artist or founder directly when studying a bronze sculpture, such as choice versus compromise, or economy and expediency over aesthetic exploration. The recurring question about how a particular process or choice might manifest in the finished cast probes for explicit descriptions of the kind of evidence these might leave on the cast. The conversation relates to the entire, large version of the figurative sculpture of a horse, but for the purposes of this exercise, the illustrations focus solely on the head and neck. The questions about 3D scanning and printing relate to a separate, reduced-size version of the sculpture that was made during an experimental stage of the entire process.

Andrew Lacey is a sculptor and founder currently living in the UK. He first gained metalworking knowledge through working in London art foundries. He later set up a number of large studios and foundries dedicated to his own sculptural practice, and studied archaeometallurgy at the Institute of Archaeology, University College London. He now carries out research on historical casting practices, often interweaving this historic knowledge into the production of his own sculptural output. It will become clear that while he follows conventional casting methods for some of a work, at times he radically subverts or alters the process to create a material aesthetic that better represents his thinking and ethos.

1. The modeling: A drawn-out experimental process

What are the overall steps you followed to design the model of the horse?

(video 3)

The modeling was a long and frustrating process, as the desired finish needed to be very refined and lifelike. I had already begun modeling the horse in clay and played around with its position but found that, because I was not happy with the way the horse was standing or its overall form, I had to create another version that I could manipulate more easily. I decided to make a rough two-piece silicone rubber mold of the clay model with which to cast a wax model of the horse. This allowed me to try out new compositions, for instance in the position of the legs.

Once I had created the wax model I liked, I had it laser scanned to preserve the greatest amount of information in case the process of making a second, more complicated silicone mold should damage the model.

How often do you use digital scanning?

(fig. 376)

Scanning in this way is very new for me, although I had read up on the subject and was fully aware of its progress in the wider industry. It is a common practice today, almost to the point of superseding other, more conventional molding methods. It's a clean, quick, nondestructive way of capturing a 3D form. All art schools can now afford it.

Were you satisfied with the scanning results?

(fig. 544)

The high-resolution scanning was carried out on-site by a specialist from Falmouth University. Acquiring the scans was simple, if time consuming, but it was immediately clear when viewing the initial data that the processing of the computer model would require some significant human interpretation due to shadows from the laser or lack of data where the laser cannot reach, and discrepancies in the overlapping of the layers of scans.

Also, as fine as the laser scan may be, the printed resin version couldn't fully replicate the fine detail of the original model. We reduced the size of the digital model so as to print a horse standing fourteen centimeters high. This resulted in a compression of the details and thus a superficially tighter shape that lacked the vitality of the original model and felt very artificial. I would still have had to rework the resin print in order to bring some life back to the modeling. It was an interesting experiment, but somewhat disappointing. In the end, it would have been better to model the smaller version by hand.

I therefore molded the wax model of the horse again with silicone rubber, but this time using greater care and consideration. I was trained to use rubber for molding sculpture early on in my career, and learned how to adapt or cheat because rubber has always been expensive and sometimes time pressures demand quicker methods. I molded the bulk of the horse in rubber. For the tail, which I had cut off of the model for the molding process to get better access to the legs, I used a plaster piece mold. I had shaped it only roughly since I planned to rework the modeling in the wax, so it did not have much complex detail. The rubber mold was designed in smaller sections that corresponded with the areas to be cast separately first in wax and then in metal—that is, the head and neck, each of the legs, and the body (video 2).

How often do you break down the model into smaller sections, and what would other foundries do?

This is something I do quite often, as it allows me to de-mold the head and neck and start casting before I have finished the rest of the mold for the horse. Again, it's a cheat; most foundries would complete the whole mold for fear of losing structural or dimensional stability, but since I'm working on something I already intimately understand, I can allow for alterations as I make the rest of the bronze.

What evidence might we find of this sectioning in the final bronze?

(fig. 545)

This choice of making the mold in small discrete sections will be invisible from the outside, as I can fix whatever evidence there is. However, thin lines can be seen between each on the inside where the sections have been welded together. The pattern of lines in the X-ray image will describe the way I planned the mold and therefore how I chose to overcome future difficulties in the casting process. Additional lines may be found where wax sections needed to be further cut down, which we see in the joining of the head and neck of the horse.

2. Methods used to fill the mold with wax: slush molding and painting

Turning specifically to wax, what process(es) did you use to produce the wax inter-model?

(fig. 21)

Making the inter-model out of wax sections is generally easy, but the technique changes from mold to mold depending on its design. For example, the head and neck of the horse were made in a two-part mold split in half along its length that forms an open vessel shape. It is therefore easy to pour in molten wax and slush it around the inside of the mold (video 2). Such a slush molding process was also used for the legs and tail. But the body was molded in two large halves divided lengthwise along the horse's back and belly and designed so that each half would be cast separately. Here I painted wax into the open half mold because that was easier in this case.

How often do you mix techniques for the same model, and how would other modern foundries do it?

It is quite common for me to use different methods of production for one object, as each specific part of the sculpture presents numerous opportunities for exploration. Other modern foundries would do so as well. Looking at how other foundries' techniques differ helped me to develop a broad range of techniques for myself.

What evidence might we find in the final bronze of these different ways of building up the wax?
(fig. 88)

These two techniques are clearly visible either by looking inside the bronze, if one has access to it, or by radiography. The short and busy patterns of the brush marks are quite distinct from the soft, smooth patterns and surfaces of the slush-molding method, which are often punctuated by tide marks.

3. Homemade recipes for casting wax

What kind(s) of wax do you work with?

I use beeswax with an addition of pine resin and earth pigments. I used to use an industrial wax made of petrochemicals and artificial fillers, which worked very well, but it was expensive. Generally, different waxes are designed to perform different tasks—for example to be malleable when modeling, or to cover sharp ridges when pouring into a mold. Over time and through experience and experimentation I began to design my own wax recipes to produce the physical characteristics I wanted. I use the same basic components but alter the ratio.

Does evidence of the kind of wax you used show up in the final bronze?

When the investment mold is eventually broken off the cast, occasionally you can see some very interesting areas of color on the surface that was in contact with the wax. These are caused by the inorganic earth pigments in the wax, which are deposited on the ceramic shell during the de-wax stage (video 11). Some of this coloration might just stain the core, and it is possible that future conservators could retrieve this material. Unfortunately, as all the wax is melted out and fully burned out during de-waxing, all of the evidence pertaining to its organic chemistry disappears with it.

4. Separately cast sections rather than cast in one shot

Why do you cast in separate sections, and how do you choose each section?
(fig. 546)

Originally, I was taught to bring all the wax sections together so that they could be cast as one object. This has some benefits: when a wax is made in this way, you can immediately see if there is anything wrong. For instance, is the shape or geometry of the body correct? Are the textures and modeled areas of the surface continuous and harmonious? My practice has developed in a different direction.

On the one hand, I know my own sculptures intimately—the surfaces, the movement, the energy, et cetera. Therefore, there is no need for me to put the whole structure together in the wax, as it is already understood. On the other hand, I work by myself and there is a limit to what I can physically handle. There are related limiting factors. For instance, the crucible I use holds sixty kilograms, so there is no point in making a wax section that would require more metal (wax is one tenth the weight of bronze). The kiln is front loading, so when setting out the pour (organizing the layout of the casting equipment), I find it difficult to lift a red-hot mold that weighs more than I can comfortably carry—maximum thirty kilograms. On reflection, the entirety of the studio is designed around these limitations, which in turn directly influence how I make sculpture. The surprising aspect is that I do not limit the scale of the final bronze sculpture. If I'm working on a very large sculpture, I simply break the form into pieces I can deal with and rebuild it in bronze. I just have to be confident of my engineering skills.

How do you choose where to section up the wax to create what we might call a casting plan?

Just to be clear, "casting plan," in my mind, relates to the organization of the sections to be cast—that is, how the model is divided. This could just mean that the top of the head is removed or that the whole model is cut into smaller sections. It might also relate to the organization of the sprues in terms of designing the direction and flow of the metal.

Prior to molding, I study the sculpture for some time. You have to imagine that you've already made the sections, then take them apart and reassemble them in your mind, seeing what the issues would be, then try a new configuration to resolve each problem in turn. By using my imagination, I can explore numerous possible dissections until I have what I consider the best casting plan for the sculpture at hand.

My reasoning for making specific dissections is threefold: namely the "economic line," the "assembly line," and the "aesthetic line." The "economic line" is the shortest line or dissection needed to remove one section. This may mean that it is better to move the line slightly higher or lower in order to shorten the length of the cut. The length of the cut will be proportional to the length of the weld. The "assembly line" considers what happens if a line goes through difficult-to-reach areas. In considering how the sections will go back together, I assess whether I can physically get the welding torch and filler rod in at the right angles to make a continuous joint. Lastly, the "aesthetic line." When making a dissection line I always avoid areas that are detailed, complex, and in frontal zones. This means, for instance, never making a line that crosses facial details or follows the direction of the modeling, so that the eye is not disturbed. If a joint must cut across the torso, which is inevitable, utilize the rib cage or muscle structure to disguise the weld seam.

What further evidence of such sectioning may show up in the bronze?

(fig. 545)

Most of the evidence will be found on the inside of the bronze, since the outer surface is worked over to disguise the seams. After many years, the bronze may corrode differentially in these areas, and the joints may become visible. Otherwise, as mentioned earlier, one can detect joints in X-rays. Thin gaps or thickening from joints between the various sections will tend to indicate how I decided to break the model down for casting. The brush marks are visibly different from the slushed sections of wax, and often the direction or orientation of these marks will be strikingly different. If you compare the inside to the modeled texture or form of the outside, you'll start to see what issues I'm trying to resolve and in turn understand something of my thinking process.

5. A sprueing system adapted to each section

What does the sprueing system look like? Do you prefer "top fed" or "bottom fed"?

(video 2)

In casting up the various wax sections, I used two very different methods of sprueing, "top fed" and "bottom fed." This is simply because some sections present better opportunities for one method over another. For example, the head, legs, and tail can be considered irregular cylinders. By pouring a plaster core into these and placing feeds (the sprues used to direct the bronze into the mold) at the top, they make for very straightforward castings. The panels of the body, on the other hand, are sprued up while still in the mold so as not to cause any distortion in the geometry of the wax. They can be sprued up from the top, but this tends to cause overheating and excessive porosity at that location. By bottom feeding them, the flow of metal is less turbulent and the heat is dispersed throughout the mold.

Is this common practice among founders?

I think most founders take such an adaptable approach when presented with different forms to cast. And really, that's only logical. Why enforce unnecessary complexities onto an already difficult task?

How does evidence of the sprue system show up in the final bronze?

(fig. 88)

In the interior of this bronze sculpture you should see some evidence of the different sprueing choices, but reconstructing the way a bronze has been sprued based on remaining sprue stubs, residual flow patterns of the dross, or oxidized metal in the cast surface, et cetera, would be difficult at best. And of course, if I had chosen to place the sprues on the outer, modeled surface of the wax, then no evidence would be left after I reworked the locations of the cut-off sprues to blend in with the surrounding areas.

6. Core supports

What technique(s) do you use to keep a core in position?

(fig. 547)

I have developed the habit of using carpet tacks as core supports. I used to use stainless steel rods, but they are expensive and can be difficult to get hold of. Carpet tacks, on the other hand, are very cheap and sold at any hardware store. The design of the tack is also perfect for the task. They are thin with sharp points, perfect for piercing the wax, and have flat heads, which assists in their removal after casting.

How do you choose where to place the pins?

I tend to arrange them in the wax with a degree of symmetry. This means that if any of the core pins get broken off during the breaking open and removal (de-molding) of the mold I should know roughly where they are. I also place them in areas that are easy to chase in the bronze, such as higher, rounded parts of the form. I tend to avoid putting core pins near faces or front-facing areas as these require more chasing later, or in hollows and crevices that only superficially conceal their presence.

For larger and more open wax sections I allow the ceramic shell to coat the outside and inside of the mold continuously. This means that the core is ceramic, too, and is locked in place by its full contact with the edge of the outer mold, and then a smaller number of core pins are needed.

How does evidence of the core pins show up in the final bronze?

(fig. 85)

Externally, it's very difficult to observe any evidence of core pins, as they are either welded over and chased clean, or excess material around the hole is hammered in and chased. On rare occasions I have lost track of a broken core pin that remained firmly locked into the surface of the bronze and showed itself by staining the patina red with rust; obviously this had to be removed. The interior of the bronze will clearly reveal either holes where the core pins have been removed or the tips of pins that have broken off and remain lodged in place (video 11).

7. Different core materials depending on the location in the statue

What material(s) do you use as core?

(fig. 2)

I was taught very early in my career how to make cores from a simple mixture of plaster and sand. Over time, experience has taught me that almost any inert grain or powder will work just as well as sand. This opens up many opportunities for experimentation, but also frees me from needing to buy prohibitively expensive commercial core materials. Also, no matter where I find myself in the world, I can now always find materials that will work for this purpose. Currently, for small bronzes I have used clay, which is the easiest material to find globally. I mix clay with inert powders and a tiny amount of plaster.

In the case of the horse head, I chose to make the core out of plaster for the nose up to the level of the eyes, as this shape is very difficult to get air-dried, and ceramic shell for the more open area of the neck, as from this point air could easily be blown into the hollow form. Using both methods of coring ensured that the combined core would be perfectly dry and also as lightweight as possible for handling and firing.

Does the core leave traces on the finished cast?

(fig. 118)

I usually remove as much core as possible, but often some small fragments of material may be left inside. In the case of the double core method, a thin line or minimal flashing would be visible inside the bronze where the two cores meet (video 11).

8. The investment: key for a good-quality and clean cast surface

What kind of investment do you usually use, and how do you prepare it?

(fig. 7)

The first studio I worked in used ceramic shell, albeit in an ad hoc way. Rather than having large tanks of liquid ceramic shell to dip the sprued wax models into, we would make up a tiny bowl of material that we painted onto the wax, which would then be coated in a fine fused ceramic grain. This sequence was repeated several times to build up the desired thickness of the ceramic shell. Small models were easy to cover that way, whereas large sculptures had to be painted one area at a time until the whole was covered. Either way, using ceramic shell in a manner outside of the industrial norm proved flexible and affordable. Ceramic shell offers great surface detail and is much lighter for handling, whereas even small molds made in a plaster-based investment can be heavy, and larger molds are so weighty they require the use of hoists, and de-waxing is very long and time consuming. The de-waxing and firing times for ceramic shell are very short, in contrast, meaning that I do not need as much fuel.

How did you work out your current method of investing the horse?

My recent experiments in re-creating one of the Rothschild Bronzes (Lacey 2018) have revealed how much care and attention are needed in applying the first layers of investment. This I already knew, but I found that I could use a historic recipe of fine clay mixed with ground inert powders for the first coats and back them up with ceramic shell (video 2), or just use ceramic shell for the less detailed parts. This made for very lightweight molds that captured detail accurately, but were easier to remove from the bronze than ceramic shell. The clay layers are also cheaper than the zirconium silicates typically used for the initial layers in ceramic shell investments. It was through these experiments that I developed a method of investing the horse that worked best for me.

Do other art foundries act likewise?

For contemporary foundries this approach would be disastrous, because they need a degree of repeatability. My method is very sensitive, but requires time to adjust mixtures and apply initial layers.

Does the investment material show up in the finished sculpture?

These new mixed techniques that I've been exploring with historical materials will rarely leave a trace on the finished bronze. Possibly some of the fine particles may remain lodged in crevices. Potentially, chemical analysis could pick this up.

9. Choice of alloys

What alloys do you use? Do you have any preferences? Why?
(video 11)

Throughout my career I have used numerous bronze alloys (for more on bronze alloys see I.2). In the early days, leaded gunmetals were common, easy to obtain, and generally compliant to work with. Over time phosphor bronze ($Cu_{90}Sn_{10}$ with phosphorus addition for de-gassing) took over. It is very clean and with a low viscosity, meaning that it casts beautifully. Silicon bronze ($Cu_{94}Si_4Mn_1Fe_1$) became ubiquitous in the late 1990s among foundries due to its great ductility and mechanical strength. It could be welded nearly flawlessly. In recent years—partly related to my historical research, but also to cost and/or the desire to control the color of the alloy—I have gone back to creating bronze alloys from raw ingredients or by recycling scrap bronze, usually from old industrial components and leftovers from previous casts. To do this I have learned how to deal with some of the more troublesome aspects of alloying. Particularly when using scrap metals, one has to be much more attuned to what happens in the melt. The surface of liquefied silicon bronze is a perfect mirror, but that of scrap bronze will have plenty of dross or slag on top and needs refining, adjusting, and cleaning.

Do commercial art foundries often experiment with alloys as you do?

I think that my long-term habit of material exploration and curiosity about historical and archaeological materials and technologies—and prescientific ways of making sense of the material world—has enabled me to deal with unknowns with moderate success, but as with the investment, this approach would be disastrous for commercial foundries.

Why do you try to avoid zinc in the alloy?

Whatever alloy I work with, I try to reduce the amount of zinc because zinc tends to make welding more difficult, and often causes porosity and occasionally differences in metal color that are exacerbated when the patina is applied. Zinc vapor is also terrible to breathe in: it can cause a strong fever and chills, all of which I have experienced in the past.

How do you deal with differences in alloy composition in separately cast sections?

In the example of the bronze horse, several mold sections might be cast in small batches from the same crucible load (video 11). Typically a week or two would pass before I cast the next batch. Inevitably, by using refined scrap bronze to cast the body and legs you will see an overall shift in composition due to continued reuse of the scrap and segregation of the alloy. For the sections that required fine chasing, such as the head and neck, I created an alloy of 90 percent copper, 10 percent tin—similar to the one prescribed for cannons and sculptures in Italian Renaissance treatises—which is receptive to sharp tooling such as chiseling and engraving. As long as the combination of alloys does not negatively affect the patina or the physical/mechanical properties of the sculpture, these differences in elemental composition do not concern me.

Do the compositional differences show up on the final bronze?
(video 16)

The surface of the sculpture should not reveal any differences in composition; ideally it should look as if it was cast in one pour. The joints will have been skillfully concealed through welding and resculpting the metal in those areas to make the surface look continuous. More time is taken to build up the patina so that the differences in alloy are disguised. It would be difficult to discern the compositional differences visually. However, if you used a handheld XRF (for more on this surface elemental analysis see II.5§2.1) to make a compositional map of the sculpture, the different alloys would show up immediately. One would have to sample quite a few areas to do that. On the basis of the data gathered, it might be possible to hazard a guess at which batches or panels were cast together, in what order, and maybe even the orientation of the sprue system in a given piece.

10. Gas furnaces versus charcoal hearths

What kind of furnace did you use for the horse?

(video 11)

It's a homemade affair, essentially a simple box of bricks. I have tried many forms of construction, materials, and dimensions over the years and found that reflective kiln brick (designed to reflect heat) held in a drum works perfectly well. There's a small aperture at the bottom for gas to enter and a cement lid on top with a hole in the center. Gas is fed into the furnace from a hand-built torch, powered by propane gas that heats a clay graphite crucible. There is no great science or magic here; it just has to heat up the crucible as fast as possible. It is a tool that needs no adjustments or sensitive management; just turn it on and leave it alone.

Do you use the same furnace for all of your castings?

No. For casting some of the small and more detailed sections I use an open charcoal hearth that contrasts greatly to the gas furnace I just described. I use a crucible no more than fifteen kilograms in capacity, in a bed of charcoal. The temperature of the charcoal can be far higher than that of gas, and the bed itself has different heat zones, meaning that you can be far more sensitive or exacting—for instance switching from oxidizing to reducing environments, accelerating the temperature or dampening it down to hold when waiting for the right time.

Do commercial foundries use charcoal hearths?

I don't think so. In contrast to a gas furnace, the hearth is full of "magic" and requires dedication and time to fully master. The hearth offers so much versatility, but from a commercial perspective it is something of a liability financially, as it requires someone to tend to it constantly.

Can the type of furnace or hearth you used be inferred from the technological investigation of the finished bronze?

I cannot imagine that much if any observable evidence is left in the bronze. There is a technique that examines the gas trapped in porosity held deep within the body of the bronze, and in theory this gas could hold information on the type of furnace used, but it is a very complicated and delicate process.

11. Welding like crochet

What are your considerations when assembling the cast sections?

(fig. 548)

I come from a long line of tailors, and there is something in the way a suit is built that echoes in the way a bronze sculpture is constructed. In both fields you work with a number of sections or panels that need to be joined together in a way that conforms to and describes the contour of the body or sculpture. Usually this works perfectly well, but occasionally you have to cut vents or seams into the bronze to better align the sections. I think this is when I started to think about how larger and more complex forms could be created in bronze.

How are the separated sections assembled?

I always weld the cast sections together using an inverter TIG welder. When using different alloys for the various component sections, tensions can form in the bronze: the art here is to join the sections to avoid "hot tearing" and porosity. With refined scrap, this can be very difficult because you never entirely know what you're dealing with, but I have learned that rather than applying more and more heat to the bronze panels, a gentler approach works better. The movement between the arc of the weld and the filler rod is also hugely important. The action looks more like crocheting: small circular movements that hook into the cooling parts of the weld literally stitch the bronze together. Typically, I use a silicon bronze welding rod to join the cast sections. It works very well, but you do have to plan where the weld line goes more carefully when using varied metals, as each of them (and the weld metal) will react differently to patination.

Is TIG welding common in most commercial art foundries?

Yes, it's the simplest way to join sections of bronze together, especially as most art foundries use silicon bronze for casting, and it welds with ease using this method. The other method commonly used is MIG welding, in which, instead of using short, thin metal welding rods, the handset has a continuous feed of wire. This is ideal for joining larger, heavier sections of bronze, but it lacks the finesse required for detailed work.

How does the welding show up in the final bronze?

(fig. 545)

As stated earlier, in an ideal world, the finishing should make the joints invisible from the outside. If you are trying to assess the location of the joints as part of technical analysis, any differences between the body metal and the weld metal will be clear with XRF. This assumes that one can take enough readings around the sculpture, but you should be able to discern from the data not only the separate cast sections, but also the layout of the weld seams. Using X-radiography, the section joints will appear as fine pale lines, occasionally with fluid shapes defined by the filler welding material.

12. Chasing: different from one edition to another

How important is chasing for you?

(video 10)

I like chasing, and strangely it's something I've always enjoyed, which is not the norm. It's a hard and very physical process. Most commercial foundries aim to reduce the amount of chasing needed by trying to cast bronzes that require little after-work. This also increases efficiency and, thus, profit. As I am principally concerned with the casting of my own work (I don't cast the work of other artists as commercial foundries do), I can judge how much extra effort I wish to put into a particular bronze. More importantly, I do not consider the artistic process as ending with the initiation of the pouring. Most artists hand over the model and after that point have little part its creation; they only see it as the final bronze. Whereas I often wish to extend the modeling or enhance a certain aspect of the sculpture well after it is cast. That's why I don't produce editions of identical casts. I see them as prohibiting artistic expression. I like to be able to alter each cast in different ways, all the way through the process.

Do you use electric tools?

(video 13)

I do use electric tools, for example grinders and pendant drills, but mostly I use hand tools like punches, chisels, and files, the majority of which I have made myself to suit my needs. My use of hand tools is not out of a sense of tradition or to uphold craft values. I'm very physical and love to express this through the bronze, and these tools allow me to communicate with the metal in an ongoing dialogue. That invigorates me, and hopefully both the sculpture and I have a near-tireless energy for such work.

How does chasing show up on the finished bronze?

(video 11)

On this sculpture, most of the tool marks are very fine and nearly impossible to see. Around the eye, for instance, I used them to refine the shape and sharpen the form without leaving marks (video 11). That said, the surfaces of most of my bronzes might be littered with marks made on the original model, as well as punch marks and traces of chiseling, texturizing, and burnishing. I enjoy this process and use it a lot to bring the surface to life. Raking light, which is easy and practical with a flashlight (or with more sophisticated photogrammetry methods), shows up the surface topography and is a descriptive way of understanding the discrete deformations made by chasing. If the surface of the bronze is distracting because of a thick or colorful application of patina, the chasing may be obscured. One could take an impression of an area with silicone rubber in order to reproduce the tool marks for study.

13. Patina: unique for each edition

What place does patina have in your creations?

(video 16)

I love patina more than chasing because it offers so much scope for exploration. I was trained to create very fine and highly repeatable patinas—the patinas produced by most commercial foundries. Over time this approach began to hold me back, and I needed to bring the energy of the model and its translation into bronze into closer alignment. Again, the feeling that the creative process does not stop when the modeling is finished is paramount!

How did you patinate the horse?

I started with a black followed by brown using ferric nitrate, deep black using oil and flame, then white with titanium dioxide, and lastly mottled grays with complex layers of chemistry. Sometimes it's easier to begin exactly where you do *not* want to end up—that is, with the traditional black or brown patina. By starting with such a contradiction, I then feel my way by deliberately destroying that patina with the application of another. I repeat the process until I reach a point of equilibrium or expression that starts to speak of the object or whatever it is that I feel is in the sculpture. Sometimes the process never ends and I melt the sculpture back down.

Is such a series of back-and-forth trials your modus operandi for patination?

Yes, I often start with an idea for the look or the finish that I want to achieve, but after starting down that route, the feeling and direction for the sculpture shifts. You need to go beyond the surface of the bronze: What is it that a particular sculpture wants to become? It's not always obvious, and this process tends to be a reflection of oneself, entirely introspective. I guess, when I find what is needed in the patina, I usually find some resolution for the bronze and ultimately myself.

Patination is such a jealously guarded secret of art foundries. Do you know of other practices like this?

There is another way of thinking about this. Instead of considering that individual areas of activity, like patina or mold making, are more shrouded in secrecy than others, in reality the entire foundry system is governed by secrecy. Although most foundries are, at least superficially, open to visitors and clients to discuss their work, there are always layers of activity behind the scenes that operate discreetly and privately, out of sight. This behavior leads to the evolution or invention of novel practice in all areas of the foundry. It is, after all, these unique inventions that give each foundry its commercial and artistic edge. Secret methods also enter the foundry when new workers or practitioners arrive from other foundries. If their methods are better than what is currently practiced, their ideas will readily be adopted. For researchers of technologies used in the arts, this is a wonderful phenomenon because these secret methods often leave evidence on the bronze that can act as a signature of an individual maker or foundry. If you follow this evidence, sometimes you can map out how ideas—or secret practices—have either been spontaneously invented or migrated and evolved from foundry to foundry.

Do you think we could find out how you patinated the horse by carrying out the kind of investigation recommended in the present volume?

One can only imagine the depth of chemical layering that forms on the surface of a bronze like this. Sometimes it appears that the chemical solutions begin their own complex set of interactions that carry on for days after the patina is applied. Chemical analysis of the surface should reveal, if not a specific recipe, the possibility of an approach by the maker that is distinct and purposeful.

14. Process duration

What insights can you offer regarding the time involved for the various steps of the process?
(fig. 33)

I can give you an outline for the working times or durations used in this casting process, although there is no such thing as a generalized time frame (table 1). And some steps relate to drying or cooling times where there is nothing to do but sit and wait.

The first stage of the process was to make a two-part silicone rubber mold from the original model, backed with plaster mother molds. This took one day to complete, but three hours of this related to time allowed for the rubber to cure. Next, I made the wax inter-model, which took one and a half hours to pour four layers of wax, the first starting at 95°C and the last ending at 65–70°C. It required a further one hour for the wax inter-model to cool down enough to be removed from the mold. The wax needed reworking to remove any imperfections such as holes and delaminations. This took around two hours.

Designing and attaching the sprue system and adding the core pins to the wax inter-model then took one hour. I poured a plaster investment core inside the wax inter-model, which took about thirty minutes, but again some of this time was just preparation and cleanup afterward: the actual pouring of the plaster took a mere ten seconds.

Applying a ceramic shell investment to the sprued wax took two and a half hours to apply for ten layers. But one needs to count a further two and a half hours between layers to ensure that each one was dry, so it took three days to complete. De-waxing is a relatively fast process, taking only twenty or thirty minutes, but because I used a plaster core rather than ceramic shell, the mold needed to be fired in a kiln at around 500–600°C for a further hour and a half. The wax reclaimed after de-waxing weighed four and a half kilograms. The weight in wax is multiplied by ten to calculate the weight of bronze needed: I had to melt forty-five kilograms of bronze to fill the mold. Both my gas and my charcoal furnaces take an hour and a half to get up to temperature. Once liquefied, it took only thirty seconds to pour the metal into the mold. I tend to leave the bronze cast to cool for three or four hours before breaking it out of its mold.

The initial fettling of the bronze cast took two hours, while the fine detailed chasing took another four hours. Finally, the patination process took one hour, but often this stage is revisited over the following days to see if any further chemical changes have occurred, in which case I usually redo the patina until I'm satisfied.

15. Closing observations

(video 16)

By following Andrew Lacey's artistic and foundry process in making *The Anatomy of Bronze* from start to finish, what has become evident is that although he was trained in London's contemporary art foundry scene, his decision-making, material considerations, and aesthetic values differ greatly from those of most commercial art foundries. He has developed a deeply idiosyncratic methodology. This may be due to a number of factors. He works in isolation in his own studio/foundry and is therefore neither influenced nor constrained by the need to collaborate with other practitioners. And thanks to his academic research, he very deliberately interweaves historic and contemporary methods of making. While conscious of costs, he is not under the tight economic constraints that most art foundries are. He is driven primarily by artistic expression and also likes the intellectual challenge that comes from investigation of materials and processes. Finally, his sculptural output and the methods behind it are clear extensions of the same artistic expression.

Further resources

https://andrewlacey.com/

Vocabulary

Edited by Jane Bassett and David Bourgarit

English: Francesca G. Bewer
French: Anne-Lise Desmas, Elisabeth Lebon
German: Joachim Kreutner, Jean-Marie Welter
Italian: Francesca G. Bewer, Lorenzo Morigi
Chinese: Yi Chen, Linda Ying-Chun Lin

Additional Contributors Ann Boulton, Jean Dubos, Irene Gunston, Patricia Harpring, Sharon Hecker, Andrew Lacey, Marjee Levine, Jeffrey Maish, Benoît Mille, Peta Motture, Uve Peltz, David Reid, Dominique Robcis, Lise Saussus, Harold Schulze, Jeffrey Springer, Nicolas Thomas, Quanyu Wang, Jeremy Warren, Frank Willer, Dimitrios Zikos

The rich terminology associated with sculpture, and bronzes and foundry work in particular, is very specialized. It can also vary to a surprising extent across different groups of specialists such as professional foundry practitioners and historians of bronze sculpture, or even across experts focused on different chronological or geographical areas within a field. One of the main goals of the interdisciplinary CAST:ING project was to shed light on some of these ambiguities, wrangle definitions, and where possible suggest a common preferred term. The selection of agreed-upon terms offered here should be useful to a range of researchers—as an aid for interpreting written sources or features on a cast, and for recording and communicating those discoveries and observations. A shared vocabulary ultimately also facilitates a wider sharing of data. Definitions for each term are provided in English and French. Translations are provided in the three other modern languages that reflect the expertise of the CAST:ING members: Italian, German, and Chinese. Early French and Italian translations are also proposed here and there. Earlier terminology is also provided for Italian and French.

It was beyond the scope of this project to produce such a thesaurus for the more than three hundred technical terms collected by its members. We agreed on a selection of fifty-eight based on their frequency of use in the study of bronze sculpture, and on the possible definitional ambiguities. For each entry, we have aimed to include a translation in the selected languages, along with references to two or more authoritative bibliographical sources, as these terms are also to be incorporated in the structured online vocabulary of the Getty Art and Architecture

aimed to include a translation in the selected languages, along with references to two or more authoritative bibliographical sources, as these terms are also to be incorporated in the structured online vocabulary of the Getty Art and Architecture Thesaurus (AAT), which is used broadly by art historians. Where necessary, multiple translations are provided with notes to explain differences between them. Entries are illustrated with relevant images drawn from the Visual Atlas of Features. And hyperlinks to the select terms are provided throughout the *Guidelines*.

English	French	German	Italian	Chinese
after-cast	surmoulé	Nachguss	sovracalco	翻铸
armature	armature	Kerneisen	armatura	塑像內部支架
as-cast surface	surface brute de coulée	Gusshaut	superficie al grezzo	毛坯铸件表面
brass	laiton	Messing	ottone	铜锌合金
brazing	brasage	Hartlöten	brasatura	硬焊
bronze	bronze	Bronze	bronzo	铜锡合金
cast (n.)	fonte	Guss	getto	铸件
cast (v.)	couler	gießen	gettare	铸造
cast-on repair	réparation par coulée secondaire	Überfangguss	getto a incastro	修补浇铸
castability	coulabilité	Fliessvermögen	colabilità	可铸性
casting defect	défaut de fonderie	Gussfehler	difetto di fusione	铸瑕
casting plan	plan de coulée	Teilungsplan	schema di fusione	分铸计划
chaplet	cale à noyau	Kernhalteplatte	chiodo distanziatore	垫片
chasing	ciselure	ziselieren	rifinitura	铸件表面加工
chef-modèle	chef-modèle	Dauermodell	modello di riferimento	主铸型
chiseling	travail au ciseau	meißeln	incisione a bulino	凿
coating	revêtement	Überzug	rivestimento	涂层
cold shut	reprise	Kaltguss	accostatura	冷界
core	noyau	Kern	anima	芯型
core pin	fer à noyau	Kernstift	distanziatore	销钉
core support	support de noyau	Kernstütze	chiodo di sostegno	芯骨
corrosion	corrosion	Korrosion	corrosione	腐蚀
edition	édition	Auflagenguss	edizione	版本
engraving	gravure	gravieren	incisione	阴刻

English	French	German	Italian	Chinese
fettling	réparure	Abgraten	sbavatura	清砂
flashing	gerce	Grat	bava	毛边
founder	fondeur	Gießer	artefice	铸工
foundry model	modèle de fonderie	Ausführungsmodell	forma da gittar di bronzo	铸型
gilding	dorure	Vergoldung	doratura	镀金
inlay	incrustation	Einlage	agemina	镶嵌
inter-model	modèle intermédiaire	Wachsmodell	modello intermedio	范制蜡型
investment	moule de potée	Formmantel	forma	熔模
life-casting	fonte sur le vif	Naturabguss	fusione dal vero	活体模铸
lost-wax casting	fonte à la cire perdue	Wachsausschmelzverfahren	fusione a cera persa	失蜡法
metal plating	placage métallique	Plattierung	placcatura	镀覆
metallurgical joint	joint métallurgique	Metallurgische Verbindung	saldatura metallurgica	冶金接合
model	modèle	Modell	modello	模
mold	moule	Form / Negativform	calco	范
mold extension	portée de noyau	Mold extension	morsa	自帶泥芯撐
overlay	incrustation	Auflage	incrostazione	包覆
patch	plaquette de réparation	Flicken	laminetta ad incastro	补修
patina	patine	Patina	patina	古色
peening	matage au marteau	Hämmern	martellatura	轻敲
piece mold	moule à pièces	Stückform	forma a tasselli	块范
plug	insert	Dübel	tappo	塞子
porosity	porosité	Porosität	porosità	气孔
pour	coulée	Guss	colata	浇注
punch	ciselet mat	Punze	cesello	冲头
refractory mold	moule réfractaire	Form	forma	耐火模具
replica	épreuve d'édition	Replik	multiplo	复制品
Roman joint	assemblage à la romaine	Steckverbindung	ghiera alla romana	罗马式接合
sand casting	fonte au sable	Sandformguss	fusione a staffa	砂型铸造
seam line	couture	Gussnaht	linea di giunzione	范线

English	French	German	Italian	Chinese
shrinkage	retrait	Schrumpfung	ritiro	收缩
soldering	brasage	Löten	brasatura	焊接
sprue	jet de coulée	Gusskanal	getto	浇铸道
variant	version	Variante	variante	变体
welding	soudage	Schweißen	saldatura	熔焊

after-cast

See figs. 1, 284.

A bronze cast that has been created from a mold taken directly from an existing bronze. After-casts are therefore made using the indirect lost-wax process, or in some instances by sand casting.

Note: An after-cast will usually replicate surface traces of alterations on its precursor model, including damage and repairs. It will also tend to be smaller and potentially less crisp than the bronze from which it was derived (see II.3).

Sources

Cultural Heritage: Dillon 2002; Motture 2019; Penny 1993

To Be Distinguished From

copy

replica

Synonyms

aftercast

surmoulage
The French term *surmoulage* is often used in English texts.

French: *surmoulé*

Reproduction en métal par moulage d'un bronze ou d'une partie d'œuvre en bronze.

Note: En français, le mot surmoulé peut être chargé d'un sens négatif (Baudry and Bozo 1978, 634); l'auteur parle aussi de contremoulage (550). Pour certains auteurs, le surmoulé n'est pas restreint aux reproductions en bronze mais peut être en plâtre, résine, etc. (Rama 1988, 375). Au 19e siècle, le « surmoulé » prend un autre sens, celui d'épreuve de série légalement obtenue à partir d'un chef-modèle (Lebon 2012). On trouve aussi « surmoulé » avec la signification de chef-modèle (Bader and Théret 1961, 615). Le surmoulage désigne l'opération consistant à réaliser un surmoulé.

To Be Distinguished From

copie

réplique

variante

version

Sources

Cultural Heritage: Baudry and Bozo 1978, 634; Lebon 2012

Art and Craft Textbook: Rama 1988, 375

German: *Nachguss*

Source: Maaz 2010, 710

Italian: *sovracalco*

Closest translation to *surmoulage*; could refer to the process as well as to the product of surmoulage; not specifically used for metal in these sources.

Sources

Cuomo di Caprio 2007, 224; Panazza 2011, 30

Alternate Translations

rifusione
Less common, related to the process of obtaining a new bronze from an already-existing sculpture
Sources: Battaglia 1961, under *rifusione* [definition 1, vol. 16, 303]; Treccani, under *rifusione*

fusione successiva
Usually has smaller dimensions due to the shrinkage of the metal while cooling

multiplo

Chinese: 翻铸

armature

See figs. 8, 31, 32, 34, 79, 221.

Assemblage of joined metal rods, tubes, and/or wires (and/or other materials such as wood) that provides a structural framework for a sculpture and usually attaches to a base. Though generally internal, it may also include external supporting components that are eventually removed. In a lost-wax bronze, the armature is created to support the model, whether it is hollow or solid. In the case of a hollow cast, it will further help support the refractory mass of the core during the pour (see GI). Armatures may also be used in the assembly of a sculpture that was cast in parts, and in the repair of sculptures that have been structurally damaged (e.g., large archaeological bronzes).

Sources

Cultural Heritage: Mattusch 1996; Boulton 2006; Dillon 2002

Copper Industry: Brunhuber 1988

Synonym

core rod

French: *armature*

Élément ou ensemble d'éléments (barres, tubes, fils) destinés à armer le noyau, et plus généralement le modèle. Ils peuvent se trouver dans le moule réfractaire pour le rendre plus résistant lors de sa manipulation et lors de la coulée. Une cire pleine peut être armée. L'armature peut, dans le cas des grandes fontes, également servir à renforcer le bronze.

Note: On parle aussi de « système d'armatures » pour décrire l'ensemble des éléments armant un noyau. Des termes plus spécialisés sont parfois employés par les fondeurs pour désigner des types spécifiques d'armatures comme la « lanterne », un tube creux perforé (Delon 1877, 144; Rama 1988, 374).

Sources

Cultural Heritage: Baudry and Bozo 1978, 626; Bewer, Bourgarit, and Bassett 2008; Azéma and Mille 2013b

Historical: Gonon 1876, 38; Guettier 1858, 300; Delon 1877, 144; Boffrand 1743, 24–25; Mariette 1768, 39–48

Art and Craft Textbooks: Lambert 2002, 266; Rama 1988, 372; Dubos 2003

Copper Industry: Association Technique de Fonderie, Commission Ingénieurs et Techniciens 1979

Alternate Translation

armature de noyau
Source: Copper Industry: Brunhuber 1988

German: *Kerneisen*

Source

Brunhuber 1988

Alternate Translations

Stützkonstruktion

Armierung
Source: Maaz 2010, 695

Italian: *armatura*

Sources

Giuffredi 2006, 19–21; Treccani, under *armatura* [definition 6]; Battaglia 1961, under *armatura* [definition 4, vol. 1, 667]

Alternate Translations

armadura
Early italian spelling of *armatura*
Source: Leonardo 1490, fols. 155r, 157v, 144v

ferramenta
Relates more specifically to the ferrous materials used for the armature
Source: Biringuccio [1540] 1977, fol. 80v

rinforzo per anime
Source: Brunhuber 1988

Chinese: 塑像內部支架

Source

Ming 2010, 1783; TNATD

Alternate Translation

雕塑骨架
Source: TNATD

as-cast surface

See figs. 5, 62, 104, 277.

Refers to the surface immediately after removal from the mold, before fettling and chasing.

Note: The quality of the as-cast surface varies depending on the characteristics of the refractory mold. It may include oxidized metal, refractory mold remains, flashes, or casting flaws and imperfections. An artist or foundry may choose to preserve the surface as-cast without much further refinishing.

Sources

Cultural Heritage: Kienlin, Bischoff, and Opielka 2006

Other: Burd and Greene 1948; Mödlinger and Sabatini 2016

To Be Distinguished From

fire scale

fire-skin

French: *surface brute de coulée*

Surface du bronze en sortie de moule réfractaire, juste après le décochage, avant la réparure.

Note: Une belle surface faisant apparaître la qualité de la « peau de pièce » peut être obtenue par un choix judicieux d'alliage, de température de coulée, de température du moule réfractaire, mais surtout par la qualité du moule réfractaire.

Sources

Art and Craft Textbook: Lambert 2002, 238

Alternate Translations

surface brute de fonderie
Source: Baudry and Bozo 1978, 632, on parle d'un état « brut de fonderie »

surface brute de fonte

German: *Gusshaut*

Sources

Alscher 1987, 555; Maaz 2010, 696

Alternate Translation

gussrauhe Oberfläche

Italian: *superficie al grezzo*

Used as part of expression "superfiecie al grezzo di fusione"

Source

Battaglia 1961, under *superficie al grezzo* [definition 10, vol. 7, 39]

Alternate Translation

grezzo

Chinese: 毛坯铸件表面

Source

Ming 2010, 998

brass

See fig. 133.

A copper alloy with zinc as the primary added element. As with bronzes, there are a wide variety of brass alloys.

Sources

Cultural Heritage: Bayley 1991; Dillon 2002, 299; Penny 1993, 297

Copper Industry: Brunhuber 1988; Campbell 2015, 270; Koch and Newell 1963

Historical: Buchanan 1903, 23

General Dictionary: https://www.merriam-webster.com/dictionary/brass [definition 1]

Other: Turner 1982

French: *laiton*

Alliage à base de cuivre dans lequel le zinc est l'élément d'addition principal.

Sources

Cultural Heritage: Baudry and Bozo 1978, 643; Garenne-Marot, Robion, and Mille 2003; Bewer, Bourgarit, and Bassett 2008; Thomas and Bourgarit 2018; Thomas 2009, 510

Copper Industry: Association Technique de Fonderie, Commission Ingénieurs et Techniciens 1979; CTIF 2022; Comité international du dictionnaire technique de fonderie 1955, 29; Brunhuber 1988; Koch and Newell 1963

Historical: Félibien 1690, 335; *Secrets concernant les arts et métiers* 1810, 224

Art and Craft Textbooks: Lambert 2002, 271; Rama 1988, 374

Alternate Translations

archal
Sources: Cultural Heritage: Thomas 2009, 505, 507–8; Historical: Lespinasse and Bonnardot 1879, 551

cuivre jaune
Sources: Historical: Félibien 1690, 335; Boffrand 1743, 54–55; Diderot et al. 1751, 4:540–47, entry « cuivre »; Mariette 1768, 75, 106; *Secrets concernant les arts et métiers* 1810, 224; Grandpierre and Avale 1867, 7; Delon 1877, 180

orichalque
Source: Cultural Heritage: Halleux 1973

German: *Messing*

Sources

Maaz 2010, 708; Gesamtverband Deutscher Metallgießereien 1982; Brunhuber 1988; Koch and Newell 1963

Alternate Translations

Messinge

Messinglegierung

Italian: *ottone*

Sources

Treccani, under *ottone*; Treccani, under *Bronzi e ottoni*; Battaglia 1961, under *ottone* [definition 1, vol. 12, 288]

Chinese: 铜锌合金

Literally "copper-zinc alloy"

Source

Ming 2010, 1440

Alternate Translation

黄铜
This is a broad, colloquial term for brass.

brazing

See figs. 106, 188, 206, 229.

In bronze sculpture, a technique for joining separately cast parts or repairs or filling casting defects by localized addition of a molten copper alloy of slightly lower melting temperature than that of the cast. Unlike in welding, the contact zones of the sections to be joined are not brought to a molten stage.

Note: In industry, brazing is often defined as a joining method using a filler metal with a melting point above 450°C, which includes both copper-based and silver-based (hard solder) alloys. In cultural heritage settings, it may be impractical to identify the alloy and melting point of the filler metal. To avoid confusion, we suggest that white filler metal be referred to as solder.

Sources

Cultural Heritage: Allen et al. 2022, 39

Historical: Buchanan 1903, 23, 60; Bolland 1894, 72

Copper Industry: Schwartz and Aircraft 1951

Other: Scott 1991, 138

To Be Distinguished From

flow fusion welding

flow welding

fusion welding

repair

soldering

welding

French: *brasage*

Technique d'assemblage de deux pièces métalliques ou plus, par addition d'un alliage à point de fusion plus bas que celui des métaux à assembler, n'entraînant pas la fusion des zones de contact, contrairement au soudage.

Note: L'anglais distingue « brazing », le procédé utilisant de la brasure tendre (métaux à bas point de fusion comme le plomb et l'étain), de « soldering », celui utilisant une brasure forte (alliage à base de cuivre). On rencontre parfois en français le terme « brasure très forte » quand de l'argent est utilisé (Baudry and Bozo 1978, 634). Par ailleurs, la brasure désigne le métal de brasure (Bader and Théret 1961, 102) ainsi que l'assemblage lui-même.

Sources

Cultural Heritage: Baudry and Bozo 1978, 634

Copper Industry: Association Technique de Fonderie, Commission Ingénieurs et Techniciens 1979

Alternate Translations

soudure indirecte

Les termes soudage ou soudure (Baudry and Bozo 1978, 634; Bader and Théret 1961) et souder (Félibien 1690, 757) sont souvent employés comme termes génériques décrivant tous les procédés d'assemblages de métaux mettant en jeu du métal à l'état liquide, brasage et soudage par fusion.

Source: Copper Industry: Bader and Théret 1961, 595

German: *Hartlöten*

Source

Fachkunde Metall: Bovenkerk, Braun, and Dörflinger 1987, 366

Italian: *brasatura*

A subset of *saldatura*

Sources

Treccani, under *brasatura*; Battaglia 1961, under *brasatura* [vol. 2, 359]

Chinese: 硬焊

Sources

Ming 2010, 1684; TNATD

Alternate Translations

钎焊

Source: Ming 2010, 1137, 1138

铜焊

Sources: TNATD; Ming 2010, 1436

bronze

See figs. 133, 134.

Depending on the user, "bronze" may designate a copper alloy that has tin as the primary added element or any other copper-based alloy. We recommend using the term "bronze" specifically for copper-tin alloys unless qualified by another term (e.g., "silicon bronze"). See I.2§1.

Note: In common parlance, most copper alloy sculptures are referred to as "bronzes" regardless of the alloy's actual elemental composition.

Sources

Cultural Heritage: Bayley 1991; Bassett 2008, 274; Dillon 2002, 299; Penny 1993, 297; Welter 2018

Historical: Buchanan 1903, 24

Art and Craft Textbook: Rome and Young 2003, 302

Copper Industry: Brunhuber 1988; Campbell 2015, 270; Koch and Newell 1963

General Dictionary: https://www.merriam-webster.com/dictionary/bronze [definitions 1a, 2]

French: *bronze*

Suivant l'utilisateur, désigne aussi bien un alliage à base de cuivre dans lequel l'étain est l'élément d'addition principal que tout alliage à base de cuivre (Félibien 1690, 333; Launay 1827, 2:251–52; Hamm 1924, 83; voir aussi I.2). Pour lever l'ambiguité, il est recommandé de restreindre le terme aux alliages cuivre-étain.

Note: Le terme désigne souvent tout objet fabriqué en alliage à base de cuivre.

Sources

Cultural Heritage: Baudry and Bozo 1978, 300, 643; Rolley 1994, 64; Bewer, Bourgarit, and Bassett 2008; Welter 2008; Thomas 2009, 510

Copper Industry: Association Technique de Fonderie, Commission Ingénieurs et Techniciens 1979; CTIF 2022; Comité international du dictionnaire technique de fonderie 1955, 7; Brunhuber 1988; Koch and Newell 1963

Historical: Diderot et al. 1751, 2:436–43, entry « bronze »; Launay 1827, 2:251; Félibien 1690, 335; Boffrand 1743, 54–55; Mariette 1768, 76, 105–6

Art and Craft Textbooks: Rama 1988, 372; Lambert 2002, 266

General Dictionary: Office québécois de la langue française n.d.

Alternate Translations

airain

Ce terme est très fréquent dans les sources écrites depuis l'antiquité comme terme générique pour décrire tout alliage à base de cuivre, bien que souvent assimilé au bronze (à l'étain) par les historiens. Terme ambigu donc.

Source: Cultural Heritage: Thomas 2009, 505, 507

bronze d'art

Tous les alliages cuivreux avec une teneur en cuivre supérieure ou égale à 65%, cf loi Lebrun 1935 dans Rama 1988, 372. Il existe de nombreuses définitions plus ou moins divergentes. Voir par exemple (Hamm 1924, 83; Launay 1827, 2:251–52).

Sources: Historical: Launay 1827, 2:251–52; Art and Craft Textbooks: Rama 1988, 372; Hamm 1924, 83

German: *Bronze*

Sources

Weihrauch 1944; Maaz 2010, 677; Brunhuber 1988; Koch and Newell 1963

Italian: *bronzo*

Sources

Treccani, under *bronzo*; Treccani under *bronzi e ottoni* and *bronzi artistici*; Battaglia 1961, under *bronzo* [definition 1, vol. 2, 395]; Biringuccio [1540] 1990, fols. 80vff; Brunhuber 1988

Chinese: 铜锡合金

Literally "copper-tin alloy"

Source

Ming 2010, 1440

Alternate Translation

青铜

This is a broad, colloquial term used for bronze.

Source: TNATD

cast (n.)

A sculpture or more generally an object that is shaped by pouring a molten material or a slurry into a mold in which it will solidify. Plaster of paris, metal, and wax are among the cast-forming sculptural materials routinely involved in bronze production. The term may also refer to the amount of homogenous molten metal resulting from a single pour.

Note: Both "cast" and "casting" can be correctly used as nouns, but the former is more common.

Synonym

casting
Source: Copper Industry: Association Technique de Fonderie, Commission Ingénieurs et Techniciens 1979

French: *fonte*

Statue ou plus généralement objet métallique obtenu par fonderie. Attention à l'homonyme fonte qui désigne la fonte de fer.

Sources

Cultural Heritage: Baudry and Bozo 1978, 624; Bewer, Bourgarit, and Bassett 2008

Historical: Boffrand 1743, 51–56; Diderot et al. 1751, 7:106, entry « fonte »; Mariette 1768, 105–10.

Alternate Translations

jet
« On dit aussi un beau jet pour dire une figure qui a esté bien jettée »
Source: Historical: Félibien 1690, 623

objet de fonderie
Probablement le terme le plus satisfaisant en français pour désigner un objet obtenu par fonderie.

œuvre fondue
Le terme est employé dans (Association Technique de Fonderie, Commission Ingénieurs et Techniciens 1979, 635) mais uniquement en titre de section, dans le sommaire. Le seul terme en relation qui y est défini est « Exemplaire (coulé) », p. 634
Source: Copper Industry: Association Technique de Fonderie, Commission Ingénieurs et Techniciens 1979

œuvre moulée
Le terme est souvent employé en histoire de l'art. Pour autant, dans Association Technique de Fonderie, Commission Ingénieurs et Techniciens 1979, le terme (défini par « moulage », p. 565) est réservé à la sculpture en plâtre, terre, cire, ciment, béton, ou matière plastique, mais pas pour la sculpture en métal.
Source: Copper Industry: Association Technique de Fonderie, Commission Ingénieurs et Techniciens 1979

pièce coulée
Source: Copper Industry: Association Technique de Fonderie, Commission Ingénieurs et Techniciens 1979

pièce d'art
Source: Copper Industry: Koch and Newell 1963

German: *Guss*

Alternate Translation

Gussstück
Source: Association Technique de Fonderie, Commission Ingénieurs et Techniciens 1979

Italian: *getto*

Foundry-specific term

Sources

Treccani, under *getto* [definition 3]; Battaglia 1961, under *getto* [definition 15, vol. 6, 724]; Baglione 1642, 324–25

Alternate Translations

colata
A more general term that can be used to specify the pouring of plaster, wax, or molten metal
Sources: Treccani, under *tecnica*; Treccani, under *procedimento di fusione del bronzo*; Battaglia 1961, under *colata* [definintion 2]

gitto
Early Italian form of *getto* that can refer to the cast object as well as the act of pouring
Source: Biringuccio [1540] 1977, fol. 81

Chinese: 铸件

Source

Ming 2010, 1828

Alternate Translation

铸品

VOCABULARY

cast (v.)

See fig. 549.

In the sculptural context, the verb refers to pouring a slurry or liquefied material (e.g., plaster, wax, metal) into a hollow matrix or mold that will determine the shape of the material in order to produce a cast (n.).

Sources

Cultural Heritage: Boulton 2006

Historical: Buchanan 1903; Bolland 1894

Copper Industry: Brunhuber 1988; Association Technique de Fonderie, Commission Ingénieurs et Techniciens 1979

General Dictionary: https://www.merriam-webster.com/dictionary/cast [definition 4]

Synonym

pour
Source: Copper Industry: Association Technique de Fonderie, Commission Ingénieurs et Techniciens 1979

French: *couler*

Opération consistant à fabriquer un objet par fonderie. Plus généralement, exécuter une opération de coulée.

Sources

Cultural Heritage: Rolley 1994, 70; Baudry and Bozo 1978, 558; Bewer, Bourgarit, and Bassett 2008; Azéma and Mille 2013b

Copper Industry: Association Technique de Fonderie, Commission Ingénieurs et Techniciens 1979; Brunhuber 1988

Historical: Diderot et al. 1751, 7:80–81, entry « fondre »

Art and Craft Textbook: Lambert 2002, 268

Alternate Translations

fondre
Sources: Historical: Boffrand 1743; Diderot et al. 1751, 7:80–81, entry « fondre »; Mariette 1768, 168; Cultural Heritage: Baudry and Bozo 1978, 624

jeter / jetter
Sources: Historical: Félibien 1690, 320; *Secrets concernant les arts et métiers* 1810

German: *gießen*

Sources

Brunhuber 1988; Gesamtverband Deutscher Metallgießereien 1982; Association Technique de Fonderie, Commission Ingénieurs et Techniciens 1979

Italian: *gettare*

Foundry-specific term that relates to the pouring of metal

Sources

Treccani, under *gettare*; Battaglia 1961, under *gettare* [definition 28, vol. 6, 719]

Alternate Translations

buttare
Literally "to throw"
Source: Battaglia 1961, under *buttare*

colare
A more general term that can be used to specify the pouring of plaster, wax, or molten metal.
Source: Treccani, under *colare*

fondere
Used in reference to the melting of the metal
Sources: Treccani, under *fondere*; Battaglia 1961, under *fondere*; Leonardo 1490, fol. 150r

gittare
Early Italian spelling of *gettare*
Sources: Leonardo 1490, fol. 143r; Biringuccio [1540] 1977, fols. 89vff

Chinese: 铸造

Sources

Ming 2010, 1831; TNATD

cast-on repair

See figs. 65, 123, 169, 177, 179, 180, 181, 182, 184, 196.

A type of repair consisting of a localized cast of molten copper alloy to fill cavities or other casting defects. Cast-on repairs may fill a void in the sculpture or secure a separately formed patch or element to the cast.

Sources

Cultural Heritage: Avery 2018, 166; Salter and Gilmour n.d.; Motture 2001, 222

To Be Distinguished From

cast-on
The shortened term "cast-on" refers to a type of joint in which molten metal is used to join parts that were intentionally cast in sections.

French: *réparation par coulée secondaire*

Type de réparation utilisé pour combler un trou ou un manque sur un bronze en y coulant un alliage de composition proche.

Sources

Cultural Heritage: Bewer, Bourgarit, and Bassett 2008; Azéma and Mille 2013b

Alternate Translations

goutte
Sources: Historical: Boffrand 1743, 59; Diderot et al. 1751, 2:442, entry « bronze »

grains de bronze
Source: Historical: Mariette 1768, 131

réparation par surcoulée

German: *Überfangguss*

Literally "cast-on." Generally used in reference to an ancient process. In German, the term describes everything related to casting-on: it designates indiscriminately a cast-on repair or a cast-on joint. *Überfanggussreparatur* might be invented to avoid confusion.

Sources

Drescher 1958; Willer 2016, 139–40

Alternate Translations

Anguss

Anguss oder

Angießverfahren
Used in reference to modern processes
Source: Brunhuber 1986

Italian: *getto a incastro*

More general term to denote casting on, could apply to joints and repairs; see German *Überfangguss*

Source

Rohnstock 1999, 402–3

Alternate Translations

rifusione
Source: Morigi and Morigi 2008, 3, 5

rigetto
Source: Morigi and Morigi 2008, 3, 5

Chinese: 修补浇铸

Source

Ming 2010, 1585

Alternate Translations

浇补
Sources: Ming 2010, 740; TNATD

补铸
Source: TNATD

castability

See figs. 437, 438.

Ability of a liquid metal to fill and pick up every detail of a mold. See I.2§2.2.

Sources

Copper Industry: Schmidt and Schmidt 1992; Brunhuber 1988; Koch and Newell 1963; Association Technique de Fonderie, Commission Ingénieurs et Techniciens 1979

To Be Distinguished From

fluidity

fusibility

viscosity

Synonym

feeding power
Source: Hanson and Pell-Walpole 1951, 151–52

French: *coulabilité*

Aptitude d'un métal à remplir un moule réfractaire dans tous ses détails, lorsqu'on le verse à l'état liquide dans le moule réfractaire et qu'il se solidifie.

To Be Distinguished From

fluidité

fusibilité

viscosité

Sources

Cultural Heritage: Baudry and Bozo 1978, 632; Mille 2017

Copper Industry: Cuénin 1997a, 5; Association Technique de Fonderie, Commission Ingénieurs et Techniciens 1979; Brunhuber 1988; CTIF 2022; Koch and Newell 1963

Art and Craft Textbooks: Rama 1988, 279; Lambert 2002, 174

General Dictionary: Office québécois de la langue française n.d.

German: *Fliessvermögen*

Source

Association Technique de Fonderie, Commission Ingénieurs et Techniciens 1979

Alternate Translations

Formfüllungsvermögen
Source: Association Technique de Fonderie, Commission Ingénieurs et Techniciens 1979

Gießbarkeit
Sources: Brunhuber 1988; Koch and Newell 1963; Brunhuber 1986

Gießeingeschaften
Source: Brunhuber 1988

Gießfähigkeit
Source: Koch and Newell 1963

Vergießbarkeit
Sources: Brunhuber 1988; Association Technique de Fonderie, Commission Ingénieurs et Techniciens 1979

Italian: *colabilità*

Sources

Brunhuber 1988

Alternate Translations

facilità di fusione

fondibilità

scorrevolezza
Source: Treccani, under *fusione di oggetti artistici*

Chinese: 可铸性

Sources

Ming 2010, 838; TNATD

Alternate Translations

铸造性
Sources: Ming 2010, 1832; TNATD

casting defect

See figs. 73, 150, 151, 153, 154, 155, 158, 159, 161.

An unintended imperfection on a bronze that occurs during casting and appears as a more or less subtle discontinuity in the desired form and is associated with either a lack or an excess of metal. See I.3.

Note: Casting defects should not be confused with other defects generated before casting (e.g., imperfection in the model) or after casting (e.g., intentional alteration, wear due to use, or damage). Some authors separate casting defects into categories as either negative or positive (Rome and Young 2003).

Sources

Art and Craft Textbook: Rome and Young 2003

Copper Industry: Ammen 1980; Campbell 2015; Neff 2011

Synonym

casting flaw

French: *défaut de fonderie*

Défaut non intentionnel dans l'objet survenant lors de la coulée.

Note: Ne pas confondre les défauts de fonderie avec toutes les imperfections qui peuvent survenir dans le modèle avant la coulée, ou sur l'objet après la coulée (voir I.3). Par ailleurs, il est fréquent que dans la littérature il soit fait mention de défauts de fonderie spécifiques, sans que le terme générique de défaut de fonderie soit employé : trous (Boffrand 1743, 59; Diderot et al. 1751, 2:436–43, entry « bronze »; Lambert 2002, 241), fentes, crevasses, déchirures (Boffrand 1743, 59; Diderot et al. 1751, 2:436–43, entry « bronze »), manques (Lambert 2002, 241).

Sources

Cultural Heritage: Baudry and Bozo 1978, 645; Azéma and Mille 2013b

Copper Industry: Mascré, Thomas, and Hénon 1952; Cuénin 1997a, 6

Alternate Translations

anomalie de fonderie
Source: Copper Industry: Association Technique de Fonderie, Commission Ingénieurs et Techniciens 1979

défaut de coulée
Sources: Cultural Heritage: Bewer, Bourgarit, and Bassett 2008; Azéma and Mille 2013b

German: difetto di fusione

Italian: *difetto di fusione*

Sources

Pecchioli 1999, 193; Treccani, under *difetto di fusione*

Alternate Translations

difetto di fonderia
Sources: Carruba 2006, 24; Pecchioli 1999, 192

manchevolezza
General term for describing defects
Source: Treccani, under *manchevolezza*

sbollitura
Term for a specific casting defect used for general reference to that defect
Sources: Treccani, under *sbollitura*; Battaglia 1961, under *sbollitura* [vol. 17, 687]

soffiatura
Term for a specific casting defect used for general reference to that defect
Source: Treccani, under *soffiatura* [definition 2a]

Chinese: 铸瑕

Source

TNATD

Alternate Translations

铸疵
Source: TNATD

铸造缺陷
Source: Ming 2010, 1832

casting plan

See fig. 367.

A methodological tool that has been developed by researchers to reverse engineer the casting sequence of a bronze sculpture and visually represent the separately cast parts. It is based on the evidence presented in the object and attempts to map the decisions made by a foundry regarding the number and position of separately cast pieces.

Note: Founders do not work with such charts or maps, although strategizing occurs at every stage (e.g., where joints in the wax or metal will occur).

Source

Cultural Heritage: Mille and Descamps-Lequime 2017

French: *plan de coulée*

Plan de découpage prévu par le fondeur d'une sculpture en plusieurs éléments coulés séparément.

Note: Ce terme est très récent, il a été proposé par Benoît Mille comme élément technique susceptible d'aider à discriminer les grands bronzes antiques.

Source

Cultural Heritage: Azéma and Mille 2013b

German: schema di fusione

Italian: schema di fusione

Chinese: 分铸计划

chaplet

See figs. 38, 39, 40, 122.

A type of metal insertion placed in the gap between the core and the outer mold as a spacer to hold the core in place during the casting operation. A number of these are placed strategically throughout the mold. They are most often made of an alloy similar to that of the surrounding metal, as they will become embedded in the cast. In modern foundries, chaplets are mainly used in sand casting, but they have been encountered in historic lost-wax casting as well.

Note: The term "chaplet" has occasionally been used in the context of technical studies to refer to nails or pins that extend from the core to the mold and serve to hold the core in place (Beale 1975, Mattusch 1996). More commonly these are referred to as core pins.

Sources

Cultural Heritage: Beale 1975, 40; Mattusch 1996, 24; Penny 1993, 298; Salter and Gilmour n.d.

Art and Craft Textbook: McCreight 1996

Copper Industry: Brunhuber 1988; Association Technique de Fonderie, Commission Ingénieurs et Techniciens 1979

To Be Distinguished From

core nail

core pin

core rod

core support

Synonyms

core plate
Source: Copper Industry: Brunhuber 1988

spacer
Source: Cultural Heritage: Strahan 2010

French: *cale à noyau*

Petite plaque métallique disposée sur le noyau ou sur la paroi interne du moule réfractaire pour maintenir le noyau en place lors de la coulée, et dont l'épaisseur déterminera l'épaisseur de la fonte.

Note: Dans la mesure ou aucun terme spécifique n'existait à notre connaissance pour décrire un élément rencontré dans plusieurs bronzes anciens, le terme « cale à noyau » a été proposé et soumis à l'approbation des membres francophones de CAST:ING. Certains auteurs (Bader and Théret 1961) précisent que « ce sont de petites cales métalliques » pour fonte au sable. D'autres qu'elles servent « à le caler dans sa position » (Association Technique de Fonderie, Commission Ingénieurs et Techniciens 1979). Tous donnent la traduction en anglais « chaplet ».

To Be Distinguished From

broche

clou

clou distanciateur

distanciateur

fer

fer de maintien

fer de soutien

fer à noyau

Alternate Translations

chapelet
Source: Historical: Diderot et al. 1751, 11:267a, entry « noyau terme d'artillerie »

plaque porte noyau
Terme réservé à la coulée sous pression.
Source: Copper Industry: Brunhuber 1988

support de noyau
Sources: Copper Industry: Bader and Théret 1961, 614; Association Technique de Fonderie, Commission Ingénieurs et Techniciens 1979

German: *Kernhalteplatte*

Source

Brunhuber 1988

Alternate Translation

Kernhalter
Source: Willer 1994

Italian: chiodo distanziatore

Chinese: 垫片

Note: This term has several translations, including "chaplet," "shim," "spacer," "gasket."

Sources

Ming 2010, 272; Zhang 2010, 85; Liu 2015, 97

Alternate Translations

撑头

Source: Ming 2010, 141

芯撑

Source: TNATD

金属撑子

Source: Ming 2010, 141

chasing

See figs. 105, 248, 249, 252, 253, 255, 511, 550.

The process of fine detailed "cold work" after casting and fettling that serves to correct or enhance the cast surface by removing and/or compressing metal using hand tools by punching, engraving, and/or chiseling, and in modern times also with power and pneumatic tools.

Note: In common art historical usage, the term "chasing," when used for bronze sculpture, refers specifically to all of the steps taken to embellish the surface of the cast after fettling. Modern founders include the detailed process of removing and repairing casting defects. In decorative arts, on the other hand, chasing of gilt bronze refers to the steps taken after filing to embellish the surface by punching (compressing the metal using punches), as opposed to engraving.

Sources

Cultural Heritage: Beale 1975; Penny 1993, 298; Dillon 2002, 299; Bassett 2008, 274

Art and Craft Textbooks: Rome and Young 2003, 273; Untracht 1982, 122

Other: Van Langh 2012

To Be Distinguished From

chiseling

engraving

fettling
In modern foundries, fettling may be carried out by the founder using power tools, whereas "chasing" refers to finer work carried out by a specialized person—the chaser—preferentially with hand tools.

punching

French: *ciselure*

Opérations servant à la fois à masquer les imperfections de fonderie et ajouter ou rehausser certains éléments décoratifs, lors desquelles une grande panoplie d'outils est susceptible d'être mise en œuvre (limes, ciseaux, ciselets, burins, etc).

Note: Le terme a un sens plus général que celui utilisé par les orfèvres et monteurs en bronze, pour qui la ciselure désigne exclusivement le travail réalisé au ciselet (Félibien 1690, 337; De Bois 1999), par opposition à la gravure (Hamm 1924, 269; De Bois 1999). Aujourd'hui en fonderie d'art, si la réparure comprend tout le gros œuvre sur le bronze au sortir du moule réfractaire (opération de décochage), souvent exécuté à l'outil tournant, la ciselure n'intervient qu'après et est le fait d'un savoir-faire spécifique.

To Be Distinguished From

ciseler

parachèvement

réparage

réparure

travail au ciseau

ébarbage

ébavurage

Sources

Cultural Heritage: Baudry and Bozo 1978, 633

Historical: Boffrand 1743, 59; Diderot et al. 1751, 3:480, entry « ciseler »

Art and Craft Textbooks: De Bois 1999, 16; Rama 1988, 328

Alternate Translation

ciselure-réparure

German: *ziselieren*

Sources

Maaz 2010, 744; Alscher 1987, 555

Italian: *rifinitura*

Not reserved specifically for metals

Sources

Treccani, under *fusione*; Battaglia 1961, under *rifinitura* [definition 2]; Formigli 2010, 17 and 20f

Alternate Translations

rinettare (v.)
Source: Cellini [1568] 1983, 1, fol. 18v, 38f

cesellare

Source: Battaglia 1961, under *cesellare* [vol. 3, 13]

ritocco a freddo (ritocco a freddo di dettagli)
Source: Formigli 2010, 20f

Chinese: 铸件表面加工

chef-modèle

See figs. 102, 103, 210, 531, 553.

French term for a replica of the artist's model (often in bronze) that is used to make molds for the production of large editions in sand casting. In rare cases, a chef-modèle has been used in lost-wax casting.

Note: Metal is often used rather than plaster because it can better withstand the wear caused by repeated sand molding. To aid in mold making, the chef-modèle is often cast in sections to allow their removal from the sand mold without damaging the piece mold sections. The edition proofs (*épreuves d'édition*) produced with a chef-modèle are a kind of after-cast.

Sources

Cultural Heritage: Grissom and Harvey 2003; Barbour and Sturman 2017, 83; Lebon 2003, 14

Synonyms

pattern

A more general term for a hard model for sand casting. The term is more commonly used in the context of industrial production of machine parts, but may be applied to sculptures as well, though in the context of European bronzes from the early twentieth century onward, "chef-modèle" has been adopted more frequently.

Sources: Cultural Heritage: Motture 2019; Historical: Bolland 1894; Buchanan 1903

French: *chef-modèle*

Modèle, souvent en bronze, servant à marquer l'empreinte dans un moule en sable dans le cas de productions en série. Terme presque exclusivement utilisé pour la fonte au sable.

Note: En sculpture, le chef-modèle est rarement sinon jamais le modèle original de l'artiste.

Sources

Cultural Heritage: Lebon 2012

Art and Craft Textbooks: Rama 1988, 372

To Be Distinguished From

modèle original

Alternate Translations

cuivre

Sources: Cultural Heritage: Baudry and Bozo 1978, 759; Lebon 2012, Glossaire: http://journals.openedition.org/inha/3448

maître-modèle

Source: Art and Craft Textbook: Rama 1988, 372

modèle

Sources: Cultural Heritage: Lebon 2012, Glossaire: http://journals.openedition.org/inha/3448; Copper Industry: Koch and Newell 1963; Bader and Théret 1961, 614; Association Technique de Fonderie, Commission Ingénieurs et Techniciens 1979; Brunhuber 1988; Art and Craft Textbooks: Hamm 1924, 43, 85; Dubos 2003

modèle mère

Source: Copper Industry: Bader and Théret 1961, 615

modèle-maître

Source: Cultural Heritage: Baudry and Bozo 1978, 624

German: *Dauermodell*

Source

Fachkunde Metall: Bovenkerk, Braun, and Dörflinger 1987, 216

Italian: *modello di riferimento*

Alternate Translation

modello

Perhaps due to the lack of consistent industrial art foundry terminology before the nineteenth century, the term is shared in all sculptural fields: marble, plaster, clay, metal.

Chinese: 主铸型

Alternate Translation

主铸模

chiseling

See figs. 226, 249, 254, 278, 281, 550.

Act of using a chisel and hammer to remove metal. With each strike of the hammer the tool jumps, often leaving a visible "step." Chiseling may be part of either fettling or chasing.

Note: This term is sometimes incorrectly used to define all steps to remove metal as well as to compress it (actions that should be referred to as "chasing"). This is an inaccuracy possibly stemming from the French term *ciselure* (which translates to "chaser," not "chiseler").

Sources

Cultural Heritage: Frel 1982, 13

General Dictionary: https://www.merriam-webster.com/dictionary/chiseling [definition 1]

To Be Distinguished From

chasing

French: *travail au ciseau*

Pour la sculpture en bronze, le terme anglais « chiseling » désigne tout travail réalisé avec un ciseau pour couper le métal, à la fois lors de la réparure (correspondrait alors principalement à l'ébarbage, voir Grandpierre and Dargent 1862, 275; Baudry and Bozo 1978, 633) et la ciselure. Il n'existe *a priori* pas de traduction directe en français, si ce n'est « travail au ciseau ».

To Be Distinguished From

ciselure

ciselure-réparure

German: *meißeln*

Source

Bol 1985, 139

Italian: *incisione a bulino*

Sources

Battaglia 1961, under *bolino*; Battaglia 1961, under *incisione* [definition 9, vol. 7, 686]

Alternate Translation

Scalpello battuto a Martello
Source: Formigli 1999d, 148

Chinese: 凿

(v., n.)

Source

Ming 2010, 1735

Alternate Translations

錾 (v., n.)
Source: Ming 2010, 1735

镌 (v.)

雕 (v.)

coating

See figs. 322, 323, 324, 326, 330.

Purposefully applied surface deposits or films on bronze sculpture that consist of materials chemically different from the metal substrate. Their composition can vary widely, ranging from natural substances such as lacquer, resin, oil, and wax to synthetic resins. Coatings may be decorative and/or protective. In the study of Renaissance bronzes, for example, applied lacquer-like coatings are often referred to as "organic patinas."

Note: Not to be confused with corrosion products or mineral compounds bonded to the metal surface, which fall under patina. Though metals are also used to coat surfaces, they are referred to here as plating.

Sources

Art and Craft Textbook: Hughes and Rowe 1989, 45–46

Other: Weil 1977; Considine et al. 2010, 94–95

French: *revêtement*

Couche volontairement appliquée en surface d'un bronze pour le protéger ou le décorer, constituée d'un matériau différent du substrat en bronze. Une grande variété de matériaux peut être utilisée, depuis des substances naturelles (laques, résines, huiles, cires) jusqu'à des produits synthétiques (résines, peintures).

Note: Ne pas confondre avec les patines chimiques, qui contrairement aux revêtements, réagissent avec le substrat en bronze pour former la couche de surface. Par ailleurs, on préférera désigner les revêtements métalliques par un terme spécifique, en l'occurrence placage métallique.

Sources

Cultural Heritage: Baudry and Bozo 1978, 656; Aucouturier et al. 2003

Copper Industry: Levy and Saurat 2005

German: *Überzug*

Source

Koller and Baumer 2000

Italian: *rivestimento*

Sources

Treccani, under *rivestimento* [definition 2]; Battaglia 1961, under *rivestimento* [vol. 16, 1058]; Giuffredi 2006, 221

Chinese: 涂层

Sources

Ming 2010, 1448; TNATD

Alternate Translation

涂料
Sources: Ming 2010, 1449; TNATD

cold shut

See figs. 6, 159.

The interface where two streams of metal come together in the mold but do not fuse properly, often due to premature cooling of the metal in the mold. A cold shut may also describe a hole or void in a cast caused by premature cooling (Rome and Young 2003, 303). The cooled metal edges will be rounded in profile.

Sources

Cultural Heritage: Bassett 2008

Historical: Bolland 1894; Buchanan 1903

Copper Industry: Ammen 1980; American Foundry Society (AFS) n.d.; Palmer 1929

Art and Craft Textbook: Rome and Young 2003, 303

General Dictionary: https://www.merriam-webster.com/dictionary/cold shut [definition 1]

To Be Distinguished From

cold tearing
Cold tearing is generated by the same phenomenon as a cold shut, but leading to a crack.

Synonyms

cold-shot (alternate spelling)

cold-shut (alternate spelling)

French: *reprise*

Défaut de fonderie dû à la fusion imparfaite de deux flux de métal liquide qui se rencontrent dans le moule réfractaire, laissant en surface un sillon. On parle de « solution de continuité ».

To Be Distinguished From

tapure à froid
Note: Même phénomène que la reprise, mais la rencontre entre les deux flux est tellement imparfaite qu'elle engendre une fissure au refroidissement.

Sources

Cultural Heritage: Baudry and Bozo 1978, 647

Copper Industry: Hénon, Mascré, and Blanc 1971, 27; Association Technique de Fonderie, Commission Ingénieurs et Techniciens 1979

Historical: Guettier 1858, notice 616, pp. 322–23

Alternate Translation

goutte froide
Attention, goutte froide désigne également une particule métallique, généralement oxydée, de même composition que la pièce, incluse en surface de la pièce (Baudry and Bozo 1978, 648).
Source: Copper Industry: Association Technique de Fonderie, Commission Ingénieurs et Techniciens 1979

German: *Kaltguss*

Source

Brunhuber 1986

Alternate Translation

Kaltschweisse

Italian: *accostatura*

Alternate Translations

giunzione fredda

solidificazione prematura

Chinese: 冷界

Source

TNATD

Alternate Translations

冷结
Source: Ming 2010, 883

冷隔
Source: Ming 2010, 883

流界
Source: TNATD

core

See figs. 8, 10, 57, 61, 69, 112, 428, 431, 473, 475.

The portion of the refractory mold that defines the internal space in a hollow bronze sculpture. It may be formed in a variety of ways and is usually (but not always) made of similar material as that used for the outer portion of the mold. The term is also used as shorthand to refer to the material it is made of. A core is generally solid but may be hollow (see GI§2.1.1).

Sources

Cultural Heritage: Bassett and Fogelman 1997

Historical Source: Bolland 1894

Copper Industry: Brunhuber 1988; Association Technique de Fonderie, Commission Ingénieurs et Techniciens 1979; Palmer 1929

Art and Craft Textbook: Rome and Young 2003

General Dictionary: https://www.merriam-webster.com/dictionary/core [definition 1e]

French: *noyau*

Partie interne du moule réfractaire déterminant le volume interne d'une fonte creuse. Le noyau est souvent (mais pas systématiquement) réalisé dans le même matériau que la partie externe du moule réfractaire (la chape). Le noyau peut-être plein ou creux.

Sources

Cultural Heritage: Rolley 1994, 70; Baudry and Bozo 1978, 625; Bewer, Bourgarit, and Bassett 2008; Azéma and Mille 2013b; Lebon 2012, Glossaire: http://journals.openedition.org/inha/3448

Copper Industry: Cuénin 1994, 5; Koch and Newell 1963; Association Technique de Fonderie, Commission Ingénieurs et Techniciens 1979; Brunhuber 1988

Historical: Mariette 1768, 61–64; Félibien 1690, 324; Launay 1827, 2:277; Guettier 1858, notice 605, pp. 311–12; Boffrand 1743, 27 et sq; Diderot et al. 1751, 11:268, entries « noyau terme de fonderie » et « noyau terme d'artillerie »

Art and Craft Textbooks: Lambert 2002, 272; Rama 1988, 374

Alternate Translations

âme

Sources: Historical: Félibien 1690, 325; Boffrand 1743, 27 et sq; Diderot et al. 1751, 11:268, entries « noyau terme de fonderie » et « noyau terme d'artillerie »

German: *Kern*

Source

Brunhuber 1988

Italian: *anima*

Sources

Treccani, under *anima* [definition 4]; Battaglia 1961, under *anima* [definition 16, vol. 1, 480]; Giuffredi 2006, 219; Biringuccio [1540] 1977, fols. 77–78; Bruni 1994, 81–82; Brunhuber 1988

Alternate Translations

anima di fusione

Source: Bruni 1994, 81–82

maschio

Early Italian term for core that literally translates as "male."

Sources: Leonardo 1490, fols. 156, 157v; Biringuccio [1540] 1977, fols. 77v, 84

nocciolo

Early Italian term for core that means "nut."

Source: Cellini [1568] 1983, 11, fols. 48, 49

Chinese: 芯型

Source

Ming 2010, 1573

Alternate Translation

芯范

Source: Zhang 2010, 82

core pin

See figs. 35, 36, 37, 64, 122, 547.

A metal rod, nail, or wire that is embedded in both the core and the outer mold and serves to secure the core in place during the pour. Core pins have traditionally been made of copper alloys, iron, or steel, and today are generally made of stainless steel.

Note: The term "transverse core pin" (or "transfixing core pin") is used to describe long core pins that extend through both sides of the core (e.g., straight through a limb) (figs. 35, 64, 122).

Sources

Cultural Heritage: Penny 1993; Lie and Bewer 2014; Mattusch 1996, 22–24

Copper Industry: Brunhuber 1988; Association Technique de Fonderie, Commission Ingénieurs et Techniciens 1979

Art and Craft Textbook: Rome and Young 2003

To Be Distinguished From

chaplet

core plate

spacer
Note that "spacer" has been used as a synonym for "core pin" in some technical studies in reference to ancient bronzes (Mattusch 1996, 22–24).

Synonyms

core nail

core rod

French: *fer à noyau*

Tige ou clou, parfois fil, généralement en fer servant à maintenir le noyau en place dans le moule réfractaire lors de la coulée.

Source

Cultural Heritage: Bewer, Bourgarit, and Bassett 2008

Alternate Translations

broche
Pour certains auteurs de l'industrie, terme spécifique à la coulée sous pression (Association Technique de Fonderie, Commission Ingénieurs et Techniciens 1979; Brunhuber 1988).
Sources: Copper Industry: Association Technique de Fonderie, Commission Ingénieurs et Techniciens 1979; Brunhuber 1988; Art and Craft Textbook: Dubos 2003

clou
Sources: Cultural Heritage: Baudry and Bozo 1978, 626; Art and Craft Textbooks: Rama 1988, 257; Lambert 2002, 181

clou distanciateur
Source: Cultural Heritage: Mille 2007, 97

distanciateur
Source: Cultural Heritage: Rolley 1994, 67

fer
Source: Art and Craft Textbook: Rama 1988, 375

fer de maintien
Source: Art and Craft Textbook: Rama 1988, 249

fer de soutien

German: *Kernstift*

Source

Association Technique de Fonderie, Commission Ingénieurs et Techniciens 1979

Italian: *distanziatore*

Source

Bruni 1994, 72

Alternate Translations

chiodo distanziatore
Sources: Giumlia-Mair 1999, 257; Pecchioli 1999, 193

chiodo di sostegno
Same as "core support"
Source: Treccani, under *fusion a cera persa*

Chinese: 销钉

Source

Ming 2010, 1559; TNATD

core support

See figs. 9, 16, 25, 42, 122, 195.

A general term for a variety of metal or other features that reinforce or support the core during the casting process. This term is used in different ways depending on the context. In industry, it is applied to metal inserts or spacers that hold the core in place during the pour; therefore, core pins, chaplets, and mold extensions are examples of core supports. In cultural contexts, the term is applied to internal wires or rods that help strengthen the core during assembly of the casting model and/or during the pour. Generally reserved for smaller wires in projecting limbs, or wires and rods used to strengthen joints between separately molded wax sections in the indirect lost-wax process. For clarity, when the second definition is intended, the term "internal core support" should be used.

Sources

Cultural Heritage: Dillon 2002; Bassett 2008; Sturman 2004; Stone 2008

Copper Industry: American Foundrymen's Society 1984

French: *support de noyau*

Terme général pour décrire tout dispositif servant à maintenir le noyau en place dans le moule réfractaire lors de la coulée (fer à noyau, cale à noyau, voire portée de noyau et armature).

Note: En anglais, à cette définition propre à la fonderie industrielle vient s'ajouter une définition plus particulière à la sculpture en bronze pour désigner également tous les systèmes d'armatures internes faits de fils de fer pour renforcer le noyau.

Sources

Copper Industry: Association Technique de Fonderie, Commission Ingénieurs et Techniciens 1979; Brunhuber 1988

Alternate Translation

support pour noyau
Source: Copper Industry: Brunhuber 1988

German: *Kernstütze*

Sources:

Brunhuber 1988; Association Technique de Fonderie, Commission Ingénieurs et Techniciens 1979

Italian: *chiodo di sostegno*

Same as "core pin"

Source

Treccani, under *chiodo di sostegno*

Alternate Translations

armatura interna

sopporto reggi anime
Source: Brunhuber 1988

sostegno di bronzo
Also found with alternative spelling: *sustegno*
Source: Biringuccio [1540] 1977, fol. 82

supporto dell'anima di fusione

Chinese: 芯骨

Source

Ming 2010, 1572

corrosion

See figs. 310, 311, 312, 314, 315, 318, 334, 348, 500.

A chemical process that causes a metal such as bronze to change from a metallic state into a chemically more stable mineral compound known as a corrosion product.

Note: Most metals undergo natural corrosion, except for pure so-called noble metals such as gold and platinum-group metals. A chemical patina refers to corrosion of the surface by either natural causes (e.g., burial) or the intentional application of chemicals. The minerals formed during the process may be similar to ones from which the metal was smelted or refined prior to manufacture. Sometimes artificially induced chemical patinas intentionally mimic natural corrosion products.

Sources

Cultural Heritage: Gettens 1970; Scott 2002; Chase 1994; Selwyn and Canadian Conservation Institute 2004

General Dictionary: https://www.merriam-webster.com/dictionary/corrosion

Other: Scott 1991, 43–47, 81

French: *corrosion*

Pour un bronze, transformation chimique du métal en un composé non métallique sous l'action d'un environnement particulier. Désigne aussi le produit résultant.

Sources

Cultural Heritage: Baudry and Bozo 1978, 648; Aucouturier et al. 2003; Robbiola 2015

Art and Craft Textbook: Rama 1988, 372

German: *Korrosion*

Source

Fachkunde Metall: Bovenkerk, Braun, and Dörflinger 1987, 128

Italian: *corrosione*

Sources

Treccani, under *corrosione*; Battaglia 1961, under *corrosione* [definition 2, vol. 3, 845]

Chinese: 腐蚀

Sources

Ming 2010, 452; TNATD

Alternate Translation

锈蚀
Sources: Ming 2010, 1586; TNATD

edition

See figs. 261, 551.

The intentional production of a sculpture in several virtually identical casts, usually from the same set of molds derived from the original model. In modern castings, item number and total number of multiples produced is often reported somewhere on the surface, as it has legal value.

Note: Casts from any given edition will be essentially identical in form and size (understanding that within an edition, different bases may be used, and there may be variations in how the edition is mounted), yet may vary owing to casting flaws or differences in chasing, patination, or deterioration of the molds/chef-modèle over time. Whereas before the nineteenth century it was rare for replicas to be made without small variations (hence the use of the term "variants"), modern bronze casting practice is more consistent, and generally predicated on the notion of editions. Starting in the early twentieth century, it became common practice to mark casts with their individual number within an edition.

Sources

Cultural Heritage: Bassett and Fogelman 1997, 32; Beale 1975; Dillon 2002

To Be Distinguished From

copy

replica

version

variant

Synonym

editioned replica

French: *édition*

Désigne les épreuves issues d'un même modèle original multiplié dans des matériaux tels que plâtre, alliage métallique, terres cuites, etc.

Note: Le nombre d'épreuves dépend de la volonté de l'artiste ou de celle du propriétaire des droits de reproduction et, depuis 1967, il est en France strictement limité par la loi pour que les épreuves puissent être qualifiées d' « originales ». Le terme se décline en de nombreuses variantes plus ou moins spécifiques: bronze d'édition (œuvres de très grande diffusion, sans numérotation, Rama 1988, 372), épreuves d'édition, exemplaires d'édition (Baudry and Bozo 1978, 545), etc.

Sources

Cultural Heritage: Baudry and Bozo 1978; Lebon et al. 2016, 330

Art and Craft Textbook: Rama 1988

German: *Auflagenguss*

Source

Maaz 2010, 671, 683

Alternate Translation

Edition
Source: Mietzsch 2009, 103

Italian: *edizione*

Source

Treccani, under *edizione*

Alternate Translations

multiplo
In the arts, specifies that the cast is not a one-off but part of a group of essentially identical sculptures.
Source: Treccani, under *multiplo* [definition 1]

serie
Correctly describes the presence of a number of issued sculptures; less used in artwork because production "in serie" has a less valuable industrial production meaning.

Chinese: 版本

Alternate Translation

翻版
Source: Ming 2010, 310; TNATD

engraving

See figs. 241, 243, 254, 276, 279, 280, 495.

In relation to bronzes, the process of decorating the surface by removing material with a chisel, burin, or graver that creates a V-shaped groove.

Note: Engraving may be undertaken in the chasing process, and/or to form or enhance signatures. See I.6§2.3.

Sources

Cultural Heritage: Beale 1975, 43

Copper Industry: Brunhuber 1988; Maskinaktiebolaget Karlebo 1982

Art and Craft Textbook: Untracht 1982, 283

General Dictionary: https://www.merriam-webster.com/dictionary/engraving

To Be Distinguished From

chasing

French: *gravure*

Technique de décoration des bronzes par enlèvement de matière à l'aide de burins.

Sources

Cultural Heritage: Baudry and Bozo 1978, 578

Copper Industry: Brunhuber 1988; Maskinaktiebolaget Karlebo 1982

Art and Craft Textbook: Hamm 1924

German: *gravieren*

Source

Brunhuber 1988

Alternate Translation

Gravur
Sources: Weihrauch 1944; Brunhuber 1988; Maskinaktiebolaget Karlebo 1982

Italian: *incisione*

Sources

Treccani, under *incisione* [definition 1]; Battaglia 1961, under *incisione* [definition 8, vol. 7, 686]; Brunhuber 1988; Maskinaktiebolaget Karlebo 1982

Chinese: 阴刻

Source

TNATD

Alternate Translation

雕刻
Sources: Ming 2010, 310; TNATD

fettling

See figs. 104, 552.

Steps carried out directly after casting to remove unwanted features, including oxidized metal, sprues, core pins, flashing, etc. Fettling may entail the use of power tools and/or hand tools such as saws, chisels, hammers, coarse files, and abrasives.

Sources

Cultural Heritage: Salter and Gilmour n.d., 23; Penny 1993

Copper Industry: Association Technique de Fonderie, Commission Ingénieurs et Techniciens 1979

Art and Craft Textbook: Untracht 1982, 526

To Be Distinguished From

chasing

French: *réparure*

Ensemble des opérations visant à nettoyer un bronze au sortir du moule réfractaire, avant ciselure et/ou patination.

Note: Suivant les périodes et l'organisation de l'atelier la réparure peut être subdivisée en plusieurs opérations distinctes et conduire, de fait, à des termes plus spécialisés. Au 19e siècle, par exemple, la réparure consiste, dans l'ordre, en l' ébarbage, le rachevage, la réparure proprement dite, et la ciselure (Baudry and Bozo 1978, 633; Grandpierre and Dargent 1862, 275; Grandpierre and Avale 1867, 8n1).

To Be Distinguished From

ciselure

ciselure-réparure

Sources

Cultural Heritage: Baudry and Bozo 1978; Bewer, Bourgarit, and Bassett 2008; Azéma and Mille 2013b

Historical: Boffrand 1743, 58; Gonon 1876, 13, 39; Grandpierre and Dargent 1862; Grandpierre and Avale 1867; Mariette 1768, 127

Art and Craft Textbooks: Rama 1988, 375; Lambert 2002, 275

Alternate Translations

ébarbage
Traduction française de « fettling » d'après la source citée ici
Source: Copper Industry: Association Technique de Fonderie, Commission Ingénieurs et Techniciens 1979

ébavurage
Traduction française de « fettling » d'après la source citée ici
Source: Copper Industry: Association Technique de Fonderie, Commission Ingénieurs et Techniciens 1979

réparage
Sources: Historical: Mariette 1768, 127; Cultural Heritage: Baudry and Bozo 1978, 762

reparure
Variante orthographique. Le terme « reparer » apparaît dans Félibien 1690, 333; Boffrand 1743, 58; Lebon 2012.
Sources: Historical: Gonon 1876, 12; Art and Craft Textbook: Rama 1988, 375

German: *Abgraten*

Source

Association Technique de Fonderie, Commission Ingénieurs et Techniciens 1979

Alternate Translations

Entgraten
Source: Association Technique de Fonderie, Commission Ingénieurs et Techniciens 1979

Fertigputzen
Source: Association Technique de Fonderie, Commission Ingénieurs et Techniciens 1979

Putzen
This term is used primarily in the modern casting industry and for mass production; in an art foundry it is not always distinguished as a separate step from chasing.
Source: Brunhuber 1986, 747

Italian: *sbavatura*

Removal of flashing or traces of sprues

Sources

Sources: Treccani, under *sbavatura* [definition 2c]; Battaglia 1961, under *sbavatura* [second entry, vol. 17, 663]

Alternate Translations

rifinitura

Same word used for chasing

Source: Treccani, under *rifinitura*; Battaglia 1961, under *rifinitura* [definition 2, vol. 16, 257]

sgrossatura

Designates more specifically the quick removal prior to chasing; see *chiseling*

Chinese: 清砂

Source

Ming 2010, 1164

Alternate Translations

修整

Source: TNATD

清理

Sources: Ming 2010, 1828; TNATD

精整

Source: TNATD

flashing

See figs. 6, 59, 61, 77, 90, 249, 545.

A ridge of excess metal that can occur when molten metal enters cracks in the refractory mold (both outer and core). Flashing most often rises perpendicularly to the inner or outer wall of bronze, although a gap in consecutive layers of the mold material may result in thin flanges of excess bronze that spread parallel to the metal wall. This type of feature can appear on the cast's internal or external surfaces (respectively called "core flashing" and "mold flashing"), and in the latter case is often removed during fettling. Flashing may also occur along seam lines of an ill-fitted piece mold.

Sources

Cultural Heritage: Beale 1975, 41

Copper Industry: Ammen 1980; Brunhuber 1988; Association Technique de Fonderie, Commission Ingénieurs et Techniciens 1979

Art and Craft Textbook: McCreight 1996

To Be Distinguished From

seam line
Note: The term "flashing" is also used to describe a seam line, but to avoid confusion, we recommend that it be used specifically in connection with casting defects.

Synonyms

finning
Source: Copper Industry: Association Technique de Fonderie, Commission Ingénieurs et Techniciens 1979

veining

French: *gerce*

Défaut de fonderie où le métal liquide pénètre dans des fissures du moule réfractaire. S'agissant de la partie interne du moule réfractaire (le noyau, on parle alors de gerces de noyau), le défaut apparaît sur la surface interne du bronze comme une excroissance en forme de veine. S'agissant de la partie externe du moule réfractaire, un défaut similaire apparaît mais sur la surface externe du bronze (notamment — mais pas seulement — à l'endroit des joints de moule quand le moule est constitué de plusieurs parties assemblées).

To Be Distinguished From

balesvre

balevre

barbe

barbure

bavochure

bavure

bavure de joint

couture

suture

toiles

ébarbure

Sources

Cultural Heritage: Baudry and Bozo 1978, 645; Bourgarit, Bewer, and Bresc-Bautier 2014

Copper Industry: Hénon, Mascré, and Blanc 1971, 17; Association Technique de Fonderie, Commission Ingénieurs et Techniciens 1979

Alternate Translations

gerçure
Sources: Historical: Boffrand 1743, 62; Gonon 1876, 6

nervure
Sources: Copper Industry: Hénon, Mascré, and Blanc 1971, 17; Association Technique de Fonderie, Commission Ingénieurs et Techniciens 1979

German: *Grat*

Source

Fachkunde Metall: Bovenkerk, Braun, and Dörflinger 1987, 219

Alternate Translation

Gussgrat

Italian: *bava*

Sources

Biringuccio [1540] 1977, fol. 78

Alternate Synonym

cresta di fusione
Sources: Carruba 2006, 24; Pecchioli 1999, 192

Chinese: 毛边

Source

Ming 2010, 997

Alternate Translations

溢料
Source: TNATD

溢边
Source: TNATD

飞边
Sources: Ming 2010, 409; TNATD

founder

See fig. 546.

Expert head of the foundry or the person who pours the metal. Person(s) responsible for the translation of the artist's sculptural model into cast metal sculptures. This may entail a variety of specialized operations, from mold making to wax chasing, investing, alloying, casting, fettling, assembling, chasing, and patination. The artist may in some cases also take on one or more of these roles.

Sources

Cultural Heritage: Boulton 2006

Historical: Buchanan 1903

Copper Industry: Brunhuber 1988; Association Technique de Fonderie, Commission Ingénieurs et Techniciens 1979

Synonyms

foundryman
Source: Copper Industry: Association Technique de Fonderie, Commission Ingénieurs et Techniciens 1979

foundrywoman

foundryworker

French: *fondeur*

Personne qui prend en charge la fonte d'une statue. Désigne souvent la personne qui dirige la fonderie.

Note: Pour lever l'ambiguïté, le terme de fondeur-fondant a été proposé (Delon 1877, 178) pour désigner spécifiquement l'opérateur (Lambert 2002, 13), le fondeur désigne plutôt un industriel, par opposition au bronzier ou bronzeur. Le terme de fondeur trouve de nombreuses déclinaisons plus ou moins spécifiques. Citons notamment fondeur d'art, fondeur d'ornement (Rama 1988, 22), sculpteur-fondeur (Bresc-Bautier 1989).

To Be Distinguished From

fabricant

mouleur
Le terme mouleur peut être ambigu. Si certains auteurs insistent sur la différence entre mouleur et fondeur (Guettier 1858 [édition originale 1844], 284), d'autres confondent les deux métiers (Launay 1827, 2:276, l'auteur précise "mouleur en cuivre"). Le fabricant est celui qui sous-traite la fonte (Lebon 2003).

Sources

Cultural Heritage: Baudry and Bozo 1978; Bresc-Bautier 1989; Rolley 1994, 72; Bewer, Bourgarit, and Bassett 2008; Azéma and Mille 2013b

Copper Industry: Brunhuber 1988; Association Technique de Fonderie, Commission Ingénieurs et Techniciens 1979; Koch and Newell 1963

Historical: Félibien 1690, 34; Boffrand 1743, 3, 4, 38, 41, and 44; Diderot et al. 1751, 7:79, entry « fondeur »; Mariette 1768, ii–iv, 72, 83, 108, 110, 128, 132–33; Guettier 1858, 284

Art and Craft Textbooks: Rama 1988, 22; Lambert 2002, 269

Alternate Translations

bronzeur
Source: Art and Craft Textbook: Lambert 2002, 13

bronzier
Pour Lambert, le bronzier ou bronzeur désigne l'ouvrier fondeur, celui qui fond et coule le bronze.
Source: Art and Craft Textbook: Lambert 2002, 13

fondeur en cuivre
Source: Chesnel de la Charbouclais 1857, notice n° 1038

fondeur en terre et sable
Ce que recouvre le terme « fondeur en terre » reste confus. Il fait probablement référence aux techniques utilisant la terre, cire perdue et technique de fonte de cloches et de canons, sachant que pour ces derniers des termes spécifiques peuvent être utilisés (fondeur de cloches, fondeur de canons ou fondeur d'artillerie). Le terme « fondeur à cire perdue » existe, mais non comme une catégorie professionnelle spécifique avant le début du 20e siècle. Par exemple, les fondeurs utilisant la cire perdue travaillant pour des orfèvres ou des bijoutiers sont dénommés « fondeurs pour orfèvres » ou « fondeurs pour joailliers » plutôt que par le procédé mis en œuvre Chesnel de la Charbouclais 1857, notice n° 1038.
Sources: Cultural Heritage: Bresc-Bautier 1989; Historical: Chesnel de la Charbouclais 1857

mouleur en terre ou en sable
Source: Historical: Chesnel de la Charbouclais 1857

German: *Gießer*

Sources

Brunhuber 1988; Gesamtverband Deutscher Metallgießereien 1982

Alternate Translations

Gießereieinhaber
Source: Association Technique de Fonderie, Commission Ingénieurs et Techniciens 1979

Gießereifachman
Sources: Brunhuber 1988; Gesamtverband Deutscher Metallgießereien 1982

Gießereileiter
Source: Association Technique de Fonderie, Commission Ingénieurs et Techniciens 1979

Gießereiunternehmer

Schmelzmeister
This designates the operator standing next to the furnace and deciding when to pour the molten metal into the mold.

Italian: *artefice*

Many sources used the general terms *artefice* or *maestro* to refer to founders.

Source

Biringuccio [1540] 1977, fols. 75–76

Alternate Translations

fonditore
Sources: Treccani, under *fonditore* [definition 1]; Morigi and Morigi 2008, 12; Bruni 1994, 133

fusore

gettatore
Source: Cellini [1568] 1983, 11, fols. 51v, 55

maestro
Sources: Biringuccio [1540] 1977, fol. 75, 107

tragettatore
Sources: Baglione 1642, 327; Marconi 2004, 71n71

Chinese: 铸工

Source

TNATD

Alternate Translation

铸造工
Source: Ming 2010, 1831

foundry model

See figs. 210, 531, 544, 553, 554.

A general term for any model or replica of the artist's model around which a mold is formed for bronze casting. It is made by the foundry in order to preserve the artist's model. A foundry model may also be used as a reference for the finishing of a bronze for the purpose of quality control.

Note: In sand casting, a foundry model is referred to as a "pattern" or "chef-modèle."

Sources

Cultural Heritage: Boulton 2006; Beale 1975; Beentjes 2019

French: *modèle de fonderie*

Terme général désignant tout modèle, modèle original de l'artiste ou réplique, servant à réaliser un moule pour une fonte, tant pour la cire perdue que pour la fonte au sable. Englobe les termes plus spécifiques à une technique de fonte donnée (e.g., chef-modèle pour la fonte au sable).

Note: Le terme de « modèle de fonderie » est rarement utilisé, jugé trop long : on lui préfère le terme de « modèle ». Cependant, pour éviter les confusions, nous recommandons vivement d'employer le terme complet.

Alternate Translations

modèle

Sources: Cultural Heritage: Baudry and Bozo 1978, 624; Historical: Launay 1827, 2:275; Art and Craft Textbooks: Hamm 1924, 74–82; Rama 1988, 36; Lambert 2002, 26

modèle servant à la fonte

Source: Cultural Heritage: Bewer, Bourgarit, and Bassett 2008

plâtre de fonderie

Comme son nom l'indique, ne concerne que les modèles en plâtre. On parle aussi de plâtre d'atelier (Jean Dubos, pers. comm., July 2021).

Source: Cultural Heritage: Lebon et al. 2016, 331

pré-modèle

Source: Cultural Heritage: Baudry and Bozo 1978, 624

German: *Ausführungsmodell*

Source

Maaz 2010, 671

Italian: *forma da gittar di bronzo*

Source

Biringuccio [1540] 1977, fol. 82v

Alternate Translation

modello di fonderia

Chinese: 铸型

Alternate Translation

铸模

gilding

See figs. 287, 290, 293, 294, 295, 296, 297, 299, 301, 555.

The application of gold to the surface of a bronze sculpture. This can be achieved by a variety of methods. Traditionally gilding was mainly undertaken using leaf gilding or mercury gilding. Starting in the nineteenth century, electrochemical plating or deposition began to be used, as well as gold powder paint or wax. See I.7.

Sources

Cultural Heritage: Drayman-Weisser 2000; Oddy 1993; Salter and Gilmour n.d.; Motture 2019, 39

French: *dorure*

Opération consistant à appliquer une couche d'or en surface, quel que soit le procédé. Désigne également le résultat.

Note: Nous n'avons pas trouvé mention du substantif « dorure » dans Félibien 1690, mais uniquement du verbe « dorer », p. 334.

Sources

Cultural Heritage: Baudry and Bozo 1978, 657; Azéma and Mille 2013b; Aucouturier et al. 2003

Historical: Diderot et al. 1751, 5:57–60, entry « dorure »

German: *Vergoldung*

Source

Lein 2004, 58

Italian: *doratura*

Sources

Treccani, under *doratura*; Battaglia 1961, under *doratura* [vol. 4, 963]; Cellini [1568] 1983, 1, fol. 41

Chinese: 镀金

Sources

Ming 2010, 328; TNATD

Alternate Translation

鎏金
Although this literally translates to "gold-plating," it is widely used to describe the technique of mercury gilding on ancient Chinese works.

inlay

See figs. 309, 340, 342, 343, 345, 346, 349, 351, 365.

Decorative element on a sculpture usually added for visual contrast through color and/or texture. The top surface of an inlay is generally flush with the surrounding metal. A wide range of attachment methods may be used, including solder, adhesives, cements, burrs, undercuts, and/or rivets, but the basic technique involves shaping the inlay and incising the ground metal to produce a cavity of the inlay shape. Inlay materials may include metals, glass, stone, or bone, among others.

Note: The terms "inlay" and "overlay" are often used indiscriminately, but they differ in one main regard: inlays are inserted into the bronze surface, whereas overlays are set on top of the surface.

Sources

Cultural Heritage: Hemingway and Abramitis 2017

Art and Craft Textbook: Untracht 1982

General Dictionary: https://www.merriam-webster.com/dictionary/inlay

To Be Distinguished From

encrustation

foil

incrustation

overlay

Synonyms

damascene
This term is often used interchangeably to describe inlays and/or overlays, especially in iron or steel arms and armor and decorative metalwork. As the term is imprecise and more applicable to those types of materials, we recommend it not be used in the context of bronze casting. The term "false damascening" is somewhat of a misnomer, as it is a different technique that renders a similar visual result: instead of cutting recesses, the metal is superficially scored and thin metal foil or wire is hammered onto the surface. Since this joint may be weak, the surface is heated to further fuse the two metals. Visually the two techniques appear similar.
Source: Art and Craft Textbook: Untracht 1982

incrustation

French: *incrustation*

Élément décoratif inséré dans la surface d'un bronze ou formant relief, fait d'un matériau différent du substrat en bronze pour un effet polychromatique. Une grande variété de matériaux et de techniques peut être mise en œuvre.

Note: il n'existe aucun terme français pour distinguer précisément un élément inséré dans la surface du bronze (« inlay » en anglais) d'un élément formant relief (« overlay »). Le japonais, au contraire, offre un vocabulaire très riche et très spécifique.

To Be Distinguished From

damassé
Acier forgé par pliages successifs pour faire apparaître un décor.

Alternate Translation

damasquinure
Pour certains auteurs (Arminjon and Bilimoff 1998, 162), désigne une sous-catégorie d'incrustations, en l'occurrence l'incrustation d'un métal sur un autre [on parle aussi de damasquinage pour décrire le procédé (Arminjon and Bilimoff 1998, 162–63; Baudry and Bozo 1978, 661). Pour d'autres auteurs, la damasquinure ne concerne que les incrustations d'or ou d'argent sur un objet en fer ou acier, typiquement une arme ou une armure (Félibien 1690, 460). Il est suggéré d'éviter d'employer ce terme pour la sculpture en bronze, on lui préférera « incrustation métallique ».
Sources: Cultural Heritage: Arminjon and Bilimoff 1998, 162–63; Historical: Félibien 1690, 460

German: *Einlage*

In principle more general (actually a direct translation of "inlay").

Source

Cüppers 1994, 1013–16

Alternate Translation

Tauschierung
Only if the inlay is of linear character, e.g. wires; you wouldn't speak of a *Tauschierung* in the case of inlaid silver eyes.
Source: Lein 2004, 63

Italian: *agemina*

A decorative process usually performed on steel and copper alloys, with thin silver wires in an undercut groove on the surface.

Source

Treccani, under *agemina*

Alternate Translations

damaschinatura

More specifically referring to inlaid decoration on bronze, brass, or iron with contrasting colored metals

inserto

intarsio

General term referring to the insertion of different materials on a decorated surface

Sources: Treccani, under *intarsio*; Battaglia 1961, under *intarsio* [vol. 8, 169]

lavori di tanccia

As in "Damascho fanno gli azzimini che commetteno in quei loro vasi pezzetti d'oro..."; see Biringuccio [1540] 1990, 373; note that translators Smith and Gnudi assume this is a misprint.

Source: Biringuccio [1540] 1977, fol. 138

Chinese: 镶嵌

Sources

Ming 2010, 1554; TNATD

inter-model

See figs. 1, 16, 23, 25, 41, 55, 72, 556.

Replica in wax obtained from the reusable mold of an original model. Inter-models are used in indirect lost-wax casting. Inter-models are often slush molded (fig. 16, Case Study 7.) One inter-model may vary from another through additions or changes made in the wax before the investment is applied.

Note: The term is subject to dispute among CAST:ING members. Depending on discipline and expertise, some initially preferred "wax working model," "wax intermediary model," or simply "wax." The current term was finally agreed upon. An inter-model is a type of casting model or foundry wax, but the latter two refer to any wax model that is destroyed during casting, including those used in direct or indirect lost-wax casting processes.

Sources

Cultural Heritage: Motture 2019; Dillon 2002; Avery 2018, 163

Synonym

intermediary model

French: *modèle intermédiaire*

Modèle en cire utilisé dans le procédé indirect de fonte à la cire perdue, obtenu dans le moule à bon creux.

Note: Terme très débattu parmi les spécialistes du bronze et en particulier les membres de CAST:ING. Il convient de bien distinguer le modèle intermédiaire de « l'épreuve en cire », en l'occurrence la cire telle que directement sortie du moule à bon creux, qui est susceptible d'être retouchée, augmentée, et de conduire ainsi au modèle intermédiaire proprement dit qui sera fidèlement traduit en bronze.

Source

Cultural Heritage: Lamouche 2021

Alternate Translations

cire

Ce terme est ambigu: une cire peut être à la fois le produit final désiré par l'artiste, le modèle en cire utilisé dans le procédé direct, l'épreuve, et le modèle intermédiaire utilisés dans le procédé indirect.

Source: Cultural Heritage: Bresc-Bautier 1989

épreuve en cire

Ce terme est ambigu: une épreuve en cire peut être à la fois le produit final désiré par l'artiste, le modèle en cire utilisé dans le procédé direct, l'épreuve et le modèle intermédiaire utilisés dans le procédé indirect.

Sources: Cultural Heritage: Baudry and Bozo 1978, 254; Bewer, Bourgarit, and Bassett 2008; Azéma and Mille 2013b; Bourgarit, Bewer, and Bresc-Bautier 2014; Art and Craft Textbook: Lambert 2002, 214

intermodèle

Source: Cultural Heritage: Lamouche 2021

modèle auxiliaire

Source: Cultural Heritage: Mille 2007

modèle de cire

Ce terme est ambigu: un modèle en cire peut être à la fois le produit final désiré par l'artiste, le modèle en cire utilisé dans le procédé direct, l'épreuve et le modèle intermédiaire utilisés dans le procédé indirect.

Sources: Cultural Heritage: Rolley 1994; Bresc-Bautier 1989

German: *Wachsmodell*

Sources

Alscher 1987, 554; Mietzsch 2009, 6

Italian: modello intermedio

Chinese: 范制蜡型

investment

See figs. 5, 13, 557, 558.

The term can refer to the refractory mold used in the lost-wax casting process or to the material used to make that mold. It also denotes the process of coating or embedding the wax model in this material and is applicable to clay-based, plaster-based, and ceramic shell molds. In all of these, the first layers have a special, fine consistency that is designed to pick up the detail and avoid problems during casting; the later layers are coarser. Clay-based investment may also be referred to as "loam." The investment is destroyed to free the case bronze.

Sources

Cultural Heritage: Motture 2019, 35; Salter and Gilmour n.d.

Art and Craft Textbook: Untracht 1982

French: *moule de potée*

Terme réservé à la fonte à la cire perdue. Désigne la partie externe du moule réfractaire, qui peut être à base de plâtre ou d'argile, notamment.

Note: La potée désigne le mélange argileux servant -à l'origine- à fabriquer le moule (Félibien 1690, 327; Lambert 2002, 216), mais peut aussi parfois désigner le moule réfractaire (Rama 1988, 375; Lambert 2002, 274). En anglais, potée et moule de potée sont désignés par le même terme (investment).

Sources

Cultural Heritage: Baudry and Bozo 1978, 625; Bewer, Bourgarit, and Bassett 2008

Historical: Boffrand 1743, 44–46; Diderot et al. 1751, 2:436–43, entry « bronze terme de fonderie »; Mariette 1768, 79–84

Art and Craft Textbooks: Rama 1988, 260; Lambert 2002, 216

Alternate Translations

moule (Terme non univoque.)
Sources: Copper Industry: Association Technique de Fonderie, Commission Ingénieurs et Techniciens 1979; Historical: Félibien 1690, 329

moule de coulée
Sources: Cultural Heritage: Rolley 1994, 69; Mille and Robcis 2012; Azéma and Mille 2013b; Historical: Gonon 1876, 6; Art and Craft Textbook: Dubos 2003

potée
Art and Craft Textbooks: Rama 1988, 375; Lambert 2002, 274

German: *Formmantel*

Source

Alscher 1987, 554

Alternate Translation

Gussmantel
Source: Maaz 2010, 706

Italian: *forma*

The variety of terms is due to regional differences in foundry terminology. *Forma* is the total of all layers of investments ready to cast. *Camicia* or *tunica* are the first fine refractory clay layer applied directly on the wax. *Mantello* or *cappa* are the second, coarser layer, often made of refractory clay, thick cloths, or iron belts to help contain the pressure developing during the pour. See refractory mold.

Sources

Morigi 1990, quaderni di restauro, vols. 1–3; Pecchioli 1999, 193; Leonardo 1490, fols. 154v, 142v; Biringuccio [1540] 1977, fols. 77v–78

Alternate Translations

camicia
Literally "shirt"
Source: Biringuccio [1540] 1977, fol. 95

cappa
Layer of the investment mold

chappa
Early Tuscan spelling of *cappa*
Source: Leonardo 1490, fol. 156v

mantello (di fusione)
Used in the way defined here under the long entry "fusione" for the casting of bells under the paragraph "formatura" in Treccani
Sources: Treccani, under *mantello (di fusione)*; Carruba 2006, 24; Formigli and Hackländer 1999, 301

refrattario
Refers to the investment material and investment mold
Source: Bruni 1994, 75

terra da forme
Refers to the investment material
Source: Biringuccio [1540] 1977, fols. 76v, 80v

tunica
Tonaca di terra, and even more often: *involucro*
Source: Treccani, under *tunica*

Chinese: 熔模

Alternate Translations

熔模材料
Source: Ming 2010, 1214

耐火材料
Sources: Ming 2010, 1036; Hua 2013, 347

life-casting

See figs. 11, 27.

Refers specifically to the reproduction of a once-living form (either plant or animal) that results in a cast characterized by its high realism and fine detail. Life-casts are made by encasing the form in a refractory mold and burning out the form, and are therefore generally solid, though there are some exceptions.

Note: The related term "burn-out method" is used for the replication of nonliving forms (such as textiles) in which the form is burned out in order to create the mold.

Source

Cultural Heritage: Smith and Beentjes 2010

To Be Distinguished From

casting from life
"Life-casting" is to be distinguished from "casting from life," in which a reusable mold is taken from a living form (e.g., a tree trunk or a body part) without harming it (e.g., a life mask). Such casts from life may be reproduced any number of times and in different materials.

Synonym

life casting (alternate spelling)

French: *fonte sur le vif*

Procédé de fonte consistant à enrober un élément végétal ou animal dans un moule réfractaire, faire disparaître le modèle par combustion, et remplir de métal liquide le creux ainsi ménagé.

Note: La seule occurrence trouvée dans la littérature ancienne de ce procédé apparaît sous la forme « mouler sur le naturel » (Making and Knowing Project et al. 2020, fol. 110v). Remarquer qu'une fonte sur le vif peut être creuse.

Source

Cultural Heritage: Making and Knowing Project et al. 2020

German: *Naturabguss*

Source

Lein 2004, 42–45

Alternate Translation

Abguss über die Natur
Source: Uhlenhuth 1920, 51–53

Italian: fusione dal vero

Chinese: 活体模铸

lost-wax casting

See figs. 13, 16, 18, 23, 25, 28, 558.

A technique in which a model made of wax is embedded in a refractory mold that is heated, thereby melting out the wax and creating a void to be filled with molten metal. Two primary variations of the technique are referred to as "direct" or "indirect" lost-wax casting, depending on whether the original model is the one sacrificed in the process. See GI§2.

Note: The French term "cire perdue" is often adopted in English as well. The wax model can be supplemented by other materials that can be burned out, such as cloth.

Sources

Cultural Heritage: Penny 1993; Beale 1975; Mattusch 1996; Motture 2019

Other: Untracht 1968

Synonyms

investment casting

lost wax casting (alternate spelling)

lost-wax molding
Source: Copper Industry: Koch and Newell 1963

French: *fonte à la cire perdue*

Procédé de fonte consistant à créer un modèle dans un matériau fusible à base de cire. Une fois entièrement recouvert par le moule réfractaire, le matériau fusible est évacué par chauffage, et l'espace vacant est rempli par du bronze. Les deux variantes principales du procédé sont le procédé direct et le procédé indirect.

Sources

Cultural Heritage: Bewer, Bourgarit, and Bassett 2008; Lebon et al. 2016, 330; Mille and Robcis 2012; Azéma and Mille 2013b

Alternate Translations

fonte à cire perdue
Sources: Cultural Heritage: Baudry and Bozo 1978, 624; Arminjon and Bilimoff 1998, 78–80

moulage à la cire perdue
Sources: Copper Industry: Association Technique de Fonderie, Commission Ingénieurs et Techniciens 1979; Bader and Théret 1961, 437; Cuénin 1994, 5; Koch and Newell 1963; Art and Craft Textbooks: Hamm 1924, 99–100; Rama 1988, 203

moulage en cire perdue
Sources: Copper Industry: Association Technique de Fonderie, Commission Ingénieurs et Techniciens 1979; Historical: Guettier 1858, notice 585, p. 300; Laboulaye 1861, 81–82

procédé à cire perdue
Source: Cultural Heritage: Rolley 1994, 65

German: *Wachsausschmelzverfahren*

Sources

Association Technique de Fonderie, Commission Ingénieurs et Techniciens 1979; Müller 2014, 167

Alternate Translation

Modellausschmelzverfahren
Source: Association Technique de Fonderie, Commission Ingénieurs et Techniciens 1979

Italian: *fusione a cera persa*

Sources

Treccani, under *fusione a cera persa* [definition 3b]; Giuffredi 2006, 61–62

Alternate Translation

fusione a cera perduta
Source: Treccani, under *fusione a cera perduta*

Chinese: 失蜡法

Source

TNATD

Alternate Translation

失蜡铸造
Sources: Ming 2010, 1286; TNATD

metal plating

See figs. 293, 304, 343, 365.

Generic term referring to the application of a different metal to the surface of a bronze sculpture by a variety of means (mechanical, chemical, electrochemical). Typically, gold and silver are used to plate sculptures, but nickel, zinc, and tin have been used for aesthetic and/or protective reasons. When the applied metal is gold or an alloy of gold, it is referred to as gilding.

Sources

Cultural Heritage: Salter and Gilmour n.d.

Copper Industry: Brunhuber 1988

French: *placage métallique*

Terme générique décrivant toute application d'une couche métallique exogène à la surface d'un bronze.

Note: La dorure, l'argenture, et dans une moindre mesure l'étamage sont les pratiques de placage les plus fréquentes rencontrées sur les bronzes.

Sources

Cultural Heritage: Baudry and Bozo 1978, 657; Darque-Ceretti and Aucouturier 2012

German: *Plattierung*

In German, *Plattierung* (the result) and *Plattieren* (the action) are specific to obtaining the adhesion of a (noble) plate on a (less noble) substrate by mechanical pressure via hammering, rolling, friction. *Plattierung* may actually not be applicable for sculpture. In German the construction *Ver* + metal + *ung* is preferred: *Vergoldung, Versilberung, Verkupferung, Verzinnung, Verzinkung.*

Source

Wallack 1840, 191–219

Italian: *placcatura*

Sources

Treccani, under *placcatura*; Battaglia 1961, under *placcatura* [vol. 13, 626]

Chinese: 镀覆

Source

Ming 2010, 327

metallurgical joint

See figs. 59, 146, 187, 188, 203, 204, 206, 207, 364.

A type of joint between two metal parts that is undertaken using molten metal. Examples of metallurgical joints include those made using welding, brazing, soldering, and interlock casting.

Note: The correct use of this term has been disputed among CAST:ING members, but it is often used generally to distinguish from mechanical joints. Diffusion welding (an expensive twentieth-century solid-state welding technique) does not involve the use of molten metal but to our knowledge is not used for bronze sculpture.

Sources

Cultural Heritage: Lechtman and Steinberg 1970; Dillon 2002, 299; Tzachou-Alexandri 2000, 92

To Be Distinguished From

mechanical joint

French: *joint métallurgique*

Joint métal-métal faisant intervenir du métal liquide (brasage ou soudage par fusion), par opposition à joint mécanique.

Note: Ce terme a été introduit par les membres de CAST:ING – et reste débattu – pour permettre de décrire un assemblage quand bien même il n'est pas possible de distinguer une brasure d'un joint obtenu par soudage par fusion.

German: *Metallurgische Verbindung*

Alternate Translation

metallische Verbindung

Italian: *saldatura metallurgica*

Sources

Sources: Formigli 2010, 20; Formigli 1999c, 318–19

Alternate Translation

giunzione metallurgica

saldatura (metallurgica) per colata
Flow fusion welding (used in antiquity)
Sources: Formigli 2010, 20; Formigli 1999c, 318–19

saldatura
Often specified by "a fusione" or "per fusione"
Treccani, under *saldatura*; Battaglia 1961, under *saldatura*; Formigli 2010, 19

Chinese: 冶金接合

Source

Ming 2010, 1643

model

See figs. 1, 52, 75, 130, 472, 559.

The creation of a bronze may involve a series of models and molds that can differ in size and material depending on the artist's design process, and ultimately also on the casting process chosen to create the bronze version. The model is a positive version of the sculpture (as opposed to the negative mold); the word may refer to any work made as a step in the preparation of a finished sculpture, at any stage. An existing artwork or a live subject may also serve as model for a bronze. See I.1.

Note: When used in reference to general sculptural practice, the term may refer to a rough sketch made in order to work out the composition (*bozzetto*, *modello*, *esquisse*, or *maquette*) or to a more complete model ("presentation model"), sometimes used to obtain approval from a patron or as a record or reference. Because a cast may be the outcome of numerous steps of models and molds, it may be best to refer to the closest model of a cast as "precursor model" (e.g., the wax model melted out of the investment to create a particular lost-wax cast) since it can be difficult to identify the relationship of a bronze cast to the earliest model in its genealogy. "Precursor model" is a term newly proposed here.

Sources

Cultural Heritage: Dillon 2002; Motture 2019

Art and Craft Textbook: McCreight 1996

General Dictionary: https://www.merriam-webster.com/dictionary/model [definition 1]

French: *modèle*

Terme générique désignant ici toute réalisation intervenant dans la fabrication d'une sculpture en bronze. Celle-ci peut mobiliser un grand nombre de modèles (et moules) successifs (modèle original, réplique, épreuve en cire, etc.).

Note: Le sens donné à modèle est ici plus large que l'« œuvre destinée à être reproduite » (Baudry and Bozo 1978, 624), et différent du modèle au sens de l'histoire de l'art qui peut revêtir une grande variété de formes (dessin, esquisse, maquette, etc.).

Sources

Cultural Heritage: Baudry and Bozo 1978, 242–47; Lebon 2012

Copper Industry: Association Technique de Fonderie, Commission Ingénieurs et Techniciens 1979

Historical: Boffrand 1743, 15–17; Mariette 1768, 23–26, 69; Diderot et al. 1751, 2:436–38, entry « bronze »

German: *Modell*

Sources

Association Technique de Fonderie, Commission Ingénieurs et Techniciens 1979; Fleming and Tilch 1993, 127

Italian: *modello*

Sources

Treccani, under *modello* [definition 3]; Battaglia 1961, under *modello* [definition 14 and 16, vol. 10, 646]; Cellini [1568] 1983, 11, fol. 46

Alternate Translations

forma
Can refer to both the model and the mold, which leads to potential confusion

forma da gittar di bronzo
This is the wax model that is used for casting, thus also possibly referring to the inter-model.
Source: Biringuccio [1540] 1990, fol. 82v

Chinese: 模

Source

Hua 2013

mold

See figs. 7, 27, 101, 112, 542, 549, 557.

A three-dimensional negative form made of one or more parts that serves as a matrix for the production of a positive by casting or pressing malleable material into it. Molds allow for the production of one or more copies of an original sculpture.

Note: Molds of varying types may be used at different stages of the bronze casting process (e.g., to make a chef-modèle, a wax inter-model, or core, or to cast a bronze). See I.1.

Sources

Cultural Heritage: Bassett and Fogelman 1997; Beale 1975

Art and Craft Textbook: Reliance Foundry n.d.

General Dictionary: https://www.merriam-webster.com/dictionary/mold [definiton 1a(1)]

Synonym

mould (UK spelling)

French: *moule*

Terme générique désignant toute forme en creux destinée à la production d'une forme en positif. Plusieurs types de moules faits de matières différentes (plâtre, argile, sable, silicone, etc.) peuvent intervenir dans la fabrication des différents modèles voire des noyaux précédant la sculpture en bronze.

Note: Dans l'industrie moderne, le moule désigne généralement le moule réfractaire.

Sources

Cultural Heritage: Baudry and Bozo 1978, 562

Copper Industry: Cuénin 1994, 5

Historical: Boffrand 1743, 17–20; Mariette 1768, 27–35; Diderot et al. 1751, 2:436–39, entry « bronze »

Art and Craft Textbook: Lambert 2002, 272

Alternate Translation

creux

Source: Cultural Heritage: Baudry and Bozo 1978, 759

German: *Form / Negativform*

Source

Müller 2014, 170–71

Alternate Translations

Negativformteile

Hilfnegative

Italian: *calco*

Sources

Treccani, under *calco* [definition 1]; Battaglia 1961, under *calco* [definition 1, vol. 2, 529]

Alternate Translations

cavo
General term for a negative impression, or hollow; Cellini uses it as "cavo di gesso," which is specifically a mold of plaster
Source: Cellini [1568] 1983, 11, fols. 45v–46

forma
General and also investment mold
Sources: Treccani, under *forma* [definition 6]; Biringuccio [1540] 1977, fols. 77v, 80v; Bruni 1994, 85–86

stampo
General term for an impression
Source: Treccani, under *stampo* [definition 2c]

Chinese: 范

Source

Zhang 2010

mold extension

See figs. 23, 41, 122, 123.

Bridge or spacer made of refractory material between the core and the outer mold. It serves a double function as core support while also providing better air flow into the core to aid in its drying before the bronze is poured.

Note: Such mold extensions may be created naturally when the core is not fully enclosed (as in the case of a bust with an open bottom) or by leaving openings in the wall of a lost-wax model. After casting, the hole in the bronze created by the mold extension will also facilitate removal of the core material and armature. The hole may later be sealed with a metal patch.

Source

Smith and Sepponen 2019

Synonyms

core extension

core print
Source: Copper Industry: Koch and Newell 1963

core seat
Source: Copper Industry: Koch and Newell 1963

French: *portée de noyau*

Partie intégrante du noyau ou partie rajoutée (fonte au sable), faite de la même matière que le noyau, reliant noyau et partie externe du moule réfractaire. Sert à maintenir le noyau en position lors de la coulée.

Note: Aujourd'hui utilisée essentiellement dans la fonte au sable, on la rencontre également dans les procédés à la cire perdue anciens.

Sources

Copper Industry: Koch and Newell 1963

Art and Craft Textbook: Dubos 2003

Alternate Translation

portée
Sources: Copper Industry: Bader and Théret 1961; Art and Craft Textbook: Rama 1988, 140, 164

German: **morsa**

Italian: **morsa**

Chinese: 自帶泥芯撐

Source

Liu 2015, 97

Alternate Translation

芯撐
Source: Ming 2010, 1572

overlay

See figs. 223, 340, 346, 347.

Decorative element resting on the surface of a sculpture via any of a variety of attachment methods, including solder, adhesives, cements, and/or rivets. Overlay materials may include a range of materials, among them metals, glass, stone, or bone.

Note: The terms "inlay" and "overlay" are often used indiscriminately, but they differ in one main regard: inlays are inserted into the bronze surface, whereas overlays are set on top of the surface.

Sources

Cultural Heritage: Hemingway and Abramitis 2017

Art and Craft Textbook: Untracht 1982, 315

To Be Distinguished From

coating

damascene

incrustation

inlay

Synonyms

encrustation
Source: Art and Craft Textbook: Untracht 1982, 315

foil

incrustation

French: *incrustation*

Élément décoratif inséré dans la surface d'un bronze ou formant relief, fait d'un matériau différent du substrat en bronze pour un effet polychromatique. Une grande variété de matériaux et de techniques peut être mise en œuvre.

Note: le seul terme français connu par les auteurs pour distinguer précisément un élément inséré dans la surface du bronze (« inlay » en anglais) d'un élément formant relief (« overlay ») est le terme « incrustation en relief » (E. Plé, Personnal communication, June 2024). Le japonais, au contraire, offre un vocabulaire très riche et très spécifique

To Be Distinguished From

damassé
Acier forgé par pliages successifs pour faire apparaître un décor.

Alternate Translation

damasquinure
Pour certains auteurs (Arminjon and Bilimoff 1998), 162, désigne une sous-catégorie d'incrustations, en l'occurrence l'incrustation d'un métal sur un autre (on parle aussi de damasquinage pour décrire le procédé [Arminjon and Bilimoff 1998, 162–63; Baudry and Bozo 1978, 661]). Pour d'autres auteurs, la damasquinure ne concerne que les incrustations d'or ou d'argent sur un objet en fer ou acier, typiquement une arme ou une armure (Félibien 1690), 460. Il est suggéré d'éviter d'employer ce terme pour la sculpture en bronze, on lui préférera « incrustation métallique ».
Sources: Cultural Heritage: Arminjon and Bilimoff 1998, 162–63; Historical: Félibien 1690, 460

German: *Auflage*

A general term closer to "coating"

Alternate Translations

Überzug
German has no noun for the technique of overlaying by mechanical attached sheet metal; there is just an adjective: *goldbeschlagen*, *silberbeschlagen*.
Source: Lein 2004, 51

Verzierung
A general term closer to "decoration" or "ornament"

Italian: *incrostazione*

Alternate Translation

riporto
Rare and not very specific
Sources: Treccani, under *riporto* [definition 2c]; Battaglia 1961, under *riporto* [definition 2, vol. 16, 689]

Chinese: 包覆

Source

Ming 2010, 36

Alternate Translation

覆盖
Source: TNATD

patch

See figs. 35, 65, 125, 152, 166, 168, 170, 172, 173, 187, 194, 198, 367.

A type of repair most often mechanically set into the bronze surface, but which may also be soldered, welded, or cast into place (see I.4). Patches are most often made of cut-out pieces of copper alloy that are the same as that of the cast metal, but they may be of a different alloy or metal (e.g., lead).

Sources

Cultural Heritage: Bassett 2008, 280, 283; Dillon 2002, 301–2; Mattusch 1996, 169–70

To Be Distinguished From

plug

French: *plaquette de réparation*

Type de réparation pour recouvrir un trou ou un manque par insertion d'une plaque faite d'un alliage à base de cuivre. L'insertion et le maintien peuvent être mécaniques, mais le recours au brasage, au soudage et même à la coulée secondaire sont également possibles.

Note: Le métal peut être très différent de celui du bronze à réparer (voir par exemple CAST:ING 2018).

Sources

Cultural Heritage: Azéma and Mille 2013b; Bourgarit, Bewer, and Dresc-Dautier 2014

Alternate Translation

pièce
Sources: Historical: Boffrand 1743, 60; Mariette 1768, 127; Diderot et al. 1751, 2:442, entry « bronze »

German: *Flicken*

Sources

Meissner, Haber, and Mach 2000, 102; Maaz 2010, 688

Italian: *laminetta ad incastro*

Sources

Pecchioli 1999, 192–93

Alternate Translations

tassellatura
Sources: Pecchioli 1999, 193; Treccani, under *tassellatura*

tassello
Sources: Treccani, under *tassello* [definition 1a]; Battaglia 1961, under *tassello* [definition 1, vol. 20, 758]

Chinese: 补修

(v., n.)

Source

TNATD

Alternate Translations

补修料
Literally "material used as a patch"
Source: TNATD

补片
Source: TNATD

补缀 (v., n.)
Source: TNATD

patina

See figs. 293, 306, 307, 308, 309, 313, 314, 323, 326, 328, 332, 338, 500.

The term has at least three different meanings: 1) a pleasing surface alteration acquired over time—whether on a bronze or marble sculpture, furniture, or a painting—that may add aesthetic value; 2) the chemical transformation of a metal surface; to a mineral layer (sometimes referred to as chemical patina, see corrosion) that usually has a different color from and reduces the bright metallic reflectance of the polished original cast surface; or 3) (as opposed to chemically induced patinas) organic coatings such as resin, lacquer, oil, wax, or synthetic resins applied to the surface of metals that can change the color, texture, saturation, and/or reflectance.

Note: Some researchers do not consider applied organic coatings part of the patina layer.

Sources

Cultural Heritage: Weil 1996, 394–414; Ward 2008

Historical: Hiorns 1907, 62–63; Littre 1873

Copper Industry: Brunhuber 1988; Association Technique de Fonderie, Commission Ingénieurs et Techniciens 1979

Art and Craft Textbook: Fishlock 1962

French: *patine*

Le terme patine recouvre plusieurs sens. 1) Désigne la belle surface aussi bien d'un bronze, d'un marbre que d'un meuble en bois, colorée par le temps. 2) Désigne, pour un bronze, les couches non métalliques de surface issues de la transformation chimique du métal par l'action de l'homme et/ou du temps (corrosion) (Baudry and Bozo 1978, 634; Bewer, Bourgarit, and Bassett 2008; Azéma and Mille 2013b; Lebon et al. 2016, 331; Robbiola 2015; Association Technique de Fonderie, Commission Ingénieurs et Techniciens 1979). 3) Peut parfois désigner (surtout en anglais), en sus des couches transformées, toute couche de surface présente sur un bronze (vernis, etc.) (Aucouturier et al. 2003).

Note: La signification de ce terme est très débattue.

Sources

Cultural Heritage: Aucouturier 2007; Baudry and Bozo 1978, 634; Robbiola 2015; Aucouturier et al. 2003; Bewer, Bourgarit, and Bassett 2008; Lebon et al. 2016, 331; Azéma and Mille 2013b

Copper Industry: Brunhuber 1988; Association Technique de Fonderie, Commission Ingénieurs et Techniciens 1979

Historical: Diderot et al. 1751, 12:173, entry « patine »

Art and Craft Textbook: Lambert 2002, 273; Rama 1988, 374

Alternate Translation

couleur
Source: Historical: Félibien 1690, 334

German: *Patina*

Source

Association Technique de Fonderie, Commission Ingénieurs et Techniciens 1979

Italian: *patina*

Sources

Giuffredi 2006, 144, 220; Treccani, under *patina*; Battaglia 1961, under *patina* [vol. 12, 822]; Bruni 1994, 126–28; Brunhuber 1988

Chinese: 古色

This term is more commonly used in art historical references and means "an aged surface accumulated over time."

Alternate Translations

铜绿
There is no general term in Chinese for patina as defined here; this term can be translated as "green corrosion on copper" and is used specifically for ancient works.
Source: TNATD

铜锈

There is no general term in Chinese for patina as defined here; this term can be translated as "corrosion on copper" and is used specifically for ancient works.

Source: TNATD

陈年色泽

More commonly used in art historical references, meaning "an aged surface accumulated over time"

Source: TNATD

peening

See fig. 246.

The act of embellishing/texturing the metal surface by repeatedly using a peen (*pein*) hammer (which has one rounded end and one flat end) to disguise porosity or other surface flaws and create a faceted effect. The term also refers to the type of tightly textured effect created by this process.

Note: The surface markings can be similar to those made by a punch and thus easily misidentified. In sand casting, the term refers to compressing the sand with the peen end of a rammer.

Sources

Cultural Heritage: Smith 2015; Avery 2018, 162; Allen et al. 2022, 254

Historical: Bolland 1894, 298

Copper Industry: Palmer 1929, 294

French: *matage au marteau*

Traitement mécanique de surfaçage directement au marteau à panne plane ou légèrement bombée et polie miroir. Utilisé à la fois pour masquer la porosité ou d'autres défauts de surface et pour créer un effet facetté. Ne pas confondre avec le planage.

Note: Il n'existe pas à notre connaissance de terme particulier en français correspondant au terme anglais « peening », qui renvoie à la fois au matage et au grenaillage.

German: *Hämmern*

Source

Bol 1985, 142

Italian: *martellatura*

Sources

Treccani, under *martellatura* [definition b]; Battaglia 1961, under *martellatura* [definition 1, vol. 9, 840]

Chinese: 轻敲

Source

TNATD

Alternate Translation

锤平
Source: TNATD

piece mold

VOCABULARY

See figs. 9, 15, 18, 25, 26, 62, 74, 111, 116, 536, 556.

A type of mold composed of two or more individually formed, interlocking sections designed to circumvent undercuts and/or be disassembled without damage to the model or to the mold sections. Most often refers to molds made of a rigid material such as plaster. Smaller piece mold sections are generally held together by a rigid outer mother mold. Starting in the nineteenth century, flexible piece molds (at first gelatin, later replaced by rubber, alginate, and silicone molds) were used instead of plaster, reducing the overall number of sections required to make a wax cast.

Note: Sometimes very large sculptures—both reliefs and in the round—need to be broken into smaller sections simply to avoid excessive suction or vacuum when removing the mold from the model. In the context of bronze sculpture, the term may denote molds used to cast inter-models in the indirect lost-wax process. Sand casting molds are also piece molds formed by compacting the sand in discrete portions around a rigid pattern. In the piece mold casting process (aka "section-mold process"), baked refractory mold clay and loess—a fine-grained refractory soil found throughout northern China—are used to make piece molds into which bronze is poured (see Case Study 3).

Sources

Cultural Heritage: Beale 1975; Boulton 2006

Art and Craft Textbook: Rich 1988

General Dictionary: https://www.merriam-webster.com/dictionary/piece mold

Synonym

piece mold (alternate spelling)

French: *moule à pièces*

Moule constitué d'au moins deux parties ou pièces assemblées pour faciliter le démoulage et souvent pour s'affranchir des problèmes de contre-dépouille.

Note: Pour la sculpture en bronze, un moule à pièces peut aussi bien désigner un moule à bon creux (pour la cire donc), un moule réfractaire pour la fonte au sable, qu'un moule réfractaire pour la fonte à pièces (voir Case Study 3 et aussi Rama 1988, 196). Il arrive que les moules des très grands bronzes doivent être conçus en plusieurs sections non pas à cause de contre dépouilles, mais pour diminuer la trop grande adhérence du moule sur la surface du modèle et ainsi faciliter le démoulage (Andrew Lacey, pers. comm., July 2021).

Source

Cultural Heritage: Baudry and Bozo 1978, 563

German: *Stückform*

Alternate Translation

mehrteilige Form

Source

Maaz 2010, 730

Italian: *forma a tasselli*

Tassello is a piece of a mold. In Treccani, the description refers uniquely to sand casting. Biringuccio describes forming a clay mold in pieces with locking parts he calls *incastrature*.

Sources

Treccani, under *tassello* [definition 2]; Biringuccio [1540] 1977, fols. 80v–81; Bruni 1994, 53–55

Alternate Translation

calco a tasselli
This process and phrase are mentioned in the Treccani entry for *fusione* under *fusione a cera persa*.
Sources: Treccani, under *calco a tasselli*; Battaglia 1961, under *calco a tasselli* [definition 1, vol. 2, 529]

Chinese: 块范

Alternate Translation

范块
Source: Zhang 2010, 80

plug

See figs. 177, 178, 193, 195, 197, 478.

A type of repair that fills a flaw that extends through the full thickness of the metal wall, specifically those that occur with the removal of core pins and armature rods. Circular plugs are often threaded in order to mechanically lock them in place. See I.4.

Note: It is not always easy to distinguish a plug from a patch. Radiography is often necessary.

Sources

Cultural Heritage: Dillon 2002; Penny 1993; Sturman 2004

Art and Craft Textbook: Rome and Young 2003, 311

General Dictionary: https://www.merriam-webster.com/dictionary/plug [definition 1a]

To Be Distinguished From

patch

French: *insert*

Type de réparation comblant un trou traversant comme ceux laissés par les fers à noyaux. Les inserts sont souvent des tiges, filetées ou non, faites d'un alliage à base de cuivre.

Note: Il n'est pas toujours aisé de distinguer un insert d'une plaquette de réparation.

To Be Distinguished From

incrustation

plaquette de réparation

Source

Cultural Heritage Publication: Bourgarit, Bewer, and Bresc-Bautier 2014

Alternate Translations

goupille
On rencontre aussi le terme goupillage, Lambert 2002, 241.
Source: Cultural Heritage: Bewer, Bourgarit, and Bassett 2008

tige
Source: Art and Craft Textbook: Rama 1988, 326

German: *Dübel*

Used primarily in speech

Alternate Translations

Flicken
Sources: Meissner, Haber, and Mach 2000, 102; Maaz 2010, 688

Plombe
Used primarily in speech
Source: Maaz 2010, 716

Stift
Used primarily in speech
Source: Bol 1985, 139

Italian: *tappo*

Alternate Translations

perno
Source: Bruni 1994, 124–25

perno filettato
Threaded plug

Chinese: 塞子

Source

Ming 2010, 1233

Alternate Translation

栓塞
Source: TNATD

porosity

See figs. 127, 152, 153, 155, 165, 184, 205, 464.

A common type of casting flaw that includes a group or area of cavities caused by shrinkage or trapped gases. Porosity may vary considerably in dimension and may or may not break through the surface of the bronze. See I.3§1.3.1.

Note: A common way to characterize the quality of a cast is to report the degree and extent of its porosity (see I.3).

Sources

Cultural Heritage: Stone 2008

Copper Industry: Campbell 2015; Brunhuber 1988; Ammen 1980

General Dictionary: https://www.merriam-webster.com/dictionary/porosity [definition 1a]

French: *porosité*

Défaut de fonderie caractérisé par des zones plus ou moins spongieuses, c'est-à-dire comportant de nombreuses cavités résultant de la contraction du métal ou de gaz piégé dans le métal lors du refroidissement.

Note: Il est fréquent de caractériser la qualité d'une fonte par son degré de porosité. On parle de porosité ouverte quand elle débouche en surface, de porosité fermée sinon. Des termes plus spécifiques sont souvent également employés : cavité (Hénon, Mascré, and Blanc 1971, 22), crique (Rama 1988, 326).

Sources

Cultural Heritage: Baudry and Bozo 1978, 646; Bewer, Bourgarit, and Bassett 2008; Azéma and Mille 2013b

Copper Industry: Hénon, Mascré, and Blanc 1971, 23; Association Technique de Fonderie, Commission Ingénieurs et Techniciens 1979

German: *Porosität*

Source

Brunhuber 1988

Italian: *porosità*

Sources

Treccani, under *tecnica*; Battaglia 1961, under *porosità* [vol. 13, 916]

Alternate Translation

spugnoso
In early Italian no direct translation has been found, but this adjective is a descriptive term that means "spongy," used in *bucato e spugnoso*, "with holes and spongy."
Source: Biringuccio [1540] 1977, fol. 90

Chinese: 气孔

Source

Ming 2010, 1128

Alternate Translations

松孔

气孔巢
Source: TNATD

砂眼
Source: Ming 2010, 1247

pour

VOCABULARY

See figs. 549, 560.

The operation of pouring or casting metal into the refractory mold.

Sources

Cultural Heritage: Mattusch 1996, 25; Maish 2017, 342

General Dictionary: https://www.merriam-webster.com/dictionary/pour

Synonyms

casting

heat
Source: Copper Industry: Koch and Newell 1963

French: *coulée*

Opération consistant à verser du métal liquide dans un moule réfractaire.

Note: Le terme a généré de nombreuses locutions liées à la fonderie : fosse de coulée, défaut de coulée, coulée de rappel (Rama 1988, 373), etc.

Sources

Cultural Heritage: Rolley 1994, 70; Bewer, Bourgarit, and Bassett 2008

Copper Industry: Bader and Théret 1961; Koch and Newell 1963

Art and Craft Textbook: Rama 1988, 371

German: *Guss*

Source

Fachkunde Metall: Bovenkerk, Braun, and Dörflinger 1987, 221

Italian: *colata*

Refers to the act of casting and to the metal contained in the crucible or furnace that is poured in one session

Sources

Treccani, under *colata*; Battaglia 1961, under *colata*; Bruni 1994, 106

Alternate Translations

getto
Also refers to the cast
Sources: Treccani, under *getto* [definition 3a]; Battaglia 1961, under *getto* [definition 15, vol. 6, 723]; Bruni 1994, 106

gitto
Early Italian spelling for *getto*; also refers to the cast
Sources: Biringuccio [1540] 1977, fols. 108–10

Chinese: 浇注

Source

Ming 2010, 741

Alternate Translation

浇铸
Sources: Ming 2010, 741; TNATD

punch

See figs. 238, 239, 240, 242, 244, 256, 272, 494.

A tool usually made of a steel rod that may be struck with a hammer at one end in order to create a pattern in the surface of the sculpture with the other end by compressing the metal.

Note: During chasing, a number of punches with a variety of custom-made textures are often used, allowing a range of possible surface patterns. The texture of some punches may be confused with peening.

Sources

Cultural Heritage: Bassett and Fogelman 1997, 74; Penny 1993, 308; Allen et al. 2022, 63

Art and Craft Textbook: Untracht 1982, 122

General Dictionary: https://www.merriam-webster.com/dictionary/punch [definition 1a]

To Be Distinguished From

chisel

cold chisel

French: *ciselet mat*

Un des deux grands types de ciselets (dit aussi « mat », l'autre grand type étant le « ciselet clair ») dont la partie active, de section quadrangulaire ou circulaire, texturée ou lisse, permet de donner à la surface du bronze une certaine texture, par matage.

Sources

Cultural Heritage: Baudry and Bozo 1978, 642

Art and Craft Textbook: De Bois 1999, 54–55

Alternate Translation

ciselet

Terme habituellement générique pour désigner aussi bien les ciselets mats que les ciselets clairs, mais dans l'Encyclopédie (Diderot et al. 1751, 3:480, entry « ciselets ») désigne spécifiquement les ciselets mats.

Sources: Cultural Heritage: Baudry and Bozo 1978, 640; Historical: Diderot et al. 1751, 3:480, entry « ciselets »; Art and Craft Textbook: De Bois 1999

German: *Punze*

Source

Maaz 2010, 718

Italian: *cesello*

Sources

Treccani, under *cesello*; Battaglia 1961, under *cesello* [vol. 6, 13]; Cellini [1568] 1983, 11, fols. 21v, 33v

Alternate Translation

punzone

Sources: Treccani, under *punzone* [definition 3]; Battaglia 1961, under *punzone* [definition 2]; Cellini [1568] 1983, 1, fol. 27

Chinese: 冲头

Sources

Ming 2010, 156; TNATD

refractory mold

See figs. 7, 27, 542, 549, 557.

A temporary, heat-resistant, cohesive, porous mass that captures the fine impression of the model to be reproduced and forms the void into which the molten metal will be cast. Investment, green sand, and ceramic shell are examples of refractory molds.

Note: When present, the core is considered part of the refractory mold. It may be designated as the inner refractory mold as opposed to the outer refractory mold. The material used for the refractory mold tends to be similar to—if not the same as—that of the core. To avoid confusion between types of refractory molds, spelling out the nature of the mold is recommended. An investment is a type of refractory mold used specifically for lost-wax casting.

Sources

Cultural Heritage: Bourgarit et al. 2003, 119; Lie and Bewer 2014

Copper Industry: Brunhuber 1988; Association Technique de Fonderie, Commission Ingénieurs et Techniciens 1979

Synonyms

fire-resistant mold

heat-resistant mold

French: *moule réfractaire*

Moule en matériau réfractaire pour la coulée.

Note: Bien que le terme de moule réfractaire soit relativement peu utilisé — « moule » lui étant préféré, il permet d'éviter les confusions avec tous les types de moule susceptibles d'intervenir dans la fabrication d'un bronze (voir terme « moule »). Noter que quand il est présent, le noyau est considéré comme faisant partie du moule réfractaire. Il peut alors être désigné comme une partie interne du moule, par opposition à la partie externe dite aussi chape.

Source

Cultural Heritage: Bewer, Bourgarit, and Bassett 2008

Alternate Translations

moule
Terme ambigu, à éviter
Sources: Cultural Heritage Publications: Baudry and Bozo 1978, 625; Copper Industry: Brunhuber 1988; Association Technique de Fonderie, Commission Ingénieurs et Techniciens 1979; Bader and Théret 1961; Koch and Newell 1963

moule de fonderie
Terme utilisé notamment en archéologie
Source: Cultural Heritage: Saussus and Thomas 2019

chape
Attention, ne désigne que la partie externe du moule réfractaire.
Sources: Cultural Heritage: Bewer, Bourgarit, and Bassett 2008; Meyer, Thomas, and Wyss 2014; Copper Industry: Association Technique de Fonderie, Commission Ingénieurs et Techniciens 1979; Historical: Félibien 1690, 321; Art and Craft Textbook: Rama 1988, 372

German: *Form*

Sources

Wallack 1840; Müller 1902; Association Technique de Fonderie, Commission Ingénieurs et Techniciens 1979

Alternate Translation

Gießform
Sources: Brunhuber 1988; Association Technique de Fonderie, Commission Ingénieurs et Techniciens 1979

Italian: *forma*

Also refers to investment

Source

Morigi 1990, quaderni di restauro, vol. 1–3; Biringuccio [1540] 1977, fols. 80v, 82; Leonardo 1490, fols. 54v, 142v; Brunhuber 1988

Alternate Translations

camicia
Also refers to the refractory investment material
Source: Biringuccio [1540] 1977, fol. 95

cappa
Also refers to investment
Source: Bruni 1994, 75

chappa
Refers to outer investment
Source: Leonardo 1490, fol. 156v

mantello

Refers to outer investment/refractory mold
Source: Formigli and Hackländer 1999, 301

refrattario
Also refers to the refractory investment material
Source: Bruni 1994, 75

stampo refrattario

Chinese: 耐火模具

Source

Ming 2010, 1037

replica

See figs. 1, 554.

Here used to describe the precise reproduction of a bronze made by the same artist or foundry as the original bronze. Also refers to same-scale reproductions of a model made at different stages in the casting process (e.g., a wax replica, and a refractory replica used to make the core in sand casting or casting piece mold [I.1§2.2]). In lost-wax casting, bronzes fashioned from inter-models made from the same piece molds taken from the master model are considered replicas of the original. In sand casting, it refers to bronzes made using the same chef-modèle. Numerous replicas of the same bronze are called "multiples."

Note: Slight differences are inevitable between replicas due to deterioration of the mold or variations in the finishing of the wax or chasing of the bronze. With regard to modern bronze production, a replica may refer specifically to a cast made by someone other than the artist, but who is under license to make the cast. An "artist's replica" is a cast made by the artist or someone sanctioned by the artist. An "authenticated replica" is certified as having been made by the artist or someone sanctioned by the artist (see Tate Papers No. 8).

Sources

Cultural Heritage: Dillon 2002, 300

General Dictionary: https://www.merriam-webster.com/dictionary/replica [definition 1]

To Be Distinguished From

copy
Note: A "copy" is understood here to refer to a cast that reproduces the features of a model or bronze, but is not created by the original artist of the model or bronze, or cast by a licensed or sanctioned foundry.

variant

version

French: *épreuve d'édition*

Epreuve en plusieurs exemplaires obtenus de la même façon et à partir d'un même modèle. Les dimensions sont conservées peu ou prou, compte tenu des retraits possibles (voir II.4§1.1.1). La série de ces épreuves constitue l'édition du modèle. Une édition peut être en nombre limité, ou illimité.

Note: Non spécifique au bronze. On parle de « multiples » pour désigner plusieurs épreuves (Lebon et al. 2016, 330).

To Be Distinguished From

copie

réplique
« Réplique » et le terme anglais « replica » sont ce qu'il convient d'appeler des faux amis. La réplique diffère de son modèle par les dimensions, le matériau ou les deux à la fois (Baudry and Bozo 1978, 549), ce qui n'est pas le cas de la « replica » (traduite par « épreuve d'édition »).

reproduction
Une reproduction est techniquement identique à une épreuve d'édition, mais différente : juridiquementé : une "épreuve d'édition" est considérée comme un original, une reproduction non.

variante

version

Source

Cultural Heritage: Baudry and Bozo 1978, 545

Alternate Translation

épreuve de série
Source: Cultural Heritage: Baudry and Bozo 1978, 545

German: *Replik*

Source

Maaz 2010, 720

Italian: *multiplo*

Alternate Translations

replica
An Italian word, but more related to later reproductions of an artwork)
Source: Treccani, under *replica* [definition 2]

riproduzione (general term)
Sources: Treccani, under *riproduzione* [definition 2]; Battaglia 1961, under *riproduzione* [definition 2, vol. 16, 719]

Chinese: 复制品

Sources

Ming 2010, 457; TNATD

Roman joint

See figs. 43, 65, 127, 211, 212, 213, 217, 218.

A type of joint between two separately cast elements in which one element slots into the hollow "sleeve" of the other. The joint is generally further secured using pins, rivets, or by soldering.

Note: "Join" is a verb and "joint" is a noun. We have chosen here to use the latter, but it is common to find references to Roman joins, sleeve joins, etc.

Sources

Cultural Heritage: Bassett 2008; Beentjes 2019; Grissom and Harvey 2003

Synonyms

mortise and tenon join
Source: Cultural Heritage: Bourgarit et al. 2003

Roman join (alternate spelling)

sleeve join

socket and tenon join

French: *assemblage à la romaine*

Assemblage par emboîtement de deux éléments coulés séparément, l'un constituant la partie mâle, l'autre la partie femelle.

Note: Des goujons (ou clavettes ou chevilles) peuvent être ajoutés. On parle alors d'assemblage « par tenon et mortaise ».

Source

Art and Craft Textbook: Rama 1988, 125

Alternate Translation

assemblage par tenon et mortaise

German: *Steckverbindung*

Source

Willer, Schwab, and Mirschenz 2016b, 122, 128, 186

Italian: *ghiera alla romana*

Source

Bruni 1994, 123

Alternate Translation

giuntura a innesto

Chinese: 罗马式接合

sand casting

See figs. 9, 61, 101, 107, 108, 542.

A casting technique in which metal is poured into a piece mold made of a specific type of sand that is bound by clay (or oil or resin in modern foundries). The piece mold is made by ramming the sand around a rigid model or chef-modèle within stacked metal frames (aka flasks). See GI§2.4.1.

Sources

Cultural Heritage: Penny 1993

Copper Industry: Brunhuber 1988

Art and Craft Textbook: Rome and Young 2003

General Dictionary: https://www.merriam-webster.com/dictionary/sand casting

---

French: *fonte au sable*

Technique de fonderie utilisant un moule refractaire en sable.

Note: Dans les textes anciens, le terme de sable est parfois ambigu et peut faire référence à d'autres matières (terre, voire cire) et donc à des procédés autres que la fonte au sable (Lebon 2012).

Sources

Cultural Heritage: Rolley 1994, 66; Baudry and Bozo 1978, 624; Bewer, Bourgarit, and Bassett 2008; Lebon et al. 2016, 330

Historical: Diderot et al. 1751, 14:353–65, entry « sable »

Art and Craft Textbook: Dubos 2003

Alternate Translations

moulage au sable
Sources: Copper Industry: Cuénin 1994, 5; Art and Craft Textbook: Rama 1988, 31

moulage en châssis
Source: Historical: Guettier 1867, 221

moulage en sable
Sources: Copper Industry: Koch and Newell 1963; Other: Guettier 1867, 221

moulage en sable vert
Sources: Copper Industry: Cuénin 1994, 5; Art and Craft Textbook: Rama 1988, 31

German: *Sandformguss*

Source

Maaz 2010, 723

Alternate Translations

Sandguss oder

Sandgussverfahren

Italian: *fusione a staffa*

Staffa is Italian for "flask," the wooden or metal frame containing the sand.

Source

Treccani, at the end of *fusione a staffa*

Alternate Translation

fusione a terra

fusione alla sabbia
A little more common than *a staffa*
Sources: Treccani, under *fusione alla sabbia*; Treccani, under *fusione a staffa*

Chinese: 砂型铸造

Source

Ming 2010, 1247

seam line

See figs. 6, 26, 43, 61, 72, 73, 74, 75.

A faintly raised line that forms at the joints between mold sections when a molten material or slurry is poured into a mold. Seam lines are found on plaster or wax casts as well as on bronzes cast in piece molds. In ancient Chinese bronzes, some seam lines were exaggerated and integrated into the design of the casts, as can be seen in the elephant-shaped vessel in Case Study 3 (fig. 26). The term also refers to the line along which the pieces of a refractory piece mold join, which is the locus of the line that forms on the bronze. Depending on how well the piece mold pieces fit together, the seam line may be more or less raised. More extreme flashing occurs with ill-fitting pieces.

Note: Compared with flashing, which generally forms in uneven, sharp, rough flanges of metal, seam lines on the bronze surface are commonly linear, rounded, and relatively smooth, and run along the high points of convex surfaces. Seam lines are generally removed as part of the fettling process. Some modern artists leave them intentionally as evidence of the working process (fig. 75).

Sources

Cultural Heritage: Lie and Bewer 2014, 47

Art and Craft Textbooks: Rome and Young 2003; Rich 1988, 410

To Be Distinguished From

finning

flashing

veining

French: *couture*

Ligne en relief en surface d'un bronze due à un joint de moule. Attention, peut provenir du moule réfractaire mais aussi de tous les moules possiblement utilisés avant la coulée pour la fabrication du modèle : moulage de l'original, moule à bon creux (joint cire-cire, cf. Mille and Robcis 2012, etc., voir I.1).

Sources

Cultural Heritage: Baudry and Bozo 1978, 565

Art and Craft Textbooks: Lambert 2002, 268; Rama 1988, 373

Alternate Translations

balesvre / balevre

Attention, fait uniquement référence à l'opération de coulée et au moule réfractaire. Terme désuet.

Source: Cultural Heritage: Baudry and Bozo 1978, 758

barbe

Attention, fait uniquement référence à l'opération de coulée et au moule réfractaire.

Source: Copper Industry: Association Technique de Fonderie, Commission Ingénieurs et Techniciens 1979

barbure

Attention, fait uniquement référence à l'opération de coulée et au moule réfractaire. Terme désuet.

Source: Cultural Heritage: Baudry and Bozo 1978, 758

bavochure

Attention, fait uniquement référence à l'opération de coulée et au moule réfractaire. Terme désuet.

Source: Cultural Heritage: Baudry and Bozo 1978, 758

bavure

Attention, fait uniquement référence à l'opération de coulée et au moule réfractaire. Les sources bibliographiques données ci-après ne concernent que la fonte au sable.

Sources: Cultural Heritage: Baudry and Bozo 1978, 632; Art and Craft Textbook: Rama 1988, 373

bavure de joint

Attention, fait uniquement référence à l'opération de coulée et au moule réfractaire.

Source: Copper Industry: Association Technique de Fonderie, Commission Ingénieurs et Techniciens 1979

ébarbure

Attention, fait uniquement référence à l'opération de coulée et au moule réfractaire.

suture

Attention, fait uniquement référence à l'opération de coulée et au moule réfractaire.

toiles

Attention, fait uniquement référence à l'opération de coulée et au moule réfractaire.

Source: Historical: Launay 1827, 2:261

German: ***Gussnaht***

Source

Maaz 2010, 696

Italian: ***linea di giunzione***

Literally "line of joint"

Alternate Translations

linea di saldatura

traccia del calco

traccia dello stampo

Chinese: 范线

Source

Zhang 2010, 80

shrinkage

See fig. 41.

The contraction of molten metal as it cools and solidifies after casting, resulting in a reduction of the overall dimensions of the cast as well as possible casting defects. See II.4§1.1.1.

Note: Defects resulting from shrinkage may also be called "shrinkage" (Association Technique de Fonderie, Commission Ingénieurs et Techniciens 1979). To avoid confusion, we recommend that such casting defects be referred to as "shrinkage porosity" (Campbell 2015).

Sources

Historical: Buchanan 1903; Bolland 1894

Copper Industry: Ammen 1980; American Foundrymen's Society 1984; Campbell 2015

General Dictionary: https://www.merriam-webster.com/dictionary/shrinkage

---

French: *retrait*

Contraction du métal lors du refroidissement conduisant à une perte dimensionnelle et/ou à des défauts de fonderie (soufflures, retassures, etc).

To Be Distinguished From

retassure
Note: La retassure est un des résultats possible du retrait.

Sources

Cultural Heritage: Baudry and Bozo 1978, 644

Copper Industry: Hénon, Mascré, and Blanc 1971, 38; Cuénin 1997a, 5; Association Technique de Fonderie, Commission Ingénieurs et Techniciens 1979

Art and Craft Textbooks: Lambert 2002, 275; Rama 1988, 375

German: *Schrumpfung*

Alternate Translation

Schwindung
Source: Brunhuber 1988

Italian: *ritiro*

Source

Treccani, under *ritiro*

Chinese: 收缩

Sources

Ming 2010, 1313; TNATD

soldering

See figs. 187, 364.

In bronze sculpture, a technique for joining separately cast parts, inlays, overlays, or repairs, or for filling casting defects, by adding a metal with a lower melting temperature than that of the primary cast. For the purpose of visual identification on sculpture, white metal can be considered solder (alloys of silver, tin, lead, etc.), whereas brazing is yellow (copper alloys). Two types of solder include soft solder (low-melting-temperature alloys such as a combination of lead, tin, and/or bismuth) and hard solder (higher-melting-temperature silver alloys).

Note: The use of hard solder (a silver alloy) is sometimes referred to as brazing, but, as not all researchers have access to methods of analysis, and to avoid confusion, we suggest it should be referred to as soldering.

Sources

Cultural Heritage: Beale 1975, 42–43

Historical: Buchanan 1903, 23, 98

Copper Industry: Schwartz and Aircraft 1951

Art and Craft Textbook: Untracht 1982, 388–423

General Dictionary: https://www.merriam-webster.com/dictionary/soldering

To Be Distinguished From

brazing

flow fusion welding

flow welding

fusion welding

welding

French: ***brasage***

Technique d'assemblage de deux pièces métalliques ou plus, par addition d'un alliage à point de fusion plus bas que celui des métaux à assembler, n'entraînant pas la fusion des zones de contact, contrairement au soudage.

Note: L'anglais distingue « brazing », le procédé utilisant de la brasure tendre (métaux à bas point de fusion comme le plomb et l'étain), de « soldering », procédé utilisant une brasure forte (alliage à base de cuivre). On rencontre parfois en français le terme « brasure très forte » quand de l'argent est utilisé (Baudry and Bozo 1978, 634). La brasure désigne aussi bien le métal de brasure (Bader and Théret 1961, 102) que l'assemblage lui-même.

Sources

Cultural Heritage: Baudry and Bozo 1978, 634

Copper Industry: Association Technique de Fonderie, Commission Ingénieurs et Techniciens 1979

Alternate Translation

soudure indirecte
Les termes soudage ou soudure (Baudry and Bozo 1978, 634; Bader and Théret 1961) et souder (Félibien 1690, 757) sont souvent employés comme termes génériques décrivant tous les procédés d'assemblages de métaux mettant en jeu du métal à l'état liquide, brasage et soudage par fusion.
Source: Copper Industry: Bader and Théret 1961, 595

German: ***Löten***

Source

Deutsches Kupferinstitut 2005, 24

Italian: ***brasatura***

A subset of *saldatura*

Sources

Treccani, under *brasatura*; Battaglia 1961, under *brasatura*

Alternate Translation

saldatura
A term also used for welding
Sources: Treccani, under *saldatura*; Cellini [1568] 1983, fols. 37v–38

Chinese: 焊接

Sources

Ming 2010, 596; TNATD

sprue

See figs. 44, 46, 71, 86, 124, 539.

Any channel that feeds metal to the mold, in contrast to a vent, which lets air escape. Both sprues and vents make up the "sprue system," which circulates bronze from the pouring cup through the refractory mold and allows air and casting vapors such as steam to be released. In lost-wax casting, "sprue" is the term used for the solid wax rods (rarely reeds or terra-cotta pipes) used to create the channels in the mold. Sprues are also the solid metal that has filled the channels upon cooling, which is generally removed during fettling.

Note: The terms "sprue," "sprueing," "sprueing system," or "sprue system" are most commonly associated with lost-wax casting. For sand casting the equivalent would be "gating" or "gating system," which is where all the other terms ("runners," "risers," "gates," etc.) come into play. In sand and piece mold casting the channels are cut directly into the refractory mold. See GI§2.7. Depending on the complexity of the cast, the sprue or gating systems may also contain other elements with special functions (e.g., reservoirs, chills, drains, traps, jets, etc.).

Sources

Historical: Buchanan 1903

Copper Industry: Campbell 1991, 36–41

Art and Craft Textbook: Rome and Young 2003

General Dictionary: https://www.merriam-webster.com/dictionary/sprue

French: *jet de coulée*

Conduit dans le moule réfractaire pour le remplissage du moule par le métal liquide. Désigne aussi les mêmes éléments remplis de métal dont il faut débarrasser le bronze après la coulée. Désigne, pour les procédés à la cire perdue, les éléments en cire servant à cette fonction.

Note: Dans la zone d'arrivée du métal dans le moule réfractaire est aménagé un « cône de coulée » pour faciliter la coulée, et qui sert également de « masselote ». Les conduits servant pour l'évacuation des gaz portent un nom spécifique, évents (Félibien 1690, 335; Boffrand 1743, 39–42; Mariette 1768, 70–75; Launay 1827, 2:264; Gonon 1876, 5; Baudry and Bozo 1978, 627; Association Technique de Fonderie, Commission Ingénieurs et Techniciens 1979; Rama 1988, 254, 373; Rolley 1994, 66; Lambert 2002, 269; Dubos 2003; Lebon 2012). Pour l'évacuation de la cire on parle de tire-cire (Baudry and Bozo 1978, 763). Certains auteurs emploient le terme d'artifices de fonderie (Saussus and Thomas 2019, 59; Lambert 2002, 27). Pour certains d'entre eux (Saussus and Thomas 2019), les artifices de fonderie désignent également, dans le cas de moules à pièces, les éléments servant au positionnement des moules (artifices de centrage).

Sources

Cultural Heritage: Baudry and Bozo 1978, 626; Saussus and Thomas 2019

Copper Industry: Association Technique de Fonderie, Commission Ingénieurs et Techniciens 1979

Alternate Translations

alimentation

Il existe tout un répertoire de termes liés dont « système d'alimentation » (Azéma and Mille 2013b; Dubos pers. comm.; Cuénin 1997a, 2; Association Technique de Fonderie, Commission Ingénieurs et Techniciens 1979). Dans l'industrie, le système d'alimentation a une définition très spécifique, c'est l'ensemble des conduits amenant du métal liquide pour compenser le retrait (Cuénin 1997a, 2; Association Technique de Fonderie, Commission Ingénieurs et Techniciens 1979).

Sources: Cultural Heritage: Azéma and Mille 2013b; Copper Industry: Association Technique de Fonderie, Commission Ingénieurs et Techniciens 1979; Cuénin 1997b, 2; Art and Craft Textbook: Dubos 2003

canal de coulée

Source: Cultural Heritage: Rolley 1994, 66

jets

Certains auteurs distinguent les jets des attaques, ces dernières désignant exclusivement les conduits reliant les jets à l'épreuve en cire (Rama 1988, 254; Association Technique de Fonderie, Commission Ingénieurs et Techniciens 1979). Il existe tout un répertoire de termes liés : attaques de coulée (Cuénin 1997a, 2), système d'attaque (Association Technique de Fonderie, Commission Ingénieurs et Techniciens 1979).

Sources: Historical: Félibien 1690, 335; Art and Craft Textbooks: Boffrand 1743, 39–42; Mariette 1768, 70–75; Gonon 1876, 5; Rama 1988, 253; Copper Industry: Cuénin 1997b, 2; Lambert 2002, 271

tranches

Spécifique à la fonte au sable.

Sources: Cultural Heritage: Baudry and Bozo 1978, 628; Art and Craft Textbook: Rama 1988, 253

German: *Gusskanal*

Source

Maaz 2010, 696

Alternate Translations

Kanalnetz
Source: Alscher 1987, 555

Versorgungskanäle

Italian: *getto*

Term also found as part of expression *getto di colata*, referring to a gate or sprue that feeds the metal into the mold. The term is also used to describe a cast (act of casting) and pour (act of pouring, as well as content of the crucible).

Sources

Treccani, under *fusione a cera persa*; Bruni 1994, 72–73

Alternate Translations

attacco
Small section of sprue that is often used between the wax model and the longer gate or vent; also referred to as *mozzicone*
Source: Bruni 1994, 73

isfiattatoio
Refers to vents; alternative modern term: *sfiato*; another alternative early Italian pronunciation/spelling: *sfiatatoro*
Source: Cellini [1568] 1983, 11, fols. 48, 50

mandata
Refers to a sprue that feeds the metal into the mold vs. a vent that helps channel the air out
Sources: Giuffredi 2006, 220; Battaglia 1961, under *mandata* [definition 8, vol. 9, 625]

mozzicone
Small section of wax sprue that is often used between the wax model and the longer gate or vent; also referred to as *attacco*
Source: Bruni 1994, 72–74

aria
Refers to vent

bocca
Refers to gate/runner
Source: Cellini [1568] 1983, 11, fols. 51v–52

canale di alimentazione

canale di colata
Sources: Reindell and Tommasi 1999, 353; Pecchioli 1999, 193

colonna di alimentazione
Source: Bruni 1994, 72–74

entrata
Refers to gate/runner
Source: Biringuccio [1540] 1977, fol. 81

esalatorio
Refers to vent
Source: Leonardo 1490, fol. 149v

gitto
Refers to a gate (sprue that feeds the metal into the mold); also used to describe a cast (act of casting) and pour (act of pouring), as well as content of the crucible
Source: Biringuccio [1540] 1977, fols. 75–78, 89, 91

sfiatatoio
Refers to vent; alternative modern term: *sfiato*; alternative early Italian: *sfiatatoro*
Sources: Battaglia 1961, under *sfiatatoio* [definition 1, vol. 18, 875]; Treccani, under *fusione a cera persa*; Leonardo 1490, fol. 149r; Biringuccio [1540] 1977, fols. 81, 89-89v

spiraculo
Refers to vent
Source: Biringuccio [1540] 1977, fol. 89

Chinese: 浇铸道

Source

TNATD

Alternate Translations

浇口
Sources: Ming 2010, 740; TNATD

竖浇道
Source: TNATD

铸口
Source: TNATD

铸道
Source: TNATD

variant

See figs. 1, 76, 544.

A bronze that is similar in form to another, but with some differences due to one having been cast from an altered or adapted wax model, or from an entirely new model. Artists can make variants of their own work, for example, by adjusting the positions of limbs between casts, or variants may be the result of others imitating the artist's work.

Sources

Cultural Heritage: Dillon 2002; Frapiccini 2017; Bassett 2008; Allen et al. 2022, 331

To Be Distinguished From

after-cast

aftercast

replica

surmoulage

Synonym

version

French: *version*

Déclinaison d'un modèle avec des variations dans la forme ou dans les dimensions.

Source

Cultural Heritage: Baudry and Bozo 1978, 549

Alternate Translations

variante

German: *Variante*

Source

Weihrauch 1967, 476

Italian: *variante*

Sources

Treccani, under *variante*; Battaglia 1961, under *variante* [definition 3, vol. 21, 672]

Chinese: 变体

Literally "modification"

Sources

Ming 2010, 65; TNATD

Alternate Translation

变型
Literally "modification"
Sources: Ming 2010, 65; TNATD

welding

See figs. 59, 146, 185, 189, 202, 203, 204, 227, 230, 545, 548, video 12.

A technique for joining separately cast parts using high temperatures resulting in partial melting of the parts. A filler metal is often applied.

Note: In a technique very specific to Greek and Roman large bronzes, a steady stream of poured molten bronze was used both to melt and to join the edges of separately cast sections or to secure repairs. This process is known as flow welding or flow-fusion welding. Modern welding processes for copper alloys include MIG (metal inert gas) and TIG (tungsten inert gas). See I.5, video 12.

Sources

Cultural Heritage: Salter and Gilmour n.d.; Beale 1975, 28–55

Copper Industry: Brunhuber 1988

Art and Craft Textbooks: Rome and Young 2003, 312; Untracht 1982

To Be Distinguished From

brazing

soldering

French: *soudage*

Technique d'assemblage de deux éléments métalliques coulés séparément, entraînant le développement, entre les deux éléments, d'une microstructure commune – on parle de solution de continuité, contrairement au brasage qui s'apparente à du collage. Le résultat est similaire à celui observé lors du soudage de deux parties d'os après fracture.

Note: L'assemblage peut se faire sans apport de métal, par soudure dite autogène (Bader and Théret 1961, par exemple au moyen d'un arc électrique ou d'un TIG). L'assemblage peut aussi se faire par apport d'un métal de composition proche de celui des parties à assembler. Pour la sculpture en bronze et en particulier les grands bronzes antiques, on parle de soudage par fusion au bronze liquide (Azéma and Mille 2013b). Dans l'industrie moderne, le soudage des métaux peut se faire à l'état solide, sans fusion, par exemple par soudage par diffusion, cf. Murry 1994.

Sources

Copper Industry: Brunhuber 1988; Murry 1994

German: *Schweißen*

Source

Deutsches Kupferinstitut 2005, 18

Italian: *saldatura*

Often specified by *a fusione* or *per fusione*

Sources

Treccani; Battaglia 1961, under *saldatura* [vol. 16, 387]; Bruni 1994, 122–23; Brunhuber 1988

Alternate Translation

saldatura metallurgica per colata
Term also used for brazing, and (in antiquity) for flow fusion welding
Sources: Formigli 2010, 20

Chinese: 熔焊

Sources

Ming 2010, 1211; TNATD

Visual Atlas of Features

The dictum "an image is worth a thousand words" is apt when trying to describe technical details, and this is particularly true for bronze sculpture. Therefore, the *Guidelines* rely heavily on illustrations. This section gathers all of the illustrations used in the publication, including radiographs, 3D models, sketches, charts, and diagrams. A fair number of these images have visual annotations to aid in their interpretation. In the online version these appear in layers that can be activated (turned on and off) by the reader. In the print-on-demand version, these are merged onto the image. In the digital publication it is possible to search for and browse through the photos and diagrams via keywords. In the print-on-demand version, the images all print together separately from the text in the order of appearance in the *Guidelines*.

The visual atlas also gathers short videos illustrating a range of fabrication processes described in the text. Links to them appear in the relevant sections of the *Guidelines*. Their inclusion is made possible by the digital format and affords readers a better sense of the gestures, timing, choreography, and sound involved in these processes—all of which are not readily conveyed in words.

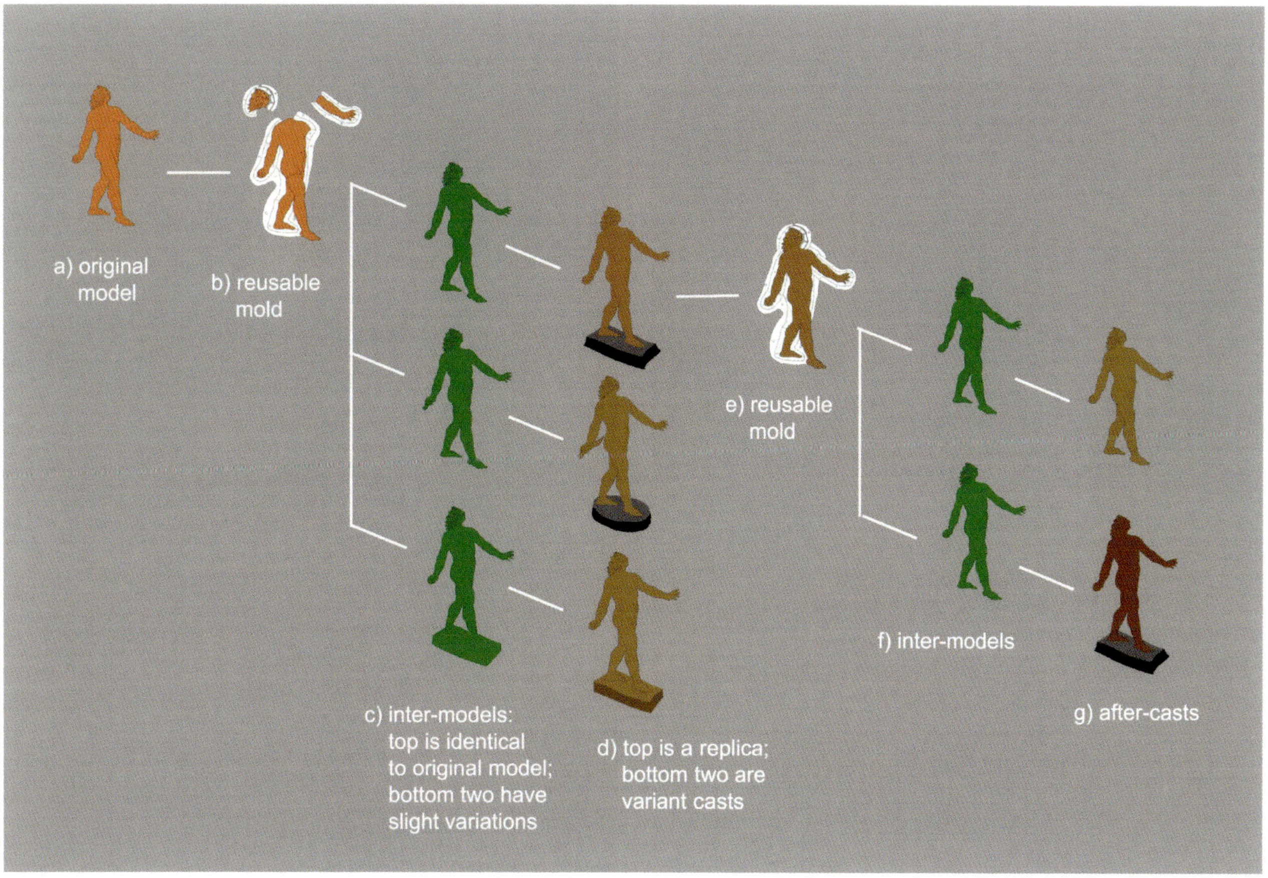

Figure 1 Diagram of generations of casts (here through the lost-wax process): a) the original model here is in clay, but could be made of any material; b) the translation of the original model into a wax inter-model may involve more than simple molding; c) a close reproduction of the original model will be referred to as a replica (a wax inter-model that is altered in details and with additional features (e.g., a base) will result in a variant); d) more alterations can be made after casting (e.g., with a base or a different patina); e) a bronze can in turn serve as model for another generation of casts by taking a mold of it; f) because wax shrinks as it cools, the inter-models are slightly smaller than the bronze model and their details softer unless reworked; g) the bronze after-cast may in turn be reworked and finished differently than its original bronze model or sister casts.

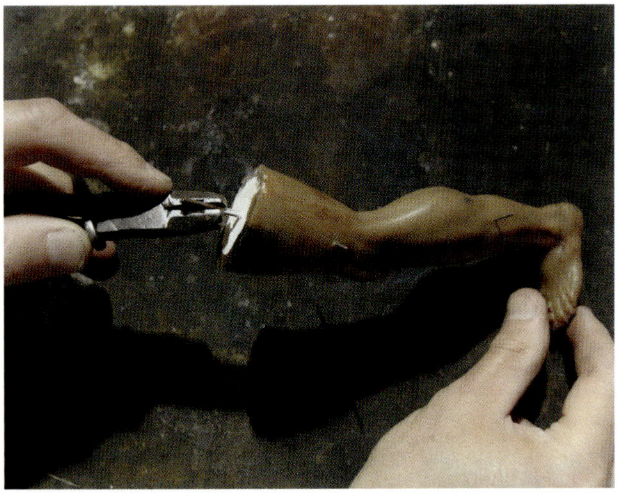

Figure 2 An internal core support is pushed into the plaster-based core of the wax leg as part of the reconstruction of *Mars* by Andrew Lacey (British, b. 1969), 1997, based on a bronze after a model by Giambologna (Flemish, active Florence, 1529–1608), ca. 1587, probably cast later, H. 40.7 cm (Victoria and Albert Museum, bequeathed by Dr W. L. Hildburgh FSA, inv. A.99-1956). See Bewer 1996a.

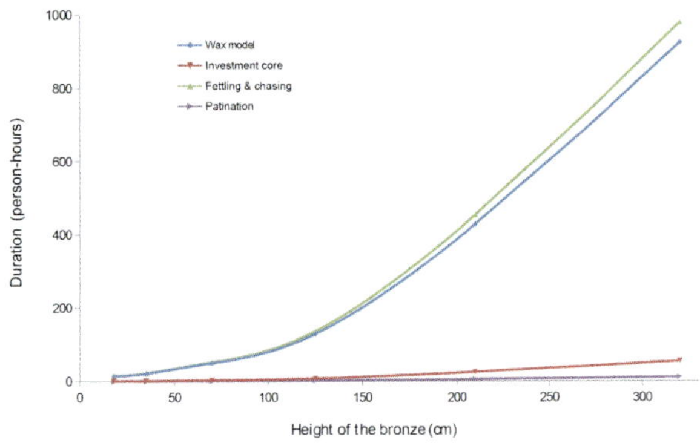

Figure 3 Graph showing the impact of a lost-wax sculpture's size on the time needed for the different steps in the fabrication process. The data are based on records gathered between 1973 and 2006 kindly provided by Jean Dubos, director of Fonderie de Coubertin, related to the casting of a standing nude man, whatever the model. The duration is given in person-hours and only factors in working hours; the time needed for drying and baking of the investment, for instance, is not taken into account here, although those steps can take a long time (as shown in table 1). Also, the more people working simultaneously on a bronze, the quicker the completion (again see table 1).

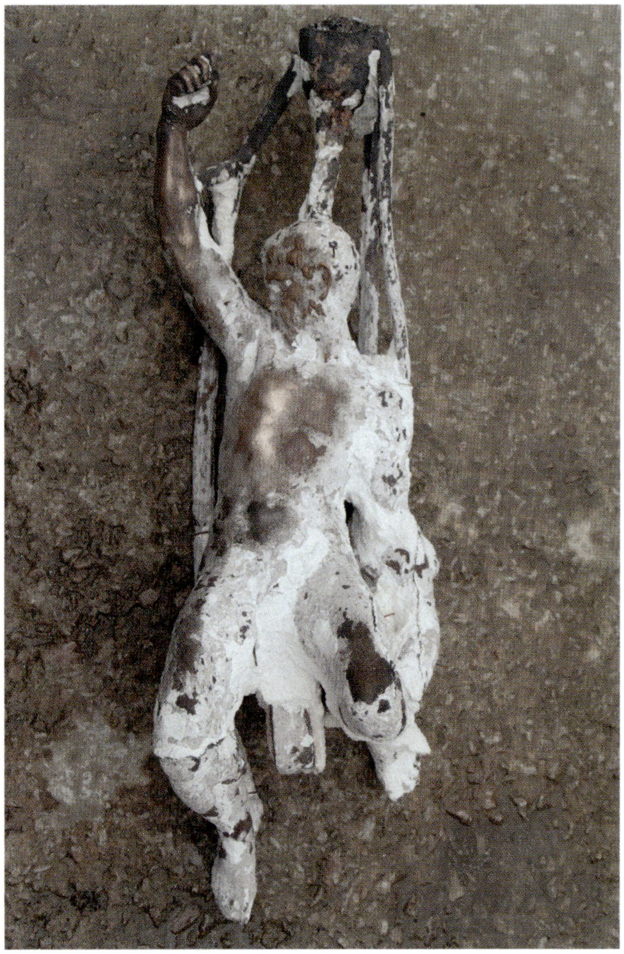

Figure 4 In-process detail of the first reconstruction by Andrew Lacey (British, b. 1969) of one of the so-called Rothschild Bronzes, attributed to Michelangelo (Italian, 1475–1564), H. 91.2 cm (private collection). The fresh casting has just been broken free from its investment. Note the marked granular texture of the bronze surface and the white plaster investment heavily bonded to it. See Lacey 2018.

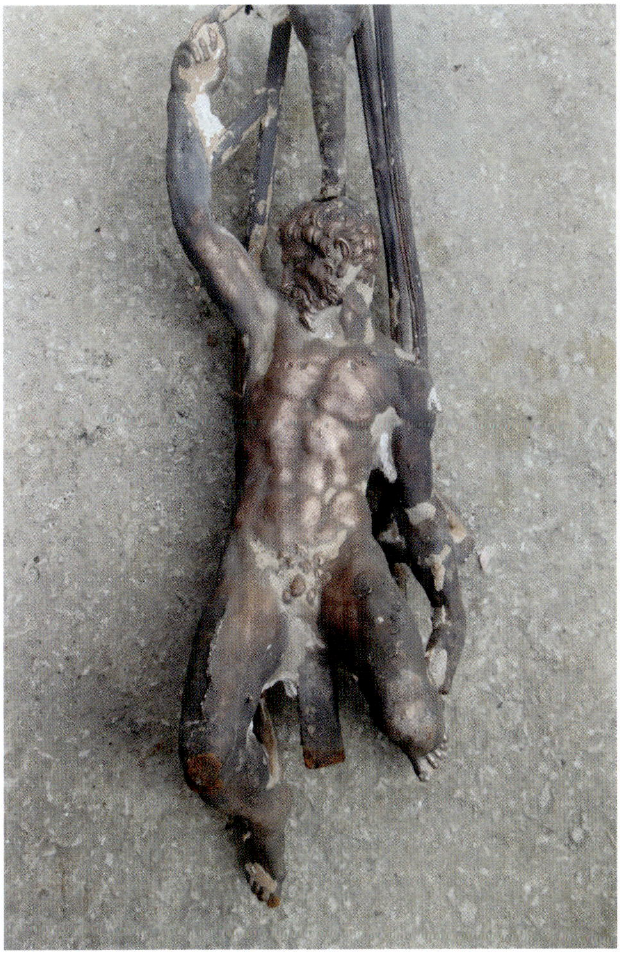

Figure 5 In-process detail of the second reconstruction by Andrew Lacey (British, b. 1969) of one of the so-called Rothschild Bronzes, attributed to Michelangelo (Italian, 1475–1564), H. 91.2 cm (private collection). The casting has just been broken out of the mold with a dark oxide layer and some of the investment bonded to the surface. Great care in the application of the first investment layers on the wax model resulted in a cleaner as-cast surface with little or no granular texture and minimal investment bonded to the surface, necessitating minimal chasing. See Lacey 2018.

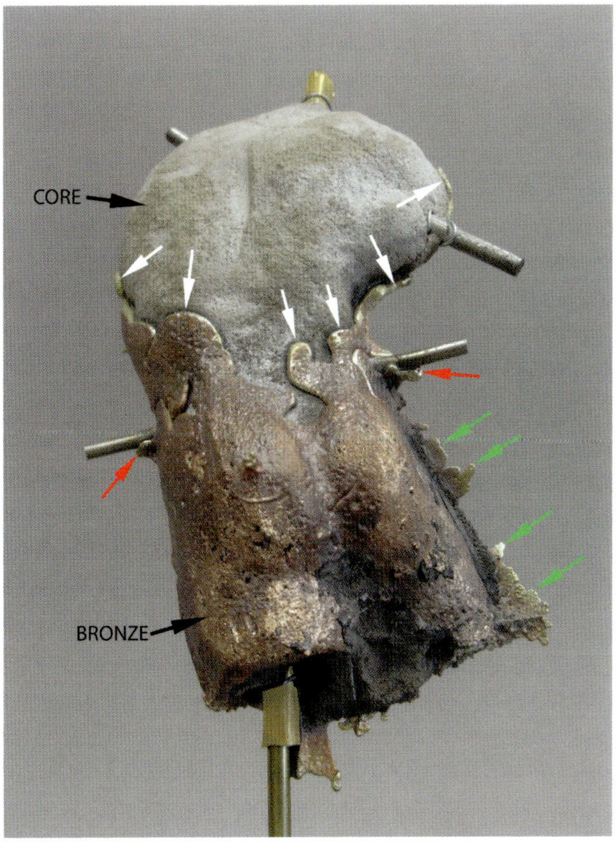

Figure 6 Experimental sand-cast torso (H. 16 cm). Yellow overlay indicates the hollow lantern made of rolled sheet metal that runs vertically through the body and has been used to mount the torso to a base. See also the area where the bronze did not fill the mold, an extreme example of a cold shut. White arrows indicate the typically rounded edge of the bronze wall where it stopped flowing. Green arrows denote the seam line along the proper right side of the torso where bronze flowed into a gap along the joint of the cope and drag. Red arrows indicate flashing at either side of the lower horizontal armature rod. Made with consultant Tonny Beentjes, Decorative Arts and Sculpture Conservation, J. Paul Getty Museum.

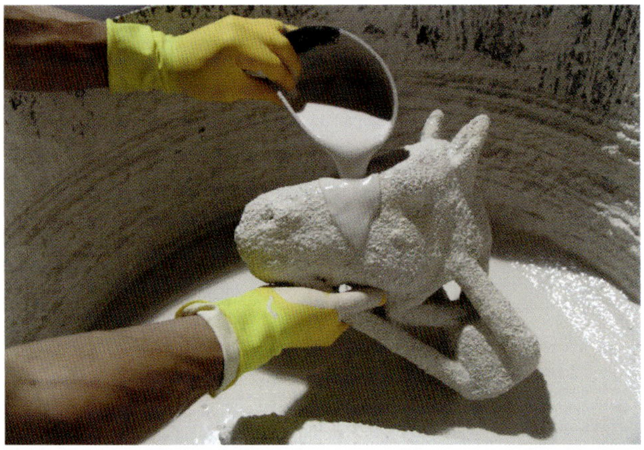

Figure 7 Building up the ceramic shell mold by repeatedly dipping the wax inter-model into colloidal silica slurry then coating that wet layer with one of larger refractory particles to build up a thick composite layer that will form a "ceramic shell" through sintering of the colloidal particles. Andrew Lacey (British, b. 1969), *The Anatomy of Bronze*, cast by the artist in 2019, Devon, UK, H. 45 cm (artist's collection).

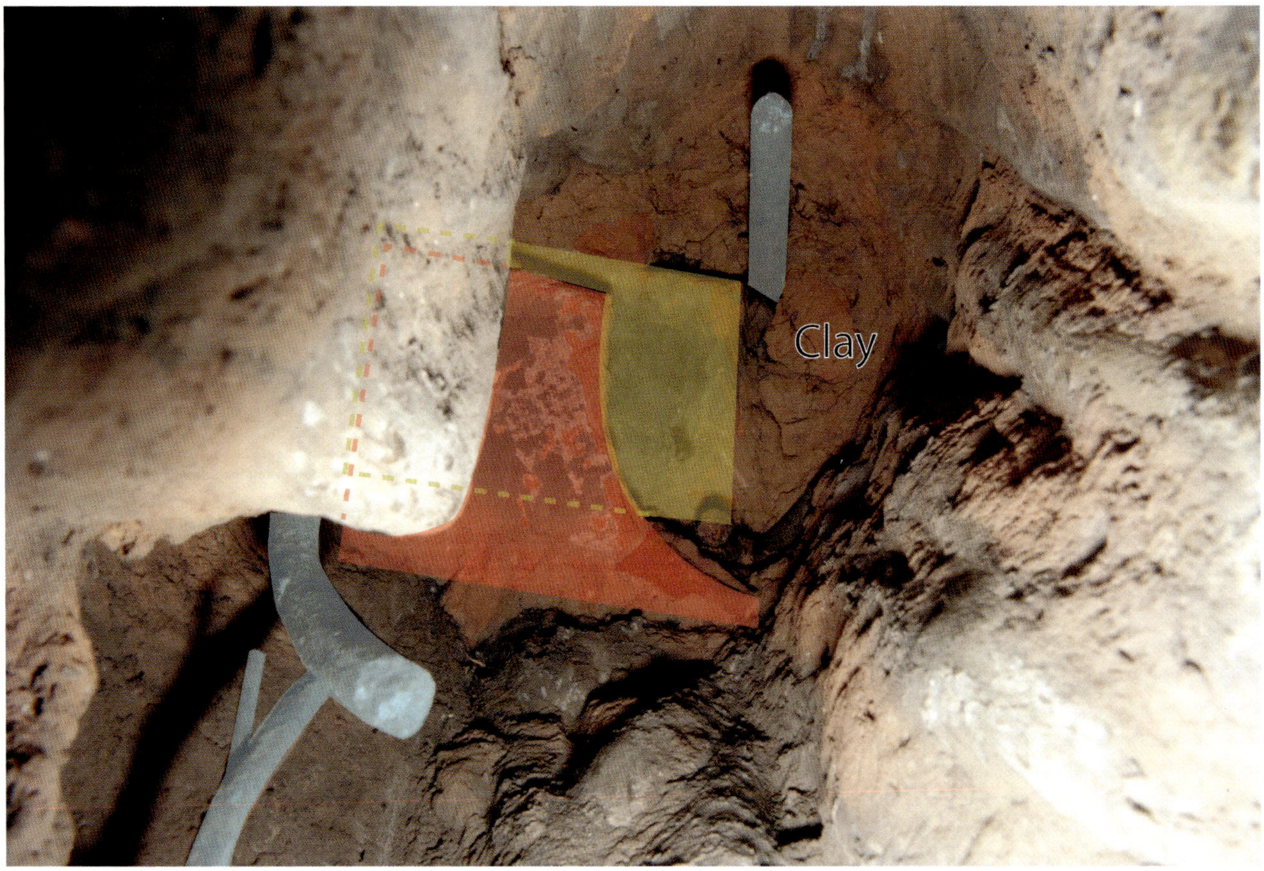

Figure 8 View of the inside of the hollow sculpture that has been partially cleaned out of its core. The cavity preserves some of the iron armature rods and wires as well as remnants of baked bricks that would have been embedded in the clay-based, refractory core mixture (image width ~50 cm). Barthélemy Prieur (French, 1536–1611), *Peace*, cast in 1571 by Nicolas Péron, H. 128 cm (Musée du Louvre, inv. MR1683). See Seelig-Teuwen, Bourgarit, and Bewer 2014; Castelle et al. 2021.

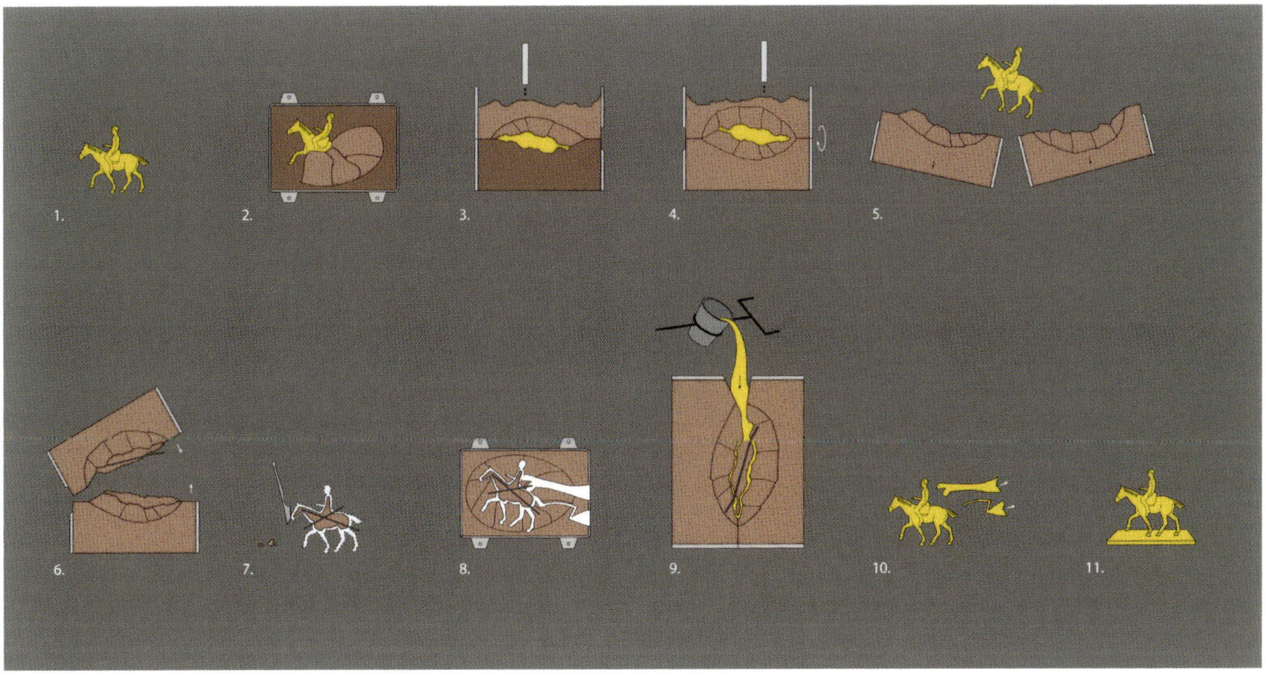

Figure 9 Diagram depicting a version of the main steps in sand casting: 1) a chef-modèle or pattern is a specially designed model for sand casting made of a hard material—often metal—to withstand the repeated handling and ramming of sand within a multipart, stacked metal or wooden frame called a casting flask; 2) the pattern is buried about halfway in backing sand in a first section of the flask called the cope; piece mold sections are built up over the model with carefully rammed special casting sand whose naturally occurring binder, clay, helps the compacted sand to hold the desired shape; 3) once the exposed part of the pattern is covered with mold pieces, the second section of the flask, or "drag," is affixed to the cope and the piece molded side is back filled with carefully rammed sand; 4) the tightly packed flask is flipped; the cope and backing sand, now on top, are removed to provide access for a similar piece molding process to be performed on the other side of the chef-modèle; 5) the two-part casting flask is parted and the mold pieces carefully disassembled to remove the pattern; 6) to make a hollow cast, a core must be created to define the thickness of the bronze walls; this is made by ramming the same special casting sand into the hollow impression left by the chef-modèle; metal core supports are incorporated into the new sand; the cope and drag are joined, thereby creating a sand replica; 7) the flask is opened again and the sand replica is removed and shaved down evenly overall to form the core; white areas indicate hollow spaces for metal to fill; the thin legs in this case will be solid, and so will not need a core; 8) the sand mold and core are baked, then carefully reassembled in the cope and drag; the core supports projecting from the core serve to suspend the core in place; channels are cut strategically into the mold to ensure the efficient distribution and flow of the metal and venting of air; 9) the cope and drag are reassembled and locked together, and the metal alloy, which has in the meantime been liquefied, is poured into the mold; 10) once the metal has solidified and cooled, the sand mold is broken open and the bronze cast removed; the metal sprues that have formed in the channels are cut off; and the surface is cleaned and repaired as needed; 11) if a bronze is cast in numerous sections, or separately from a base, it will be joined at this stage. The final steps may consist in refinishing of the metal surface and coloration through the application of a chemical patina or some form of coating. Diagram based on Charles Marion Russell (American, 1864–1926), *Medicine Whip*, modeled 1911, sand cast 1912–16, H. 14.8 cm (Gilcrease Museum, Tulsa, 0837.14). See Case Study 6.

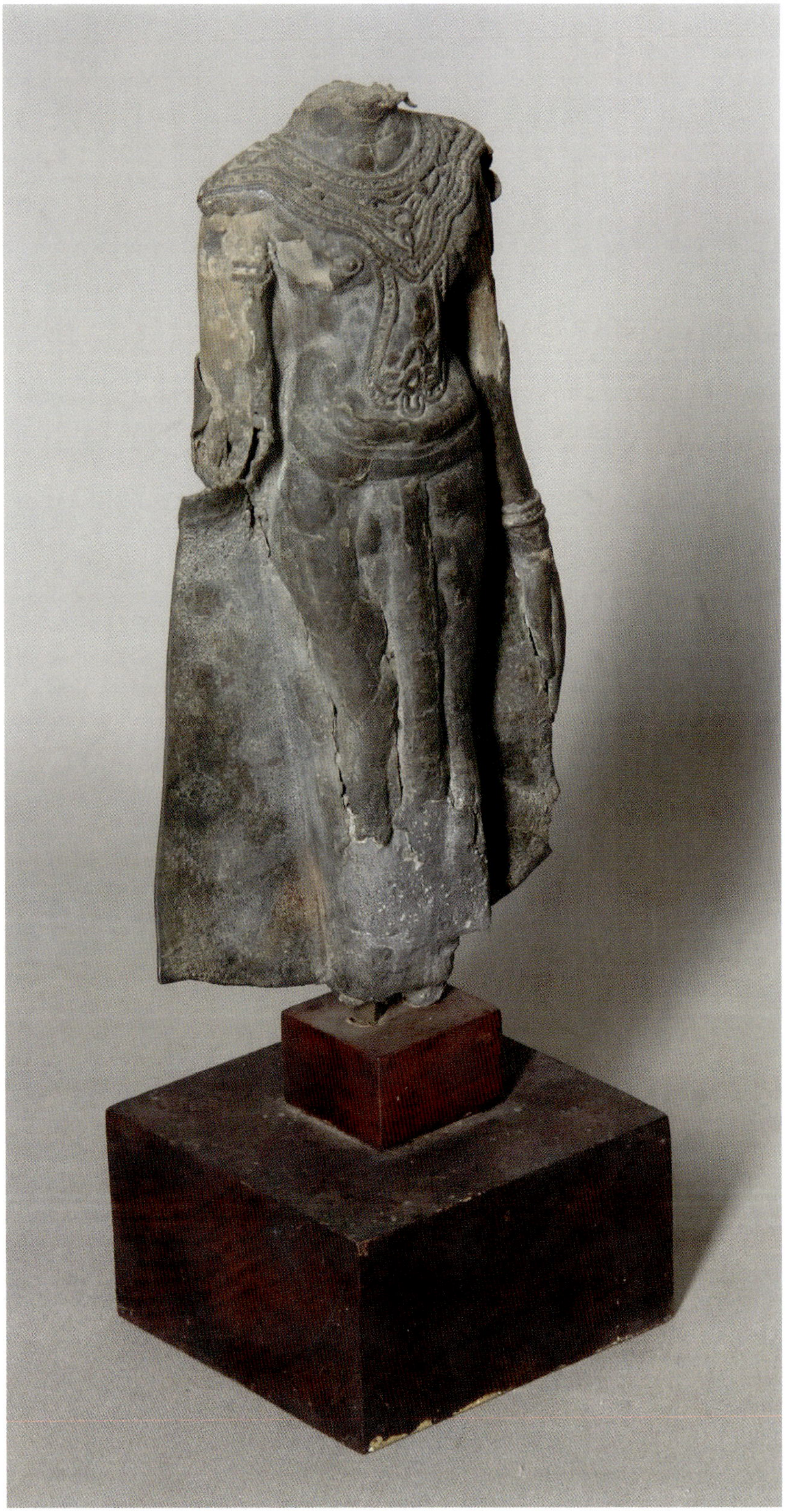

Figure 10 Large casting flaws in the arms of a Khmer divinity reveal the tan-colored, clay-based core that remains inside the bronze. Note the very thin walls of the bronze at the edges of the losses. Standing Buddha in Abhayamudrā, post-Angkorian, Prasat Bakan (Pursat), 16th century, H. 24.2 cm (National Museum of Cambodia, inv. Ga.3330).

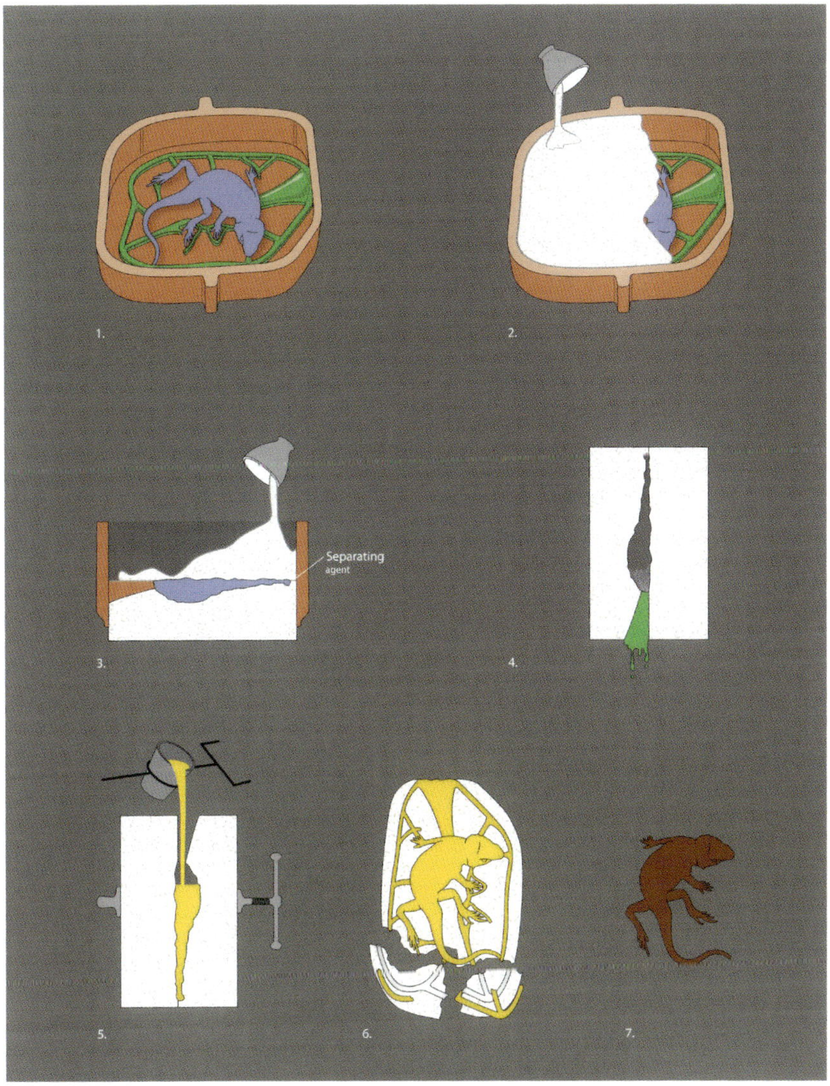

Figure 11 Diagram depicting an example of life-casting: 1) the lizard is placed in a bed of clay; rolls of wax are strategically joined to the lizard to form the sprue system; a clay wall is then built around it; 2) a plaster-based slurry is poured over the lizard and sprue system, embedding the lizard and sprues, forming the top half of the refractory mold; 3) when set, the mold is turned over, the bottom clay slab is removed, and new clay walls are built up to contain the other half of the mold and additional refractory material is poured in; 4) the mold is heated until dry; the wax and as much of the lizard as possible are burned away; the bone remnants can be removed by opening the mold; 5) the two sides of the mold are clamped together and filled with molten bronze; 6) the refractory mold and sprue system are removed from the cast; 7) the finishing of the solid bronze includes sharpening of details as needed and patination. Diagram based on research related to the Making and Knowing Project at Columbia University. See Lacey and Lewis 2020.

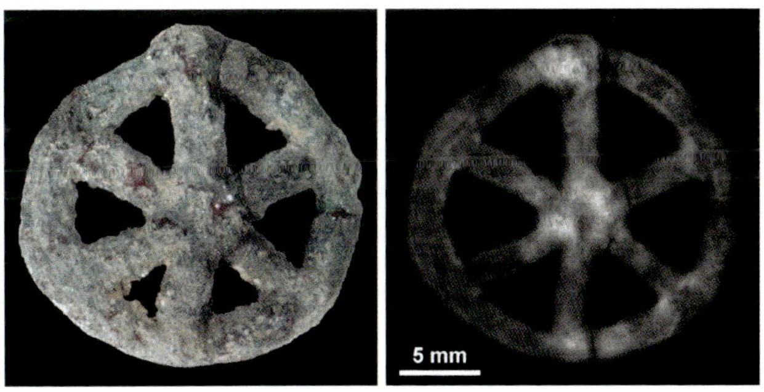

Figure 12 Overall view and X-ray radiograph of the earliest known object made by lost-wax casting. Copper Amulet, Mehrgarh, Baluchistan, mid-5th–mid-4th millennium BCE, diameter 2 cm (Mission Archéologique Française du Bassin de l'Indus, inv. MR.85.03.00.01). See Thoury et al. 2016; Mille 2017.

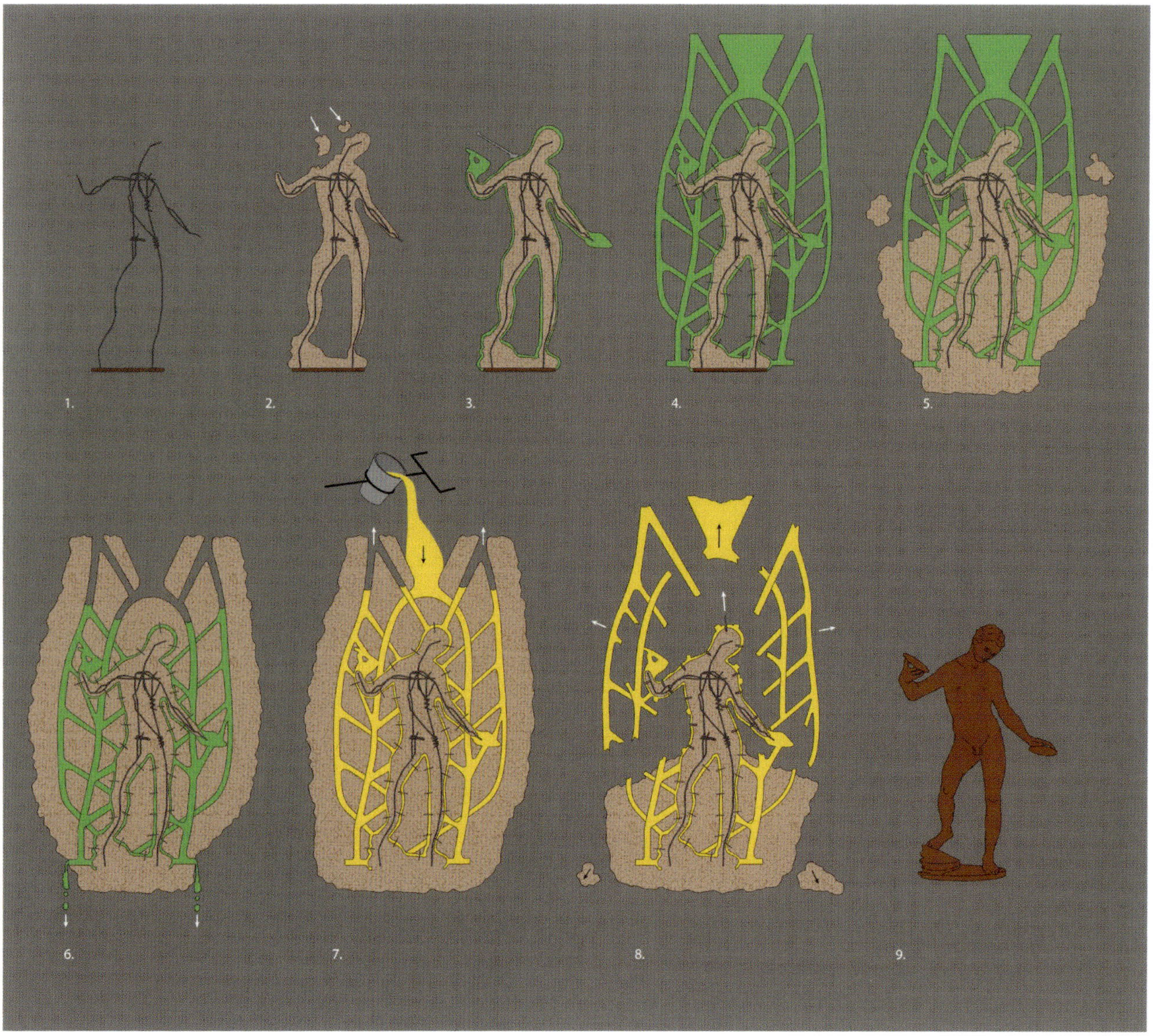

Figure 13 Diagram depicting a version of the direct lost-wax casting technique: 1) an armature is constructed of iron rods and wires; 2) a refractory clay investment is built up over the armature to form the core; 3) a wax layer is then modeled over the core and details refined in order to create a unique wax model; 4) the wax sprue system is joined to the model and core pins inserted through the wax and into the core; 5) the wax is invested in a refractory mold; 6) the mold is heated until dry and all traces of wax are melted and burned out; 7) the mold is filled with molten bronze; 8) when the metal is cool, the investment, core pins, and sprue system are removed; 9) chasing includes polishing, burnishing, and the addition of texture and sharpening of details as needed. As a final step, the surface is patinated. Diagram based on X-radiographs of Adriaen de Vries (Netherlandish, 1556–1626), *Juggling Man*, ca. 1615, H. 76.8 cm (J. Paul Getty Museum, inv. 90.SB.44). See fig. 79, Bewer 1999.

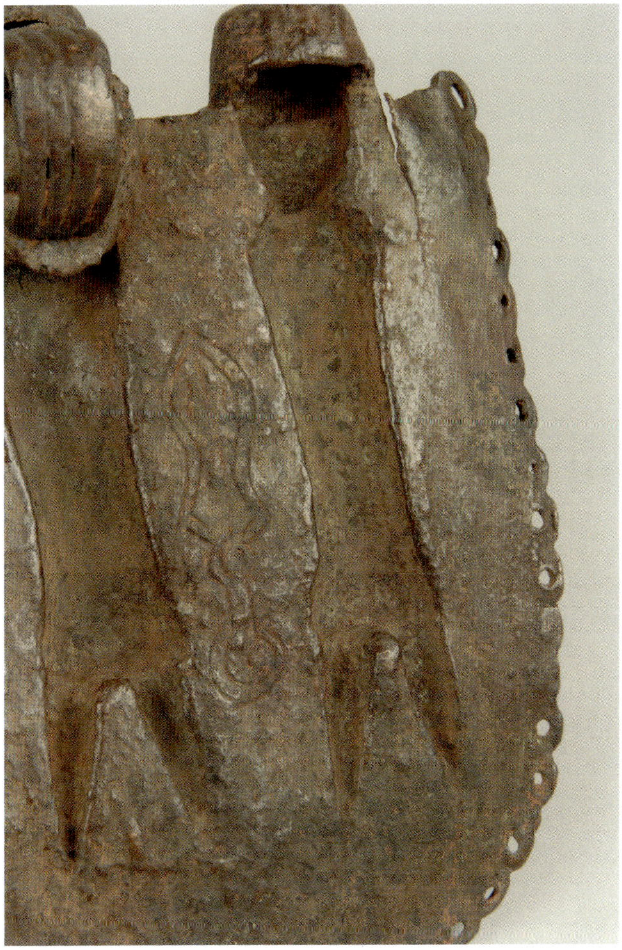

Figure 14 Detail of an Eben ceremonial sword on the back of a pendant, possibly a maker's mark. Pendant with Royal Triad, Benin kingdom, Nigeria, 18th century, H. 15.4 cm (Weltmuseum Wien, Vienna, inv. VO 64271). At least twenty pendants are known to bear this mark on their backs.

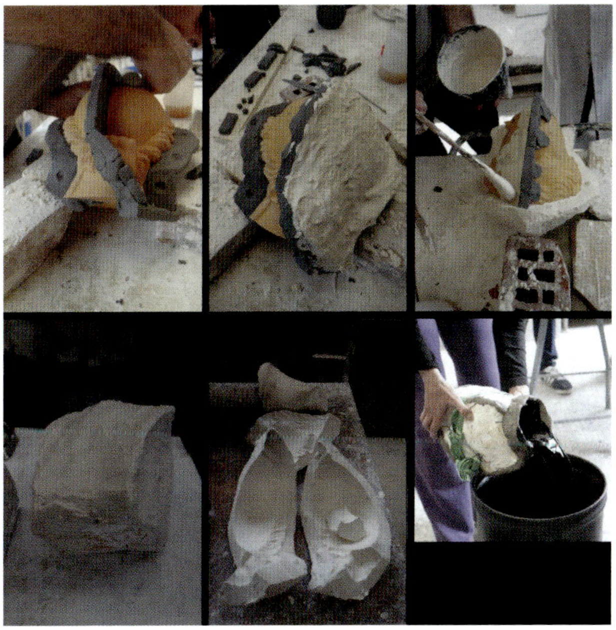

Figure 15 Plaster molding of the 3D resin print of the head of *Apollo of Lillebonne* during the 2016 CAST:ING meeting, Coubertin foundry, France. Top left: plasticine (black) is used to mark the sections of the piece mold on the resin head (yellow). Top middle and right: plaster is applied to the model in sections. Bottom left: the completed piece mold. Bottom middle: the sections of the mold are separated. Wax or clay will be applied on the internal surface by different methods (slabs, painting, balls, see figs. 17, 19, 24). Bottom right: the reassembled piece mold is used for slush molding. The height is 18 cm (two-thirds of the original size). Original: *Apollo of Lillebonne*, France, 1st century BCE–1st century CE, H. 193 cm (Musée du Louvre, inv. Br37 [NIII65]). See fig. 288.

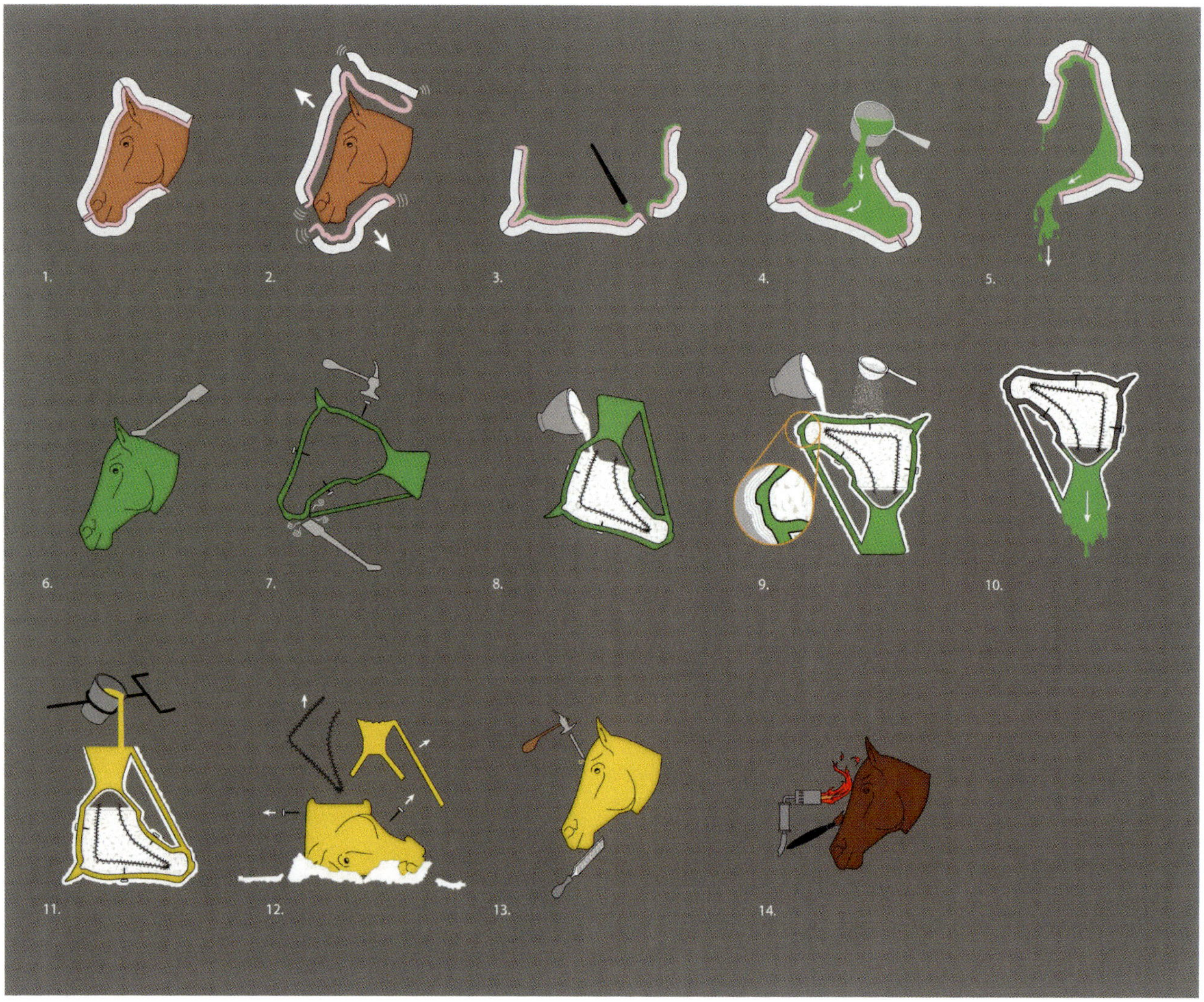

Figure 16 Diagram depicting a version of the lost-wax slush-molding process combined with ceramic shell casting: 1) a flexible silicone two-piece mold is created from the model (in this example made of clay); the layer of silicone is backed with a rigid mother mold; 2) this reusable mold is disassembled and the model removed; 3) to form the inter-model, molten wax may first be brushed into the interior of the piece mold to ensure that the wax reaches all areas and captures the fine details; 4) the mold is reassembled; the molten wax is poured in and slushed around in order to coat the interior; 5) excess molten wax is poured out; this "slush molding" is repeated until the desired thickness is achieved; 6) the hollow wax inter-model is removed from the mold, imperfections repaired, and details refinished; 7) core pins are pushed through the wax walls and the wax sprue system is fused to the inter-model; 8) an internal core support (armature) is inserted into the hollow wax, which is then filled with refractory core material (note: in ceramic shell casting, the core can be a plaster-based investment or the same ceramic shell materials as the outer mold); 9) the inter-model is then coated with a liquid ceramic slurry followed by grains of fused ceramic to provide bulk and a textured surface that will ensure that the next layer of slurry adheres well; once this first layer is dry, the process is repeated (eight to ten times) to create the desired thickness of ceramic shell; 10) when the refractory mold has dried, it is placed upside down and heated rapidly to burn out all traces of wax; it is then placed in a kiln to dry the plaster investment core and sinter the ceramic shell into a sturdy, unified body that can withstand the pressure of the molten bronze; 11) the hot mold is removed from the kiln, turned with the casting cup facing up, and the molten bronze poured into the mold; 12) when the metal has cooled, the ceramic shell casing is broken off of the cast and remains are removed by sandblasting; the bronze sprue system is cut off and the core pins and plaster core removed along with the internal armature; 13) chasing includes polishing, burnishing, and the addition of textures and sharpening of details as needed; 14) as a final step, a chemical patina and/or a coating may be applied. Diagram based on Andrew Lacey (British, b. 1969), *The Anatomy of Bronze*, cast by the artist in 2019, Devon, UK, H. 45 cm (artist's collection). See Case Study 7.

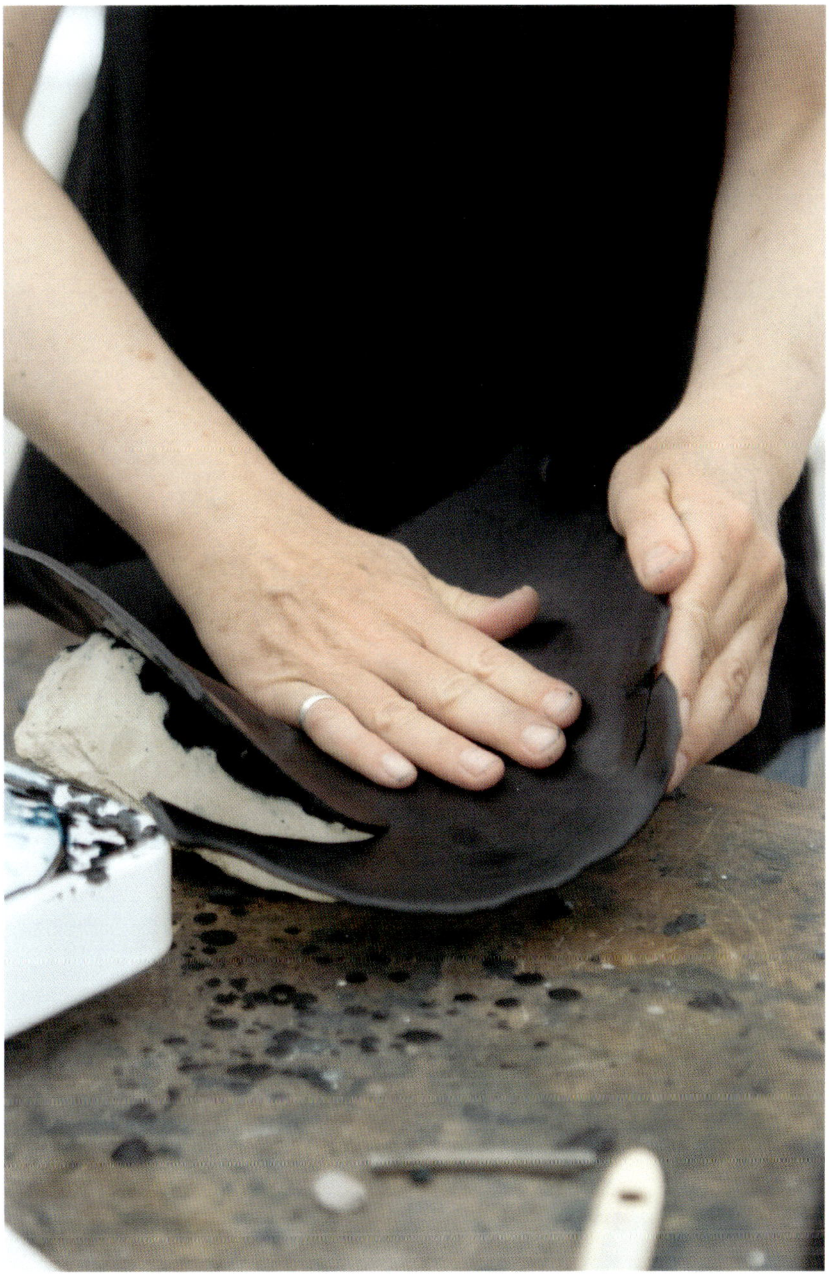

Figure 17 Application of wax slabs on the internal surface of one half of a plaster mold used for a bronze reproduction of the head of the *Apollo of Lillebonne* during the 2016 CAST:ING meeting, Coubertin foundry, France. The height is 18 cm (two-thirds of the original size). Original: *Apollo of Lillebonne*, France, 1st century BCE–1st century CE, H. 193 cm (Musée du Louvre, inv. Br37 [NIII65]). See fig. 288.

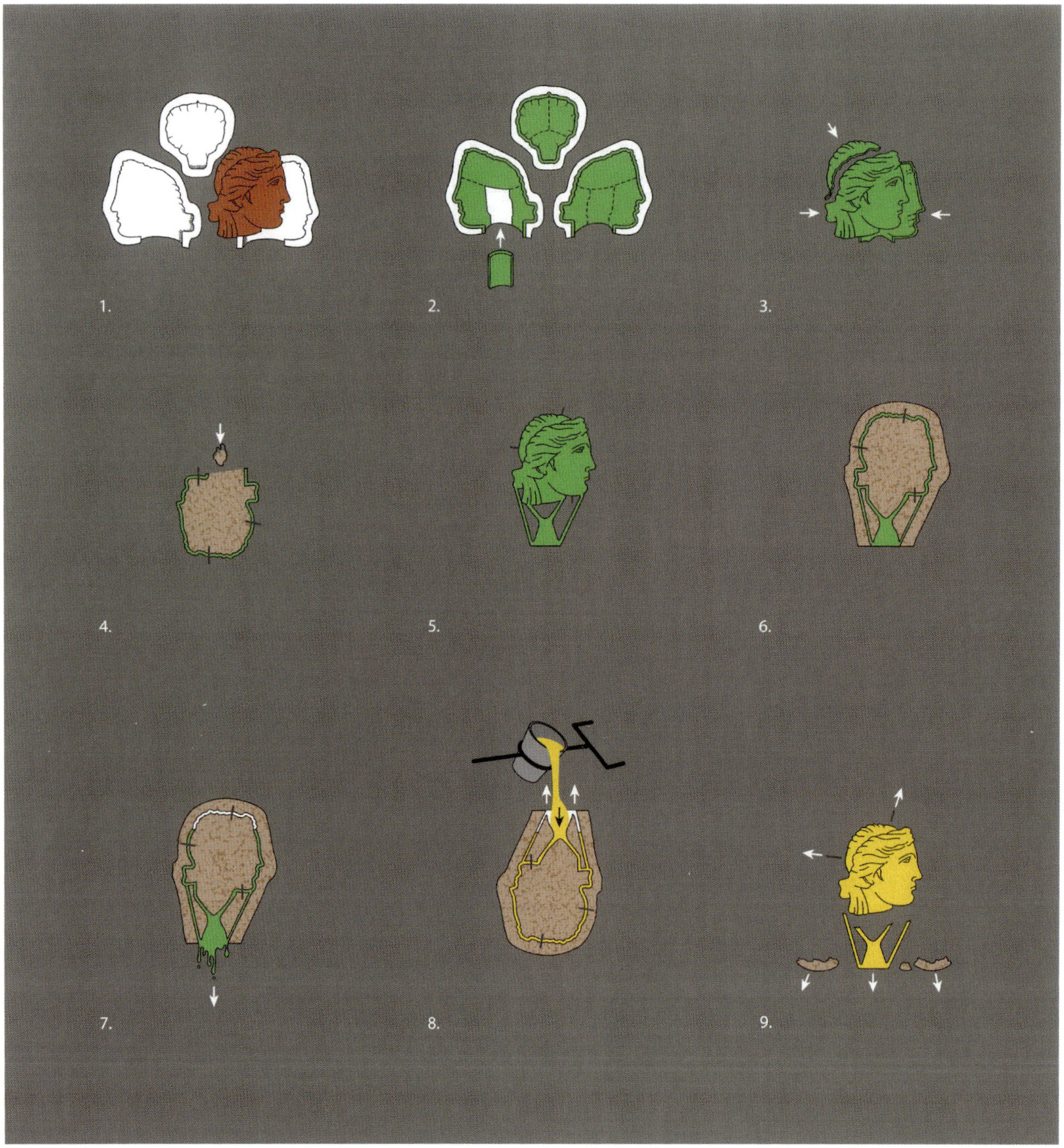

Figure 18 Main steps of the indirect wax slab process: 1) a piece mold is formed on the model, composed of smaller pieces to account for the undercuts; 2) the piece mold is removed and sections are lined with thin, warmed sheets (slabs) of wax; 3) the wax parts are joined with a hot tool to form a hollow inter-model, and the surface is reworked as needed; 4) the hollow wax inter-model is filled with refractory core material and core pins are inserted through the wax and into the core; 5) the wax sprue system is fused to the inter-model in strategic locations with a hot tool; 6) the inter-model is invested in a refractory mold; 7) the mold is then heated until dry and all traces of the wax burned out; 8) molten bronze is poured into the mold; 9) fettling includes breaking off the refractory mold, removing the core pins, and cutting away the sprue system. Chasing may include polishing, burnishing, and the addition of details as needed; the surface color may also be enhanced with inlays, plating, and/or a patina.

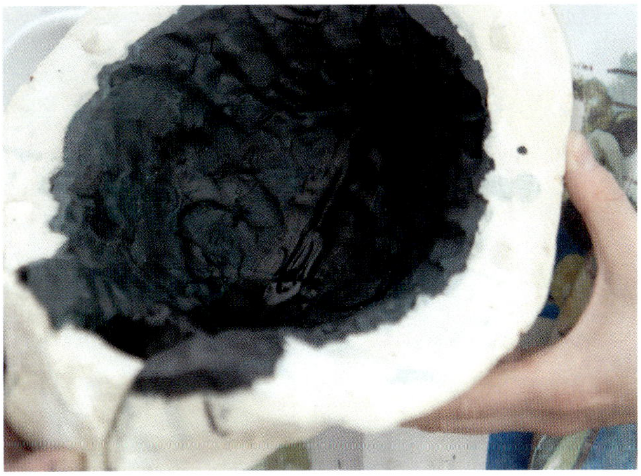

Figure 19 Application of balls of wax on the internal surface of one half of a plaster mold for the bronze reproduction of the head of the *Apollo of Lillebonne* during the 2016 CAST:ING meeting, Coubertin foundry, France. The height is 18 cm (two-thirds of the original size). Original: *Apollo of Lillebonne*, France, 1st century BCE–1st century CE, H. 193 cm (Musée du Louvre, inv. Br37 [NIII65]). See fig. 288.

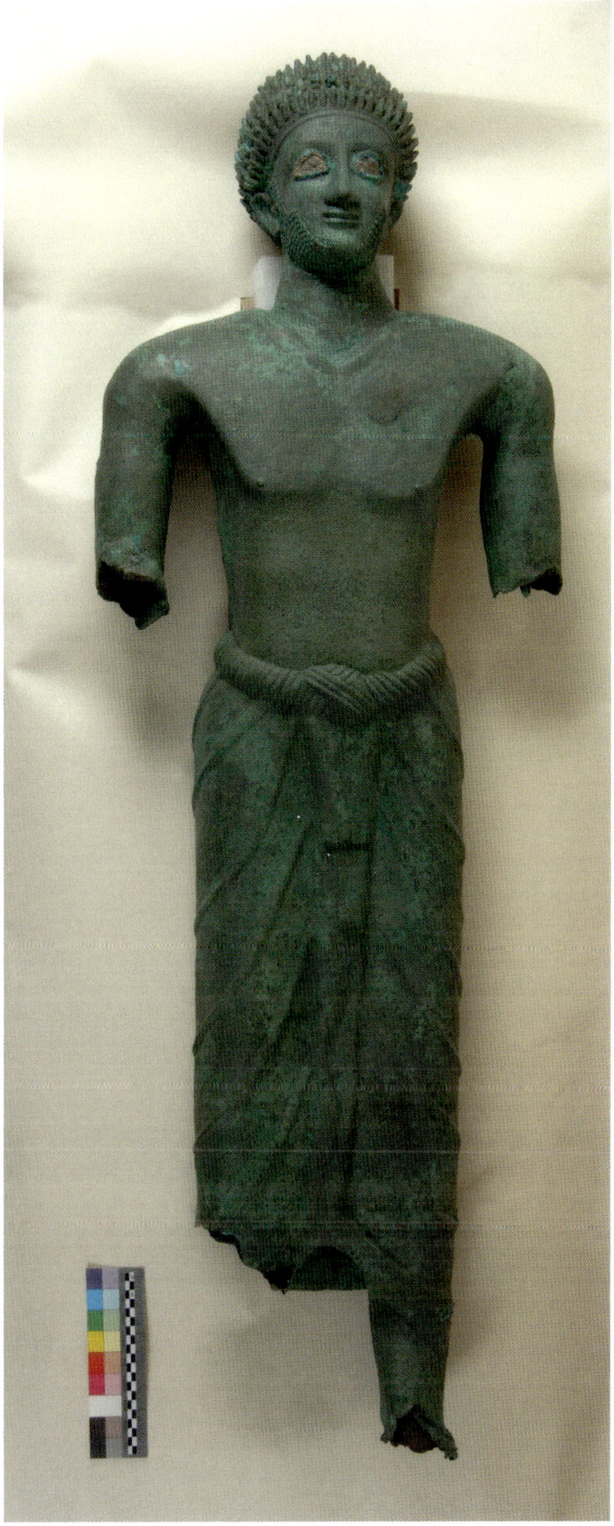

Figure 20 Example of a very regular and thin metallic wall (~2 mm) on a large bronze. See also the hair made on the wax model via the addition of rolled wax threads and rods. Bronze Statue of Hawtar'athat, Yemen, 1st millennium BCE, H. 140 cm (National Museum of Sana'a, Yemen, inv. YM 23206). See Mille et al. 2010.

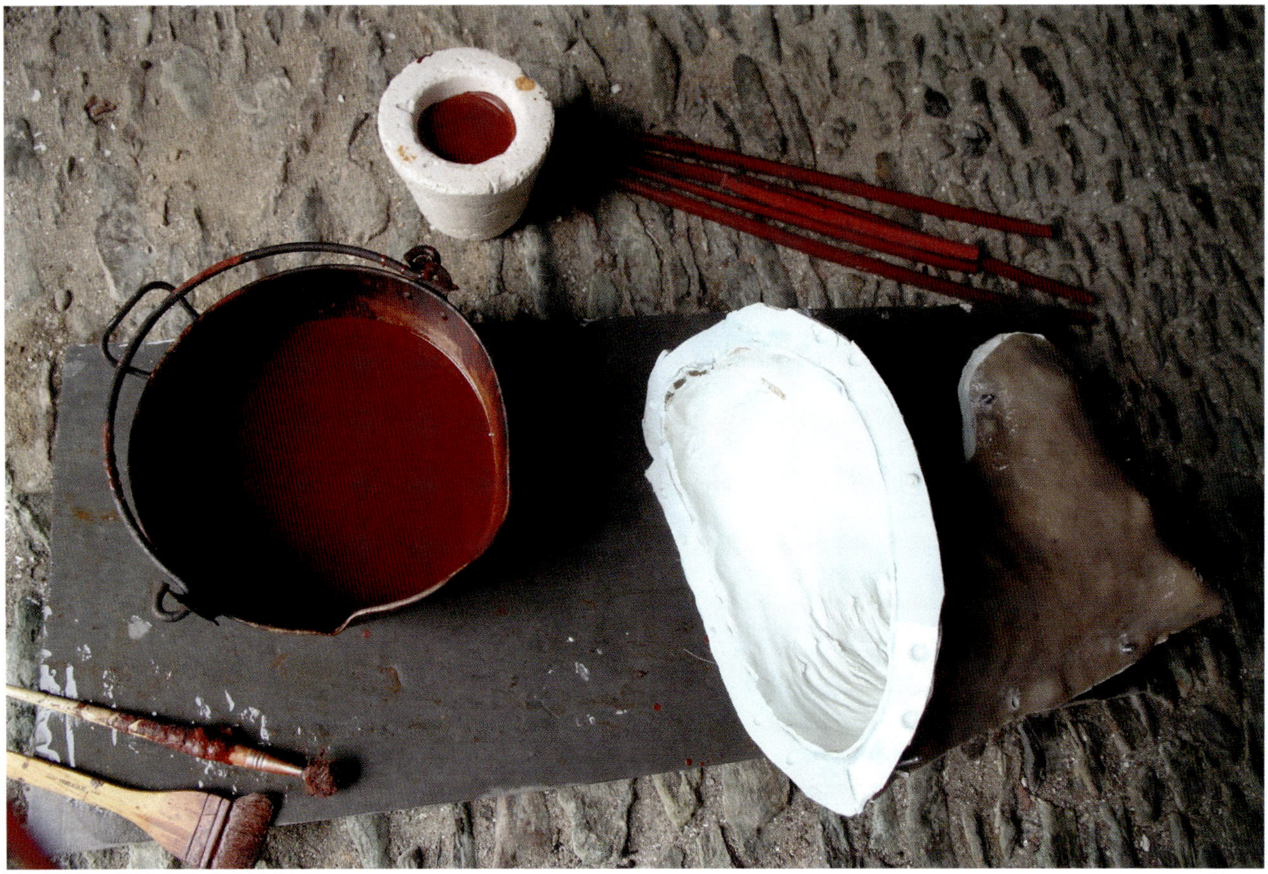

Figure 21 Assembled silicone rubber mold of the head and neck of the horse awaiting wax application along with bucket of red casting wax, wax sprues, a casting cup freshly formed in its mold, and brushes for wax application. Andrew Lacey (British, b. 1969), *The Anatomy of Bronze*, cast by the artist in 2019, Devon, UK, H. 45 cm (artist's collection).

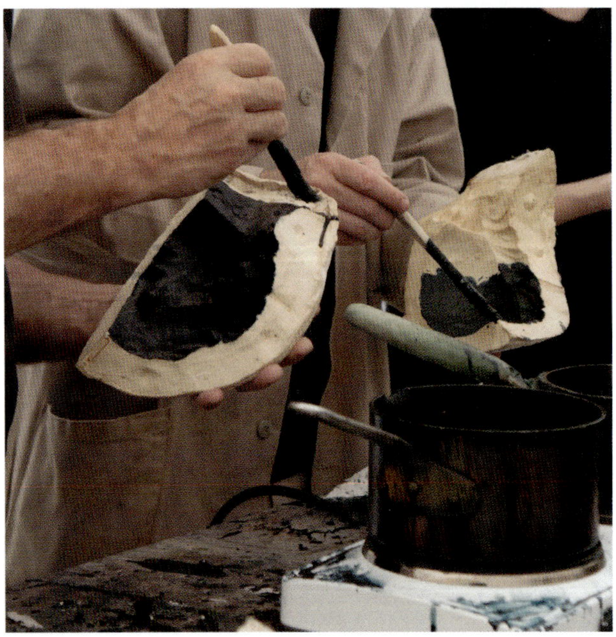

Figure 22 Hot molten wax is brushed onto the internal surface sections of a plaster mold as part of an indirect lost-wax casting process carried out during the 2016 CAST:ING meeting, Coubertin foundry, France.

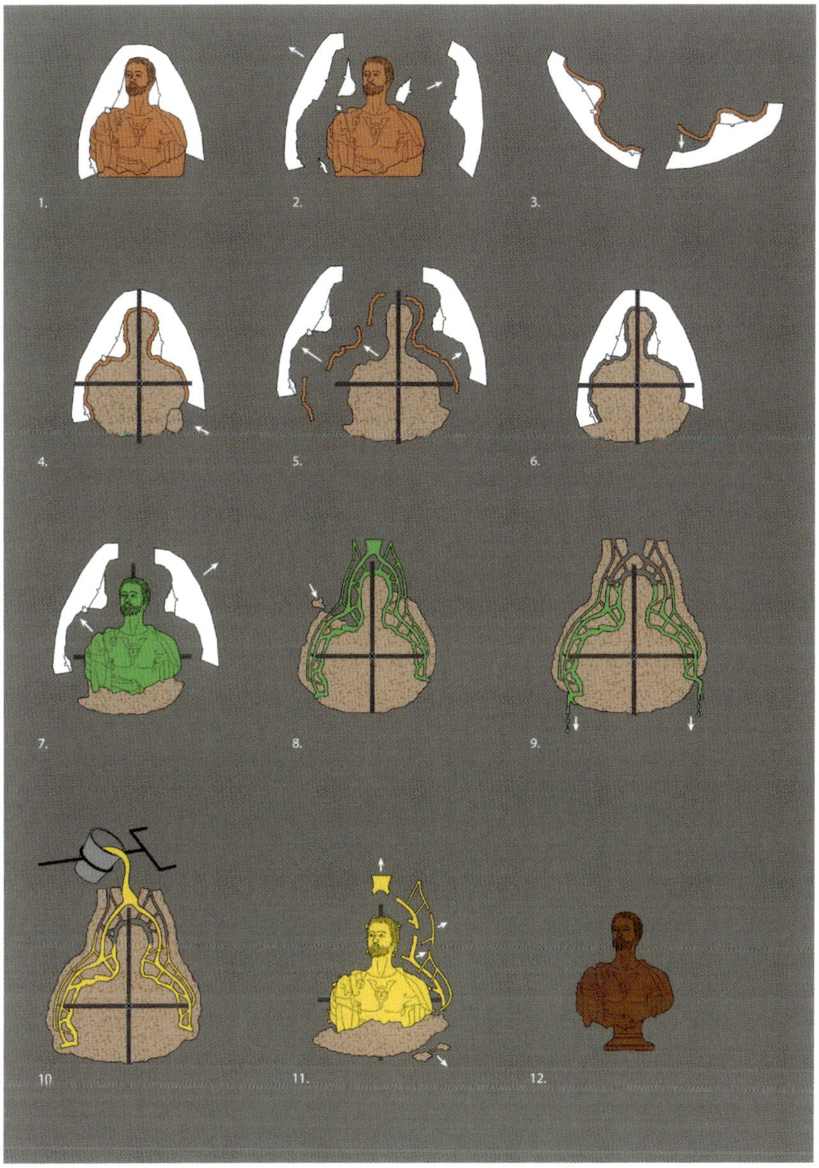

Figure 23 Diagram of a version of the lasagna technique. 1) a plaster piece mold is built up in sections around the clay model; 2) the reusable mold is disassembled; 3) sections of the mold are lined with sheets of clay that Cellini referred to as "lasagna" and that create the space to be filled eventually by bronze; 4) a fire-resistant clay core is built up around an iron armature to fit snugly into the lasagna-lined mold and then dried; 5) the piece mold is disassembled in order to remove the lasagna; 6) the core is baked and reinforced by binding with wires; the fired core and plaster piece mold are assembled; the armature and mold extension at the base help to preserve the space created by the lasagna; 7) molten wax poured into the space between the core and the piece mold creates the inter-model; when freed from the mold it is reworked to the artist's liking; 8) the sprue system is joined to the inter-model and invested with refractory mold material; 9) the mold is then heated until dry and all traces of wax are burned out; 10) molten bronze is poured into the baked mold until it is full; 11) when the metal has cooled, the refractory mold is broken away, the armature and core removed, and the sprue system cut off; 12) separately cast parts, such as a base, may be added at this stage, and the surface color may be enhanced by various means, including patination or gilding. Sketch based on Benvenuto Cellini (Italian 1500–1571), *Bust of Bindo Altoviti*, 1549, H. 85.2 cm (Isabella Stewart Gardner Museum inv. S26e21). See Dewer and MacNamara 2012.

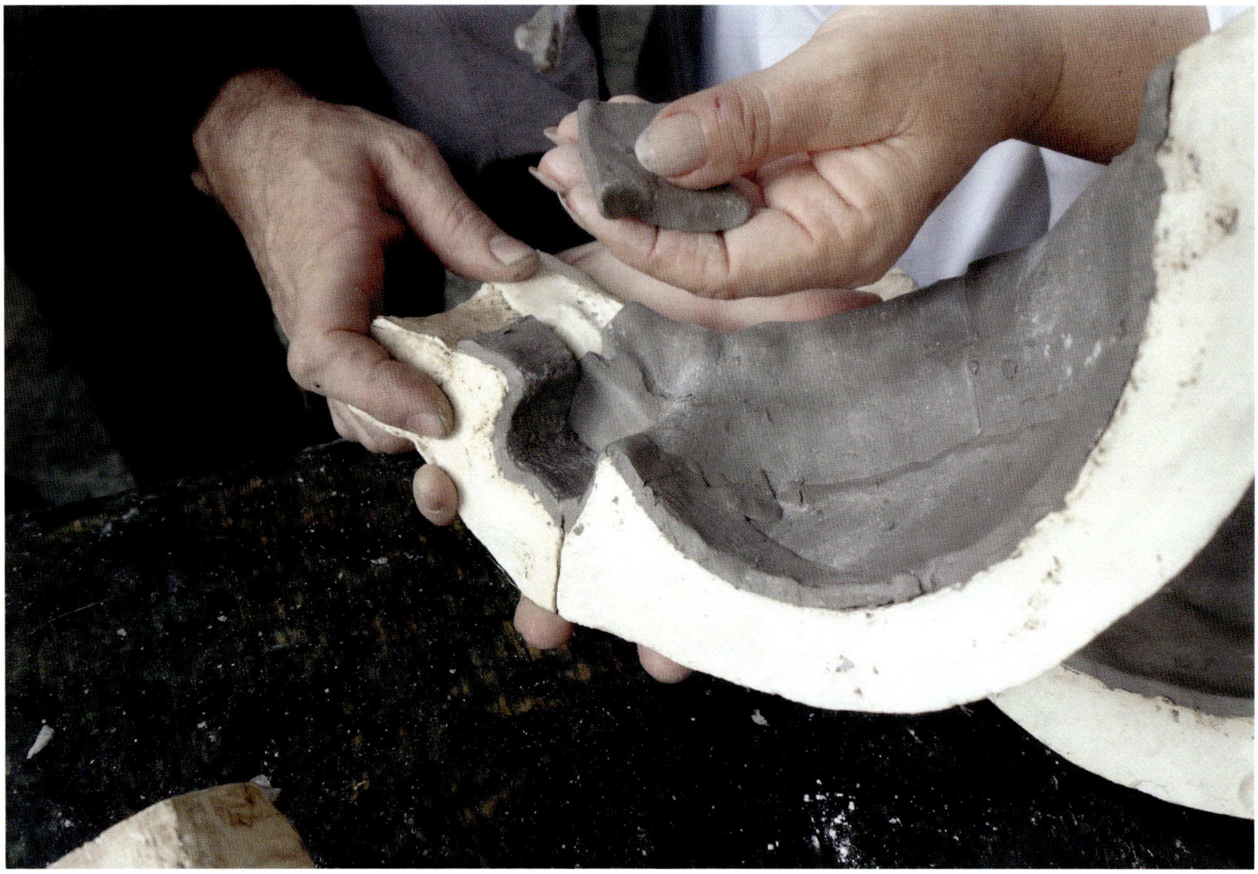

Figure 24 Application of clay sheets (lasagna) on the internal surface of one half of a plaster mold for the bronze reproduction of the head of the *Apollo of Lillebonne* during the 2016 CAST:ING meeting, Coubertin foundry, France. The height is 18 cm (two-thirds of the original size). Original: *Apollo of Lillebonne*, France, 1st century BCE–1st century CE, H. 193 cm (Musée du Louvre, inv. Br37 [NIII65]). See figs. 23 and 288.

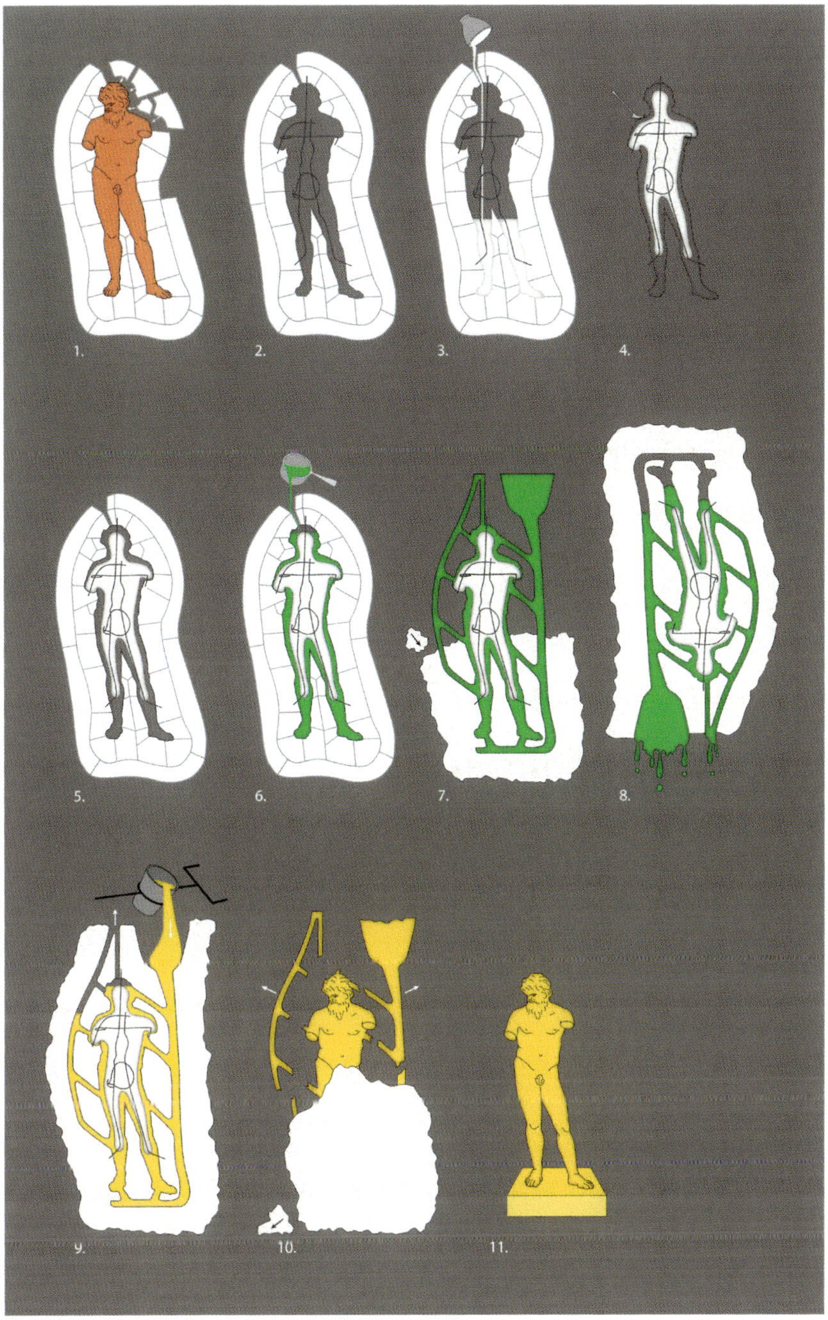

Figure 25 Diagram depicting a version of the cut-back-core process: 1) a plaster piece mold composed of smaller pieces to account for the undercuts is formed on the model; the sections are held together by an outer mother mold; 2) the piece mold is removed from the model; core support rods and wires are positioned as needed in the partially reassembled empty mold; 3) the mold is fully reassembled and filled with refractory material (here plaster-based) to create a replica of the original model; 4) the surface of the replica is pared down (cut back) to form the core of the mold, thereby creating the space that will be filled first by wax and later by metal; 5) the cut-back core is placed back into the piece mold, suspended in place by the core support rods that also serve as core pins in this example; 6) molten wax is filled into the cavity around the core, thereby creating the inter-model; 7) the inter-model is removed from the mold and reworked as needed; the sprue system is joined to it with a heated tool and the assemblage is invested in a refractory mold; 8) the mold is baked to dry it thoroughly and to burn out all traces of the wax; 9) the mold is filled with molten metal; 10) the bronze is freed of the investment, sprues, and protruding core supports; 11) in the final stages the bronze surface may be refinished with details, then joined to the separately cast arms and base and a patina applied. Diagram based on Michel Anguier (French 1612/14–1686), *Melancholy Pluto*, before 1699, H. 23.9 cm (Staatliche Kunstsammlungen, Grünes Gewölbe, Dresden, inv. IX.37). See Bassett and Bewer 2014.

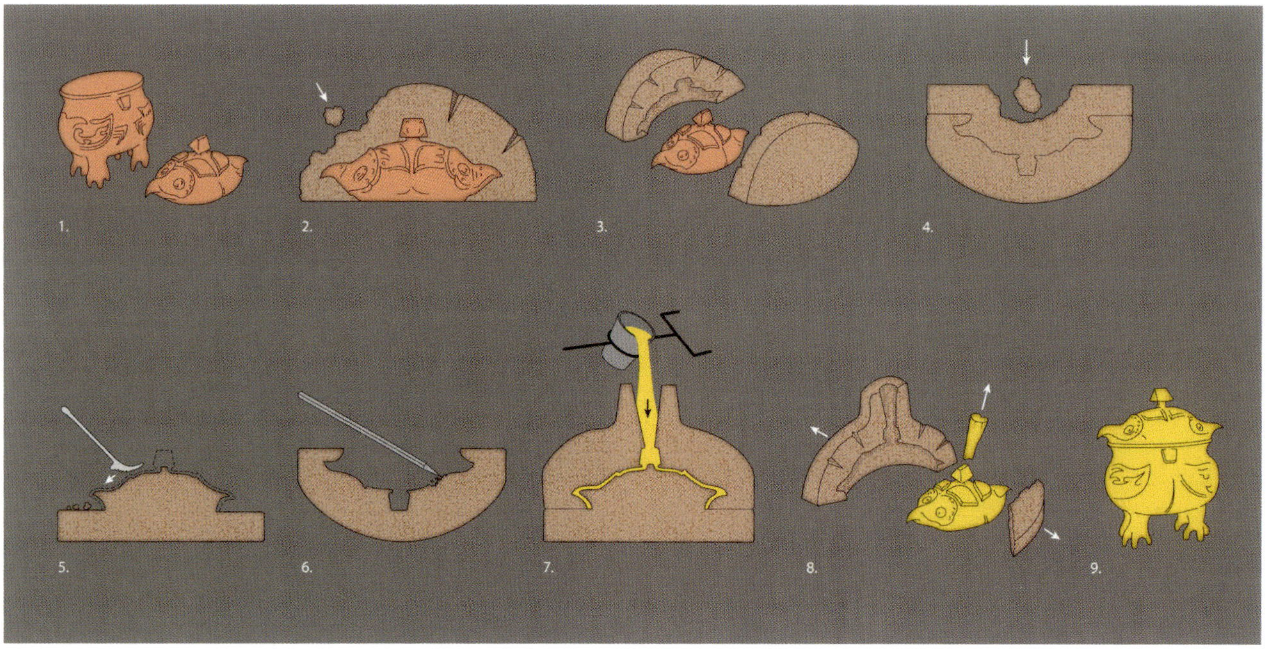

Figure 26 Diagram depicting a speculative version of the ancient Chinese piece mold casting process: 1) a model of the You vessel with lid is prepared out of loess (a fine-grained refractory soil found throughout northern China) and baked dry; 2) the piece mold segments need to account for undercuts in the model; half of the upper part is covered in loess to form the first section of the mold; it is smoothed and keyed to ensure alignment of the next section; 3) the second section of the mold is built up alongside the first and allowed to dry and harden; the mold segments are carefully removed from the model; 4) in order to make the core, the top two pieces of the mold are reassembled and packed tightly with loess to form a replica of the original model and to add a base; 5) the mold is opened and the outer surface of the loess replica is cut back (pared down) to create a core that, once repositioned in the outer mold, will help define the space to be filled with molten metal; 6) at this stage additional details can be carved into the face of the mold; 7) the mold sections are reassembled around the core and a funnel/gate is cut into the mold; the assembled mold is baked until dry and filled with molten bronze; 8) the mold is removed and the gate cut off; except for some basic removal of flashing and sprues, the surface would merely be polished; 9) the bronze lid and body of the vessel (produced by the same process) are fine-tuned to fit perfectly.

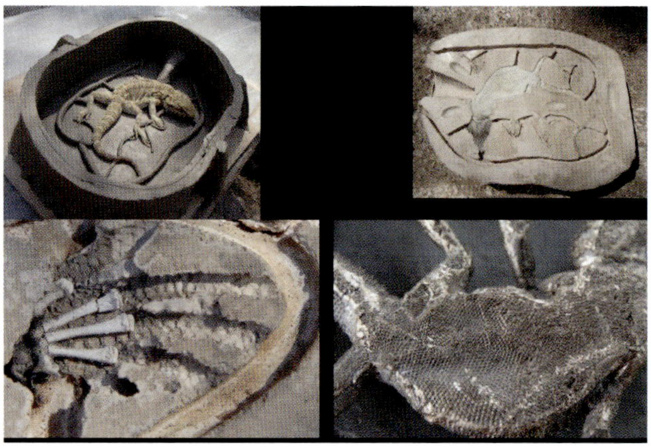

Figure 27 Andrew Lacey's life-cast of a lizard. Top left: the lizard is sprued in the bottom of the mold. Top right: top part of the plaster refractory mold with the lizard still in. Bottom left: detail of the plaster mold once the lizard was partially removed by burning. Bottom right: detail of the cast bronze. The cast of the lizard was made in a mold designed in parts to remove any remaining bones or ash. Had it not been, some of these remains might have become lodged in the metal and formed an inclusion that could result in a lacuna in the cast. See I.3. Andrew Lacey (British, b. 1969), *Life Cast of a Lizard*, cast by the artist in 2015, Devon, UK, L. 15 cm (Victoria and Albert Museum, inv. NCOL.517-2015).

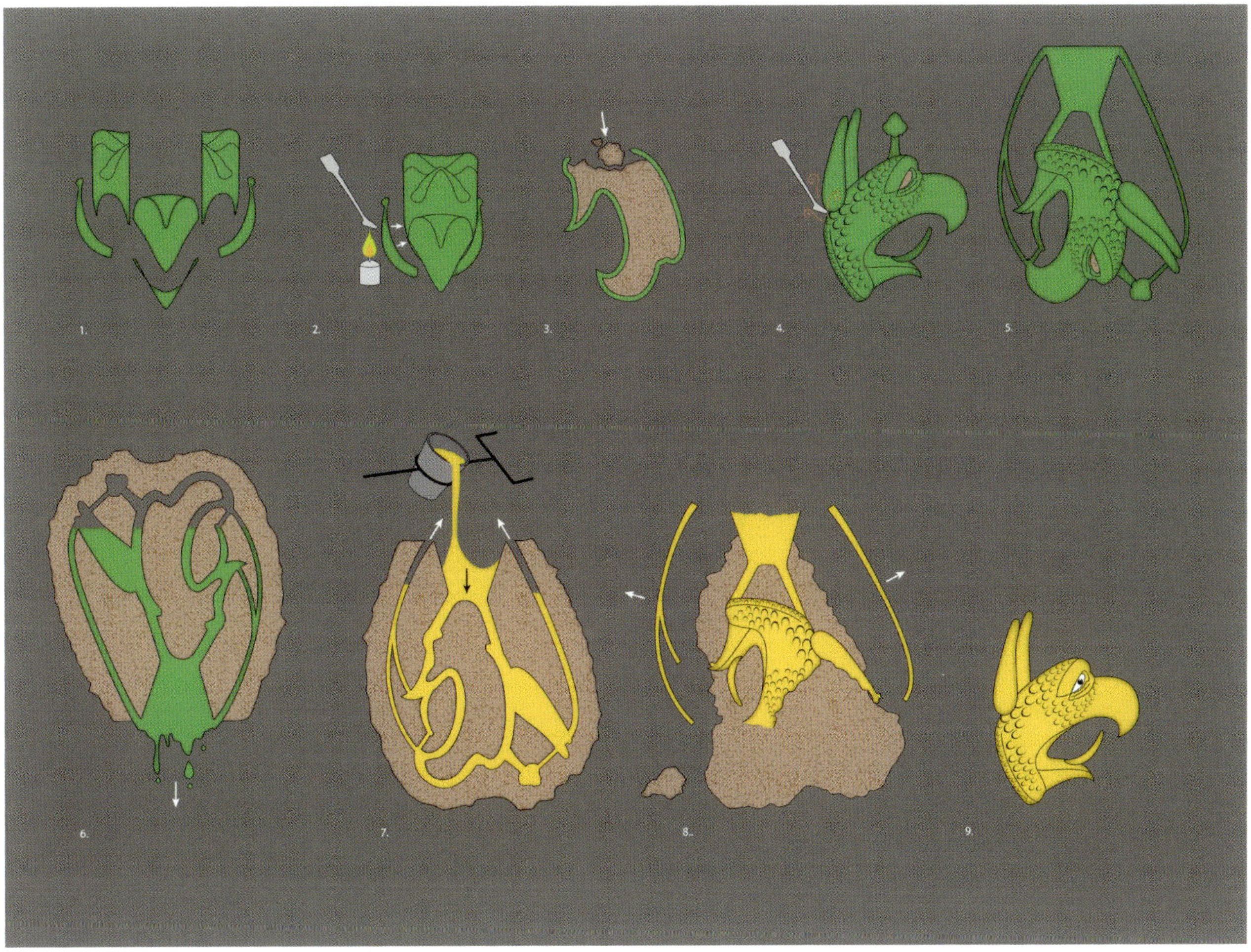

Figure 28 Diagram depicting a version of the direct wax-slab process: 1) the main shape of the griffin head casting model is formed of wax slabs that are cut into their proper shapes; the raised shapes of the upper and lower eyelids are formed by pressing the slabs out, producing recesses on the reverse; 2) the wax elements are joined with a heated tool to form a hollow head; 3) the hollow wax griffin head is filled with a refractory clay core material; the eye cavities are cut out and ledges formed on the interior to support inlays that will later fill the voids; 4) details are added to the wax; these include incised lines around the eyes and a scaly pattern formed with punches; the solid wax tongue, ears, and top knob are joined to the head; 5) the sprue system is joined to the model; 6) the wax model is invested in a refractory mold that is heated until dry and all traces of wax are burned out; the openings in the eyes and the neck act as core extensions (see I.1§3.2), preserving the space between the core and the outer mold; 7) the mold is filled with molten bronze; 8) when cool, the cast is broken out of the mold, and the sprue system is removed; 9) the bronze is then chased and in this example the eyes inlaid. Diagram based on Mattusch 1990.

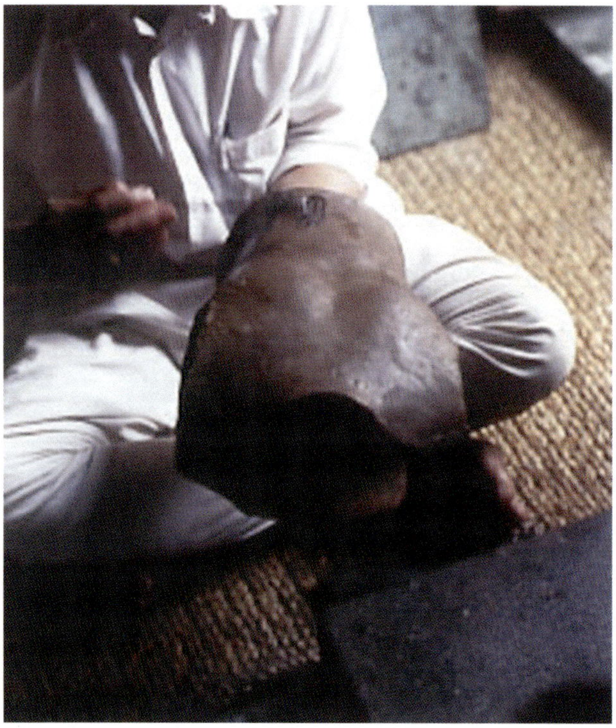

Figure 29 Example of the direct wax slab process as seen in Patan, Gujarat, India, in the 1980s. The wax torso of a large statuette is shaped by modeling wax slabs, before adding the core. See Craddock 2015; Furger 2017.

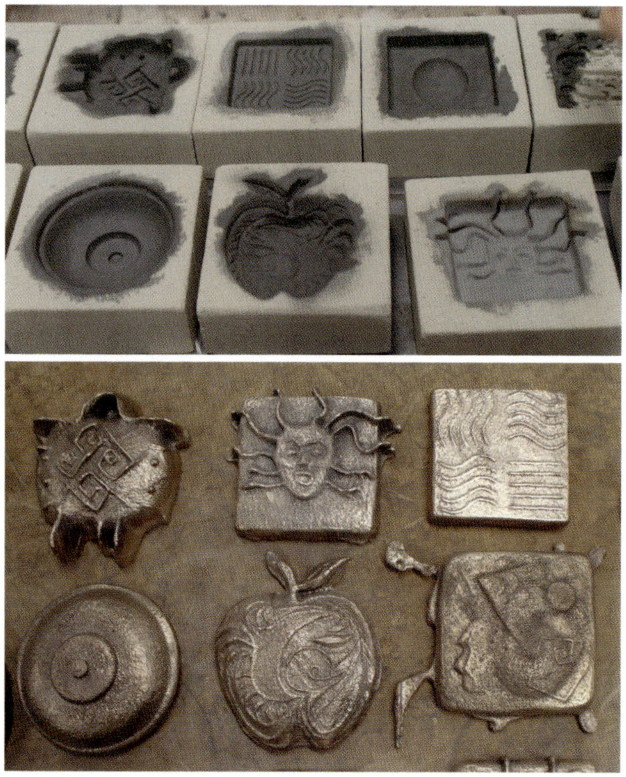

Figure 30 Images from a 2012 Harvard University January-term bronze-casting course led by Francesca G. Bewer showing the participants' carved, open-faced, resin-bonded sand molds that were painted with graphite powder before casting and the resulting bronze reliefs.

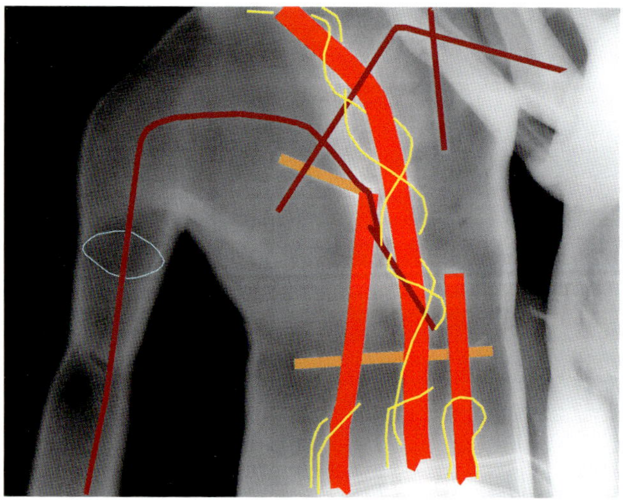

Figure 31 Radiographic detail of the torso of Apollo clearly shows a wax-to-wax joint as a dark ellipse at the junction of the arm to the shoulder (blue). See also the complex armature and notably the transverse rods (orange) one of which goes through the main armature rods (red). *Apollo and Daphne*, probably France, early 18th century, H. 85 cm (Musée Fabre, Montpellier, France, inv. 836.4.86). See C2RMF Internal Report 2018a.

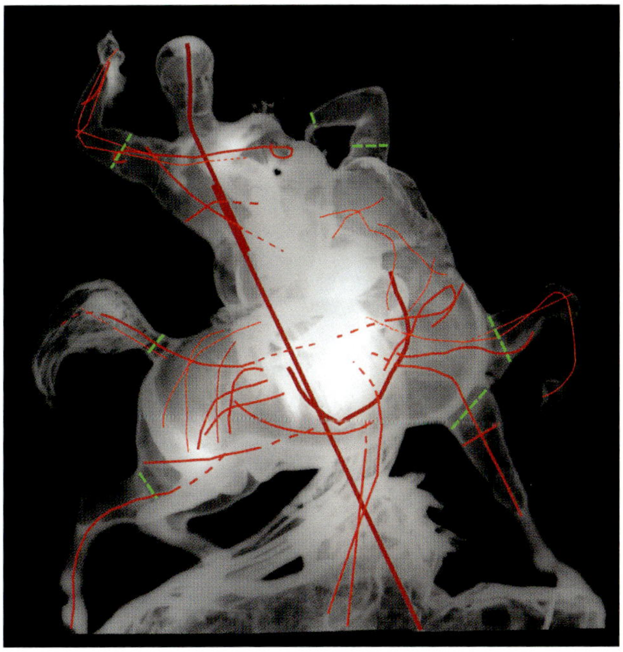

Figure 32 X-radiograph showing a complex armature made of different sizes of rods and wire (11 mm, 7 mm, and 4 mm in diameter). See also the wax-to-wax joints. A 15 mm Cu filter was needed to keep the same exposure parameters for all the different areas of the sculpture, despite the significant variations in thickness (400kV, 8mA, 15 min. exposure, 3 m source-to-object distance, AA400PB film). Antoine-Louis Barye (French, 1795–1875), *Theseus and the Centaur Bienor*, cast by Eugène Gonon (1814–1892) in 1877, H. 1.3 m (Musée du Louvre, inv. RF 3882). For daylight photography of the statue see fig. 395. See C2RMF Internal Report 2016b.

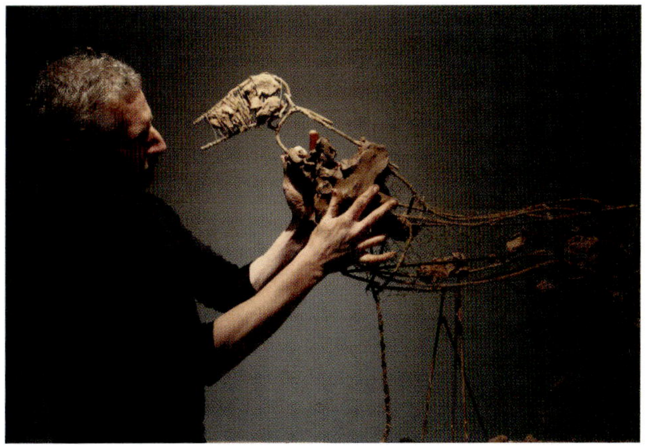

Figure 33 The artist beginning the modeling process on a metal armature. Andrew Lacey (British, b. 1969), *The Anatomy of Bronze*, cast by the artist in 2019, Devon, UK, H. 45 cm (artist's collection).

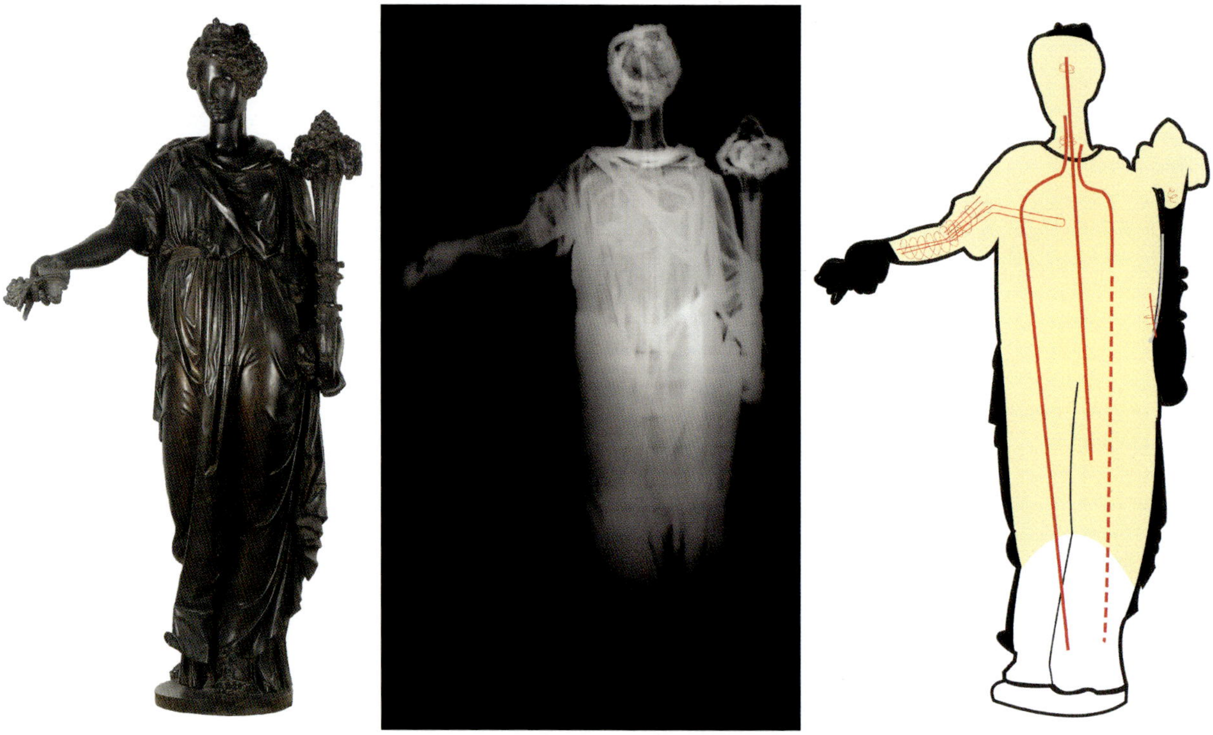

Figure 34 Overlaying the daylight photograph and the X-radiograph of a sculpture helps to separate out forms that are part of the visible composition (e.g., folds of the drapery or overlapping body parts) from information about the internal structure (such as the joint in the neck, the armature, and the wire wrapping the core of the right arm) shown in the layer with the interpretive diagram. Barthélemy Prieur (French, 1536–1611), *Abundance*, cast in 1571 by Nicolas Péron, H. 128 cm (Musée du Louvre, inv. MR1681). See Case Study 5.

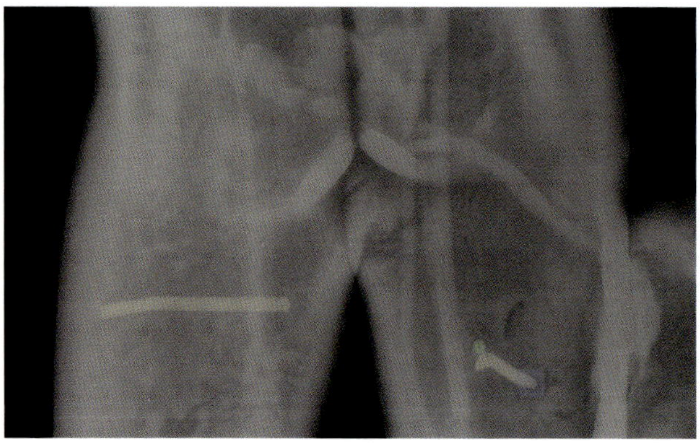

Figure 35 Detail of an annotated digital X-radiograph showing transverse core pins in both thighs (yellow overlays), one running side to side, the other front to back. A circular plug (green overlay) and rectangular surface patch (blue overlay) repair the holes left in the bronze wall when the front-to-back core pins were cut off at the surface and pushed into the hollow bronze cavity. Giovanni Battista Foggini (Italian, 1652–1725), *Dancing Faun*, ca. 1700, H. 54 cm (J. Paul Getty Museum, inv. 2000.8).

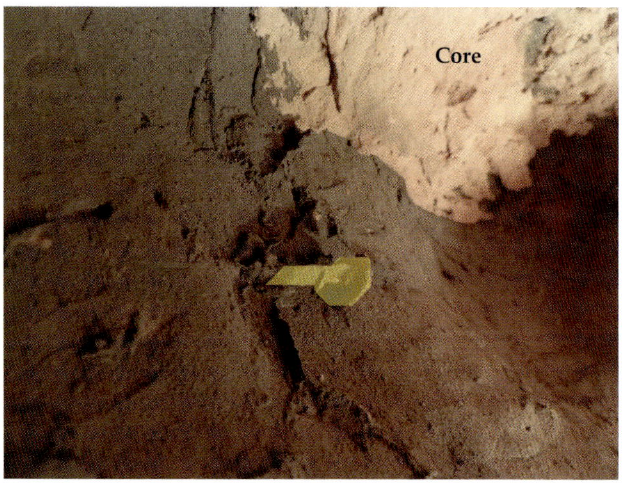

Figure 36 Detail of internal surface showing a core pin with the head in the interior (yellow overlay), testifying to the use of a process in which the core was set after the wax in the mold. Remains of clay core are visible (top right). Giovan Francesco Rustici (Italian, 1475–1554), *Tomb Effigy of Alberto III Pio, Prince of Carpi*, ca. 1535, L. 167 cm (Musée du Louvre, inv. MR 1680). See C2RMF Internal Report 2016a.

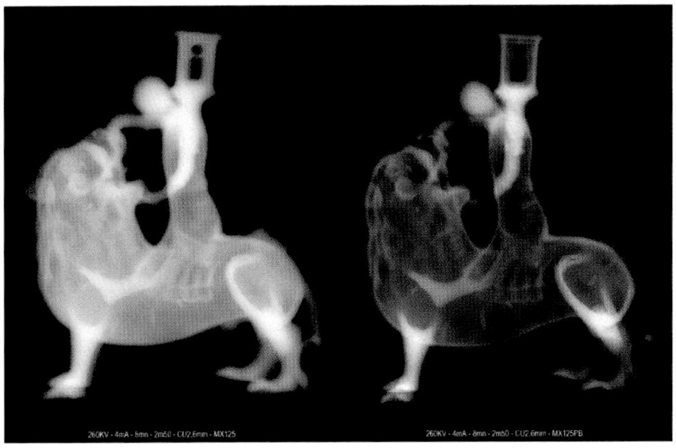

Figure 37 X-radiographs showing the influence of lead intensifying screens placed directly in front of and behind the film. Without lead screens (left), the contrast is low. On the right, the addition of the lead reinforcing screens (MX125PB film) reveals technical features, including core pins (some marked with red overlays). Operating conditions for both images are: 260kV, 4mA, 8 min. exposure, 2.5 m source-to-object distance, 2.6 mm Cu beam filtering. Candlestick representing Samson and the Lion, Lower Saxony, ca. 1175–1225, H. 22 cm (Musée départemental des antiquités de Rouen, France, inv. R.91.23). See C2RMF Internal Report 2015b.

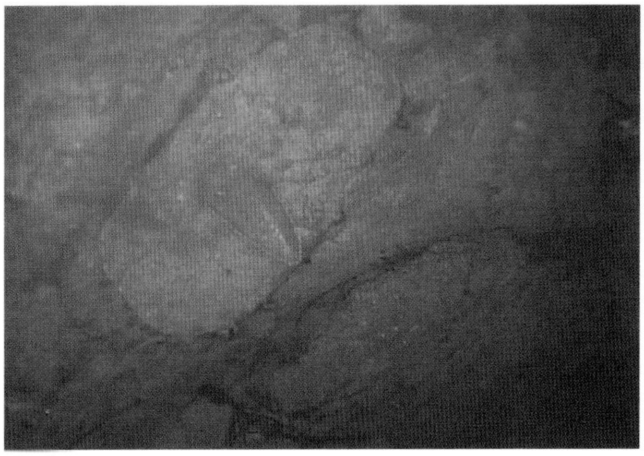

Figure 38 Endoscope image of a detail from the inner surface of the figure showing a distinctive kind of chaplet that sports a tip that helped to anchor it in the core during casting (approx. H. 1 cm). Martin Lefort (French, dates unknown), *Justice*, 1571, H. of chaplet 128 cm (Musée du Louvre, inv. MR1682). See Castelle et al. 2021.

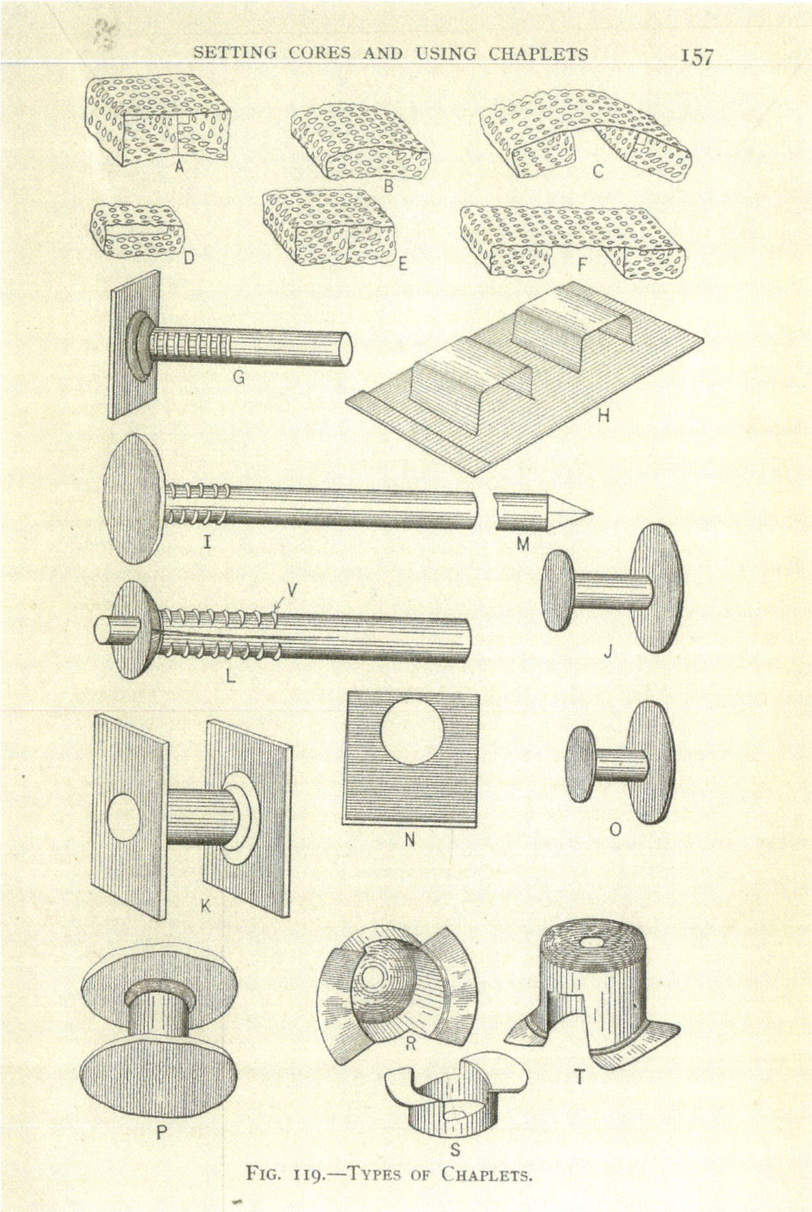

Figure 39 Different types of chaplets for twentieth-century industrial iron sand casting. Each type has a special use: A–F are perforated sheet tin for heavy loads; I–M are for the heavier castings (tons); G–I are the most common types. From Palmer 1929, 157, fig. 119.

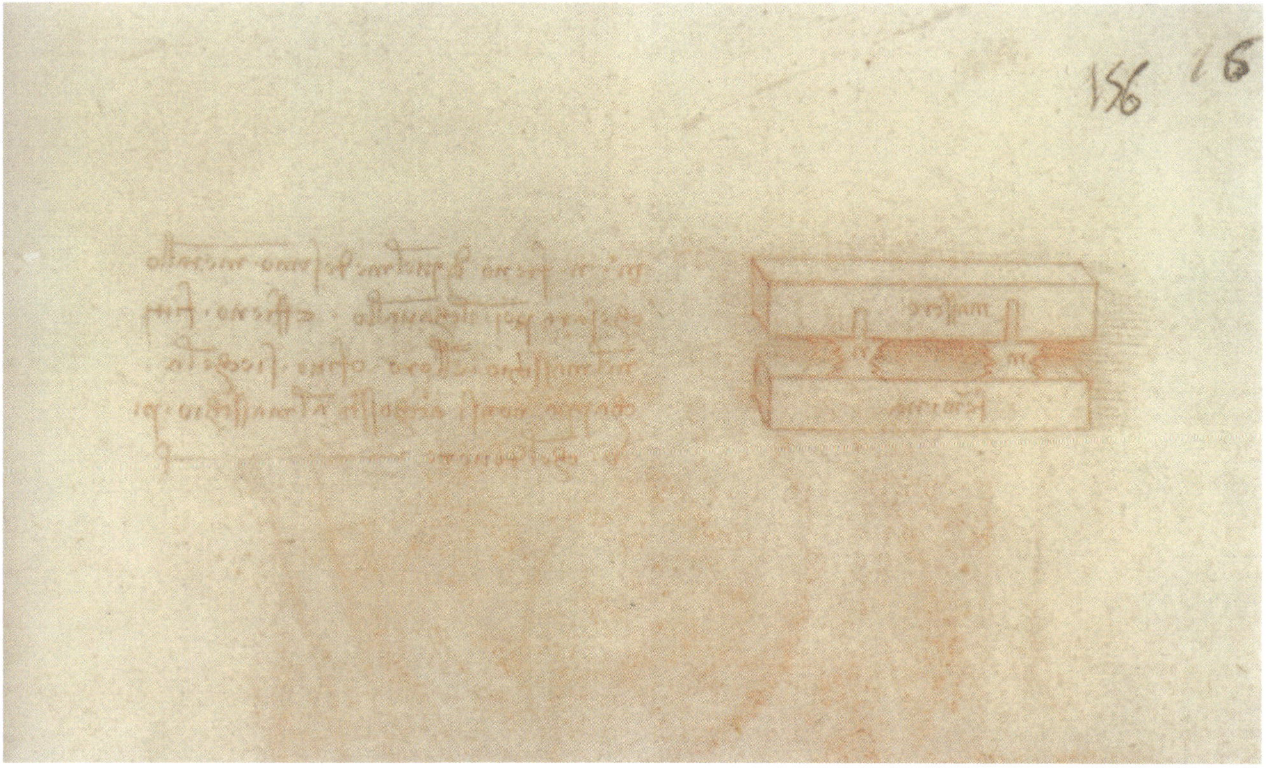

Figure 40 Sketch of a chaplet firmly wedged between the core and the outer mold, with a protruding point securing it in place in the clay on one side. This is one of many notes that Leonardo made in connection with the casting of the *Sforza Horse*. Leonardo da Vinci (Italian, 1452–1519), *Madrid Codex II*, 1503–5, fol. 156r (Biblioteca Nacional de España, Ms. 8936).

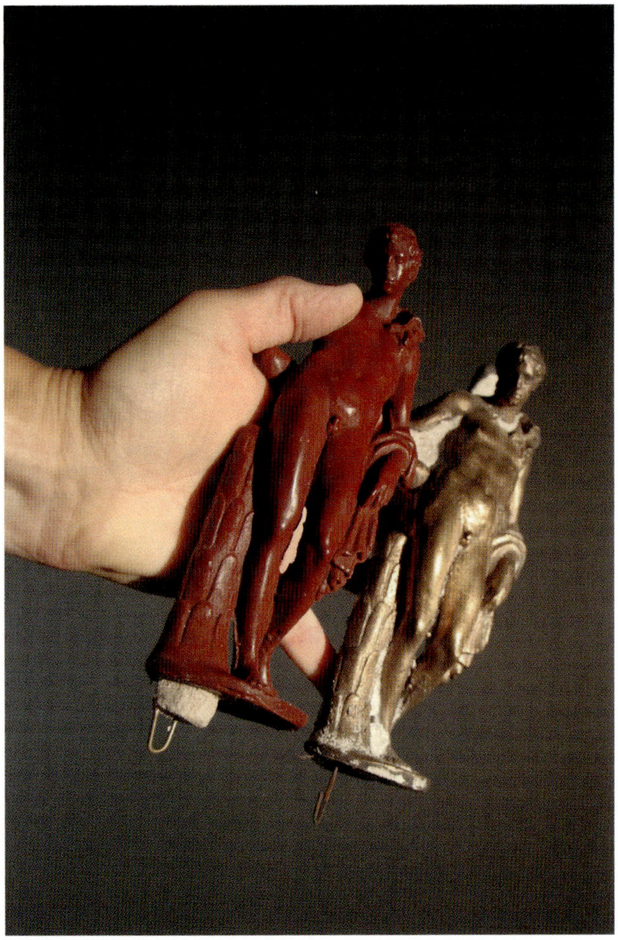

Figure 41 Wax inter-model and bronze copy of *Antinous* (French School) made by Andrew Lacey in 2014. The core extending out of the base will function as a mold extension, tying the core to the investment and reducing the number of core pins needed. In the indirect process, both the wax inter-model and the casting alloy will contribute to the overall shrinkage of the sculpture. This is most noticeable when comparing the original model to the final bronze sculpture. The wax inter-model will shrink by ~2.5–3% in linear dimension and the casting alloy by ~1–1.5%; therefore the final sculpture will shrink by ~4% in total.

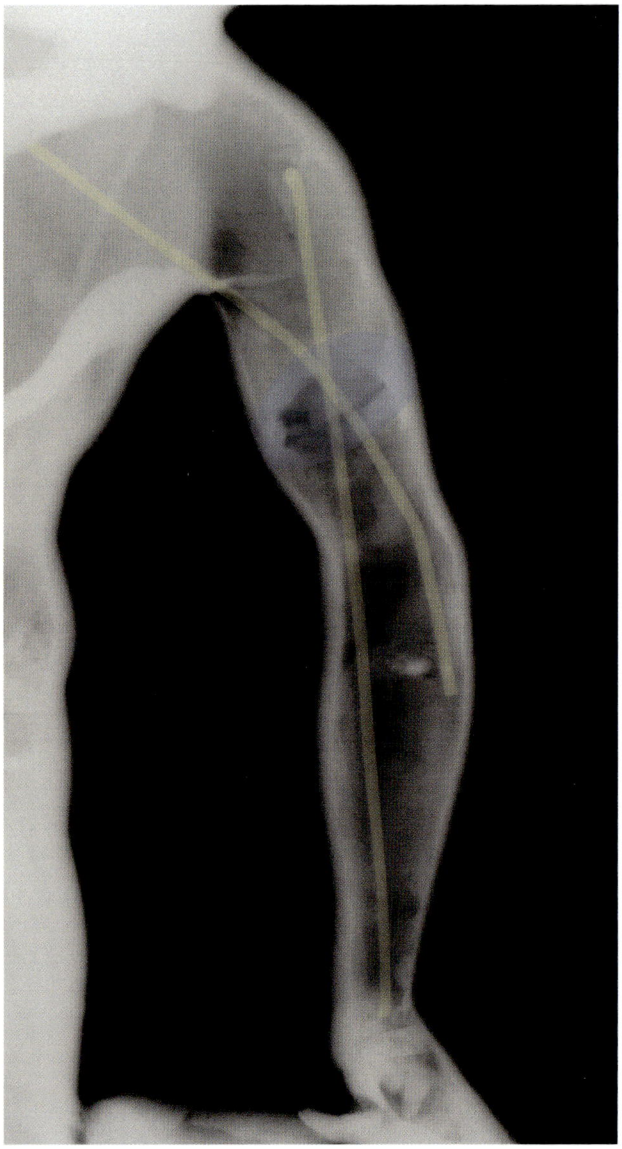

Figure 42 Annotated detail of an X-radiograph showing core support rods (yellow overlay) supporting a wax-to-wax joint in the arm (blue overlay). Giovanni Battista Foggini (Italian, 1652–1725), *Dancing Faun*, ca. 1700, H. 54 cm (J. Paul Getty Museum, inv. 2000.8). See Fogelman and Fusco 2002, 238, 357.

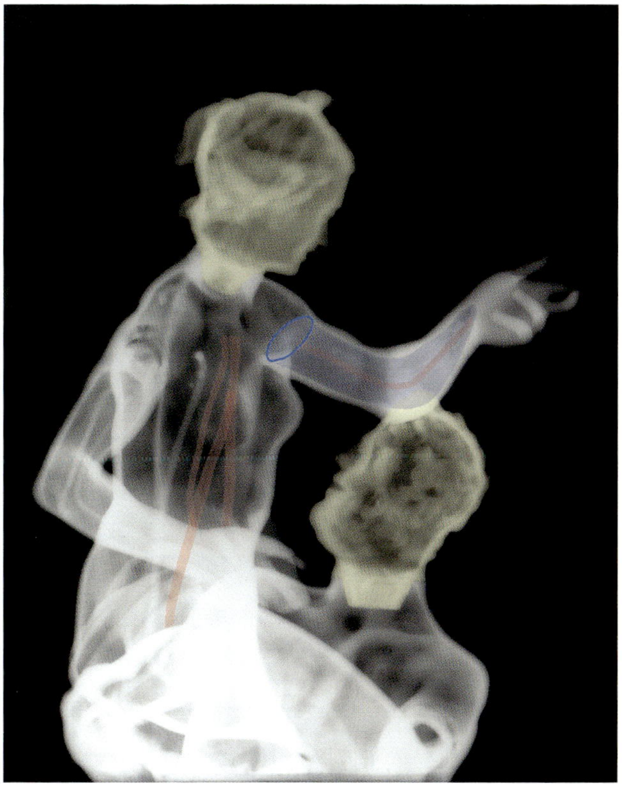

Figure 43 Annotated X-radiograph of a sand cast. Both Dejanira's and Hercules's heads were cast separately and secured at the neck with Roman joints (yellow overlay). Roman joints also secure Dejanira's proper right arm as well as Hercules's proper left arm. There is an internal seam line in Dejanira's left shoulder where the separately formed core in the arm abuts the core used in the torso (blue line). Lanterns (or metal core vents) can be seen in the torso and the proper left arm (red overlays). Charles Crozatier (French, 1795–1855), *Hercules Freeing Dejanira from the Centaur Nessus*, Paris, 19th century, after Adriaen de Vries (Netherlandish, 1556–1626), H. 78.5 cm (Rijksmuseum, Amsterdam, inv. BK 1957-2).

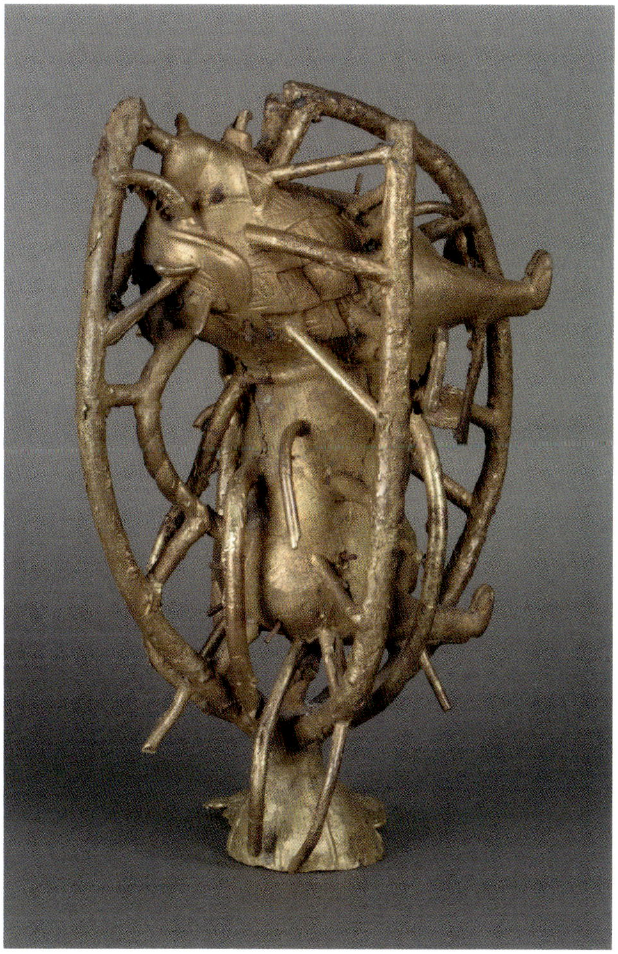

Figure 44 Cast reproduction of an aquamanile after removal of its investment, still retaining its core pins and its sprue system translated into metal. Ubaldo Vitali (American, b. 1944), Reproduction of a Lion Aquamanile, Maplewood, New Jersey, 2006, H. 19 cm, after Aquamanile in the Form of a Lion, probably northern Germany, 12th century, H. 19.5 cm (Metropolitan Museum of Art, gift of Irwin Untermyer, 1964, inv. 64.101.1491). See Barnet and Dandridge 2006, and video 8.

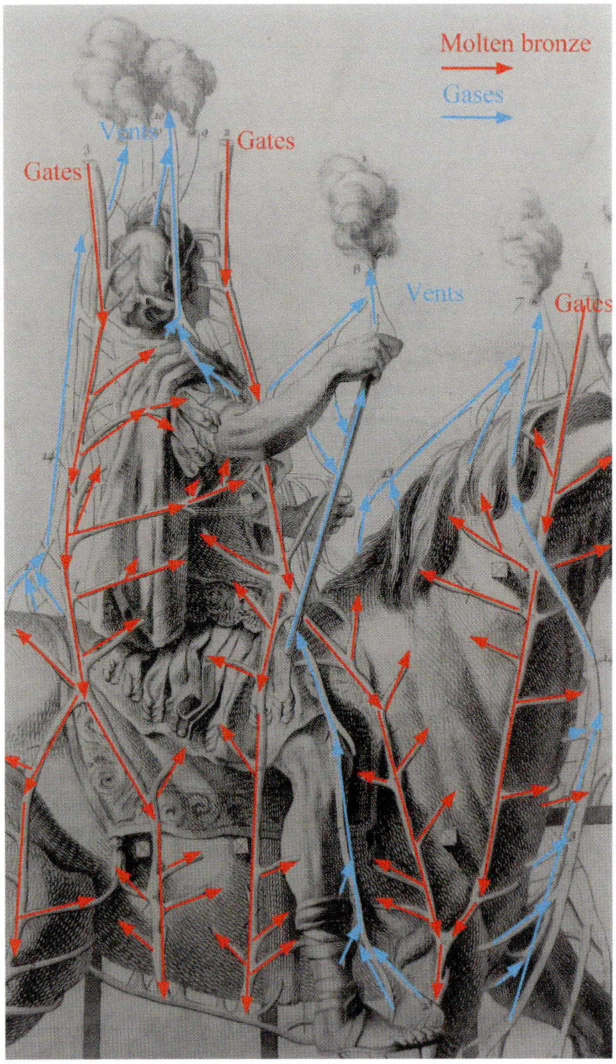

Figure 45 Representation of the complex sprue system planned for casting the equestrian monument in one piece. The overlay shows the initial downward direction of the molten metal through the gates (red) and upward movement of displaced air along the vents as the bronze fills the mold (blue). Mariette 1768, 78, plate II detail, showing Edme Bouchardon (French, 1698–1762), *Louis XV*, cast in 1758 by Pierre Gor (French, 1720–1773), H. 520 cm. After Desmas 2014.

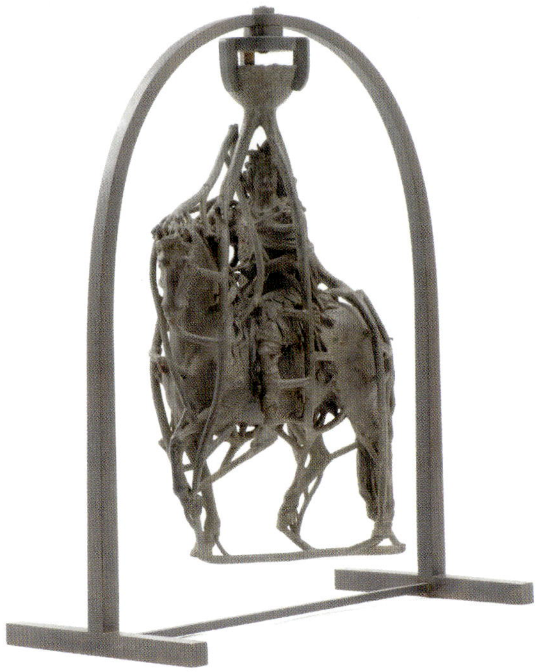

Figure 46 Bronze model suspended from the pouring cup within a modern mount. The sculpture depicts the complex sprue system designed for casting the figure in one piece. After Martin Van Den Bogaert (Desjardins) (Dutch, 1637–1694), *Louis XIV on Horseback*, cast by Roger Schabol (Dutch, d. 1727), early 1700s, H. 52.4 cm (Statens Museum for Kunst, Copenhagen, inv. KMS 5403).

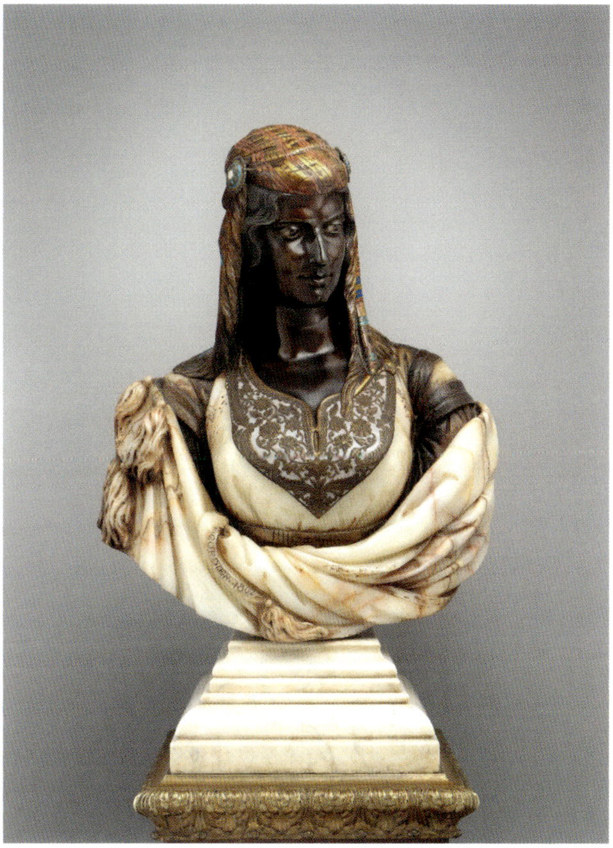

Figure 47 Sculpture made of Algerian onyx-marble, bronze, gilt bronze, enamel, and amethyst eyes, and white marble socle. Sculptors such as Cordier took advantage of enamel to add color, as in this draped head covering. Charles-Henri-Joseph Cordier (French, 1827–1905), *The Jewish Woman of Algiers*, Paris, 1862, H. 90.2 cm (Metropolitan Museum of Art, European Sculpture and Decorative Arts Fund, 2006, inv. 2006.113a–c).

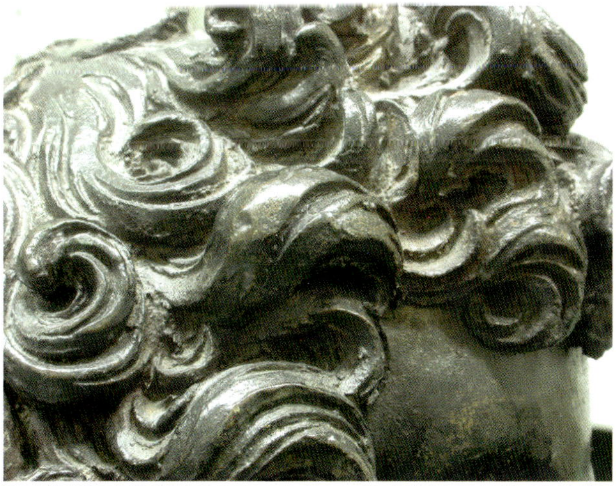

Figure 48 Detail of head showing how the vivacious modeling of the curls in the wax model has been captured in the bronze. Barthélemy Prieur (French, 1536–1611), *Funerary Genius*, 1583–85, L. 109 cm (Musée du Louvre, inv. MR 1685). See Bewer, Bourgarit, and Bassett 2009; Seelig-Teuwen, Bourgarit, and Bewer 2014. See fig. 526 for an overall view.

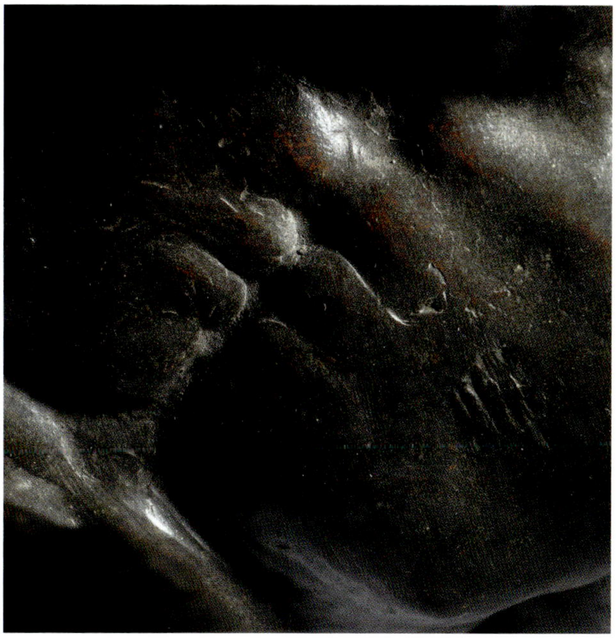

Figure 49 Detail of chest showing the discrete additions of wax in the modeling of the ribs. Barthélemy Prieur (French, 1536–1611), *Funerary Genius*, 1583–85, L. 109 cm (Musée du Louvre, inv. MR1685). See Seelig-Teuwen, Bourgarit, and Bewer 2014; Castelle 2016. See fig. 526 for an overall view.

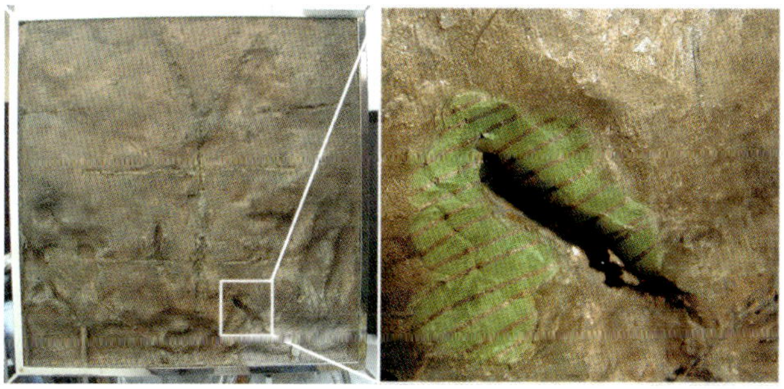

Figure 50 Reverse of panel showing evidence of sprues and working from the back of the wax model. The detail shows reinforcement of the edges of a hollow portion with small masses of soft wax. They correspond to the area where the hollow protruding figure of David meets the background of the relief. The malleable material also captured partial fingerprints. Lorenzo Ghiberti (Italy, 1378–1455), David and Saul panel in the *Gates of Paradise*, left door, design begun after 1425, installed 1452, H. 518 cm (Museo dell'Opera del Duomo, Florence, inv. 2005/905). See Bewer, Stone, and Sturman 2007.

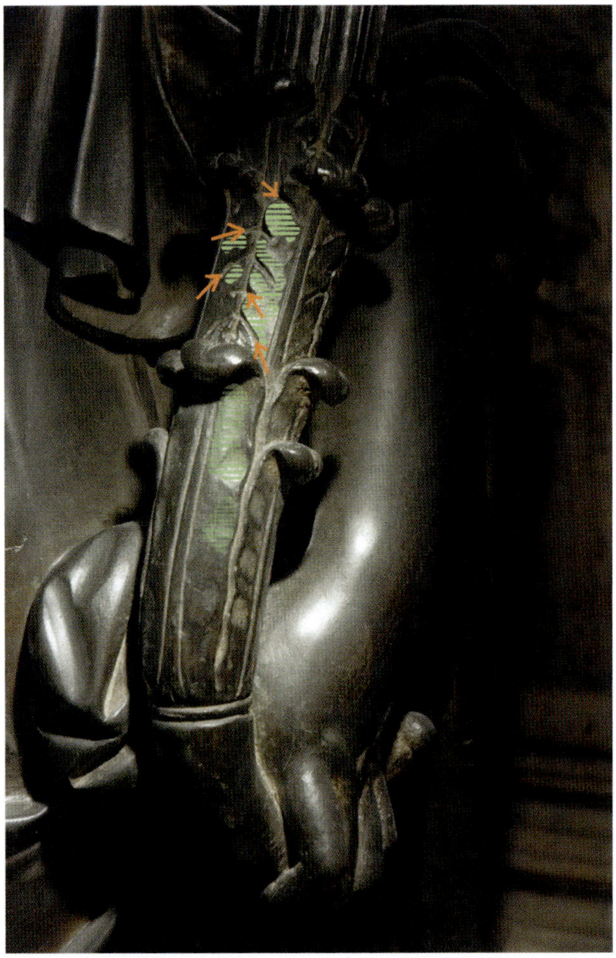

Figure 51 Detail of cornucopia showing tool marks in the wax (green overlay). Marks of metal hammering are also visible (orange arrows). Barthélemy Prieur (French, 1536–1611), *Abundance*, cast in 1571 by Nicolas Péron, H. 128 cm (Musée du Louvre, inv. MR 1681). See Seelig-Teuwen, Bourgarit, and Bewer 2014. See fig. 34 for an overall view.

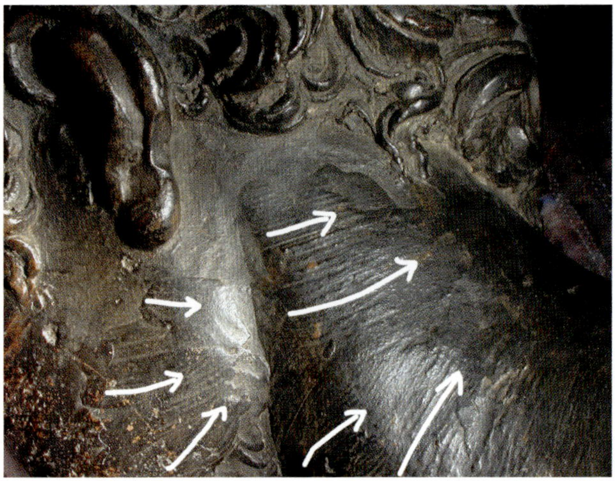

Figure 52 Detail of neck that preserves the raw tool marks made on the wax model (arrows indicate the direction of the stroke). The finer striated marks in the bottom right corner may have been made in the metal with a file. Barthélemy Prieur (French, 1536–1611), *Funerary Genius*, 1583–85, L. 109 cm (Musée du Louvre, inv. MR 1685). See Seelig-Teuwen, Bourgarit, and Bewer 2014. See fig. 526 for an overall view.

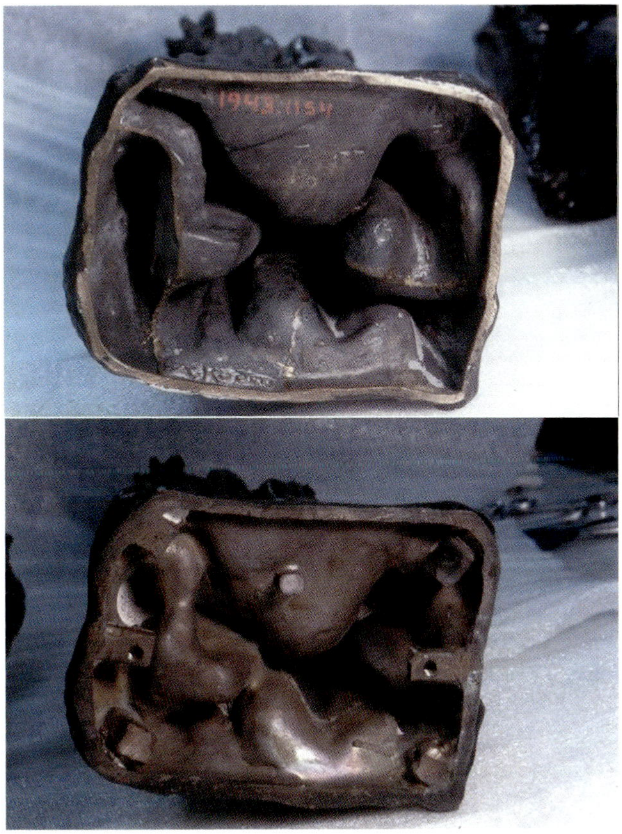

Figure 53 Underside of two reduced-size casts showing the characteristic, more angular inner surface of a sand cast produced by the pared-down core (top), and the softer forms of a lost-wax cast (bottom). Auguste Rodin (French, 1840–1917), *Burgher of Calais: The Man with the Key, Jean d'Aire*, after 1895, H. 47 cm (Harvard Art Museums / Fogg Museum, Bequest of Grenville L. Winthrop, inv. 1943.1154 [top], and modern cast from private collection [bottom]).

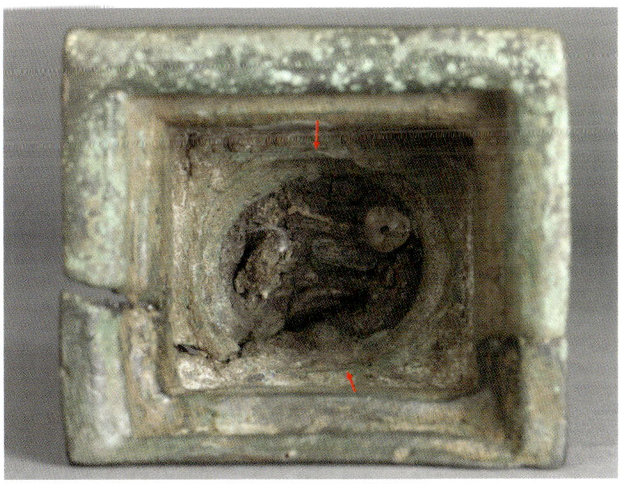

Figure 54 View of the underside showing consecration deposits. See also the raised ridges (arrows) indicating wax-to-wax joints. *Kubera/Jambhala*, Java, first half of the 10th century, H. 18 cm (Musée national des arts asiatiques – Guimet, Paris, inv. MG 3625). See Mechling et al. 2018. See fig. 508 for an overall view.

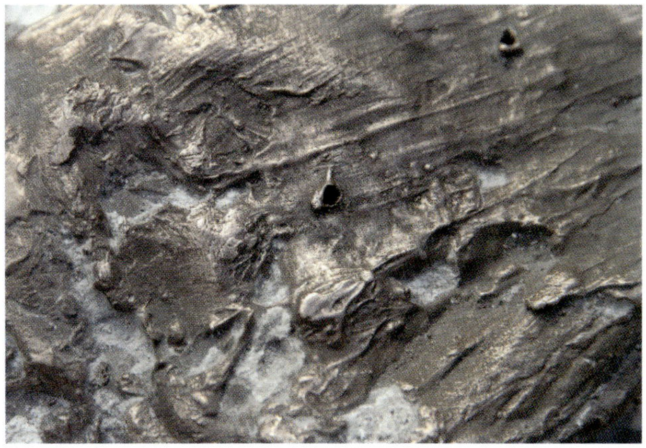

Figure 55 Detail of the inside of the bronze showing a core pin hole (diameter 2.5 mm) made in the wax inter-model. The softened rim of the hole was created as the warmed core pin was pushed through the wax. Brush marks and fingerprints are further evidence of the production of the wax inter-model. Andrew Lacey (British, b. 1969), *Ashen*, 2021, H. 192 cm (artist's collection).

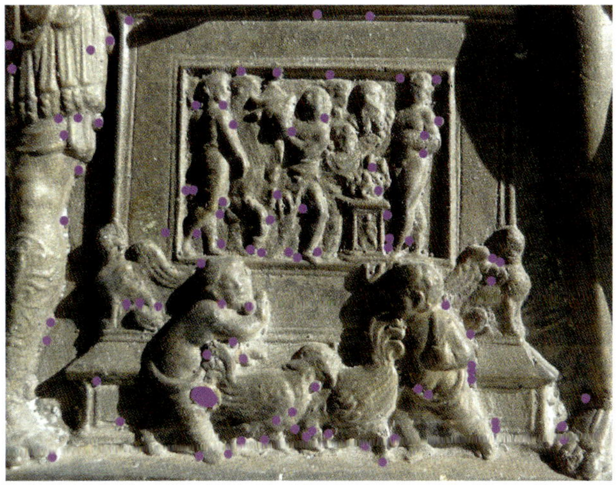

Figure 56 Detail of the lower portion of a lost-wax after-cast of a Renaissance partially gilded silver plaquette as cast. The myriad small bronze spherules (see overlay) result from air bubbles in the poorly applied plaster-based investment slurry, which settled on the surface of the wax inter-model. Francesca G. Bewer after-cast (private collection) made from an electrotype copy (date unknown) after Galeazzo Mondella (Italian, 1467–1528), *Sacra Conversazione*, ca. 1510, H. 13.9 cm (Kunsthistorisches Museum, Vienna, inv. Kunstkammer, 1107).

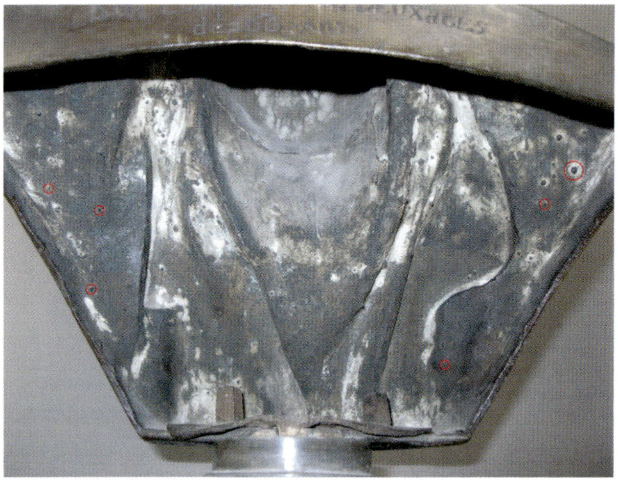

Figure 57 Detail of the open back of the bust, where spheres of bronze are the result of air bubbles from the plaster core trapped against the wax casting model (some examples circled). Jean-Baptiste Pigalle (French, 1714–1785), *Bust of Denis Diderot*, 1777, H. 52.2 cm (Musée du Louvre, inv. RF 1396).

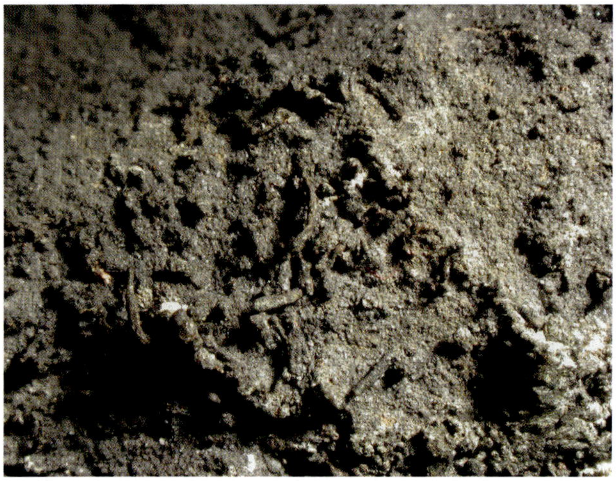

Figure 58 View of internal cavity showing details of the translation into bronze of porosities created by organic materials incorporated in the core that were close to the surface and replaced by the metal. Barthélemy Prieur (French, 1536–1611), *Funerary Genius*, Christophe de Thou Monument, 1583–85, L. 109 cm (Musée du Louvre, inv. MR 1685). See fig. 526 for an overall view.

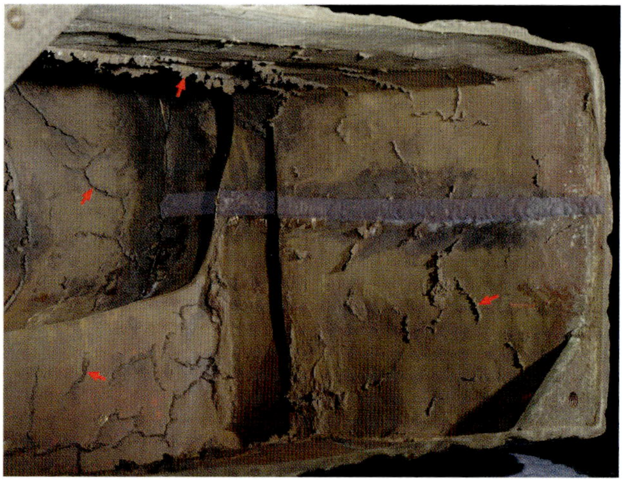

Figure 59 Interior detail of a welding line that joins two large sections of a lost-wax cast base and part of the figure (blue overlay). Note the characteristically jagged-edged flashing due to cracks in the core that formed perpendicularly to the bronze walls (red arrows). Henry Moore (British, 1898–1986), *Seated Woman*, designed 1958–59, cast 1975, H. 203.2 cm (J. Paul Getty Museum, inv. 2005.117.3). See fig. 203 for an overall view.

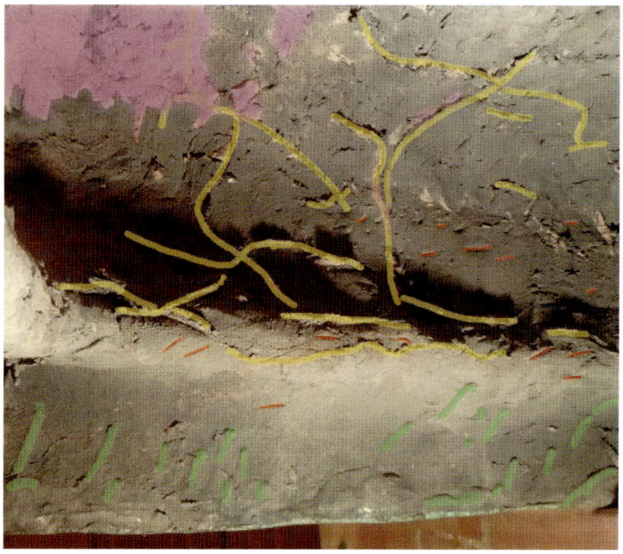

Figure 60 Detail of the underside of the tomb effigy showing drips on the inner edge that were formed through use of a hot tool on the wax model (green marks), the network of flashing due to cracks in the core (yellow marks), and short, straight, crisp chisel marks (red marks) created during removal of flashing and of the clay core (pink overlay). Giovan Francesco Rustici (Italian, 1475–1554), *Tomb Effigy of Alberto III Pio, Prince of Carpi*, ca. 1535, L. 167 cm (Musée du Louvre, inv. MR 1680). See C2RMF Internal Report 2016a.

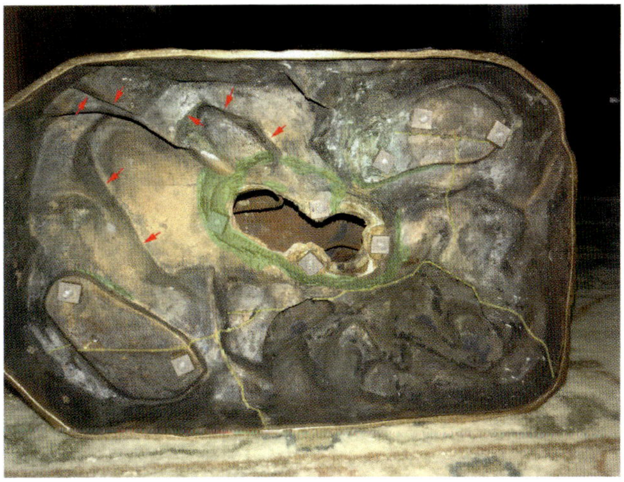

Figure 61 The underside of the base reveals numerous features indicative of sand casting: the overall angular and geometric contours of the base (red arrows); seam lines likely formed between sections of the core (yellow lines); the larger seam lines (flashing) appear to have been chiseled down); nuts on threaded rods used to secure the separately cast figures (white overlay); hammering to better align the joined sections (green overlay). Traces of the sand mold are present (examples overlaid in blue). François-Auguste-Hippolyte Peyrol (French, 1856–1929), *Indian Fighting Cougar* (French title according to the inscription plaque is *La Lutte pour la vie*), cast in France, marked Peyrol foundry, inscription date 1933 (probably cast prior to 1927, when the foundry closed), H. 75 cm (collection of George Syrmos).

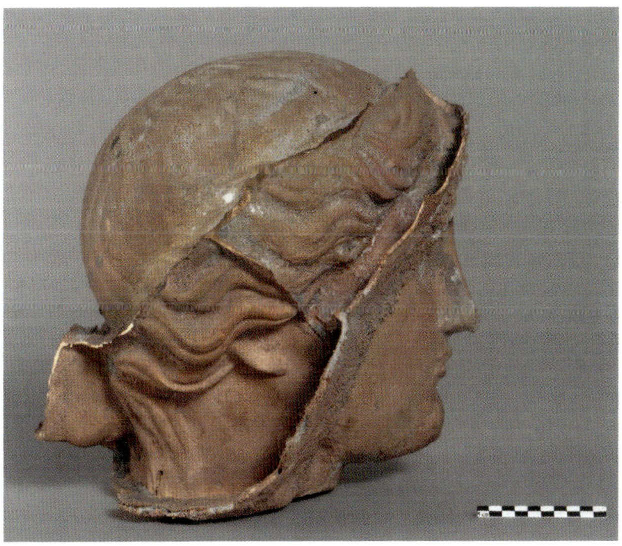

Figure 62 Bronze reproduction of the head of the *Apollo of Lillebonne* made by sand casting during the 2016 CAST:ING meeting, Coubertin foundry, France, before any repairing. See the rough as-cast surface representing the texture of the sand mold, and the stepped linear feature following the piece mold joint line. The height is 18 cm (two-thirds of the original size). Original: *Apollo of Lillebonne*, France, 1st century BCE–1st century CE, H. 193 cm (Musée du Louvre, inv. Br37 [NIII65]). See fig. 288.

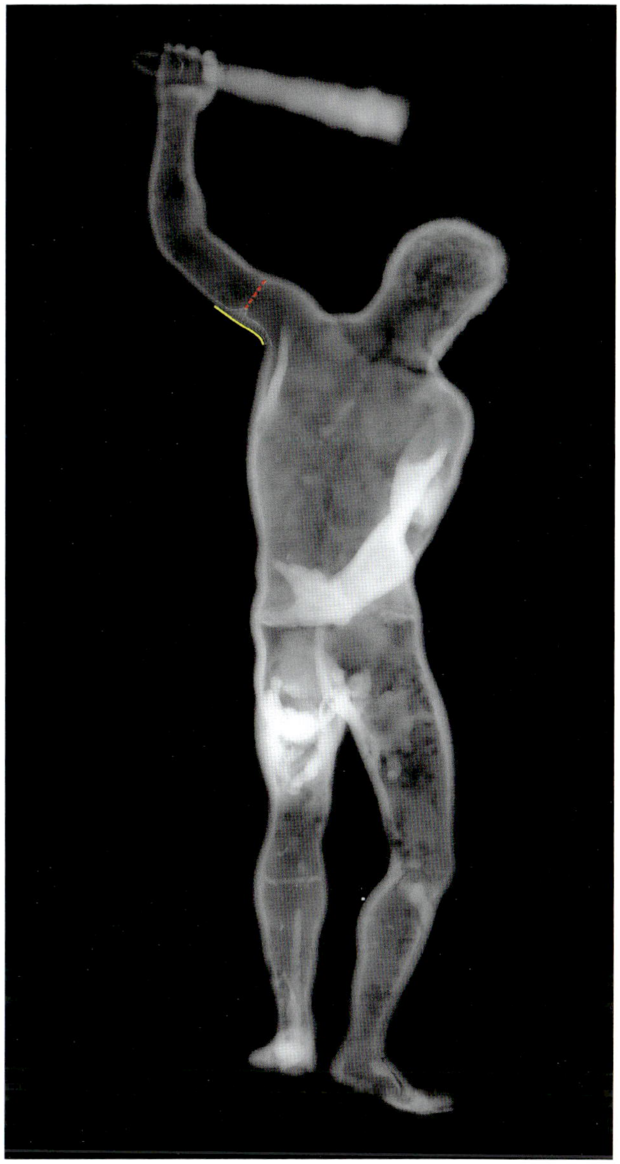

Figure 63 X-radiograph showing a wax-to-wax joint on the right arm (red dotted line), as revealed by the discontinuity in the metal wall thickness (blue overlay). Note that making precise measurements of wall thickness from radiographs may be difficult, as the external edge may not be visible if overexposed (yellow line). As a consequence, the metal wall thickness may be up to 30% greater than that inferred from the radiograph (blue overlay). After Giambologna (Flemish, active Florence, 1529–1608), *Hercules*, ca. 1580, cast in Italy, Germany, or the Netherlands, 17th century or later, H. 43 cm (Musée des Beaux-Arts d'Angers, France, inv. 2003-1-180). See C2RMF Internal Report 2018b.

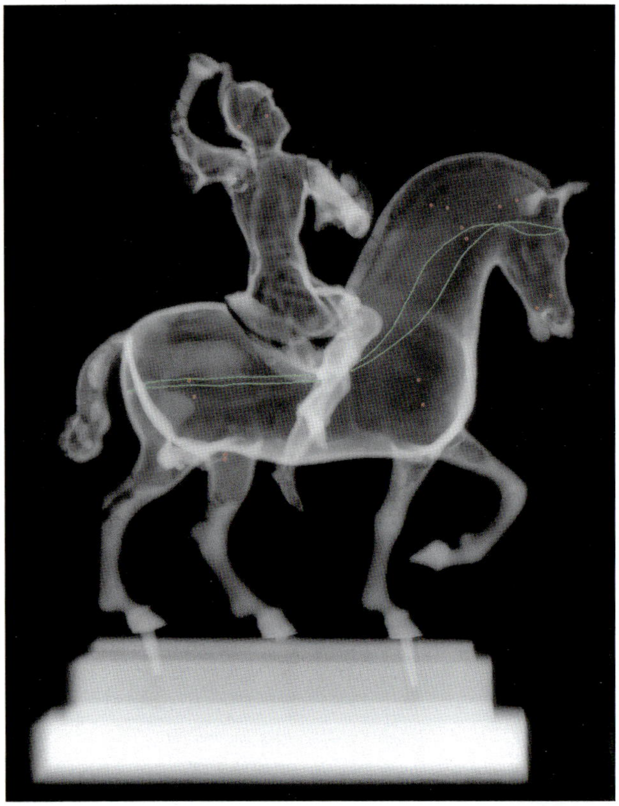

Figure 64 Annotated X-radiograph showing numerous small circular holes (red dots) testifying to the use of transverse core pins passing through the horse's body from one side to the other. See also the long wax joint running the length of the horse (green lines). *Horse and Rider*, Italian, 16th century, H. 33 cm (Musée des Beaux-Arts d'Angers, France, inv. 2003-1-182). See C2RMF Internal Report 2018b.

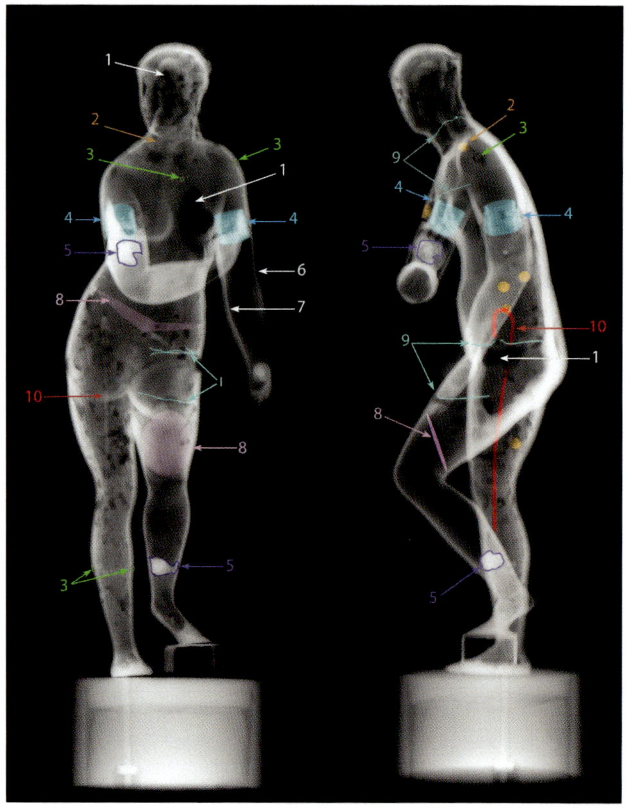

Figure 65 Annotated X-radiograph illustrating a number of technical features: 1) gaps in the core; 2) threaded plug repair; 3) repair patches; 4) mechanical assembly (Roman joints); 5) cast-on repairs; 6–7) variation of thickness of the metal wall due to core shift; 8) core juncture; 9) wax flashing; 10) internal core support. *Venus déhanchée*, northern Europe, 17th century, H. 45 cm (Musée des Beaux-Arts d'Angers, France, inv. 2003-1-200). See C2RMF Internal Report 2018b.

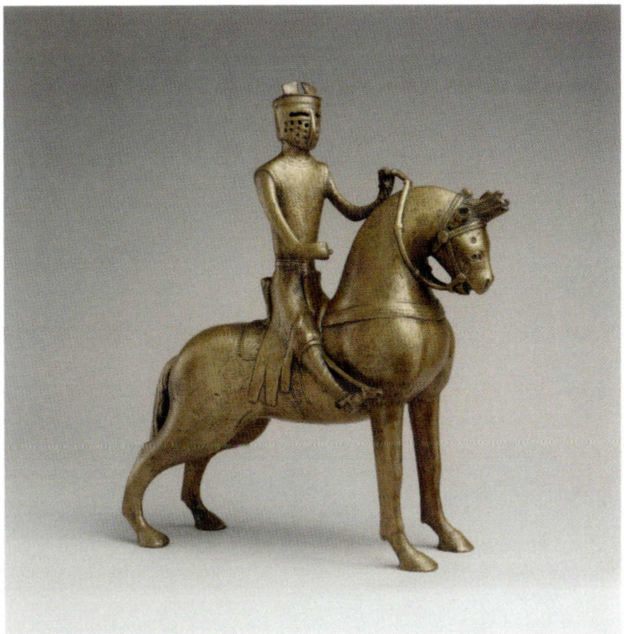

Figure 66 The horse fittings were added as wax rods on the wax model. Aquamanile in the Form of a Knight on Horseback, Lower Saxony, Germany (probably Hildesheim), ca. 1250, H. 37.3 cm (Metropolitan Museum of Art, gift of Irwin Untermyer, 1964, inv. 64.101.1492). See Dandridge 2006.

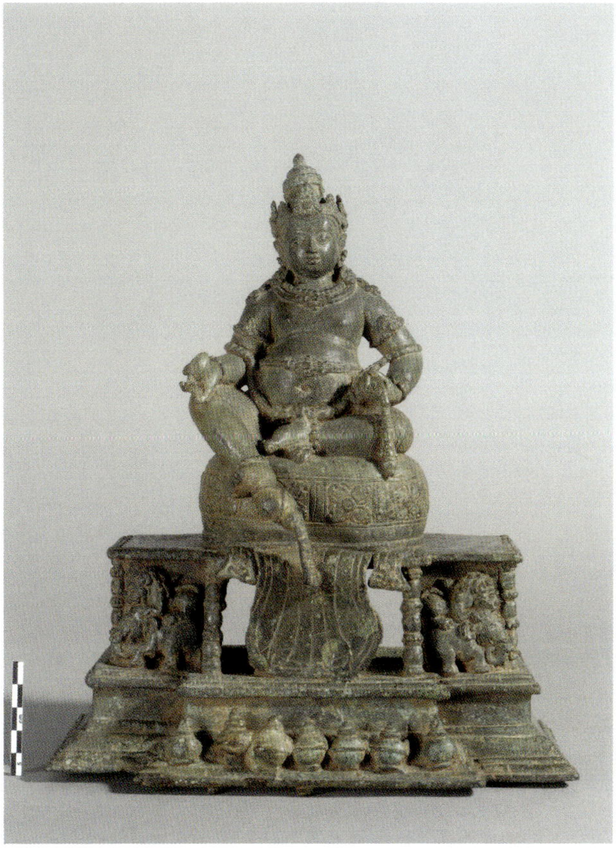

Figure 67 Example of direct wax additions on a wax model, notably the attribute in the left hand. Statuette of Jambhala, Java, first half of the 9th century, H. 28 cm (Musée national des arts asiatiques - Guimet, Paris, inv. MG 3814). See Mechling et al. 2018.

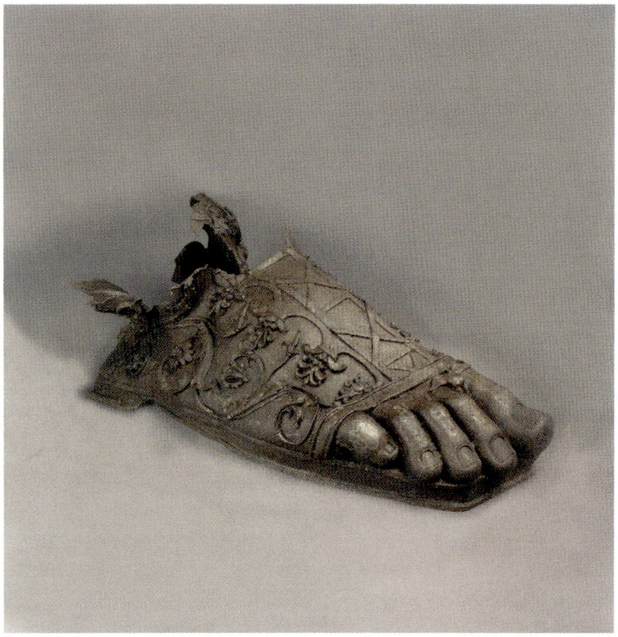

Figure 68 Remaining foot of a colossal statute. The height of the statue (probably an emperor) is 3.3–3.5 m. While the overall shape was produced using the indirect lost-wax process, the decoration was created by adding wax rods and threads directly to the wax model. Colossal Foot of Clermont-Ferrand, Roman (Gaule), early 2nd century CE, L. 58 cm (Musée de Clermont-Ferrand, Gaule). See Darblade-Audoin and Tavoso 2008.

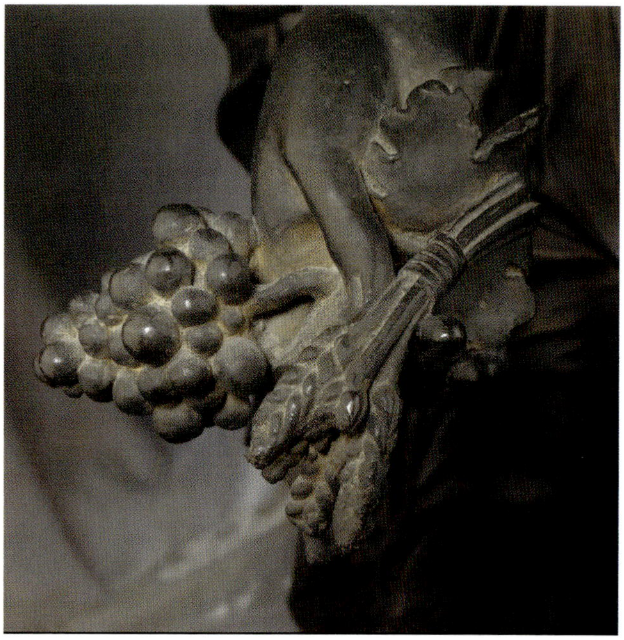

Figure 70 Detail of vine leaf and grapes in the figure's right hand that bear signs of having been modeled in wax. Barthélemy Prieur (French, 1536–1611), *Abundance*, cast in 1571 by Nicolas Péron, H. 128 cm (Musée du Louvre, inv. MR 1681). See Seelig-Teuwen, Bourgarit, and Bewer 2014; Castelle et al. 2021. See fig. 34 for an overall view.

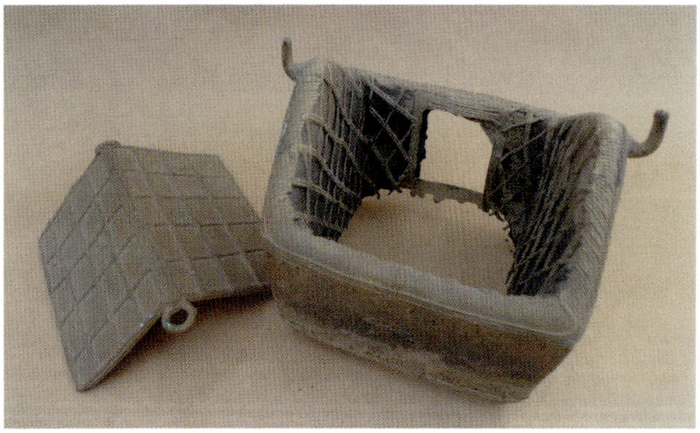

Figure 69 The open lattice patterns that decorate the inner walls of this small, directly cast, lost-wax structure reproduce the carefully applied wax threads that were laid over the carbon-rich core. Artist unknown, partial grouping of architectural structures on bronze base, Dhoka, Indian (?), 20th century, H. 13.3 cm (private collection).

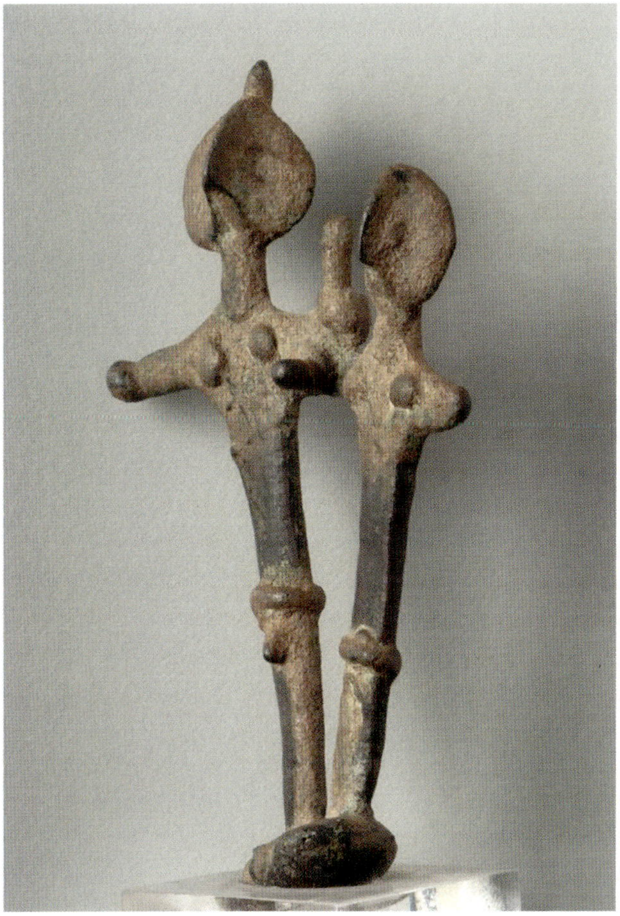

Figure 71 This small pair of human figures were modeled directly in wax. The heads were fashioned out of flattened balls of soft wax from which the noses were pinched out into thin, vertical ridges. The breasts are small, flattened balls of wax pressed onto the torsos. The base was formed by the remains of the metal that puddled at the bottom of the casting cup—evidence that the group was cast head-down in the mold. The figures could thus essentially be seen as decorated sprues. *Group of Two Human Figures*, Syro-Hittite, 15th–13th century BCE, H. 7.4 cm (Harvard Art Museums / Arthur M. Sackler Museum, gift of Benjamin and Lilian Hertzberg, 2004.205). See Lie and Bewer 2014, 46.

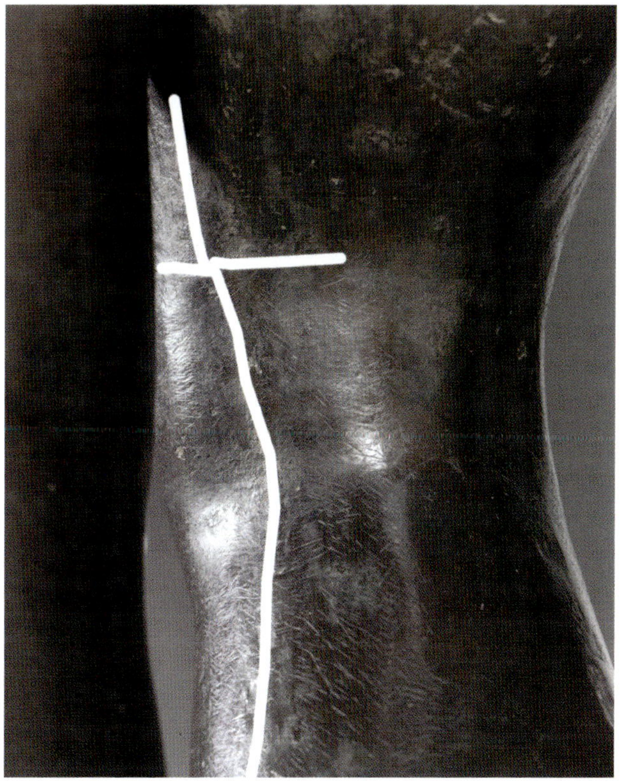

Figure 72 Detail of the back of the right knee showing the remains of seam lines from the piece molds when forming the wax intermodel (white overlay). Attributed to Hubert Le Sueur (French, ca. 1580–1658), *Medici Venus*, H. 150 cm (Musée du Louvre, inv. MR 3278). See Castelle 2016.

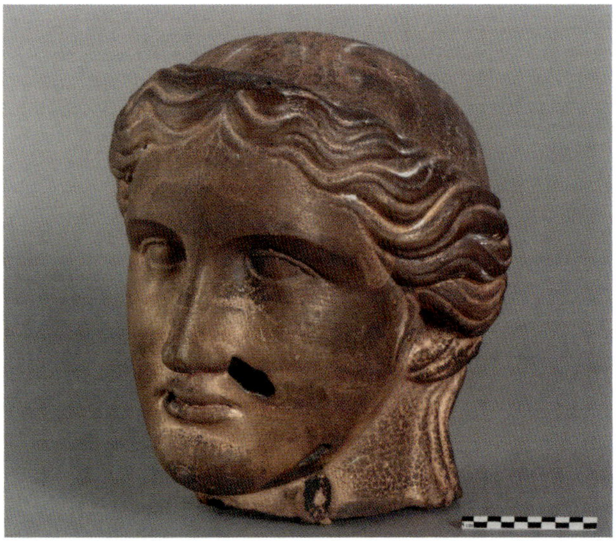

Figure 73 Bronze reproduction by slush molding of the head of the *Apollo of Lillebonne* made during the 2016 CAST:ING meeting, Coubertin foundry, France. Despite its complex shape, the cavity on the left cheek does not stem from any exogenous material fallen in the melt, but rather from the thinness of the wax model in this area, which caused air to be trapped, thereby preventing the mold from filling. The vertical seam lines in the middle of the forehead are a direct reproduction of the mold lines on the wax model (see fig. 74). The height is 18 cm (two-thirds of the original size). Original: *Apollo of Lillebonne*, France, 1st century BCE–1st century CE, H. 193 cm (Musée du Louvre, inv. Br37 [NIII65]). See fig. 288.

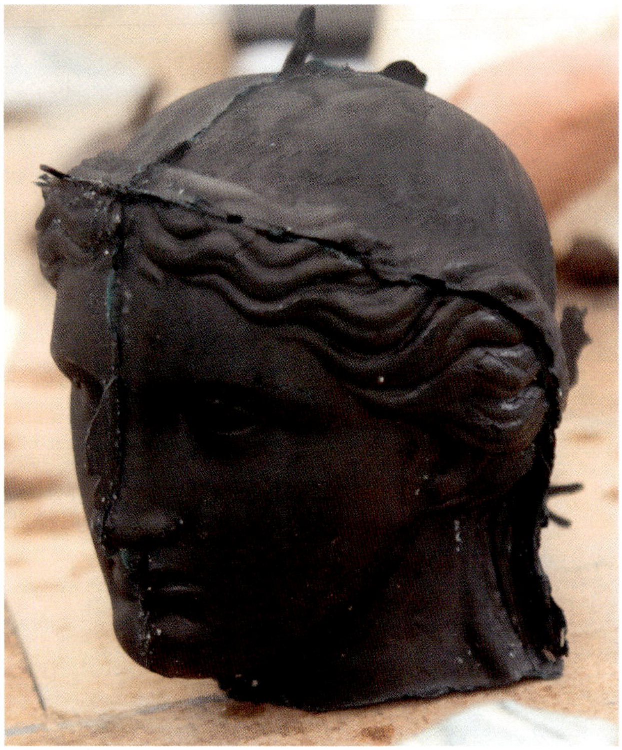

Figure 74 Wax reproduction of the head of the *Apollo of Lillebonne* made during the 2016 CAST:ING meeting, Coubertin foundry, France. This wax was obtained by slush molding in a plaster piece mold. All the mold lines (seam lines) are clearly visible. Note the distortion of the model due to imperfection in the piece mold assemblage. The height is 18 cm (two-thirds of the original size). Original: *Apollo of Lillebonne*, France, 1st century BCE–1st century CE, H. 193 cm (Musée du Louvre, inv. Br37 [NIII65]). See fig. 288.

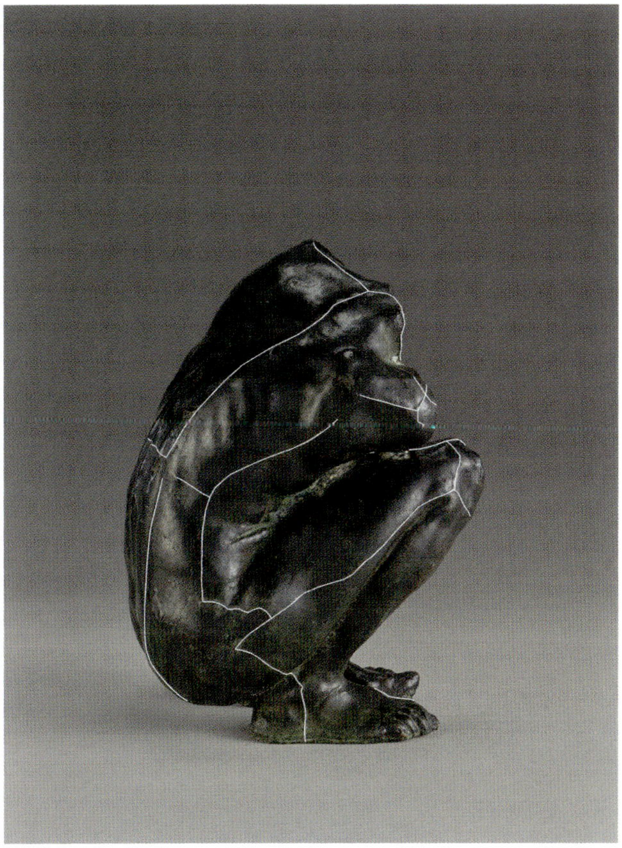

Figure 75 Raised seam lines appear to correspond to traces of the piece mold used to create a plaster model that was then used to cast the bronze using the indirect lost-wax process. Camille Claudel (French, 1864–1943), *Torso of a Crouching Woman*, model ca. 1884–85, cast by 1913, H. 35 cm (J. Paul Getty Museum, inv. 2018.32).

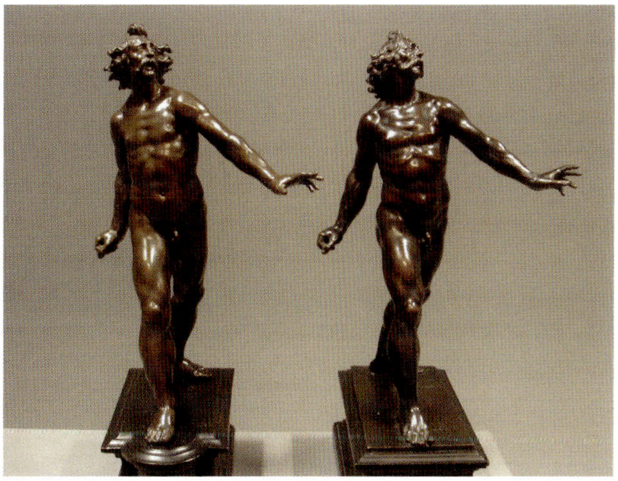

Figure 76 Juxtaposition of these two indirect lost-wax casts shows how their assemblage from separate wax parts—in this case, the limbs and torso (as confirmed by X-radiography)—can result in slight variations in their postures. Willem van Tetrode (ca. 1525–1580), *Striding Warrior*, 1562–65, H. 40 cm (Collection of Mr. & Mrs. J. Thomilson Hill, New York [left]) and H. 39.4 cm (Hearn Family Trust, New York [right]). See Bewer et al. 2003.

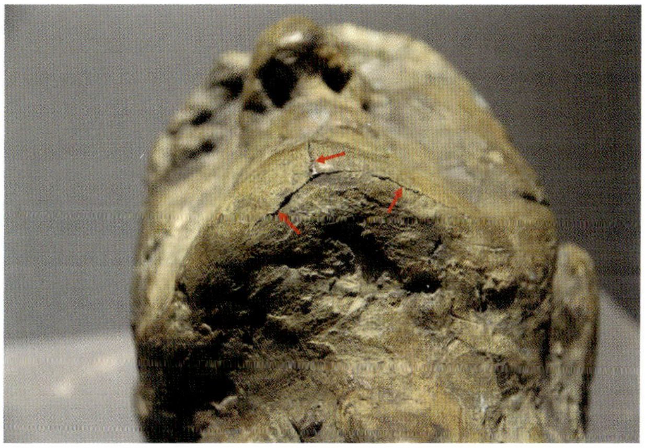

Figure 77 Detailed image of a small area of the chin showing flashing in the refractory mold that has been filed down. The lines are so angular, they look like the intersecting lines of a mold. Andrew Lacey (British, b. 1969), *Head of Henry Tonks*, cast by the artist in 2015, H. 17 cm (artist's collection).

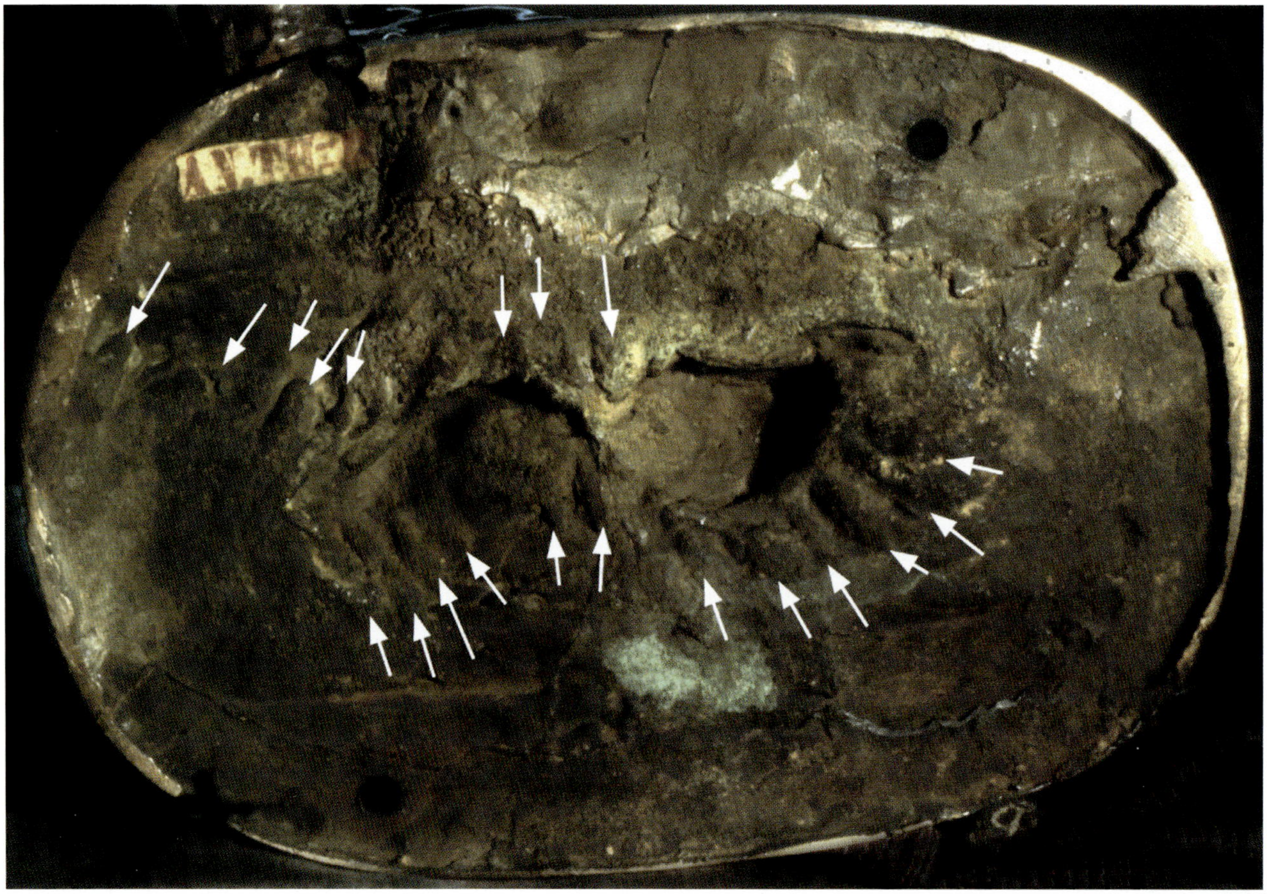

Figure 78 Underside of the *Antiope and Theseus* group showing the long marks that the artist's fingers produced in pulling up a mound of fresh clay from the slab that formed the core for the base. Adriaen de Vries (Netherlandish, 1556–1626), *Antiope and Theseus*, ca. 1600–1601, H. 95 cm (Royal Collection, Buckingham Palace, London, inv. RCIN 57961). See Bewer 1998, 70; Bewer 2001.

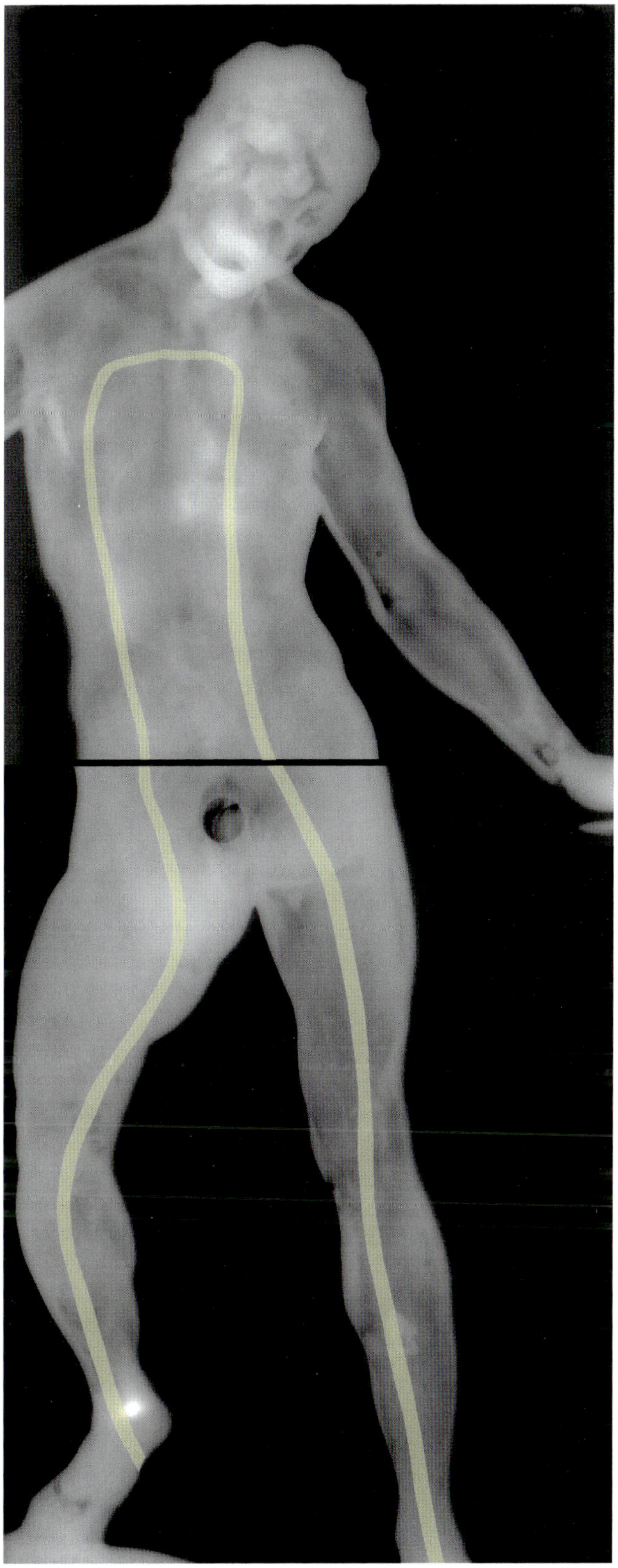

Figure 79 Annotated composite X-radiograph. Note in particular the continuous armature that extends throughout the direct lost-wax cast. Adriaen de Vries (Netherlandish, 1556–1626), *Juggling Man*, ca. 1615, H. 76.8 cm (J. Paul Getty Museum, inv. 90.SB.44).

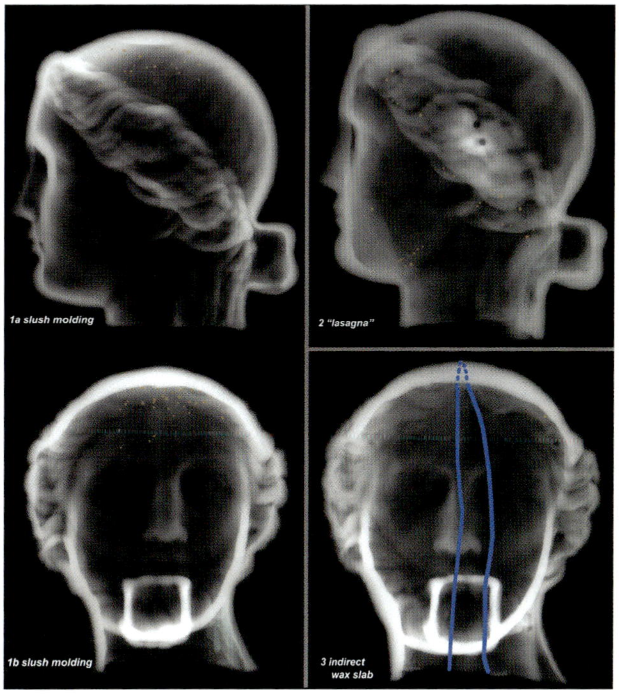

Figure 80 X-radiographs of experimental casts after the same model of Apollo using three different lost-wax casting processes. The heads were produced by beginners during the 2016 CAST:ING meeting at the Coubertin foundry, France, and can therefore not claim to be representative of features in expertly cast historical bronzes, but present a few noteworthy features. 1a–b) Side and frontal views of the same head produced by the slush molding process. They show the relatively even thickness of metal wall that follows the contours of the molding. The frontal view shows the drips that formed in the neck (turquoise overlay) on the internal surface when the excess liquefied wax was poured out of the reusable mold. The solid spheres on the top of this and the other casts are from air bubbles trapped along the inner surface when the fresh plaster-based core slurry was poured into the mold (orange overlay dots). 2) Unlike the head in 1a, this cast of the Apollo head is more uneven, as seen in particular in the facial area, where the flattened shape of the inner surface is the result of a rather thick clay slab (the "lasagna") that was initially laid into the reusable mold to create the negative space to be filled by the bronze. 3) The frontal view of the third experimental cast shows the overall even thickness of the metal wall that results from the use of the thin wax slabs pressed into the mold. The joint of several of the wax slabs runs vertically down the face and back of the head (blue overlay lines). See also fig. 121. The height of the casts is 18 cm (two-thirds of the original size). Original: *Apollo of Lillebonne*, France, 1st century BCE–1st century CE, H. 193 cm (Musée du Louvre, inv. Br37 [NIII65]). See fig. 288.

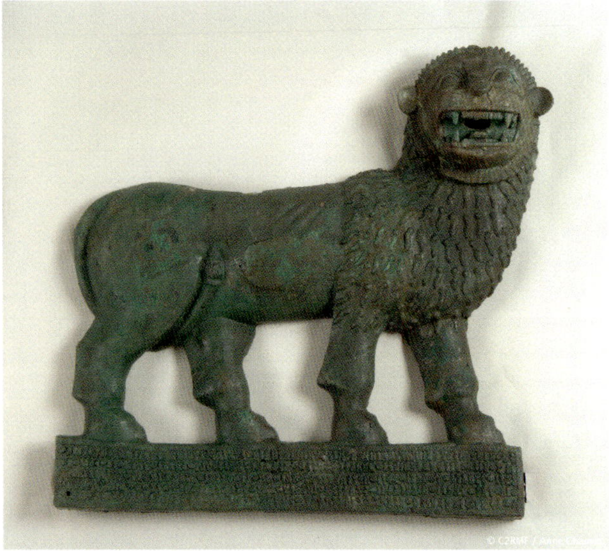

Figure 81 *Lion Appliqué*, probably discovered in Tamna, Yemen, third quarter of the 1st century BCE, H. 67 cm (National Museum of Sana'a, Yemen, inv. YM36526). See Mille et al. 2010.

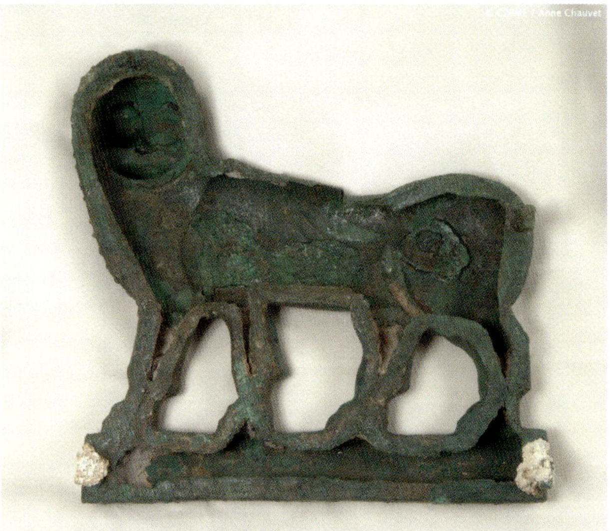

Figure 82 Rear view showing the bent metallic walls at the edge that are considered evidence of the use of wax slabs. *Lion Appliqué*, probably discovered in Tamna, Yemen, third quarter of the 1st century BCE, H. 67 cm (National Museum of Sana'a, Yemen, inv. YM36526). See Mille et al. 2010.

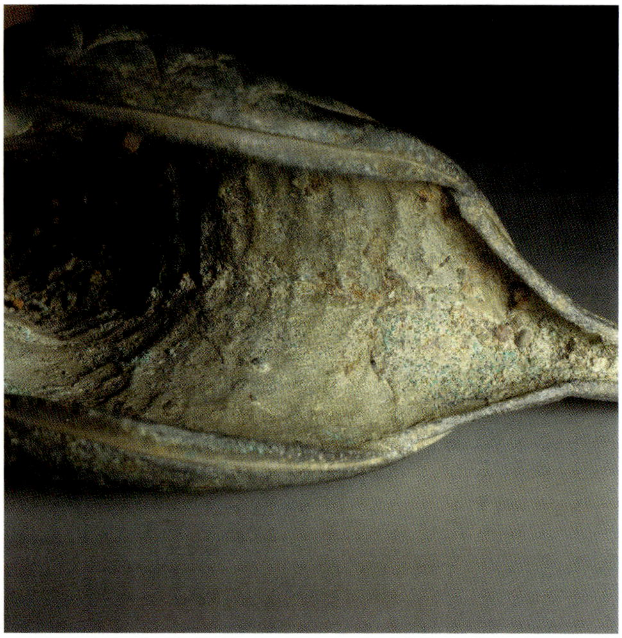

Figure 83 Detail showing the marks of a tool used to work the interior surface of the wax inter-model. *Zoomorphic Figure Evoking a Griffin*, France, early-late antiquity, H. 26 cm (Musée de la préhistoire du Grand-Pressigny, France, inv. 2008.001.0001). See C2RMF Internal Report 2008a.

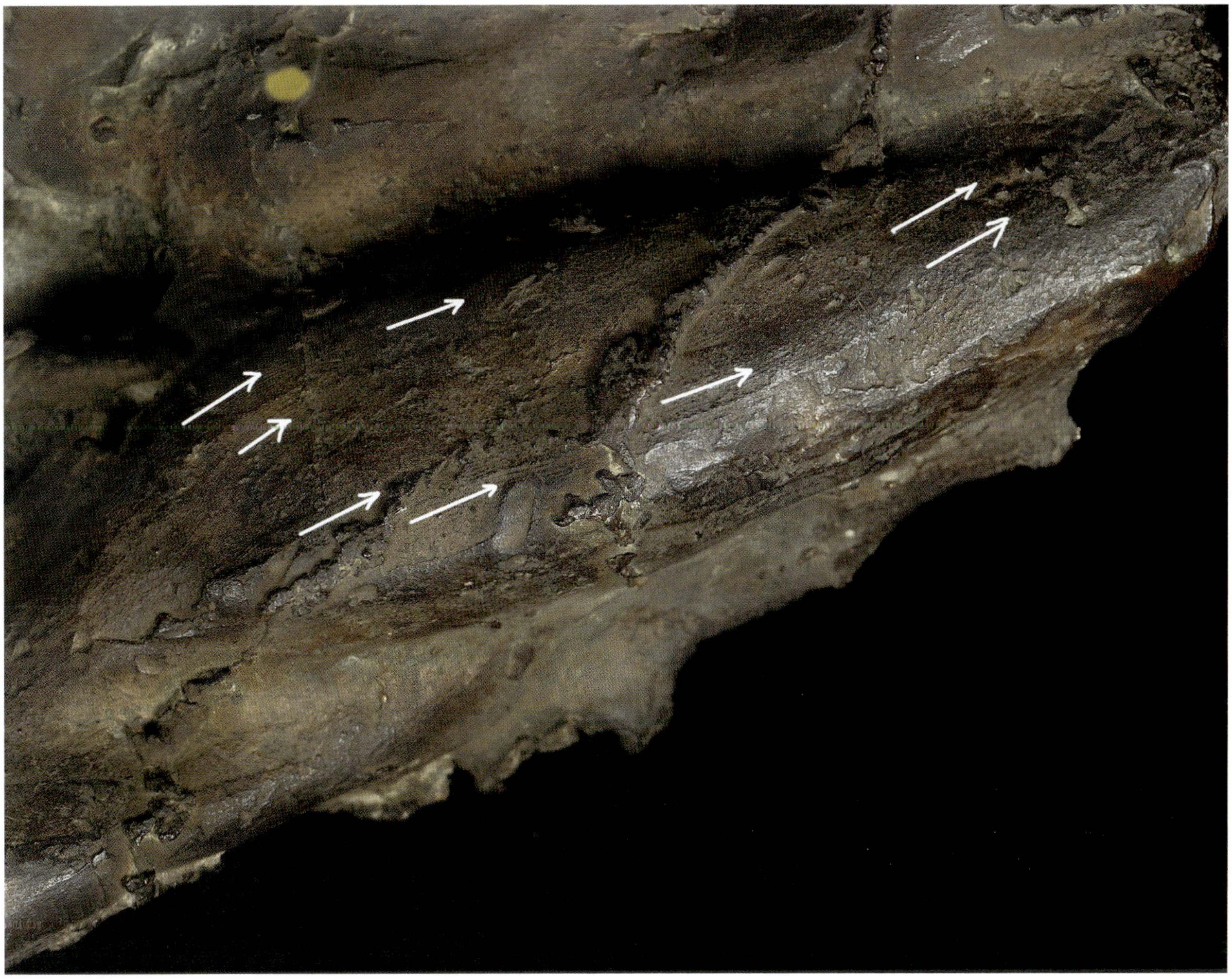

Figure 84 Rear surface of a fragment of the dolphins showing a fingerprint (yellow overlay) and the elongated traces made by a fine-toothed tool or finger (white arrows indicate the direction of the strokes) drawn along the inside of the wax model. *Great Dolphins of Vienne*, Roman period, Vienne, France, 2nd century CE, max L. 260 cm (Musée des Beaux Arts et d'Archéologie de Vienne, France, inv. R.1998.2.43). See technical sketch, fig. 367.

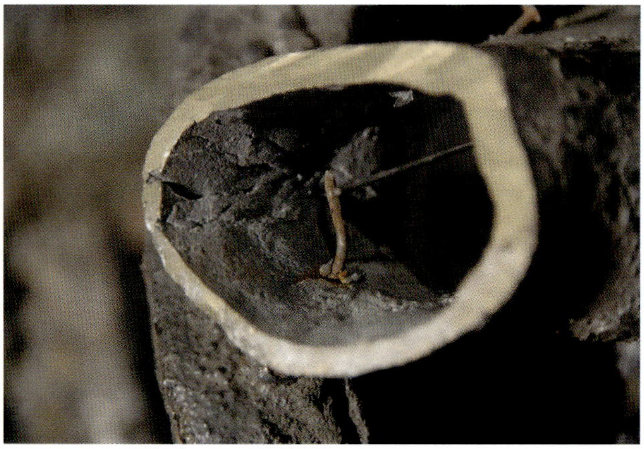

Figure 85 This internal view of a small arm shows an extant wire core pin and the uneven blobby accretions at its point of contact with the inner surface. These occurred during casting, with the molten-metal-filled gaps created in the core when the wire was pushed through the core-filled wax model, causing some localized breakdown. Andrew Lacey (British, b.1969), *Hand*, 1998, H. 18 cm (private collection).

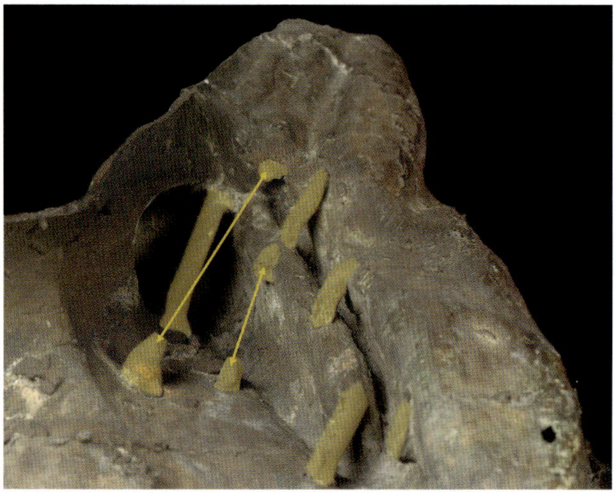

Figure 86 Rear surface of the head of the front dolphin showing sprues added on the wax model (yellow overlay). The wax model of the head was made of several pieces joined together at the juncture of the upper and lower lips, and also at the rostrum. *Great Dolphins of Vienne*, Roman period, Vienne, France, 2nd century CE, max L. 260 cm (Musée des Beaux-Arts et d'Archéologie de Vienne, France, inv. R.1998.2.43). See technical sketch, fig. 367.

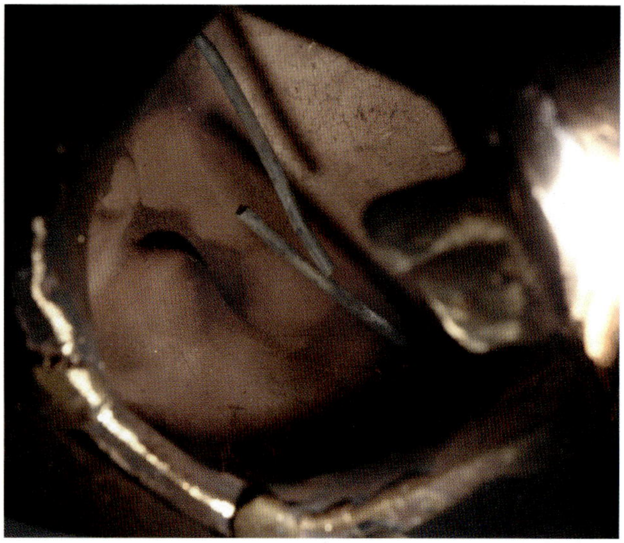

Figure 87 View from the inside of a bronze reproduction of the head of the *Apollo of Lillebonne* made during the 2016 CAST:ING workshop, Coubertin foundry, France, showing the typical smooth flowing surface of slush-molding casts. The height is 18 cm (two-thirds of the original size). Original: *Apollo of Lillebonne*, France, 1st century BCE–1st century CE, H. 193 cm (Musée du Louvre, inv. Br37 [NIII65]). See fig. 288.

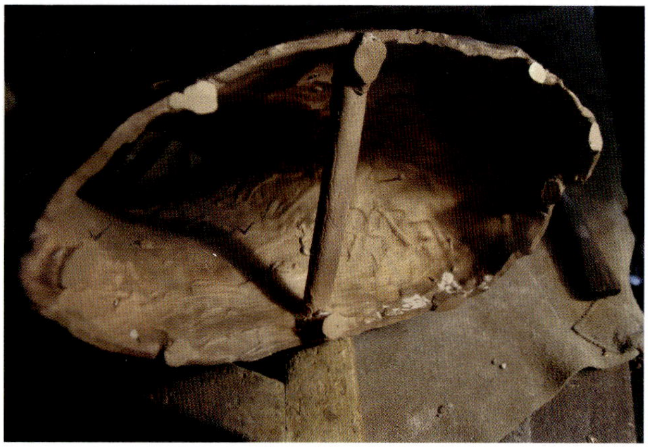

Figure 88 Inside view of the horse head showing tide lines from the flow of liquid wax captured on the inner surface of the bronze. A fully preserved internal sprue is seen here cutting across the inside at the base of the neck. And several circular areas along the rim of the cast are where sprues connecting to the casting cup were cut off during fettling. Andrew Lacey (British, b. 1969), *The Anatomy of Bronze*, cast by the artist in 2019, Devon, UK, H. 45 cm (artist's collection).

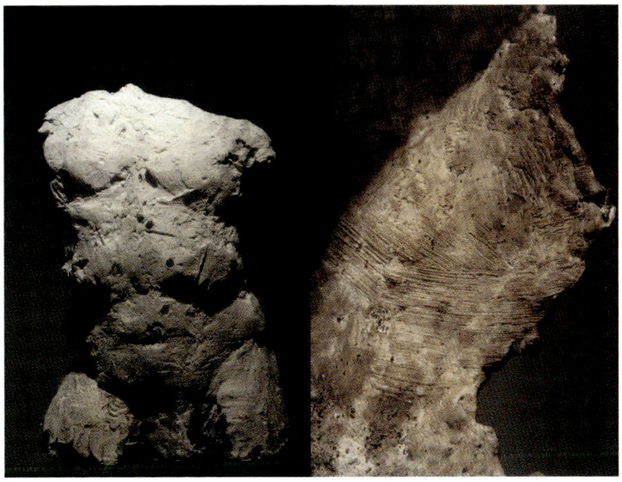

Figure 89 General view and detailed image of the work's reverse showing the brushstrokes from painting in liquefied wax. Andrew Lacey (British, b. 1969), *Study for Saint Sebastian with Bullet Holes*, cast by the artist in 2017, H. 60 cm (artist's collection).

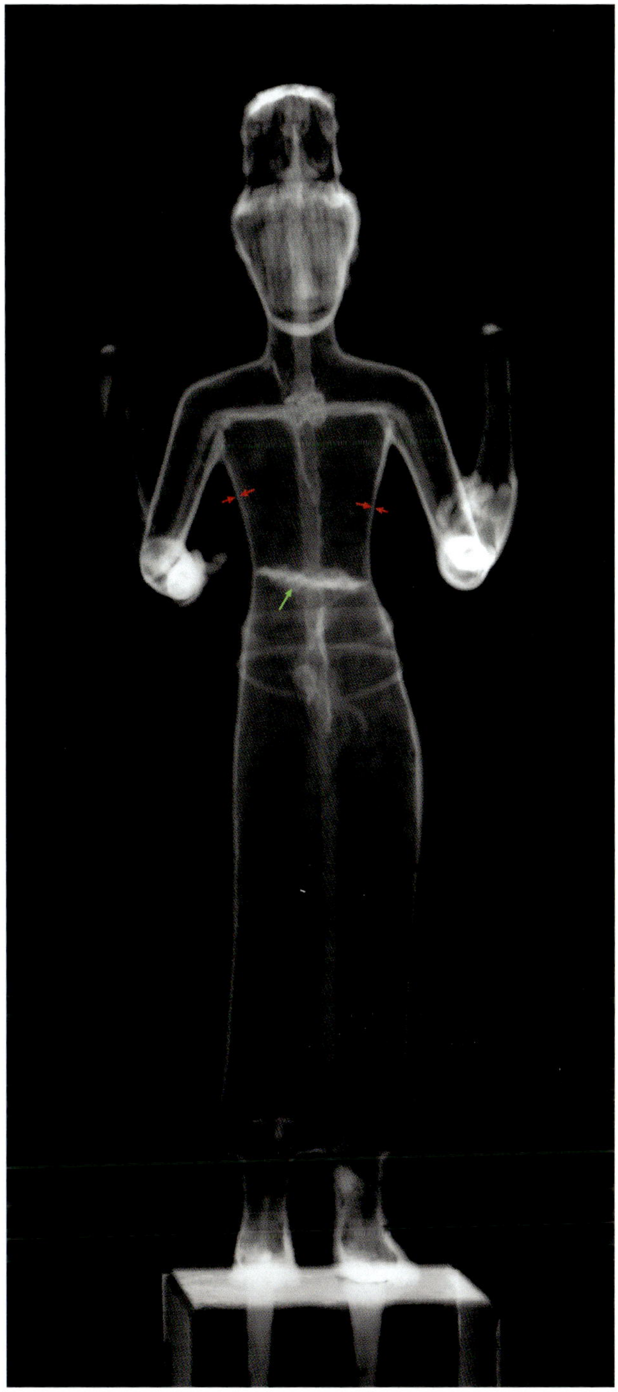

Figure 90 Annotated X-radiograph showing the very thin and even metal walls (1–2 mm, red arrows). See also the large core flashing at the level of the belly (white feature shown by the green arrow). *Maitreya*, Khmer, pre-Angkor, 8th century, H. 46 cm (Musée national des arts asiatiques – Guimet, Paris, inv. MA 3321). See Bourgarit et al. 2003.

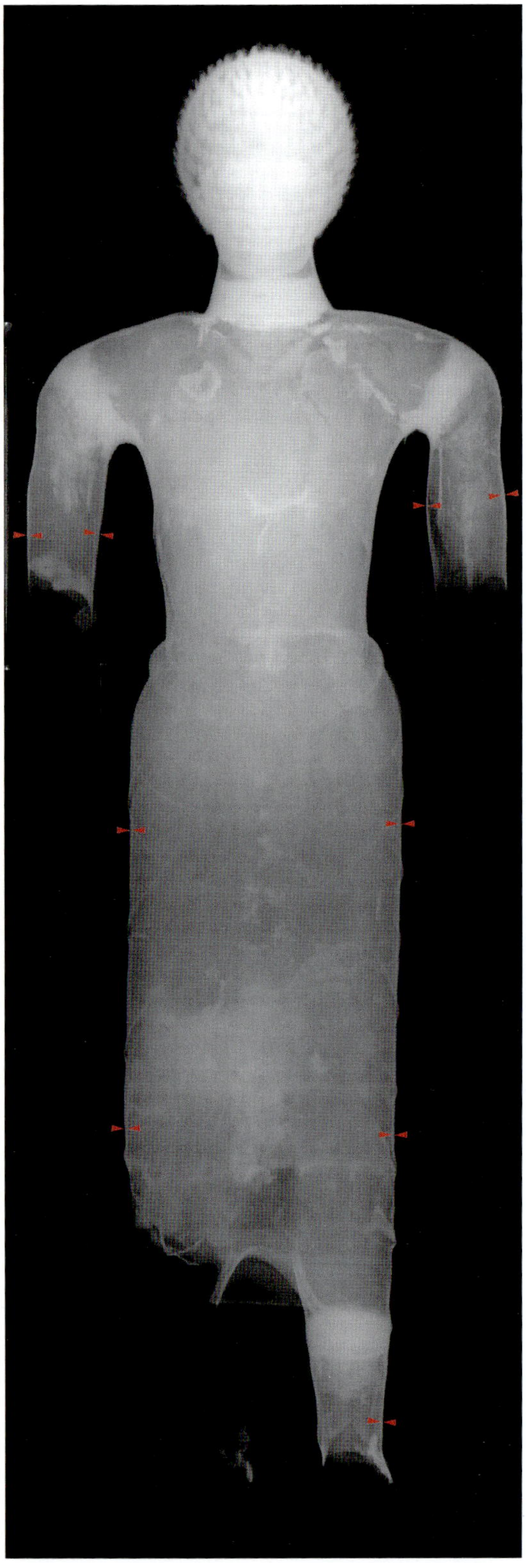

Figure 91 X-radiograph showing the very thin and even metal walls (2 mm). Bronze Statue of Hawtar'athat, Yemen, 1st millennium BCE, H. 140 cm (National Museum of Sana'a, Yemen, inv. YM 23206). See Mille et al. 2010.

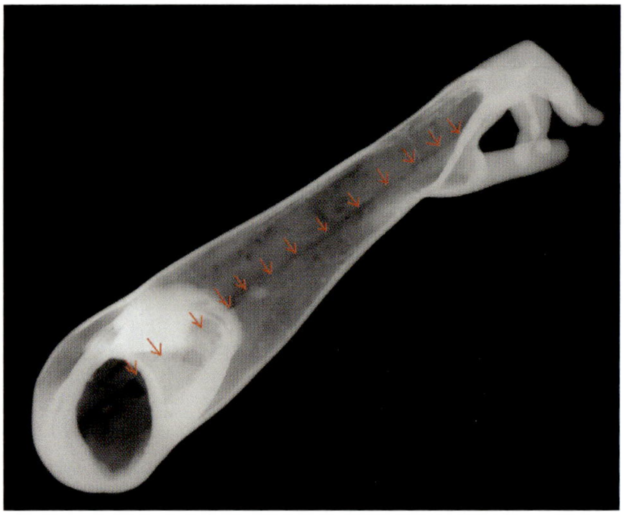

Figure 92 X-radiograph of the forearm of a monumental Roman statue cast by the lost-wax process. The long, dark, linear area corresponds to a thinning in the original wax model where wax slabs were not joined all the way through. The area would have been filled with core material and the imperfection was thus translated into bronze. *Arm of Essegney*, Roman, Essegney, France, probably 2nd–3rd century CE, L. from hand to elbow 43 cm (Musée départemental d'art ancien et contemporain d'Epinal, France, inv. M0536_2005.2.1). See Caumont et al. 2006.

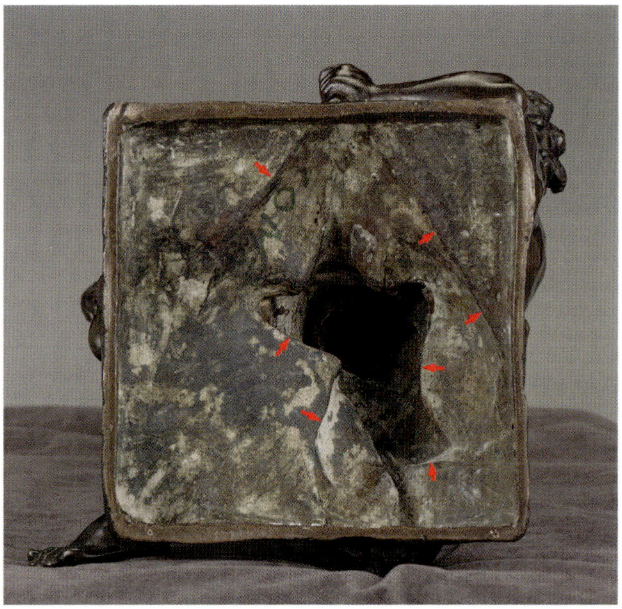

Figure 93 View of underside showing the angular contours of the interior, typical of the cut-back-core process. Gaspard Marsy (French, 1624–1681), *Boreas Abducting Orithyia*, probably cast late 18th century, H. 55.2 cm (J. Paul Getty Museum, inv. 74.SB.18). See Bassett and Bewer 2014.

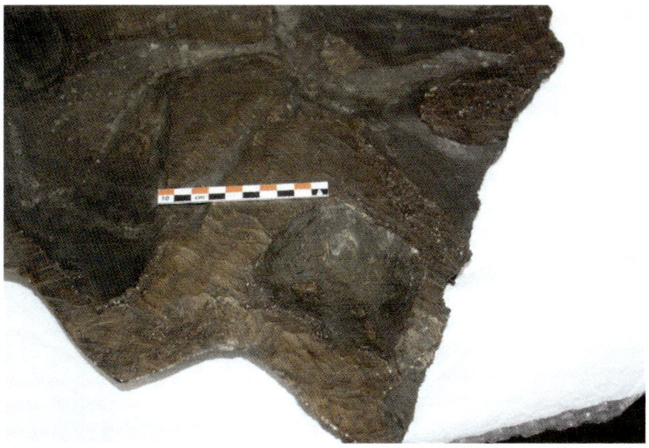

Figure 94 Tool marks testifying to the cutting back of the core. Shown here is the back view of a bronze fragment of the Vendôme Column. This may be related to what the founder of most of the Vendôme Column bronzes, Jean-Baptiste Launay (French, 1768–1827), described as *amaigrissement du noyau* (core thinning). Various French sculptors, *Vendôme Column*, 1805–10, dismantled 1871, re-erected 1873–75, H. 44 m (Place Vendôme, Paris). See Launay 1827, 1:xv.

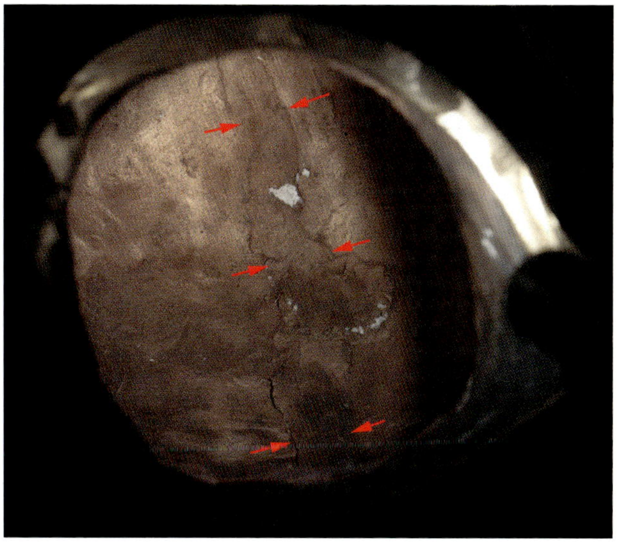

Figure 95 Inside view of a bronze reproduction of the head of the *Apollo of Lillebonne* made by the lasagna process during the 2016 CAST:ING meeting at the Coubertin foundry, France. The outer contours of the hair are not at all visible on the simplified interior of the head. See the working marks of the clay sheet (lasagna), and particularly the thick joint between the two main sheets crossing the length of the head (red arrows). The height is 18 cm (two-thirds of the original size). Original: *Apollo of Lillebonne*, France, 1st century BCE–1st century CE, H. 193 cm (Musée du Louvre, inv. Br37 [NIII65]). See fig. 288.

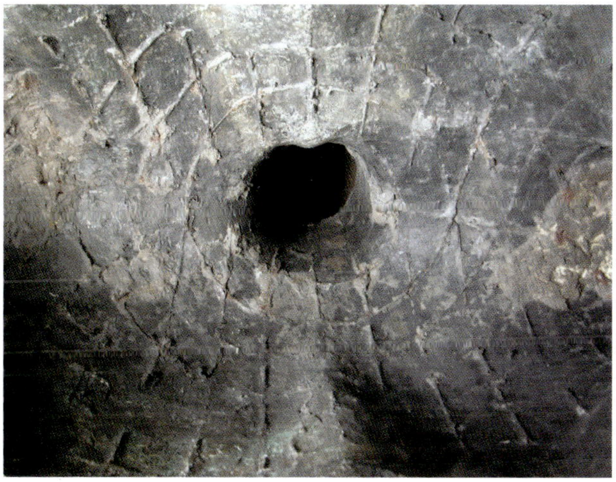

Figure 96 View of the interior of a bust showing the wire mesh that would have bound the core and is now embedded in the inner metal surface of the cast. Benvenuto Cellini (Italian, 1500–1571), *Bust of Cosimo I*, 1546–47, H. 110 cm (Museo Nazionale del Bargello, Florence, inv. 358 Bronzi). See Bewer and MacNamara 2012.

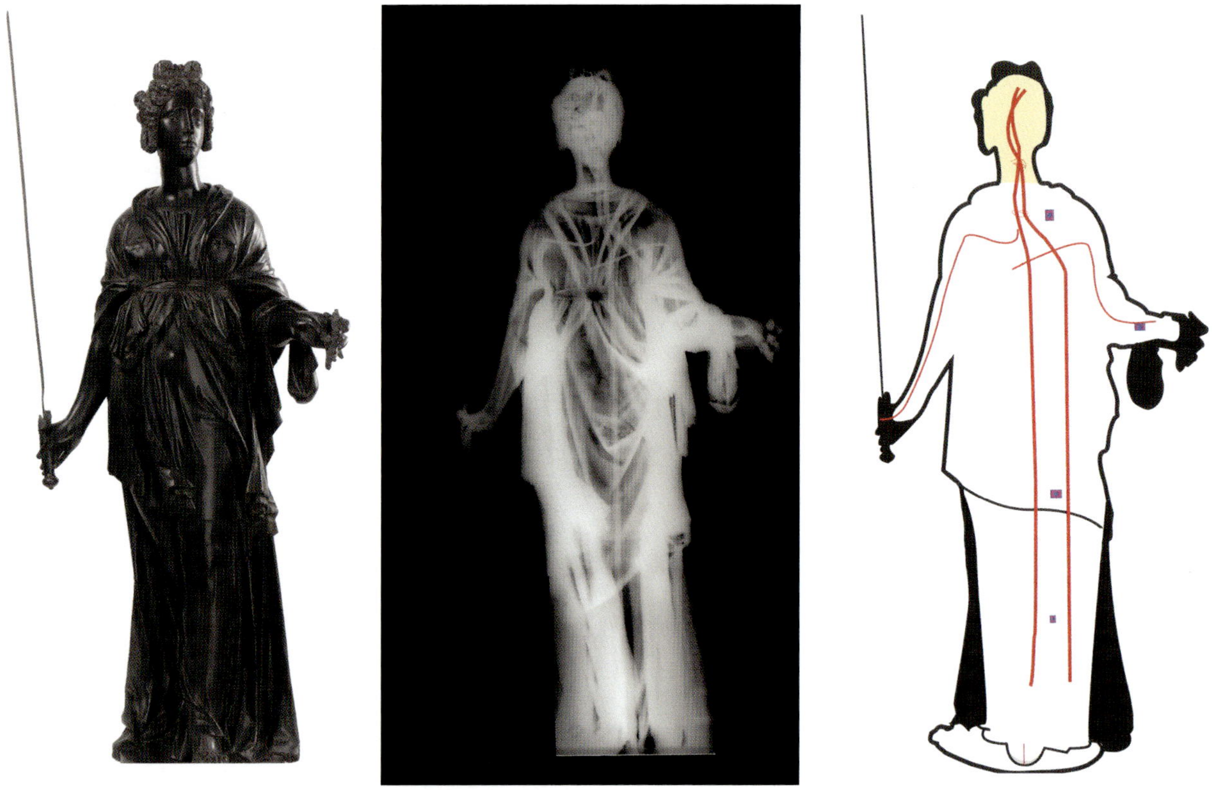

Figure 97 Overlay of interpretive technical diagram and both X-radiograph and daylight photography. Martin Lefort (French, dates unknown), *Justice*, 1571, H. 128 cm (Musée du Louvre, inv. MR1682). See Case Study 5.

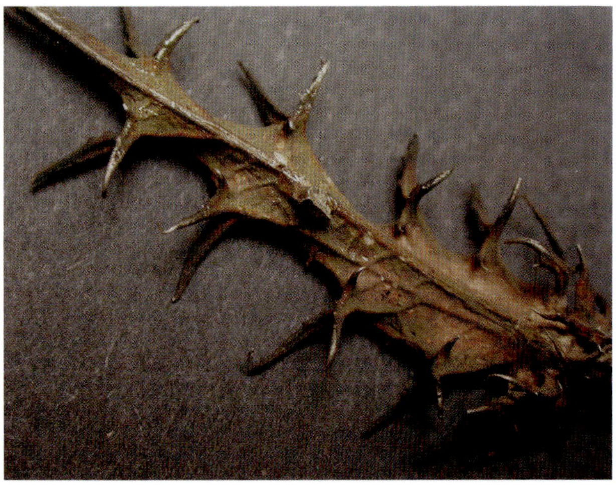

Figure 98 An example of life-casting in which the model was entirely burned out in the process. Andrew Lacey (British, b. 1969), *A Modern-Day Thistle*, cast for the artist's own experimental research, 2015, H. 3.5 cm (artist's collection).

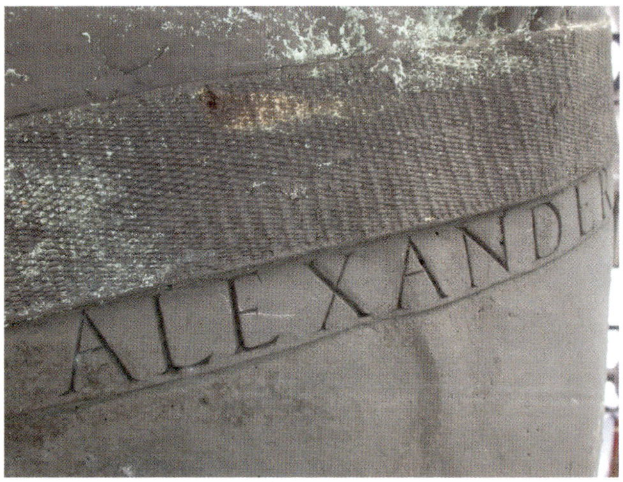

Figure 99 The saddle belt is cast "from life" from a woven band that was added onto the wax model's belly. Andrea del Verrocchio (Italian, ca. 1425–1488) and Alessandro Leopardi (Italian, ca. 1465–1512), *Equestrian Monument of Bartolomeo Colleoni*, Venetian, 1479–96, over life size (Piazza SS Giovanni e Paolo, Venice). See general view of the monument, fig. 100.

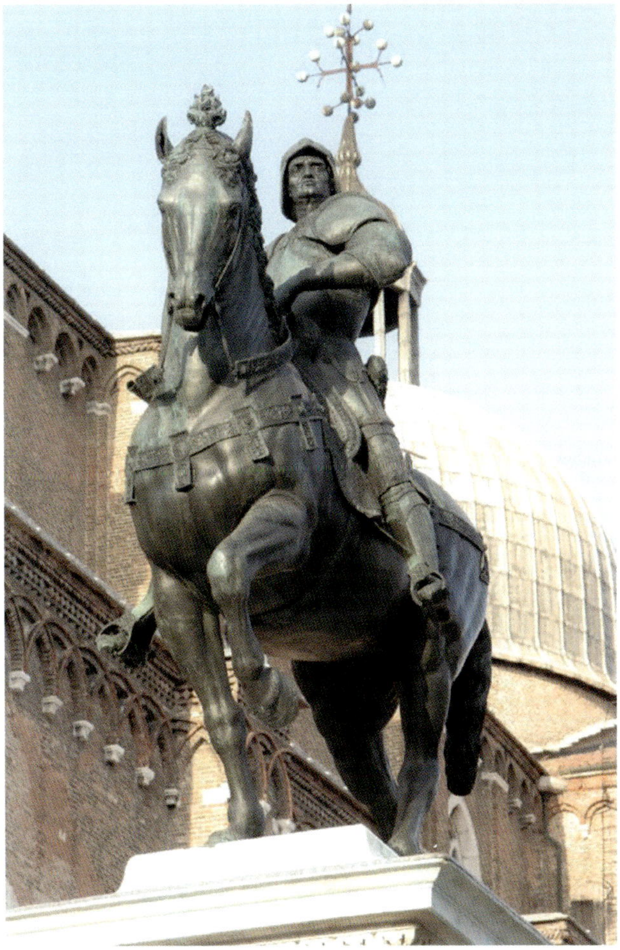

Figure 100 Andrea del Verrocchio (Italian, ca. 1435–1488) (model of horse) and Alessandro Leopardi (Italian, ca. 1465–1512 (finished model and cast), *Equestrian Monument of Bartolomeo Colleoni*, 1479–96, over life size (Piazza SS Giovanni e Paolo, Venice), after 2006 conservation by Giovanni and Lorenzo Morigi.

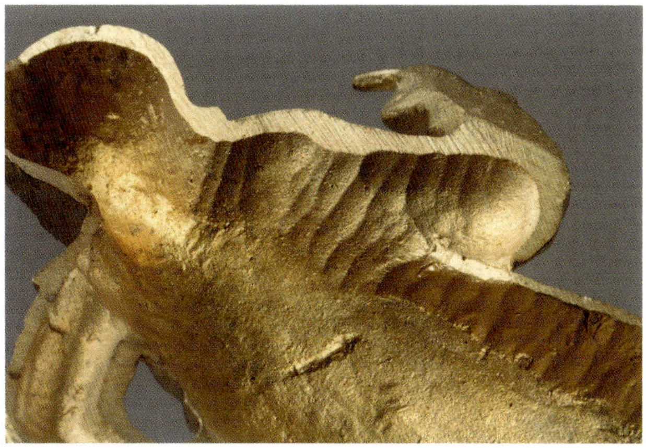

Figure 101 Detail of the reverse of a gilded brass relief figure showing gouge marks from the carved, hollow wooden pattern and the granular surface of the sand mold. Because this is a relief, the reverse should be considered the external surface as well. Relief figure of Benjamin Franklin, purportedly cast by Paul Revere (American, ca. 1735–1818) after a wood model by Simon Skillen, from Joseph Pope's (American, 1750–1826) *Grand Orrery*, 1776–87, H. 31 cm (Collection of Historical Scientific Instruments, Harvard University, inv. 0005).

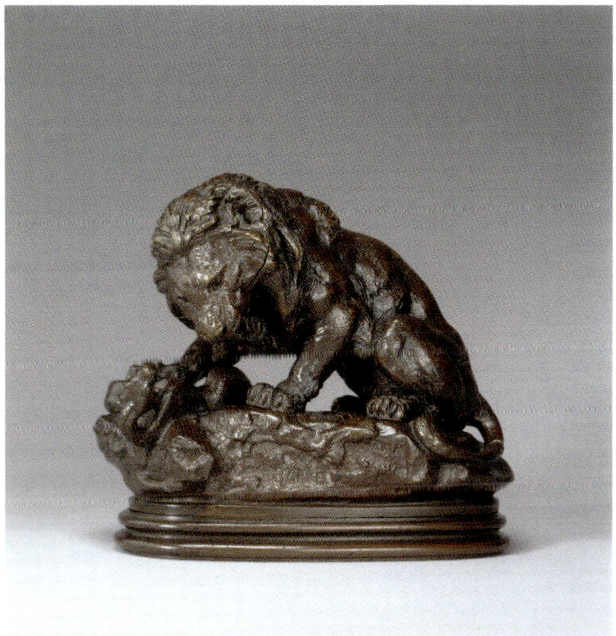

Figure 102 This chef-modèle has been cast in four sections (here assembled by way of interlocking pins, ridges, and slots) so as to provide more manageable segments to be sand molded and cast separately. Antoine-Louis Barye (French, 1795–1875), *Lion and Serpent* (No. 2), chef-modèle cast by Victor Paillard, ca. 1858, H. 18.4 cm (Walters Art Museum, Baltimore, inv. 27.92).

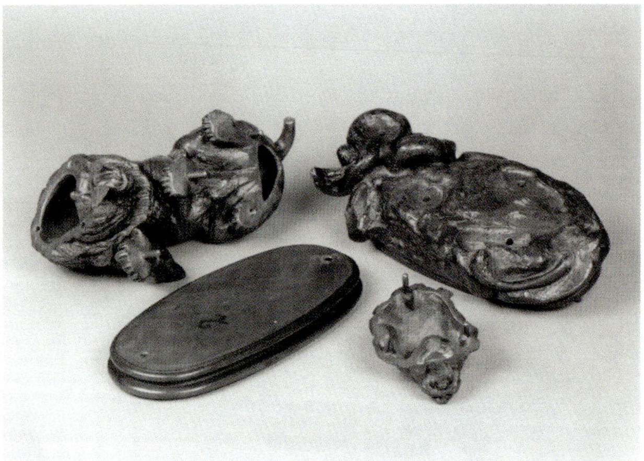

Figure 103 Dismantled chef-modèle. The piecing is characteristic of sand casting of complex forms. Antoine-Louis Barye (French, 1795–1875), *Lion and Serpent* (No. 2), chef-modèle cast by Victor Paillard, ca. 1858, H. 18.4 cm (Walters Art Museum, Baltimore, inv. 27.92).

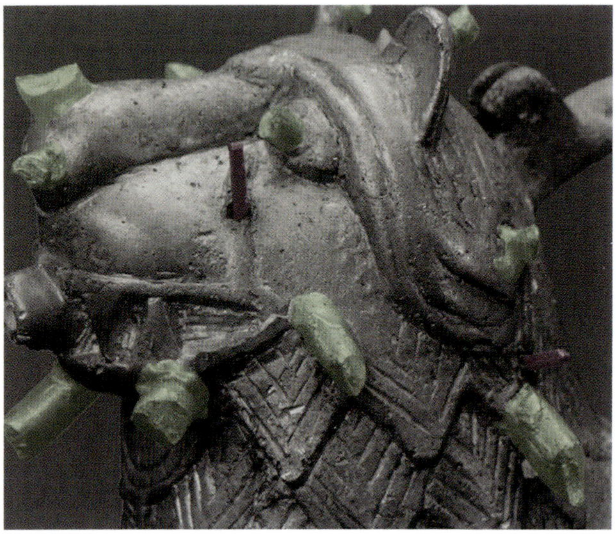

Figure 104 Detail of an experimental casting during fettling showing sprue ends and core pins. It also shows the coarse, as-cast surface that preserves the V-shaped incised lines of the mane made in the wax model. Ubaldo Vitali (American, b. 1944), Reproduction of a Lion Aquamanile, Maplewood, New Jersey, 2006, H. 19 cm, after Aquamanile in the Form of a Lion, probably northern Germany, 12th century, H. 19.5 cm (Metropolitan Museum of Art, gift of Irwin Untermyer, 1964, inv. 64.101.1491). See Dandridge 2006.

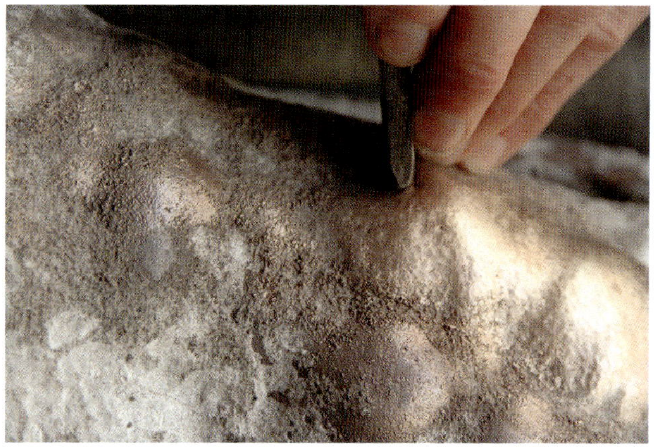

Figure 105 Chasing by punching over casting imperfections. Note the surface that has bonded with the white investment. In-process detail of a reconstruction by Andrew Lacey (British, b. 1969) of one of the so-called Rothschild Bronzes, attributed to Michelangelo (Italian, 1475–1564), H. 91.2 cm (private collection). See Lacey 2018.

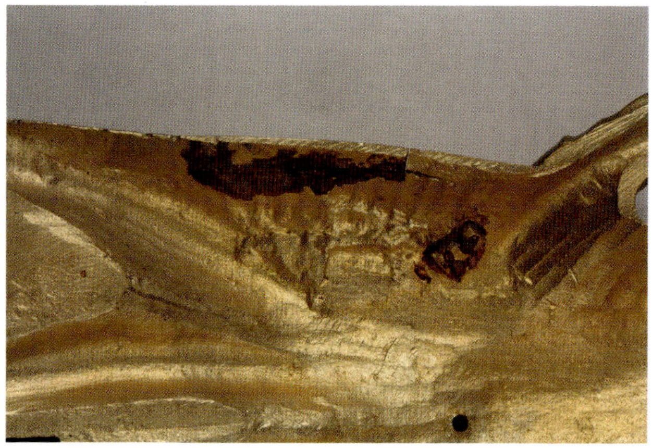

Figure 106 Detail of the reverse of a gilded brass relief figure shows the grainy texture of the sand mold, the gouge marks of the original carved model that were replicated in the bronze, as well as the patched, brazed repairs with flux damage. Relief figure of Benjamin Franklin, purportedly cast by Paul Revere (American, ca. 1735–1818) after a wood model by Simon Skillen, from Joseph Pope's (American, 1750–1826) *Grand Orrery*, 1776–87, H. 31 cm (Collection of Historical Scientific Instruments, Harvard University, inv. 0005).

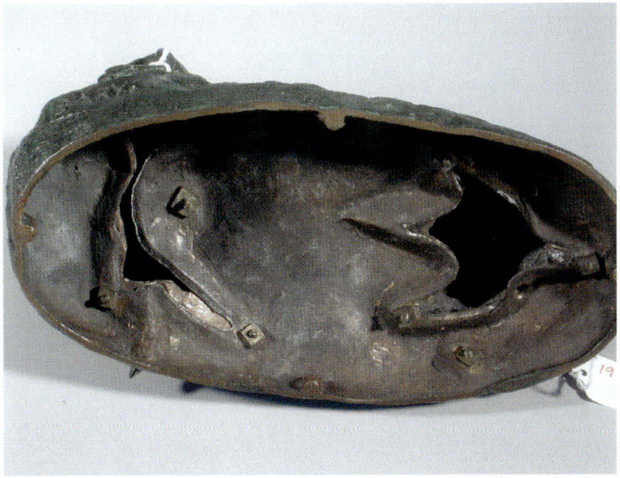

Figure 107 View from underside showing the angular shapes of the inner surface that are characteristic of pared-down sand-cast cores. As is also typical with sand casts, the group was made in pieces and assembled in the metal. None of these joints are visible from the outer surface. Antoine-Louis Barye (French, 1795–1875), *Lion Devouring a Boar*, ca. 1865, H. 16.9 cm (Harvard Art Museums / Fogg Museum, The Henry Dexter Sharpe Collection, inv. 1956.150). For a general view see fig. 108.

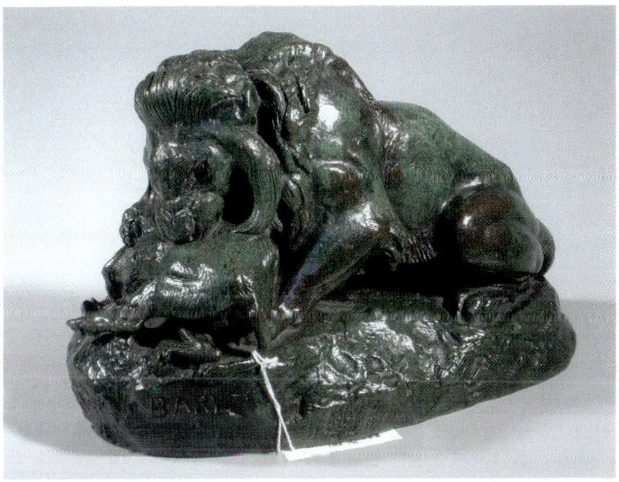

Figure 108 This work was cast in several parts and then bolted together in the metal. The chasing perfectly conceals any trace of the assemblage, which only becomes apparent from the underside (see fig. 107). Antoine-Louis Barye (French, 1795–1875), *Lion Devouring a Boar*, ca. 1865, H. 16.9 cm (Harvard Art Museums / Fogg Museum, The Henry Dexter Sharpe Collection, inv. 1956.150).

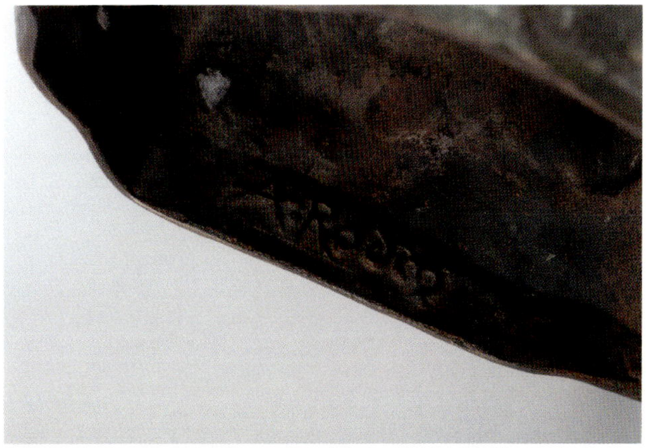

Figure 109 Inscription on the inner surface of the bust: "A. Rodin" stands for Auguste Rodin. It has been marked on the sand core and thus appears in relief on the bronze. Auguste Rodin (French, 1840–1917), *Victor Hugo*, reduction sand casting by Eugène Rudier (French, 1875–1952), 1883–92 for the reduction, cast in November 1916, H. 38.4 cm (Galerie Univers du Bronze, Paris).

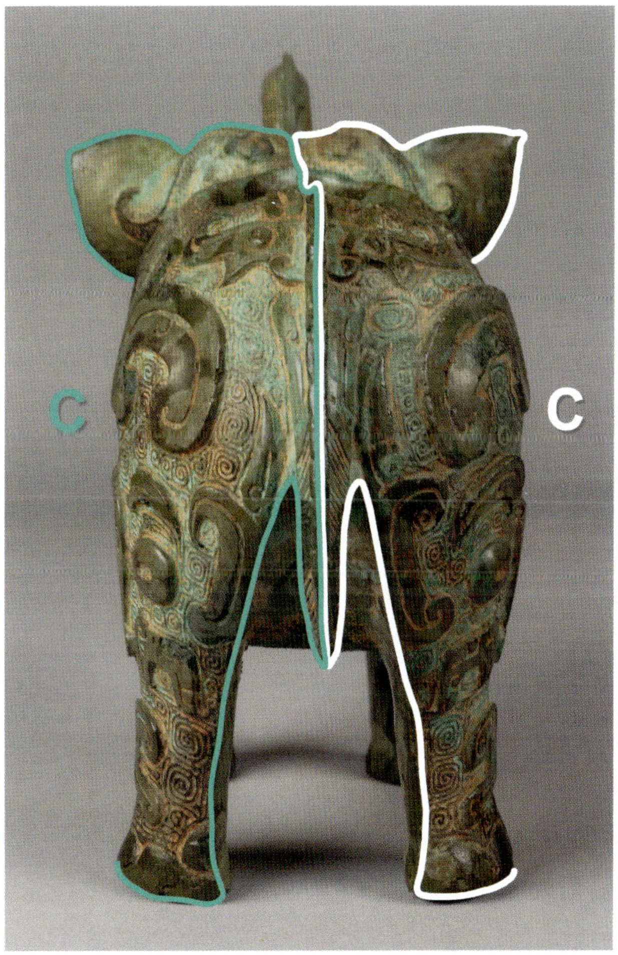

Figure 110 Annotated back view without the lid indicating the location of the seam lines. *Huo Ritual Vessel*, 11th century BCE, Middle Yangtze Valley, China, H. 17.2 cm (Freer Gallery of Art, Smithsonian Institution, Washington, DC, purchase—Charles Lang Freer Endowment, inv. F1936.6a–b / Department of Conservation and Scientific Research).

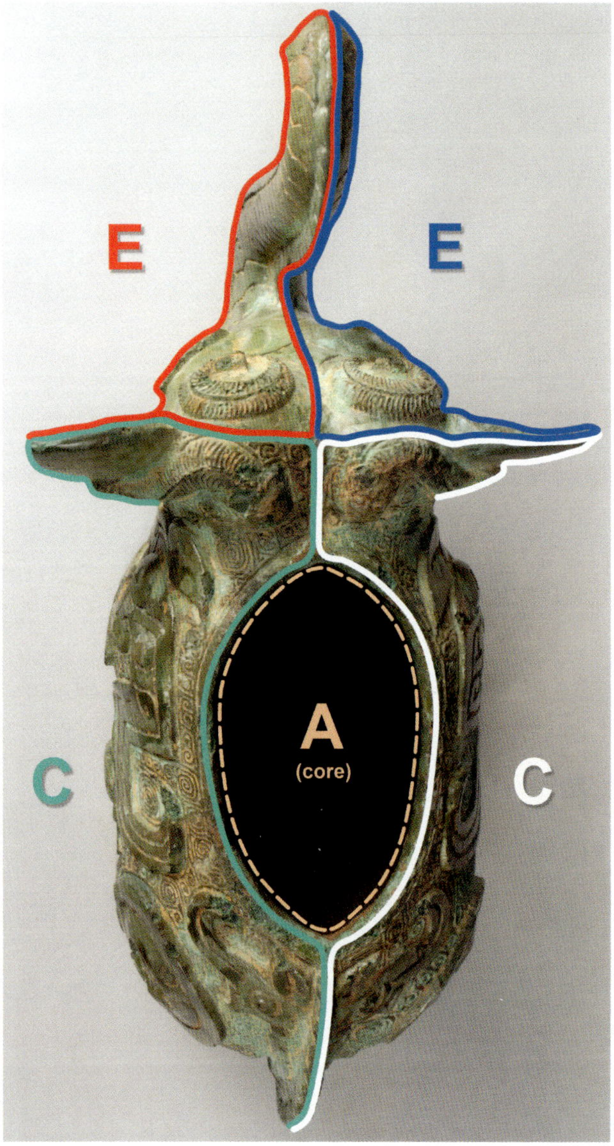

Figure 111 Annotated top view without the lid, indicating the location of the seam lines. *Huo Ritual Vessel*, 11th century BCE, Middle Yangtze Valley, China, H. 17.2 cm (Freer Gallery of Art, Smithsonian Institution, Washington, DC, purchase—Charles Lang Freer Endowment, inv. F1936.6a–b / Department of Conservation and Scientific Research).

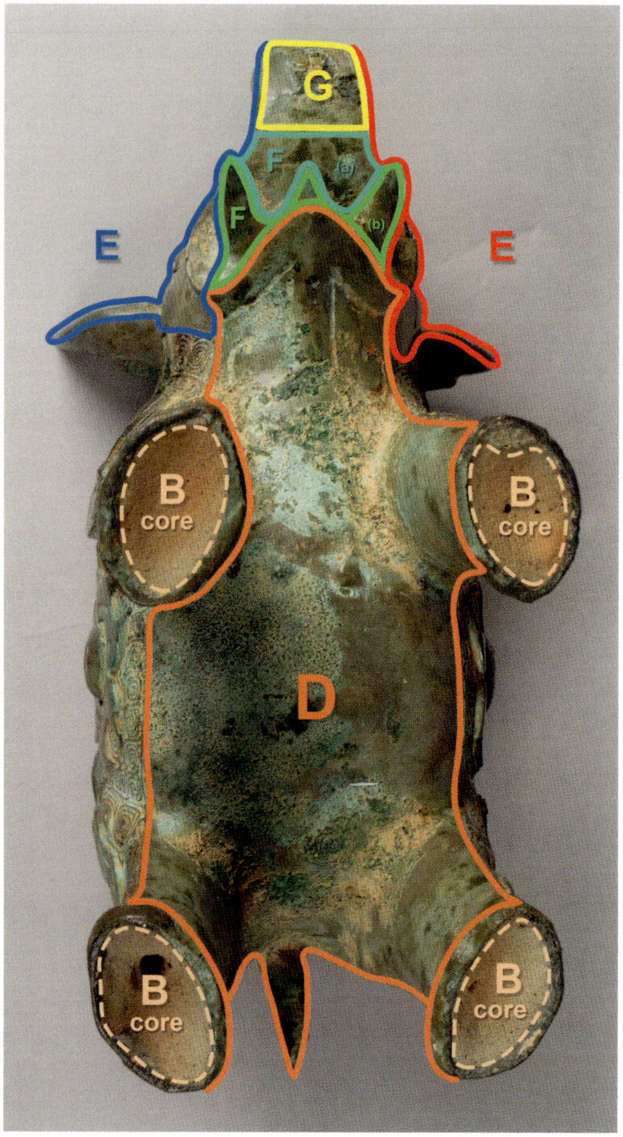

Figure 112 Annotated view of the underside indicating the location of the seam lines created at the joints of the loess piece mold sections. Note remnants of core material in the legs. *Huo Ritual Vessel*, 11th century BCE, Middle Yangtze Valley, China, H. 17.2 cm (Freer Gallery of Art, Smithsonian Institution, Washington, DC, purchase—Charles Lang Freer Endowment, inv. F1936.6a–b / Department of Conservation and Scientific Research).

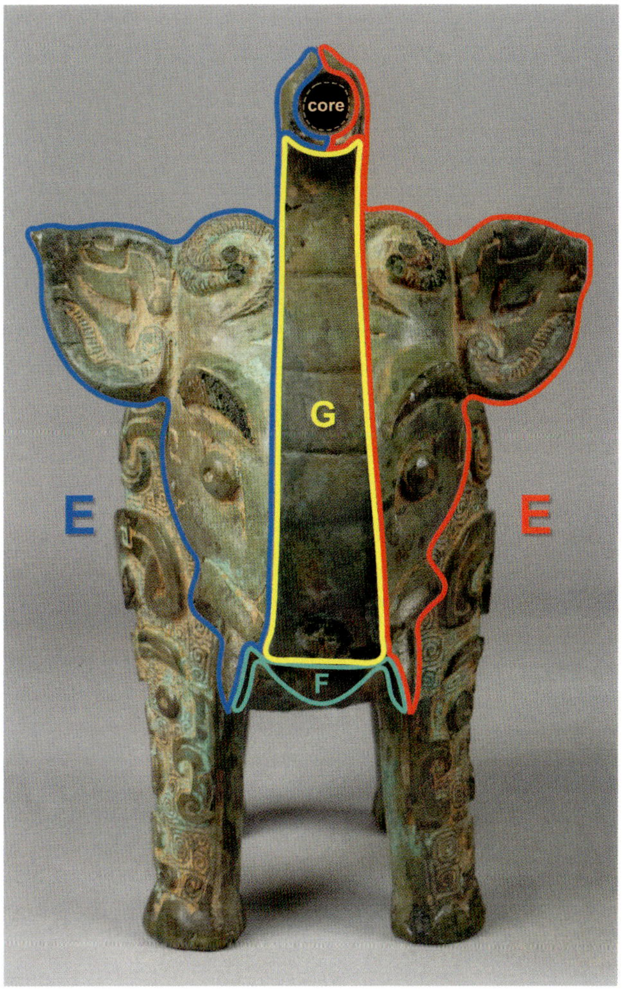

Figure 113 Annotated frontal view indicating the location of the seam lines. *Huo Ritual Vessel*, 11th century BCE, Middle Yangtze Valley, China, H. 17.2 cm (Freer Gallery of Art, Smithsonian Institution, Washington, DC, purchase—Charles Lang Freer Endowment, inv. F1936.6a–b / Department of Conservation and Scientific Research).

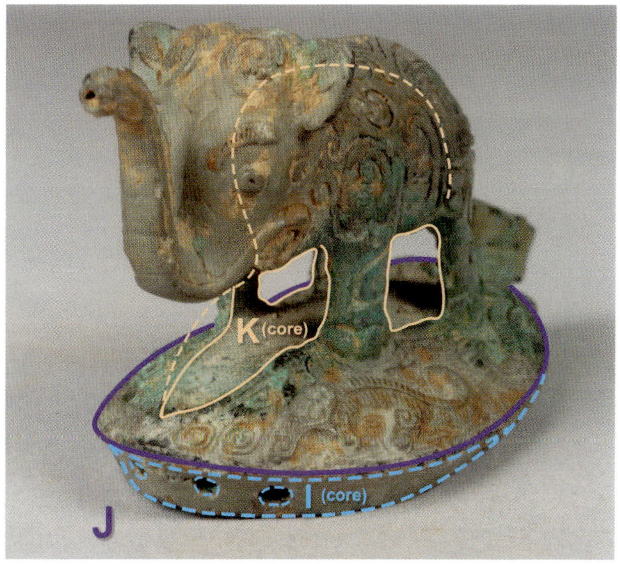

Figure 114 Annotated three-quarter view of the lid indicating the location of the seam lines. *Huo Ritual Vessel*, 11th century BCE, Middle Yangtze Valley, China, H. 17.2 cm (Freer Gallery of Art, Smithsonian Institution, Washington, DC, purchase—Charles Lang Freer Endowment, inv. F1936.6a–b / Department of Conservation and Scientific Research).

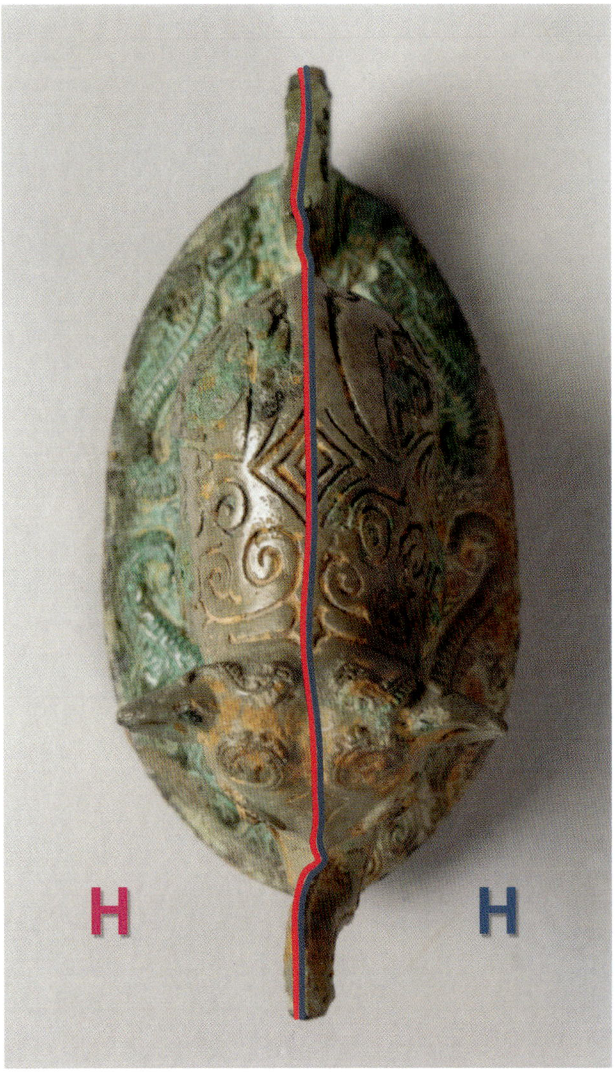

Figure 115 Annotated view from above of the lid indicating the location of the central seam line. *Huo Ritual Vessel*, 11th century BCE, Middle Yangtze Valley, China, 17.2 cm (Freer Gallery of Art, Smithsonian Institution, Washington, DC, purchase—Charles Lang Freer Endowment, inv. F1936.6a–b / Department of Conservation and Scientific Research).

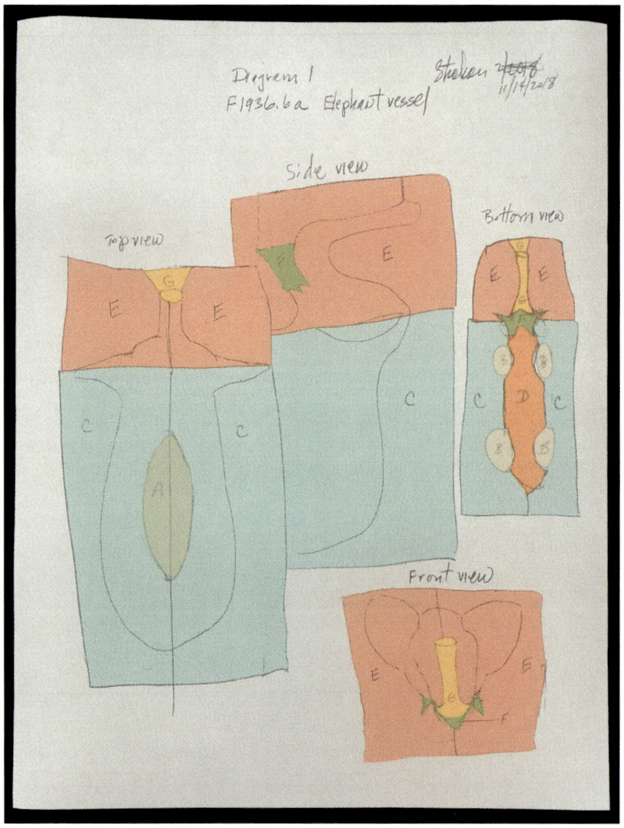

Figure 116 Conservator's working drawings mapping the piece mold sections used to create the *huo*. *Huo Ritual Vessel*, 11th century BCE, Middle Yangtze Valley, China, H. 17.2 cm (Freer Gallery of Art, Smithsonian Institution, Washington, DC, purchase—Charles Lang Freer Endowment, inv. F1936.6a–b / Department of Conservation and Scientific Research).

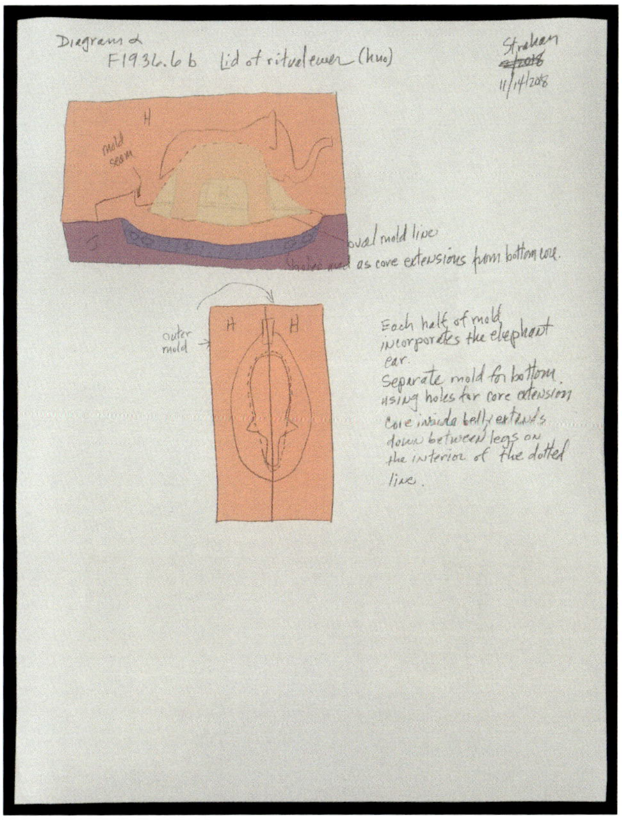

Figure 117 Conservator's working drawings mapping the piece mold sections used to create the *huo*. *Huo Ritual Vessel*, 11th century BCE, Middle Yangtze Valley, China, H. 17.2 cm (Freer Gallery of Art, Smithsonian Institution, Washington, DC, purchase—Charles Lang Freer Endowment, inv. F1936.6a–b / Department of Conservation and Scientific Research).

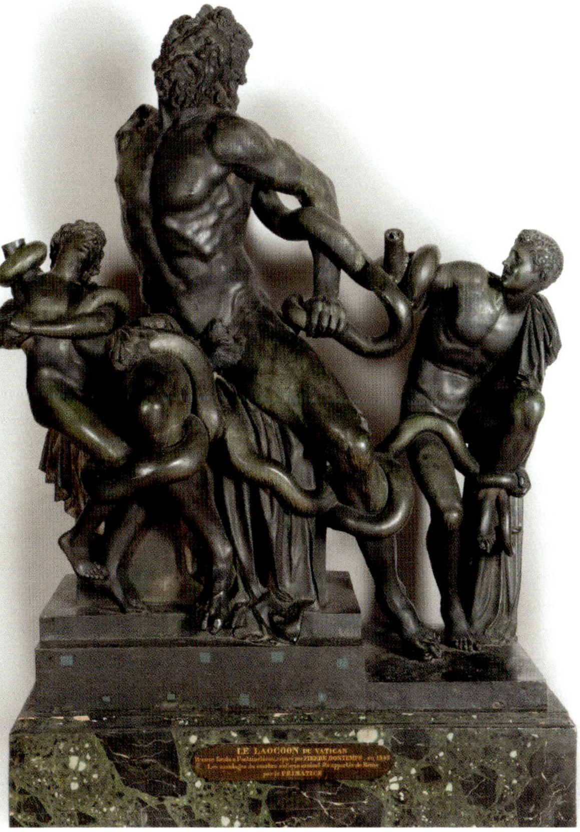

Figure 119 The repair patches linked to the armatures used for casting the main section of the bronze pedestal are visible on the external surface (blue squares). Francesco Primaticcio (Italian, 1504–1570), *Laocoön and His Sons*, 1542, H. 191 cm (Musée national du château de Fontainebleau, France, inv. MR 3290). See Castelle 2016.

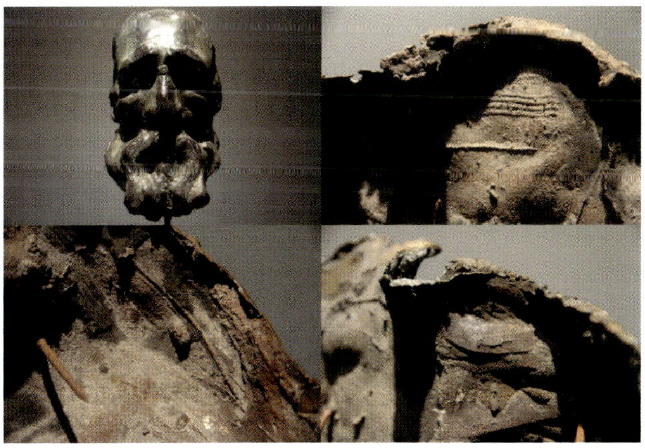

Figure 118 Overall view and details of the inside of the sculpture showing tool marks that indicate the core has been shaved down. This was done using a coarse-toothed tool, hence the parallel lines (top right). Other singular lines were used either to define the depth that was needed to cut back or to sketch the design or form of the head/features directly onto the core (bottom right). A flat spatula was also used to scrape back the surface of the core (bottom left). Andrew Lacey (British, b. 1969), untitled head, cast by the artist in 2017, H. 31 cm (artist's collection).

VISUAL ATLAS

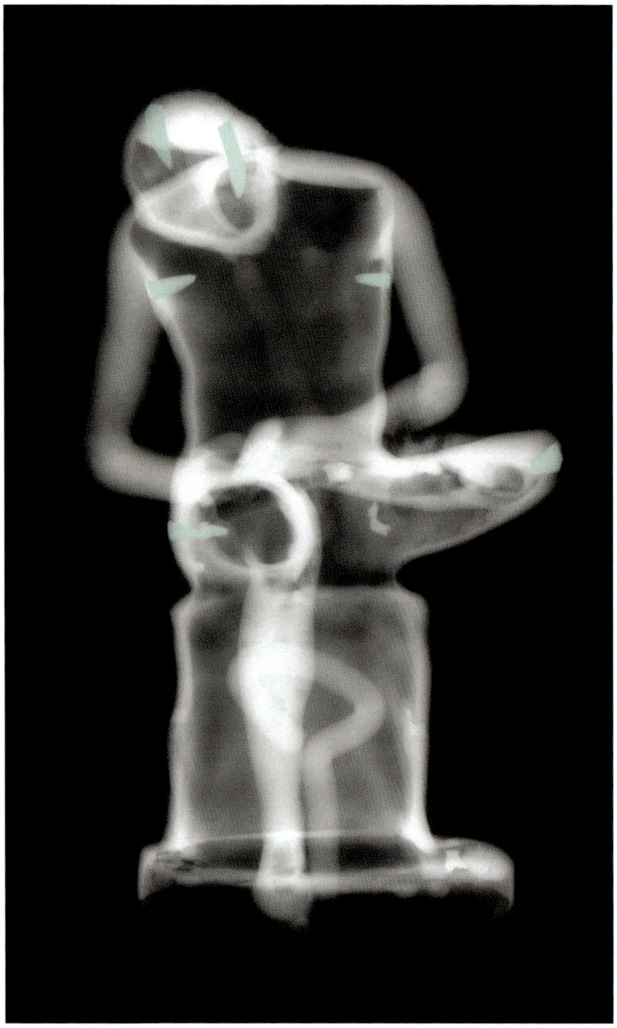

Figure 120 Annotated X-radiograph shows the core pins under the shoulders (turquoise overlays). *Spinario*, northern Italy, first quarter of the 16th century, H. 15.5 cm (Musée des Beaux-Arts d'Angers, France, inv. 2003.1.188). See C2RMF Internal Report 2018b.

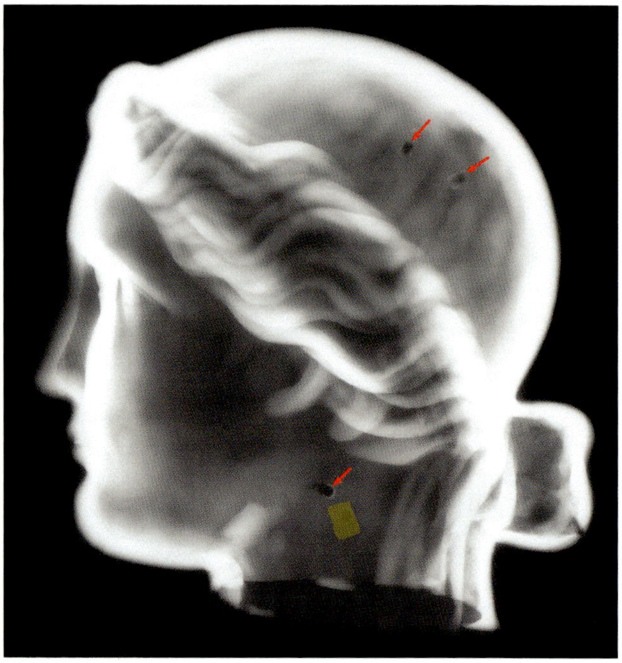

Figure 121 X-radiographic sequence (viewable in online edition) of a two-thirds-scale bronze reproduction of the head of the *Apollo of Lillebonne* showing the core-pin holes that have been left unplugged (black spots) and an attempt to repair one of these holes: a rectangular shape has been carved around the hole in the metal but has been left unpatched (dark-gray rectangular shape on the left side of the neck). Two-thirds-scale reproduction using the indirect wax-slab process during the 2016 CAST:ING meeting, Coubertin foundry, France. Original: *Apollo of Lillebonne*, France, 1st century BCE–1st century CE, H. 193 cm (Musée du Louvre, inv. Br37 [NIII65]). See fig. 288.

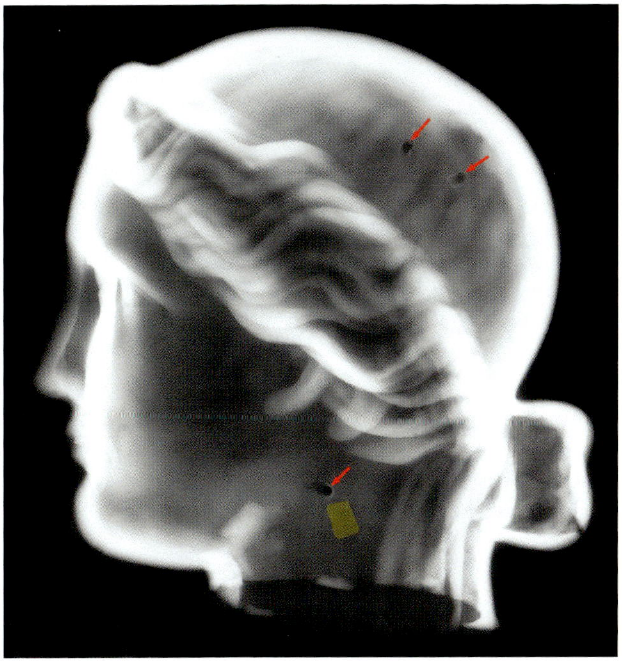

Figure 121 X-radiographic sequence (viewable in online edition) of a two-thirds-scale bronze reproduction of the head of the *Apollo of Lillebonne* showing the core-pin holes that have been left unplugged (red arrows) and an attempt to repair one of these holes: a rectangular shape has been carved around the hole in the metal but has been left unpatched (yellow overlay). Two-thirds-scale reproduction using the indirect wax-slab process during the 2016 CAST:ING meeting, Coubertin foundry, France. Original: *Apollo of Lillebonne*, France, 1st century BCE–1st century CE, H. 193 cm (Musée du Louvre, inv. Br37 [NIII65]). See fig. 288.

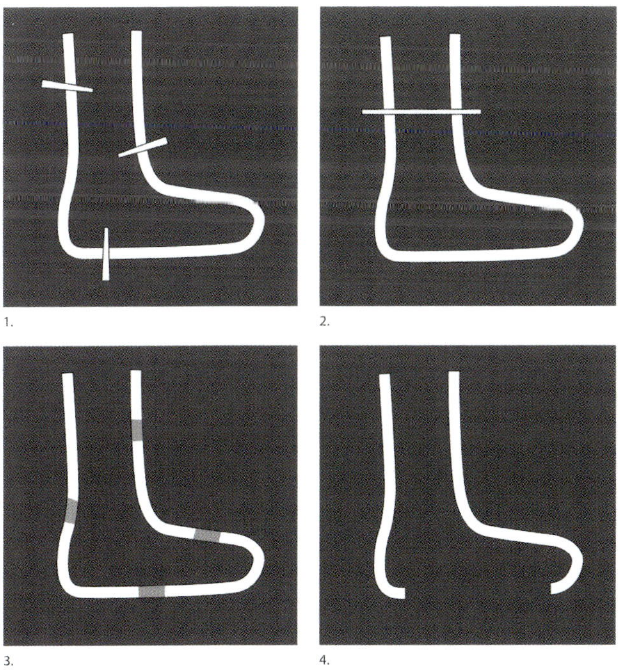

Figure 122 Four types of core supports used to hold the core in place during the pour: 1) core pins; 2) transverse core pin; 3) chaplets; 4) mold extension.

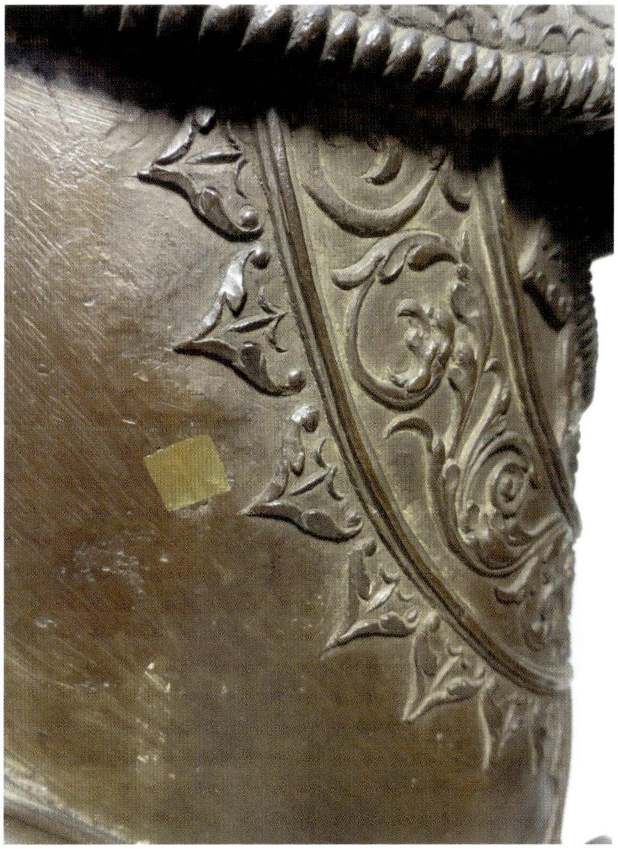

Figure 123 Detail of proper left side of torso. The square cast-on patch (yellow overlay) is an original repair of a hole left during casting by the presence of a mold extension. After Leone Leoni (Italian, 1509–1590), *Emperor Charles V, Innsbruck*, second half of the 16th century, H. 109 cm (National Gallery of Art, Washington, DC, Samuel H. Kress Collection, inv. 1952.5.104). See Smith and Sepponen 2019.

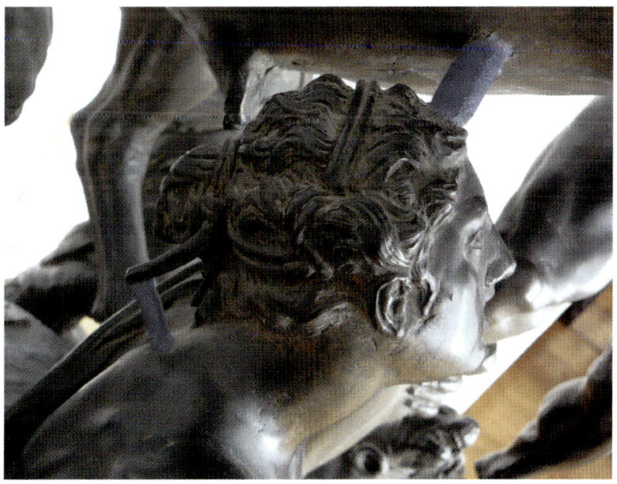

Figure 124 Detail showing the sprue remains shaped as rolls of wax preserved on the top and back of the head (see overlay). The artist left these intact on the finished bronze as subtle evidence that the complex sculpture was cast in one piece—an extraordinary feat. Adriaen de Vries (Netherlandish, 1556–1626), *Farnese Bull*, 1614, H. 103.5 cm (Schlossmuseum, Gotha, Germany, inv. P 50). See Bewer 2001.

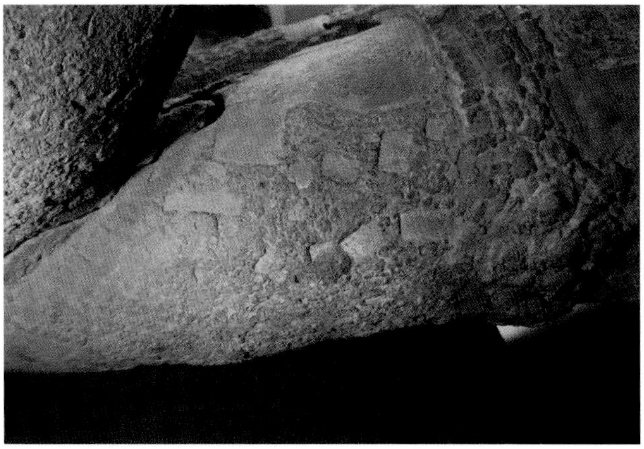

Figure 125 Patches used to repair casting flaws on the right arm are made of unalloyed copper. They were less corroded during burial than the surrounding tin bronze. This phenomenon is probably due to differences of metallurgical state and subsequent porosity: the patches are hammered and much less porous than the primary cast. *Reclining Vishnu*, West Mebon (Khmer), 12th century, L. 220 cm (National Museum of Cambodia, inv. Ga 5387). See CAST:ING 2018.

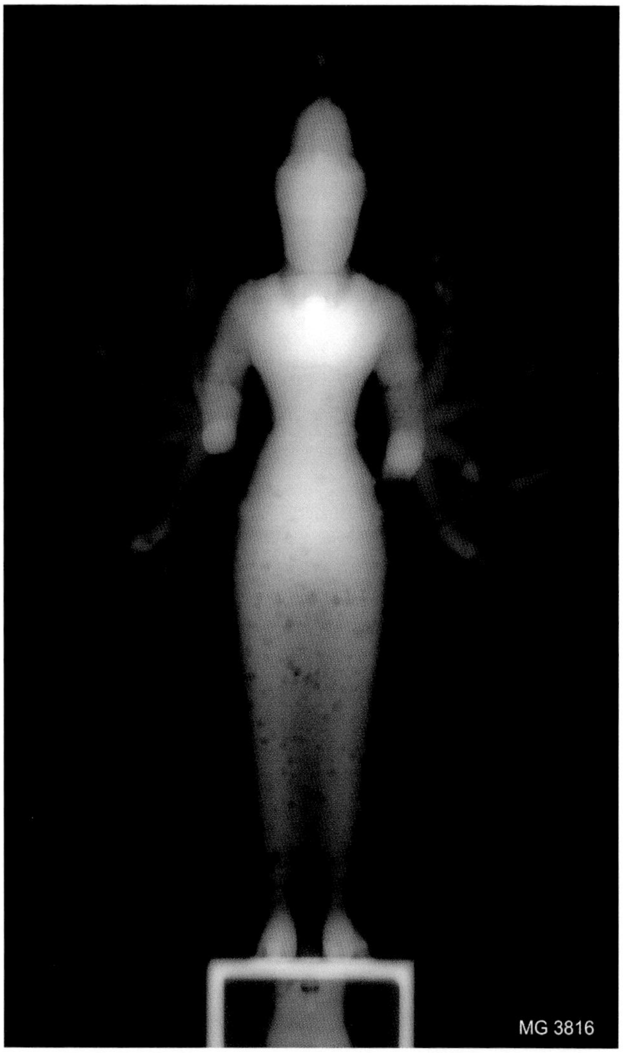

Figure 126 X-radiograph of frontal view, showing that the figure is cast solid. The concentration of porosity in the legs supports the hypothesis of an upside-down casting. *Ten-Armed Avalokiteśvara*, Java, first half of the 9th century, H. 34 cm (Musée national des arts asiatiques - Guimet, Paris, inv. MG 3816). See Mechling et al. 2018.

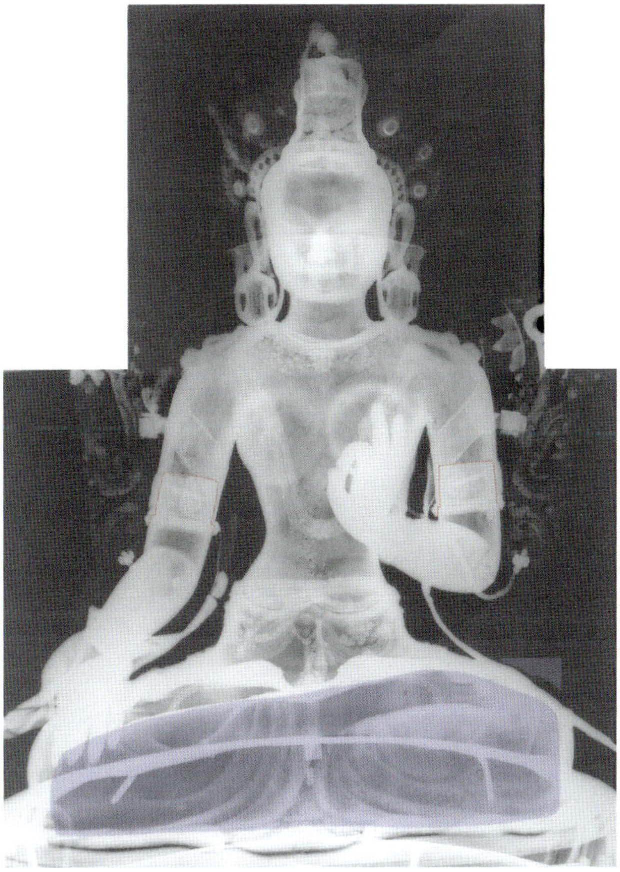

Figure 127 Annotated X-radiograph. Porosity at the bottom may indicate that the sculpture was cast upside down (blue overlay). Note also the Roman joints (red lines). *White Tara*, Tibet, 17th century, H. 65 cm without base (Musée national des arts asiatiques - Guimet, Paris, inv. MA 12495). See C2RMF Internal Report 2012.

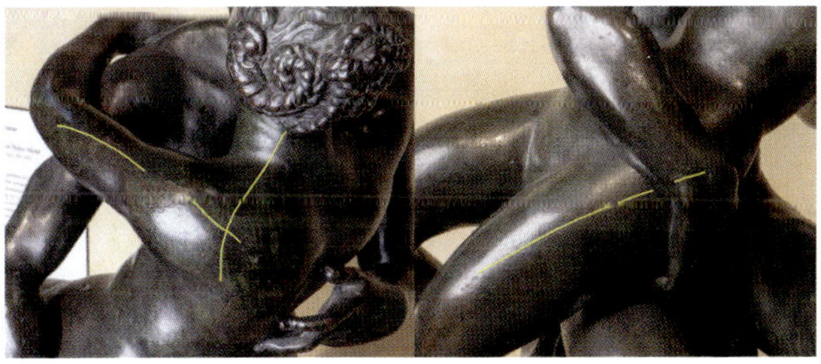

Figure 128 Fine lines scratched through the paint-like patina are from mold making, so-called *surmoulage* lines. Attributed to Hans Mont (Flemish, ca. 1545–after 1585), *Mars and Venus*, 1580, H. 53.9 cm (J. Paul Getty Museum, inv. 85.SB.75).

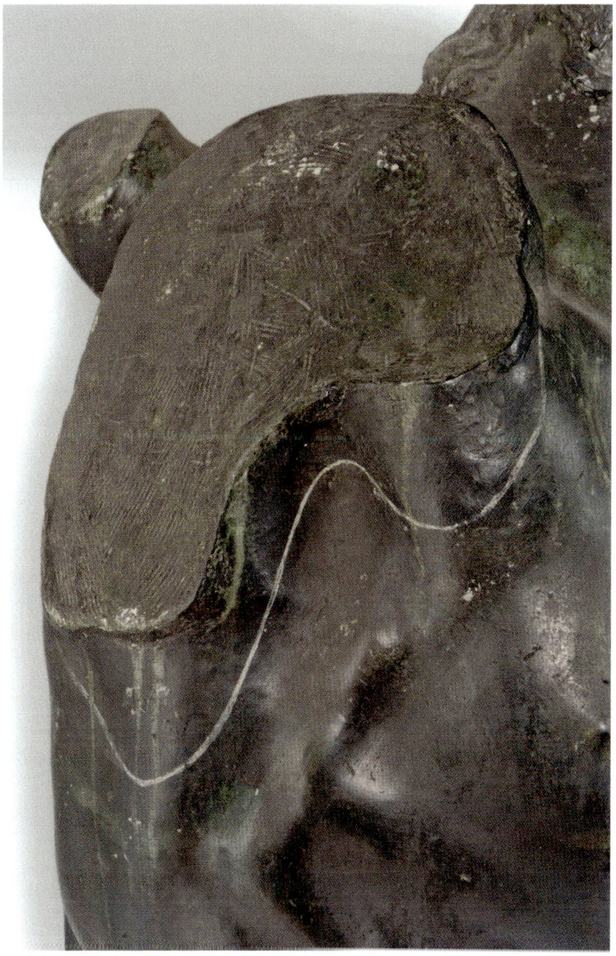

Figure 129 Detail of right shoulder showing white lines, probably drawn before molding. Francesco Primaticcio (Italian, 1504–1570), *Laocoön and His Sons*, 1542, H. 191 cm (Musée national du château de Fontainebleau, France, inv. MR 3290). See Castelle 2016. See fig. 119 for an overall view.

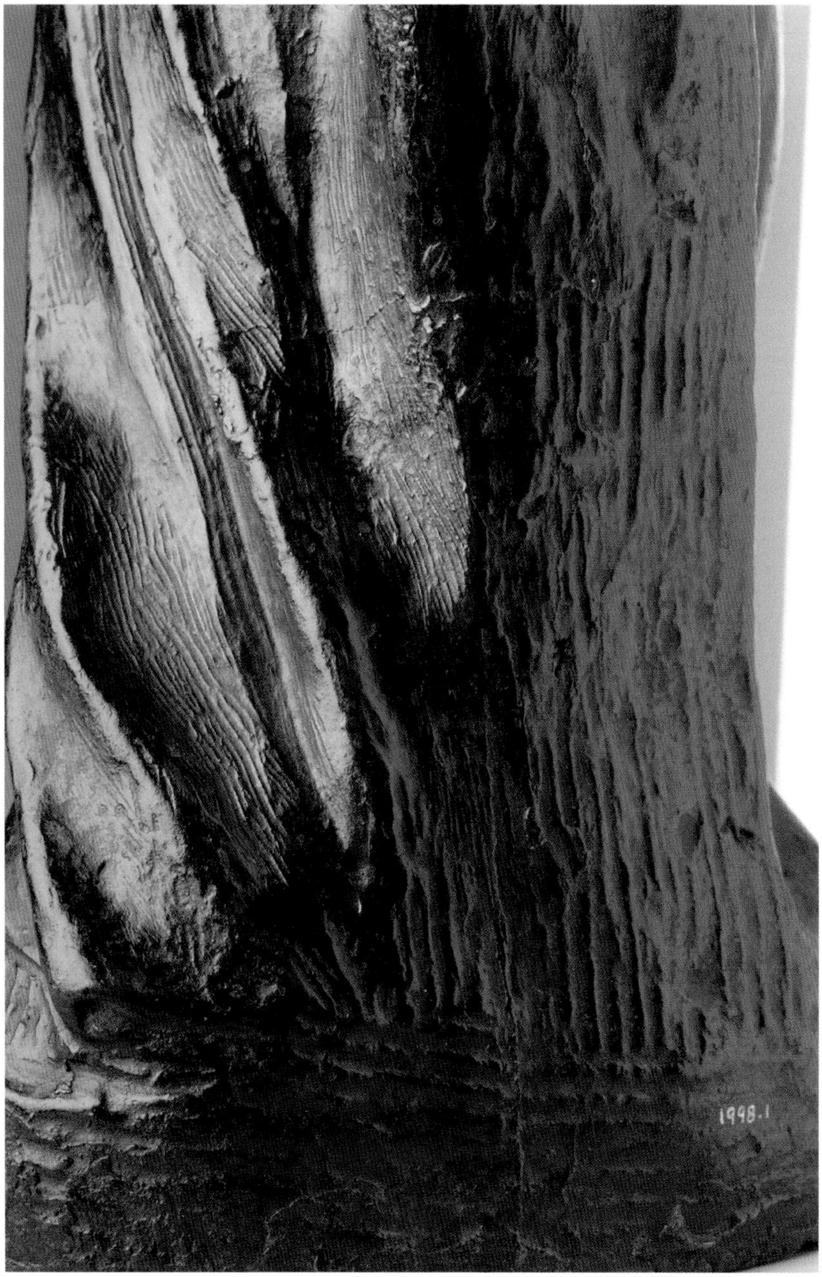

Figure 130 Detail of the rear of the Matilda statuette showing different tool marks. The wider striations on the right and bottom were most likely made in the clay model. The finer parallel lines coming down on the left at a diagonal in the fold appear to have been made at the wax stage with a finer-toothed tool. After a model by Gian Lorenzo Bernini (Italian, 1598–1680), *Countess Matilda of Tuscany*, ca. 1633–34, H. 40 cm (Harvard Art Museums / Fogg Museum, partial gift of Max Falk and partial purchase through the Director's Acquisition Fund, 1998.1). See Bewer 1999. See fig 258 for an overall view.

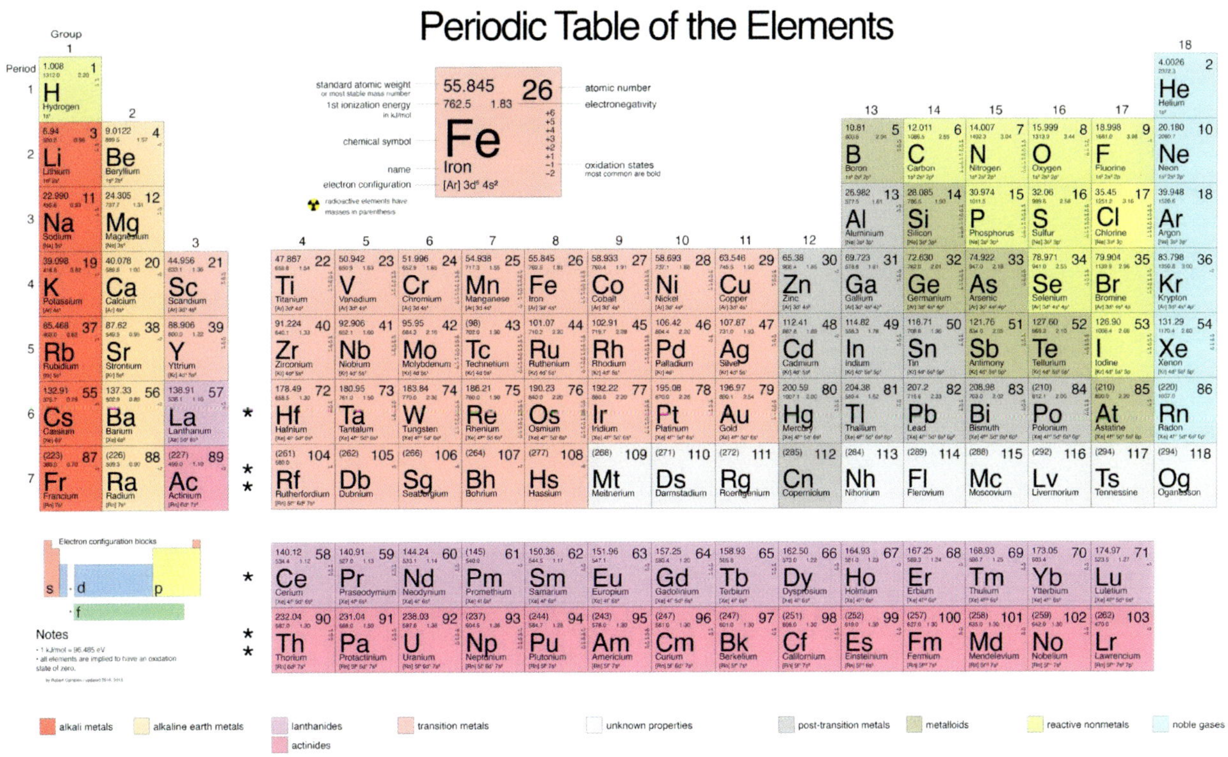

Figure 131 Periodic table of the elements.

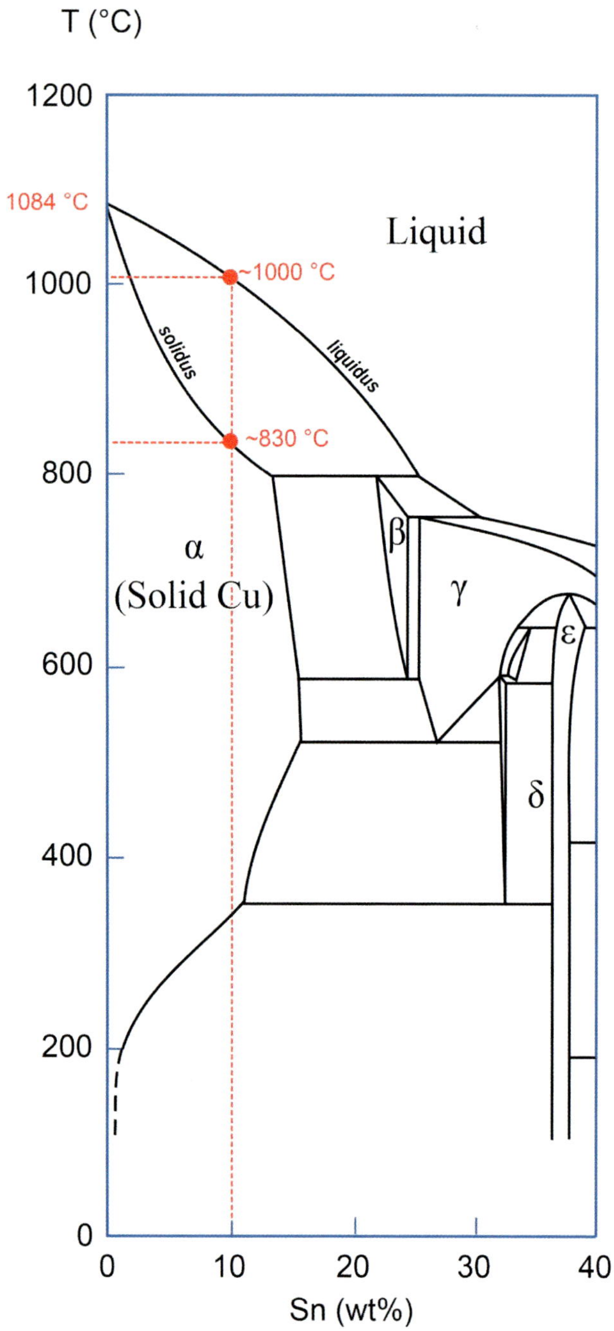

Figure 132 A detail of the equilibrium phase diagram of the copper-tin (Cu-Sn) system shows the different phases that may exist in the microstructure of such an alloy depending on the temperature (vertical axis), and the weight percentage of tin in the alloy (horizontal axis). In pure copper, Cu atoms are aligned in a face-centered cube: this is the alpha (α) phase (see fig. 139). The delta (δ) phase forms when there is more than ~20 wt% tin. This face-centered cubic phase is responsible for the specific acoustic properties of bells. The diagram also indicates melting temperatures depending on the alloy composition. For example, an alloy with 10% tin melts fully at ~1000°C, once the liquidus line is passed, as shown by the red dotted lines. It solidifies fully at ~830°C, once the solidus line is crossed.

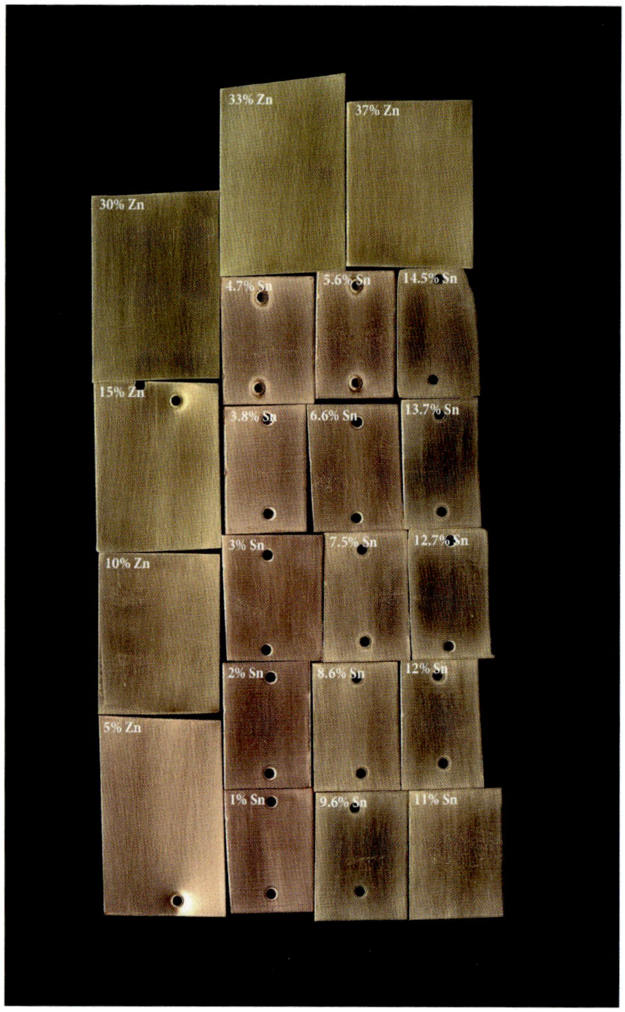

Figure 133 Cast coupons made of tin bronzes (smaller coupons) and brasses showing that zinc has a greater impact on the color than tin. The coupons were prepared at KME Italy in Fornacidi Barga. The image of the polished samples was created on a flatbed scanner.

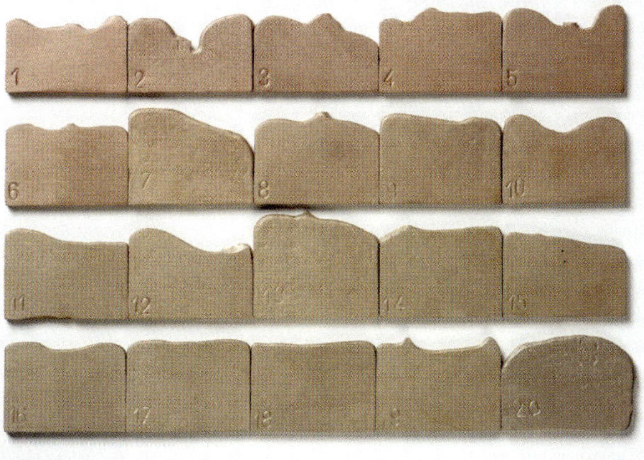

Figure 134 Synthetic samples of tin bronzes from 1 wt% tin (top left) to 20 wt% tin (bottom right), showing the evolution of color. Bronze plates, 5 × 6 cm, were cast by Frank Willer/LVR-Landesmuseum Bonn as part of the experimental archaeology for a research project on Roman large bronzes (see Willer, Schwab, and Mirschenz 2017). The pieces were polished with pumice stone to enable a clearer perception of the actual color of the alloy with as little reflection as possible. Photograph using neutral studio strobes.

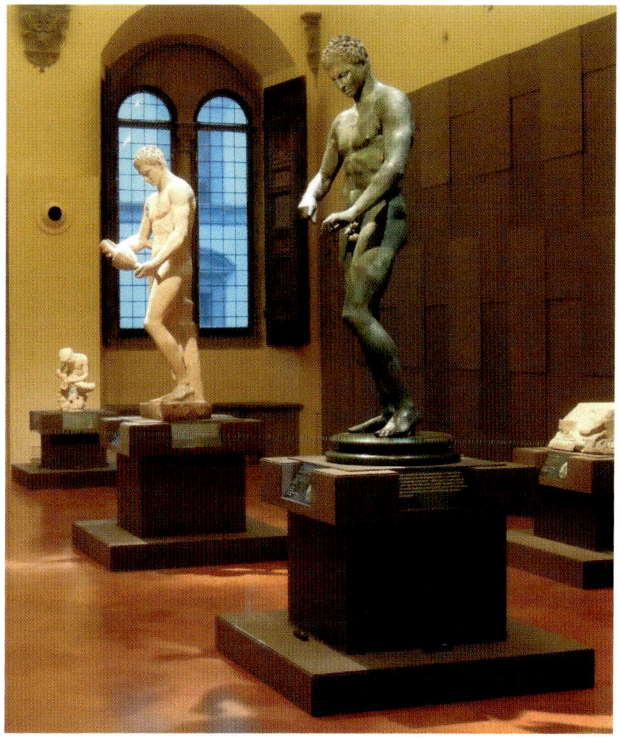

Figure 136 Side-by-side view of the Ephesian bronze Athlete, Roman, copy after Greek original, last quarter 4th century BCE, H. 193 cm (Kunsthistorisches Museum Vienna, Collection of Greek and Roman Antiquities, inv. VI 3168), and the Florentine marble Apoxyomenos, 2nd century CE, H. 193 cm (Gallerie degli Uffizi, Florence, inv. 100). Both appeared the 2015 exhibition *Power and Pathos: Bronze Sculpture of the Hellenistic World*, Palazzo Strozzi, Florence. Although the extension of the limbs is less dramatic than on other statues, the two arms are secured by a strut on the marble statue but not on the bronze.

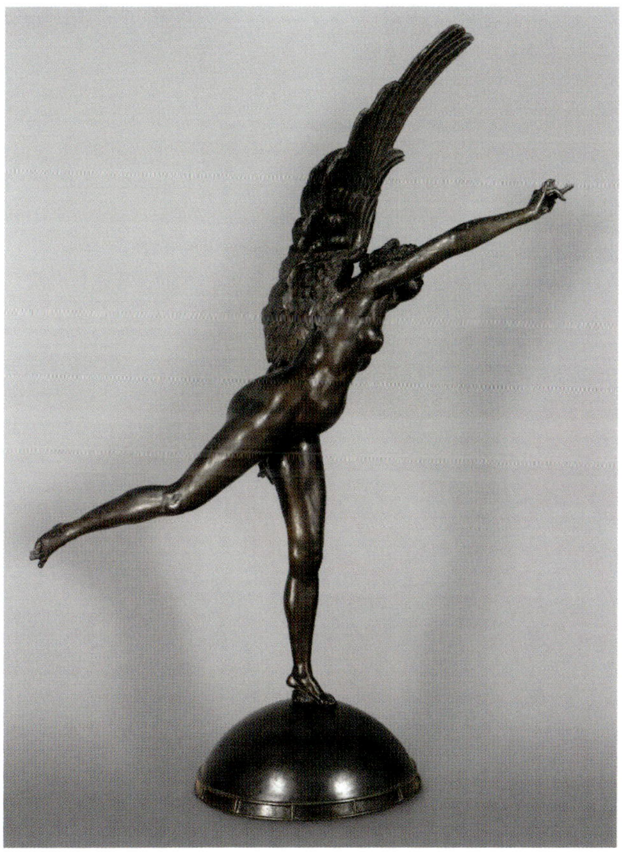

Figure 135 Example of how the mechanical strength of bronze allows large offsets (here, the right leg and arm). Pierre Biard I (French, 1559–1609), *Fame*, ca. 1607, H. 134 cm (Musée du Louvre, inv. LP361). See C2RMF Internal Report 2017.

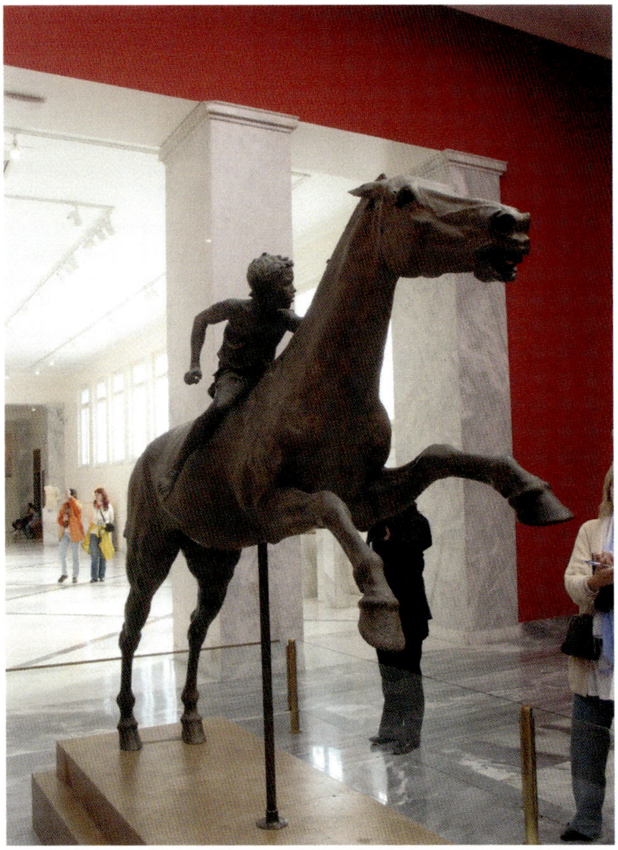

Figure 137 The armature for this sculpture was removed after casting. The Horse and Jockey from Artemision, found at Cape Artemision, Greece, ca. 150–146 BCE, H. 210 cm (National Archaeological Museum of Athens, inv. X15177).

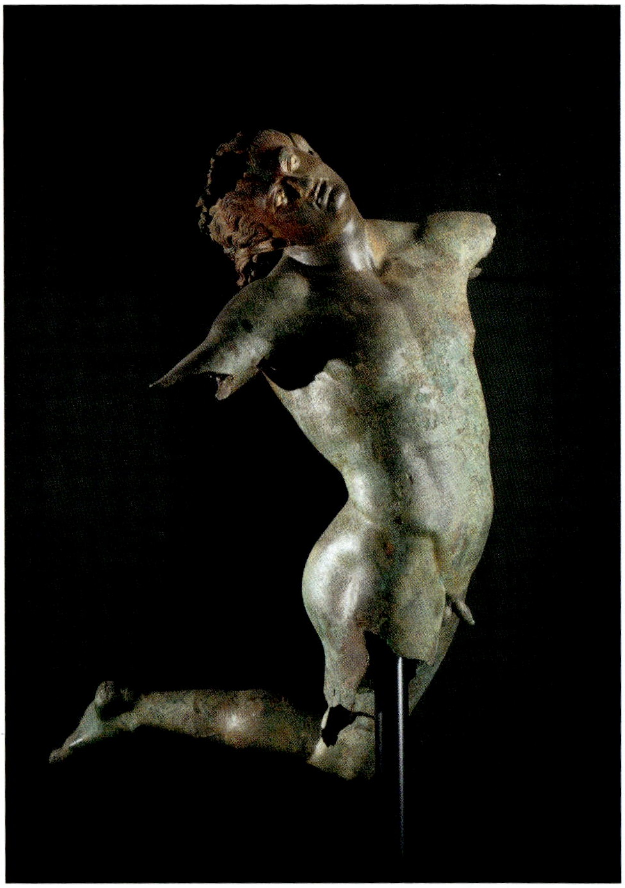

Figure 138 The original armature for this sculpture was removed after casting. A modern structural one has been added to support the fragmentary figure. *Dancing Satyr*, origin unknown, late 4th–mid 1st century BCE, H. 200 cm (Museo del Satiro, Mazara del Vallo, Italy).

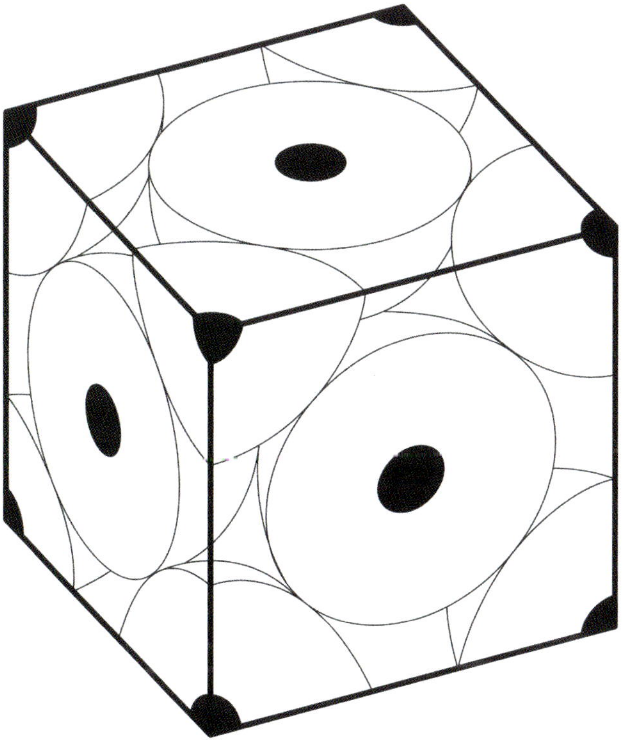

Figure 139 Diagram showing how atoms of copper are stacked within a cube. The black dots represent the cores of the atoms; they are located on each corner of the cube and at the center of each face. This crystalline structure is called face-centered-cubic (FCC). After Scott 1991, 1, fig. 2.

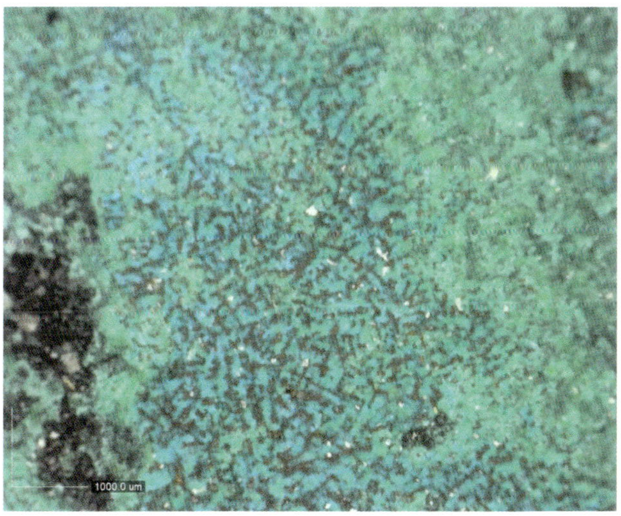

Figure 140 Digital micrograph showing a dendritic microstructure as seen on the corroded surface. Auguste Rodin (French, 1840–1917), *The Burghers of Calais*, 1885–89, cast in 1926 at the Alexis Rudier foundry by Eugène Rudier (French, 1875–1952), H. 217 cm (Musée Rodin, inv. S.00450).

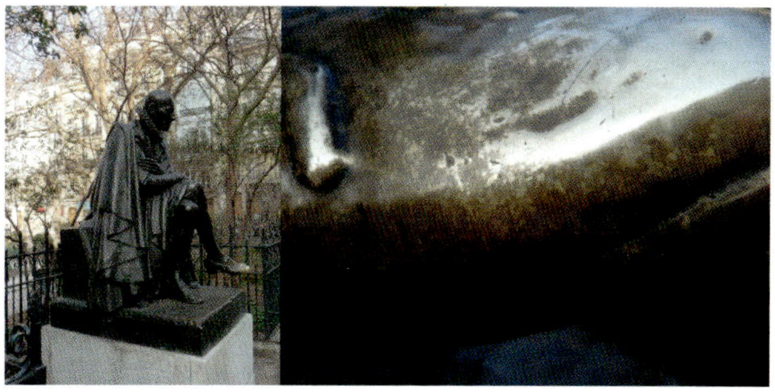

Figure 141 Touching the surface has removed the patina and revealed the grains of the microstructure of the cast metal by natural etching, as seen in the detail on the right. Paul Landowski (French, 1875–1961), *Michel de Montaigne*, ca. 1931, cast by Didier Landowski (French, b. 1940) at the Blanchet-Landowski foundry in 1989, H. 200 cm (Ville de Paris, place Paul-Painlevé).

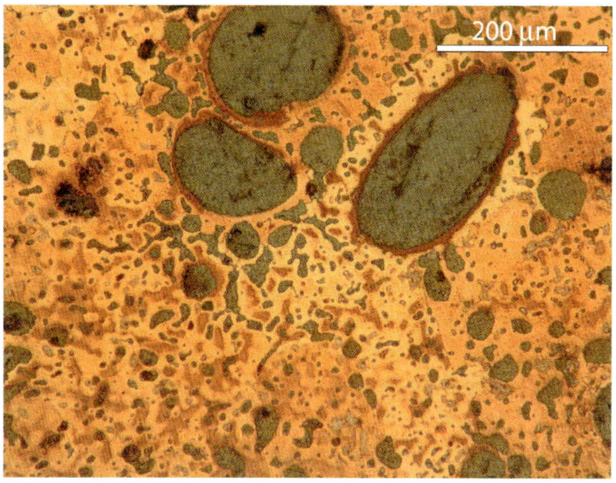

Figure 142 Bright field photomicrograph of an etched cross section showing the dendritic microstructure of the primary casting. The outlines of dendrites appear brown: the interdendritic space is filled with phase alpha + delta (eutectoïd composition) together with nodules of lead (gray oval shape) and various inclusions. *The Youth of Agde*, France, 1st century CE, H. 140 cm (Musée de l'Ephèbe et d'Archéologie Sous-Marine de la Ville d'Agde, France, inv. 839). See fig. 561 for a general view. See Mille 2012.

VISUAL ATLAS

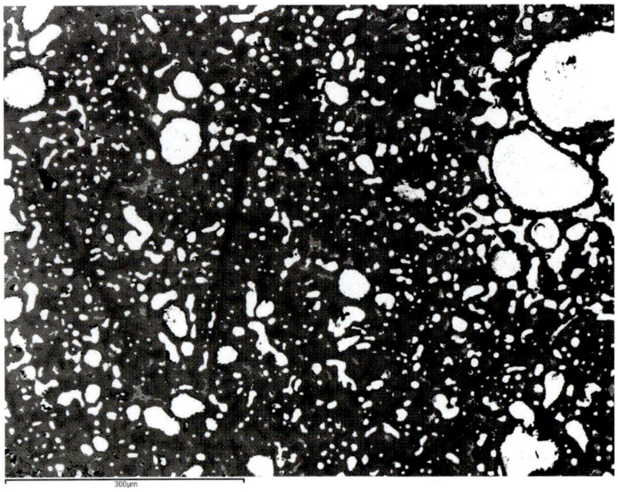

Figure 143 Backscattered electron micrograph of a polished cross section showing the dendritic microstructure of the primary casting. The outlines of dendrites appear dark gray; the interdendritic space is filled with phase alpha + delta (eutectoïd composition) and various inclusions (light gray) as well as nodules of lead (white oval shape). *The Youth of Agde*, France, 1st century CE, H. 140 cm (Musée de l'Ephèbe et d'Archéologie Sous-Marine de la Ville d'Agde, France, inv. 839). See fig. 561 for a general view. See Mille 2012.

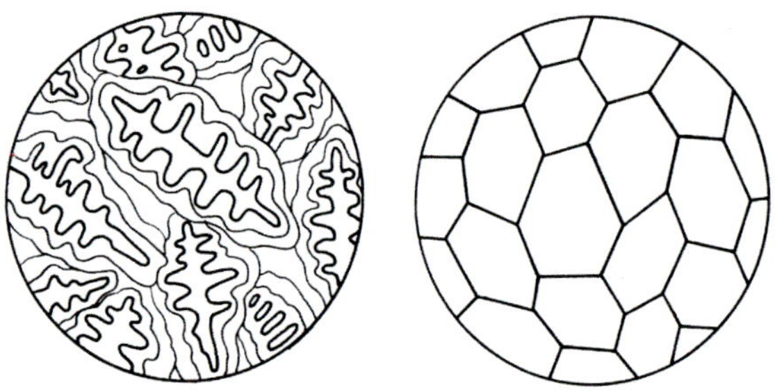

Figure 144 Diagrams of two types of common metallographic microstructures of copper alloys. Left: in the dendritic microstructure characteristic of a cast alloy, the dendrites are enclosed by the boundaries of forming grains; right: in the fully recrystallized microstructure, the hexagonal grains are equiaxed. From Scott 1991, 7, figs. 11a and 11c.

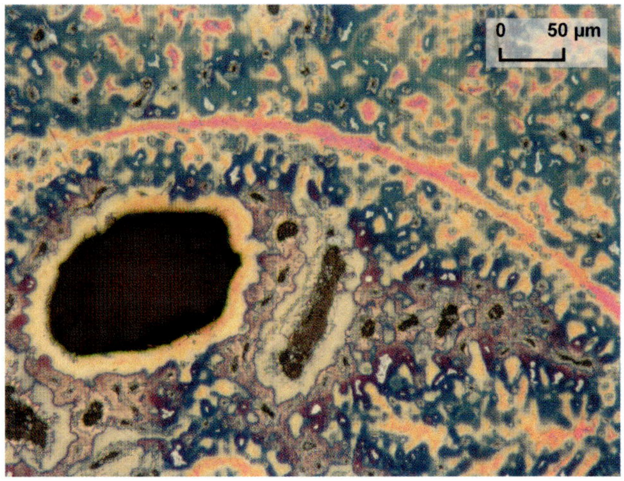

Figure 145 Bright field photomicrograph of an etched polished cross section of a repair filling a core-pin hole. The dendritic microstructure shows that the repair was cast on. The large curved dendrites of the alpha phase (yellow to pink) are marked by a strong primary chemical segregation (known as zoning) between the center of the dendrite (pink) and its periphery (yellow). *Colossal Foot of Clermont-Ferrand*, Roman (Gaule), early 2nd century CE, L. 58 cm (Musée de Clermont-Ferrand, France). See Darblade-Audoin and Tavoso 2008.

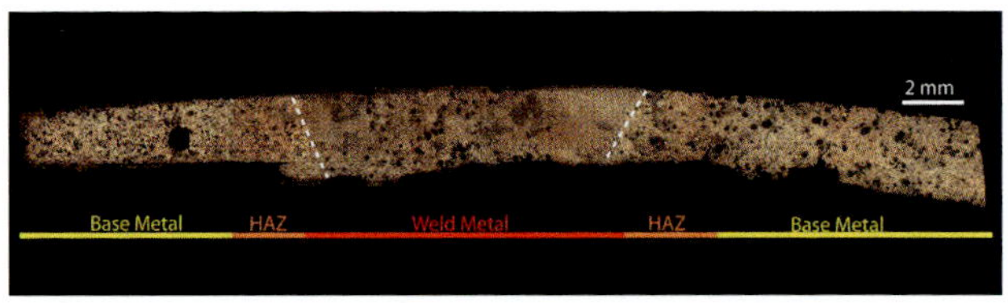

Figure 146 Bright field photomicrograph of an etched polished cross section of a weld on a large bronze fragment. Three zones may be distinguished: the primary cast (Base Metal), the heat-affected zone (HAZ), and the welding zone (Weld Metal). The joining zone is indicated (white dotted line). Fragment from the Hoard of Evreux, France, 1st century CE, L. 14 cm (Musée d'Evreux, France, inv. 4864). See Azéma et al. 2012; Azéma and Mille 2013a.

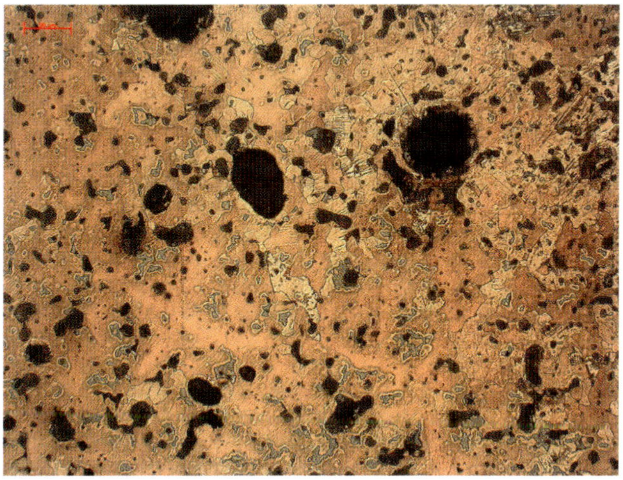

Figure 147 Bright field photomicrograph of an etched polished cross section of a weld on a large bronze fragment. Detail of the heat-affected zone (HAZ) due to flow fusion welding showing the recrystallized microstructure. Fragment from the Hoard of Evreux, France, 1st century CE, L. 14 cm (Musée d'Evreux, France, inv. 4864). See Azéma et al. 2012; Azéma and Mille 2013a.

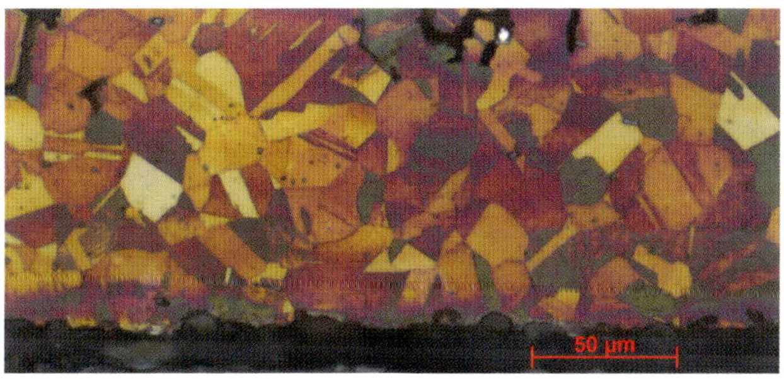

Figure 148 Bright field photomicrograph of an etched polished cross section of a Roman *cornu*. Grains are quite large (50–100 µm equivalent diameter) and homogeneous in size with a regular polygonal morphology. Thermal twins are frequent, as seen by the thick straight dark lines crossing the grains from one boundary to another. A number of features, including metal composition and microstructure, have proved to be very similar across the five investigated *cornua* from Pompeii, strongly suggesting that they were made by a single workshop. Very advanced technical skills (notably for tuning) are evident. The sample P20 studied here comes from the pavilion area of Cornu B1, Pompeii, 1st century CE, developed length of the Cornu 366 cm (Museo Archeologico Nazionale di Napoli, registered under #114261 in the January 1884 Pompeii excavation registry). See Caussé, Mille, and Tansu 2020.

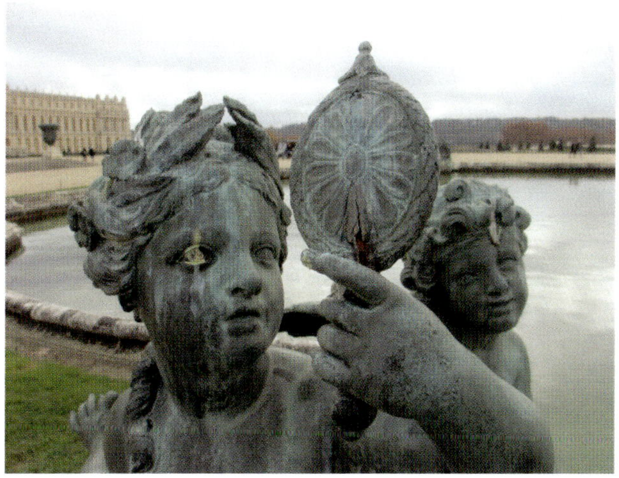

Figure 149 Detail showing how the expansion of the iron armatures due to corrosion has generated a crack in the mirror. Jean-Baptiste Poultier (French, 1653–1719), *Two Putti and a Girl Holding a Mirror*, ca. 1685, cast by Aubry, Bonvallet, Schabol, and Taubin in 1685–87, Paris, H. 145 cm (Musée National des châteaux de Versailles et de Trianon, France, inv. 1850.8925). See Maral, Bourgarit, and Amarger 2014 for the metal composition of the cast.

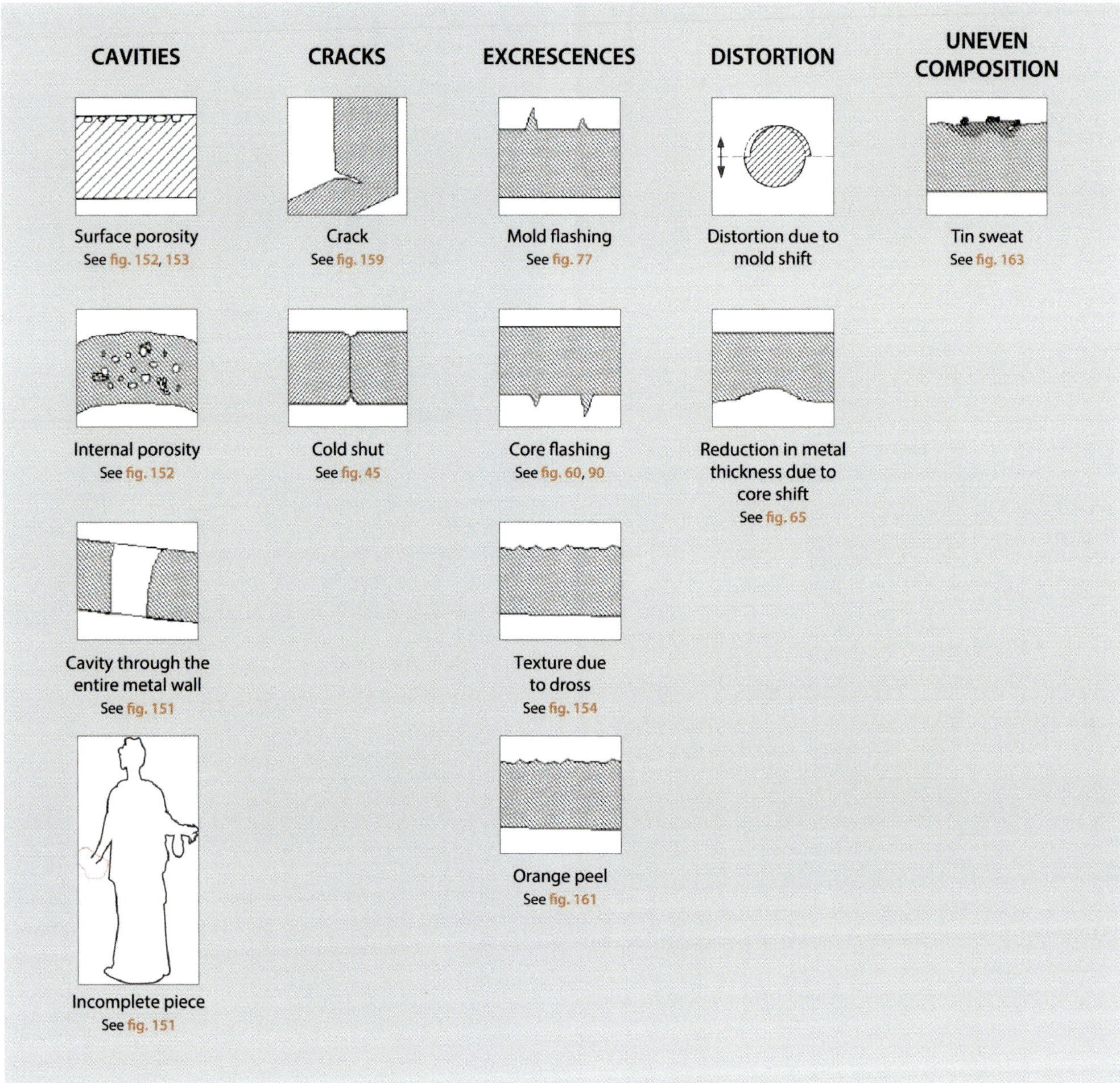

Figure 150 The five types of casting defects commonly encountered on bronze sculpture.

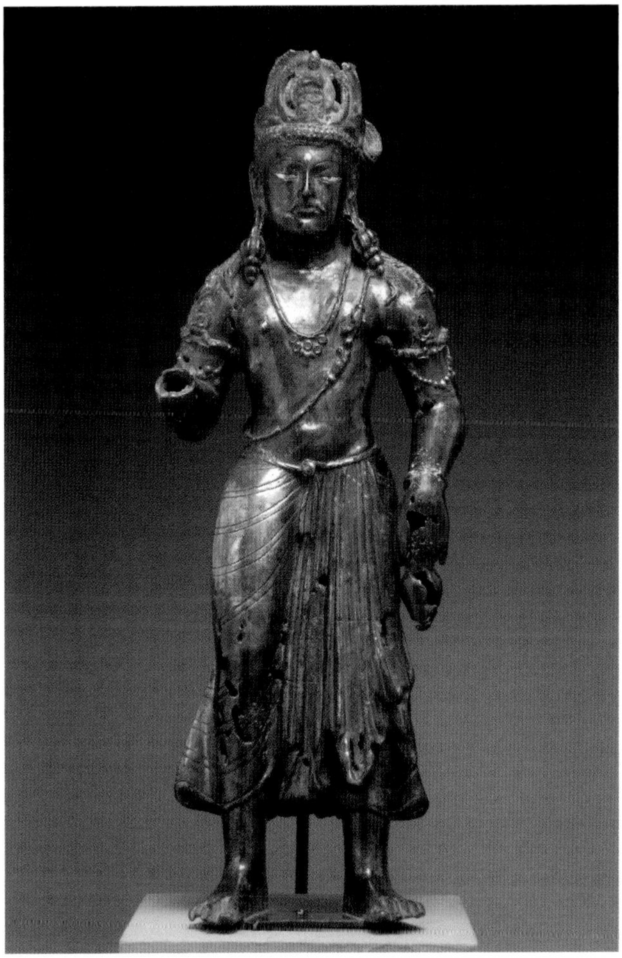

Figure 151 Example of heavy losses in an incomplete high-tin bronze cast: the right hand is missing, and the left hand and attribute are full of defects. *Bodhisattva Avalokitesvara*, Gandhara, 3rd century CE, H. 37 cm (Musée national des arts asiatiques – Guimet, Paris, inv. MAO 12128). See forthcoming C2RMF Internal Report.

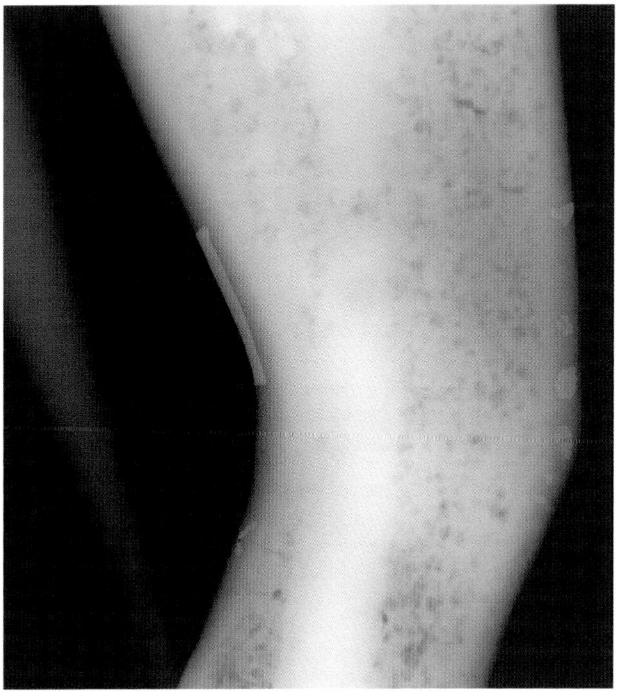

Figure 152 Annotated X-radiograph revealing extensive porosity. A large casting defect that extends partway through the surface of the figure's proper left leg has been repaired with a set-in patch (yellow overlay). In many areas, porosity is contained within the wall (some are indicated in green). Detail of X-radiograph of fig. 171, Jean-Antoine Houdon (French, 1741–1828), *Diana*, 1782, H. 208.3 cm (Huntington Art Museum, San Marino, California, inv. 27.186). See Bassett and Scherf 2014; Bennett and Sargentson 2008, 492–93. See fig.171 for an overall view.

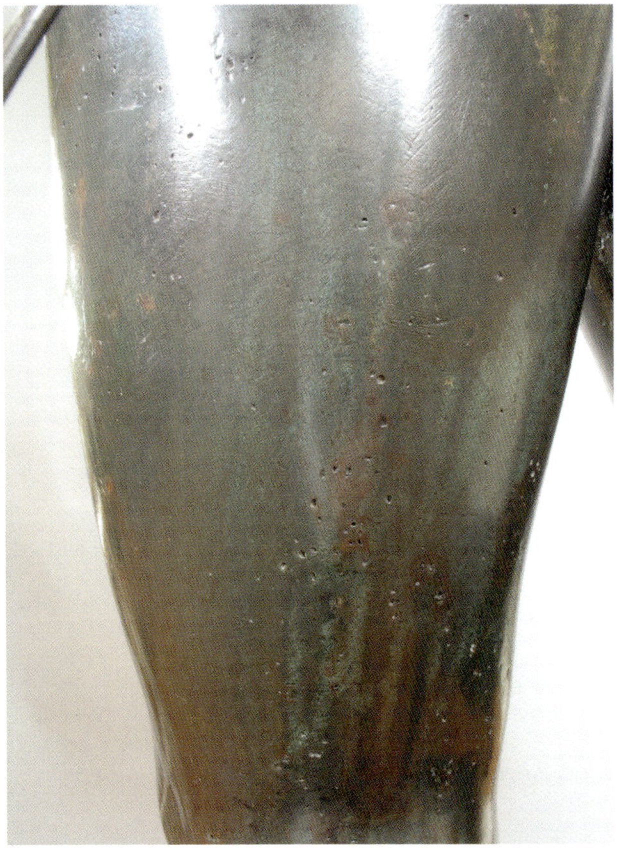

Figure 153 Porosity on the surface of the bronze has been left unrepaired. Detail of fig. 171, Jean-Antoine Houdon (French, 1741–1828), *Diana*, 1782, H. 208.3 cm (Huntington Art Museum, San Marino, California, inv. 27.186). See Bassett and Scherf 2014; Bennett and Sargentson 2008, 492–93.

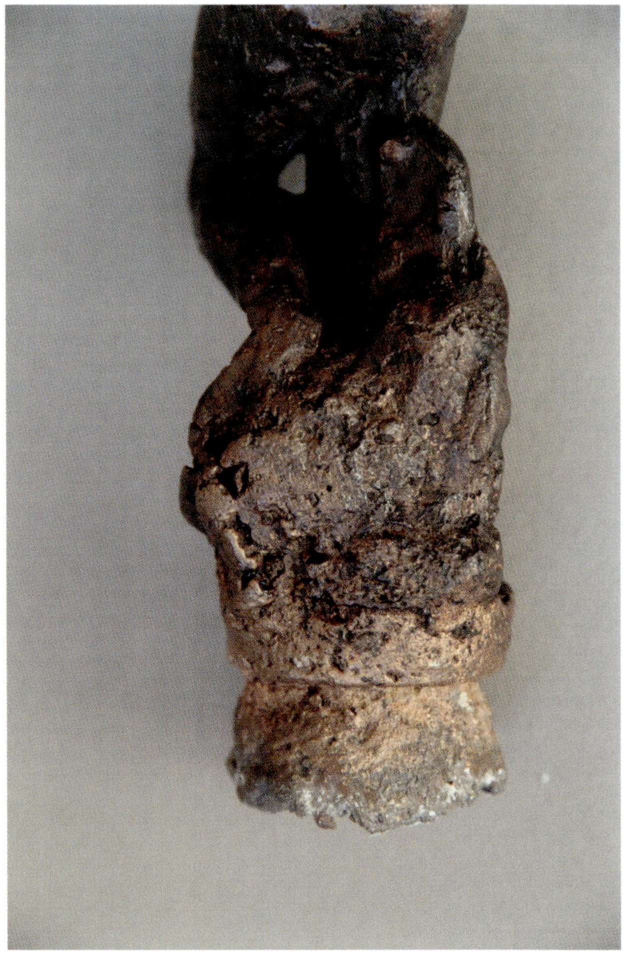

Figure 154 Example of defects on the surface near the base of a figure due to dross flowing into the mold with the melt. Andrew Lacey (British, b. 1969), hand-modeled copy in wax of Michelangelo, *Slave*, ca. 1516–19, H. 17.6 cm (Victoria and Albert Museum, inv. 4117-1854), used for private research.

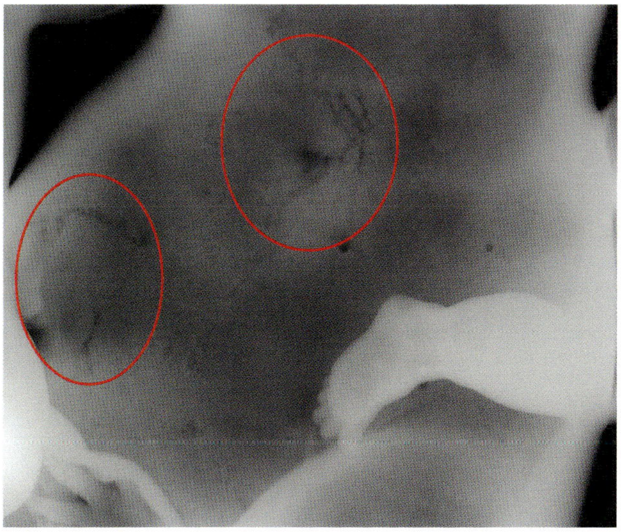

Figure 155 Annotated X-radiograph reveals shrinkage porosity in the stomach and proper right hip (circled). Adriaen de Vries (Netherlandish, 1556–1626), *Psyche Borne Aloft by Putti*, 1590–92, H. 187 cm (Nationalmuseum, Stockholm, inv. NM Sk352).

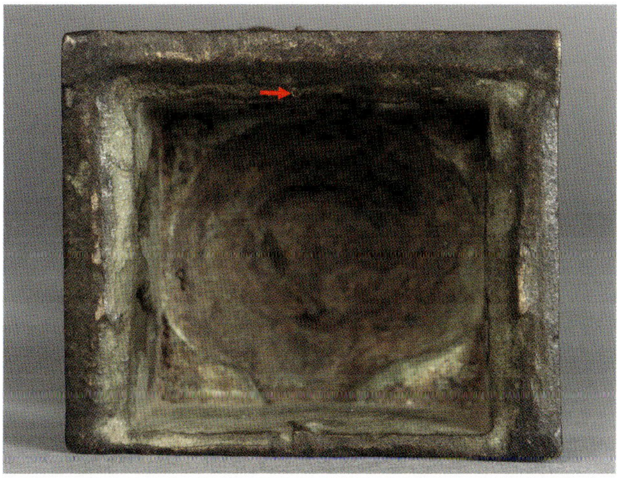

Figure 156 Bottom view of a statuette showing the 1 mm-diameter hole (red arrow) after metal sampling (drillings) for elemental analysis by ICP-AES. Kubera/Jambhala, Java, ca. late 9th–early 10th century, H. 17 cm (Musée national des arts asiatiques - Guimet, Paris, inv. MG 3619). See Mechling et al. 2018. See fig. 507 for an overall view.

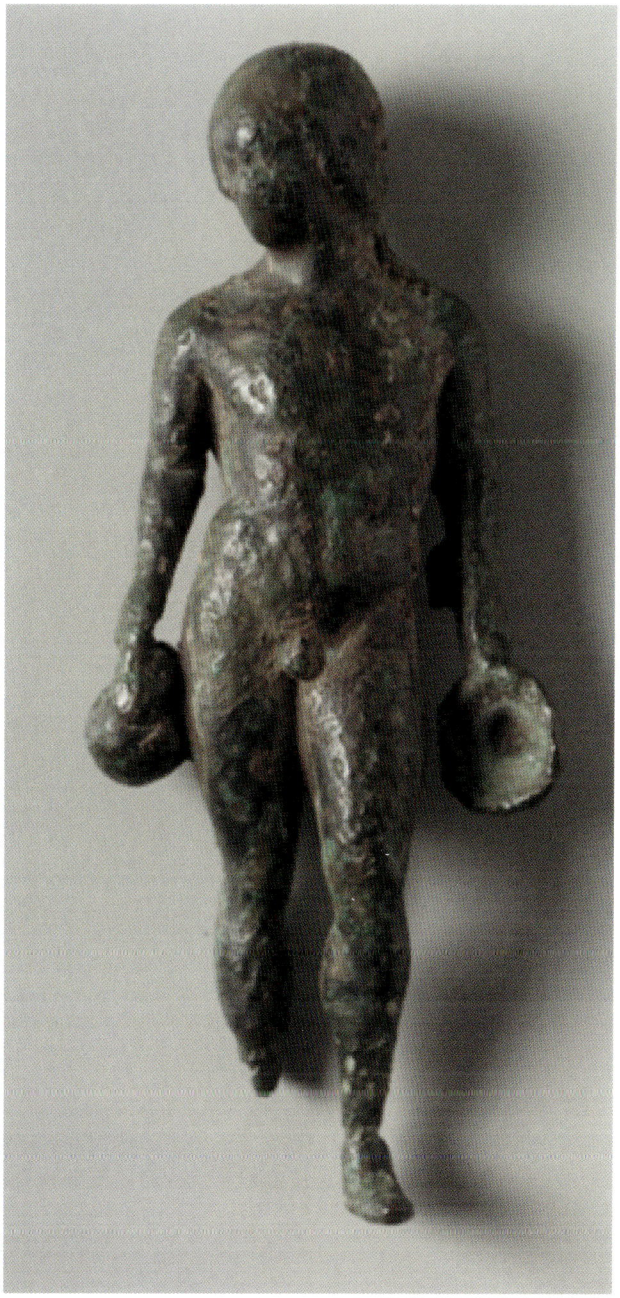

Figure 157 The many losses and worn appearance of the statuette are probably a result of ill-controlled electrolytic cleaning. *Youth Holding an Oinochoe and Strainer*, Greek, second quarter of the 5th century BCE, H. 11.8 cm (Harvard Art Museums / Arthur M. Sackler Museum, gift of Mrs. Edward Jackson Holmes, inv. 1952.20).

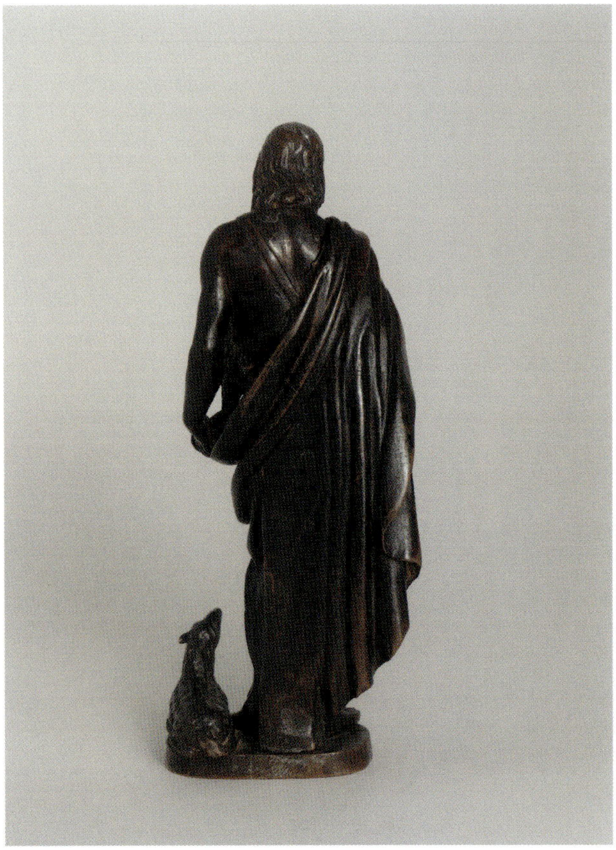

Figure 158 Back view of statuette showing numerous cracks. *Saint John the Baptist*, Italian (Venice), 16th century, H. 25 cm (Musée des Beaux-Arts d'Angers, France, inv. 2003-1-192). See C2RMF Internal Report 2018b.

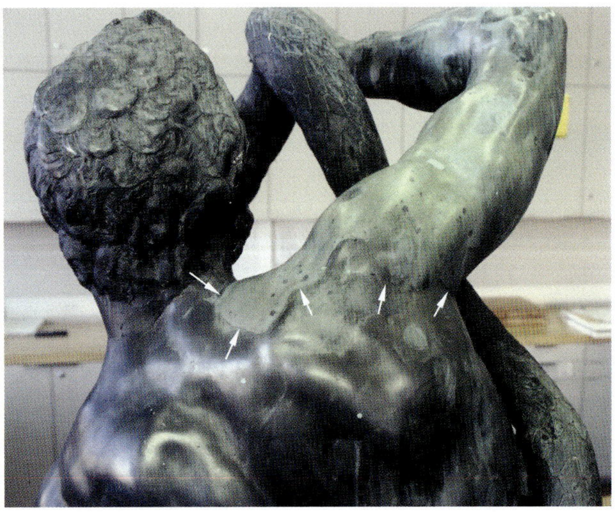

Figure 159 Detail of a cold shut on the proper right shoulder of the figure of Laocoön caused by premature cooling of the metal during the pour. Corrosion has accentuated the line where the cooling metal entered the mold from two different locations. Adriaen de Vries (Netherlandish, 1556–1626), *Laocoön and His Sons*, 1623, H. 169 cm (Nationalmuseum, Stockholm, inv. Drh Sk 68).

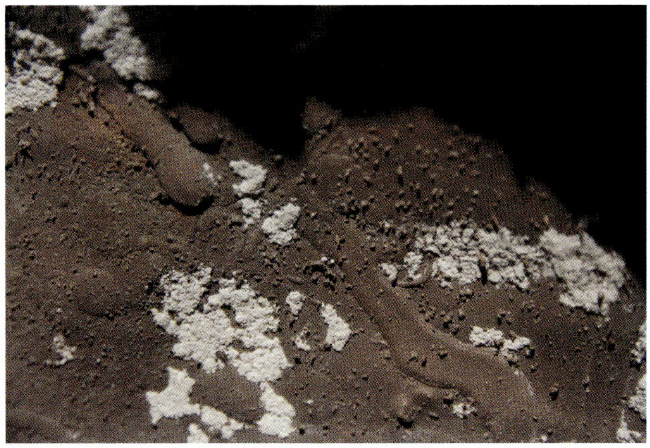

Figure 160 The breakdown of the core shows the surface of the investment being lost and causing a granular encrustation inside the bronze. Experimental reconstruction of a torso by Andrew Lacey (British, b. 1969), 2012 (private collection).

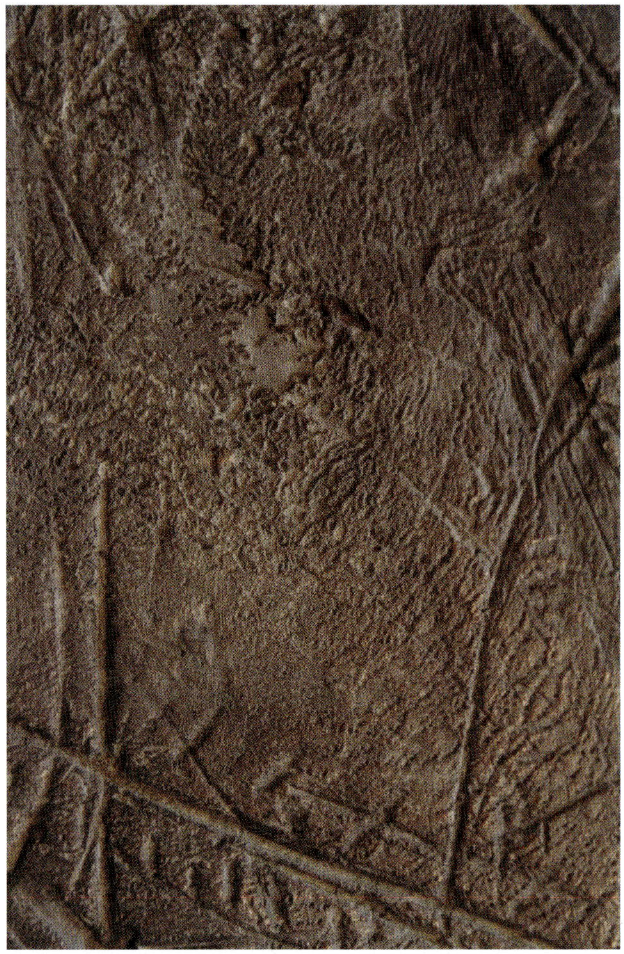

Figure 161 "Orange-peel" effect on the bronze surface of the base. The size of the detail is ~50 mm in the vertical direction. Siân Lewis (Welsh, b. 1964), *Tree Tableaux II*, cast by Andrew Lacey (British, b. 1969), 2017, H. 180 cm (artist's collection).

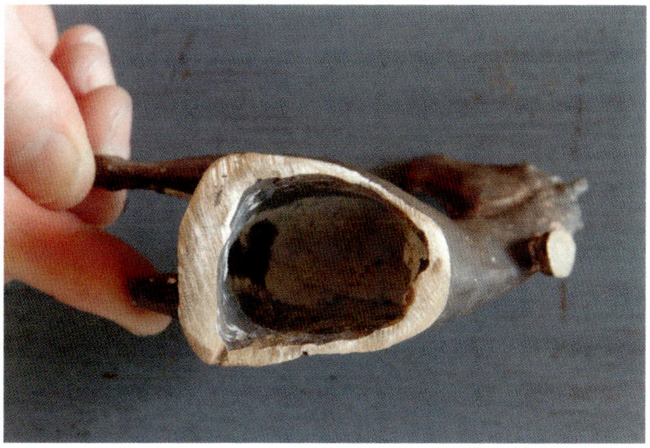

Figure 162 View of the cut-through section of a torso. During the metal pouring, the plaster core has floated upward, causing one side to be considerably thicker than the other. Andrew Lacey (British, b. 1969), maquette for *Iconoclash*, 2020, H. 13.5 cm (artist's collection).

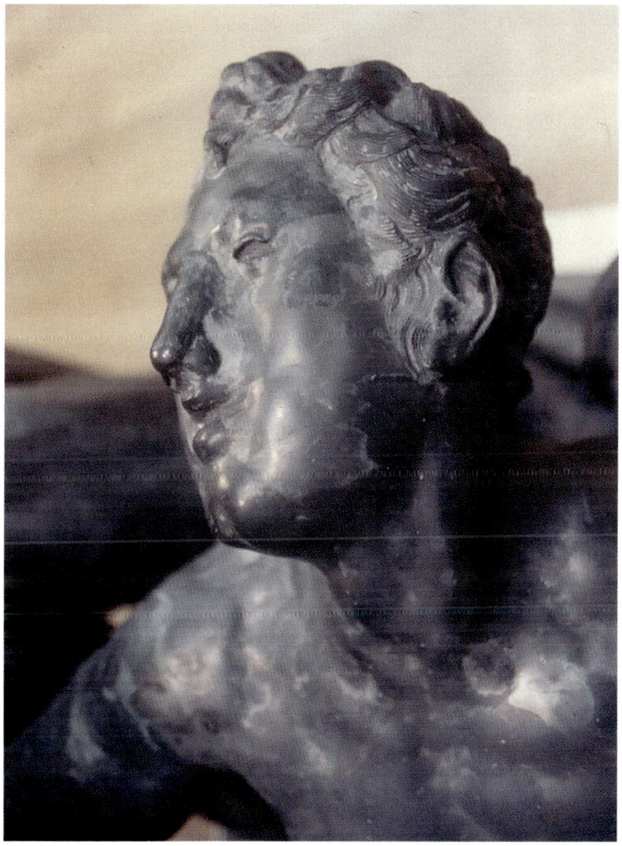

Figure 163 Detail of inverse segregation (tin sweat) on the figure's cheek (one of the sons), appearing as a silvery color due to elevated surface tin content. Adriaen de Vries (Netherlandish, 1556–1626), *Laocoön and His Sons*, 1623, H. 169 cm (Nationalmuseum, Stockholm, inv. Drh Sk 68). See Bassett 2008.

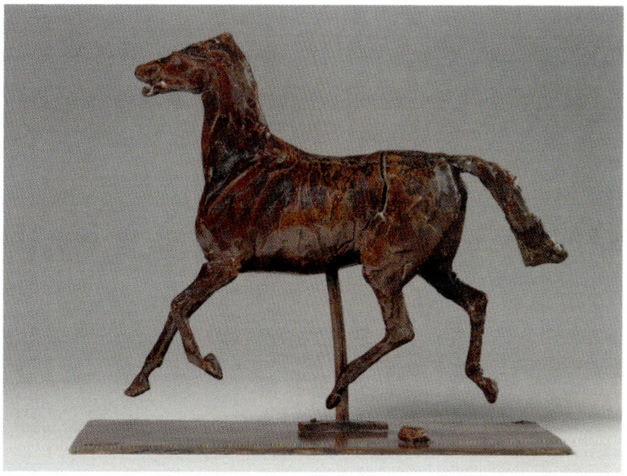

Figure 164 The cast of the horse records large cracks that developed over time in the wax model, but are no longer visible on the sculpture, as the damage was later filled during restoration. Hilaire-Germain-Edgar Degas (French, 1834–1917), *Horse Trotting, Feet Not Touching the Ground*, ca. 1881–90, H. 23.5 (Harvard Art Museums / Fogg Museum, Bequest from the Collection of Maurice Wertheim, Class of 1906, inv. 1951.79).

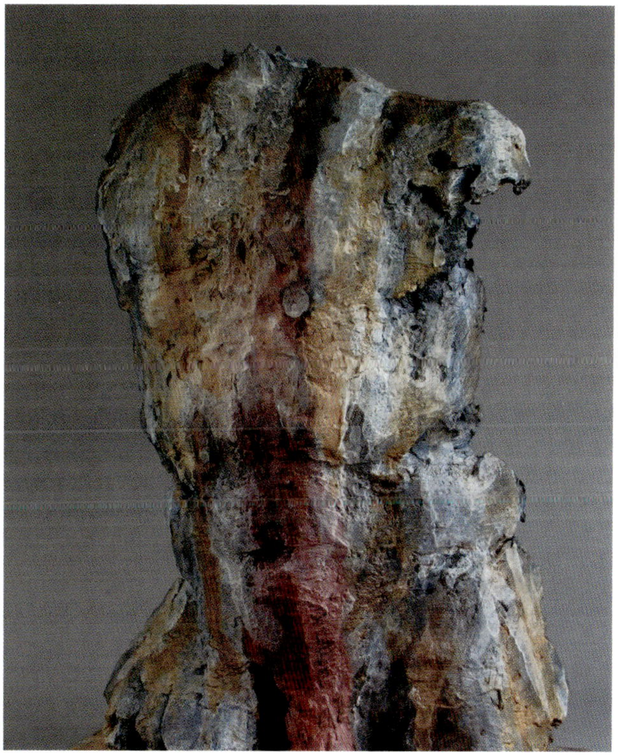

Figure 165 Artist Andrew Lacey states: "The area between the breast or middle of the chest near the heart was moistened just prior to casting. This subjected the bronze to violent steam gassing, porosity, and a small vertical hole in the casting. It is highly organic in form and texture but leaves a mark that suggests weakness or wounding, or at least as a question in itself." Personal communication to the authors, 2019. Andrew Lacey (British, b. 1969), *Mud Ash Ochre*, 2009, life-size torso (artist's collection).

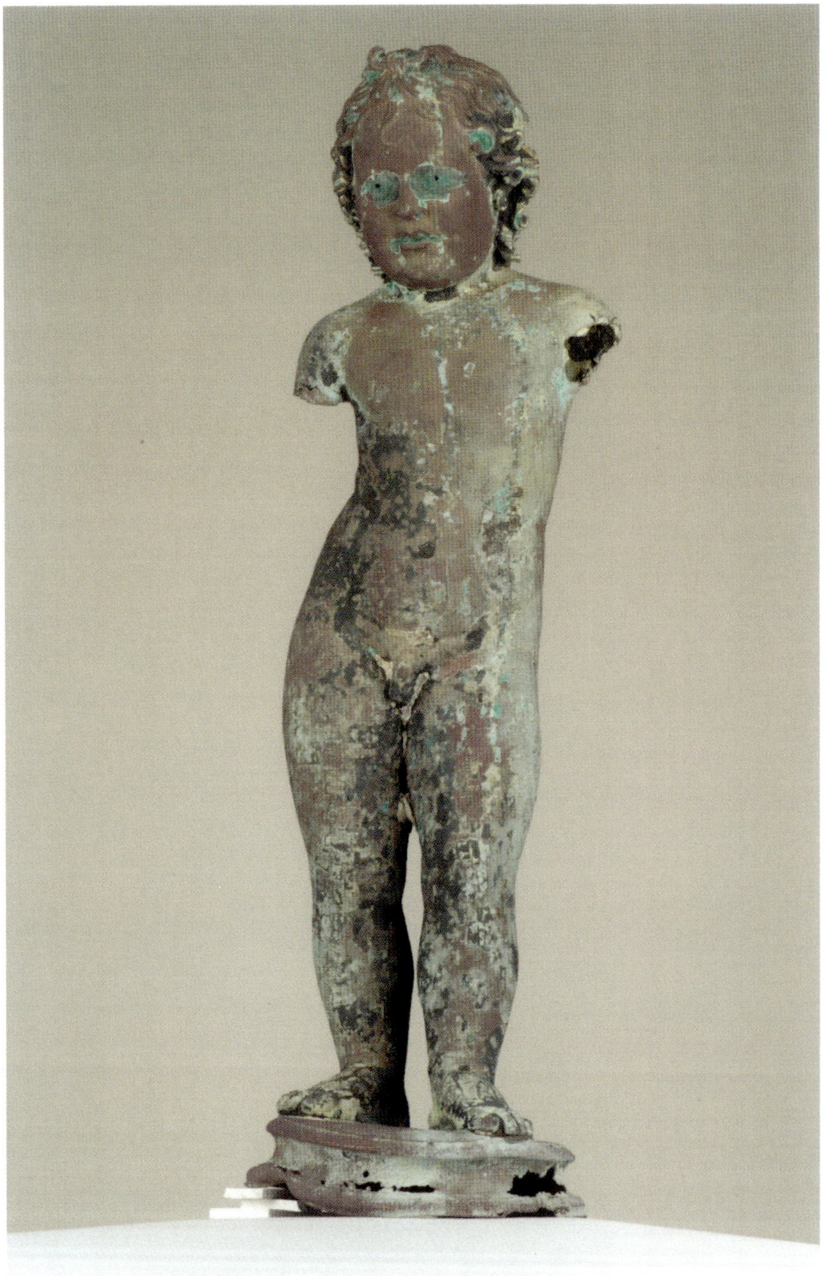

Figure 166 360-degree overall view (viewable in online edition) before restoration. Despite the high quality of the cast, several hundred repair patches have been observed. *Eros of Agde*, France, 1st century BCE, H. 63.5 cm (Musée de l'Ephèbe et d'Archéologie Sous-Marine de la Ville d'Agde, France, inv. 2888). See Mille 2012.

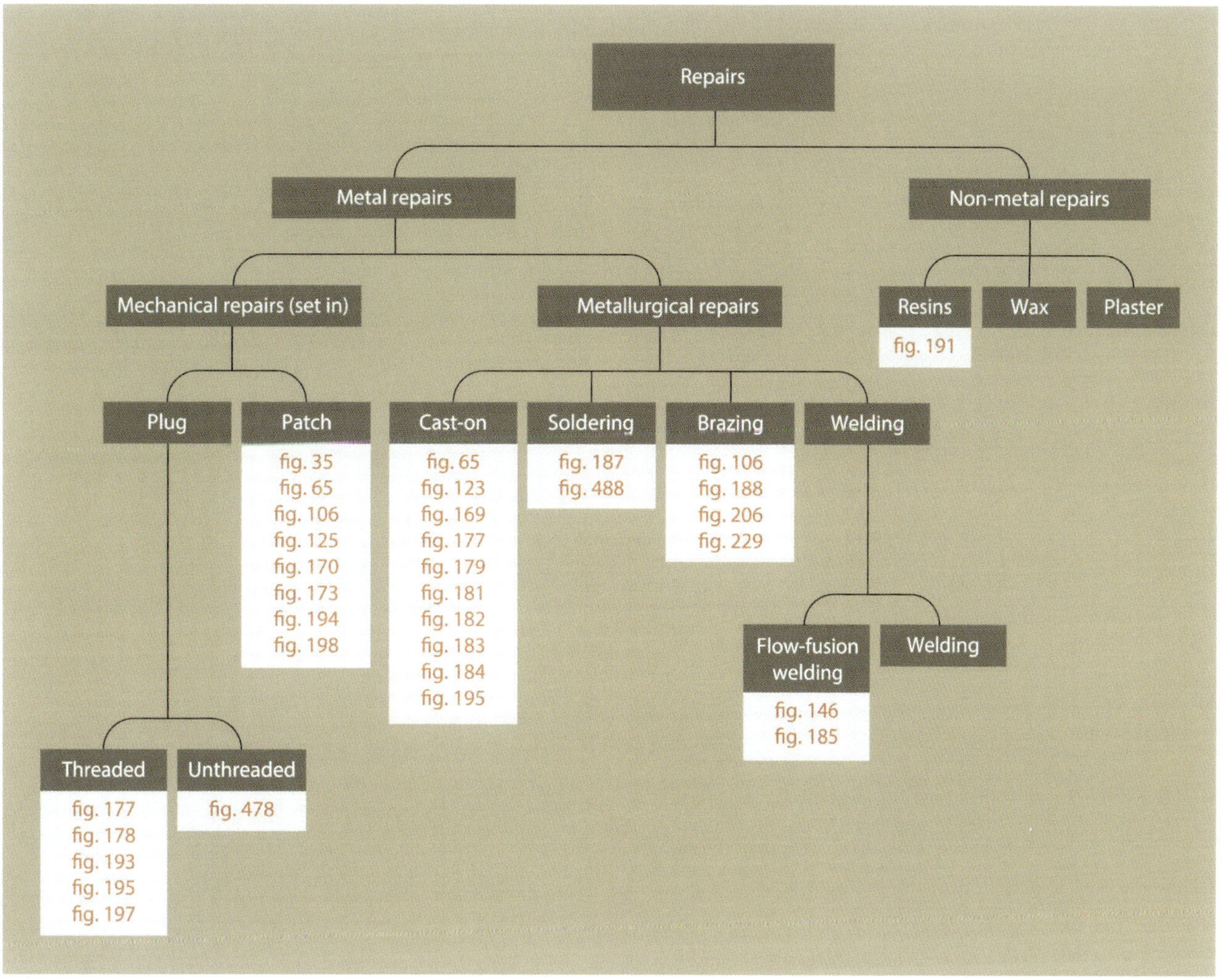

Figure 167 Diagram of the common types of repairs found on bronze sculpture.

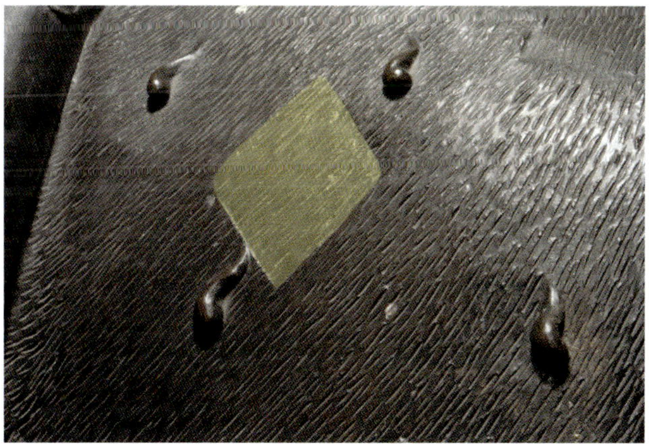

Figure 168 Detail of mantle showing a pattern of cold-worked decorative striations applied over a rectangular patch that covers a casting defect. Germain Pilon (French, ca. 1525–1590), *Kneeling Figure of Henri II*, ca. 1567, H. 150 cm (Saint-Denis Basilica, France). See Castelle 2016.

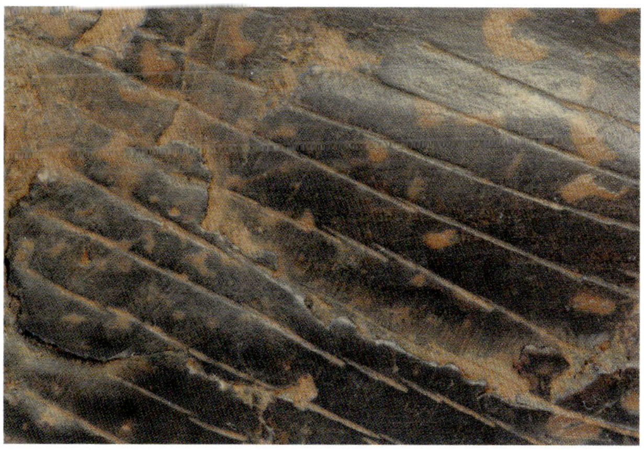

Figure 169 Detail of a cast-on repair. Decorative lines were struck and/or engraved after the repair was cast in place. Modern copy of an Ife head, Nigeria, H. 32 cm (former private collection of Simon du Chastel, gift from Nathalie du Chastel to the Université Libre de Bruxelles).

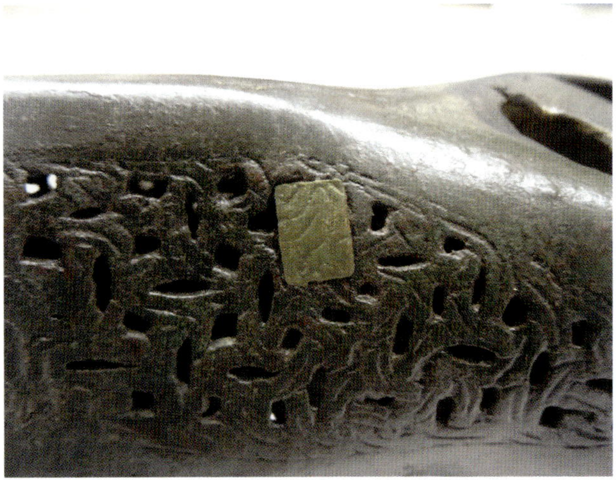

Figure 170 Casting flaw with ancient patch repair. Lynx Incense Burner, Iran (Khorasan province), ca. 12th century, H. 28.5 cm (Musée du Louvre, inv. AA 19). See Collinet 2021.

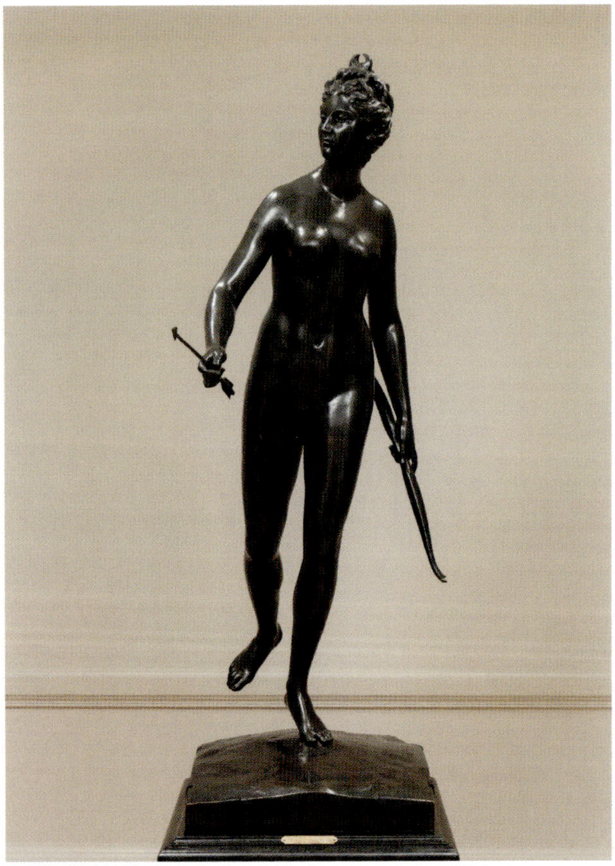

Figure 171 Jean-Antoine Houdon (French, 1741–1828), *Diana*, 1782, H. 208.3 cm (Huntington Art Museum, San Marino, California, inv. 27.186). See Bassett and Scherf 2014; Bennett and Sargentson 2008.

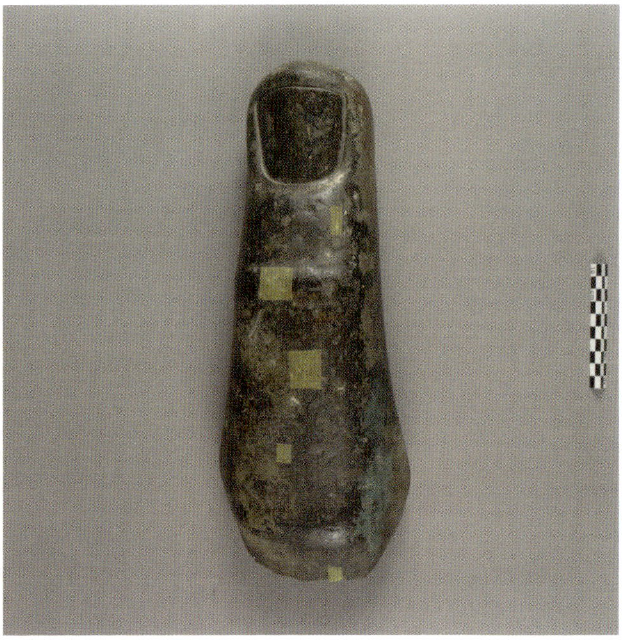

Figure 172 Monumental bronze finger recently shown to belong to the bronze statue of the Roman Emperor Constantine held at the Capitoline Museum. The outer edges of the patch repairs are clearly visible (yellow overlay). Monumental bronze finger, Rome, ca. 330 CE, H. 38 cm (Musée du Louvre, inv. Br78). See Azéma, Descamps-Lequime, and Mille 2018.

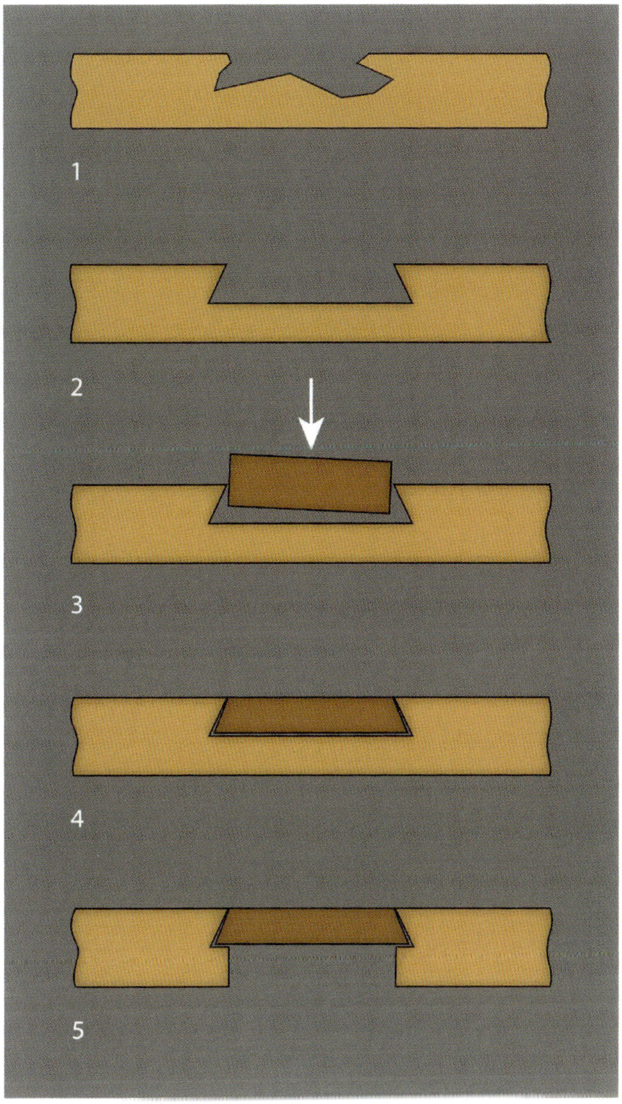

Figure 173 The process of setting in a patch to mask a superficial casting flaw: 1) casting flaw noted on the surface of the bronze (cross section); 2) the flaw is chiseled to create a geometric shape that is easily duplicated; an undercutting bevel is cut into the sides of the wall; 3) a metal patch is set into the flaw; 4) the patch is hammered and chased to fill the flaw and match the adjacent surface as much as possible; the repair is locked in place by the undercut of the bevel; 5) a similar process can be used to insert a plug to repair a flaw that extends through the entire thickness of the wall.

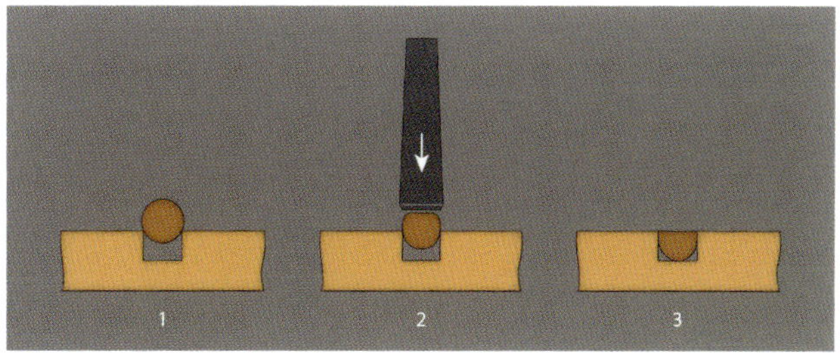

Figure 174 Diagram of a patch repair of a type seen on the statue of the Xanten Youth, Roman, last quarter of the 1st century BCE, H. 143 cm (Staatliche Museen zu Berlin, Antikensammlung, Neues Museum, inv. SK 4). 1) A copper alloy "granule" is positioned over a surface flaw; 2) a punch is used to force the granule into the flaw; and 3) the surface is further chased to conceal the flaw. After Peltz 2011, 127.1, fig. 12.

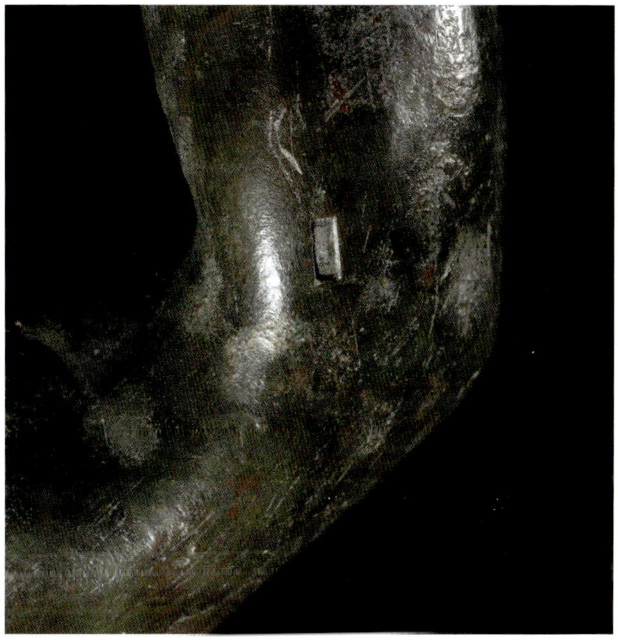

Figure 175 Detail of a casting defect in the right leg, showing where the patch repair has been lost. *Jupiter*, Roman, 1st century CE, H. 92 cm (Musée d'Art, Histoire et Archéologie d'Evreux, France, inv. FZ1520). See Azéma et al. 2012.

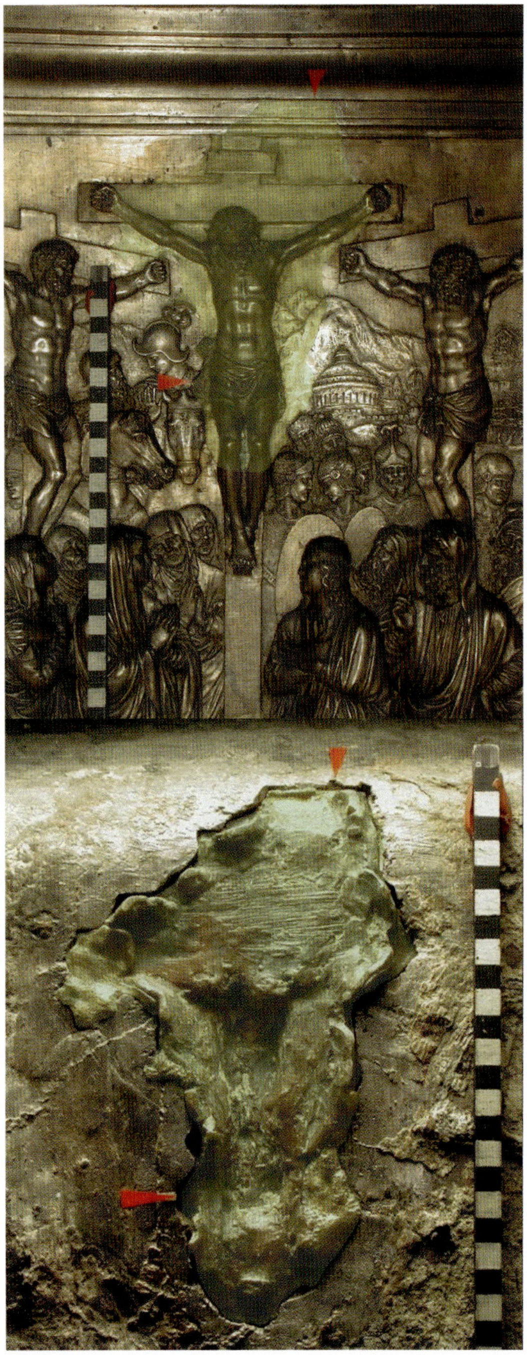

Figure 176 Detail of a tabernacle relief (H. 25 cm). The relief was cast with a large flaw. The figure of Christ was entirely remodeled in wax and cast separately in bronze. The cast repair was then inserted into the casting flaw and chased. Due to differences in the color of the alloy, the repair became visible while cleaning the surface (upper image). As seen at the back (lower image), once inserted in the primary cast, the newly cast repair was hammered to secure it. Note: the lower image has been rotated 180 degrees so that the two images are visually aligned. Girolamo (Italian, ca. 1510–1584/89) and Ludovico Lombardo (Italian, 1507/8–1575), Tabernacle, Chapel of the Holy Sacrament, Fermo Cathedral, Marche, Italy, by 1570–71, H. 230 cm.

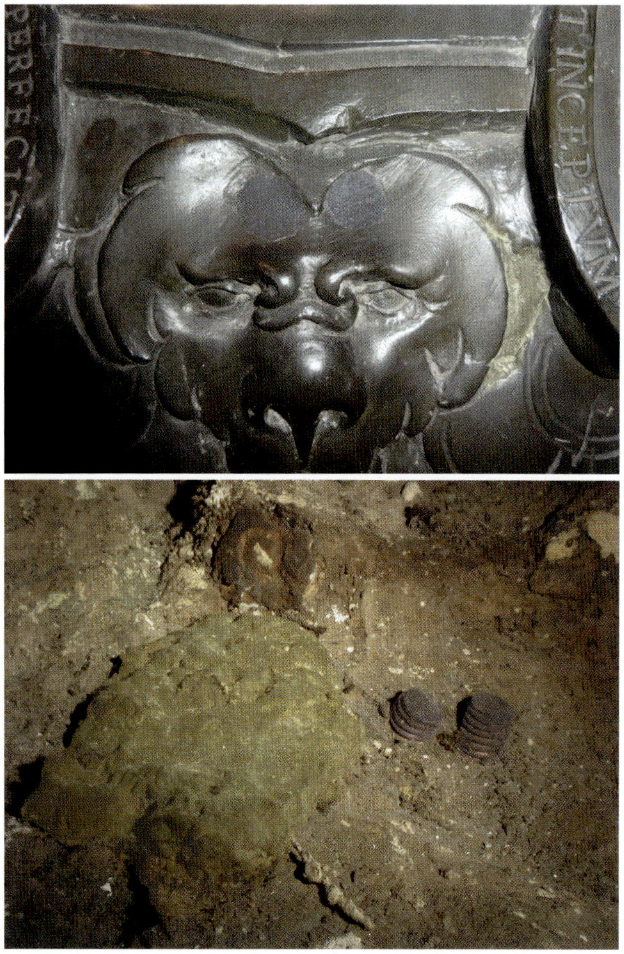

Figure 177 Two types of repairs shown both inside and outside the cuirass that supports a figure. The blue overlay highlights two threaded plug repairs; the yellow overlay highlights a cast-on repair. Francesco Bordoni (Italian, ca. 1576–1654), *Young Captive*, 1618, H. 160 cm (Musée du Louvre, inv. MR 1668). See Bourgarit, Bewer, and Bresc-Bautier 2014.

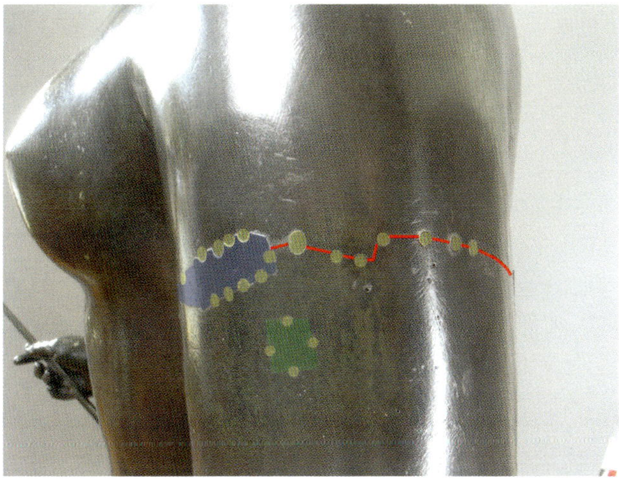

Figure 178 The separately cast arm is attached with a Roman sleeve joint (red line). A large patch fills a flaw along the joint (blue overlay). A second patch is unrelated to the joint (green overlay). Threaded circular plugs (yellow overlays) help lock the arm joint and patches in place. Jean-Antoine Houdon (French, 1741–1828), *Diana*, 1782, H. 208.3 cm (Huntington Art Museum, San Marino, California, inv. 27.186). See Bassett and Scherf 2014; Bennett and Sargentson 2008, 492–93.

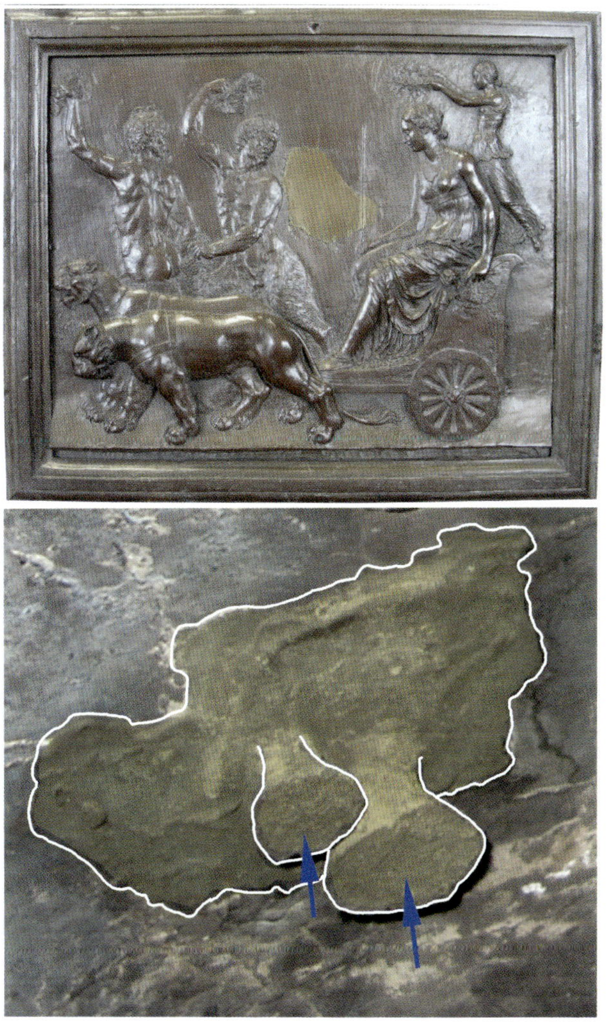

Figure 179 Detail of a cast-on repair in the center of a relief panel that has been set into the side of a Renaissance pedestal on which a Hellenistic bronze is mounted (yellow overlay). Also in yellow, the lower image shows the back side of the cast-on repair. The sprues, including a vent and a gate, were left in place and not removed (blue arrows). The repair was made with the relief face-down before being mounted to the pedestal. *Idolino* (Ephebe from Pesaro), ca. 30 BCE, base by the workshop of Girolamo (Italian, ca. 1510–1584/89), Aurelio (Italian, ca. 1501–1563), and Ludovico (Italian, 1507/8–1575) Lombardo, perhaps after a drawing by Sebastiano Serlio (Italian, 1475–1544), 1530–38/40, H. 148 cm (300 cm with base) (Museo Archeologico Nazionale, Florence, inv. 1637).

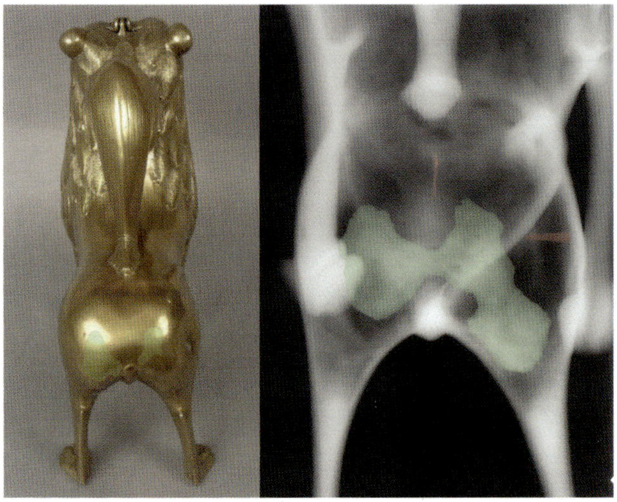

Figure 180 Annotated detail of a cast-on repair (left). The cast-on metal is surrounded by a thin gap due to shrinkage of the repair as it cooled. The annotated X-radiograph (right) reveals the gap around the repair and the excess metal cast into the interior of the hollow bronze as well as two tapering core pins. See fig. 491 for a detail of the backside of the horse. *Aquamanile in the Form of a Lion*, probably northern Germany, ca. 1200, H. 21.2 cm (Metropolitan Museum of Art, The Cloisters Collection, 1947, inv. 47.101.52). See Dandridge 2006.

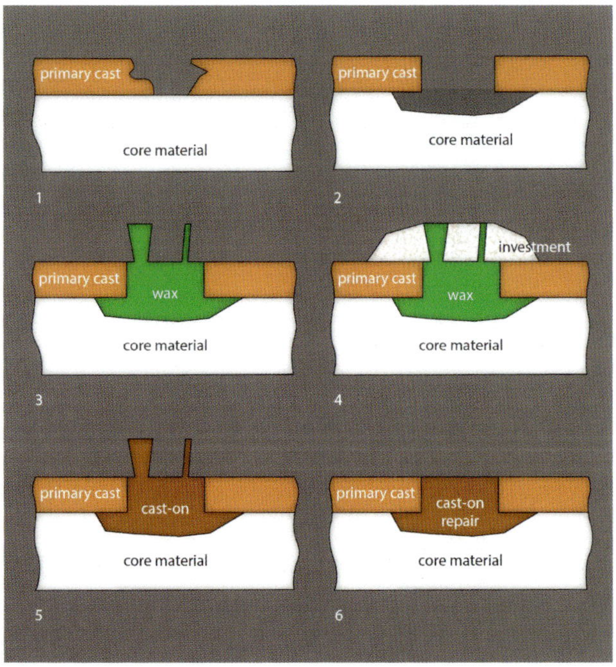

Figure 181 Cast-on repair process: 1) casting flaw that extends through the wall of the bronze; 2) the edges of the flaw are cleaned up and core under the flawed area is partially excavated; 3) the flawed surface is modeled in wax; a pouring cup and vent are added; 4) the wax is invested; heat is applied locally until the wax melts out; 5) bronze is poured into the temporary mold; when cool, the investment is removed; 6) the pouring cup and vent are removed; the surface is chased to minimize the thin gap between the bulk metal and the cast-on repair caused due to shrinkage, and to unify the surface of the bronze.

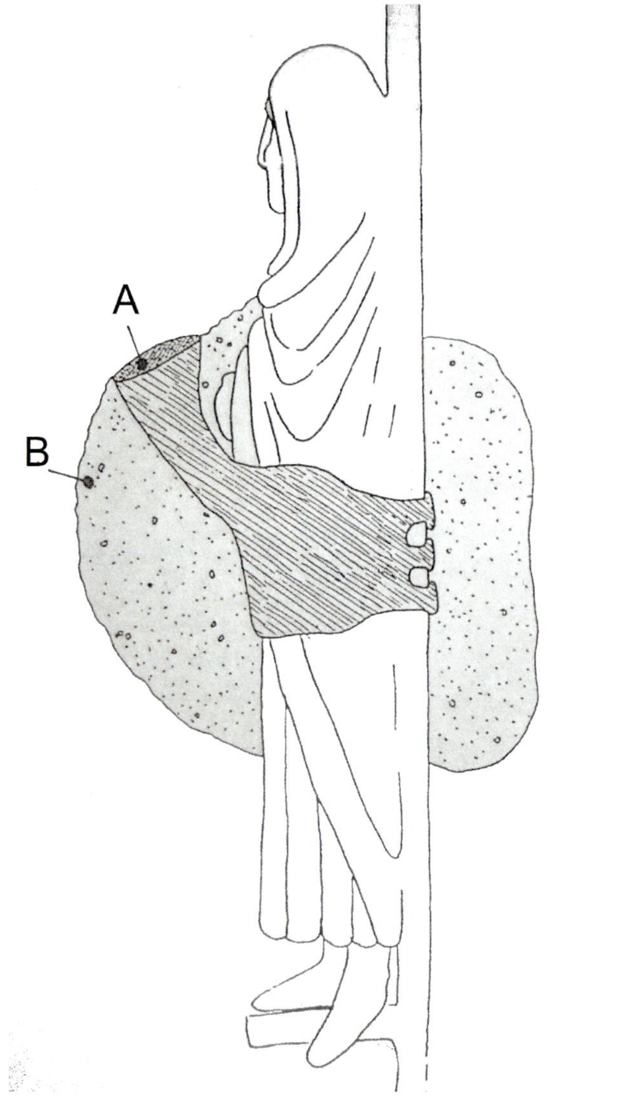

Figure 182 Diagram illustrating a cast-on repair. A) wax is applied to the loss and modeled in situ with a pouring gate; B) investment is applied to cover the wax completely (for clarity, the investment in the sketch has been cut through to show the applied wax). Based on Bonanno Pisano (Italian, active late 12th century), *Porta di San Ranieri*, Pisa Cathedral, Italy, ca. 1180, H. 470 cm. See Morigi 1999.

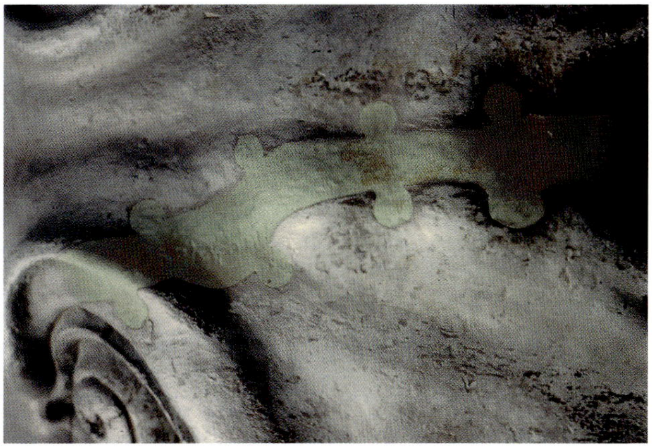

Figure 183 Surface detail (35 × 25 cm) showing a substantial cast-on repair. Circular holes cut along the borders of the gap allow the cast-on section to securely lock into place like a jigsaw. Pietro Tacca (Italian, 1577–1640), Fountain of the Marine Monsters, Florentine, 1626–37, H. 320 cm (Piazza Santissima Annunziata [northwest], Florence).

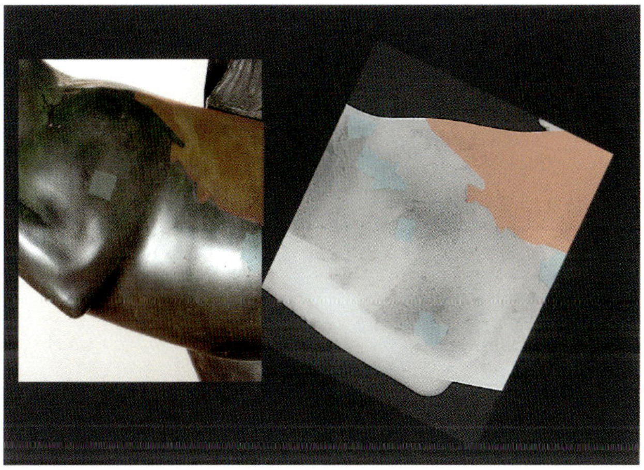

Figure 184 Detail of a cast-on repair on the back of the doe with dovetails that lock the repair in place (red overlay). The color variation between the repair and the primary casting (left) is due to the diverse microstructures and resulting corrosion of these sections. The corresponding gamma radiograph (right) helps to clarify the location of the different repairs (blue overlay) as well as the heavy porosity in the cast-on metal (dark spots). Barthélemy Prieur (French, 1536–1611), *Diana the Huntress*, 1603, H. 200 cm (Musée national du château de Fontainebleau, France, inv. RF 261). See Castelle 2016.

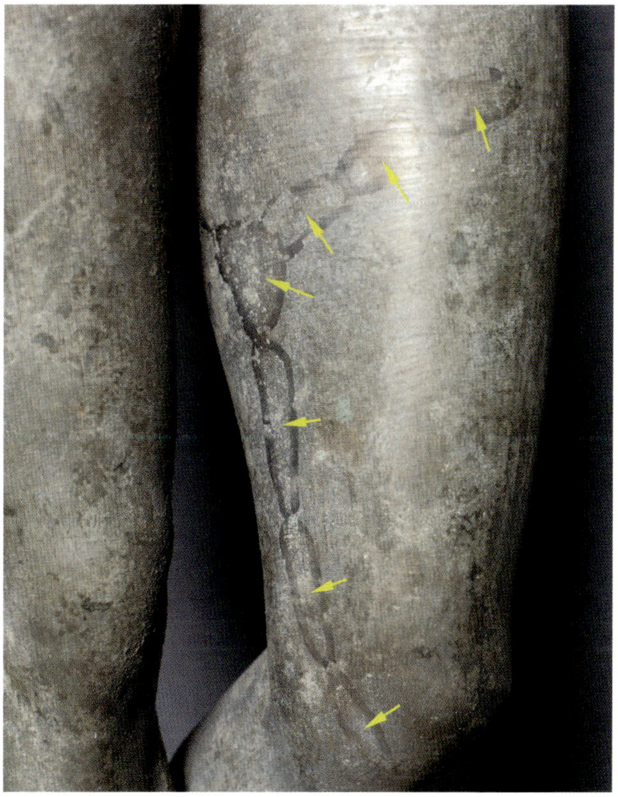

Figure 185 Detail showing flow fusion welding repair. Each darkly outlined ovoid shape represents a fusion welding pour. The dark line is created at the melted zone between the fresh metal and the original bronze wall. It would not have been visible originally, but differences in the microstructure have led to corrosion variations. Statue of a Youth, ancient Rome, probably Asia Minor, ca. 140 CE, H. 142.6 cm (Toledo Museum of Art, purchased with funds from the Libbey Endowment, gift of Edward Rummond Libbey, inv. 1966.126). See Mattusch 1996.

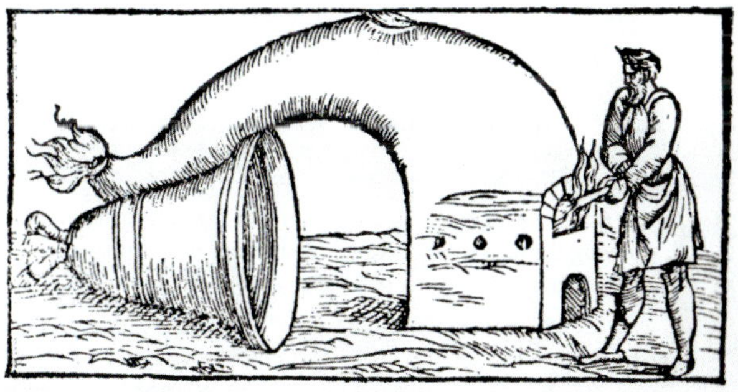

Figure 186 Illustration of a specially designed furnace used to repair a cracked bell by welding. From Vannoccio Biringuccio (Italian, 1480–before 1539), *The Pirotechnia*, 1540. See Biringuccio [1540] 1942, 276.

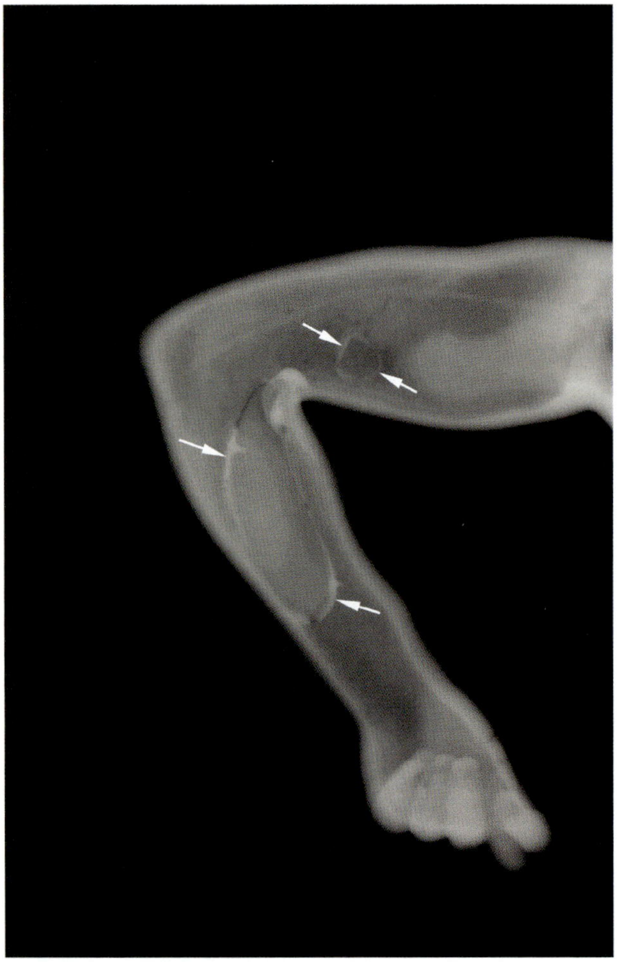

Figure 187 Annotated detail of an X-radiograph of the higher-density lead-based soldering metal that surrounds the set-in patch on the nymph's right arm (arrows). Antonio Susini (Italian 1558–1624), *Faun and Nymph*, 1580–90, H. faun 48.2 cm, H. nymph 34.6 cm (Staatliche Kunstsammlungen, Grünes Gewölbe, Dresden, inv. IX 36). See Bassett 2008, 284, fig. A.27.

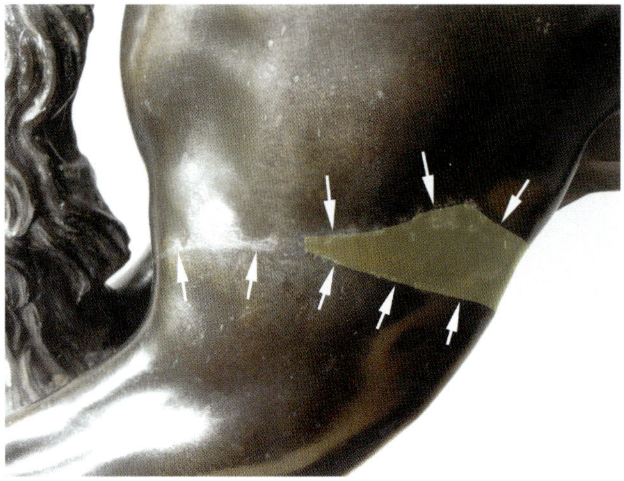

Figure 188 Brazing metal (arrows) secures the separately cast back right leg and patch repair (yellow overlay). Brazing is confirmed by the elevated levels of zinc detected by XRF along the joint. Raised corrosion along the joints was probably caused by the remains of flux on the surface. Unidentified foundry, after Antoine Coysevox (French, 1640–1720), *Fame*, 1700–1710, H. 60.3 cm (Huntington Art Museum, San Marino, California, inv. 14.14). See Bennett and Sargentson 2008, 479.

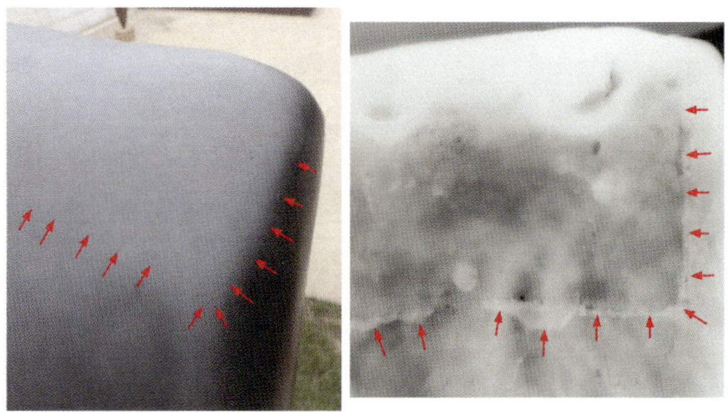

Figure 189 Detail of a weld repair. In the photograph, the surface porosity indicates the location of the weld line on the carefully chased and polished surface; note that the color along the weld does not vary from that of the adjacent surfaces. The X-radiograph (right) reveals the uneven line of excess welding metal that remains on the interior along the join. Joan Mirò (Spanish, 1893–1983), *Personnage*, designed 1976, cast 1985, H. 213.4 cm (J. Paul Getty Museum, inv. 2005.116).

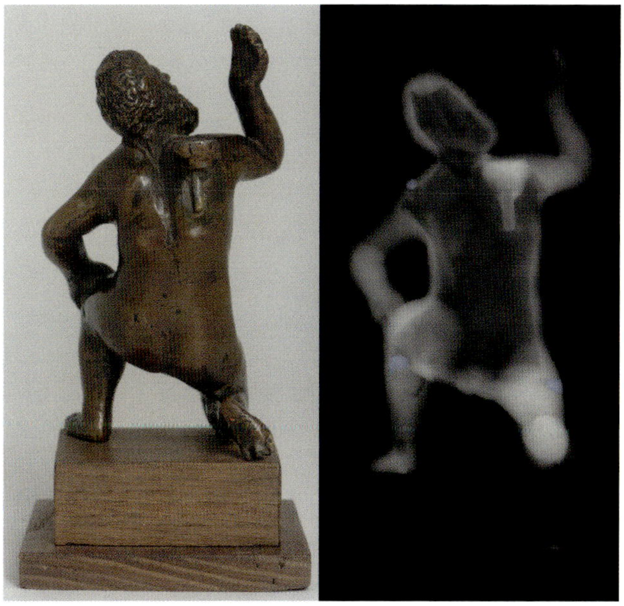

Figure 190 X-radiograph showing lead repairs filling surface flaws (overlay). A vertical bar behind the right shoulder was soldered in place as part of the support for Atlas to carry his burden (now missing). Workshop of Severo da Ravenna (Italian, active 1496–before 1538), *Atlas*, Italy, second quarter of the 16th century or later, H. 14 cm (Musée des Beaux-Arts d'Angers, France, inv. 2003.1.198). See C2RMF Internal Report 2018b.

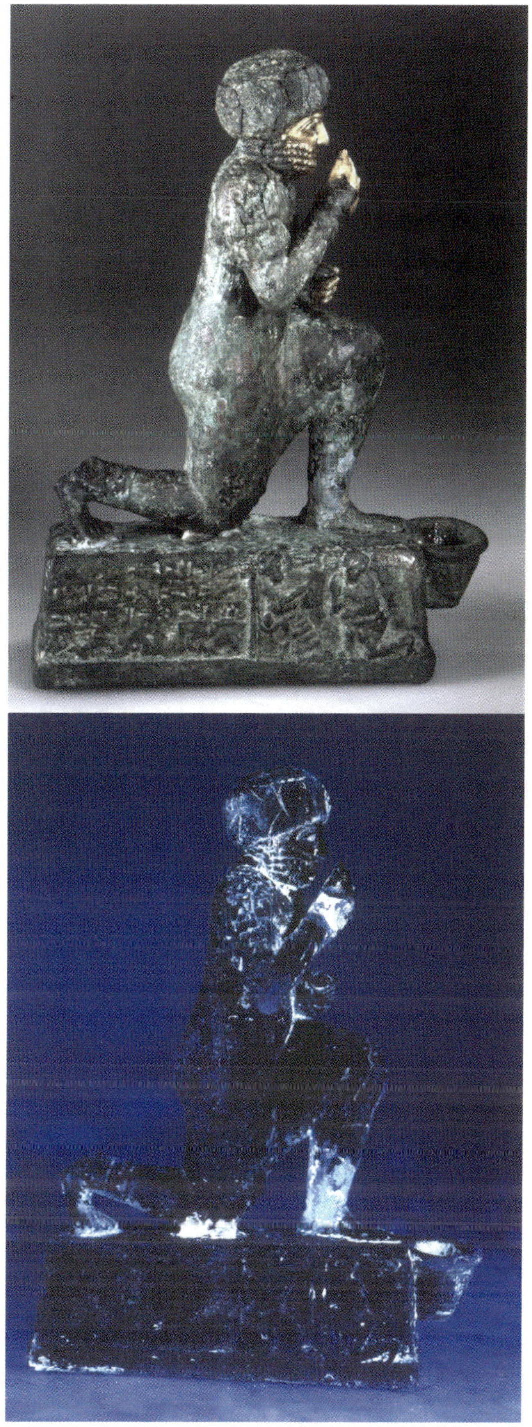

Figure 191 Some restorations made of organic materials (here likely made of a type of resin) fluoresce under ultraviolet illumination. *The Worshipper of Larsa*, Mesopotamia, 18th century BCE, H. 19.6 cm (Musée du Louvre, inv. AO15704). Views in daylight and ultraviolet light. See C2RMF Internal Report 2008b.

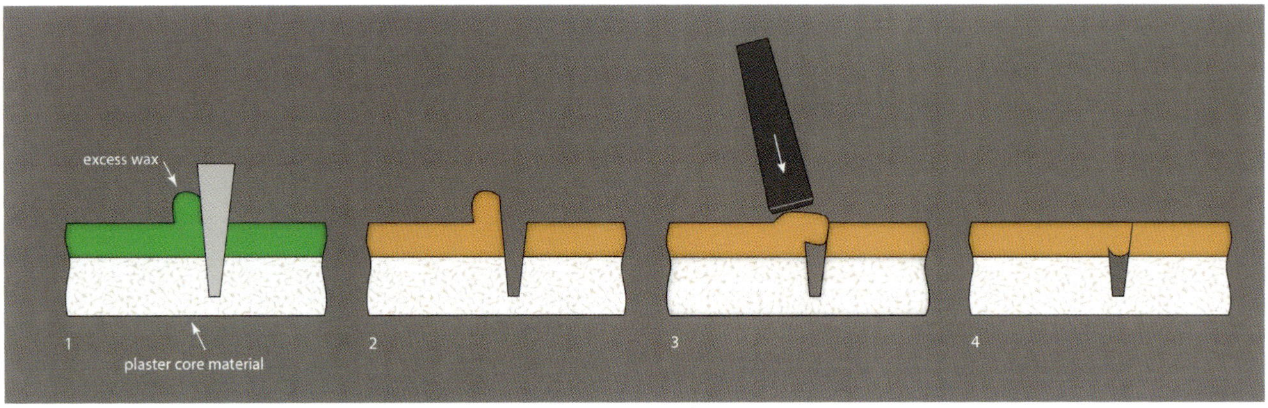

Figure 192 Diagram of a simple core pin repair: 1) a ball of excess wax is applied adjacent to a core pin that has been inserted into the wax model (cross section); 2) the bronze is poured and the core pin is removed; 3) the resulting hole is filled by hammering the excess bronze into the void; 4) the metal that fills the hole is chased to match the surrounding surfaces.

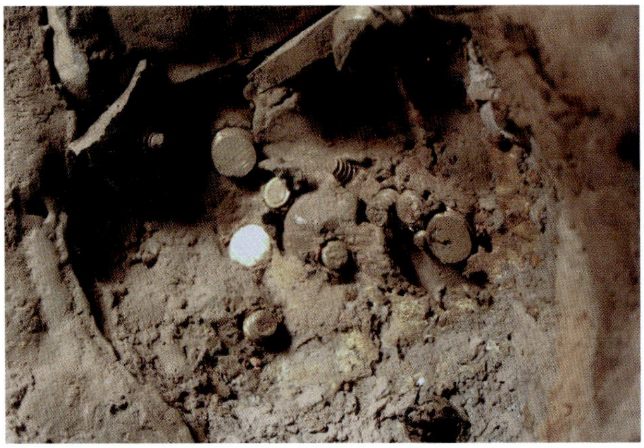

Figure 193 Interior shot showing a seriously flawed area that has been repaired by inserting threaded plugs. Note the variation in their diameter. Giovanni Battista Bianco (Italian, d. 1657), after a design by Domenico Fiasella (Italian, 1589–1669), *Virgin as Queen of Genoa*, High Altar, San Lorenzo, Genoa, Italy, 1651.

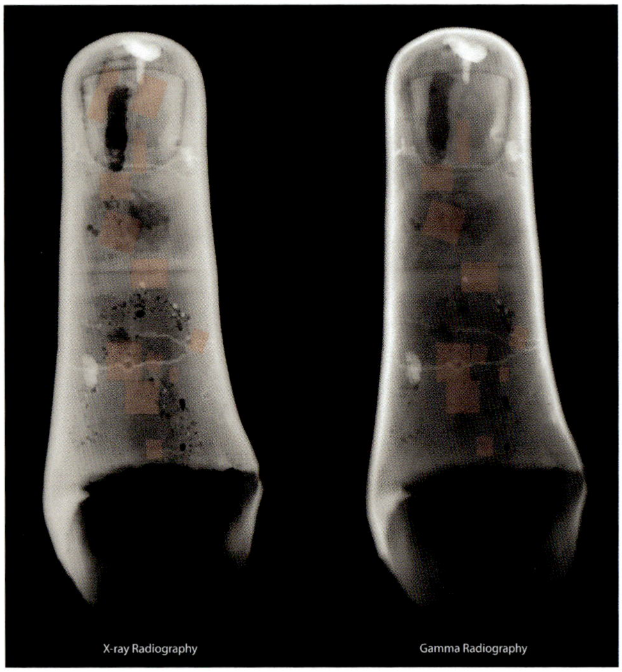

Figure 194 Comparison of X-radiography (left) and gamma radiography (right) of a monumental bronze finger—a fragment of the monumental statue of Emperor Constantin held at the Capitoline Museum. The polygonal patch repairs are highlighted in red. The operating conditions for gamma radiography were: ~2 mm ^{192}Ir 30Ci (1,1 TBq) source, 1 m source/object distance, 20 min. exposure, ERLM Dürr NDT H CR plate (resolution 50 µm), 0.1 mm Pb filtering before and after detector. Monumental bronze finger, Rome, ca. 330 CE, H. 38 cm (Musée du Louvre, inv. Br78). See fig. 172 for an overall view. See Azéma, Descamps-Lequime, and Mille 2018.

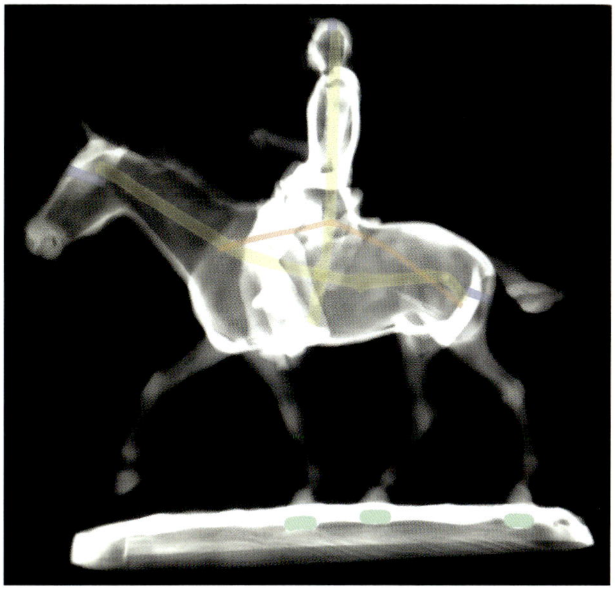

Figure 195 Annotated X-radiograph showing internal features of a sand cast: threaded repair plugs (blue overlay), core vents (yellow overlay), core support (orange overlay), and nuts that secure the horse to the base (green overlay). Charles Marion Russell (American, 1864–1926), *Medicine Whip*, modeled 1911, sand cast 1912–16, H. 14.8 cm (Gilcrease Museum, Tulsa, 0837.14).

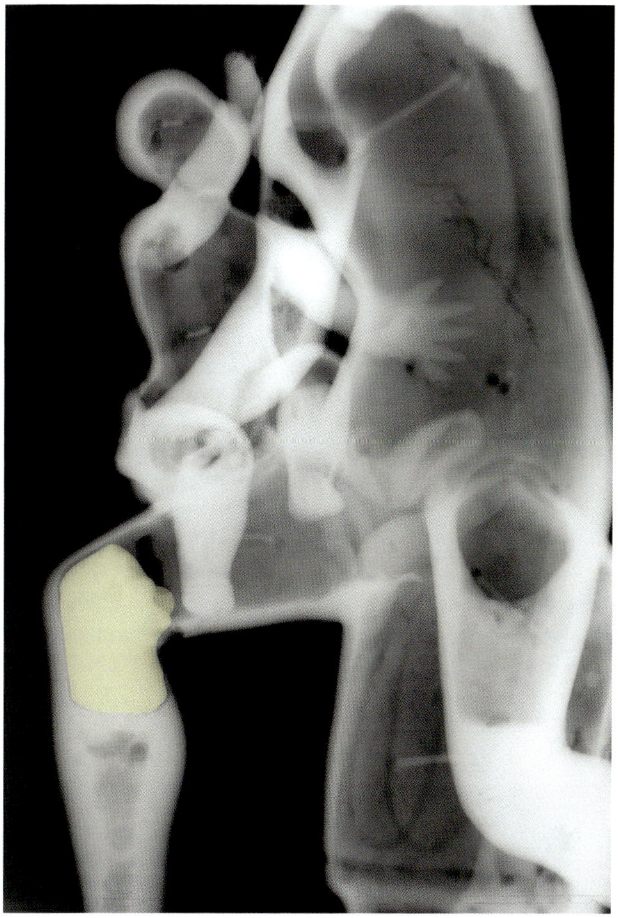

Figure 196 Annotated detail of an X-radiograph showing a cast-on repair in the seated figure's right knee that shrank as it cooled, revealing the distinctive gap between the original wall of the bronze and the later cast-in metal. Francesco Bertos (Italian, 1678–1741), *Group of Eleven Figures (Allegory of Autumn)*, ca. 1725, H. 79.5 cm (J. Paul Getty Museum, inv. 85.SB.74).

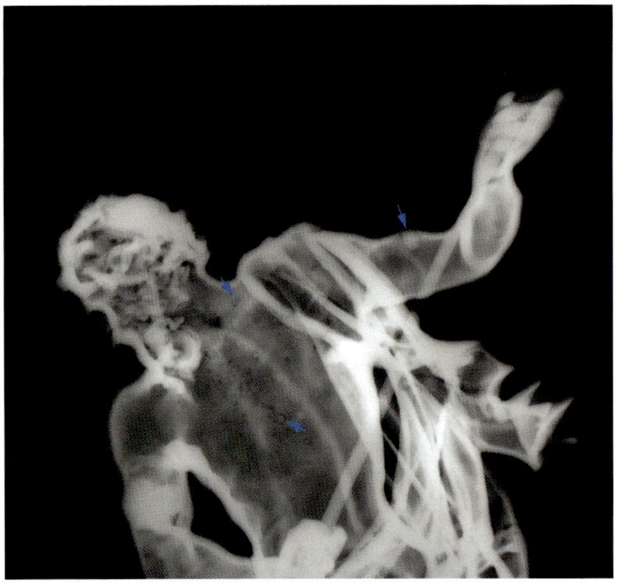

Figure 197 Threaded plugs are visible in the X-radiograph throughout the torso and in the raised proper left bicep (some marked with blue arrows). Attributed to Gian Lorenzo Bernini (Italian, 1598–1680), *Neptune and Dolphin*, 17th century (probably after 1623), H. 56 cm (J. Paul Getty Museum, inv. 94.SB.45).

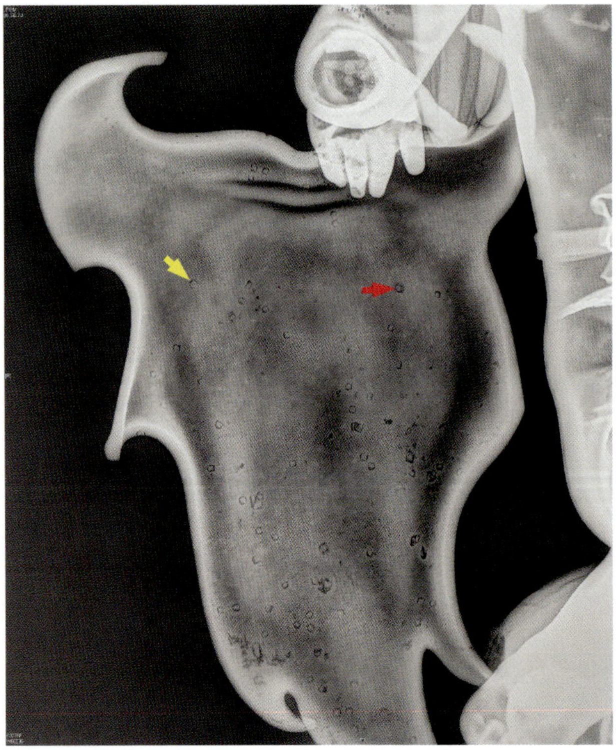

Figure 198 Annotated digital X-radiograph of a single-walled shield showing extensive small surface patches (one example indicated with a yellow arrow) and a small number of threaded plugs (one indicated with a red arrow). Ferdinando Tacca (Italian, 1619–1686), *Putto Holding a Shield*, Florence, ca. 1650–55, H. 64.5 cm (J. Paul Getty Museum, inv. 85.SB.70). See Fogelman and Fusco 2002, 217–23.

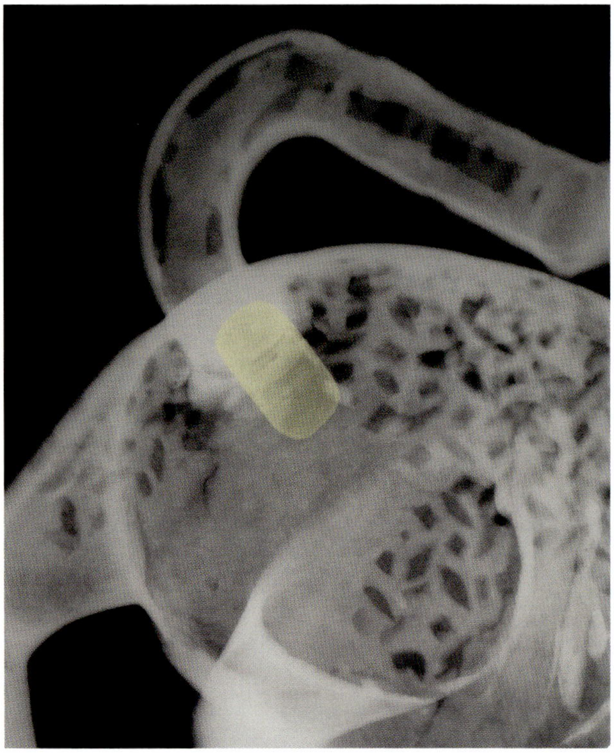

Figure 199 Detail of X-radiograph revealing an internal metal cylinder added during restoration to join the body and tail. Lynx Incense Burner, Iran (Khorasan province), ca. 12th century, H. 28.5 cm (Musée du Louvre, inv. AA 19). See Collinet 2021.

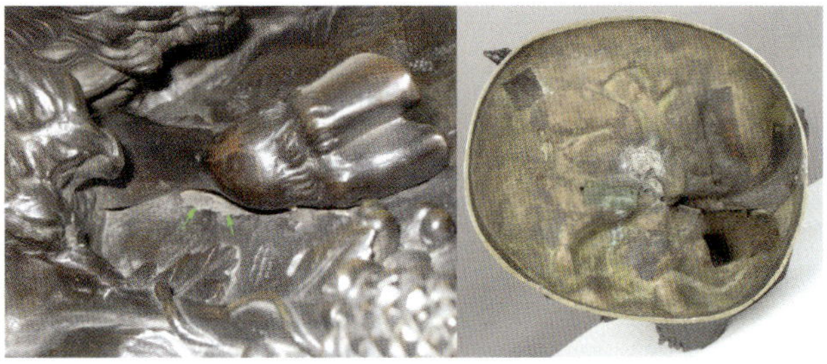

Figure 200 A casting flaw in a relatively hidden and difficult-to-reach location under the satyr's sand-cast-bronze right leg (left image) has been repaired from below with sheet metal pinned under the base (right image). In the style of Claude Michel, called Clodion (French 1738–1814), *Nymph and Satyr*, ca. 1850–75, H. 35.6 cm (Huntington Art Museum, San Marino, California, inv. 78.20.54). See Bennett and Sargentson 2008, 474–75.

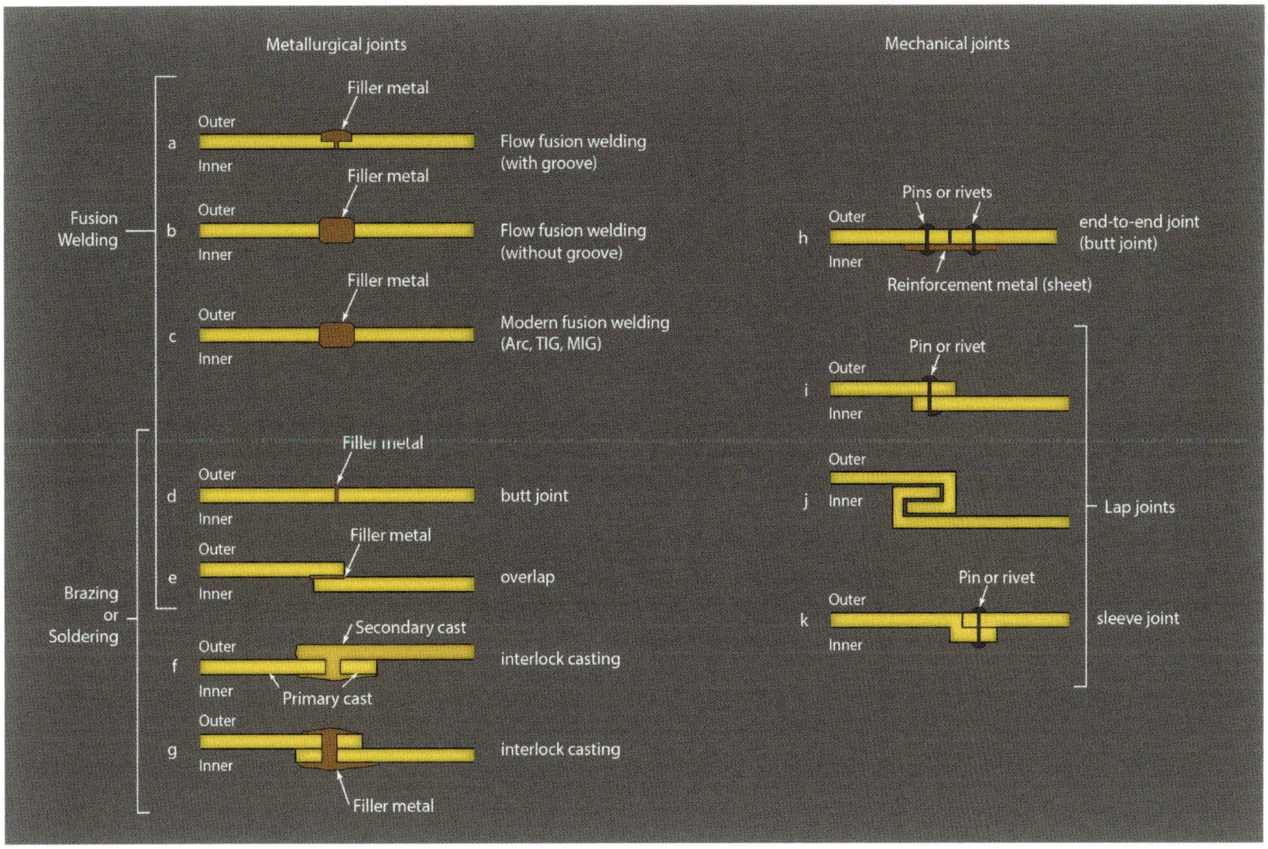

Figure 201 Main assembly techniques encountered on bronze sculpture.

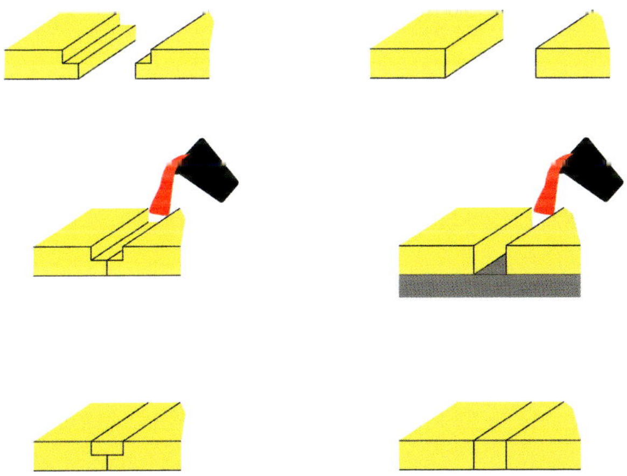

Figure 202 Two ways of preparing the two parts to be assembled by flow fusion welding as witnessed on various antique large bronzes. Left side: the base metal is cut away to half its thickness to make a channel through which the molten bronze can run; right side: a space is left between the two parts to be joined and the welding metal is directly poured through this space. See Azéma et al. 2011.

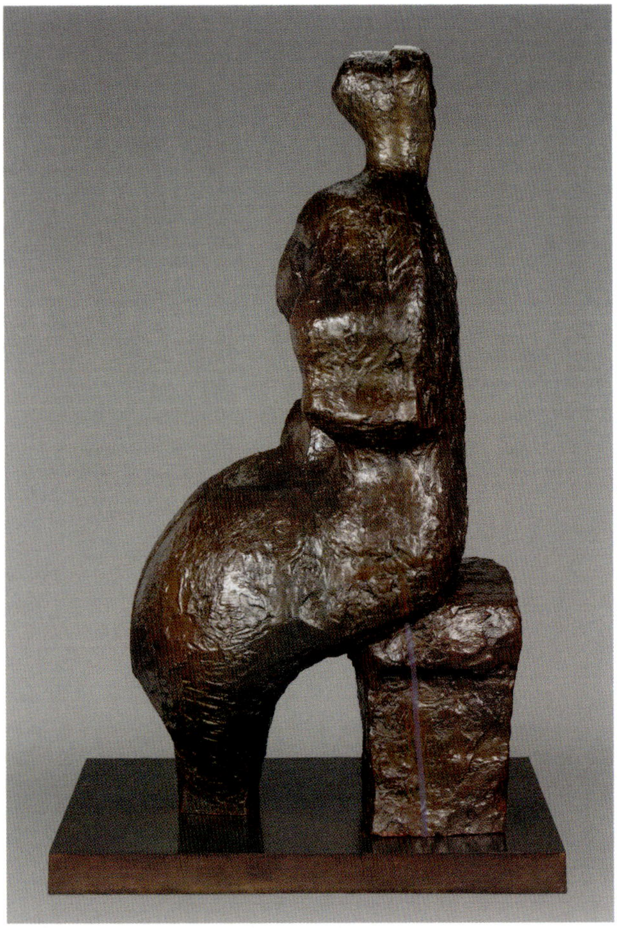

Figure 203 A welding line along separately cast parts of the base is visible on the surface due to variations in the patina (blue overlay). Henry Moore (British, 1898–1986), *Seated Woman*, designed 1958–59, cast 1975, H. 203.2 cm (J. Paul Getty Museum, inv. 2005.117.3).

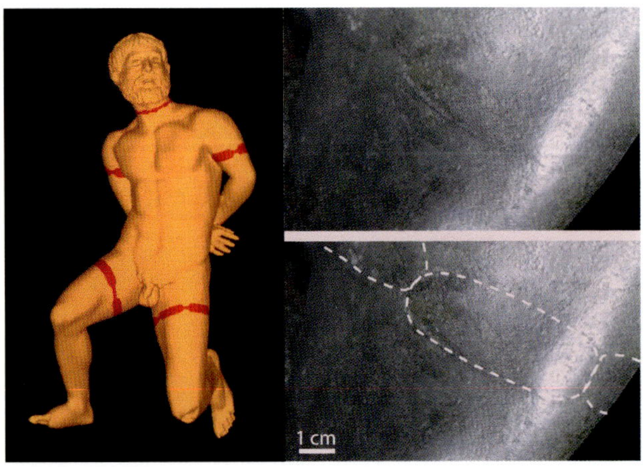

Figure 204 Statue made of six separately cast pieces assembled by flow fusion welding (left). The welds appear as a sequence of adjacent ovals corresponding to a welding in basins (right). *Captive Gaul*, Gaul, last quarter of the 1st century BCE, H. 63 cm (Musée départemental Arles Antique [MDAA], France, inv. Rho.2007.06.1962). See Azéma et al. 2013.

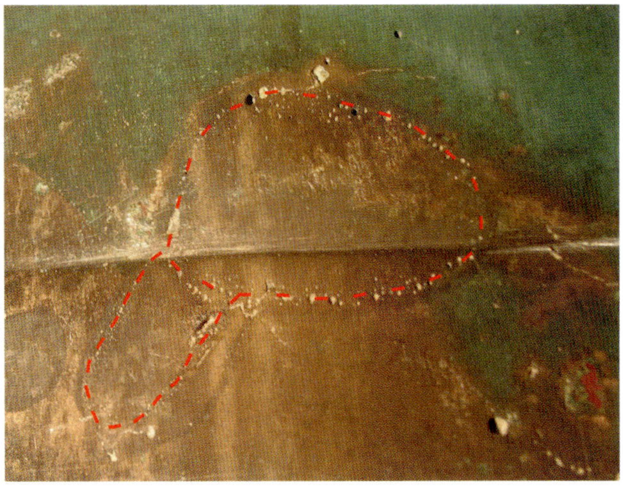

Figure 205 Characteristic basin of a flow fusion weld defined at its outer edge by a discrete line of porosity (dotted red line). *Horse of Neuvy-en-Sullias*, France, 1st century BCE–1st century CE, H. 113 cm (Musée historique et archéologique de l'Orléanais, Orléans, France, inv. A6286). See Mille 2007.

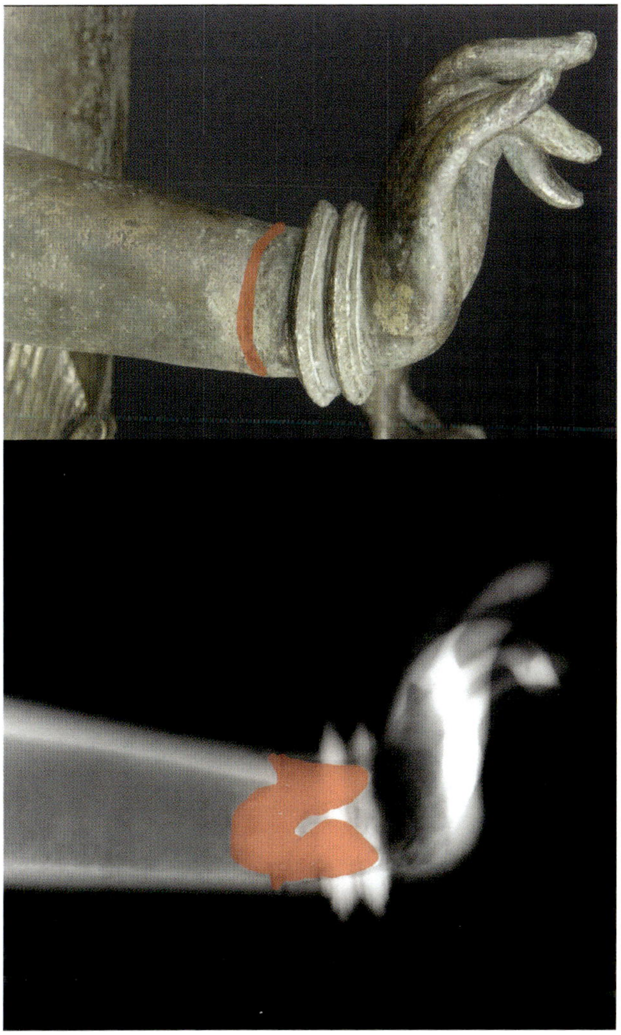

Figure 206 Daylight photograph and X-radiograph of a detail of a right arm showing the brazing between the hand and the wrist (red overlay). The area was finished, removing any exterior signs of the joining method. *Gilded Standing Divinity*, Cambodia, Angkor, Siem Reap province, 11th century, H. 130.8 cm (Metropolitan Museum of Art, from the Collection of Walter H. and Leonor Annenberg, inv. 1988.355).

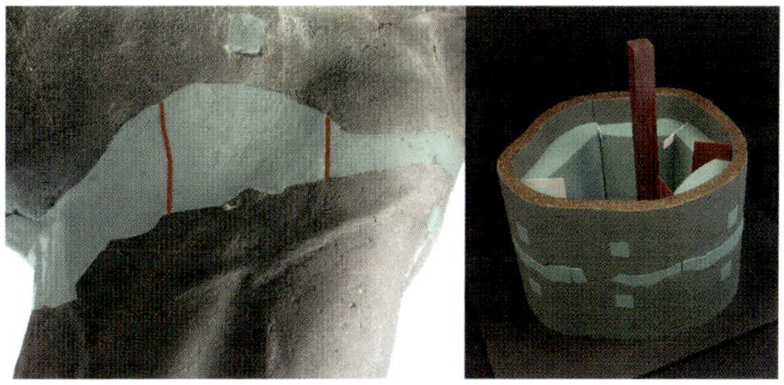

Figure 207 Detail of horse's leg showing the location of the joint made by interlock casting (left). The joint mechanism is visualized in the diagram (right). The separately cast leg and body were assembled using a series of localized bronze pours (teal overlay). Iron shims (red overlay) and core material packed around the internal iron armature were used to contain each section of that pour. The joint was further secured by interlock casting, which entailed creating small openings on either side of the joint (teal squares) for the metal to key into. Andrea del Verrocchio (Italian, ca. 1425–1488) and Alessandro Leopardi (Italian, ca. 1465–1512), *Equestrian Monument of Bartolomeo Colleoni, Venetian*, 1479–96, over life size (Piazza SS Giovanni e Paolo, Venice). See general view of the monument, fig. 100.

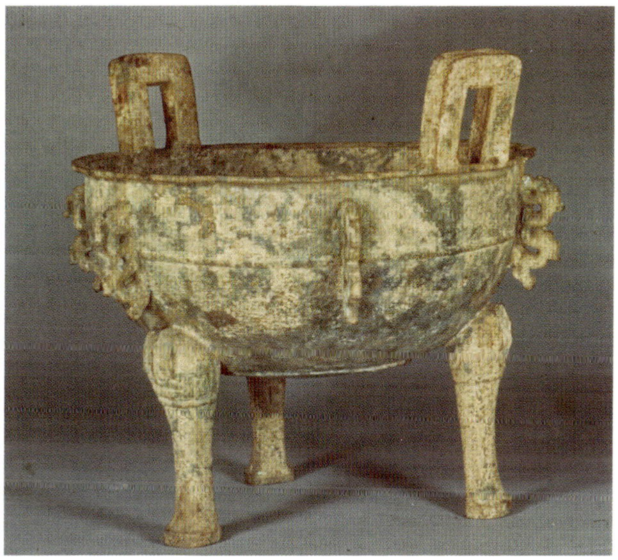

Figure 208 An example of casting-on in which the body has been cast onto the legs. See fig. 209 for a detailed view. Ritual vessel, China, Zhou dynasty, ca. 830 BCE, H. 36.5 cm (Freer Study Collection, Freer Gallery of Art and Arthur M. Sackler Gallery, Smithsonian Institution, Washington, DC, purchase—Charles Lang Freer Endowment, inv. FSC-B-56).

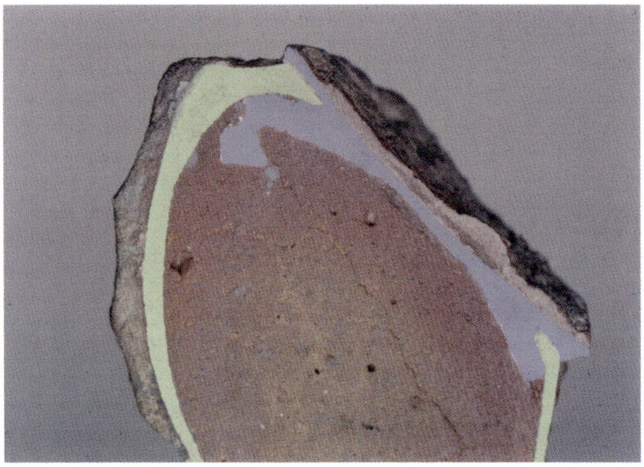

Figure 209 Cross section of a vessel fragment. The leg (green overlay) was cast first, then put into the mold. The body of the vessel (lilac overlay) was then cast onto the leg, mechanically locking them together. The reddish material in the center is the core preserved in the leg; the widest dimension of the current section is 5.5 cm. Fragment of ritual vessel, China, Zhou dynasty, ca. 830 BC, H. 36.5 cm (Freer Study Collection, Department of Conservation and Scientific Research, Freer Gallery of Art and Arthur M. Sackler Gallery, Smithsonian Institution, Washington, DC, purchase—Charles Lang Freer Endowment, inv. FSC-B-56). See fig. 208 for an overall view.

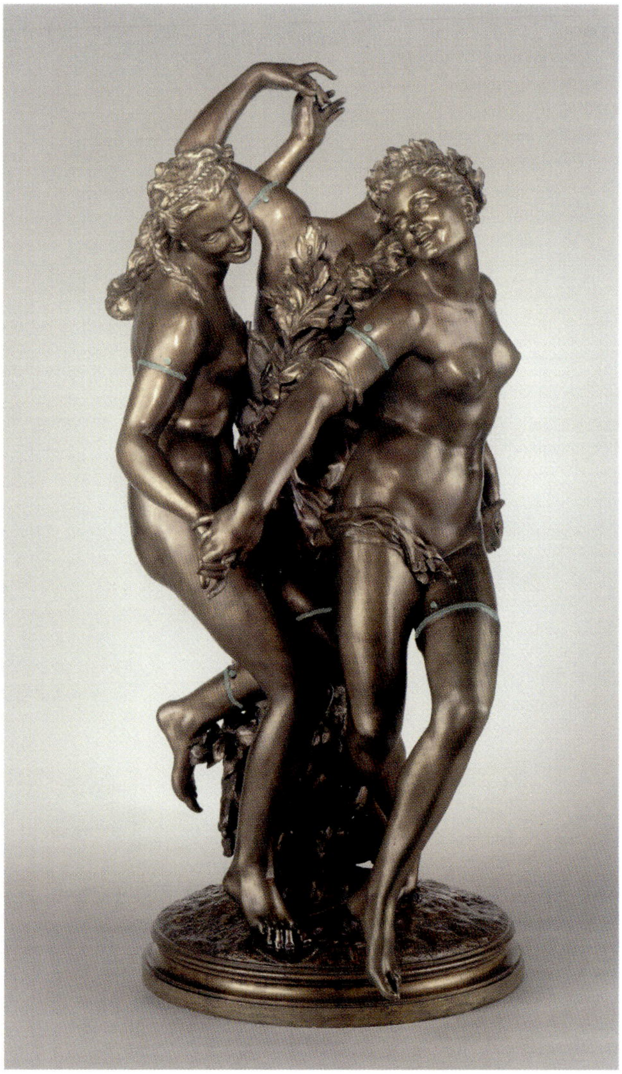

Figure 210 The joints between the separately cast parts of this bronze chef-modèle are clearly visible. The group was designed to be disassembled and the sections used as patterns for sand molding. Jean-Baptiste Carpeaux (French, 1827–1875), chef-modèle of *The Three Graces*, 1872, H. 81.3 cm (Clark Art Institute, Williamstown, Massachusetts, inv. 1955.976).

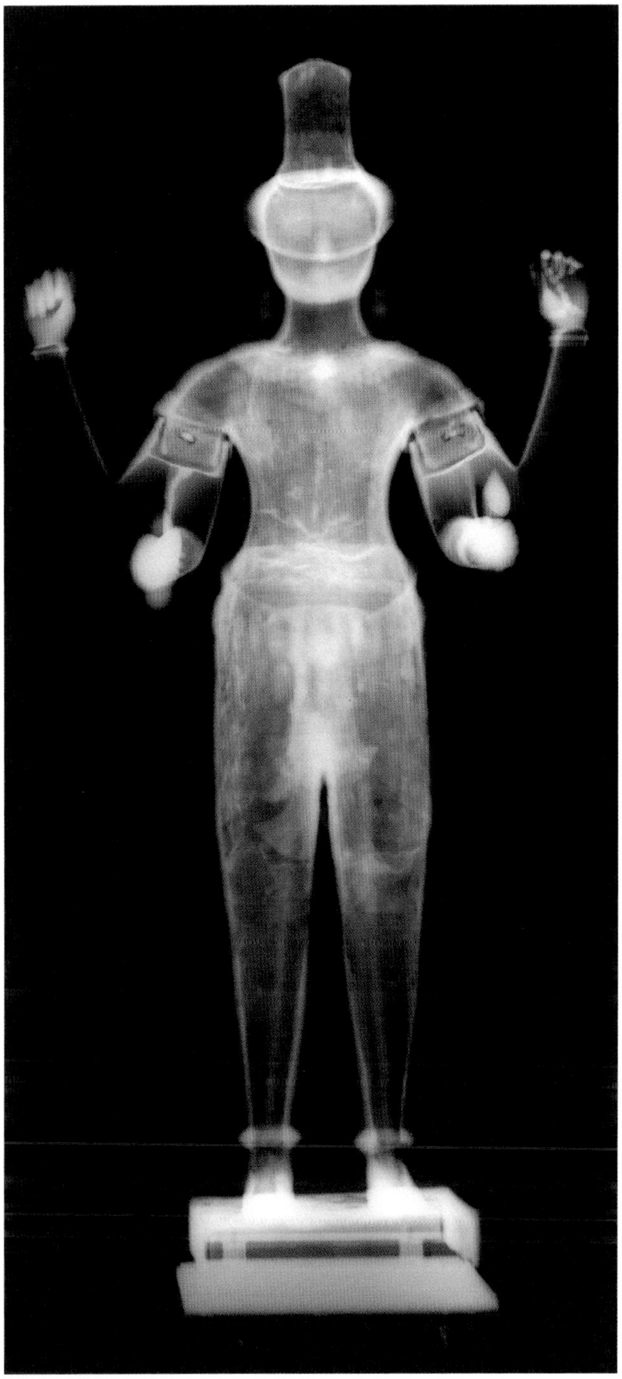

Figure 211 X-radiograph showing the assembly of the arms by mortise and tenon (Roman joint). *Lokeshvara*, Khmer, late 12th century, H. 63 cm (Musée national des arts asiatiques - Guimet, Paris, inv. MA 5960). See Bourgarit et al. 2003.

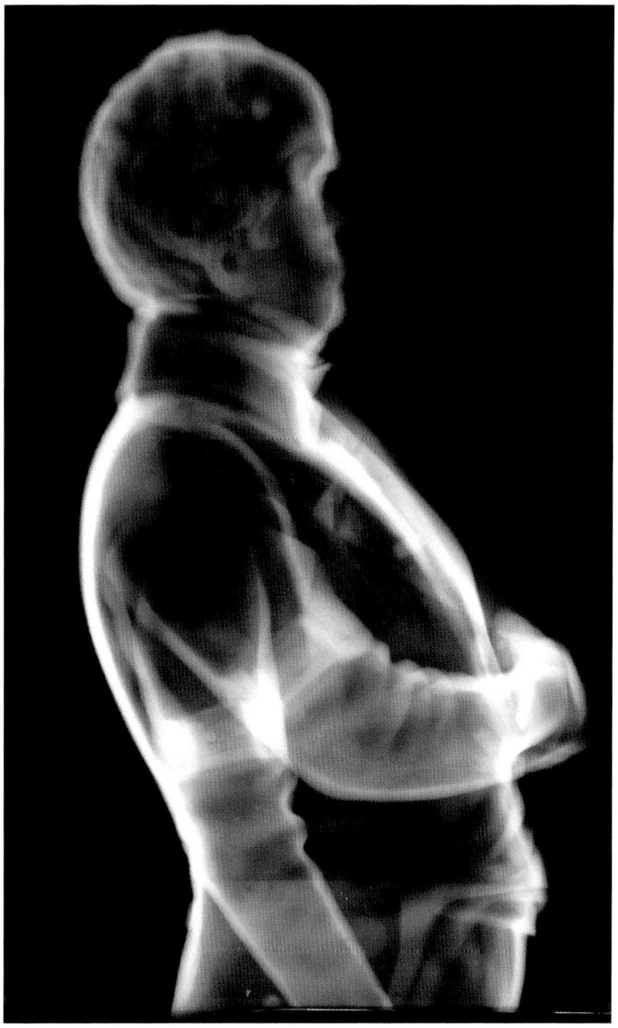

Figure 212 Gamma radiograph showing Roman joints in the arms of a figure. The type of joint used to attach the separately cast head cannot be determined in this radiograph. Gamma radiography was used due to the lack of available high-kV X-radiography. Thomas Ball (American, 1819–1911), *Daniel Webster*, 1853, sand cast probably after 1858, probably by the Ames Manufacturing Company, Chicopee, Massachusetts, H. 75.6 cm (Gilcrease Museum, Tulsa, inv. 0826.93). See fig. 213 for an overall view.

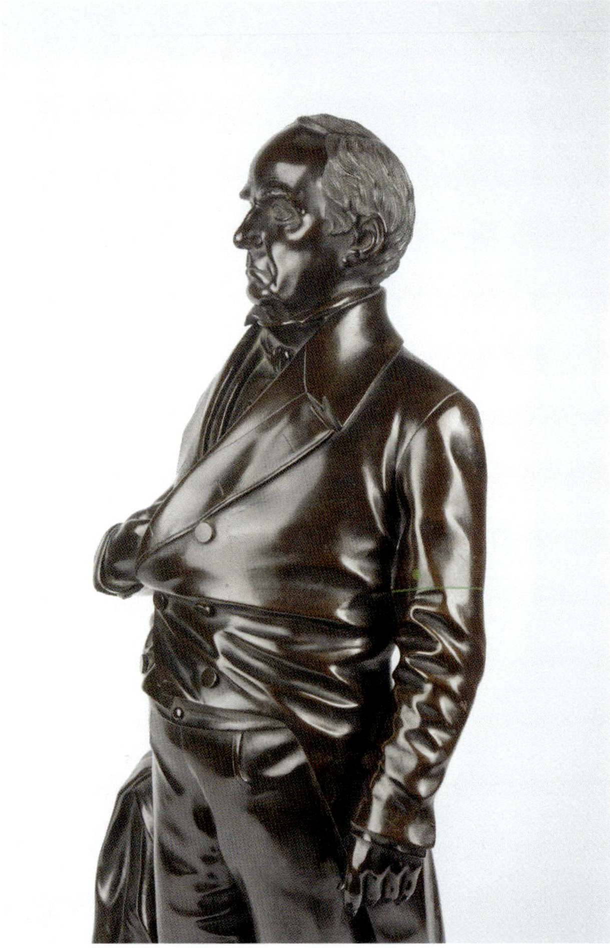

Figure 213 Exterior detail of the figure's straight arm showing the fine join line and circular outline of the pin of a Roman joint. A joint in the neck is obscured by the collar. Thomas Ball (American, 1819–1911), *Daniel Webster*, 1853, sand cast probably after 1858, probably by the Ames Manufacturing Company, Chicopee, Massachusetts, H. 75.6 cm (Gilcrease Museum, Tulsa, inv. 0826.93).

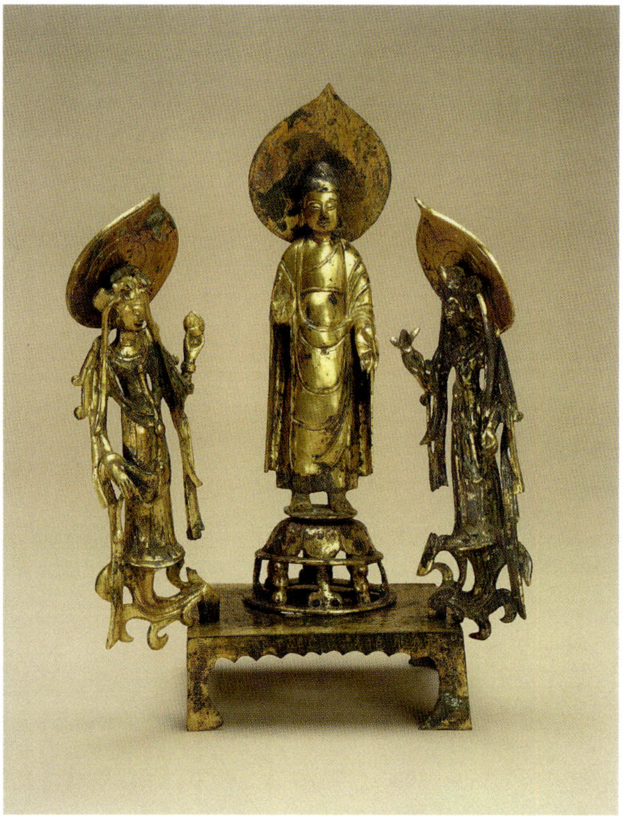

Figure 214 Figures attached to a base using tangs and pins. In addition, a tang at the back of the gilt bronze Buddha figure has a hole for a pin to be inserted once the halo has been placed over the tang. Historical Buddha and Bodhisattvas, Buddhist altarpiece, northern China, Sui dynasty, 597 CE, H. 32.1 cm (Freer Gallery of Art and Arthur M. Sackler Gallery, Smithsonian Institution, Washington, DC, purchase—Margaret and George Halderman, and Museum Funds, gift of Charles Lang Freer, inv. F1914.21a–h).

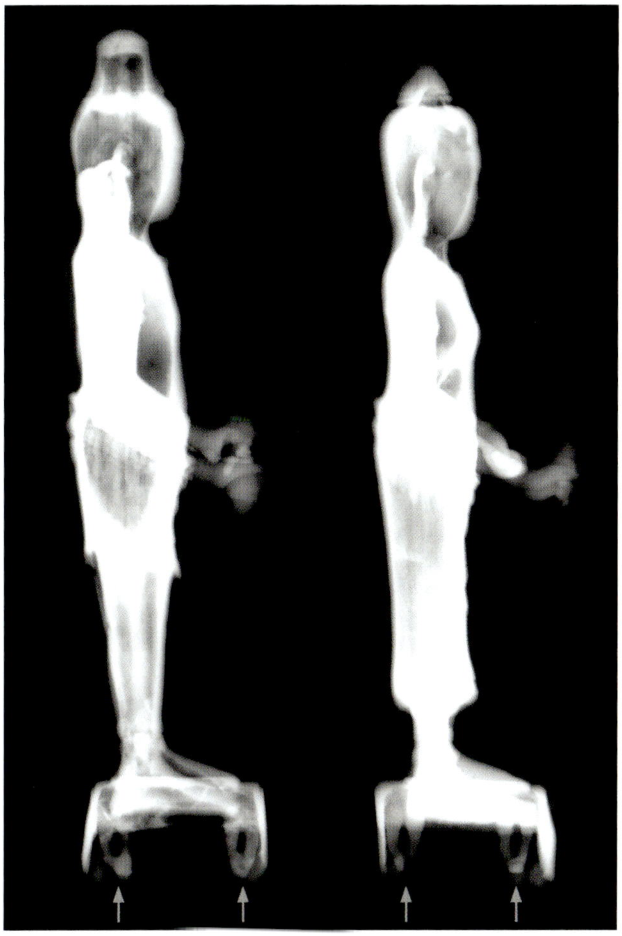

Figure 215 Annotated X-radiograph of Prajnaparamita and Lokeshvara showing the solid ring-shaped tangs cast with the statuettes to attach the figures to their base (arrows). *Khmer Buddhist Triad*, late 12th–early 13th century, H. 49.5 cm (National Museum of Cambodia, inv. Ga 2424). See Bourgarit et al. 2003.

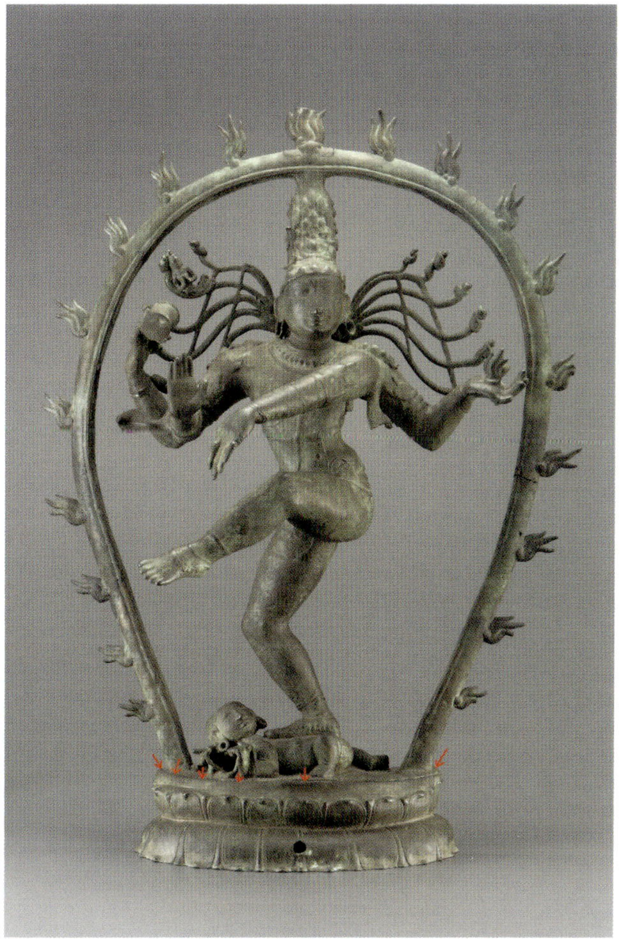

Figure 216 The statue was cast in two sections—the figures with halo, and the base—which were joined by hammering the top edge of the base over the sculpture's footplate to secure it. The red arrows in the overlay point to areas of overlap. *Shiva Nataraja (Lord of the Dance)*, India, state of Tamil Nadu, Chola dynasty, ca. 990, H. 70.8 cm (Freer Gallery of Art and Arthur M. Sackler Gallery, Smithsonian Institution, Washington, DC, purchase—Margaret and George Halderman and Museum Funds, F2003.2).

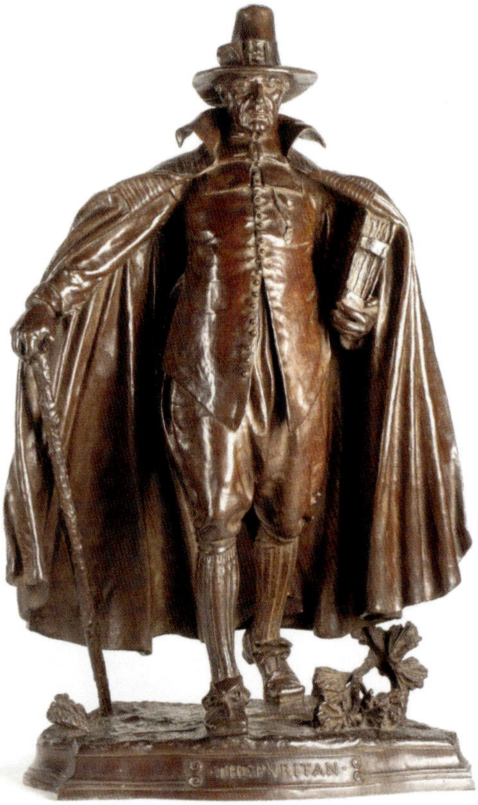

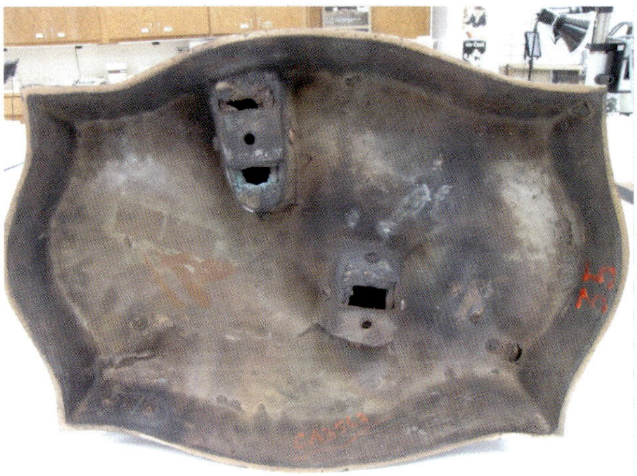

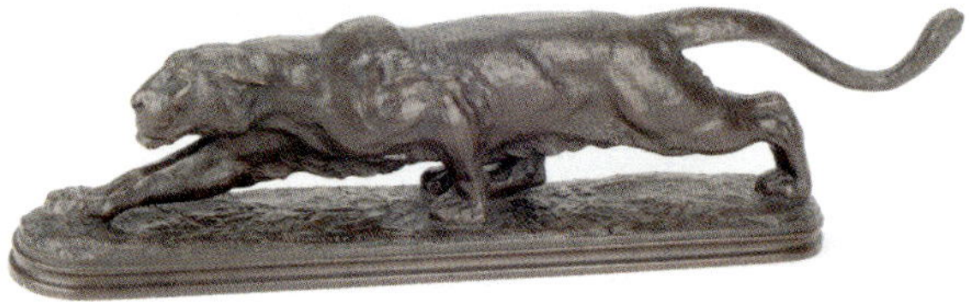

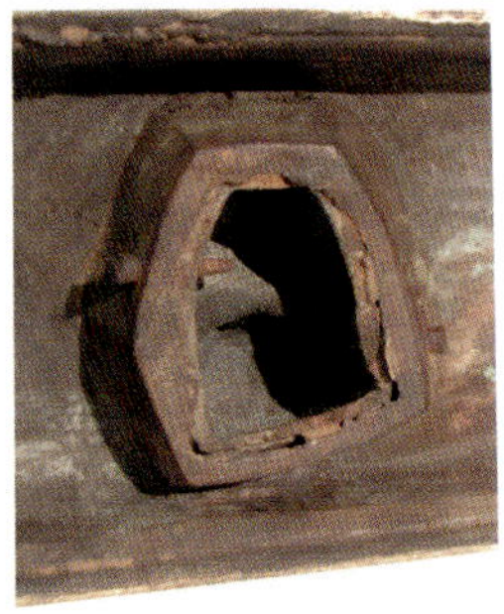

Figure 218 Three of the panther feet are set into recesses in the base and secured with pins (top), and detail showing tang securing one of the feet (bottom). Alexander Phimister Proctor (American, b. Canada, 1862–1950), *Prowling Panther*, 1891–92, sand cast 1905–12 at the John Williams Foundry, New York, H. 24.8 cm (Gilcrease Museum, Tulsa, inv. 0876.80).

Figure 217 On the underside of the base, two tangs secure the feet with a type of pinned Roman joint. The figure is a lost-wax cast and the base is sand cast. Augustus Saint-Gaudens (American, b. Ireland, 1848–1907), *The Puritan*, 1883–84, cast after 1899 at the Roman Bronze Works, New York, H. 78.1 cm (Gilcrease Museum, Tulsa, inv. 0826.114).

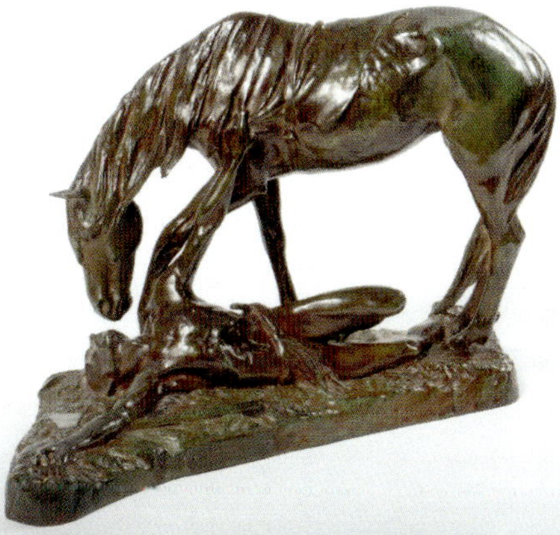

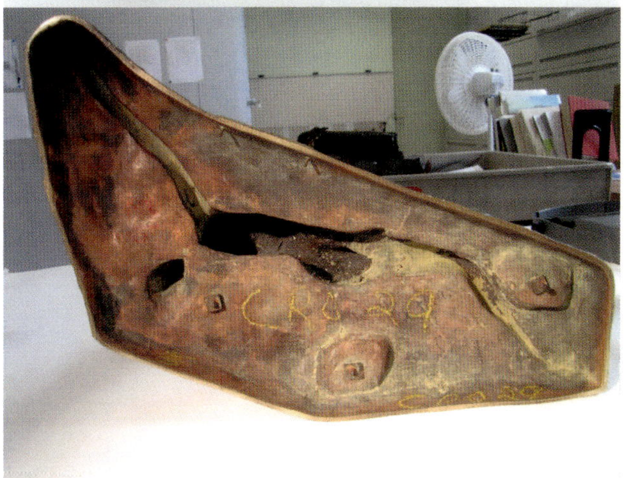

Figure 219 Detail underneath the base shows tang attachments made with threaded rods that were cast integrally with the figure. Nuts are used to secure the base to the threaded rods. Gutzon Borglum (American, 1867–1941), *The Fallen Warrior*, ca. 1891, sand cast probably around that time in France, foundry unknown, H. 27.6 cm (Gilcrease Museum, Tulsa, inv. 0876.125).

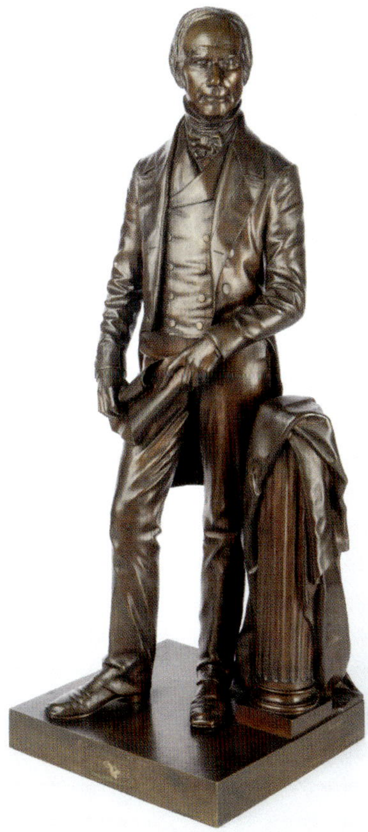

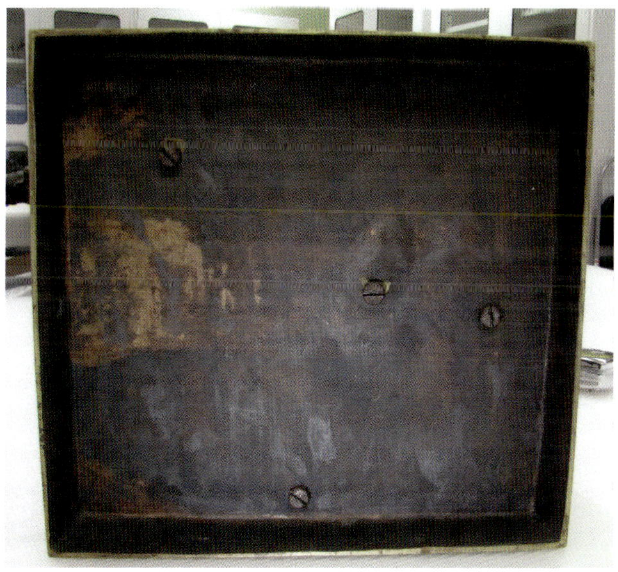

Figure 220 Detail underneath side of a base shows screw attachments without Roman joints. Thomas Ball (American, 1819–1911), *Henry Clay*, 1858, sand cast after 1858, probably by the Ames Manufacturing Company, Chicopee, Massachusetts, H. 78.1 cm (Gilcrease Museum, Tulsa, inv. 0826.94).

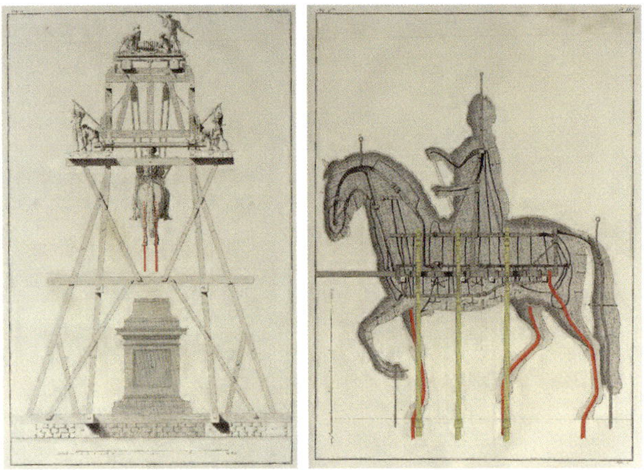

Figure 221 Sketches showing how armature rods included in the construction of the model are used to mount the over-life-size equestrian figure of Louis XV to its pedestal. Left: overall view showing the armature rods extending beyond the hooves; right: cross section showing the internal armature in the legs (red overlay) and some of the external structural supports removed after casting (yellow overlay). Mariette 1768, ch. 4, plate IX, p. 56 (left), ch. 14, plate III, p. 160 (right), showing Edme Bouchardon (French, 1698–1762), Louis XV, Paris, cast in 1758 by Pierre Gor (French, 1720–1773), H. 520 cm. After Desmas 2014.

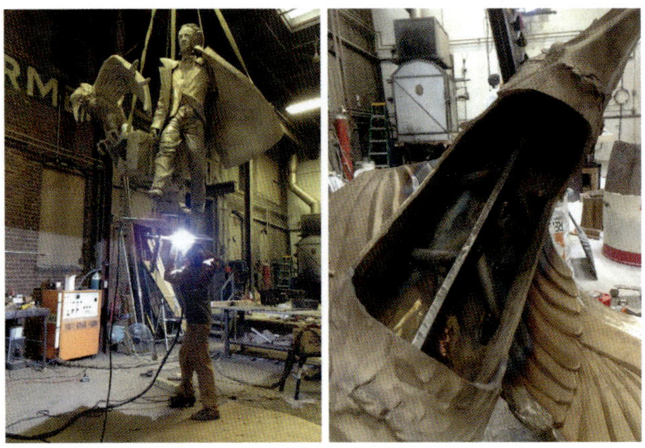

Figure 222 Dan Kendall of New England Sculpture Service fabricated the stainless-steel armature while assembling the separately cast bronze parts of the monument (left). It was welded in to provide structural support and to anchor the work to its base. The detail inside the raven (right) seen here from the bottom was taken mid-process. Stefanie Rocknak (American, b. 1966), *Poe Returning to Boston*, 2014, H. 173 cm (City of Boston).

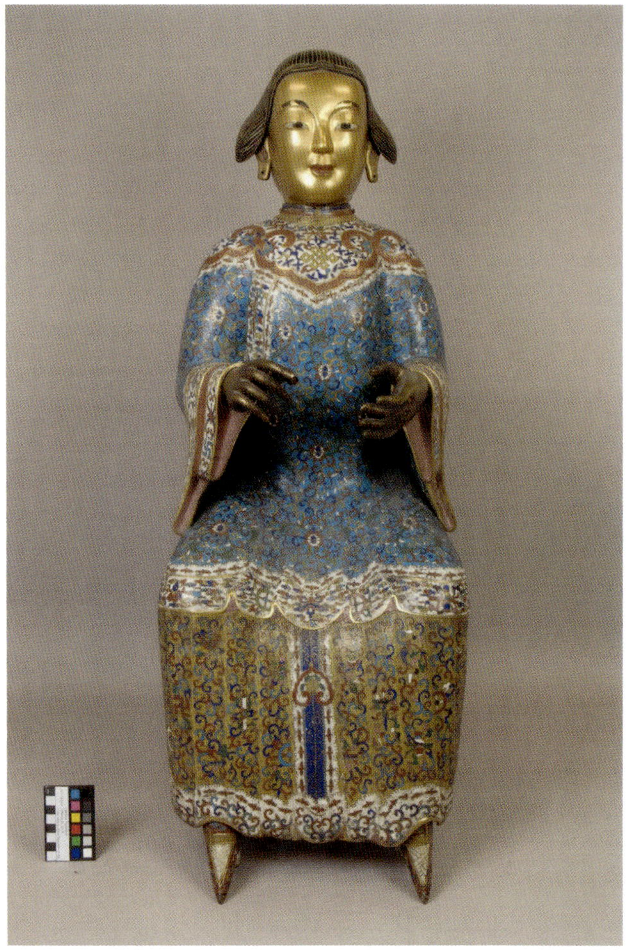

Figure 223 One of a pair of cloisonné enamel figures combining both cast sections (hands) and hammered sections (body and head) using mechanical methods for joining, namely bent tabs between head and body and rivets for the hands. Note that the enamel overlays are only on the hammered sections. *Seated Figure*, China, Qing dynasty, Qianlong period (1736–95), H. 88.3 cm (Metropolitan Museum of Art, gift of A. W. Bahr in memory of his wife, Helen Marion Bahr, 1954, inv. 54.124.2a–b.).

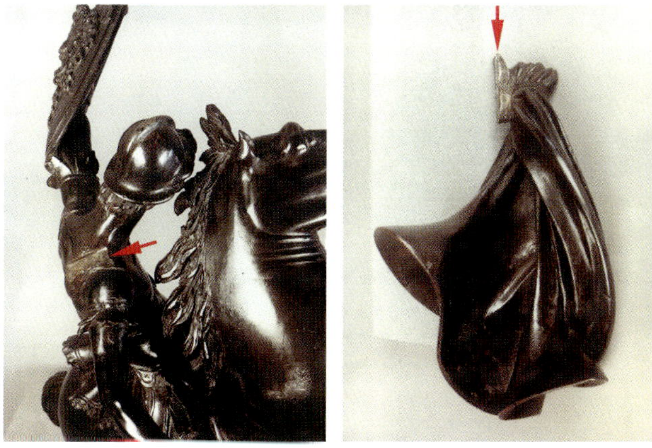

Figure 224 The separately cast cape secures to the proper right shoulder with a dovetail joint. The nearly identical alloys of the figure and the cape suggest that they were cast at the same time. After Willem Danielsz. van Tetrode (Dutch, ca. 1525–1580), *Warrior on Horseback*, 1562–65, H. 39.7 cm (J. Paul Getty Museum, inv. 85.SB.90).

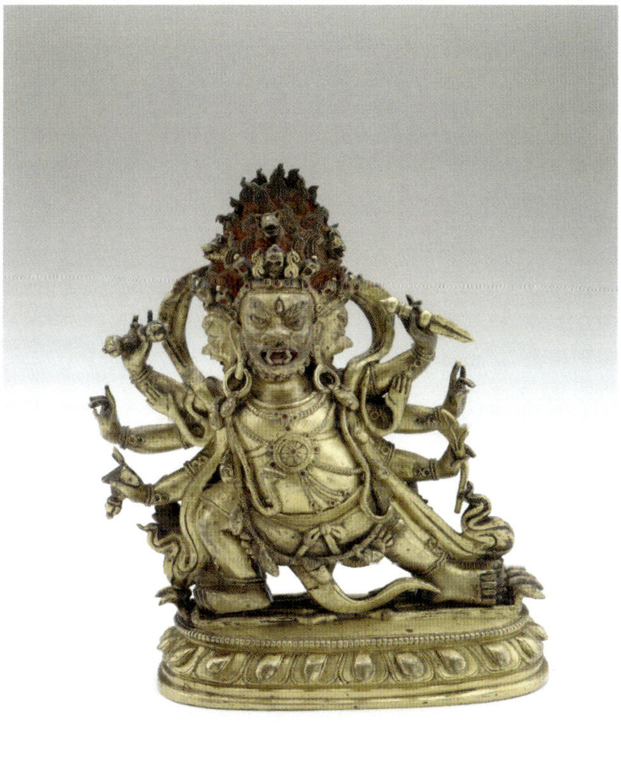

Figure 225 Example of a figure mechanically attached to a separately cast base, which is closed and holds sacred relics. Six-Armed Hayagriva, China, Qing dynasty, second half of the 18th century, H. 16.5 cm (Freer Gallery of Art and Arthur M. Sackler Gallery, Smithsonian Institution, Washington, DC, The Alice S. Kandell Collection, inv. S2018.32).

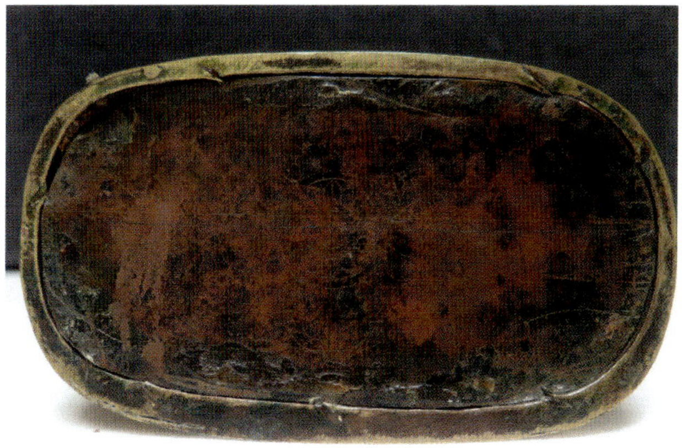

Figure 226 Bottom view of a statuette showing the cover being held on the bottom of the base by chiseling burrs from the base of the sculpture over the copper cover. The remains of resins to aid in sealing the joint can also be seen around the outer edges of the cover. Six-Armed Hayagriva, China, Qing Dynasty, second half of the 18th century, H. 16.5 cm (Freer Gallery of Art and Arthur M. Sackler Gallery, Smithsonian Institution, Washington, DC, The Alice S. Kandell Collection, inv. S2018.32). See fig. 225 for an overall view.

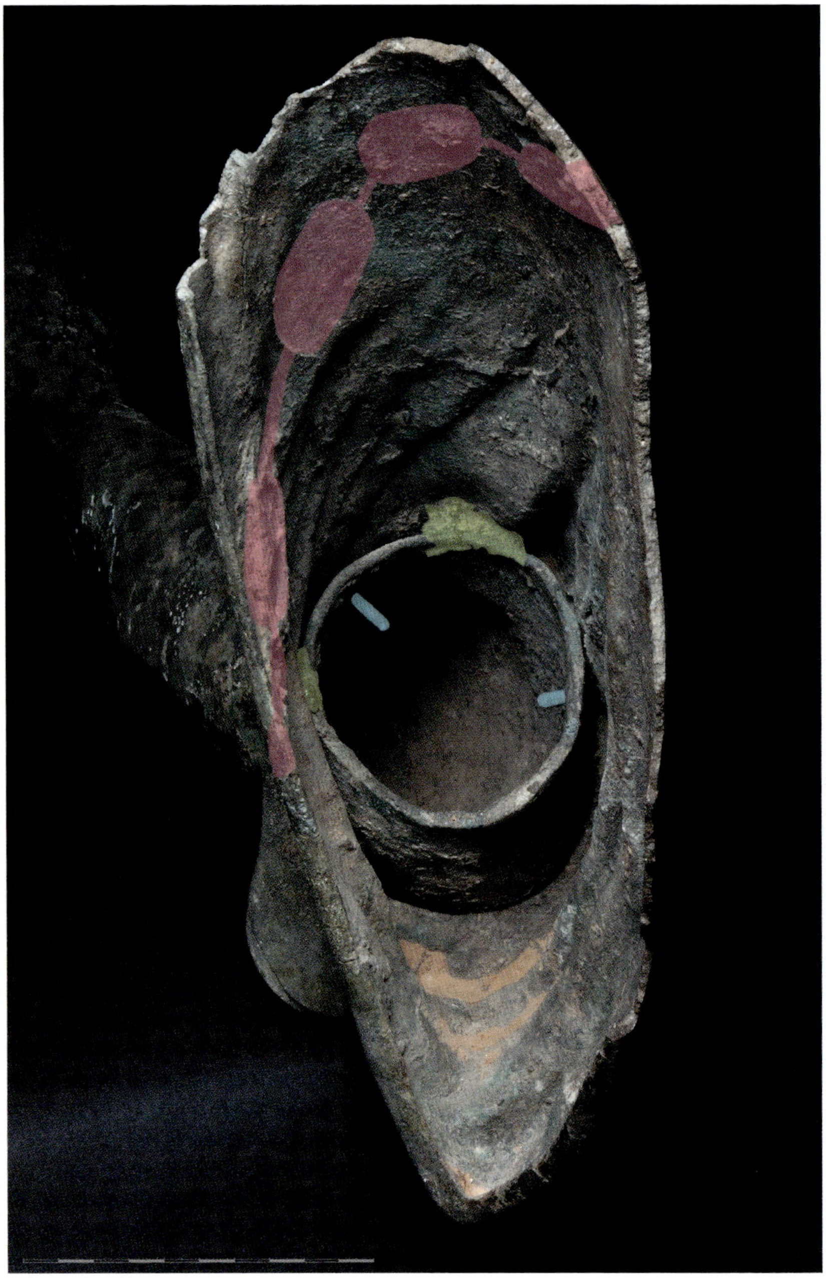

Figure 227 The drapery of the bronze arm was cast in sections and assembled using a flow fusion weld (pink overlay) that is only visible from the interior. The welding appears as a chain of oval shapes. The arm was then joined to the drapery by welding (green overlay) with securing pins (blue overlay). Arm of a statue found in the Villa of Arconciel, Switzerland, end of the 1st–3rd century CE, over life size (Service Archéologique de l'Etat de Fribourg, Switzerland, inv. Arconciel/Es-Nés FR 85). See Mille and Serneels 2012.

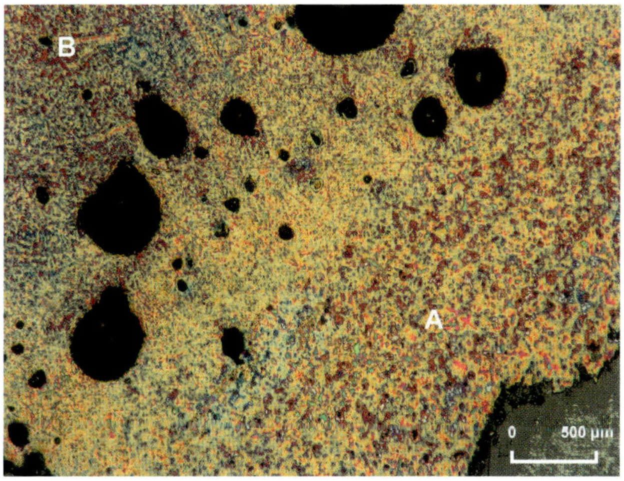

Figure 228 Bright field photomicrograph of an etched polished cross section of the welded zone between the body and the neck of a bronze horse. Zone B corresponds to the filler metal (oval area of a basin), zone A to the neck of the horse (primary casting). The yellowish area between A and B is the melted zone (MZ) and the heat-affected zone (HAZ), which consist of a mixture of the filler metal and the primary cast. This area is also marked by heavy porosity (black holes). Note that the elemental composition of the weld metal is strictly the same as that of the primary cast—that is, the neck (leaded bronze with 9.6 wt% Sn and 7.2 wt% Pb). *Horse of Neuvy-en-Sullias*, France, 1st century BCE–1st century CE, H. 113 cm (Musée historique et archéologique de l'Orléanais, Orléans, France, inv. A6286). See Mille 2007.

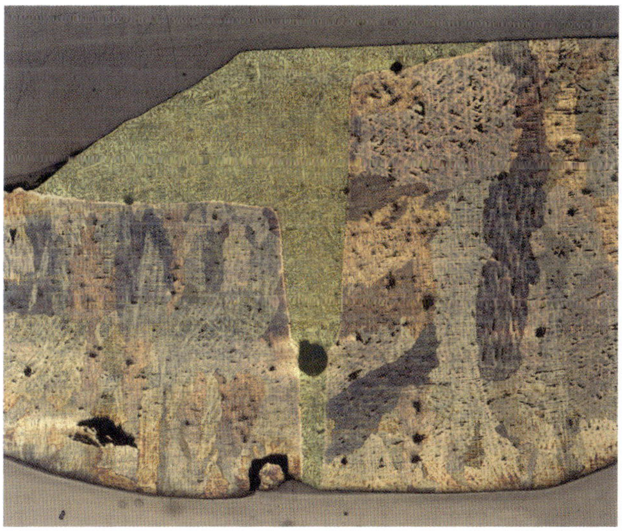

Figure 229 Bright field photomicrograph of an etched cross section across a French 18th-century brazing joint between two cast brass elements. The section shows the typical large grains and dendritic structure of the two cast elements as well as the so-called Widmanstätten microstructure of the quickly cooled, high-zinc brass brazing metal (green overlay). Note that the two metals did not melt into each other. Etched with a solution of 25% hydrochloric acid and 8% iron chloride in industrial methylated spirit. Cross-section prepared by Yannick Chastang.

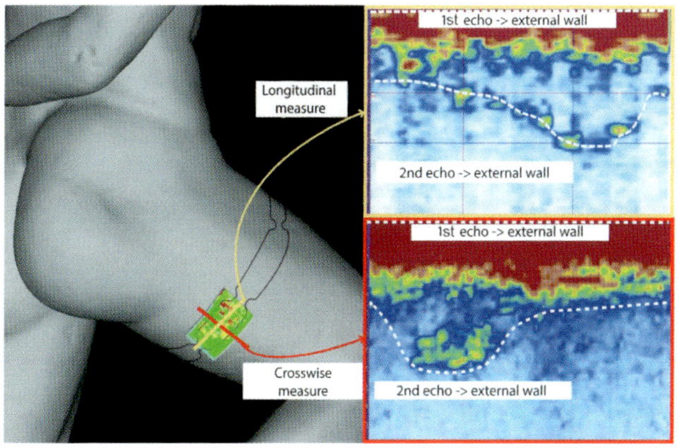

Figure 230 Ultrasonic phased array applied to the study of the flow fusion welding of an antique large bronze. The C-scan represents the variations in thickness of the wall of bronze at the right thigh to upper leg junction (left). The profilometry of a welding basin (longitudinal and cross profiles) is shown (right). *Captive Gaul*, Gaul, last quarter of the 1st century BCE, H. 63 cm (Musée départemental Arles Antique [MDAA], France, inv. Rho.2007.06.1962). See Azéma et al. 2013.

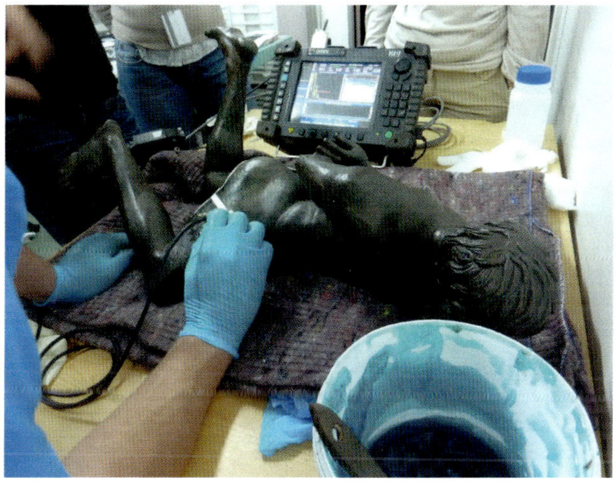

Figure 231 Ultrasonic phased array applied to the study of the welding. *Captive Gaul*, Gaul, last quarter of the 1st century BCE, H. 63 cm (Musée départemental Arles Antique [MDAA], France, inv. Rho.2007.06.1962). See Azéma et al. 2013.

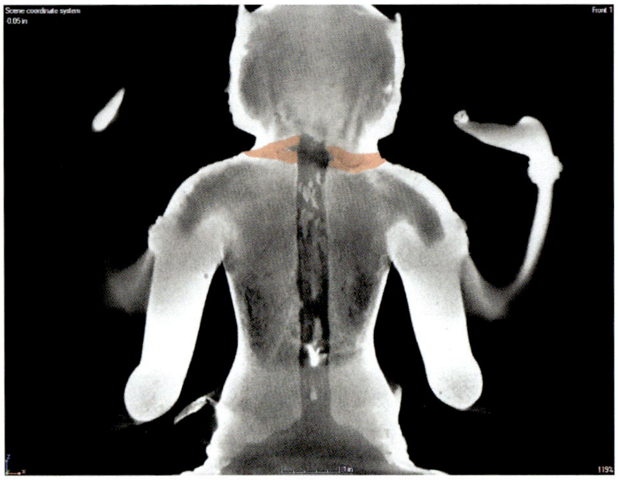

Figure 232 This slice from a computer tomography (CT) scan captured at the widest area of the neck shows a layer of metal separating the head from the body (see overlay). This is evidence that the head was cast onto an already-cast body by a localized lost-wax process. Excess metal dripped down a hollow channel along the vertical wooden armature in the torso. Elemental mapping by X-ray fluorescence (XRF) analysis revealed that the body is made of a different alloy than the head, confirming that the head was joined to the body at a separate stage. *Seated Brahma*, Cambodia, late 12th or early 13th century, H. 31.5 cm (Walters Art Museum, Baltimore, inv. 54.2734). See Strahan 1998.

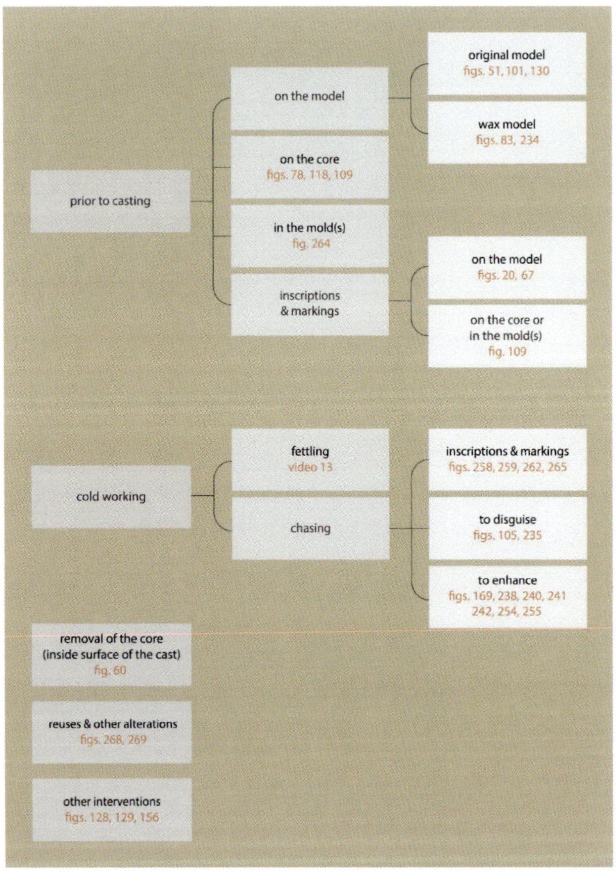

Figure 233 Diagram of characteristic tool marks produced at various stages of the production of a bronze sculpture.

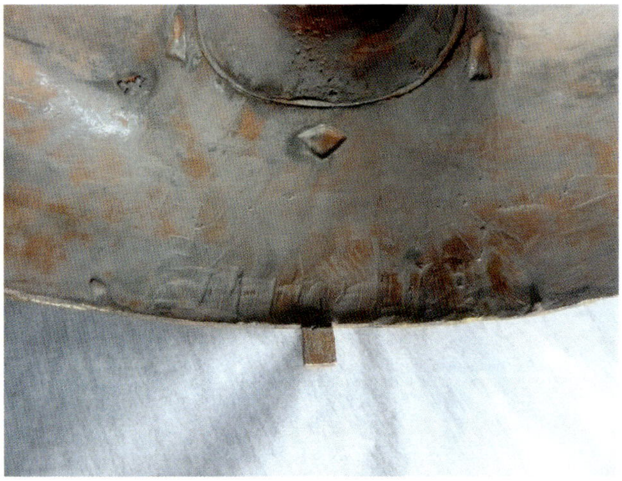

Figure 234 Tool marks made in the wax as seen under the base of a Tibetan divinity. *White Tara*, Tibet, 17th century, H. 65 cm without base (Musée national des arts asiatiques - Guimet, Paris, inv. MA 12495). See C2RMF Internal Report 2012. See fig. 240 for an overall view.

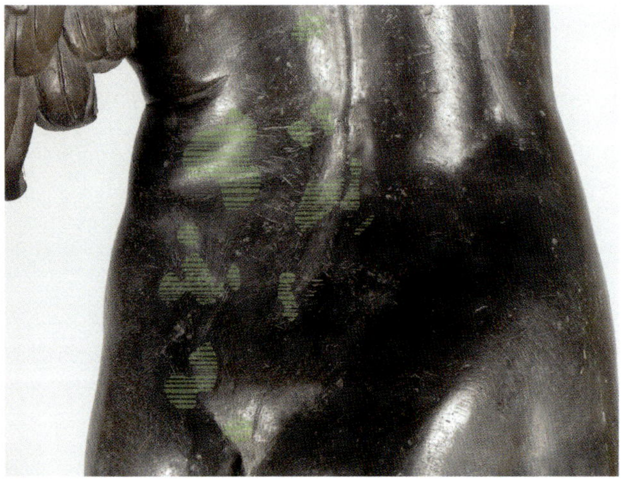

Figure 235 Detail of the figure's back showing scattered clusters of rectangular-shaped matting marks disguising the repairs of casting defects (see overlay). Pierre Biard I (French, 1559–1609), *Fame*, ca. 1607, H. 134 cm (Musée du Louvre, inv. LP361). See fig. 135 for an overall view.

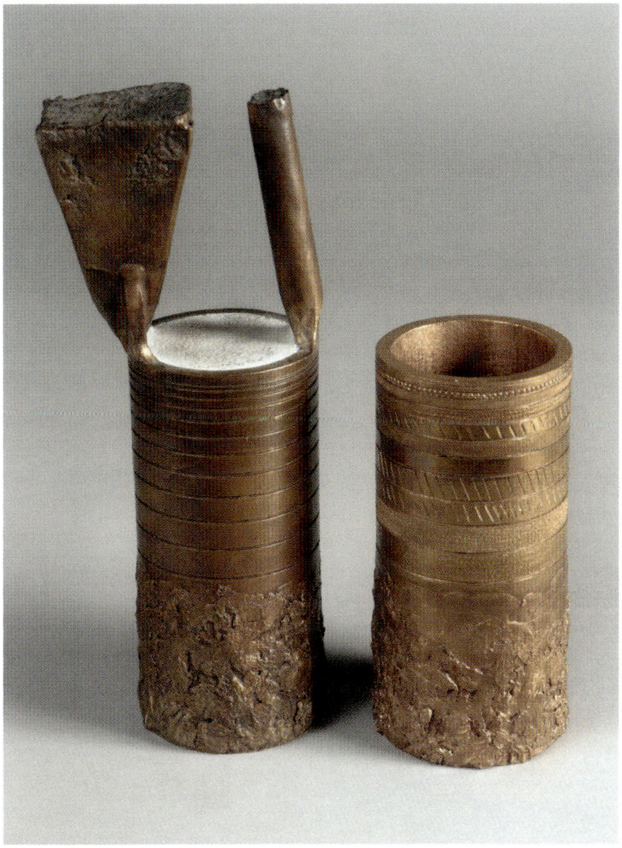

Figure 236 Two bronze cylinders (H. ~30 cm each) made by the Coubertin foundry, France, in the 1990s for the Musée du Louvre to show the different steps of casting and finishing. The as-cast state at the bottom of both cylinders is somewhat exaggerated. Note the triangular sprue and cylindrical vent in the left example, and their removal at right.

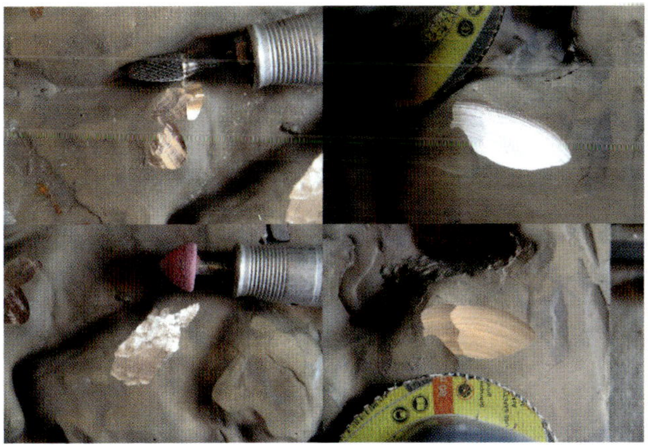

Figure 237 Marks of air tools and electric grinders on a cast bronze.

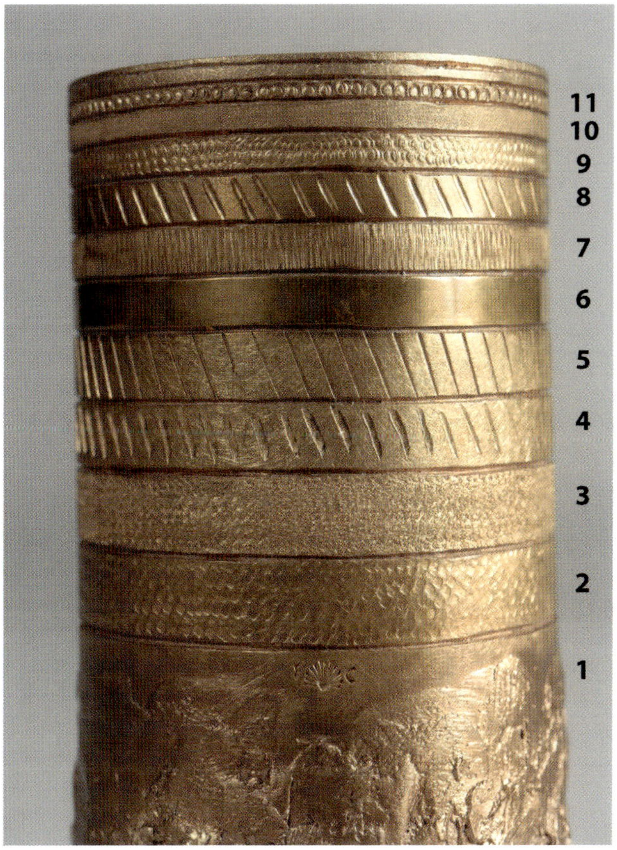

Figure 238 Bronze cylinder (the portion visible is ~12 cm high) demonstrating a range of chasing marks created using the chasing tools shown in fig. 239. From bottom to top: 1) raw, as-cast texture modeled in wax; at the center is the foundry mark struck with a steel punch (*frappé*) flanked by the punched letters "F" and "C"; 2) repetitive marks of a diamond-shaped matting tool with a smoothly textured tip (in French *ciselet*, and more specifically *mat à bout lisse carré*); 3) repetitive marks of a matting punch with a textured, granular tip (*mat sablé moyen*); 4) oblique marks of a veining punch with a coarse profile/contour (*mat à tracer large* or *traçoir gros*); 5) oblique marks of a veining punch with a refined, "sharpened" profile/contour (*mat à tracer fin* or *traçoir fin*); 6) polishing cloths of increasingly finer grit (200 to 1000) followed by electrical polishing wheel; 7) repetitive marks of a very fine veining punch (*mat à tracer très fin*); 8) oblique marks of a veining punch with a rounded profile/contour (*mat à tracer large* or *traçoir gros*); 9) repetitive marks of "dot" punch with a hemispherical, indented profile/contour (*perloir* or *mat à perler*); 10) repetitive marks of a matting punch with a finely textured tip (*mat sablé fin*); 11) repetitive marks of a "circle" punch with a rounded, circular perimeter and a recessed center profile/contour (*perloir à touche* or *mat à perler*). Bronze cylinder, 1990s, H. ~30 cm, made by the Coubertin foundry, France, for the Musée du Louvre, Département des sculptures.

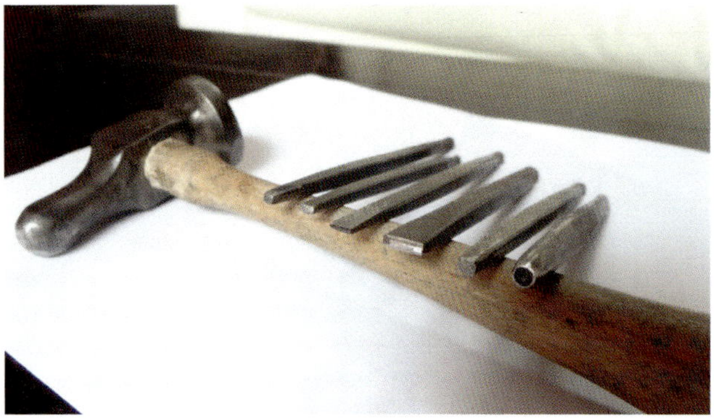

Figure 239 Chasing tools belonging to Jean Dubos, Coubertin foundry, France, used notably to work the demonstration cylinder (fig. 238). From left to right: punch with smooth texture (in French *ciselet*, and more specifically *mat à bout lisse carré*); punch with a textured, granular surface (*mat sablé moyen*); tracer or scribe (*mat à tracer fin* or *traçoir fin*); veining tool (*mat à tracer large* or *traçoir gros*); fine-textured matting punch (*mat sablé fin*); and domed "dot" punch (*perloir* or *mat à perler*).

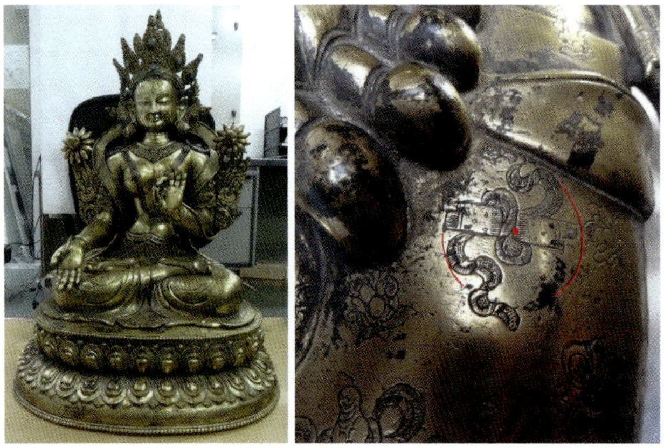

Figure 240 Tibetan divinity front view with a detail on the right showing the chased decoration in the fabric. Note the preparatory circle made with a compass (red lines) and the punch marks texturing the snake. *White Tara*, Tibet, 17th century, H. 65 cm without base (Musée national des arts asiatiques - Guimet, Paris, inv. MA 12495). See C2RMF Internal Report 2012.

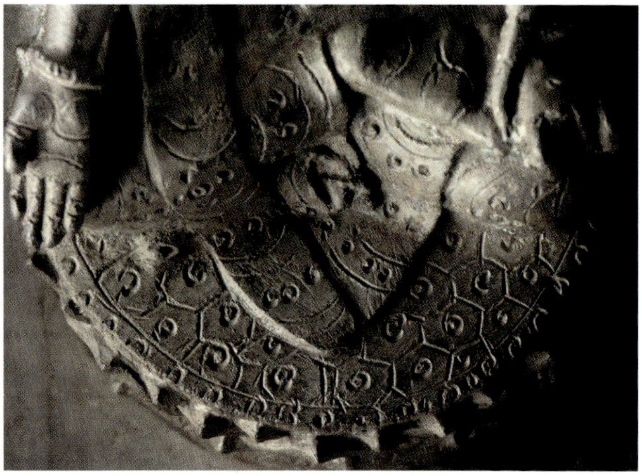

Figure 241 Reflectance transformation image (RTI) of the upper surface of the lotus of a lost-wax-cast statuette. This imaging technique allows examination of engraving marks made in the metal under variable lighting conditions. *Statuette of Bodhisattva*, Java, late 9th–early 10th century, H. 17 cm (Musée national des arts asiatiques - Guimet, Paris, inv. MG 3822). See Mechling et al. 2018.

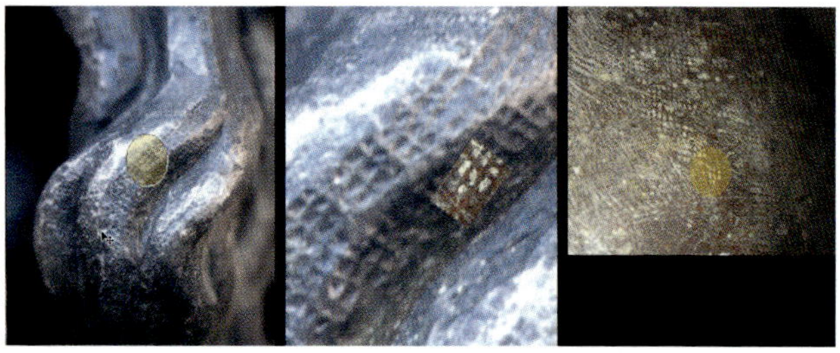

Figure 242 The same punch marks with the same sequential strikes were found on two statues by Benvenuto Cellini (Italian, 1500–1571). The right image from the forehead of the *Bust of Bindo Altoviti*, 1546–50, H. 105.5 cm (Isabella Stewart Gardner Museum, Boston, inv. S26e21) and the left and middle images from Medusa's neck on *Perseus with the Head of Medusa*, 1545–54, H. 320 cm (bronze group) (Loggia dei Lanzi, Piazza della Signoria, Florence). Micro-hardness tests were undertaken using a calibrated prism penetration durimeter, showing values very similar to that of a cold-worked copper surface. See I.6, note 17; Morigi and Morigi 2004.

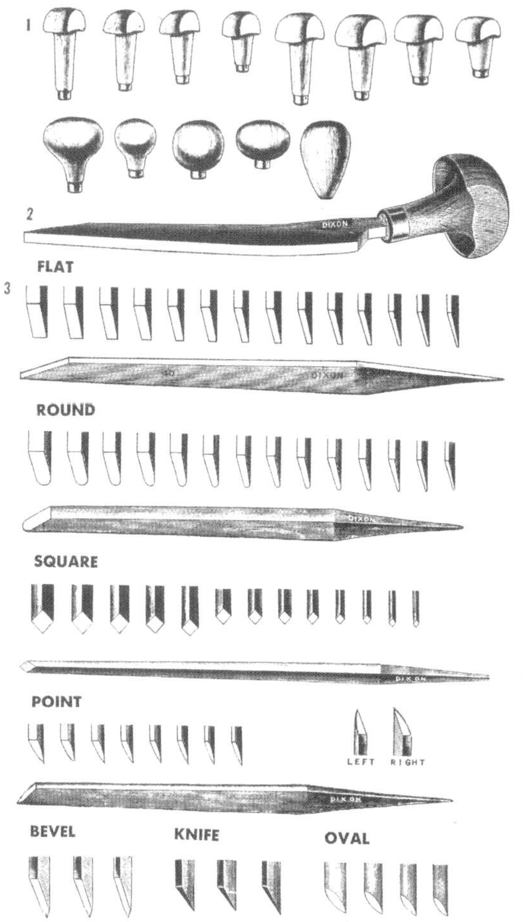

Figure 243 Range of different cutting faces possible for engraving a surface: 1) wooden handles; 2) burin blade in a handle; 3) variety of shapes and sizes of cutting faces and burins. After Untracht 1968, 112.

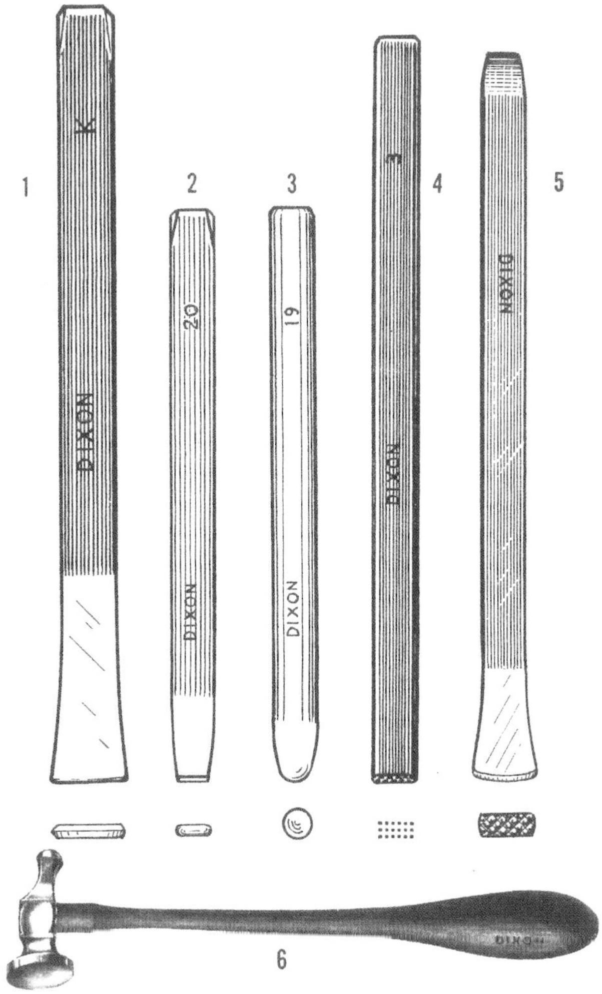

Figure 244 Sampling of punches a chaser might use to cold work a metal surface. After Untracht 1968, 85.

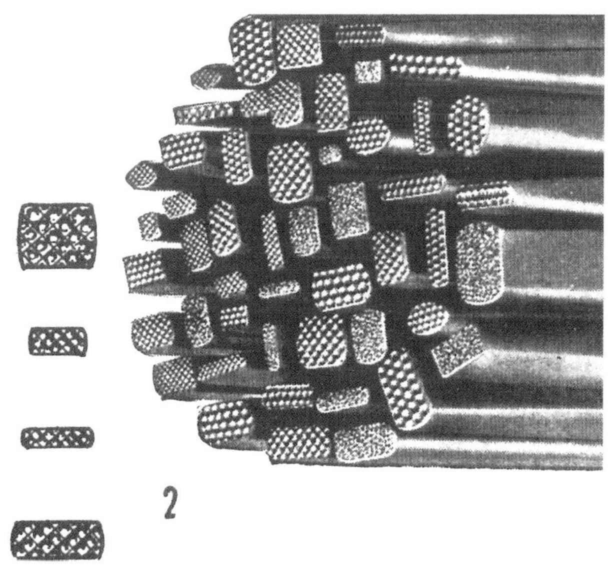

Figure 245 Sampling of matting tool faces used in cold working a surface. After Untracht 1968, 96.

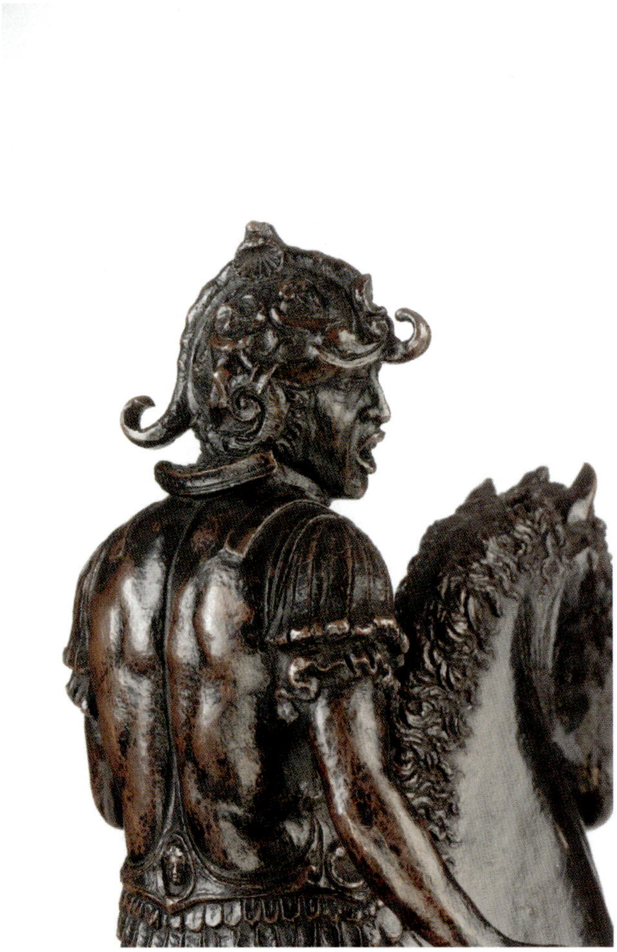

Figure 246 Detail showing peening on the back and arm of the warrior. This technique allows the light to bounce off the facets on the surface, creating a vibrancy when compared to a smooth, shiny surface reflection. Andrea Riccio (Italian, 1470–1532), *Shouting Horseman*, ca. 1510–15, H. 34.2 cm (Victoria and Albert Museum, Salting Bequest, inv. A.88:1-1910). See Stone 1981; Stone 2008; Motture 2019.

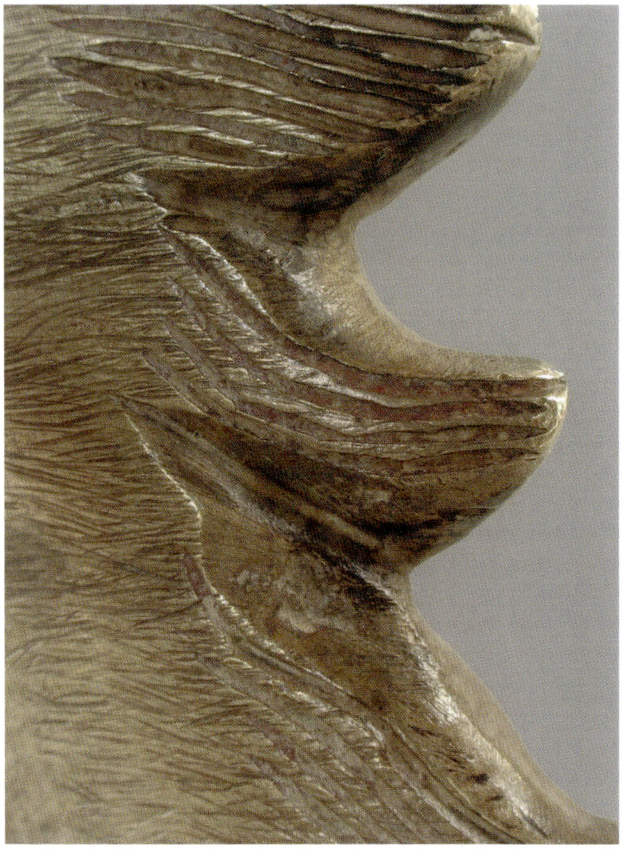

Figure 247 Detail of a flame handle with the marks of two different files visible across the surface. The red accretions within the incised lines are residues of a polishing compound utilized at some point in the object's history. Aquamanile in the Form of a Lion, Nuremberg, Germany, ca. 1400, H. 31.9 cm (Metropolitan Museum of Art, The Cloisters Collection, 1994, inv. 1994.244). See Dandridge 2006.

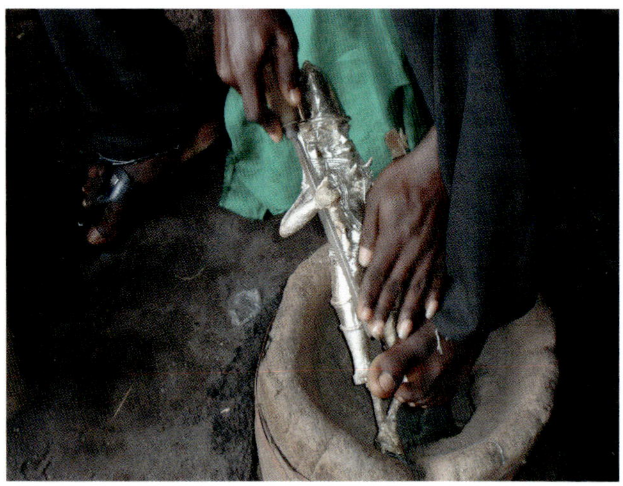

Figure 248 Filing the surface of a contemporary Benin figure in a brass casting workshop, Igun Street, Benin City, Nigeria, 2002.

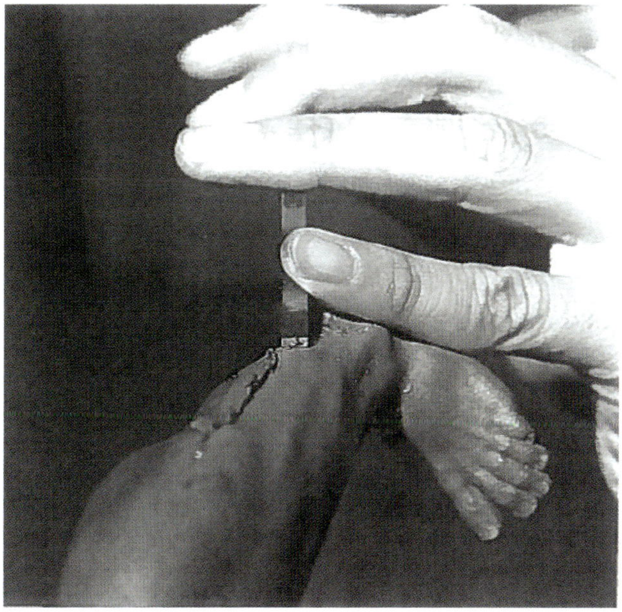

Figure 249 Francesca G. Bewer chisels a metal flash off the leg of her experimental reproduction cast of Mars, 1989, after a model by Giambologna (Flemish, active Florence, 1529–1608), ca. 1587, H. 40.5 cm. See Bewer 1996b.

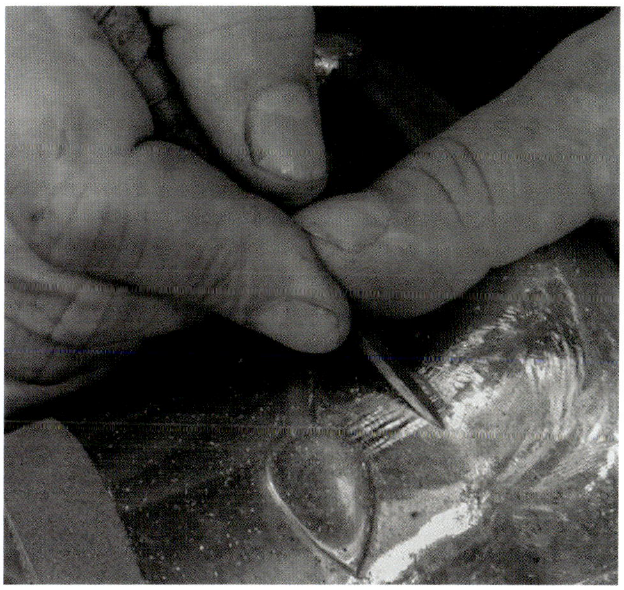

Figure 250 Detail of experimental casting showing a scraper being pushed and pulled across the previously filed surface. The action creates a series of parallel, shallow depressions that are retained throughout the finishing process, creating a subtle modulation across the surface. Ubaldo Vitali (American, b. 1944), Reproduction of a Lion Aquamanile, Maplewood, New Jersey, 2006, H. 19 cm, after Aquamanile in the Form of a Lion, probably northern Germany, 12th century, H. 19.5 cm (Metropolitan Museum of Art, gift of Irwin Untermyer, 1964, inv. 64.101.1491). See Dandridge 2006.

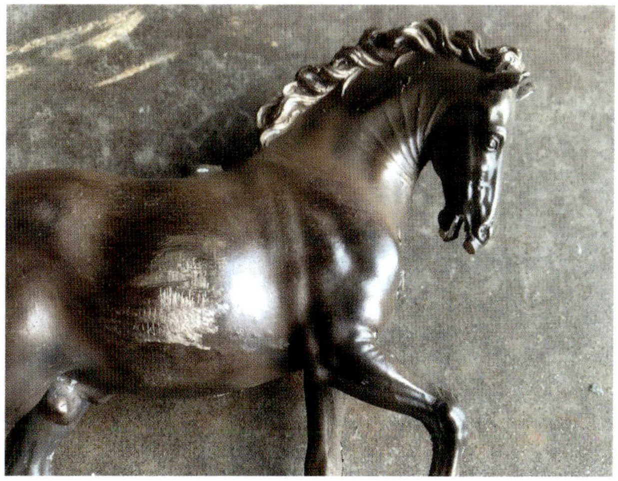

Figure 251 During the initial fettling stages the removal of the main feed on the side of the horse was carried out using a course file and large scraper to reduce or rectify the surface. The scraper produced chatter lines—here the short, parallel, vertical cuts—and the file the longer, undulating horizontal marks that reflect brightly against the surrounding darker, as-cast oxidized surface. Finer files and scrapers are used subsequently to smooth out these marks. Andrew Lacey (British, b. 1969), *Pacing Horse*, 2016, H. 23 cm, aftercast of bronze attributed to Gianfrancesco Susini (Italian, 1585–1653) after a model by Giambologna (Flemish, active Florence, 1529–1608) (private collection).

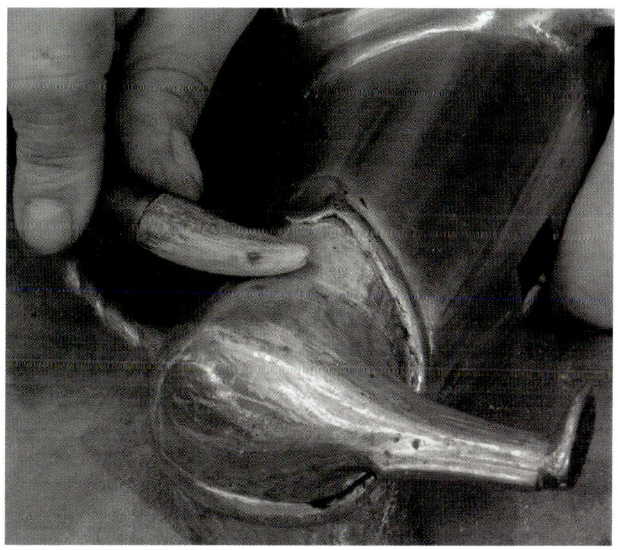

Figure 252 Burnishing compresses the metal surface to reduce the distortion left by fine files or scrapers. This detail shows the haunch burnished with a highly polished and rounded antler tool. Ubaldo Vitali (American, b. 1944), Reproduction of a Lion Aquamanile, Maplewood, New Jersey, 2006, H. 19 cm, after Aquamanile in the Form of a Lion, probably northern Germany, 12th century, H. 19.5 cm (Metropolitan Museum of Art, gift of Irwin Untermyer, 1964, inv. 64.101.1491). See Dandridge 2006.

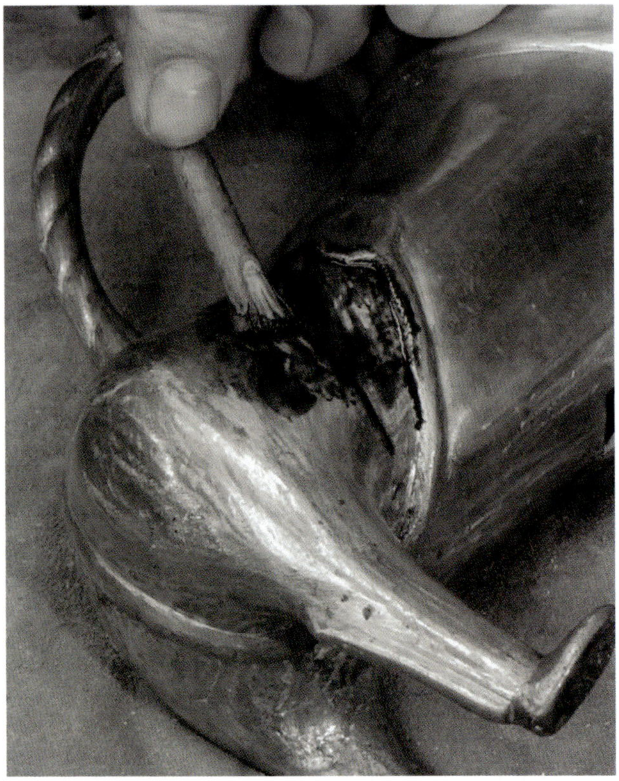

Figure 253 Detail illustrating finishing of the surface with a shredded stick and a series of graded polishing compounds in a water or oil slurry. Ubaldo Vitali (American, b. 1944), Reproduction of a Lion Aquamanile, Maplewood, New Jersey, 2006, H. 19 cm, after Aquamanile in the Form of a Lion, probably northern Germany, 12th century, H. 19.5 cm (Metropolitan Museum of Art, gift of Irwin Untermyer, 1964, inv. 64.101.1491). See Dandridge 2006.

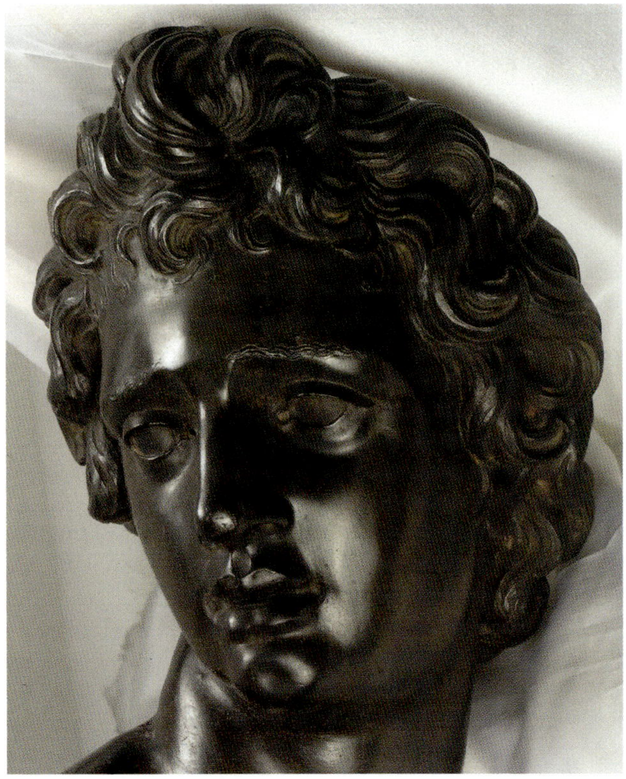

Figure 254 Detail of head showing the crisp and faceted shape of the hair, partially reworked in the metal, and the use of a V-shaped graver to incise the eyebrows. Francesco Bordoni (Italian, 1580–1654), *Young Captive*, 1618, H. 160 cm (Musée du Louvre, inv. MR 1668). See Bourgarit, Bewer, and Bresc-Bautier 2014.

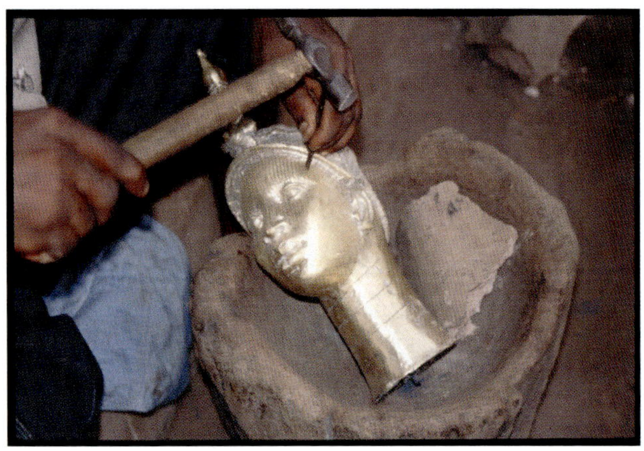

Figure 255 Chasing the surface of a copy of an Ife head in a brass casting workshop, Igun Street, Benin City, Nigeria, 2002.

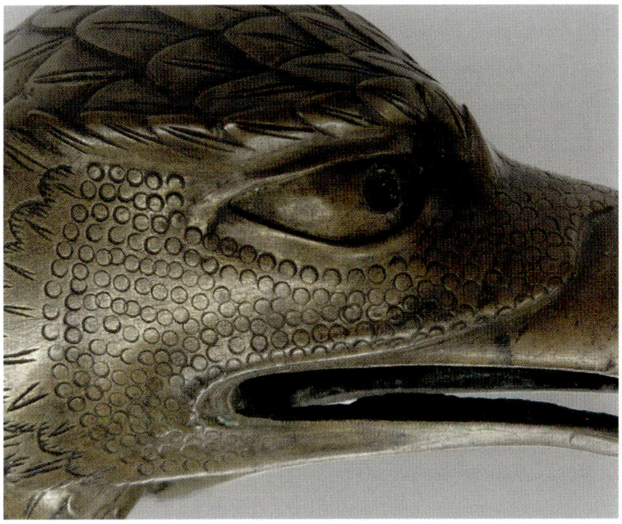

Figure 256 Dense pattern of hollow circular punch marks created in the metal. Eagle Lectern, Hildesheim, Germany, ca. 1220, H. 57.5 cm (Dom Museum Hildesheim, Germany, inv. D 1984-2).

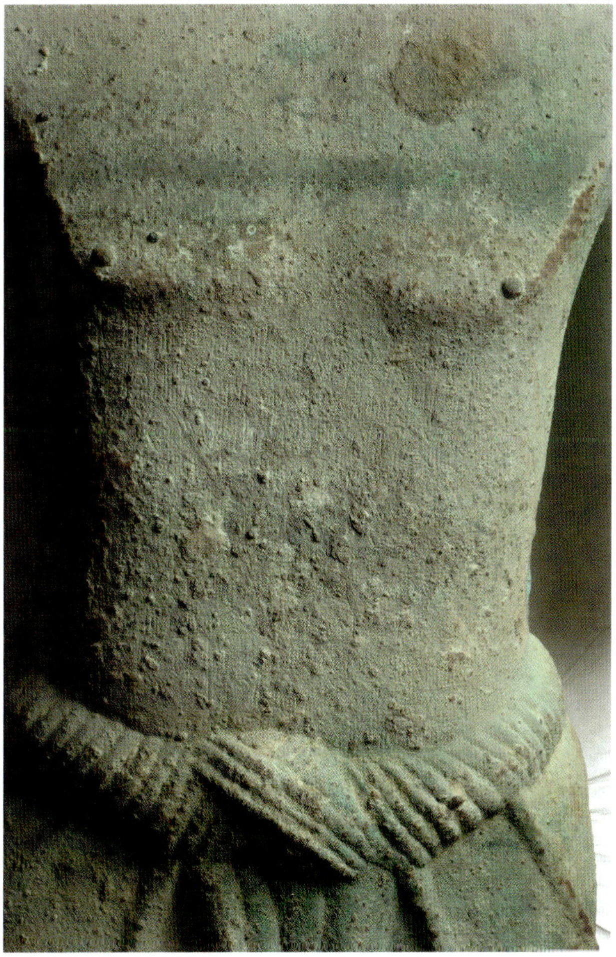

Figure 257 Detail of breast showing a lengthy inscription in the area across the diaphragm/solar plexus. The letters are quite irregular in shape and not deeply chased, which might testify to their cold working in the metal rather than in the wax. Bronze Statue of Hawtar'athat, Yemen, 1st millennium BCE, H. 140 cm (National Museum of Sana'a, Yemen, inv. YM 23206). See Mille et al. 2010. See fig. 20 for an overall view.

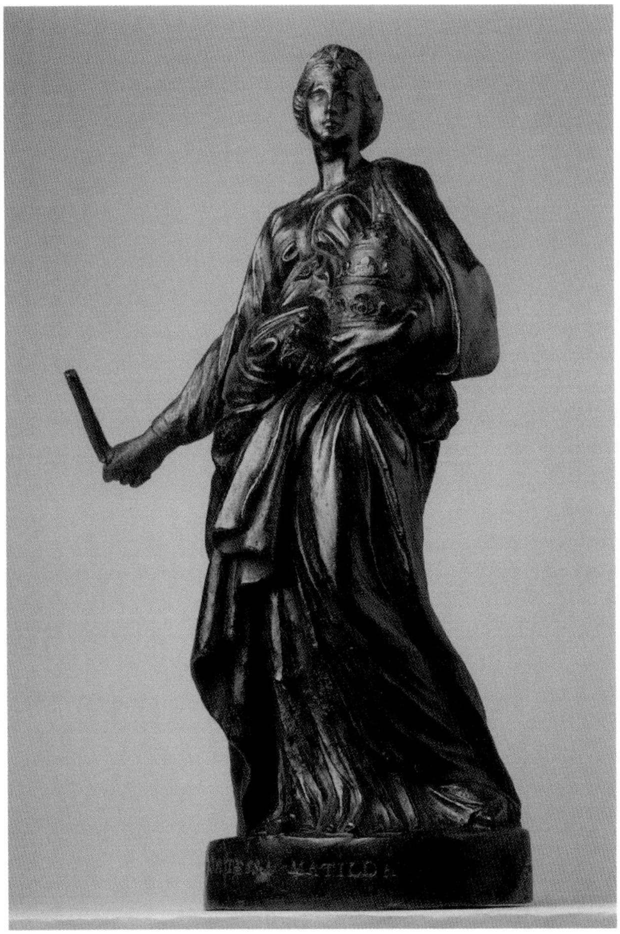

Figure 258 This lost-wax statuette has an inscription on the base that is stamped into the metal. After a model by Gian Lorenzo Bernini (Italian, 1598–1680), *Countess Matilda of Tuscany*, ca. 1633–34, H. 40 cm (Harvard Art Museums / Fogg Museum, partial gift of Max Falk and partial purchase through the Director's Acquisition Fund, 1998.1). See Bewer 1999.

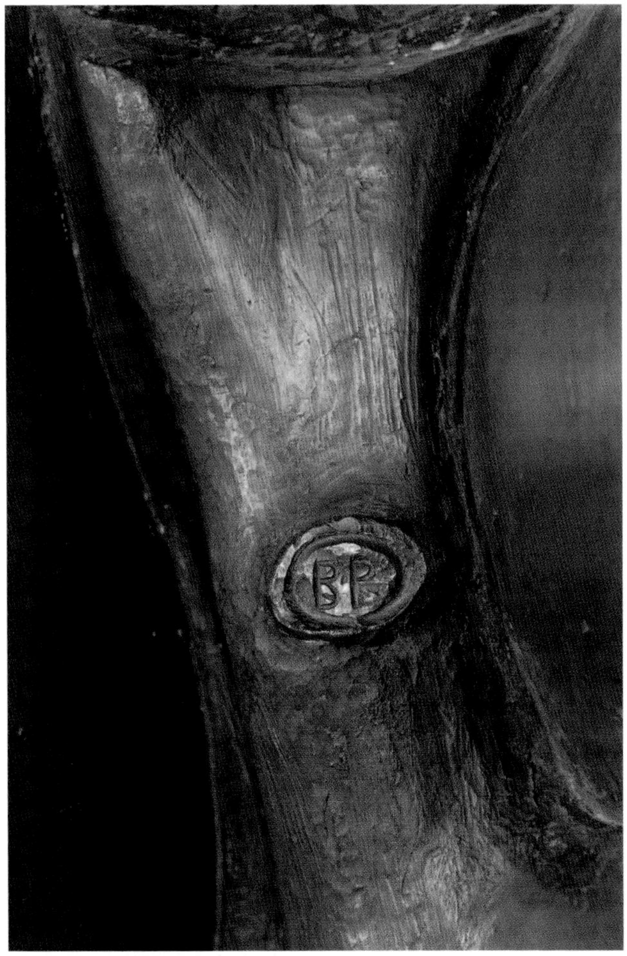

Figure 260 Inscription showing the initials of the sculptor (BP). Barthélemy Prieur (French, 1536–1611), *Diana the Huntress*, 1603, H. 200 cm (Musée national du château de Fontainebleau, France, inv. RF 261). See Castelle 2016.

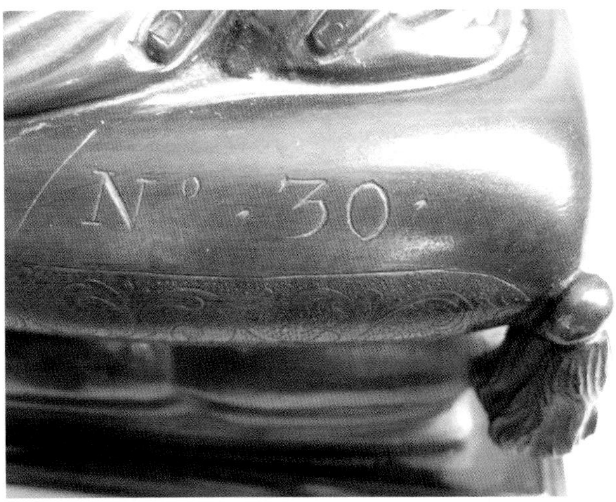

Figure 259 Detail showing the number in the "Bronzes de la couronne" collection chased in the metal in the couch. Hermaphrodite, Florence, 1640–60, L. 41 cm (Musée national des châteaux de Versailles et de Trianon, France, inv. MV 7778, collection de la couronne n°30, on deposit at Musée du Louvre).

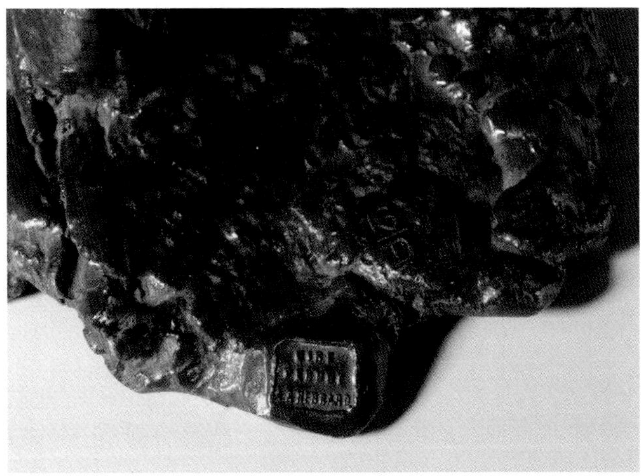

Figure 261 Detail of the dancer showing the "cire perdue" stamp of the Hebrard foundry applied in the wax model, and edition number "16/D" engraved in the bronze. Edgar Degas (French, 1834–1917), *Grande Arabesque, Third Time*, ca. 1885–90, H. 40.6 cm (Harvard Art Museums / Fogg Museum, bequest from the Collection of Maurice Wertheim, Class of 1906, inv. 1951.78).

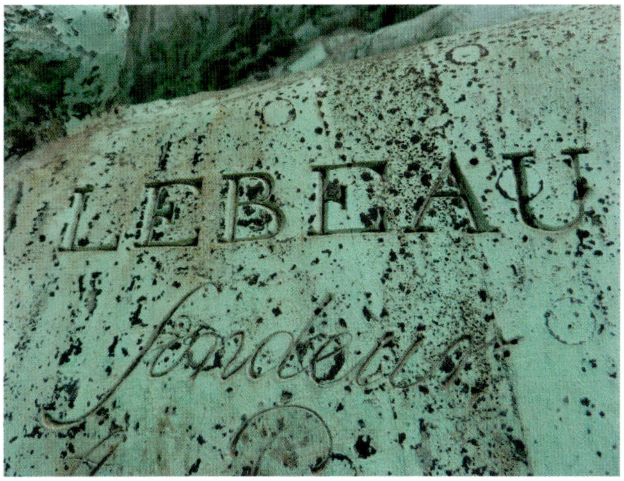

Figure 262 Inscription of the founder's name chased into the cast. Antoine-Louis Barye (French, 1795–1875), cast by Lebeau (French, active 1850), Lion at the Porte des Lions, Palais du Louvre, Paris, commissioned 1847, H. 160 cm (Musée du Louvre, inv. LP 3483 bis). See Lebon 2016; http://www.fontesdart.org/fontes-n100-elisabeth-lebon.

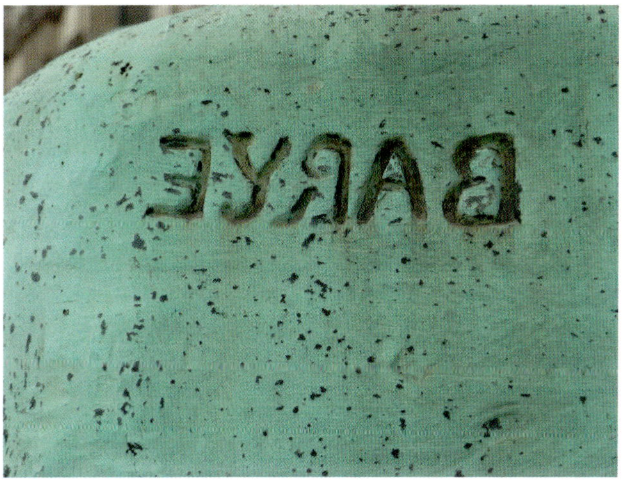

Figure 263 Inscription chased on the base of the metal copy showing the inverted name of the artist. Copy of the Lion by Antoine-Louis Barye (French, 1795–1875), ordered in 1878, today at the Porte des Lions, Palais du Louvre, Paris, H. 160 cm (Musée du Louvre, no inv. number). See http://www.fontesdart.org/fontes-n100-elisabeth-lebon/.

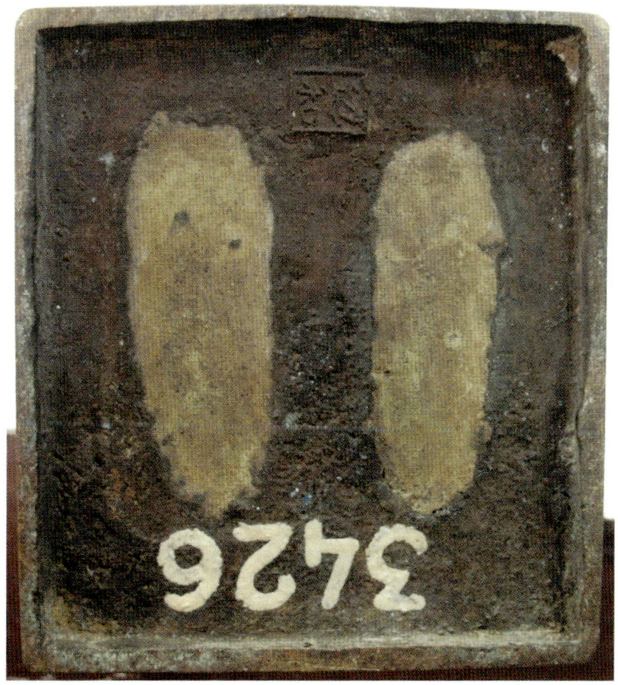

Figure 264 Raised profile of the Bien Hoa foundry mark under the base of the Shiva (top of the image) indicates it was incised into the mold taken from the original before casting. Copy of Khmer Shiva, 20th century, H. 54 cm (Museum of Vietnamese History, Ho Chi Minh City, inv. BTLS.640).

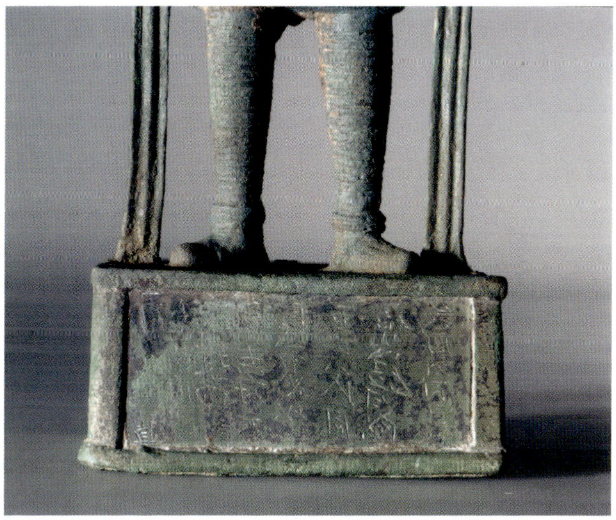

Figure 265 Detail of the chased inscription. Roi Gardien, Korea, 12th century, H. 41 cm (Musée national des arts asiatiques – Guimet, Paris, inv. MA 8153). See C2RMF Internal Report 2004.

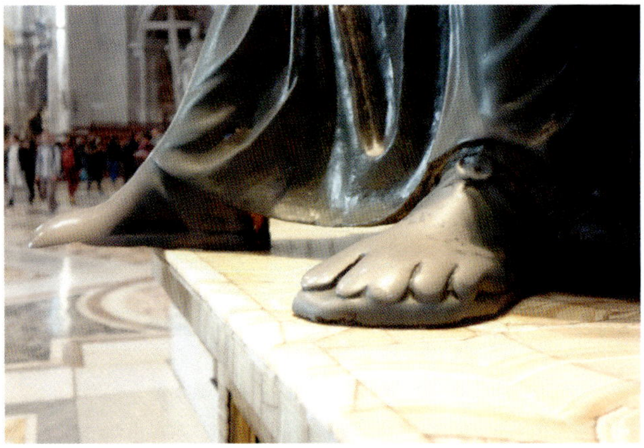

Figure 266 Detail of left foot showing wear as a result of touching. Attributed to Arnolfo di Cambio (Italian, ca. 1240–ca. 1302/10), *Saint Peter*, ca. 1296, life size (Basilica of Saint Peter, Rome). See Carruba 2006.

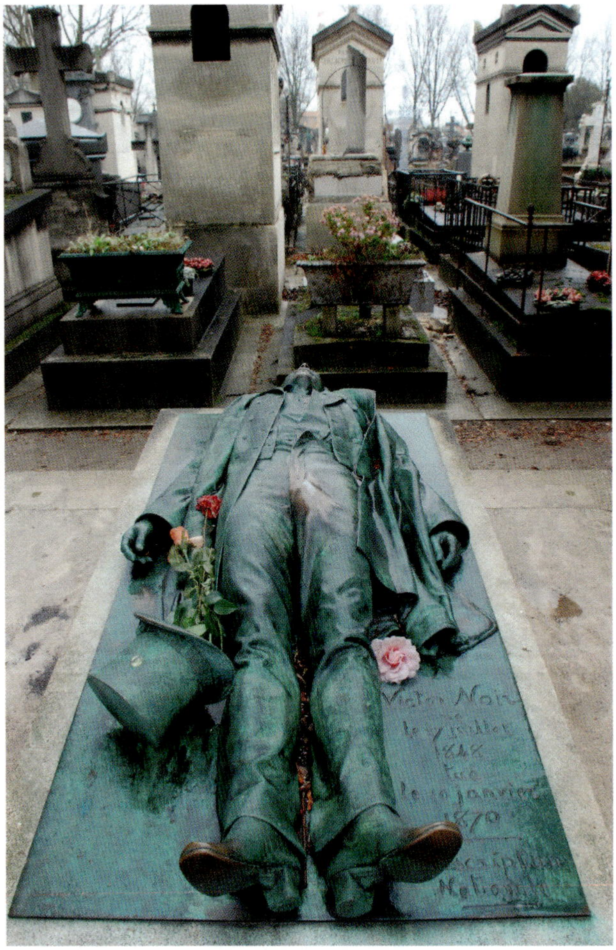

Figure 267 Repeated touching has worn down the patina in certain areas, including the shoes. Jules Dalou (French, 1838–1902), *Effigy on the Tomb of Victor Noir*, 1890, life size (Père Lachaise Cemetery, Paris).

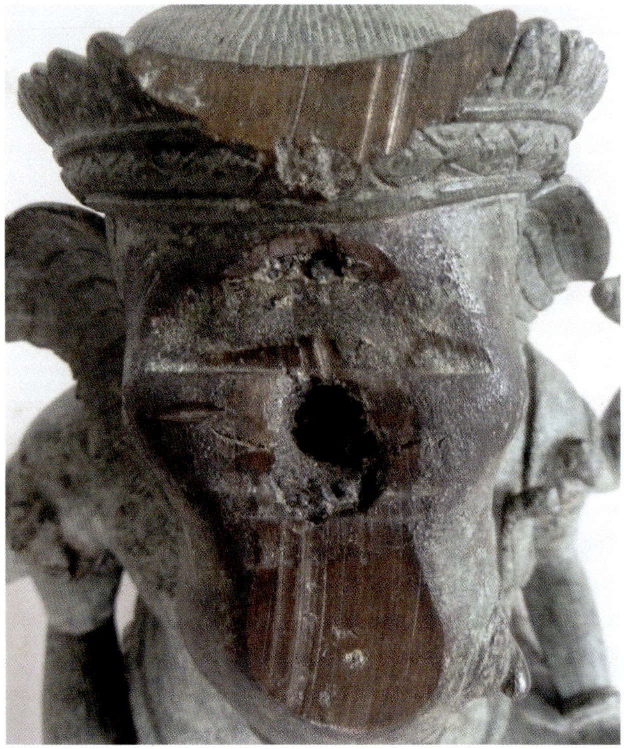

Figure 268 A section of the head and the trunk of a Ganesha, cut off by a modern tool. Khmer Ganesha, 13th century, H. 26 cm (National Museum of Cambodia, inv. Ga 5437).

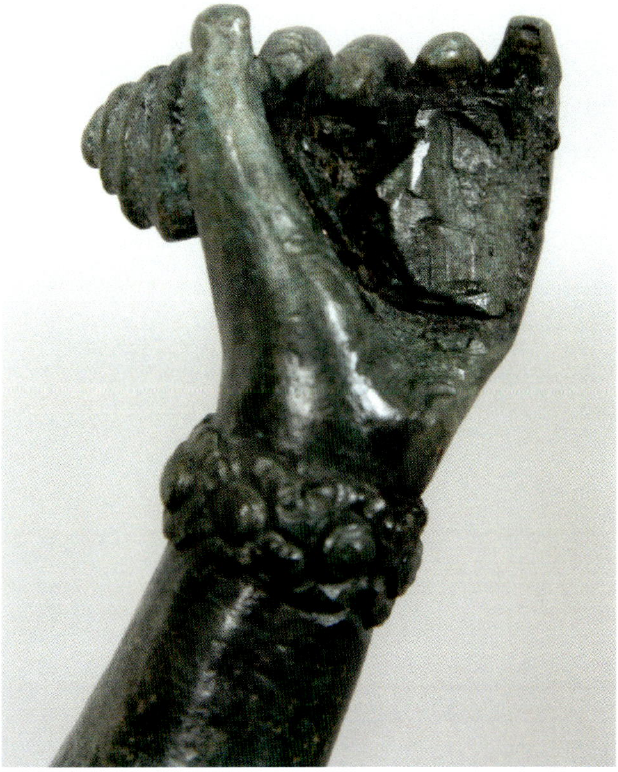

Figure 269 Example of deconsecration of a Khmer divinity by cutting off the end of the attribute. Khmer Lokeshvara, late 12th–early 13th century, H. 42 cm (National Museum of Cambodia, inv. Ga 5340).

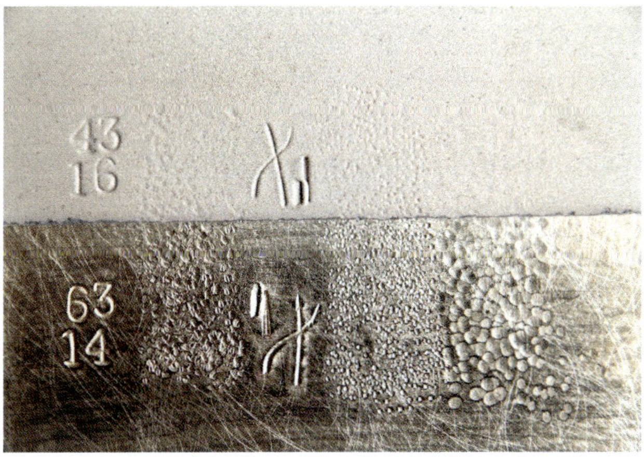

Figure 270 Comparison of select punch marks created on a copper alloy surface in which the upper half has been abraded (blasted with glass beads), to demonstrate how this dulls their crisp contours and makes them appear as if they were done in the wax rather than the metal. Didactic sheet by Andrew Lacey, May 2019.

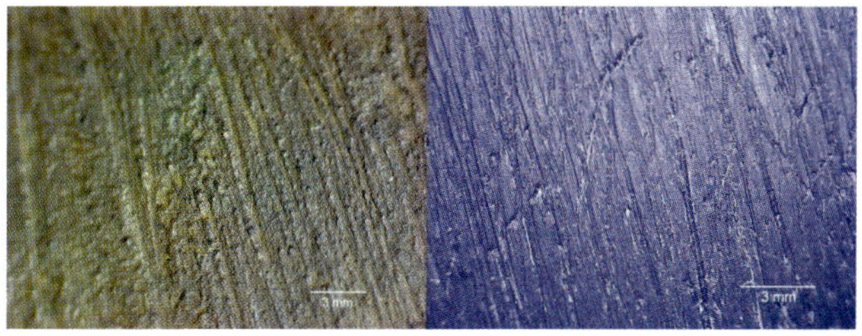

Figure 271 Experimental simulation of engraving on wax (right) to demonstrate that the tool marks observed on the surface of the bronze bust of Antoninus Pius (left) were made in the wax model rather than in the cast bronze. The size of the area in the picture is about 2 × 2 cm. Attributed to Giovanni Bandini (called Giovanni dell'Opera, Italian, 1540–1599), *Bust of Roman Emperor Antoninus Pius*, Urbino, before 1582, H. 68 cm (Museo Nazionale del Bargello, Florence, inv. 441B). See Morigi 2018.

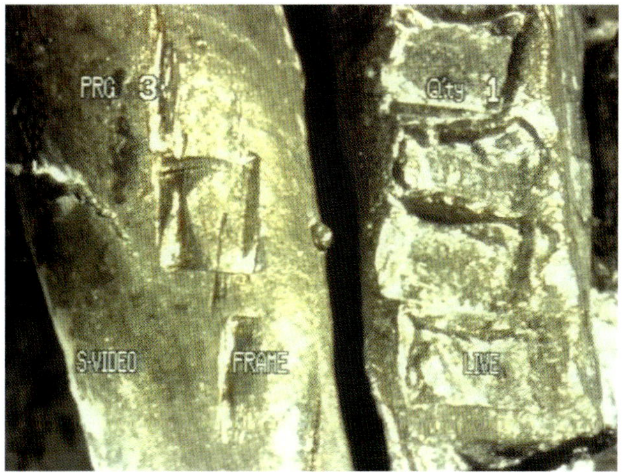

Figure 272 Photomicrograph of punch marks in experimental cast produced during the 2004 ancient materials and technologies workshop "The Technical Analysis of Renaissance Bronze Casting" (AMTeC), Chatham Dockyard, UK. The same punch was used in the metal after casting (left), where it preserves its crisp outline, and in the wax before casting (right), where the distortion of the punch mark, the ridge of displaced malleable material along the edge, and the loss of crispness are all more common in marks made in the wax.

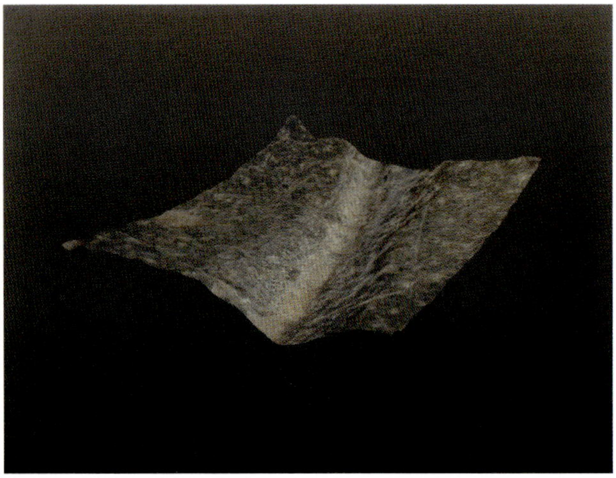

Figure 273 3D model of an engraved line (approximately 1 mm across) in a lost-wax cast statuette. The blurred V shape of the imprint results from the chasing tool following along a line already present in the wax model. The image was generated by a digital microscope using automated focus stacking. *Statuette of Jambhala*, Java, first half of the 9th century, H. 28 cm (Musée national des arts asiatiques - Guimet, Paris, inv. MG 3814). See Mechling et al. 2018.

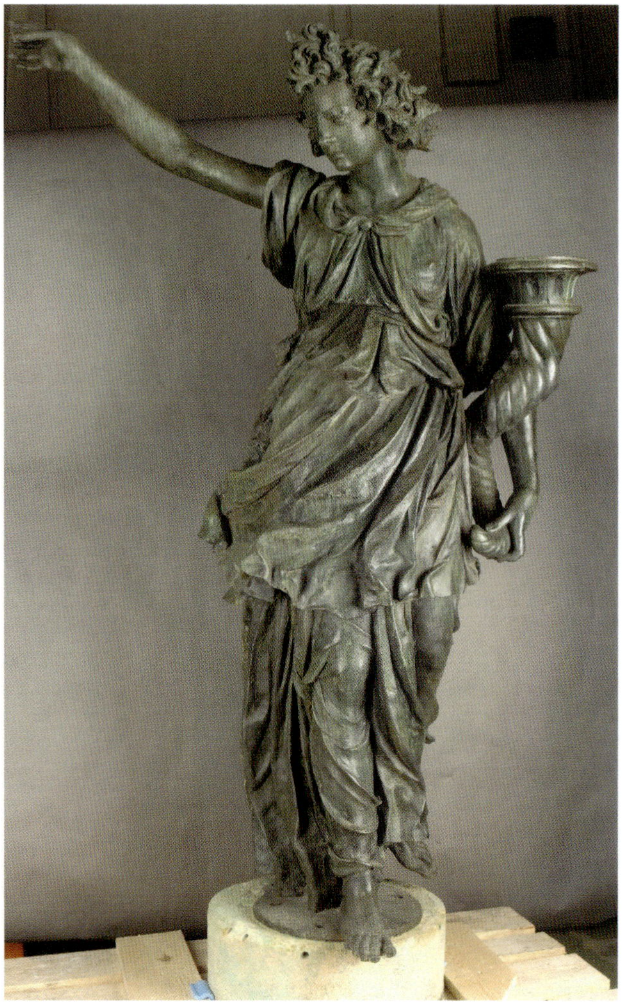

Figure 274 Francesco Bordoni (Italian, 1580–1654), *Torchbearing Angel*, 1613, H. 150 cm (École Nationale Supérieure des Beaux Arts, Paris, inv. WB38). See Bourgarit, Bewer, and Bresc-Bautier 2014.

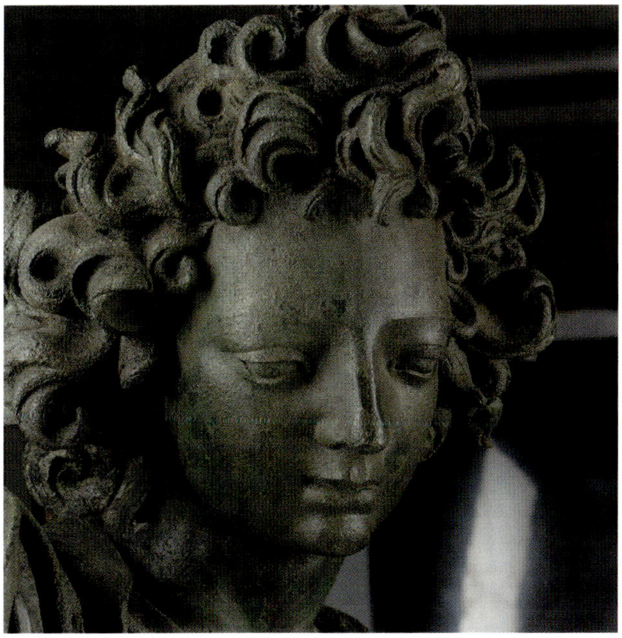

Figure 275 Detail of head whose surface has been severely weathered, destroying evidence of its cold working. Francesco Bordoni (Italian, 1580–1654), *Torchbearing Angel*, 1613, H. 150 cm (École Nationale Supérieure des Beaux Arts, Paris, inv. WB38). See Bourgarit, Bewer, and Bresc-Bautier 2014.

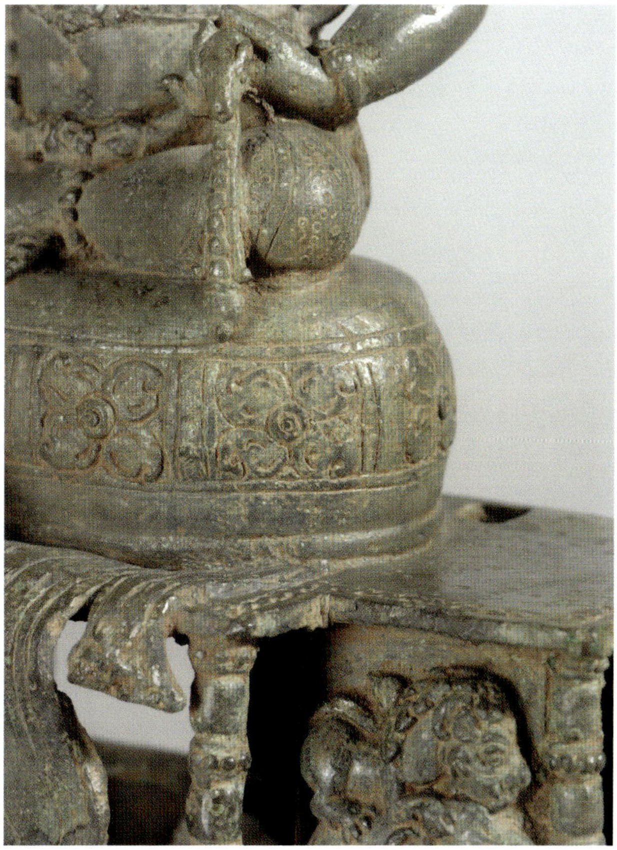

Figure 276 Detail of the surface of a lost-wax cast statuette. Only digital microscopy (see fig. 279) proved efficient in overcoming the wear and corroded surface, thus revealing that the linear details were V-shaped and sharply defined, and thus indicative of having been made or reinforced by a graver. *Statuette of Jambhala*, Java, first half of the 9th century, H. 28 cm (Musée national des arts asiatiques - Guimet, Paris, inv. MG 3814). See Mechling et al. 2018.

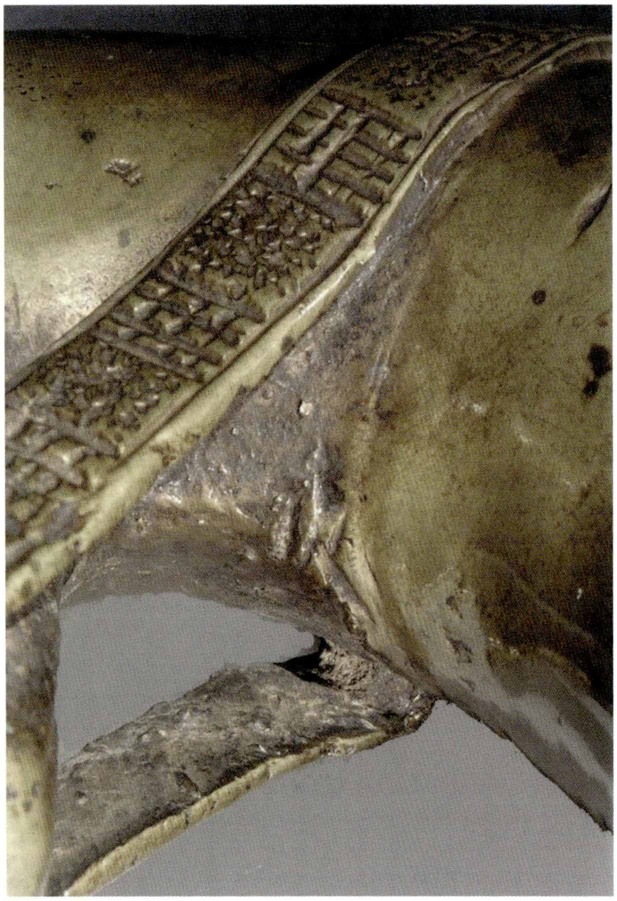

Figure 277 Detail showing the transition between the retained as-cast surface beneath the horse's neck and the adjacent chased surface of the metal. *Aquamanile in the Form of a Knight on Horseback*, Germany (Lower Saxony?), ca. 1350, H. 45.2 cm (Metropolitan Museum of Art, Robert Lehman Collection, 1975, inv. 1975.1.1409). See Dandridge 2006.

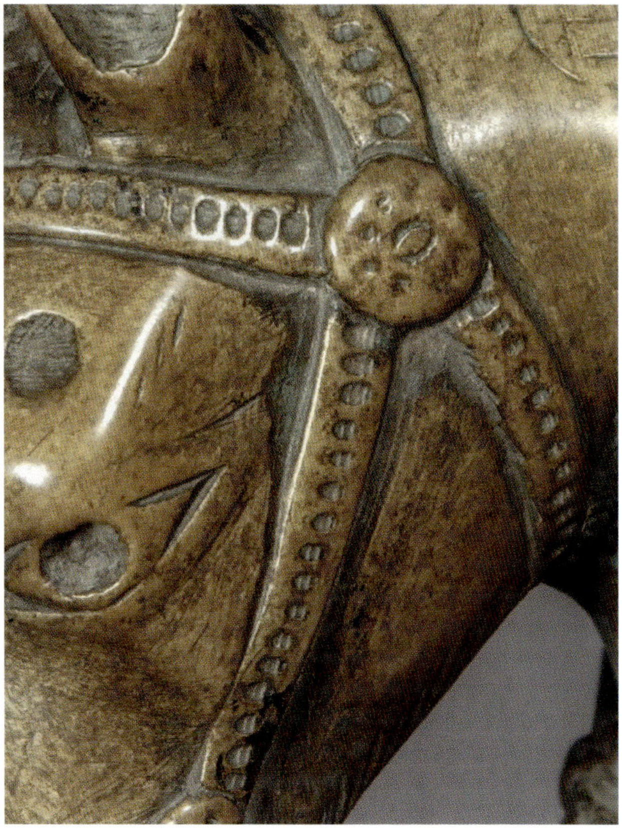

Figure 278 Features and textures of the bridle and eyes of the horse's head were enhanced in the metal with chisel, engraver, and punch. *Aquamanile in the Form of a Knight on Horseback*, Lower Saxony, Germany (probably Hildesheim), ca. 1250, H. 37.3 cm (Metropolitan Museum of Art, gift of Irwin Untermyer, 1964, inv. 64.101.1492). See Dandridge 2006.

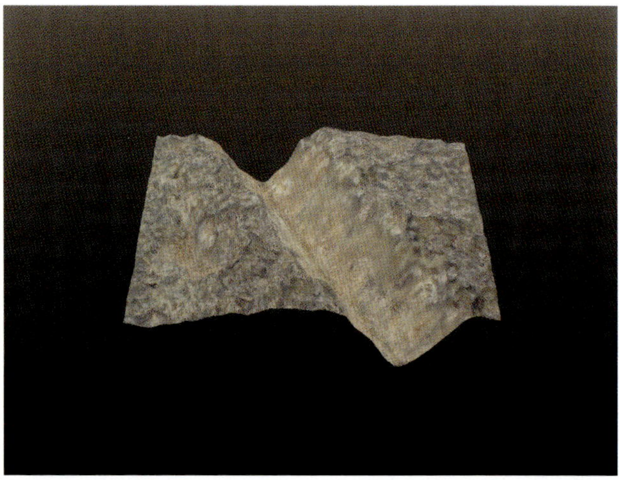

Figure 279 3D reconstructed image of a chased line made in the metal (detail of left knee). The V shape of the imprint demonstrates the use of engraving. The image was generated with a digital microscope using automated focus stacking. *Statuette of Jambhala*, Java, first half of the 9th century, H. 28 cm (Musée national des arts asiatiques - Guimet, Paris, inv. MG 3814). See Mechling et al. 2018.

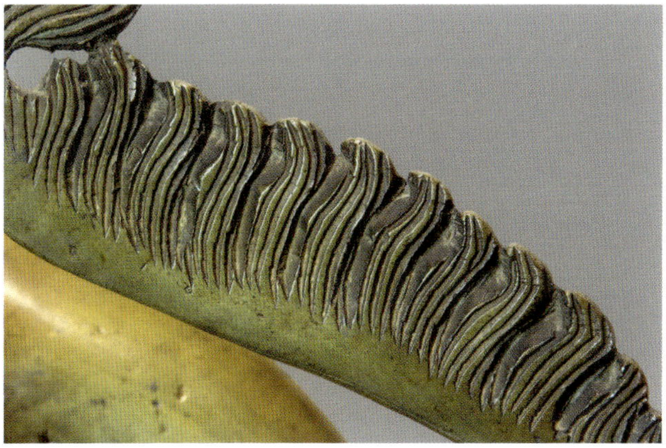

Figure 280 Detail from the handle that takes the form of the animal's tail whose as-cast details have been aggressively reinforced with a V-shaped graver. With each strike of the graver, the tool jumps, creating the visible steps within the engraved line. *Griffin Aquamanile*, Nuremberg, Germany, 1425–50, H. 32 cm (Metropolitan Museum of Art, Robert Lehman Collection, 1975, inv. 1975.1.1413). See Dandridge 2006.

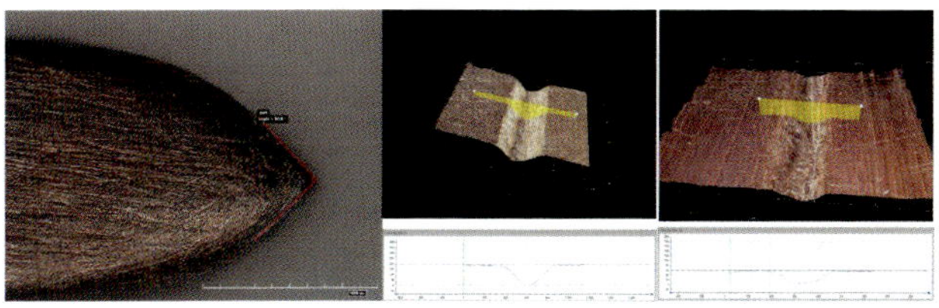

Figure 281 Digital micrographs comparing the results of experimental chasing using a tracer on two different metals in two different metallurgical states. View of the chisel (left), the mark left in the as-cast tin-bronze and its dimensions (middle), and the mark in the laminated sheet of unalloyed electrolytic copper and dimensions (right). The copper sheet, having not been annealed, is much harder than the as cast bronze, and consequently the mark is much shallower. Chasing by Dominique Robcis during the CAST:ING chiseling workshop, C2RMF, Paris, May 2017.

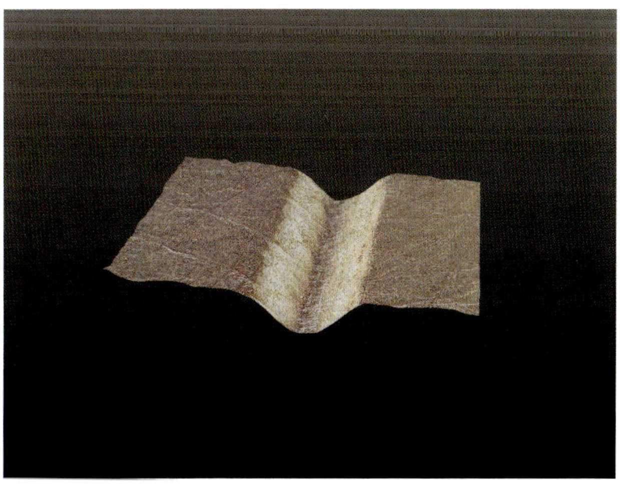

Figure 282 3D model of a chisel mark in an as-cast tin-bronze. Model created with a digital microscope using automated focus stacking during the CAST:ING chasing workshop, C2RMF, Paris, May 2017.

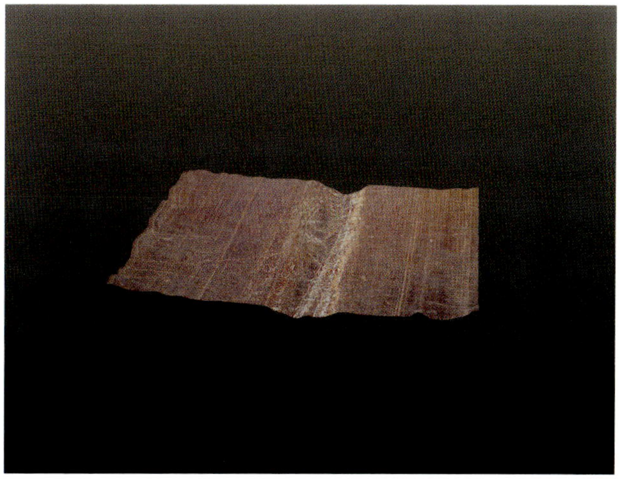

Figure 283 3D model of a chisel mark in an unalloyed electrolytic copper. Model created with a digital microscope using automated focus stacking during the CAST:ING chasing workshop, C2RMF, Paris, May 2017.

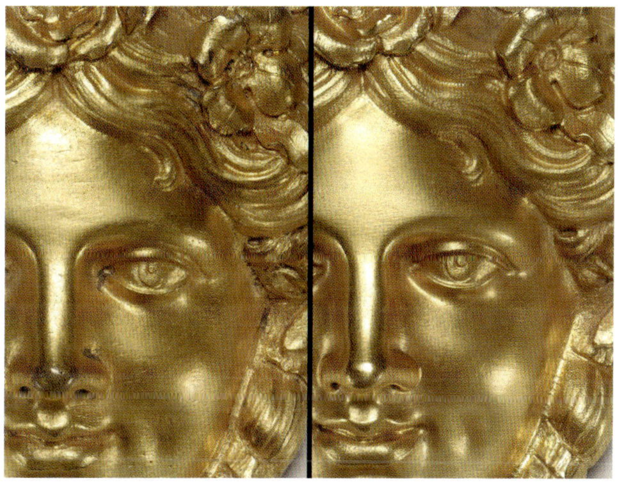

Figure 284 Comparison of an original gilt bronze mount (left) and an after-cast, nineteenth or twentieth century copy (right). The copy is more finely and "mechanically" chased in comparison to the original. The scale of the image is 9 cm from top to bottom. Attributed to André Charles Boulle (French, 1642–1732), pair of pedestals, ca. 1700 (J. Paul Getty Museum, inv. 88.DA.75.1–2). After Heginbotham 2014, fig. 1.

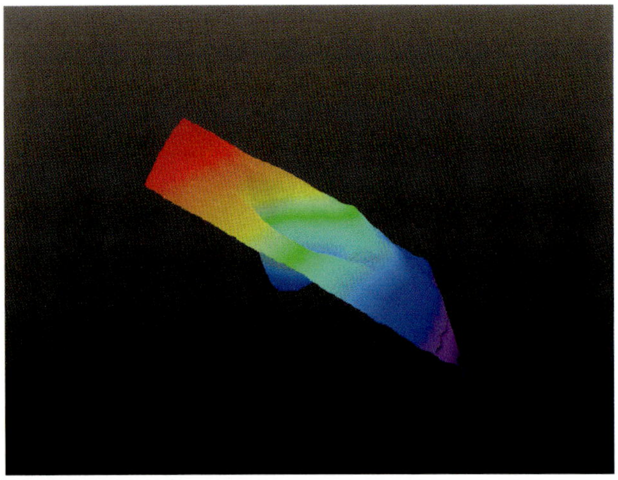

Figure 285 Digital microscopy detail of a punch mark (0.8 mm in diameter) on the lotus of a Buddha Vairocana. Note the smooth, compressed surface produced by the punch and the ridge of displaced metal around the edge. *Buddha Vairocana*, Indonesia, first half of the 10th century, H. 12 cm (Musée national des arts asiatiques - Guimet, Paris, inv. MA 3475). See Mechling et al. 2018.

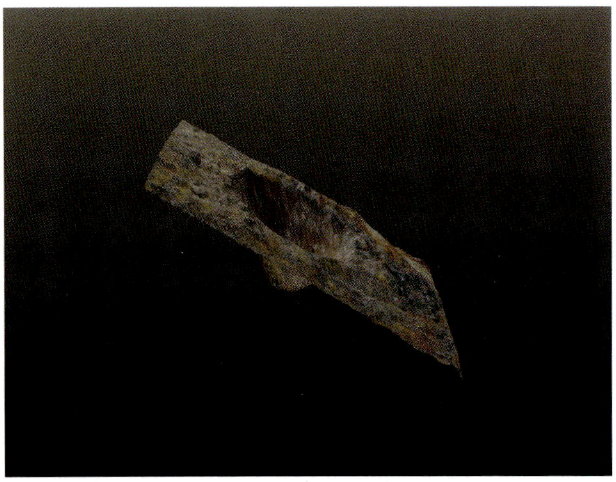

Figure 286 3D model of a punch mark (0.8 mm in diameter) from the lotus of a Buddha Vairocana (see fig. 285). The model was generated by a digital microscope using automated focus stacking. Note the asymmetry in the punch mark resulting from a defect in the face of the punching tool. *Buddha Vairocana*, Indonesia, first half of the 10th century, H. 12 cm (Musée national des arts asiatiques - Guimet, Paris, inv. MA 3475). See Mechling et al. 2018.

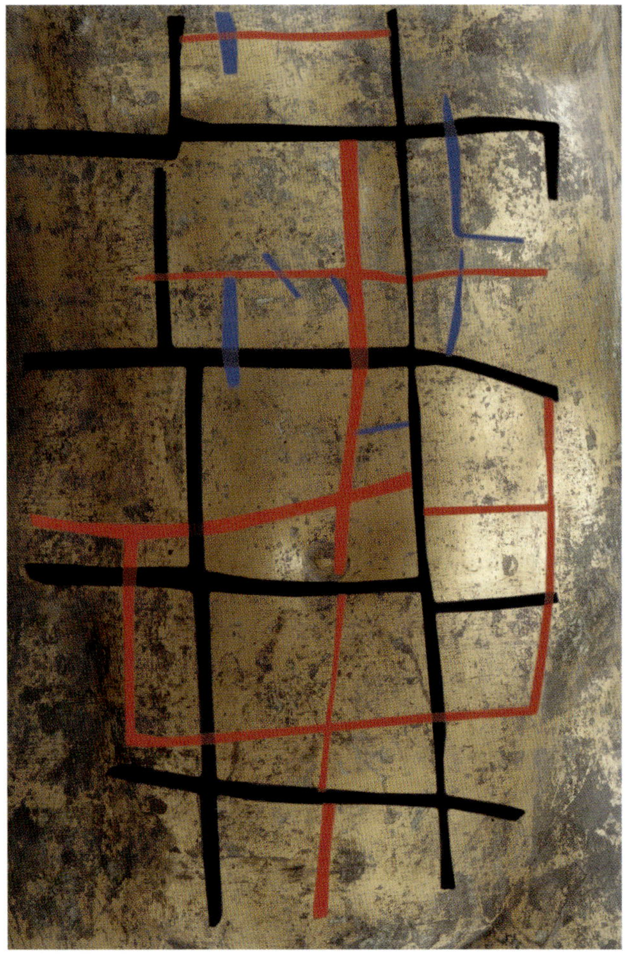

Figure 287 Annotated detail of fig. 288 showing three different phases of gilding as shown by the three different networks made by overlapping square gold leaves. *Apollo of Lillebonne*, France, 1st century BCE–1st century CE, H. 193 cm (Musée du Louvre, inv. Br37 [NIII65]). See Robcis et al. 2017.

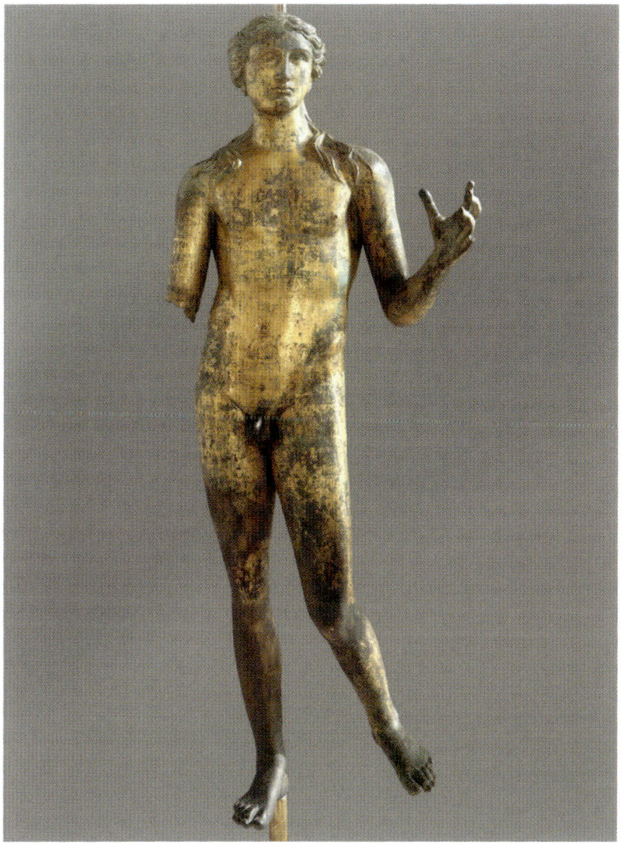

Figure 288 A Roman bronze gilded with gold leaf. *Apollo of Lillebonne*, France, 1st century BCE–1st century CE, H. 193 cm (Musée du Louvre, inv. Br37 [NIII65]). See Robcis et al. 2017.

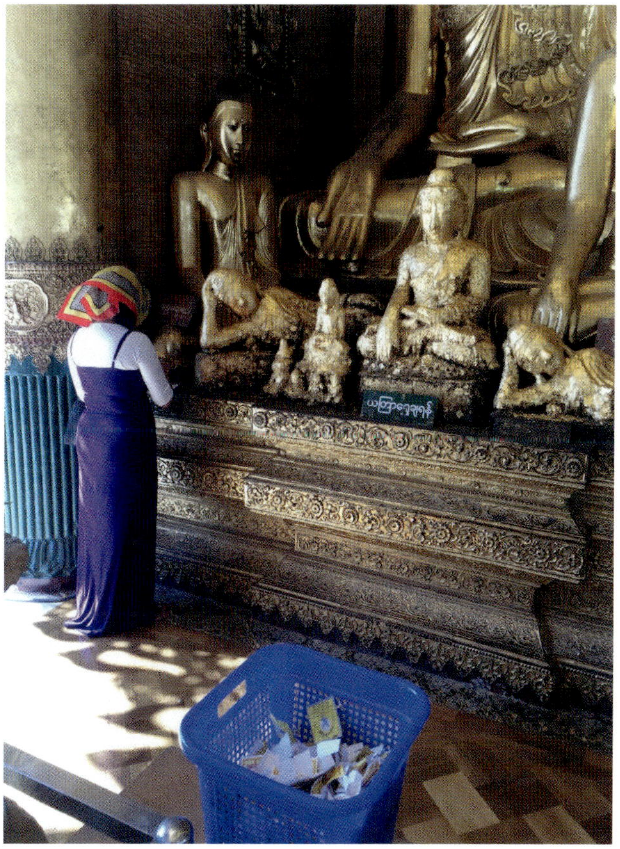

Figure 289 Woman applying gold leaf to Buddha statues as an act of worship, Shwedagon Pagoda, Yangon, Myanmar, 2018. Behind her in the blue basket are empty booklets of gold leaf.

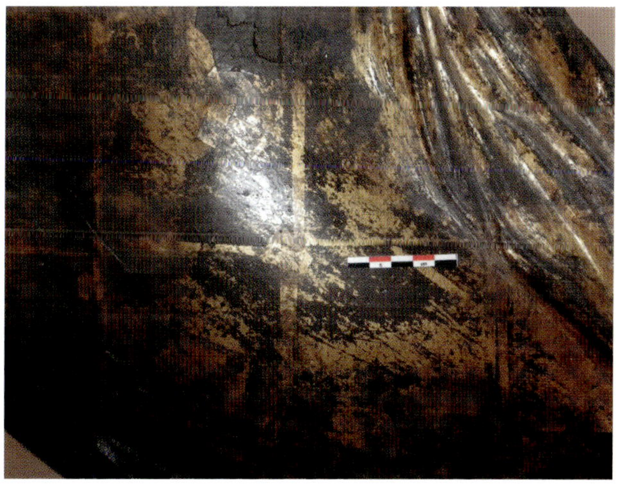

Figure 290 Detail of the body showing the more prominent overlap of the gold leaf squares. *Great Dolphins of Vienne*, Roman period, Vienne, France, 2nd century CE, max L. 260 cm (Musée des Beaux-Arts et d'Archéologie de Vienne, France, inv. R.1998.2.43).

VISUAL ATLAS

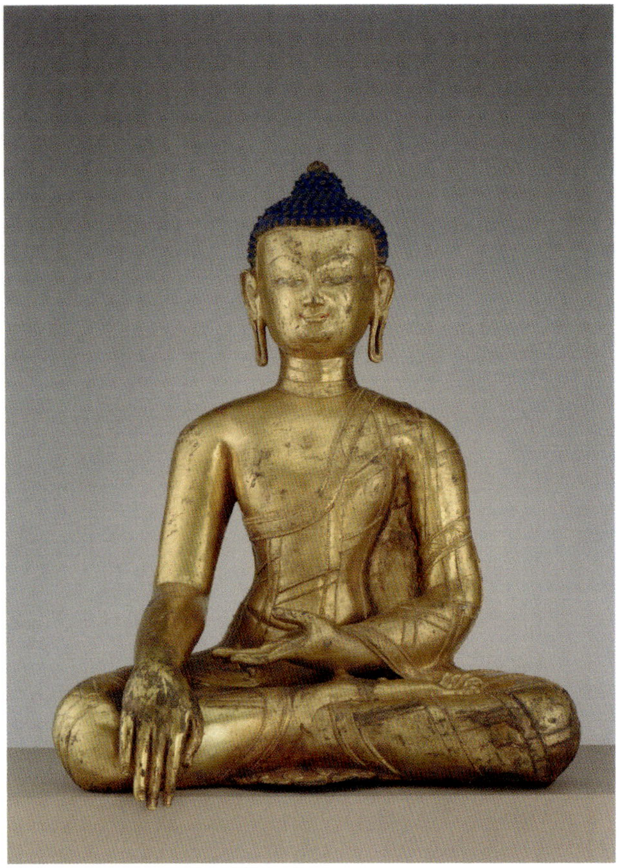

Figure 291 The surface of this cast-brass seated Buddha is mercury gilded except for the hair, which is pigmented blue. *Seated Buddha*, Central Tibet, 14th century, H. 45 cm (Freer Gallery of Art and Arthur M. Sackler Gallery, Smithsonian Institution, Washington, DC, purchase—Friends of the Freer and Sackler Galleries, S1997.28).

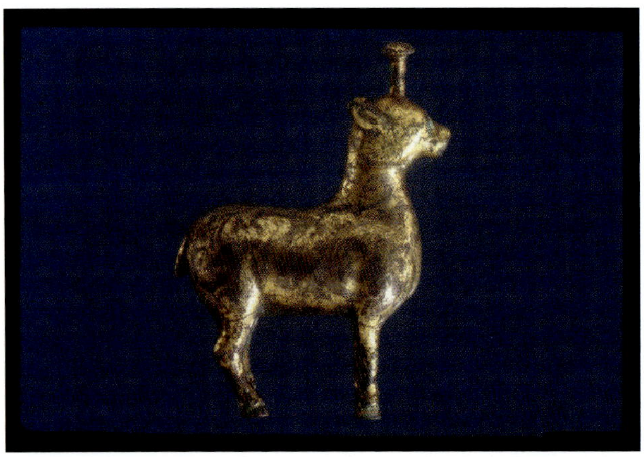

Figure 292 *Bronze Statuette of a Quadruped* (decorated with mercury gilding and silvering), China, Han dynasty, 202 BCE–220 CE, H. 7.5 cm (British Museum, reg. 1936,1118,257). See La Niece 1990.

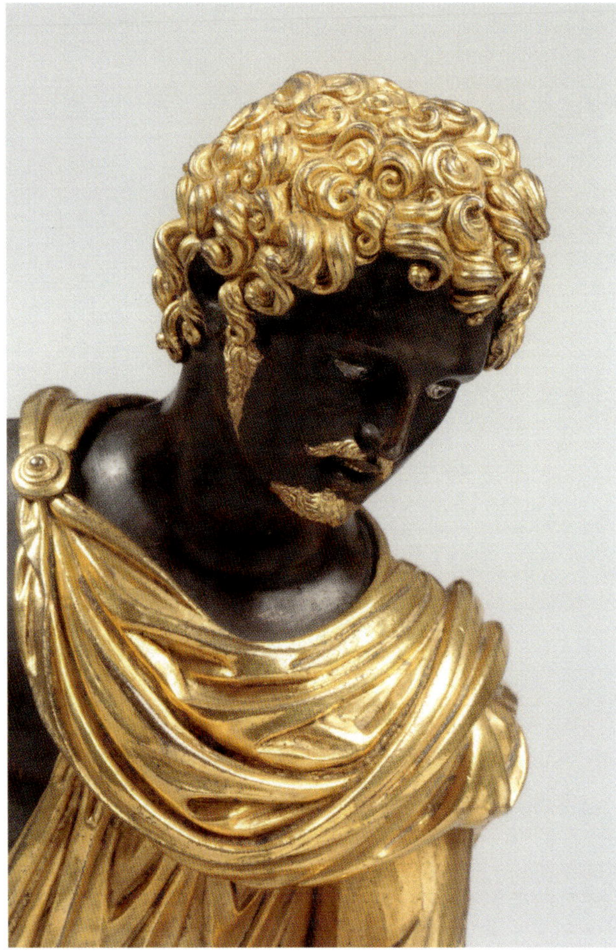

Figure 293 Detail showing the eyes inlaid with silver, and the bridge of the nose, lips, and sideburns for evidence of where the mercury gilding dripped while being applied, or did not stay within intended boundaries. Note the black patina typical of Antico and the particularly glossy surface. Pier Jacopo Alari Bonacolsi, known as Antico (Italian, ca. 1455–1528), *Meleager*, ca. 1484–90, H. 32.2 cm with original base (Victoria and Albert Museum, purchased with support from the Horn Bequest, the Bryan Bequest, and Art Fund, inv. A.27-1960). See Stone 1981; Smith and Sturman 2011; Motture 2019.

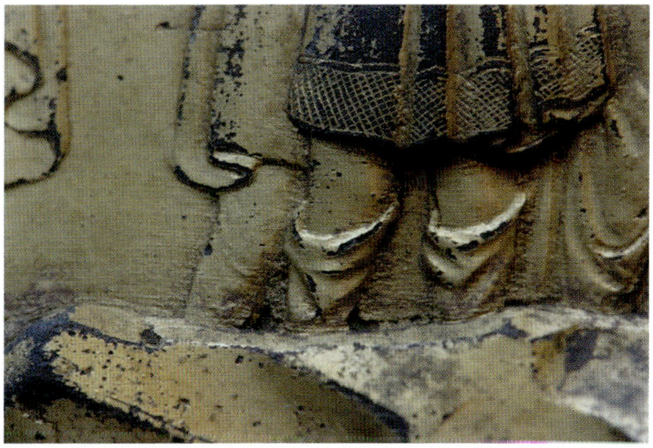

Figure 294 Detail showing horizontal brushstrokes from amalgam gilding application (detail size 4.4 × 2.9 cm). The horizontal brushstrokes are particularly visible in the background to the proper right of the legs. Lorenzo Ghiberti (Italian, ca. 1381–1455), Joseph panel in the *Gates of Paradise*, right door, H. 518 cm (design begun after 1425; installed 1452) (Museo dell'Opera del Duomo, Florence, inv. 2005/905). See Bewer, Stone, and Sturman 2007.

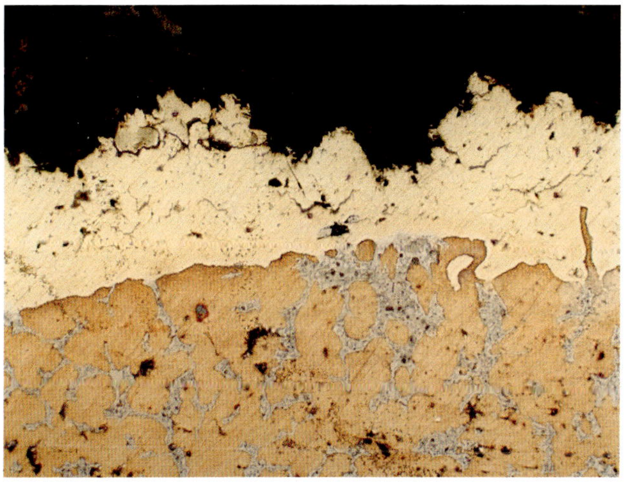

Figure 295 Bright field photomicrograph of an etched polished cross section of a modern mercury-amalgam gilded cast bronze. Typically the mercury gilding layer (top of image, ~5 μm thick) appears bright. Within the gilding layer it is common to see irregular lines that are residual outlines of the solidifying grains of gilding. The surface of the gilding is often smoother than seen in this modern example because extensive burnishing is carried out when a shiny finish is intended. The lower surface of the gilding layer can be uneven but well bonded to the substrate because heating causes some reaction of the gilding with the bronze. (Freer Department of Conservation and Scientific Research, Washington, DC, metallographic reference collection ME90069D).

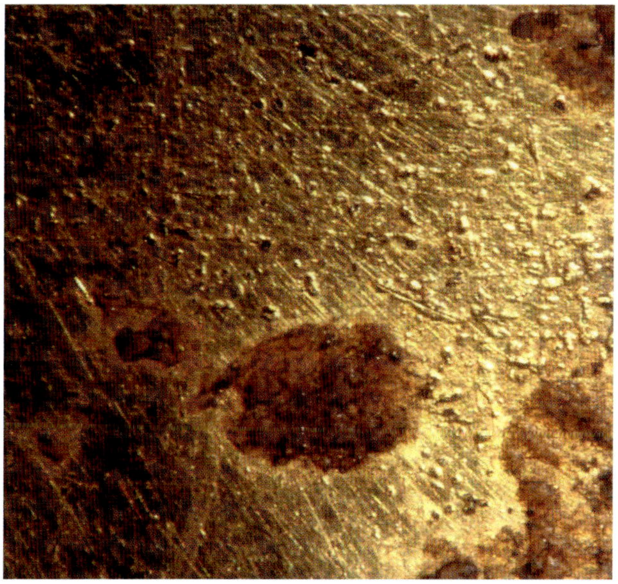

Figure 296 Detail of the surface of a fake Han dynasty bronze showing the blistering electro-gilding. The width of the field of view is 1.5 cm. The blisters are tiny raised gold spots in the gilding that indicate the bronze was poorly cleaned to prepare for the gilding process. The visible red waxy deposits were applied by the forger to "age" the finish. Modern forgery of a Chinese Han period (202 BCE–220 CE) cast bronze in the form of a bear, H. 17.2 cm (British Museum, reg. no. 1947,0712.382). See Jones 1990, 257.

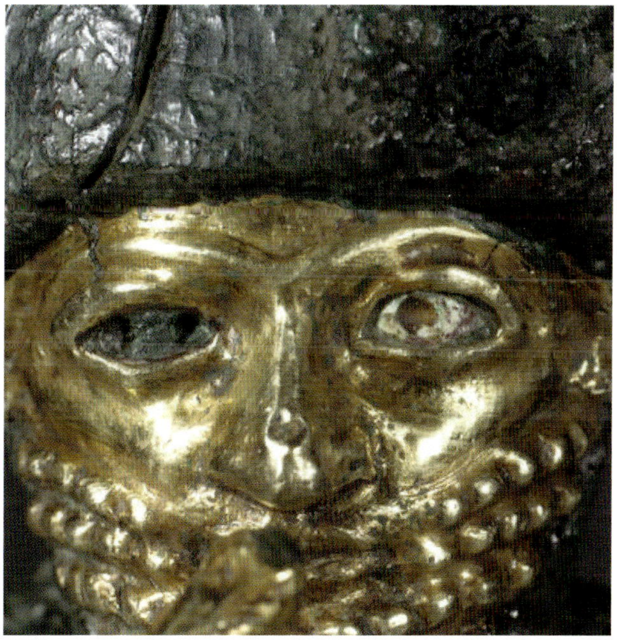

Figure 297 Detail showing foil gilding of face and hands using 0.1–0.2 mm gold foil (30–40% Ag) on an unalloyed copper cast. See also the aragonite in the left eye. *The Worshipper of Larsa*, Mesopotamia, 18th century BCE, H. 19.6 cm (Musée du Louvre, inv. AO15704). See C2RMF Internal Report 2008b.

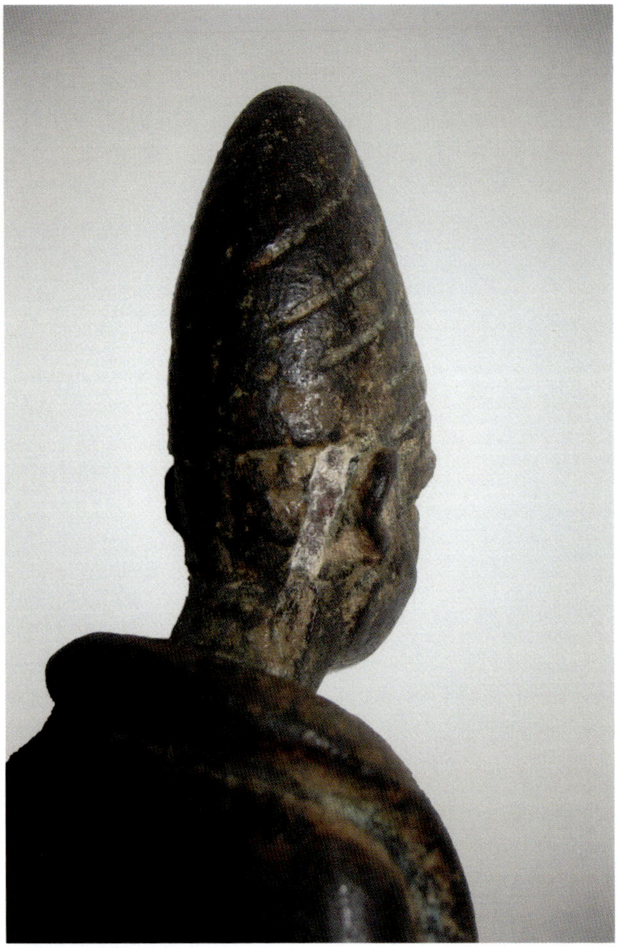

Figure 298 Two vertical grooves in the neck (one is visible in the photograph) are filled with lead, possibly to hold gold foil. *Enthroned God*, Mishrife, Syria, 17th century BCE, H. 17.2 cm (Musée du Louvre, inv. AO3992). See C2RMF Internal Report 2009.

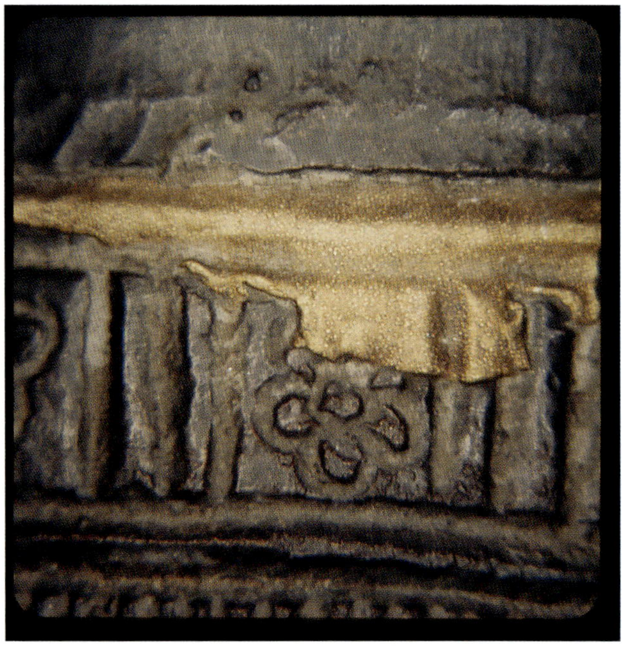

Figure 299 Detail of the surface of an ancient Anatolian bronze showing mechanically applied foil gilding. The width of the field of view is 3.5 cm. Note that the foil masks the sharpness of the surface details on the bronze. Cast bronze, Uratu, northern Syria/Turkey, 7th–5th century BCE, H. 21 cm (British Museum, reg. no. 1880,1216.9).

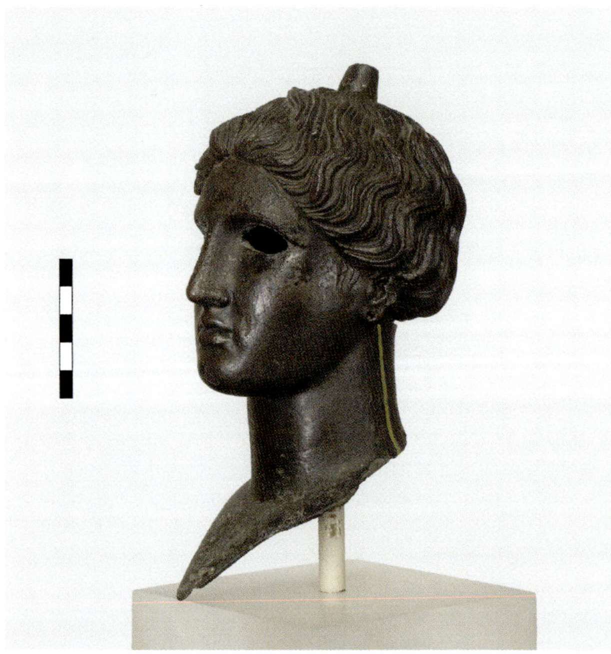

Figure 300 Detail of the grooves on the neck and hairline (overlay) for the mechanical attachment of metal foils on the bronze head. *Nike*, Athens, 420–415 BCE, H. 20 cm (Ephorate of Antiquities of Athens City, Ancient Agora Museum [ASCSA, Agora Excavations B 30]).

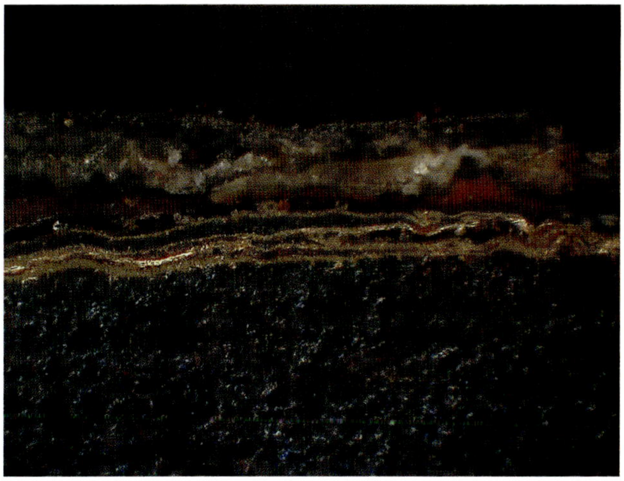

Figure 301 Digital micrograph of a cross section showing several layers of gold due to regilding at the surface (H. of image approx. 100µm). *Apollo of Lillebonne*, France, 1st century BCE–1st century CE, H. 193 cm (Musée du Louvre, inv. Br37 [NIII65]). See Robcis et al. 2017.

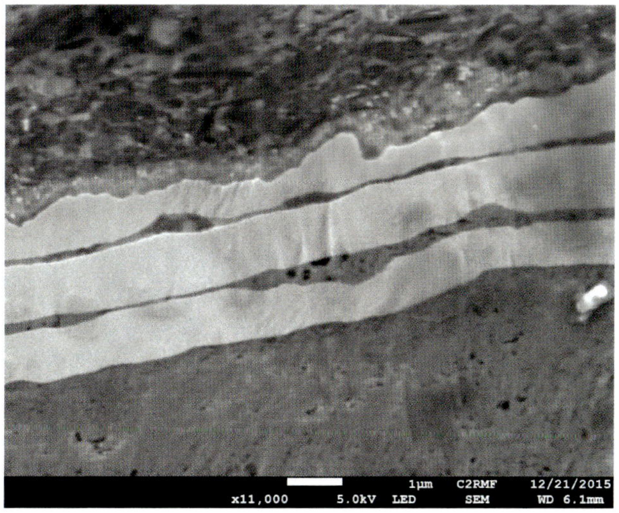

Figure 302 Backscattered electron micrograph of a polished cross section (ionic polishing) showing several layers of gold (white) due to regilding at the surface. *Apollo of Lillebonne*, France, 1st century BCE–1st century CE, H. 193 cm (Musée du Louvre, inv. Br37 [NIII65]). See Robcis et al. 2017.

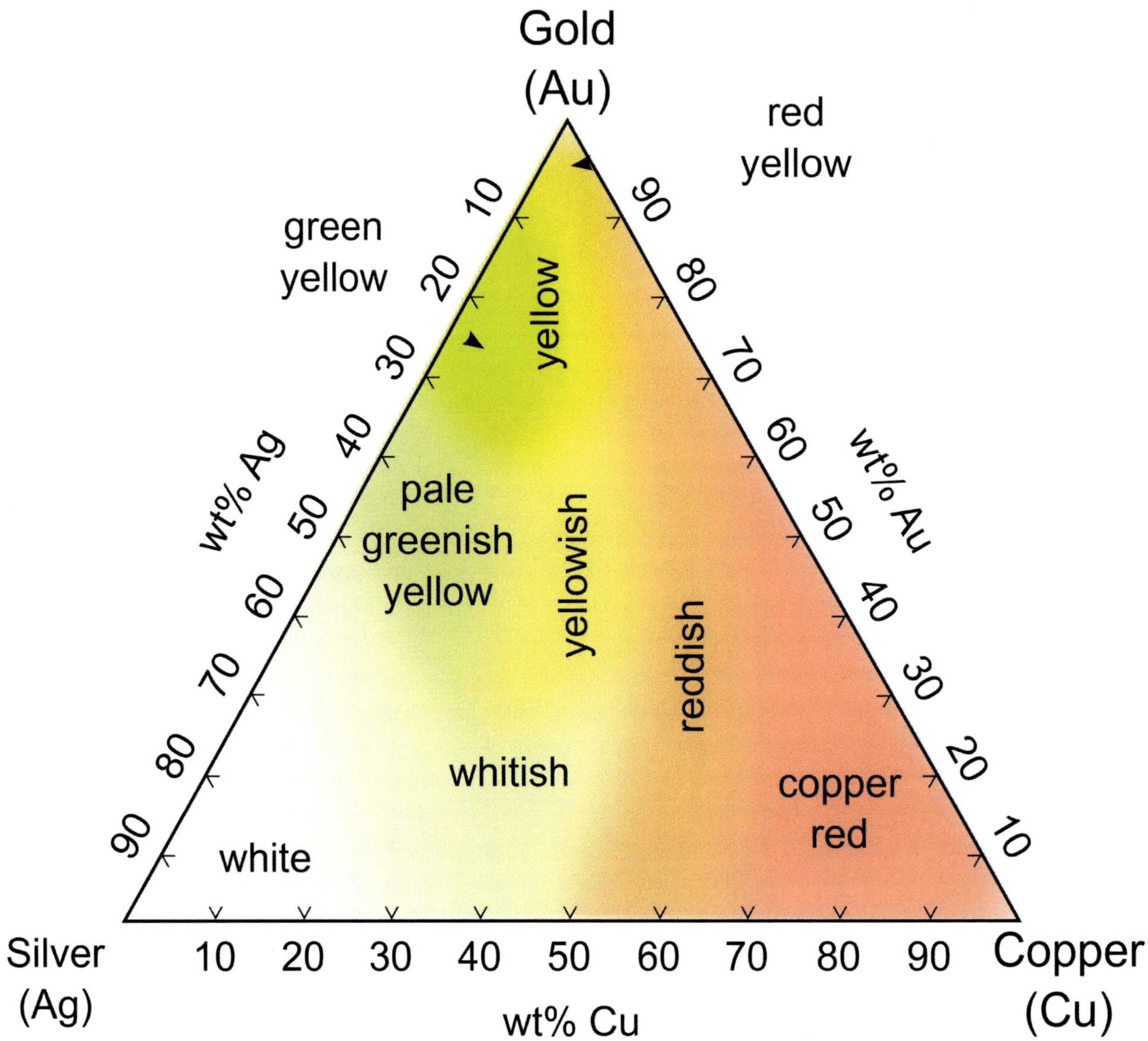

Figure 303 Ternary diagram for gold-silver-copper alloys indicating the color of different alloy compositions.

PRELIMINARY INVESTIGATION OF PLATING BY VISUAL EXAMINATION

(use a magnifying lens and good raking light; see text for details of the plating techniques mentioned)

Identify the color of the body metal of the sculpture. Look for metallic areas of a different color than the body metal and/or bright metal within corrosion layers.

Note: There may be more than one type of plating present. Use elemental analysis if available (e.g., XRF or spot tests) to identify the plating material and the sculpture metal (see table 10).

If there is plating inside a hollow sculpture: suspect electroplating (but small splashes or runs may be amalgam plating if mercury is detected, or fusion plating if it has a cast structure and no mercury).

If there is evidence of rivets, grooves, or other fixings: suspect mechanical foil gilding. Look for fragments of foil, and/or try X-radiography.

If there is scoring on the sculpture: suggests close plating with silver foil or foils added as local repairs.

If there is a gesso coating on the sculpture: look for leaf gilding.

If the plating has straight edges and/or square shapes: leaf gilding likely. Undertake elemental analysis: Is it gold, or copper + zinc (Dutch leaf)? Silver or tin?

If the plating is flaking off: indicative of deterioration of adhesive used for leaf gilding.

If the plating is blistering: a problem of electroplating, but also caused by corrosion under most plating.

If the plating has a granular texture: suspect paint; also look for brushstrokes, drips, etc. (see I.8).

If there are metal runs and drips: suspect amalgam plating (check for mercury) or fusion plating.

If only selected areas are plated: leaf, paint, amalgam, or electroplating (using a masking material) are possible.

If a gilding layer is worn on high points but not flaking: suggests amalgam (test for mercury) or electroplating (look for blistering).

If gilding is well bonded but thin: suggests amalgam (test for mercury; if none, diffusion bonding or electroplating are possible).

The next stage of investigation would be to take a sample that provides a cross section of the plating in relation to the substrate metal. See II.6.

Figure 304 Decision tree for the preliminary investigation and characterization of plating by visual examination.

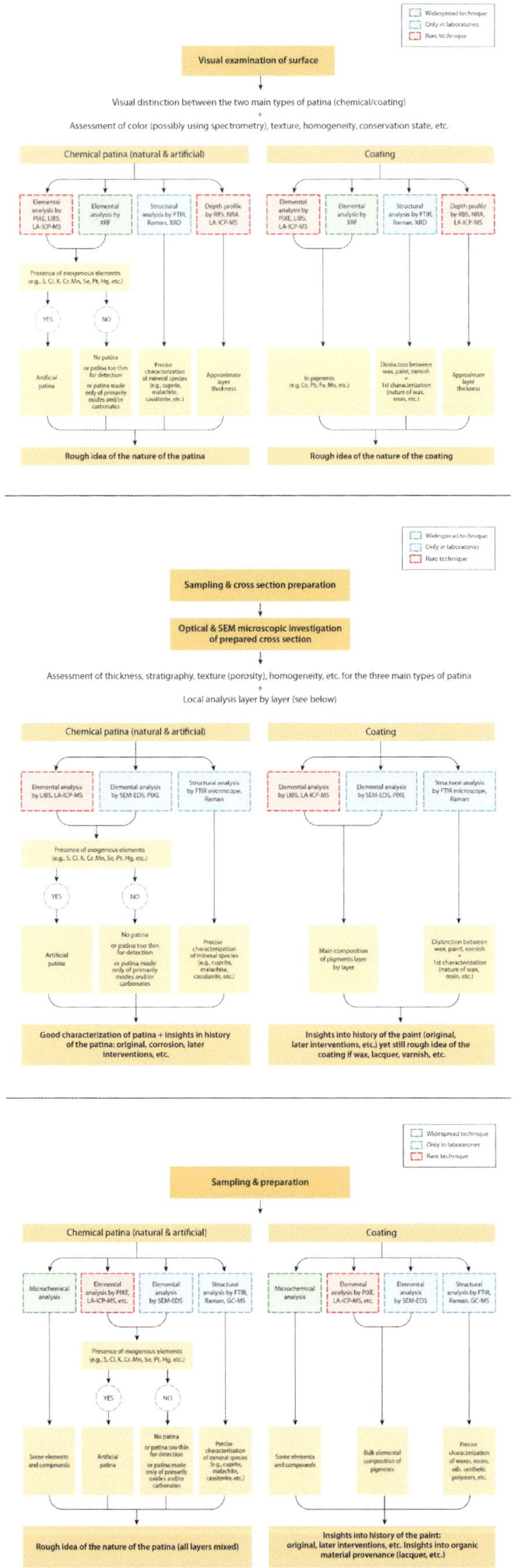

https://www.getty.edu/publications/bronze-guidelines/visual-atlas/305/

Figure 305 Decision chart for assessing the condition and characterizing an archaeological patina layer (in a museum laboratory setting). The protocol can be useful for the study of other surface layers as well, including metal plating and inlays. Three scenarios are covered: Top of chart: sampling is not possible; Middle of chart: cross section sampling is possible; Bottom of chart: powder or flake samples are available. The expected outcome of each approach is indicated at the bottom of the chart.

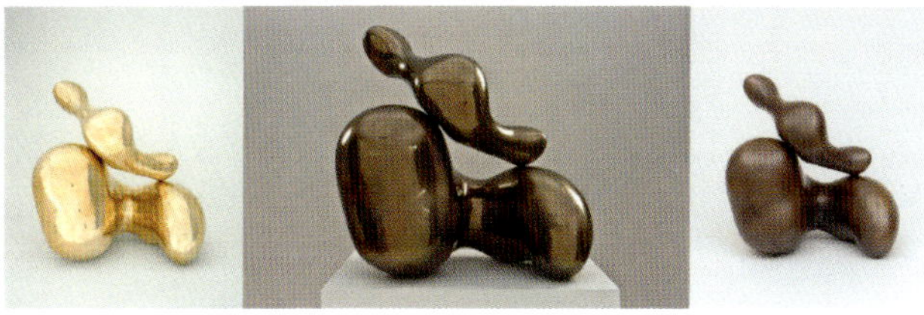

Figure 306 Research indicated that, although brightly polished when acquired (left), this sculpture was intended by the artist to have a thin brown patina. It was therefore allowed to darken by exposure to air for some years (middle). This appearance was still considered unacceptable, so repatination was deemed necessary to return it to its original appearance (right). Jean Arp (French German, 1886–1966), *Human Concretion without Oval Bowl*, 1933, cast 1961, H. 58.4 cm (San Francisco Museum of Modern Art, William L. Gerstle Collection, William L. Gerstle Fund purchase, inv. 62.2431). See Hamilton 2018.

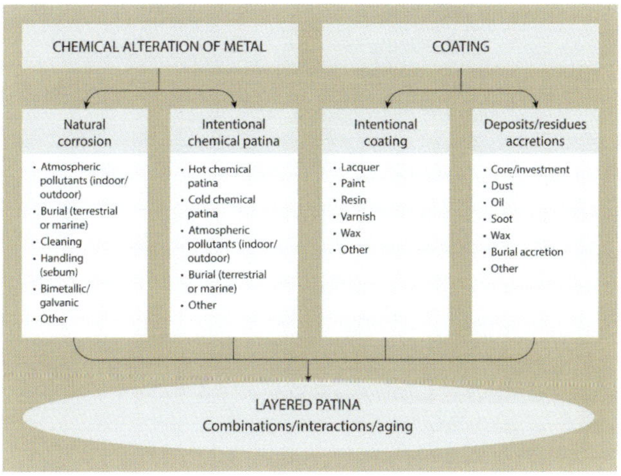

Figure 307 Diagram showing the various factors involved in patina formation.

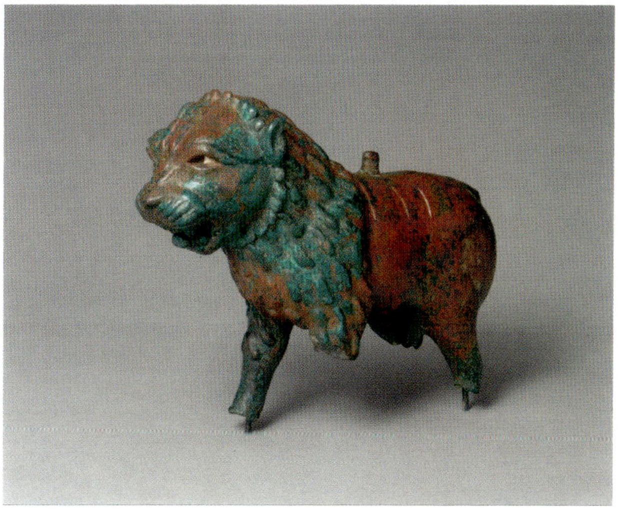

Figure 308 The well-adhered red and green patina formed naturally during centuries of burial. Partial cleaning has revealed the multicolored surface. Figure of a Lion, Anatolia or Syria, 1st millennium BCE, H. 18.7 cm (Metropolitan Museum of Art, purchase, Joseph Pulitzer Bequest, gift of Dr. Mortimer D. Sackler, Theresa Sackler and Family, and funds from various donors, inv. 2002.457 a–b).

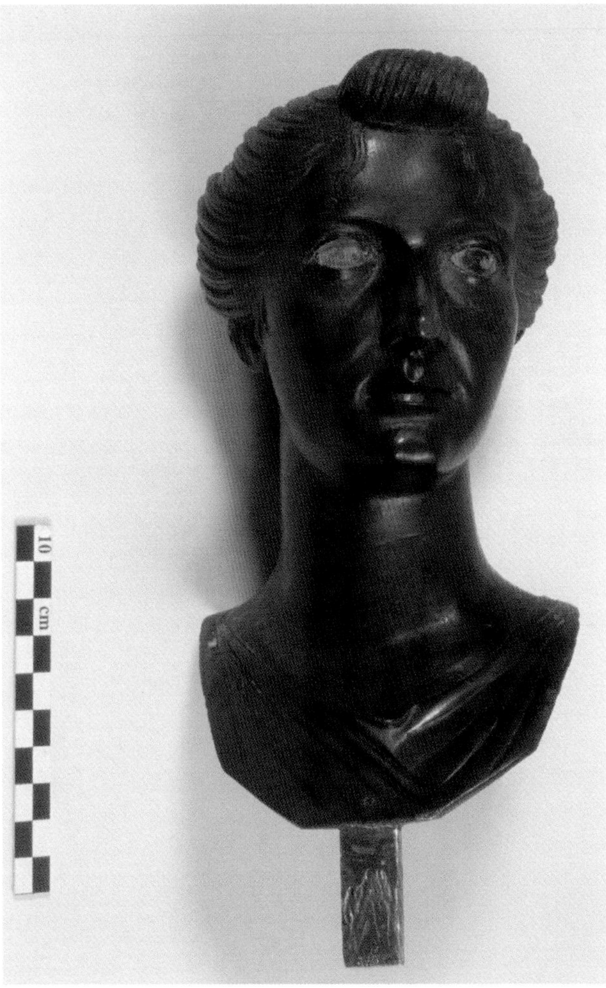

Figure 309 Metallographic investigation of a cross section has shown that the dark-green patina on this bust formed naturally during burial and is of the type referred to as "noble patina" (Gettens 1970; Robbiola and Hurtel 1997). Heated glass-paste inlay is used to create solid and translucent colors within depressions or elevated chambers. Here, elementary chemical analysis indicates the eyes were produced using luxury glassmaking techniques, including colored glass (Descamps-Lequime, Biron, and Langlois 2017). *Livia*, Neuilly-le-Réal (Allier, France), 1st century CE, H. 21 cm; head H. 10 cm (Musée du Louvre, inv. Br 28).

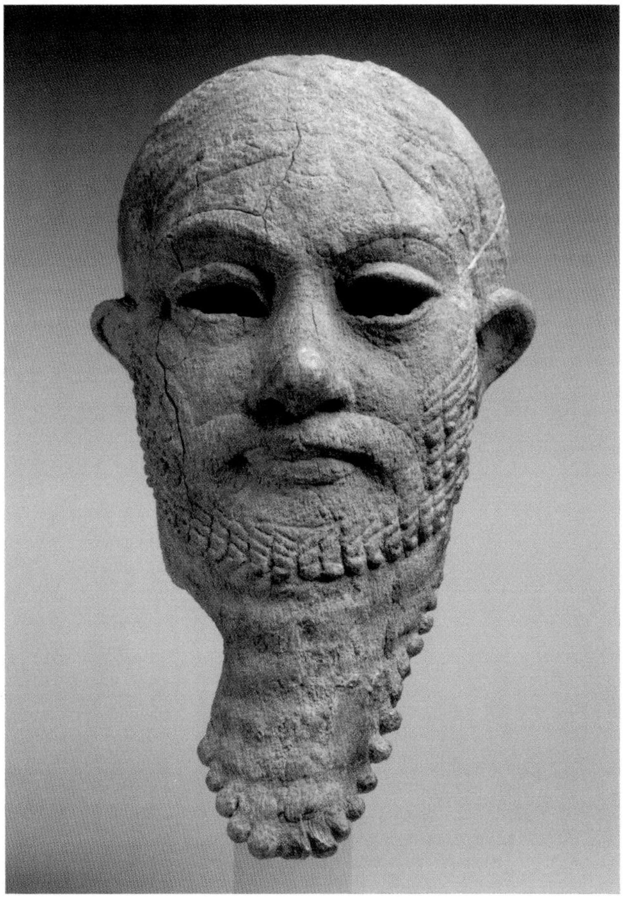

Figure 310 The large cast-bronze head of a ruler displays typical corrosion found on bronzes from archaeological origin, with deep cracks formed by the advanced mineralization of the metal. *Head of a Ruler*, Iran or Mesopotamia, 2300–2000 BCE, H. 34.3 cm (Metropolitan Museum of Art, inv. 47.100.80).

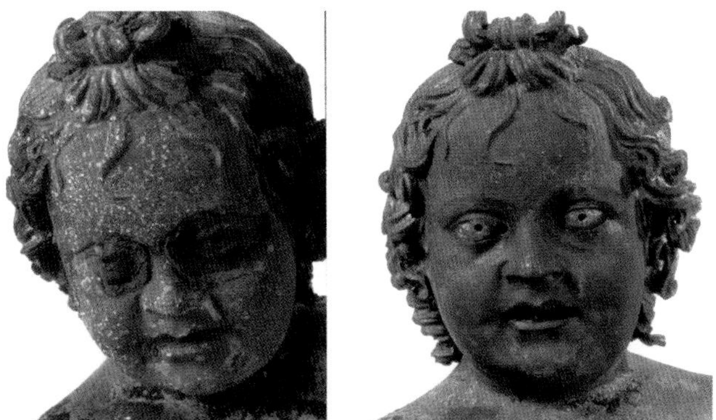

Figure 311 Detail of face before restoration, showing the heavy corrosion due to long-term immersion in seawater (left) and after restoration (right). Electrolytic dechlorination was achieved via the total immersion of the figure in a basic solution over ninety days. Note the silver eye inlays revealed at right. *Eros*, Roman, 1st century BCE, H. 63.5 cm (Musée de l'Ephèbe et d'Archéologie Sous-Marine de la Ville d'Agde, France, inv. 2888). See Mille et al. 2012.

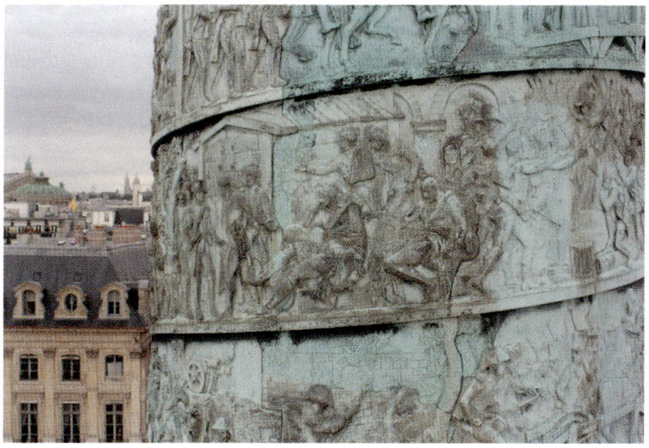

Figure 312 Detail of the Vendôme Column before the 2015 restoration, showing how exposure to urban air pollution has caused corrosion. Various French sculptors, *Vendôme Column*, 1805–10, dismantled 1871, re-erected 1873–75, H. 44 m (Place Vendôme, Paris). See Texier et al. 2016.

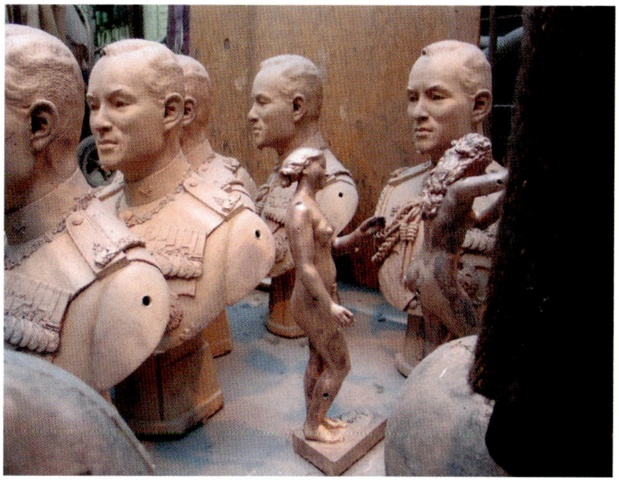

Figure 313 Recently cast bronze replica busts, acid cleaned and awaiting patination at the Susse foundry outside Paris, 2012.

Patina type	Cross section	SEM analysis	X-ray fluorescence analysis (portable)	X-ray diffraction analysis
Blue tinted 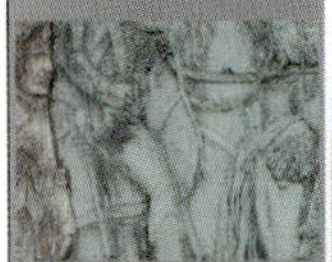	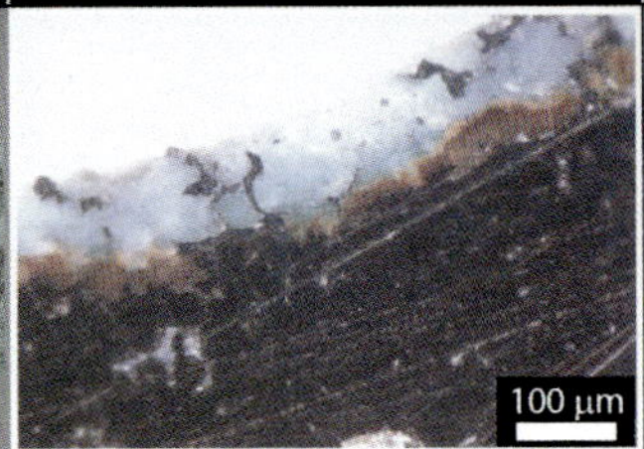	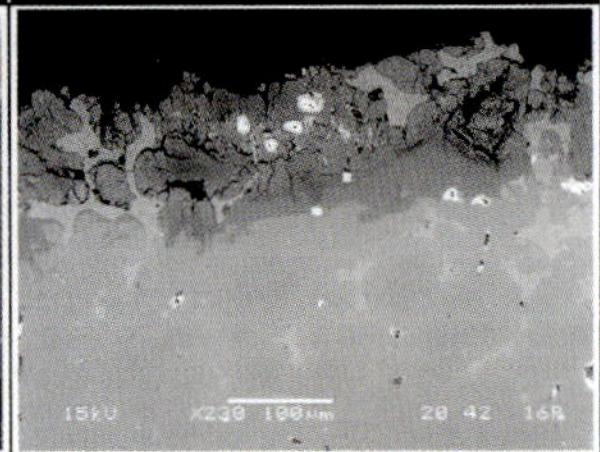	Sn = 7.2 %wt Pb = 1.2 %wt Zn < 1%wt	Brochantite $[Cu_4SO_4(OH)_6]$ Antlerite $[Cu_3SO_4(OH)_4]$
Green 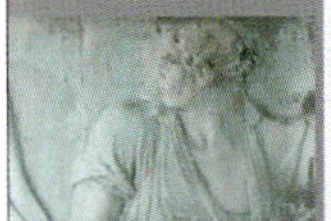	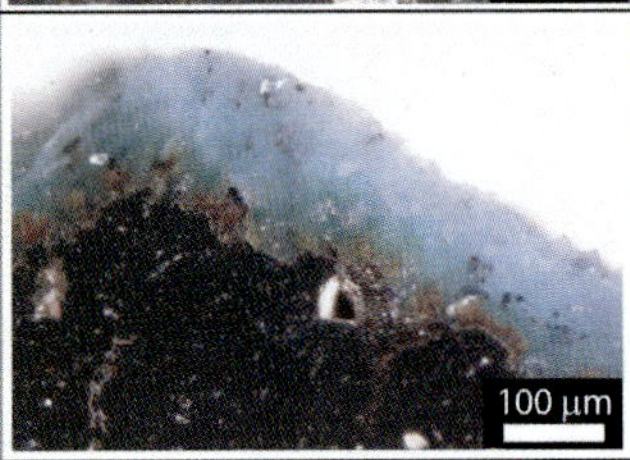	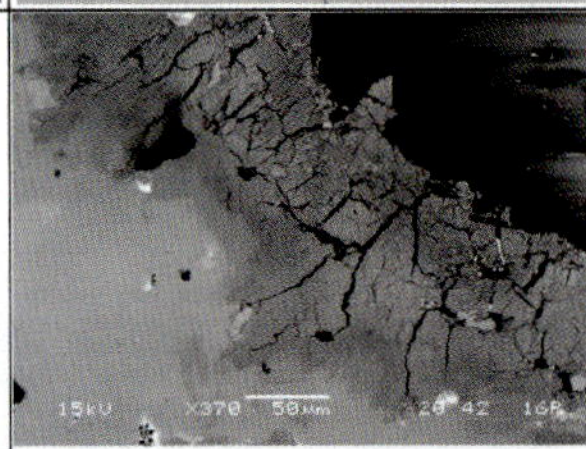	Sn = 7.5 %wt Pb = 5.7 %wt Zn < 1 %wt	Brochantite $[Cu_4SO_4(OH)_6]$ Antlerite $[Cu_3SO_4(OH)_4]$
Greenish-white 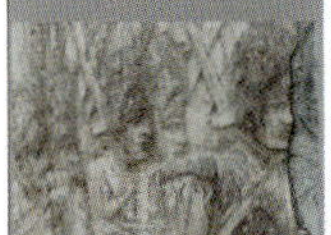	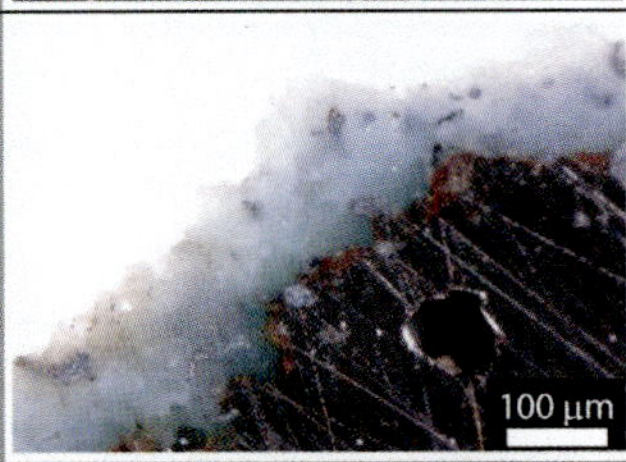	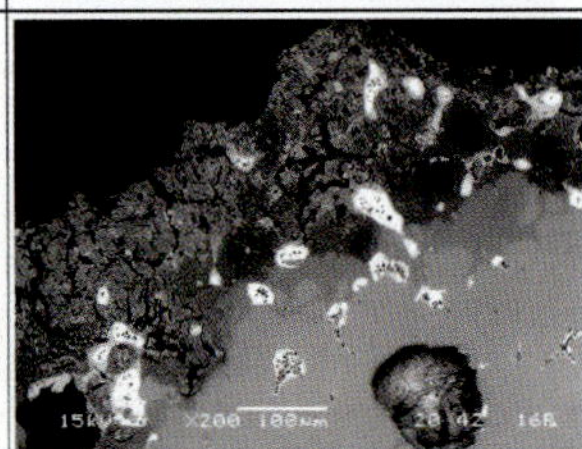	Sn = 5.6 %wt Pb = 12 %wt Zn < 1 %wt	Anglesite $[PbSO_4]$ Minor Compounds: Brochantite $[Cu_4SO_4(OH)_6]$ Antlerite $[Cu_3SO_4(OH)_4]$
Repair 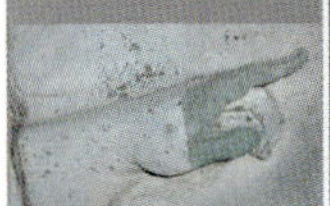	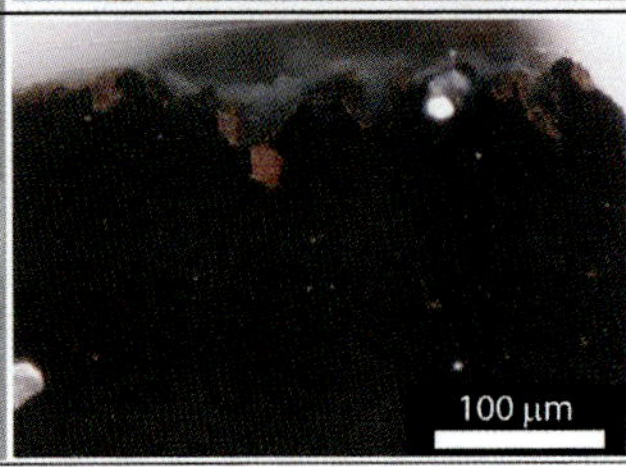	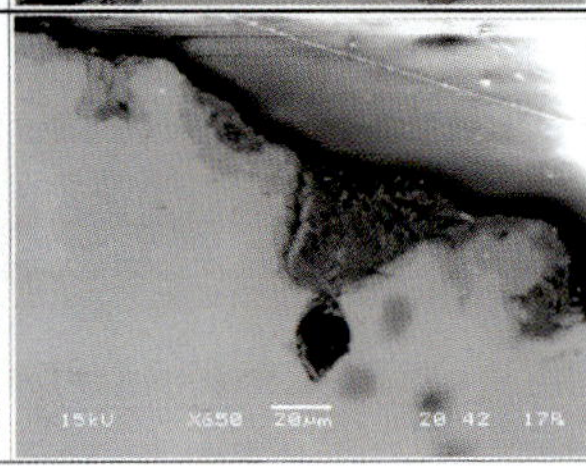	Sn ~ 2 %wt Pb < 1 %wt Zn ~ 2 %wt	No Data

Figure 314 Details of four of the bronze plates from the Vendôme Column showing the influence of metal composition on the patina shade. All plates are made of leaded tin-bronze, except the repairs (zinc and tin-copper alloy). Tin content is steady, and lead controls the color: the higher the lead content, the higher the lead in the corrosion products, and the whiter the patina. In each plate, the different patina layers are revealed by polished cross sections observed under optical microscopy (center images) and scanning electron microscopy (SEM-BSE mode) (right images). The layers are, from bottom to top: metal, red cuprite, outer layer. The composition of the metal (XRF analyses) and of the patina (XRD) is indicated. Various French sculptors, *Vendôme Column*, 1805–10, dismantled 1871, re-erected 1873–75, H. 44 m (Place Vendôme, Paris). See Texier et al. 2016.

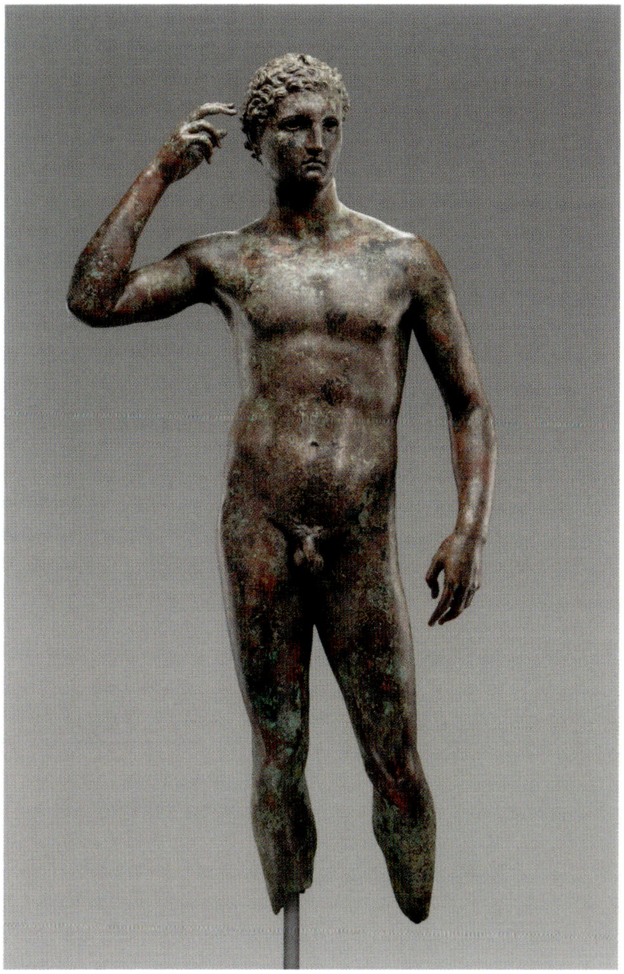

Figure 315 This statue was retrieved from the bottom of the ocean, heavily encrusted with marine deposits and corrosion layers. It is pictured here after mechanical cleaning, revealing a fairly smooth, continuous layer that is assumed to be close to the original surface. In the process, embedded copper chloride was also exposed, rendering the sculpture vulnerable to outbreaks of occasional active corrosion. For its long-term preservation, it is displayed and constantly monitored in an entirely desiccated gallery, which is continuously fed dry air. *Statue of a Victorious Youth*, Greek, 300–100 BCE, H. 151.5 cm (J. Paul Getty Museum, inv. 77.AB.30).

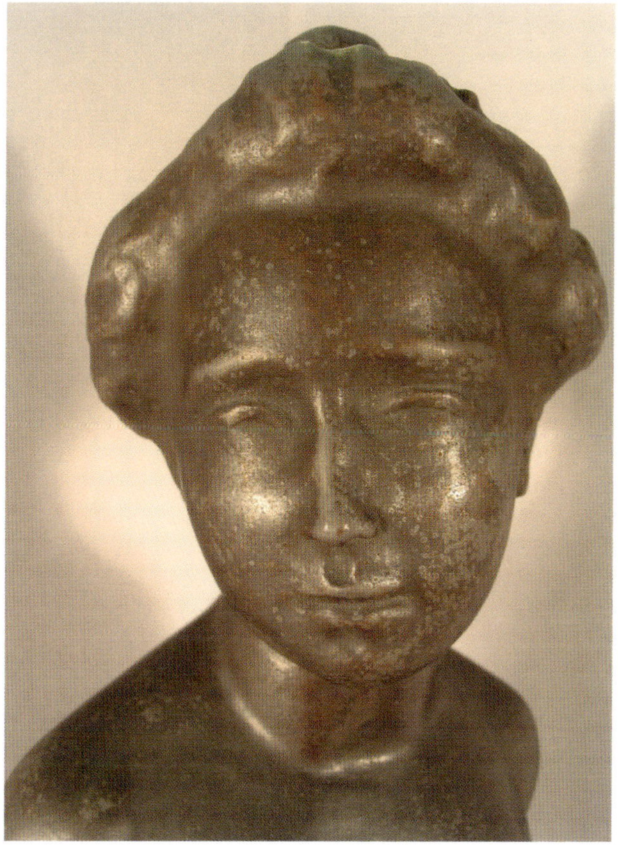

Figure 316 Detail of face, showing how storage near a furnace for many years caused damage in the form of dark spots to the patina from fumes. Aristide Maillol (French, 1861–1944), *Standing Bather*, 1899, H. 66 cm (Baltimore Museum of Art, gift of Blanche Adler, 1941.120).

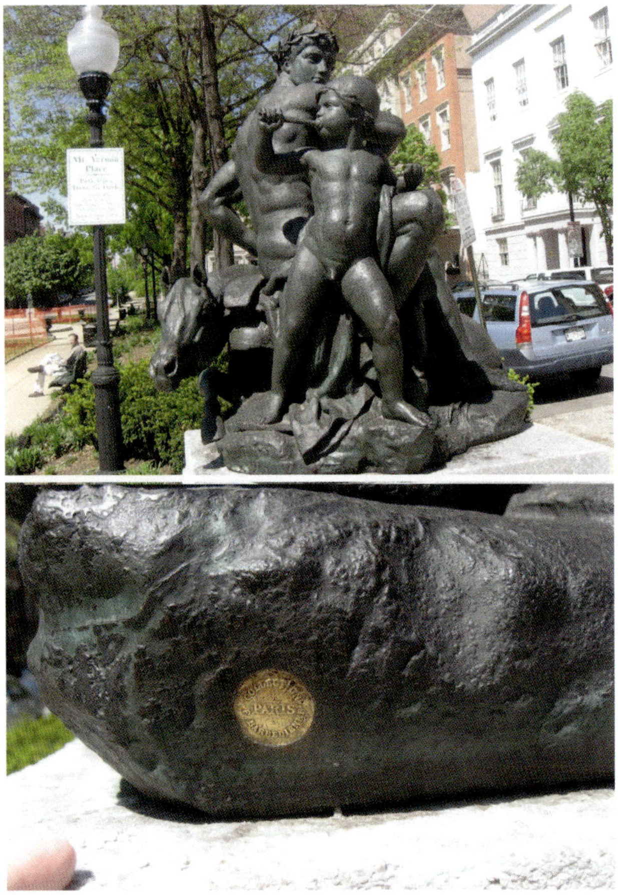

Figure 317 Raised black islands on the surface of this badly corroded sculpture are due to aggressive urban air pollution over more than a century. A recent maintenance treatment of dark-tinted wax has minimized the surface color variation, although, as seen in the detail of the base with the foundry mark, the irregular surface texture caused by the corrosion is still quite apparent. Antoine-Louis Barye (French, 1795–1875), *La Guerre* (War), 1884 cast from 1855–56 model, H. 102 cm (Mount Vernon Place, Baltimore, given to the City of Baltimore by William Thompson Walters).

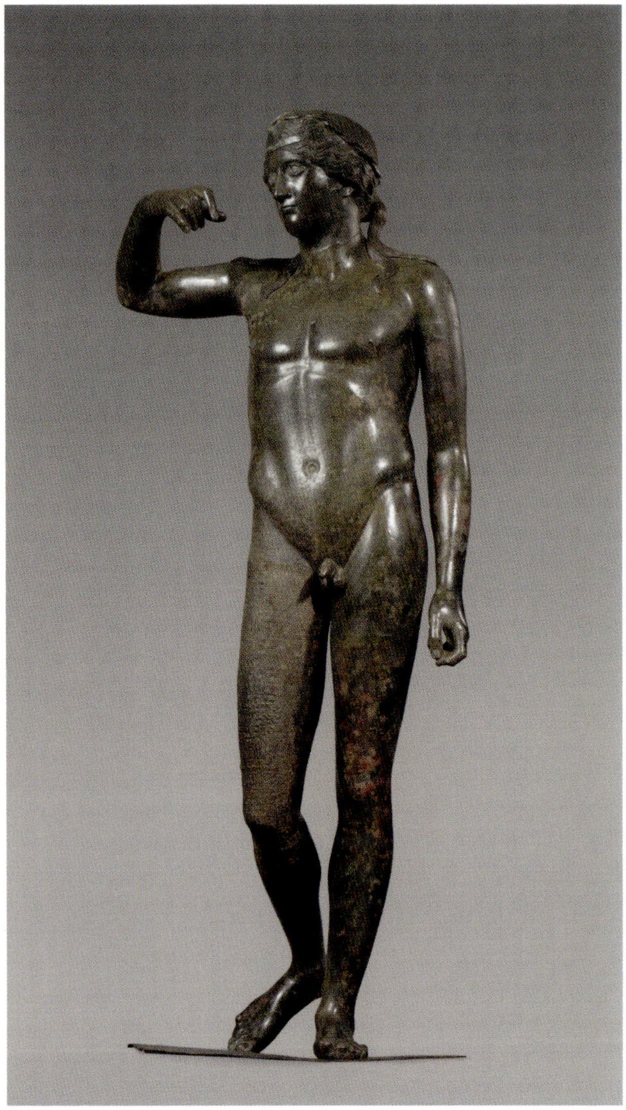

Figure 318 The rough texture on the front of the right thigh is the trace of a coarse textile that was once in contact with the bronze surface and formed into a pseudomorph during the statue's prolonged burial. *Dionysos*, Roman, Late Hellenistic, 1st century BCE–1st century CE, H. 135.8 cm (private collection, on loan to the Art Institute of Chicago). See Mattusch 1996, no. 23; Ekserdjian 2012, no. 44.

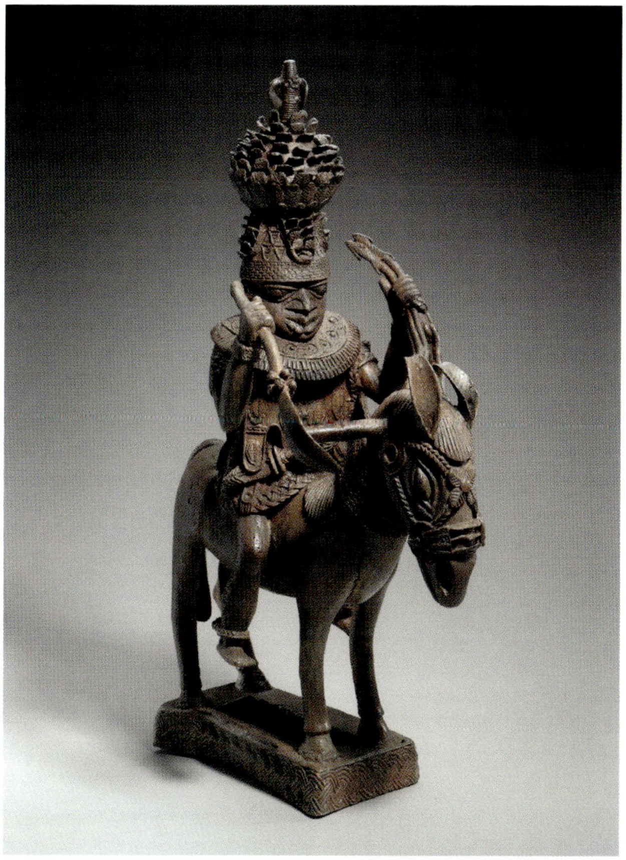

Figure 319 Bronze sculpture of horse and rider clearly made by a lost-wax casting process. The work exhibits a modern coating; the original surface appearance is unknown (and may never be known) due to repeated surface treatments it experienced after leaving Africa. *Mounted Ruler*, Edo Peoples, Benin, Nigeria, 16th century, L. 45.7 cm (Museum of Fine Arts, Boston, inv. L-G 7.12.2012).

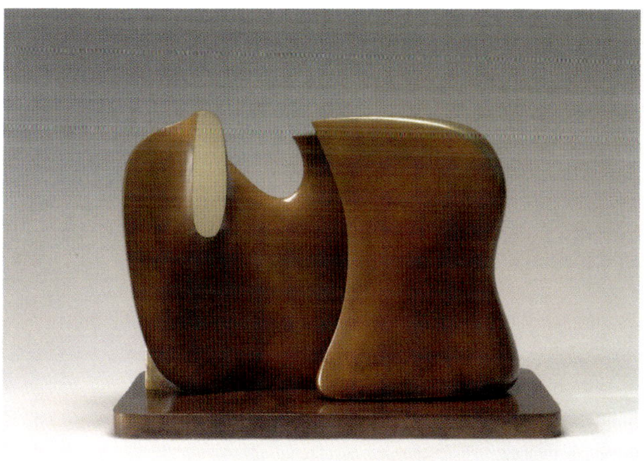

Figure 320 The highly polished, unpatinated, oval bronze plane contrasts with the darker patinated surface of the rest of the sculpture. Henry Moore (British, 1898–1986), *Two Piece Mirror Knife Edge*, 1978, H. 47.6 cm (National Gallery of Art, Washington, DC, Adolph Caspar Miller Fund, gift of the Morris and Gwendolyn Cafritz Foundation, inv. 1978.13.1).

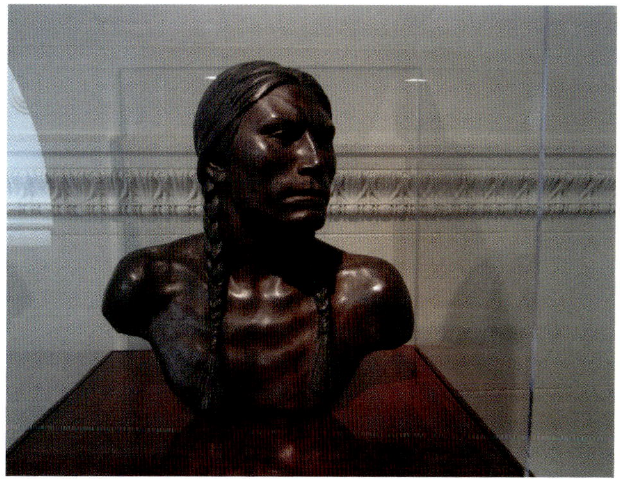

Figure 321 Malvina Hoffman chose different colors for skin and hair of *Apache Man* and worked closely with her chosen foundries to create realistic images of her subjects, as seen here. Malvina Hoffman (American, 1885–1966), *Apache Man*, 1934, H. 50.8 cm (Field Museum of Natural History, Chicago, inv. 337093).

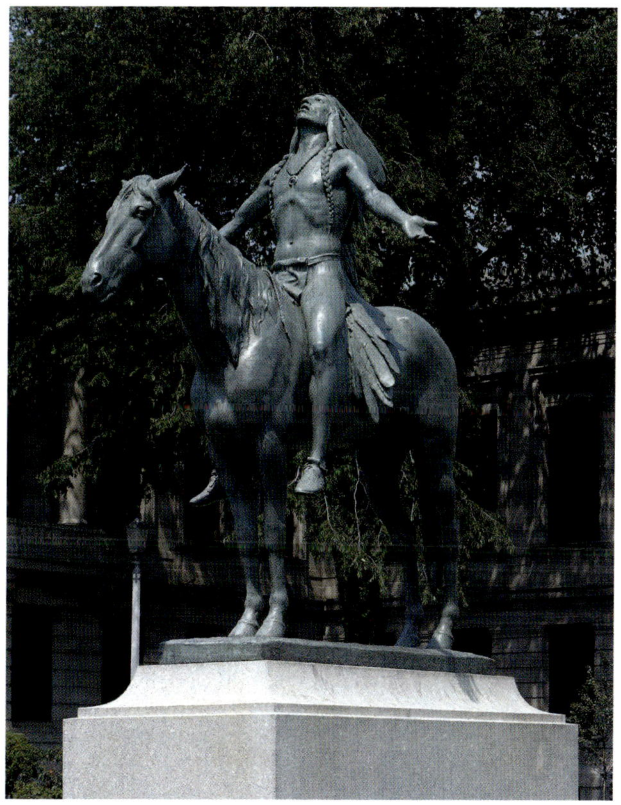

Figure 322 After several trials in the early twentieth century to preserve the original dark-brown patina, this sculpture that stands in front of the MFA Boston was painted green at the artist's behest to suggest even, natural corrosion. Cyrus Dallin (American, 1861–1944), *Appeal to the Great Spirit*, 1908, H. 309.9 cm (Museum of Fine Arts, Boston, inv. 13.380). See Newman 2011, 36.

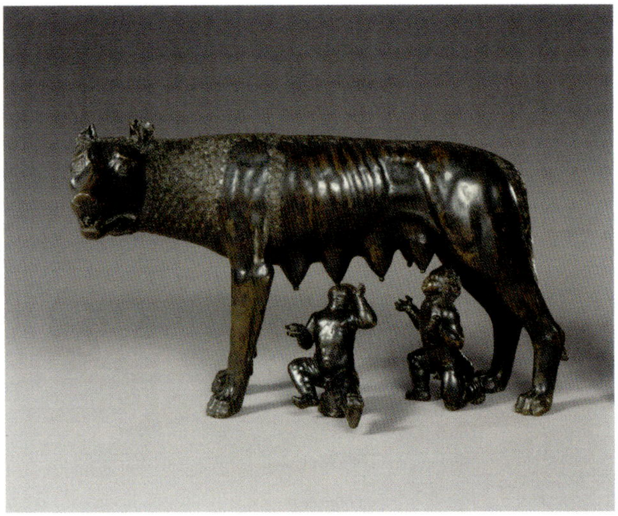

Figure 323 Example of darker applied coating patina that has been worn through handling, revealing the lighter brown oxidized surface on the high spots. *The Capitoline Wolf Suckling Romulus and Remus*, Florentine, 15th century, H. 6 cm (National Gallery of Art, Washington, DC, Samuel H. Kress Collection, inv. 1957.14.8).

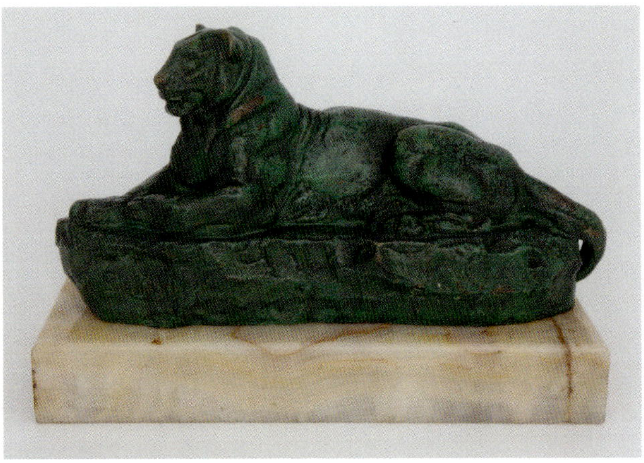

Figure 324 The bright bottle-green patina is not a chemical patina, but rather a coating that was painted on. It is worn through on several high points and chipped off in a few areas. After Antoine-Louis Barye (French, 1795–1875), *Panther of Tunis*, ca. 1930–39, H. 13.3 (Harvard Art Museums / Fogg Museum, The Henry Dexter Sharpe Collection, inv. 1956.169).

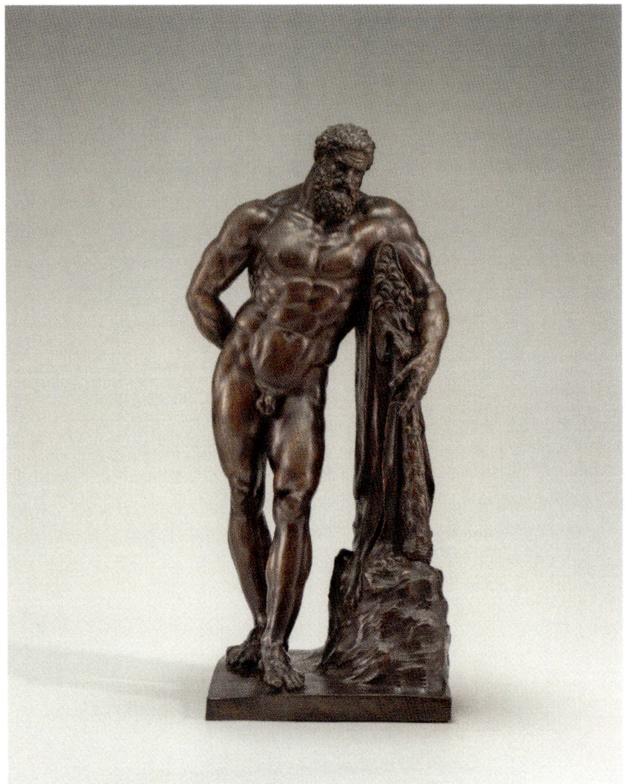

Figure 325 Example of translucent reddish-gold patina typical of some Florentine Renaissance bronzes. *Farnese Hercules*, Florentine, ca. 1550–99, H. 56.8 cm (National Gallery of Art, Washington, DC, gift of Stanley Mortimer, inv. 1960.10.1).

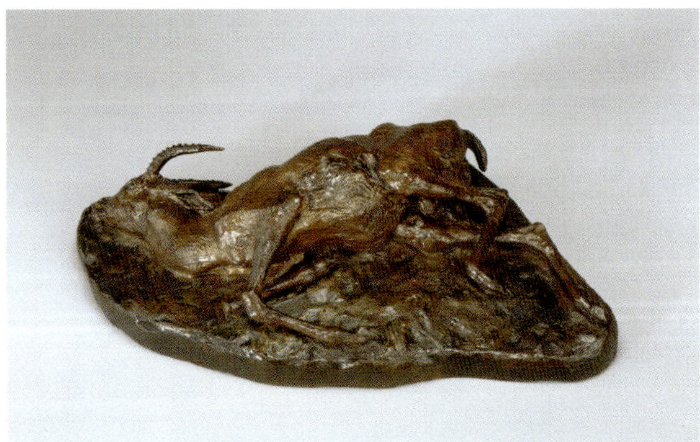

Figure 326 Bronze cast with patina partly made with organic coating mixed with metallic flakes. Antoine-Louis Barye (French, 1795–1875), *Dead Gazelle*, modeled 1832, cast 1833–34 by Honoré Gonon (French, 1780–1850), H. 7.6 cm (Walters Art Museum, Baltimore, inv. 27.96).

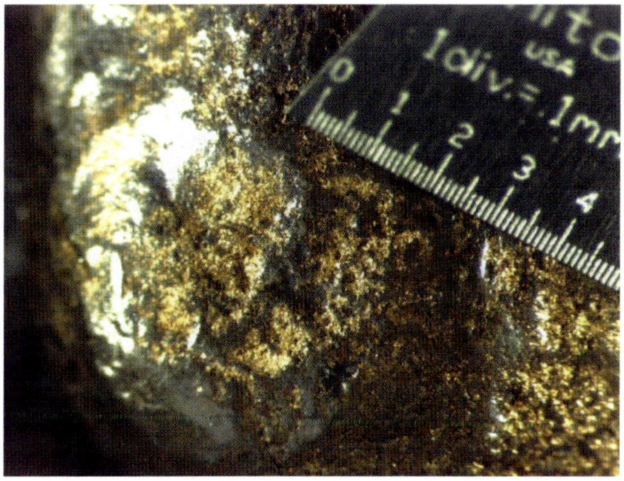

Figure 327 Photomicrograph detail showing metallic flakes in the organic patina. Antoine-Louis Barye (French, 1795–1875), *Dead Gazelle*, modeled 1832, cast 1833–34 by Honoré Gonon (French, 1780–1850), H. 7.6 cm (Walters Art Museum, Baltimore, inv. 27.96).

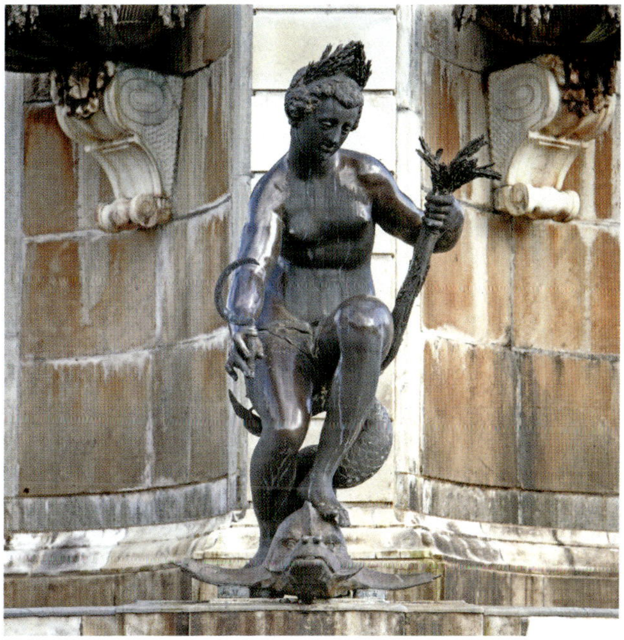

Figure 328 This piece has lived outdoors for more than three centuries as part of the Neptune Fountain at Drottningholm Palace in Sweden. Its natural patina is varied due to different types of exposure. It ranges from an attractive, glossy, warm brown to a more matte and streaky green and black. Adriaen de Vries (Netherlandish, 1556–1626), *Naiad (Ceres)*, 1615–18, H. 142 cm (Nationalmuseum, Stockholm, inv. NMDrh.Sk50). See Scholten 1998, 218.

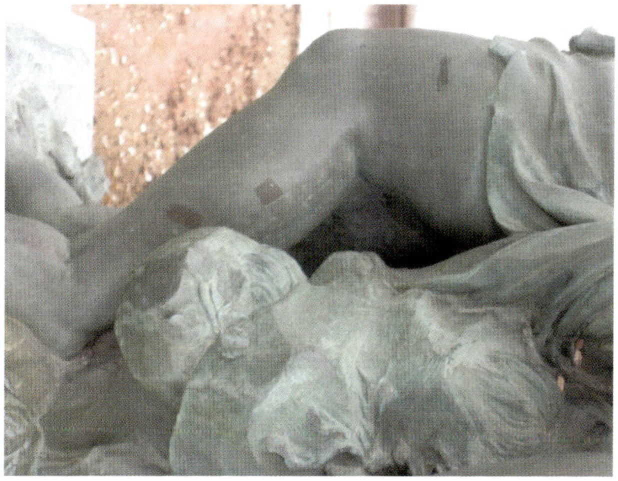

Figure 329 Detail of leg showing darker rectangular repairs. It is likely that the numerous repaired casting flaws were originally intentionally hidden by a dark or opaque coating. These repairs are now visible due to exposure to the weather and loss of original coating. Simon Mazière (French, 1649–1722), *Un Amour tenant un oiseau et deux enfants*, ca. 1685, cast by Aubry, Bonvalet, Schabol, and Taubin 1686–90, H. 161 cm (Water Parterre, Musée National des châteaux de Versailles et de Trianon, France, inv. 1850.8931). See Maral, Bourgarit, and Amarger 2014.

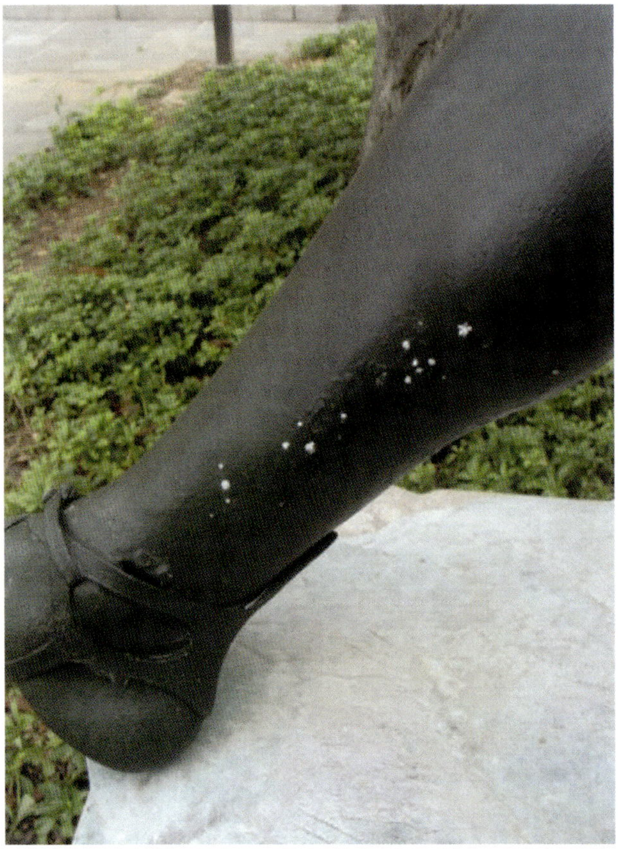

Figure 330 Detail of leg, showing core material from inside the sculpture that migrated to the surface of this porous bronze when the weather was particularly wet. The surface has been covered with a tinted wax coating to reduce porosity, water incursion, and irregularities in the surface color. *Mercury*, 19th-century Italian reproduction of a bronze from Herculaneum, H. 131 cm (Baltimore Museum of Art, gift of the City of Baltimore, Department of Recreation and Parks, inv. 1948.46).

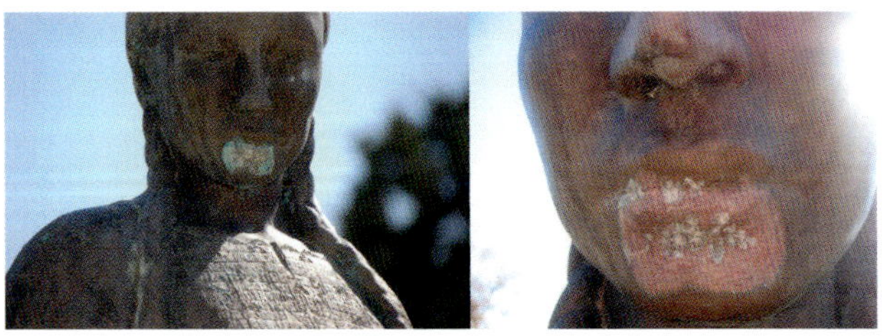

Figure 331 Core material from inside has migrated to the surface of the chin in this porous bronze during wet years, causing green corrosion. The left image shows it before treatment. After removal of the green corrosion by power washing, the raw pink bronze was revealed because the corrosion had destroyed the brown patina (right). More core material has since migrated (white areas) from the interior. Jo Mora (American, b. Uruguay, 1876–1947), *Indian Maiden*, 1928, cast by the California Art Bronze Foundry, Los Angeles, H. 182.8 cm (Woolaroc Museum and Wildlife Preserve, Bartlesville, Oklahoma, inv. SCT-47).

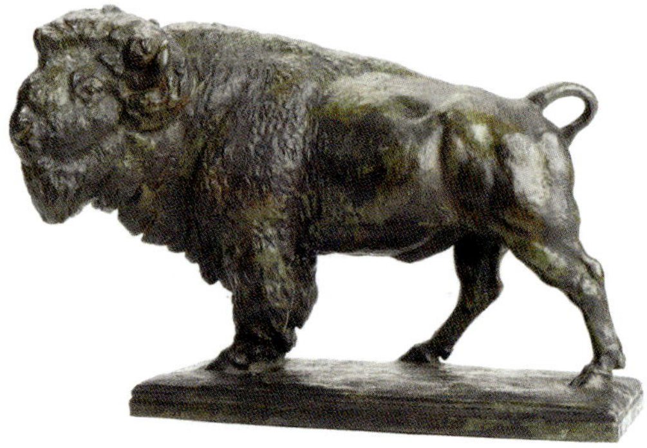

Figure 332 Machine oil was applied in a misguided attempt to spruce this work up, and years later, disfiguring black patches have emerged on the surface of the patina due to interaction with the oil. Alexander Phimister Proctor (American, b. Canada, 1862–1950), *Q-Street Buffalo*, 1912, sand cast in 1912–13 in Brussels at Verbeyst Foundry, H. 33.7 cm (Gilcrease Museum, Tulsa, inv. 08.840).

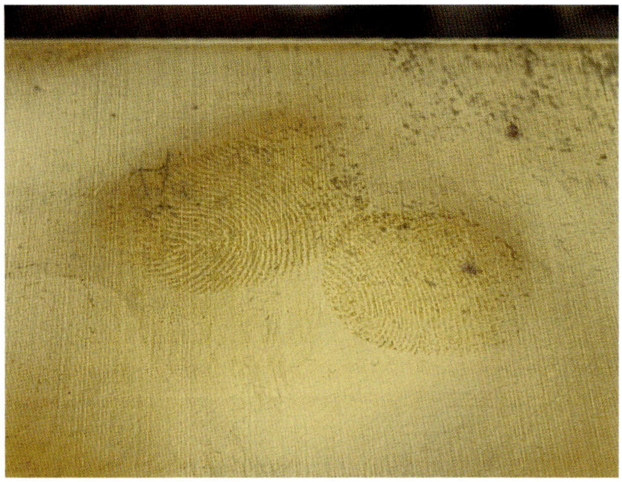

Figure 333 Fingerprint on polished brass caused by a reaction of the metal with salts and acids on human skin.

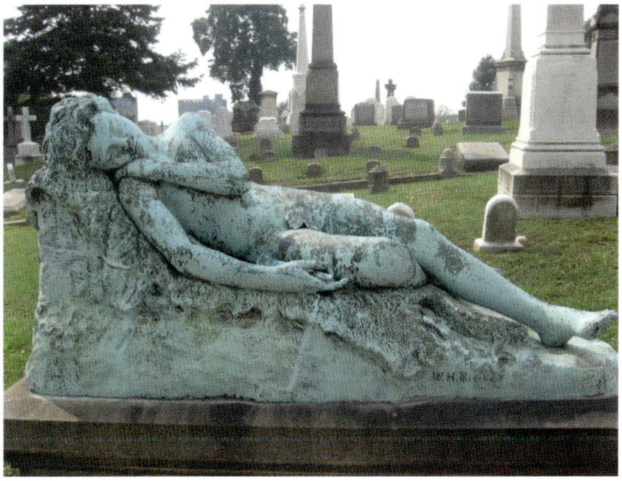

Figure 334 Exposure to a harsh urban environment has altered any original patina into a mottled, discolored, and somewhat powdery corroded surface that is very different from the original appearance. William Henry Rinehart (American, 1825–1874), *Endymion*, 1874, L. 182.9 cm (marker for the sculptor's own grave at Greenmount Cemetery, Baltimore).

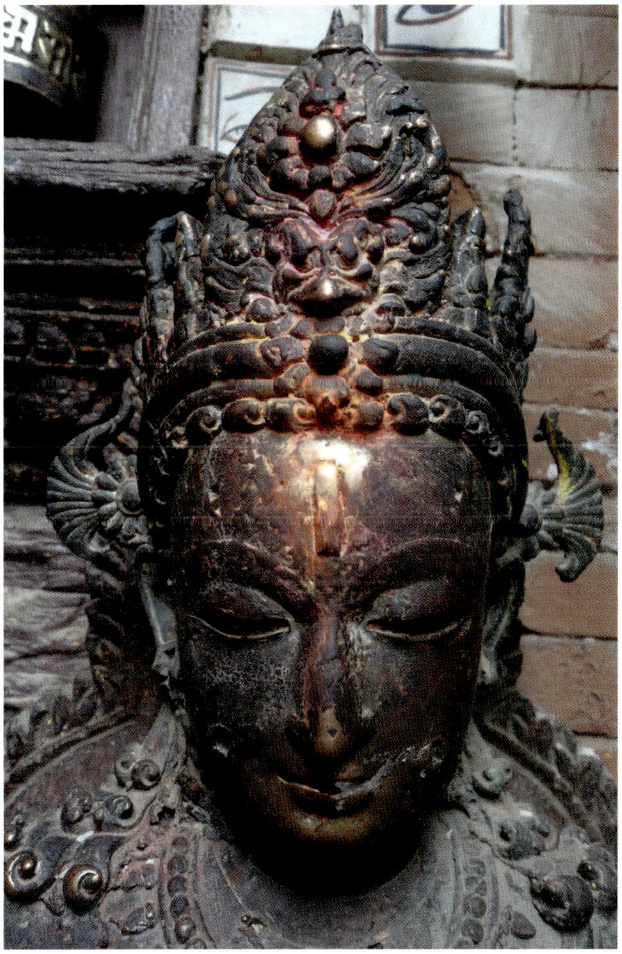

Figure 335 The application of libation and holy pigment to a copper alloy sculpture forms multilayered encrustations. Divinity in the Courtyard of the Golden Temple, Patan, Nepal, date and dimensions unknown.

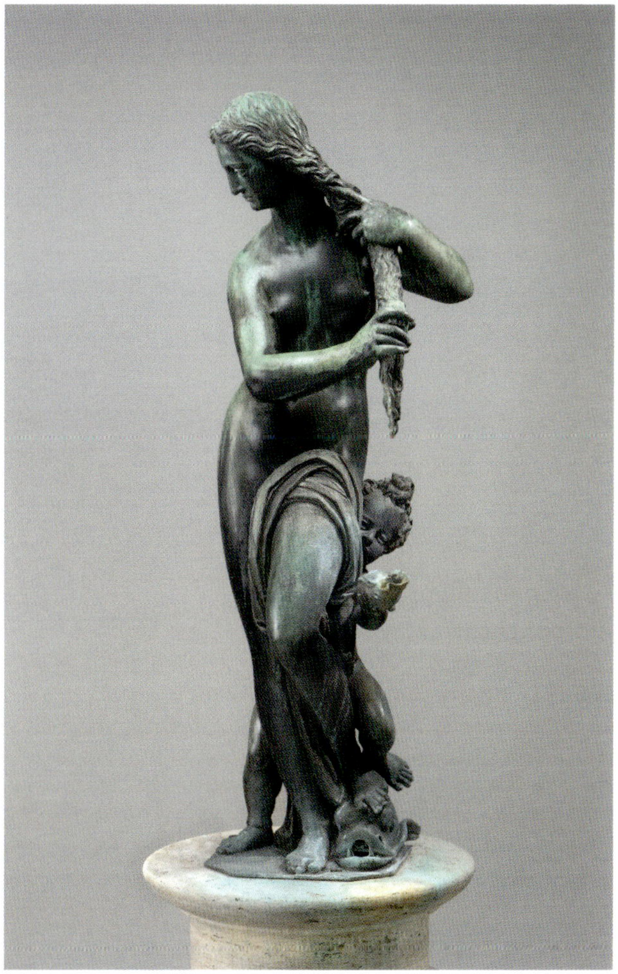

Figure 336 Bronze fountain sculpture that developed a natural patina over time from water runoff in an outdoor environment. Follower of Giovanni Bologna (called Giambologna, Flemish, active Florence, 1529–1608), *Venus and Cupid*, Italian, 16th century, H. 124.5 cm (National Gallery of Art, Washington, DC, gift of John and Henrietta Goelet, in memory of Thomas Goelet, and Patrons' Permanent Fund, inv. 1991.242.1).

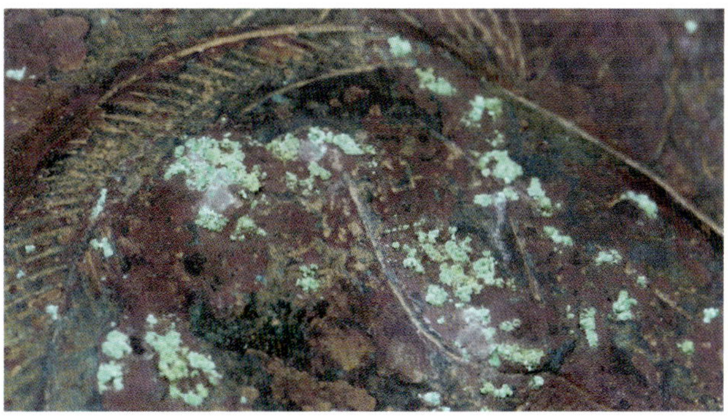

Figure 337 Detail of a thin bronze plate showing "bronze disease"—light-green-colored eruptions of copper chlorides. Breastplate with Relief Decoration of a Four-Horse Chariot, southern Italy, ca. 480 BCE, L. 107 cm (J. Paul Getty Museum, inv. 83.AC.7.3).

A. SCULPTURE THAT IS NOT LIKELY TO HAVE BEEN BURIED

If there is a suspected organic coating present such as wax or lacquer:

- Examine with strong raking and UV light, with and without magnification. Make note of any brush marks, brush fibers, drips, spray patterns, etc., and/or layering that may be revealed in areas of damage or loss.
- If multiple coating layers are found, note the order.
- Fluorescence color under UV may help with preliminary characterization of the coating(s) (Measday, Walker, and Pemberton 2017). Discrete solvent testing by a trained conservator may also help with preliminary characterization.
- Consider whether the coating(s) may be part of original manufacture, from a historically significant restoration, or from a more recent conservation/restoration treatment.

If a coating appears to be original to the manufacture of the object:

- Do not remove (except for minute sample for analysis if possible).
- Microscopic examination is useful to identify the best-preserved areas of original coating.
- Look for the presence of pigment particles or transparent colorants. Note surface condition and texture (craquelure, etc.).
- Comparative surface elemental analysis of the coating with bare metal areas may reveal inorganic elements in the coating (fig. 305 A, II.6).
- If sampling is possible, attempt identification of organic and inorganic components with the appropriate analytical techniques (fig. 305 B and C).
- Consider whether an old coating may have changed over time to become darker or more opaque.

If a coating appears to be not original:

- Examine carefully, particularly in areas of damage or loss, to determine if (possibly fragmentary) coatings are present below. A minute cross section may be taken for microscopy to aid in this determination.
- Consider whether the non-original coating(s) may significantly alter or conceal the underlying appearance.
- Consider whether identification of the non-original material is warranted to better understand the history of the object.

Be aware that a chemical patina layer can sometimes be mistaken for an organic coating:

- Chemical layers are generally thinner than organic coating layers and may not show edges in losses. Use magnification to study edges of patina application or loss if possible.
- Consider whether the original color of a chemical patina may have changed over time. Compare to any areas protected from air and light if possible.
- Comparative surface elemental analysis of patina layer(s) with bare metal areas may reveal differences (fig. 305 A, II.6). The presence of certain elements may point to either a pigmented coating or to a chemical patina.

B. ARCHAEOLOGICAL PATINA LAYER

If the surface is metallic and free of corrosion:

- Examine porosities and surface texture under magnification; look for any residual traces of corrosion.
- Consider possible reasons for the lack of corrosion (burial environment, previous treatments).

If the surface shows corrosion products:

- Examine with strong raking and UV light, as well as under magnification/stereo microscope. Make note of any:
  - inclusions
  - accretions
  - pseudomorphs
  - embedded original surface elements (gilding, gesso, paint, inlays, patination)
  - drip marks/patterns
  - layers of different colors
  - layer order
  - degree of mineralization (X-radiography may be required)
  - disfigurement of original decoration
- Consider whether corrosion is due to burial, post-excavation treatment, possible residual chemicals, or environmental causes.
- If sampling is possible, attempt identification of organic and inorganic components with the appropriate analytical techniques (fig. 305 B and C).
- Map the corrosion and any differences in color, drip marks, and any patterns that may indicate original use or orientation during burial.

A minute cross section may be sampled using a micro scalpel in order to reveal layers and information about formation and corrosion mechanisms, as well as original manufacture.

It is relatively common that materials used as inlays (metals, niello, etc., see I.9) may have become degraded by deposition and interfere with corrosion layers. Also, be aware of the possibility of staining from close contact with other materials such as iron or lead oxides.

Figure 338 Set of examination and testing protocols for the characterization of a patina layer both for sculpture that is not likely to have been buried and for archaeological bronzes. For the latter, these are very general notes; the question of burial might be more nuanced than presented here and can lead to much more complex situations (e.g., archaeological bronzes may be subsequently coated). See fig. 305 for more details on the relevant techniques of analysis.

Figure 339 Two patinated copper-alloy plates (red and black) showing the influence of surface texture on the perception of color. For each plate (~20 × 20 cm), the upper half of the surface has been left as-cast, while the lower half has been polished before patination. Samples provided to the C2RMF by the Coubertin foundry, France, in the late 1990s.

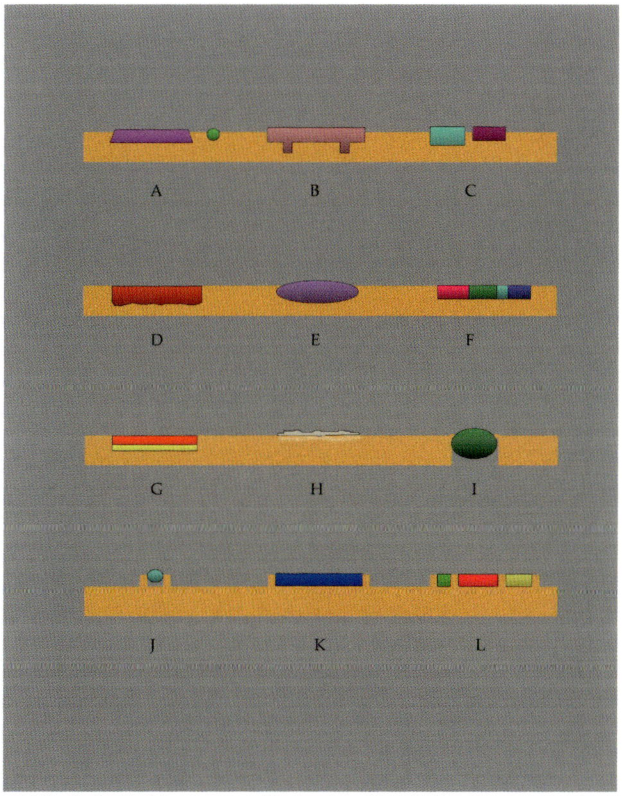

Figure 340 Schematic representation of general types of inlay (A–I) and overlay (J–L). A) cavity with undercuts and chased line with undercut; B) cast cavity with secondary receiver holes to reinforce the primary element with pins; C) cast depressions left in the surface to receive inlays; D) complex preformed inlay inserted into mold and cast in place; E) formed metal inlay in chased recesses with undercuts; F) pre-shaped abutting inlay elements in single chamber; G) translucent inlay backed by metallic foil; H) hammered metal foil on roughened surface; I) through-hole cavity with insertion from reverse; J) cloisonné-type chamber surface mount for gemstones; K) cloisonné type chamber on surface; L) cloisonné-type chambers on surface with multicolored design.

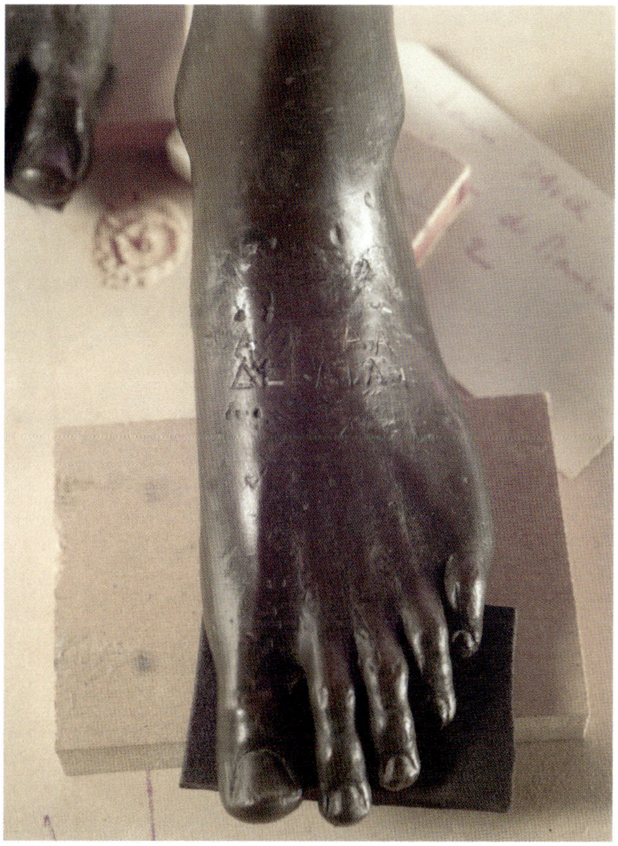

Figure 341 Detail of left foot showing silver inlay on the inscription. *Piombino Apollo*, Greek, 120–100 BCE, H. 115 cm (Musée du Louvre, inv. Br. 2). See Descamps-Lequime and Mille 2017.

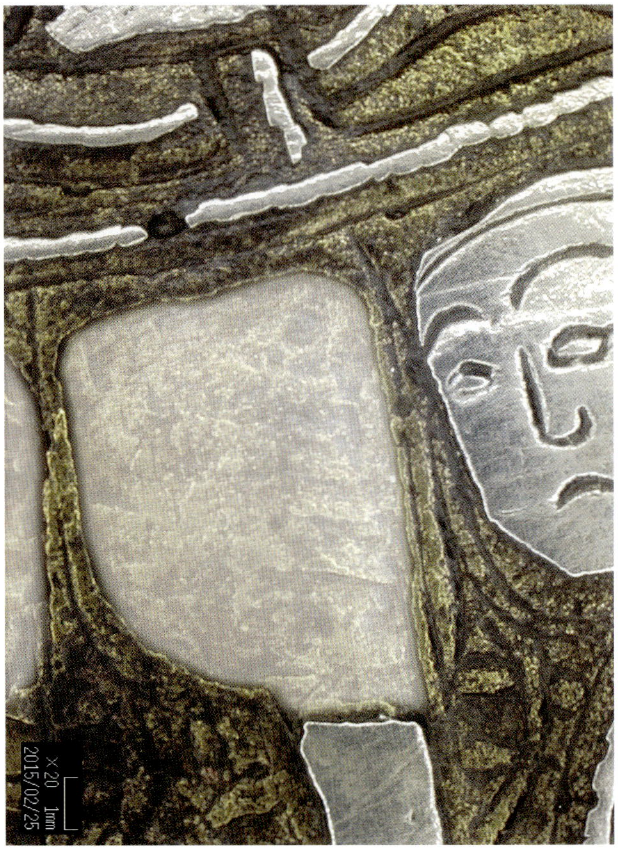

Figure 342 Digital photomicrograph showing areas of preserved silver wire and of a larger silver-plate inlay in hammered brass that has been chased to draw the face (right). Some of the silver has been lost, revealing the champlevé preparation for the larger silver-plate inlay (white overlay at center and left). A number of the chased lines would also have been inlaid with silver wire. *Ewer*, Afghanistan (Herat), ca. 1200, H. 39 cm (Musée du Louvre, inv. OA 5548). See Collinet 2021.

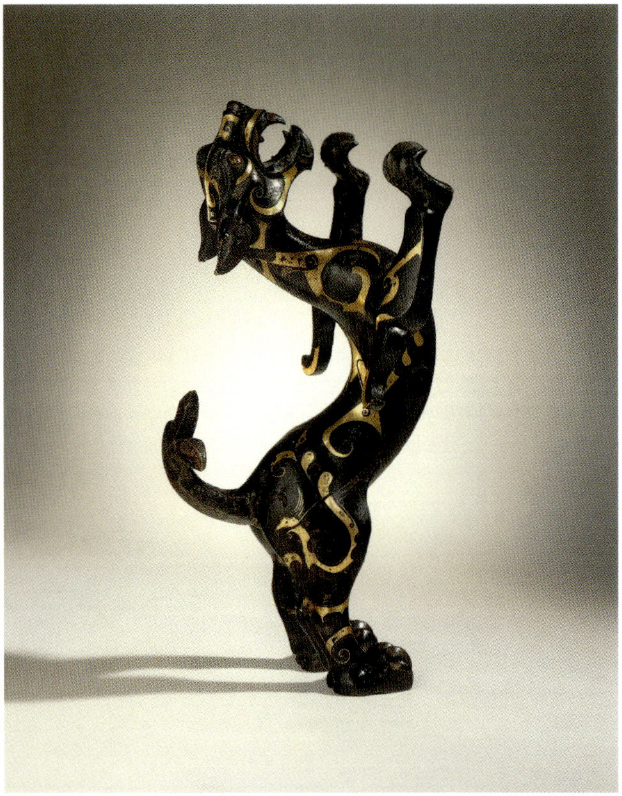

Figure 343 Stylized gold and silver inlays depict snakes and birds on a figure that was probably used as a tray support. *Figure of a Leaping Feline*, China, Eastern Zhou dynasty, 4th–3rd century BCE, H. 23 cm (British Museum, 1883, donated by Sir Augustus Wollaston Franks, inv. 1020.5).

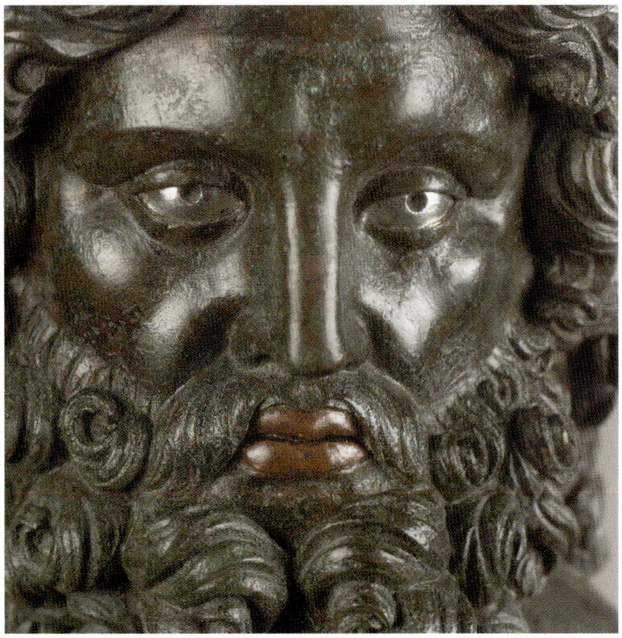

Figure 344 Detail showing inlaid eyes (silver) and lips (copper). *Jupiter*, Roman, 1st century CE, H. 92 cm (Musée d'Art, Histoire et Archéologie d'Evreux, France, inv. 5404). See Azéma et al. 2012.

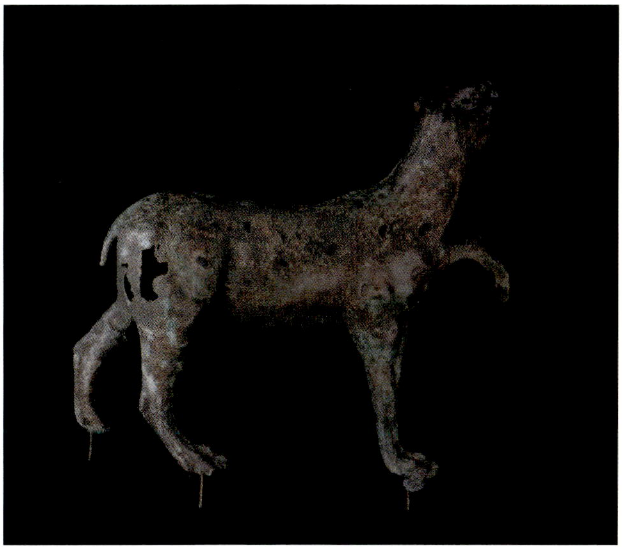

Figure 345 Eyespots made of copper inlays as shown by PIXE analysis (black on the photograph, probably due to corrosion) in a leaded tin bronze. *Panther*, Gaul, 1st century CE, H. 50 cm (Musée des Antiquités Nationales, Saint-Germain-en-Laye, France, inv. 79767). See C2RMF Internal Report 1997.

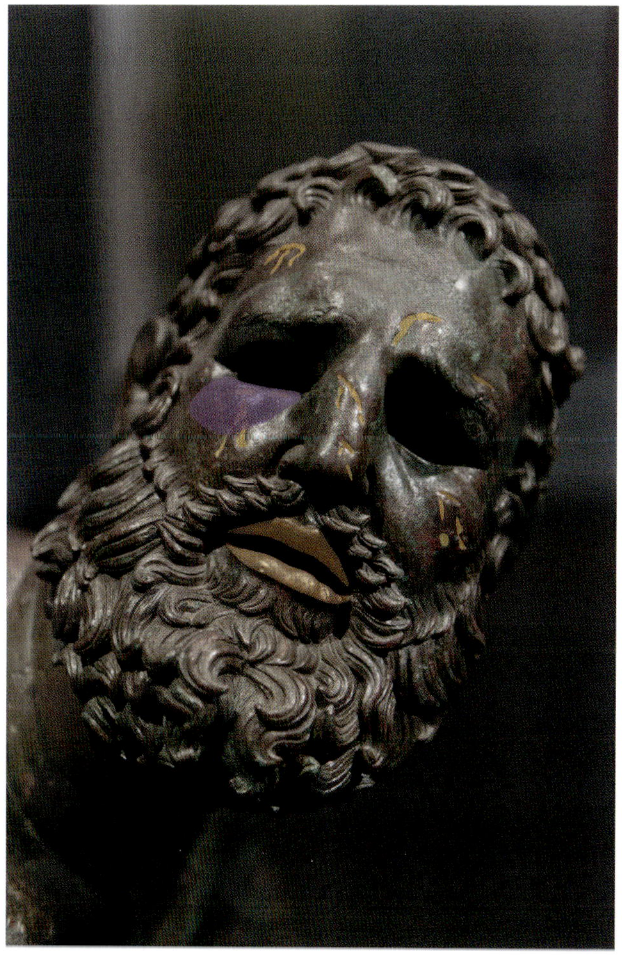

Figure 346 Annotated detail of copper-alloy inlays. The lips and blood drips are inlaid with a reddish unalloyed copper (orange overlays). A lead-rich copper alloy overlay simulates a swollen cheek contusion of a different color (purple overlay). *Boxer at Rest*, Greek, Hellenistic period, late 4th–2nd century BCE, H. 128 cm (Museo Nazionale Romano di Palazzo Massimo, Rome, inv. 1055). See Alessandri et al. 2013.

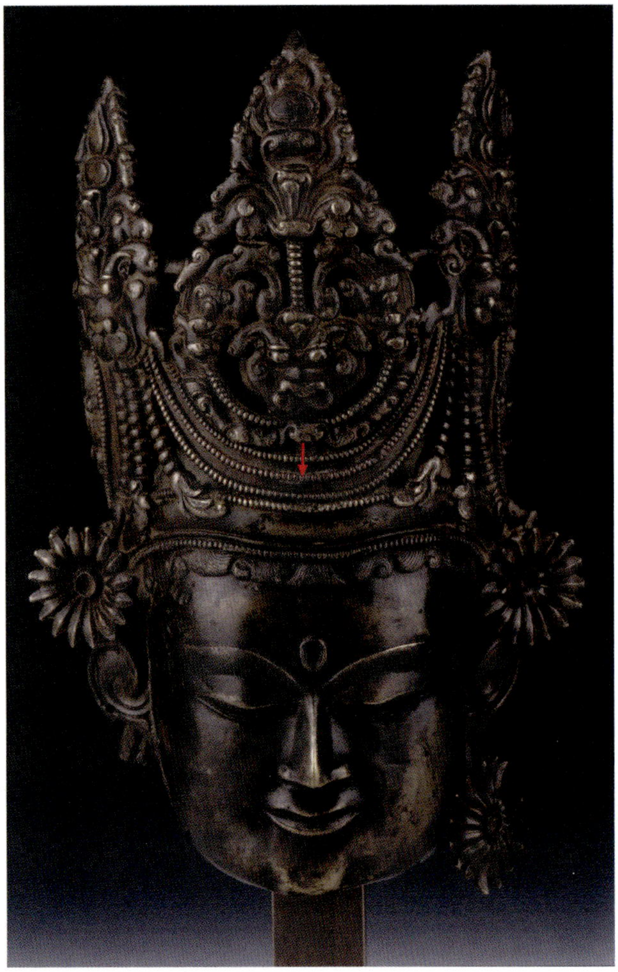

Figure 347 Lost-wax cast of a head of a Bodhisattva showing the brass and copper bead decoration in the tiara. The brass beads belong to the primary cast (both contain 21 wt% Zn), whereas one strand of copper beads was shaped from a wire and overlaid on the cast (red arrow). *Bodhisattva*, Tibet, 11th–13th century, H. 34.5 cm (Musée national des arts asiatiques - Guimet, Paris, inv. MA 6273). See C2RMF Internal Report 1998.

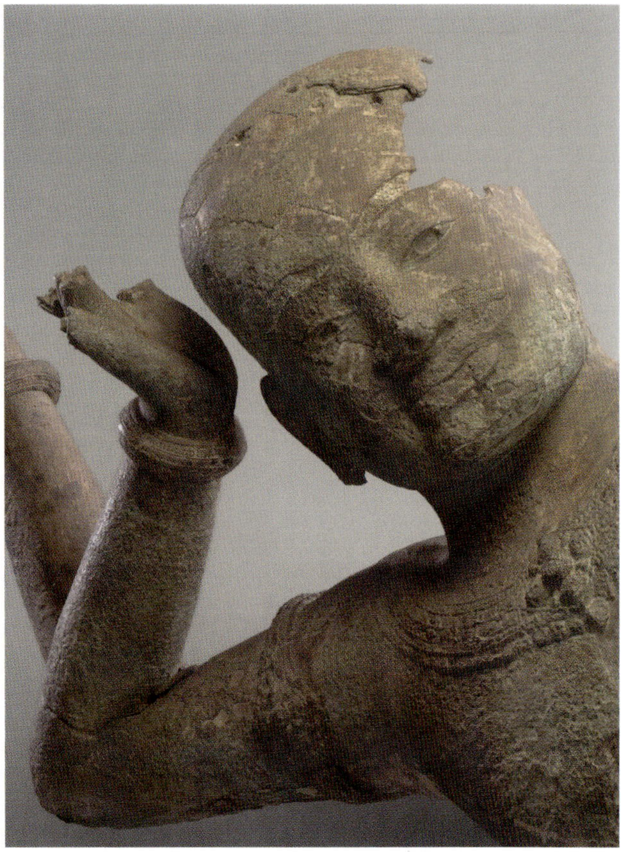

Figure 348 Detail of the face of the Reclining Vishnu. Because of heavy corrosion, the inlays are no longer visible, but only known through surface analysis: lead has been detected in the eyebrows, moustache, and beard. Composition of the eye inlays is unknown. Note texture in the eyebrows of material probably used to attach the inlay. *Reclining Vishnu*, West Mebon (Khmer), 12th century, L. 220 cm (National Museum of Cambodia, inv. Ga 5387). See CAST:ING 2018.

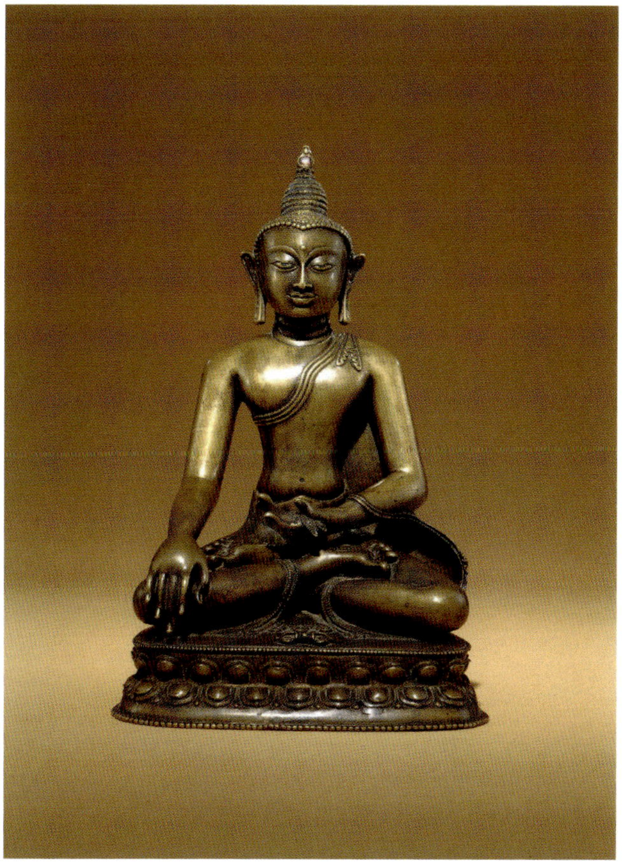

Figure 349 Buddhist deity represented in cast bronze with silver and copper inlays to enhance the eyes, and a stone set in the *usnisa* (top). The inset stone emphasizes the symbolism of the attainment of reliance on a spiritual guide. *Seated Statue of Dhyanibuddha Akshobhya*, Tibet, 12th–13th century, H. 32 cm (British Museum, Trustees of the British Museum, inv. 1981,1109.1).

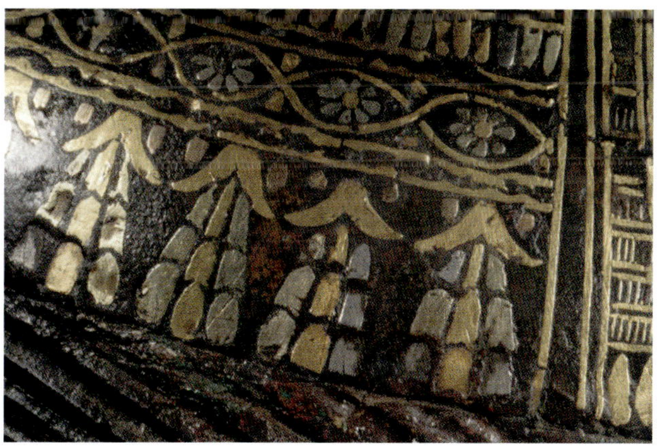

Figure 350 Detail of the chest of an ancient Egyptian statuette showing different color inlays: yellow, pink, and white-gray. PIXE analysis identified four different gold alloys: pure gold, electrum (gold and silver in equal proportions), gold with 25 wt% Cu, and gold with 34 wt% Cu and 6 wt% Ag. *Karomama*, Egyptian, Twenty-Second Dynasty, first quarter of the 1st millennium BCE, H. 59.5 cm (Musée du Louvre, inv. N500). See Delange, Meyohas, and Aucouturier 2005.

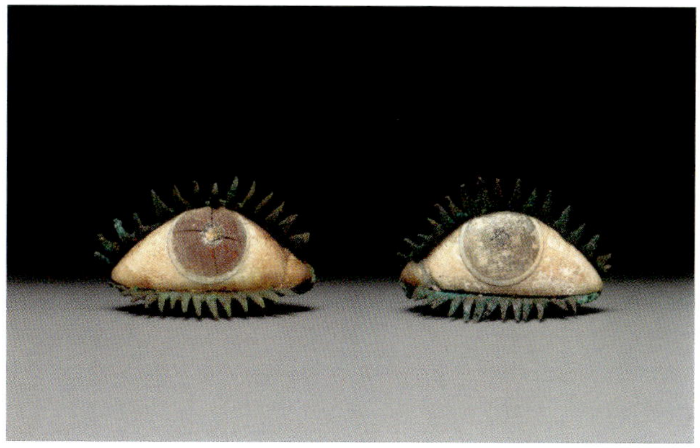

Figure 351 Eye inlays held in place within folded copper sheets, often with eyelashes as shown here (bronze/copper lashes, marble sclera, glass frit iris, quartz and obsidian pupil). Originally, the eyes would have been inserted into a cast-through cavity. Eye inlays, Greek, 5th century BCE or later, H. 3.8 cm (Metropolitan Museum of Art, purchase, Mr. and Mrs. Lewis B. Cullman Gift and Norbert Schimmel Bequest, inv. 1991.11.3a–b).

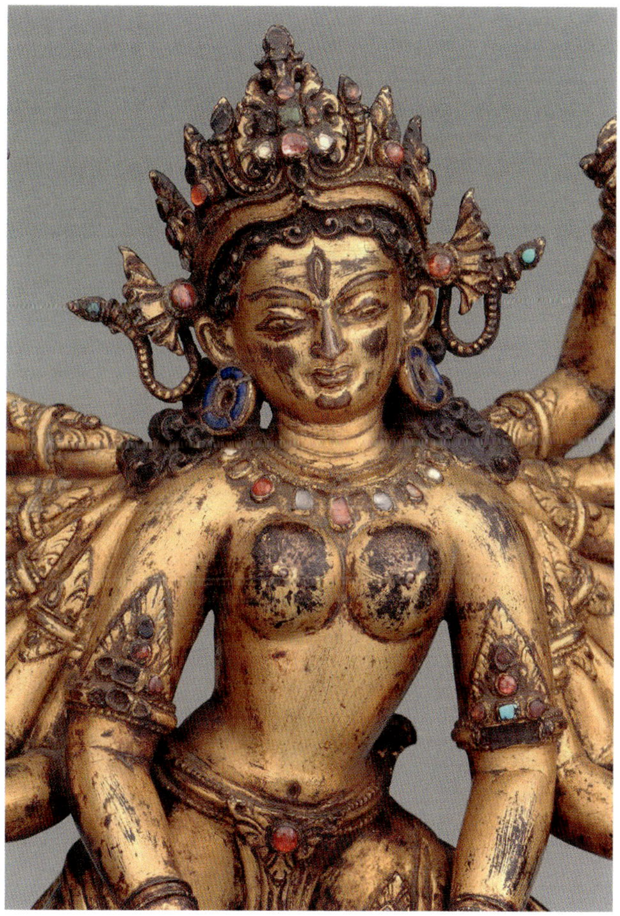

Figure 352 Gilt statuette overlaid and bejeweled with a range of semiprecious stones (possibly lapis lazuli, turquoise). Note the cavities and loss of stones in the proper right armband. *Durga as Slayer of the Buffalo Demon Mahishasura*, Nepal, 14th–15th century, H. 22.2 cm (Metropolitan Museum of Art, gift of Alice and Nasli M. Heeramaneck, 1986, inv. 1986.498).

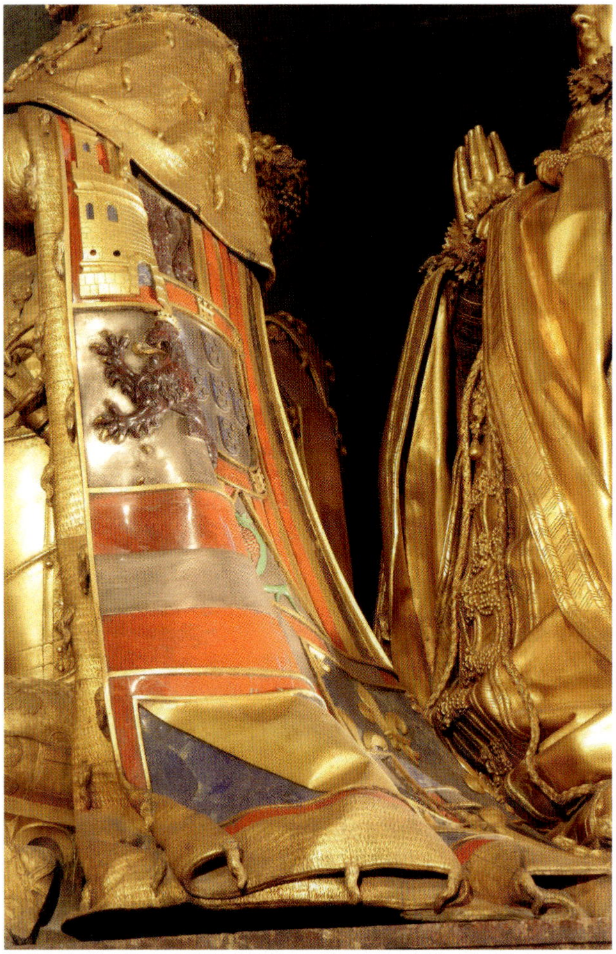

Figure 353 Detail of cloak. The royal arms are created from lapis lazuli and silver, with the heraldry representing the king's various territories. Pompeo Leoni (Italian, 1533–1608) and others, *Cenotaph of Philip II of Spain*, 1597–1600, H. 170–190 cm (Basilica of the Royal Monastery of San Lorenzo of El Escorial, Spain, inv. 10034776). See Arias 2012.

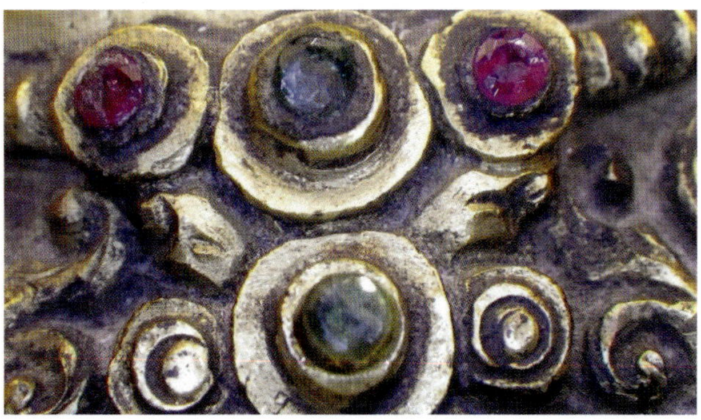

Figure 354 Detail of overlaid stones and gems. *White Tara*, Tibet, 17th century, H. 65 cm without base (Musée national des arts asiatiques - Guimet, Paris, inv. MA 12495). See C2RMF Internal Report 2012. See fig. 240 for an overall view.

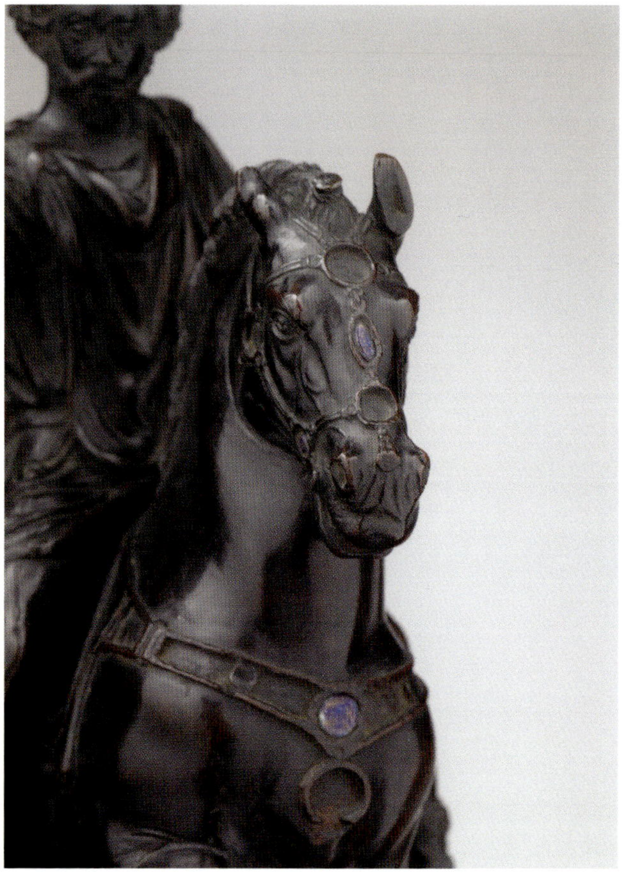

Figure 355 Detail of front inlay. Previously thought to be enamel, under microscopic examination by Michael Wagner and Rainer Richter (Grünes Gewölbe) all the blue inlays were shown to be blue glass with oil-bound gold pigment on top. Filarete (Antonio di Pietro Averlino, Italian, ca. 1400–1469), *Marcus Aurelius*, 1440–45, H. 38 cm (Grünes Gewölbe, Dresden, inv. H4 155/037). Personal communication with Claudia Kryza-Gersch, 2018.

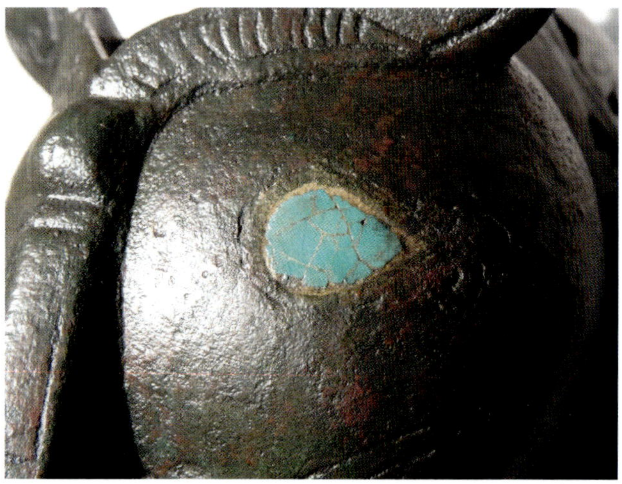

Figure 356 Detail of turquoise-blue glazed fritware inlay in the empty eye socket of a high-lead brass incense burner. *Lynx Incense Burner*, Iran (Khurasan province), ca. 12th century, H. 28.5 cm (Musée du Louvre, inv. AA 19). See Collinet 2021.

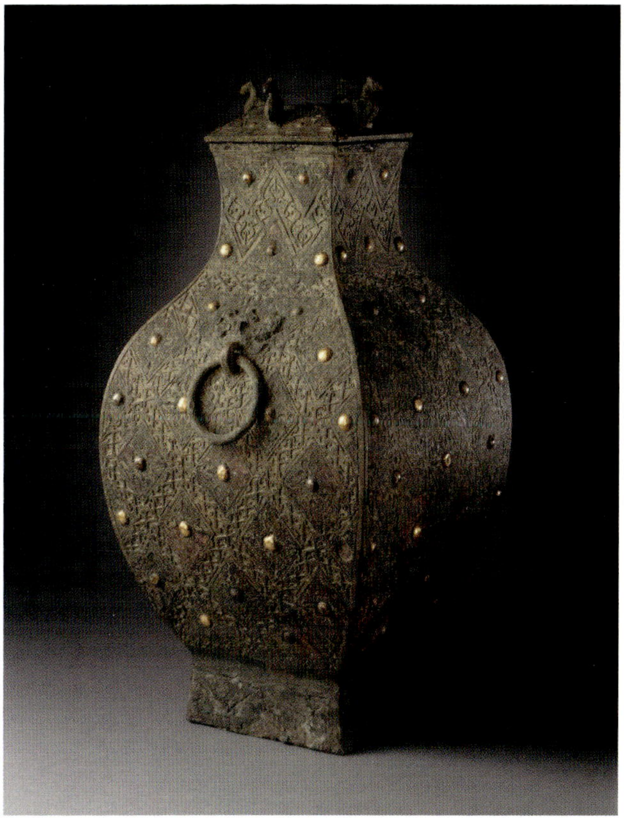

Figure 357 Cast bronze with gold appliqués and traces of powdered mineral inlay. *Lidded Square Wine Jar (Fanghu)*, China, Late Eastern Zhou dynasty or Middle Warring States period, ca. 481–300 BCE, H. 48.3 cm (Los Angeles County Museum of Art, gift of Mr. and Mrs. Lidow, inv. M.76.109a–b).

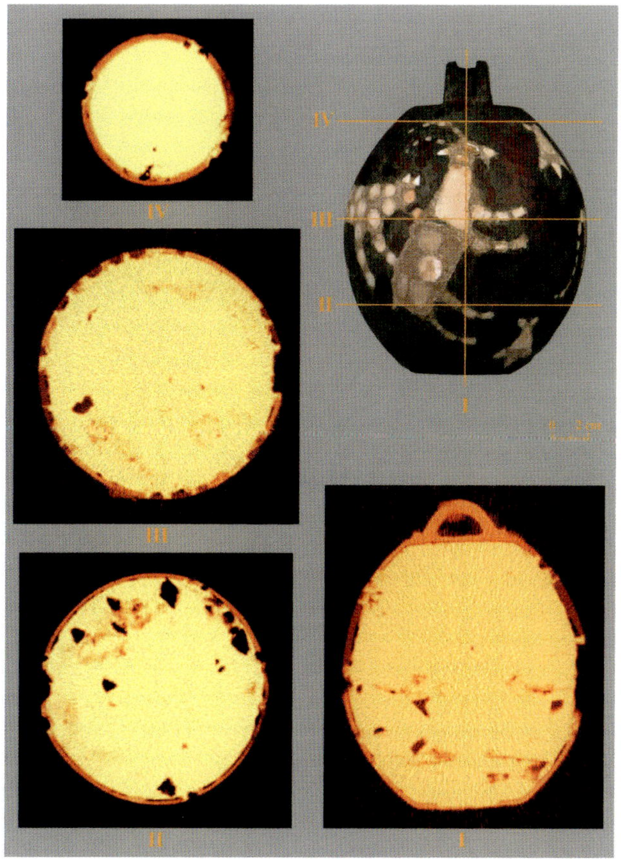

Figure 358 An early example of a hollow cast bronze (weight 15 kg); the leaded copper-arsenic jacket is filled with lead. The jacket is decorated with seashell inlays depicting a hunting frieze in which a leopard attacks an ibex. The scene is repeated twice, separated by stylized flies. High energy X-ray tomography at 8 MeV was conducted at the linear accelerator at the Laboratoire d'électronique des technologies de l'information (CEA-Leti). *Leopard Weight*, Pakistan, end 4th–early 3rd millennium BCE, H. 16.7 cm (Mission archéologique française au Makran, Shahi-Tump, inv. 298II402PO644). See Mille, Besenval, and Bourgarit 2004; Mille 2017.

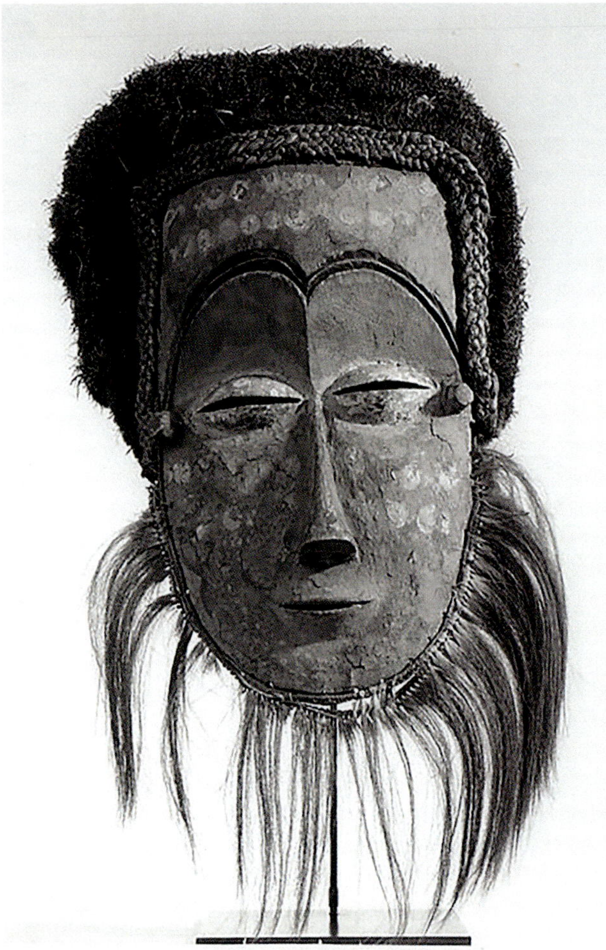

Figure 359 Wooden mask with inlays and attachments made of raffia, fur, and pigment. Inlays and attachments to wood and ivory masks and sculptures may provide clues as to lost elements in bronze masks and statues. *Mask*, Democratic Republic of the Congo, Kasai River region (Leele peoples), 19th–20th century, H. 51.4 cm (Metropolitan Museum of Art, the Michael C. Rockefeller Memorial Collection, gift of Mr. and Mrs. Gustave Schindler, 1967, inv. 1978.412.540).

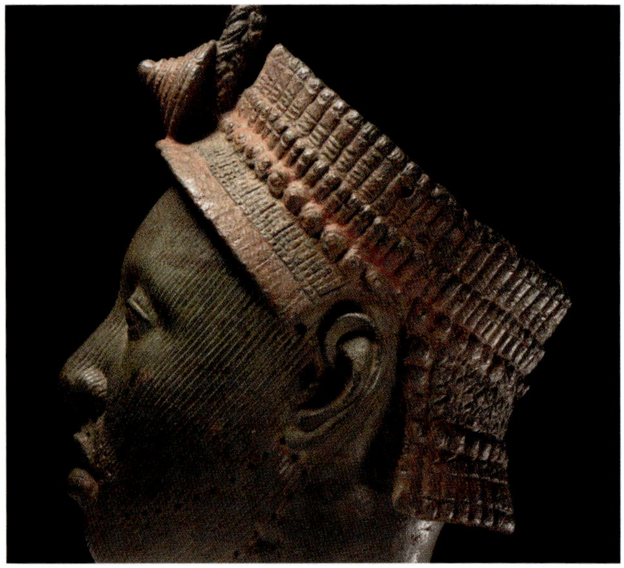

Figure 360 Lines of holes run between the lobes of the ears and across the upper lip of a brass Ife head. Based on ethnographic parallels, these may possibly be for the attachment of added elements. Red paint is used to add color to feathers; tubular beads and rosettes and traces of black paint appear on incised elements of the crown. *Ife Head*, Yoruba, probably 14th–early 15th century, H. 35 cm (British Museum, inv. Af1939,34.1).

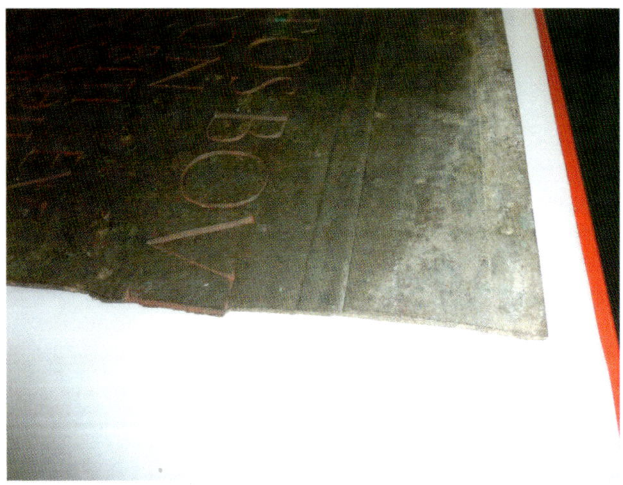

Figure 361 Detail of an engraved bronze plate showing red pigment inside the letters, identified as cinnabar by XRD and PIXE. Plate, 1st–3rd century CE, 34 × 50 cm (Musée d'Art, Histoire et Archéologie d'Evreux, France, inv. 4891). See Azéma et al. 2012.

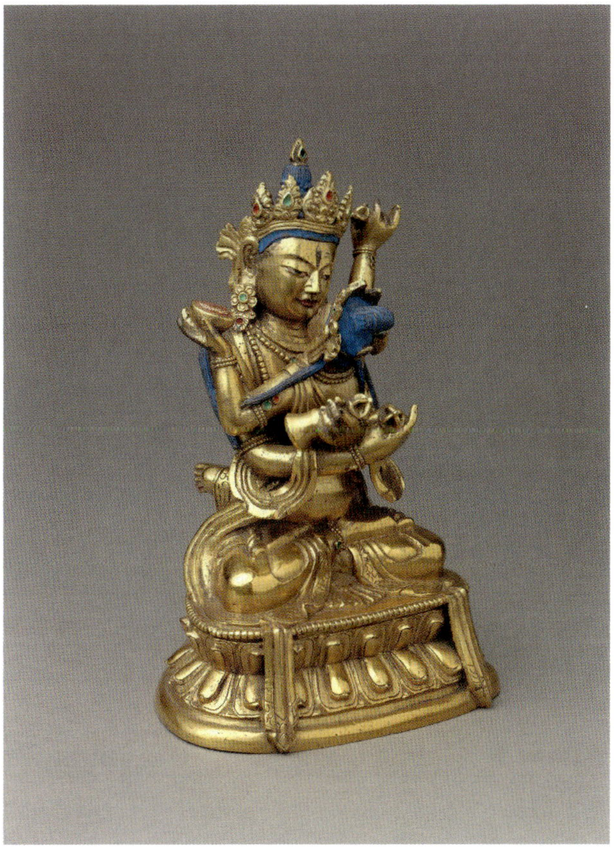

Figure 362 Bronze, mercury-gilded figure with overlaid stones and painted pigments. *Buddhist Deity Vajradhara in Union with His Consort Prajnaparamita*, probably Chinese or Tibetan, 1403–24, H. 28.6 cm (Metropolitan Museum of Art, Robert Lehman Collection, inv. 1975.1.1442).

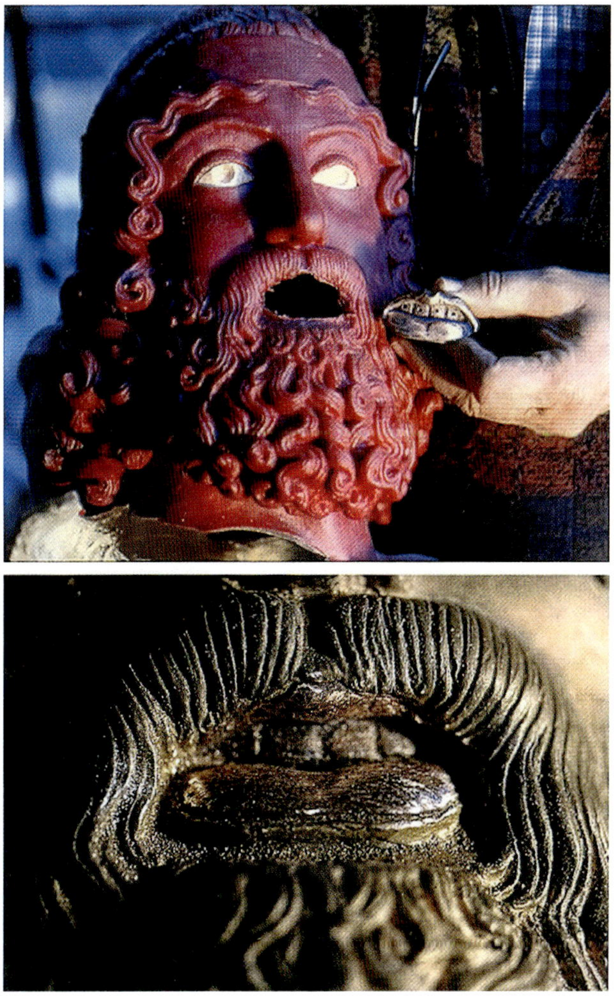

Figure 363 Top: reproduction wax; bottom: detail showing copper lips, which were precast inlays, manufactured and set into the wax model prior to bronze casting. This method is suitable for more complex inlay shapes and profiles. *Riace Bronze Statue A*, Greek, ca. 460 BCE, H. 198 cm (Museo Nazionale della Magna Grecia, Reggio Calabria, Italy). See Bertelli et al. 2013.

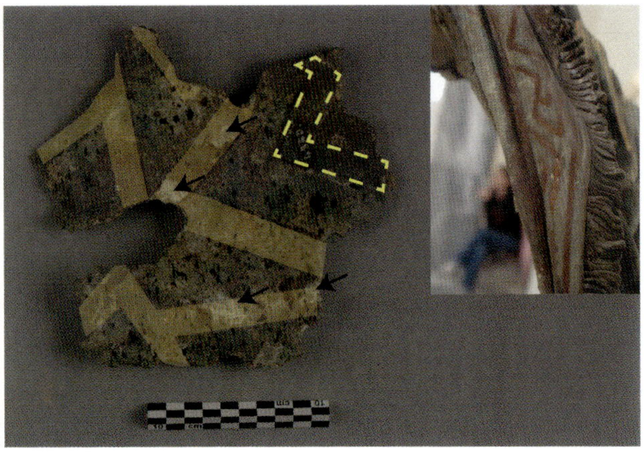

Figure 364 Left: remains of soldering used to attach inlays on the fragment of an unidentified large bronze statue, Roman, 1st–3rd century CE, H. 21 cm (Musée d'Art, Histoire et Archéologie d'Evreux, France, inv. 4860). See Azéma et al. 2012. Right: similar inlays of similar sizes have been observed on numerous antique large bronzes, as can be seen on the cloth (left hand), *Statue of the Emperor Augustus*, Roman, 29 BCE–14 CE, life size (National Archaeological Museum of Athens, inv. X23322).

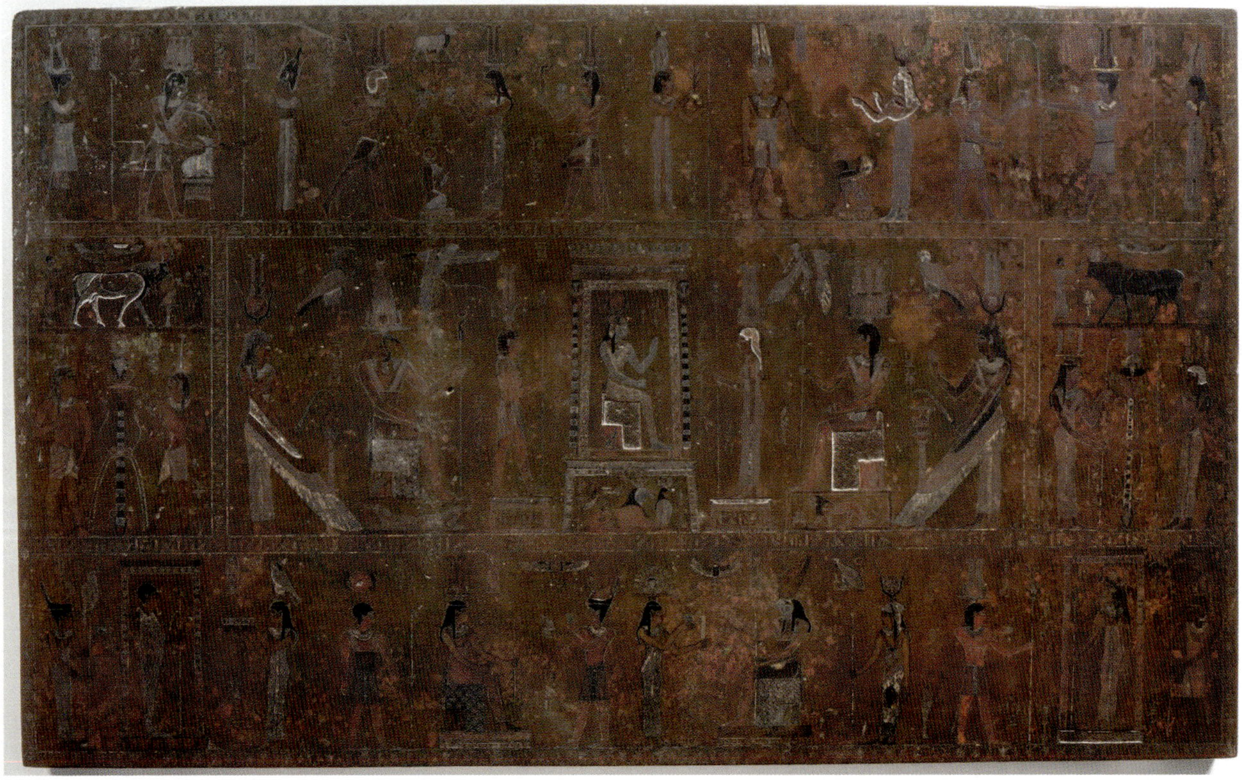

Figure 365 The *Mensa Isiaca*, a bronze table or altar, is decorated with three registers of Egyptian-style images, the goddess Isis in its center, and with pseudo hieroglyphs. All imagery and script were created by inlays of various metals and alloys, often multipart, some of which may have been intentionally patinated in antiquity to create a complex, polychrome surface. *Mensa Isiaca*, Roman, 1st century CE, H. 75.5 cm (Turin Museo Egizio, Italy, inv. 715).

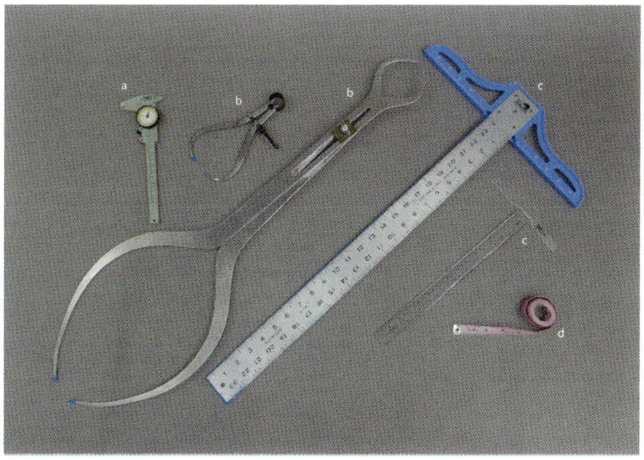

Figure 366 Some basic tools for measuring: a) dial vernier caliper; b) outside (or external) calipers; c) T-squares; d) cloth tape measure.

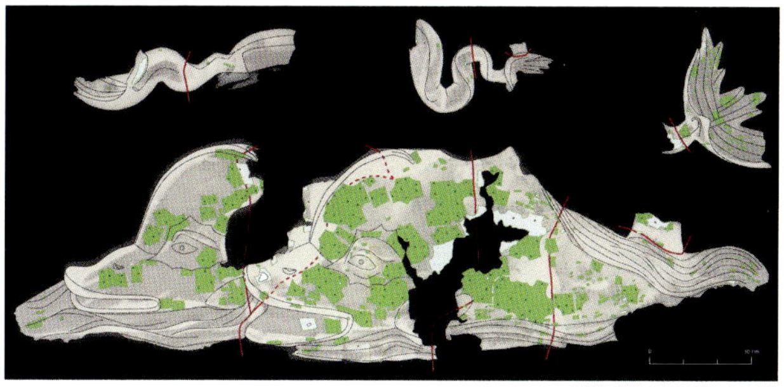

Figure 367 Annotated X-radiograph showing technical features observed on the front surface (refer to fig. 467 for mapping of evidence visible on the reverse). *Great Dolphins of Vienne*, Roman period, Vienne, France, 2nd century CE., max L. 260 cm (Musée des Beaux-Arts et d'Archéologie de Vienne, France, inv. R.1998.2.43).

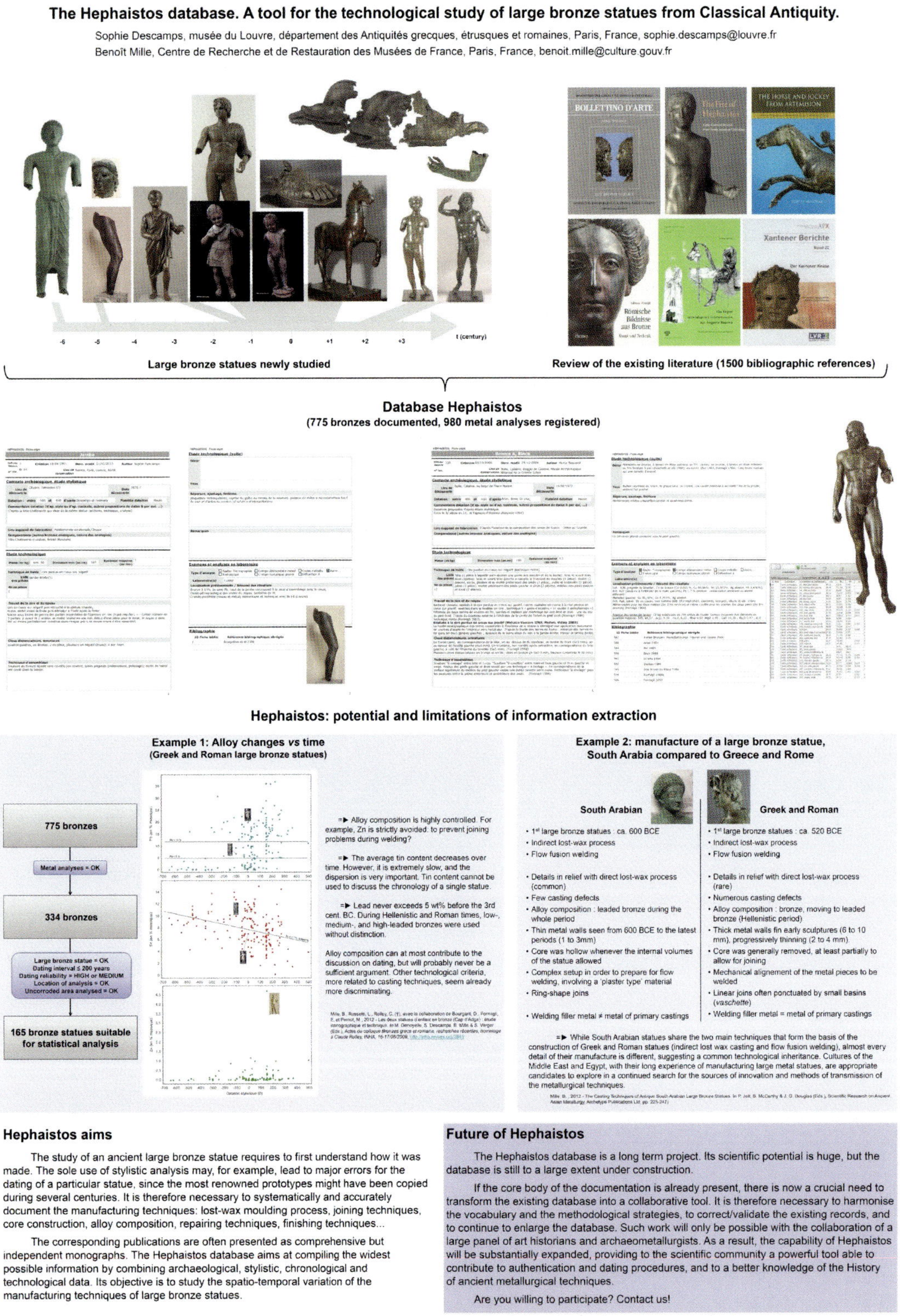

https://www.getty.edu/publications/bronze-guidelines/visual-atlas/368/

Figure 368 Poster titled "The Hephaistos database. A tool for the technological study of large bronze statues from Classical Antiquity," presented at the XVIIIth congress on ancient bronzes in Zurich, 2013. See Descamps-Lequime and Mille 2017.

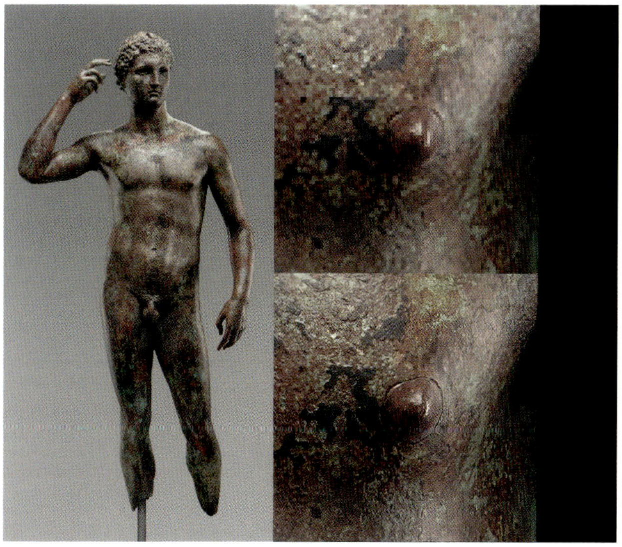

Figure 369 An image of a sculpture 150 cm high, photographed at a resolution of 2048 × 1536 pixels, will not resolve details at the mm scale (see inset at top right). Macro photography is necessary to resolve fine features such as tool marks and signatures (see inset at bottom right). *Statue of a Victorious Youth*, Greek, 300–100 BCE, H. 151.5 cm (J. Paul Getty Museum, Villa Collection, inv. 77.AB.30).

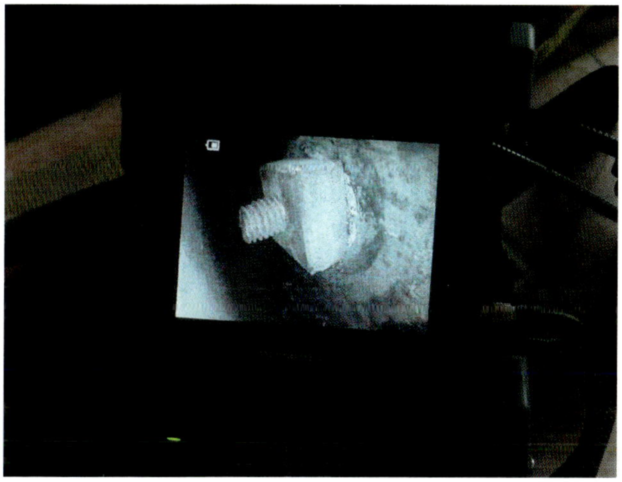

Figure 370 Examination of the inside of the base of Venus with a video endoscope, which presents the images live on screen. Francesco Primaticcio (Italian, 1504–1570), *Venus*, 1542, H. 192 cm (Musée national du château de Fontainebleau, France, inv. MR 3277). See Castelle 2016.

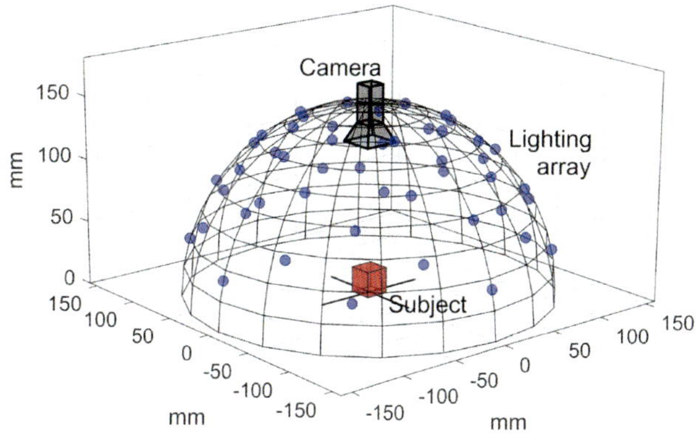

Figure 371 The principle of reflectance transformation imaging (RTI).

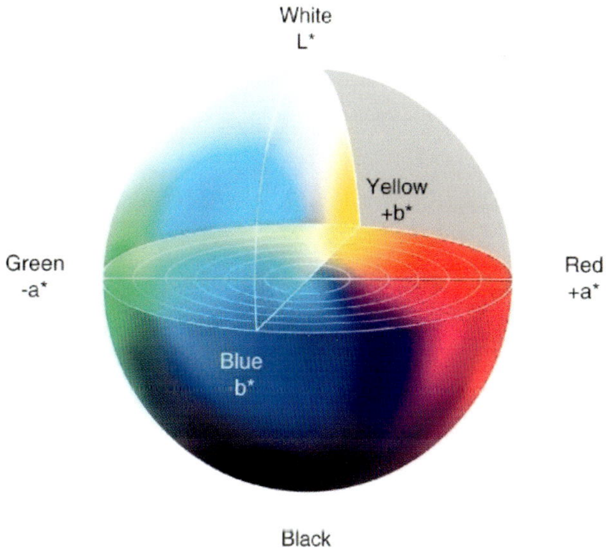

Figure 372 Color measurements on bronze sculpture typically utilize the CIELAB color space, where colors are defined by the three variables (L*, a*, and b*) representing light-dark, red-green, and yellow-blue, respectively.

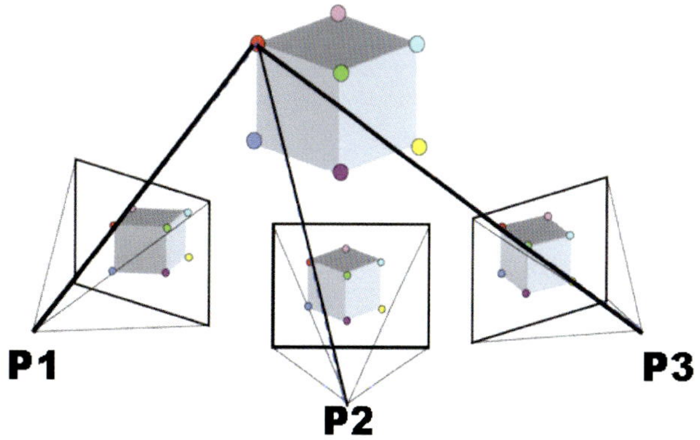

Figure 373 Principle of photogrammetry.

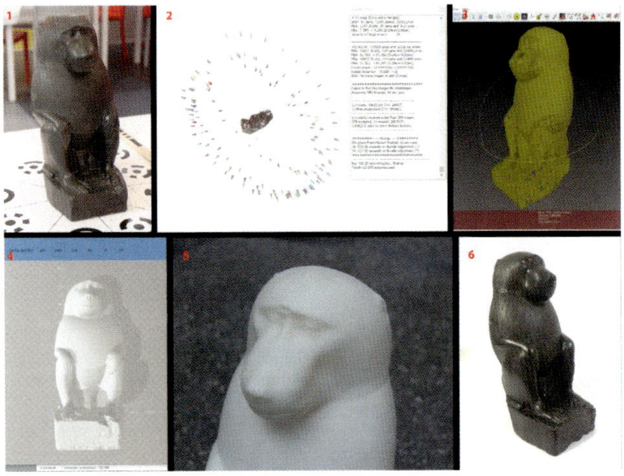

Figure 374 The main steps to design a resin model of a sculpture using photogrammetry: 1–2) digital photography of the original sculpture is undertaken at many angles to cover the whole object; 3) a digital model made of a dense cloud of points is made via reconstruction software; 4) the digital model is meshed and cleaned; 5) a 3D model is printed (here in resin) using the digital model; 6) details are refined by hand and the model is painted.

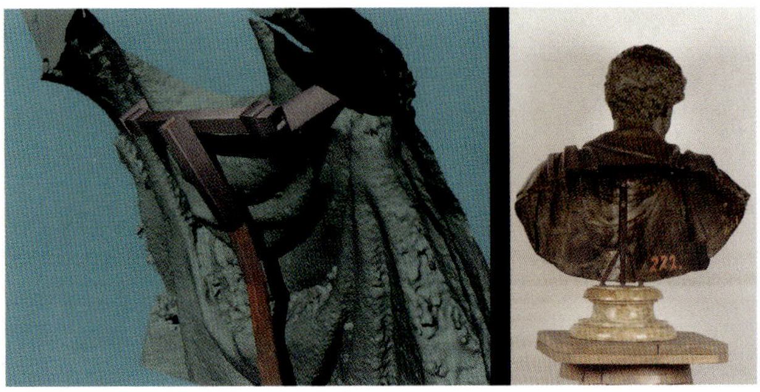

Figure 375 3D model of the new stainless steel mount for the bust (left), and back view of the bust once mounted (right). The mount was created using a 3D model designed using photogrammetry. Attributed to Giovanni Bandini (called Giovanni dell'Opera, Italian, 1540–1599), *Bust of Roman Emperor Antoninus Pius*, Urbino, before 1582, H. 68 cm (Museo Nazionale del Bargello, Florence, inv. 441B). See Morigi 2018.

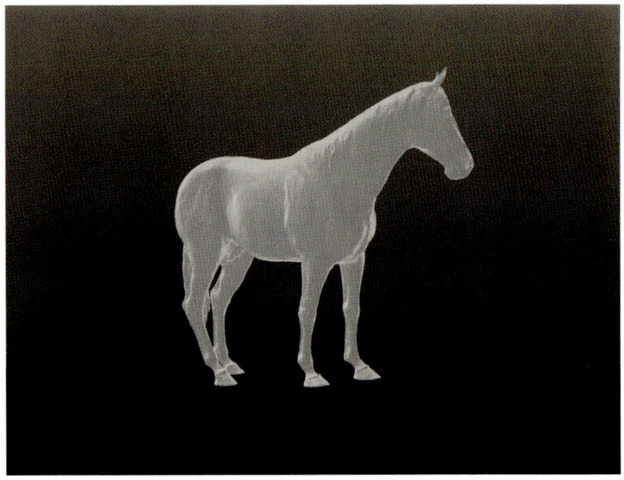

Figure 376 Digital 3D structured light scan. Andrew Lacey (British, b. 1969), *The Anatomy of Bronze*, cast by the artist in 2019, Devon, UK, H. 45 cm (artist's collection).

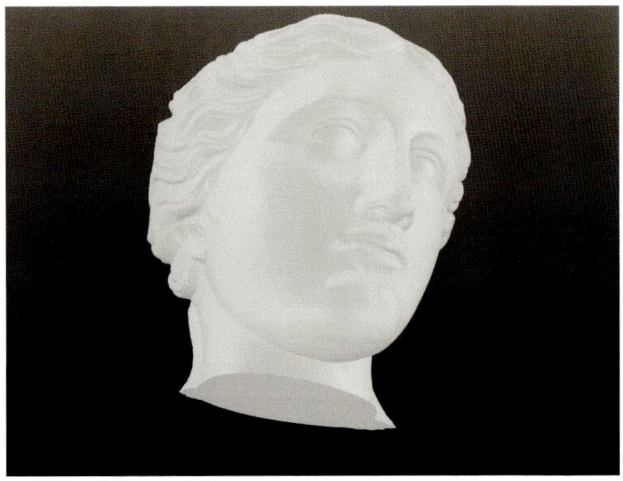

Figure 377 3D structured light scan. *Apollo of Lillebonne*, France, 1st century BCE–1st century CE, H. 193 cm (Musée du Louvre, inv. Br 37 [NIII65]). See fig. 288.

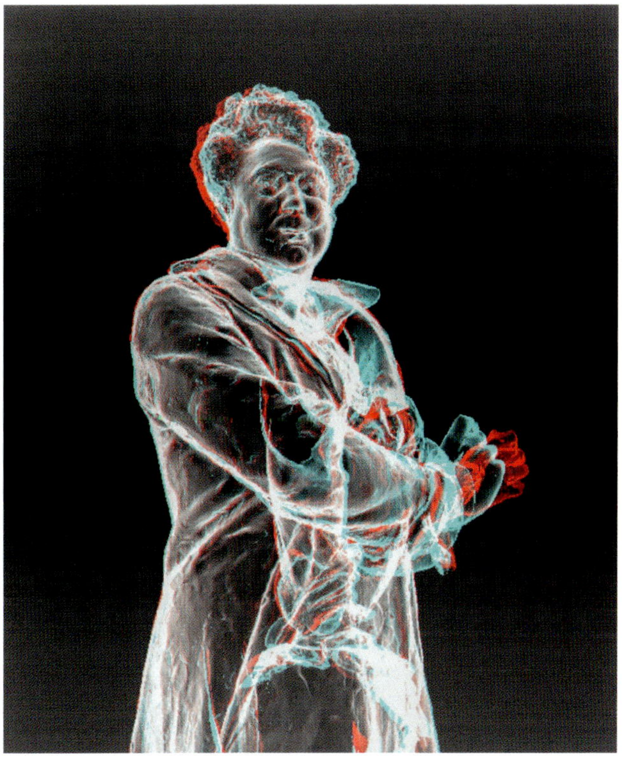

Figure 378 Two 3D models of different versions of a terra-cotta sculpture were made using a structured light scanner. By overlaying the scans, it is clear that the torsos are closely aligned. The hands, hair, and lapels are misaligned, suggesting possible joint locations for the separately mold-made parts. Albert-Ernest Carrier-Belleuse (French, 1824–1887), *Model for a Monument to Alexandre Dumas*, 1885, H. 76 cm (Musée Carnavalet, Paris, inv. S1893; Musée Alexandre Dumas, Villers-Cotterêts, France, inv. 91.2.54). See Carré 2005.

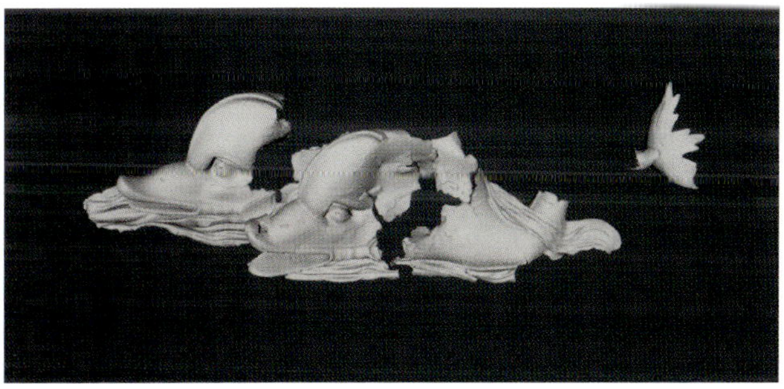

Figure 379 3D laser scan of the reassembled surviving sections of the dolphin relief produced by P. Mora and L. Espinasse and presented on the Archéovision website. *Great Dolphins of Vienne*, Roman period, Vienne, France, 2nd century CE, max L. 260 cm (Musée des Beaux-Arts et d'Archéologie de Vienne, France, inv. R.1998.2.43).

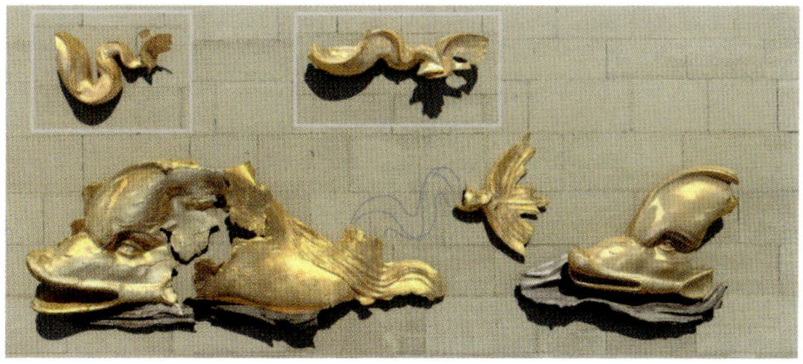

Figure 380 A reconstruction of the frieze based on 3D laser scans. *Great Dolphins of Vienne*, Roman period, Vienne, France, 2nd century CE, max L. 260 cm (Musée des Beaux-Arts et d'Archéologie de Vienne, France, inv. R.1998.2.43).

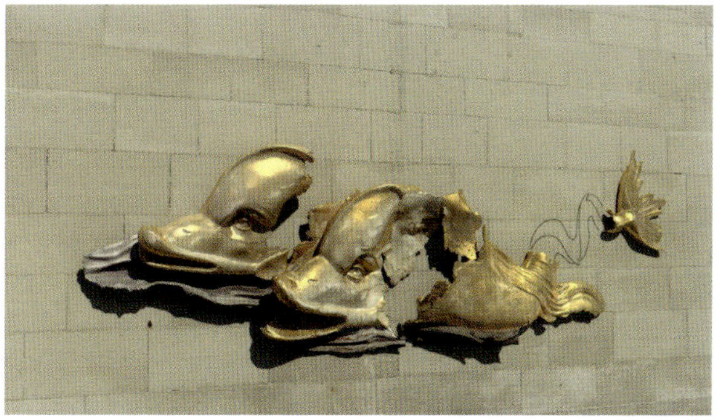

Figure 381 Detail of the virtual 3D reconstruction of the frieze showing what may have been its original dichromatic appearance. *Great Dolphins of Vienne*, Roman period, Vienne, France, 2nd century CE, max L. 260 cm (Musée des Beaux-Arts et d'Archéologie de Vienne, France, inv. R.1998.2.43).

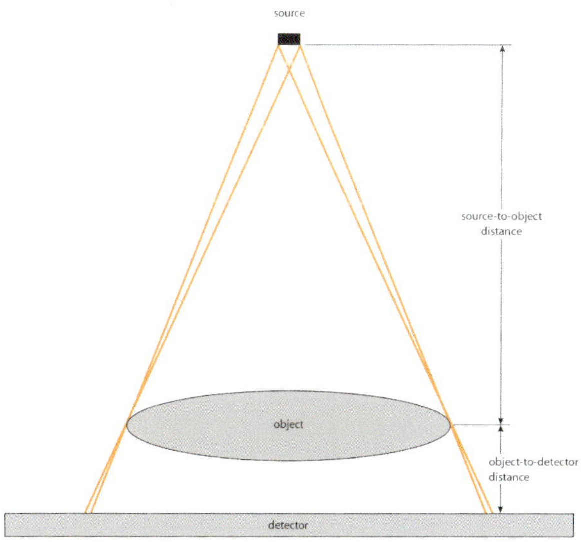

Figure 382 Schematic diagram of an X-radiography setup.

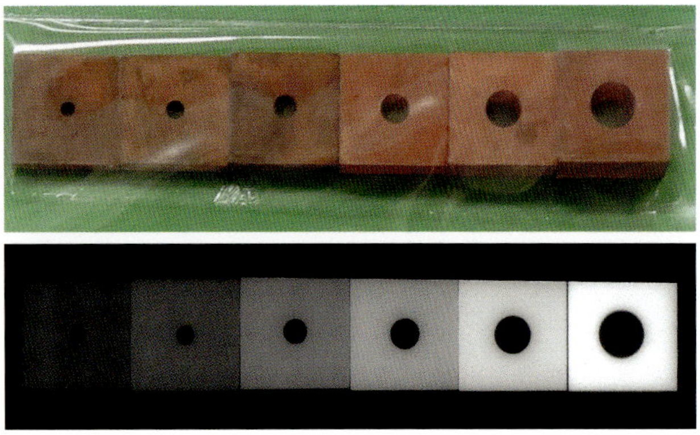

Figure 383 The influence of the thickness of copper samples (upper image) on the attenuation of X-rays (lower image). The thicker the sample, the greater the attenuation, and the lighter the resulting radiograph. The thickness is indicated by the diameter of the hole: 2 / 2.5 / 3.2 / 4 / 5 / 6.3 mm.

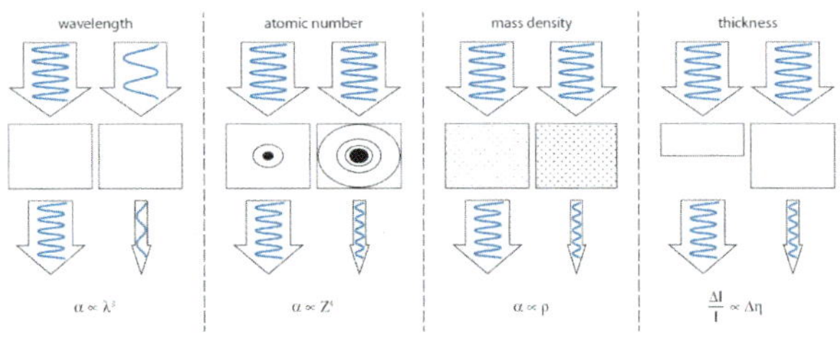

Figure 384 The four different parameters controlling the attenuation of X-rays in matter: the wavelength λ of the X-ray beam (the attenuation factor α [alpha] is proportional to λ^3); the atomic number Z of the element the beam passes through (α is proportional to Z^4); the mass density ρ of the matter the beam passes through (α is proportional to ρ); and the thickness η of the object to be radiographed (the variation of intensity of the transmitted X-ray $\Delta I/I$ is proportional to the difference of thickness $\Delta\eta$). The thinner the arrow below the object (the cube), the greater the attenuation of X-rays. See CETIM 2017.

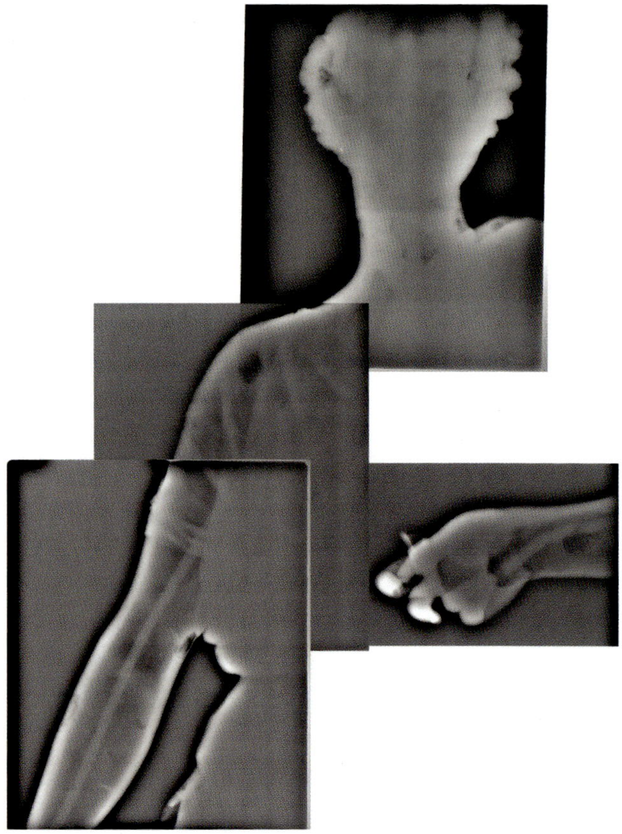

Figure 385 Assembled gamma radiographs. Ponce Jacquiot (French, 1515–1571), *Prudence*, Henri II and Catherine of Medici Funerary Monument, 1567, H. 190 cm (Saint-Denis Basilica, France) See Castelle 2016.

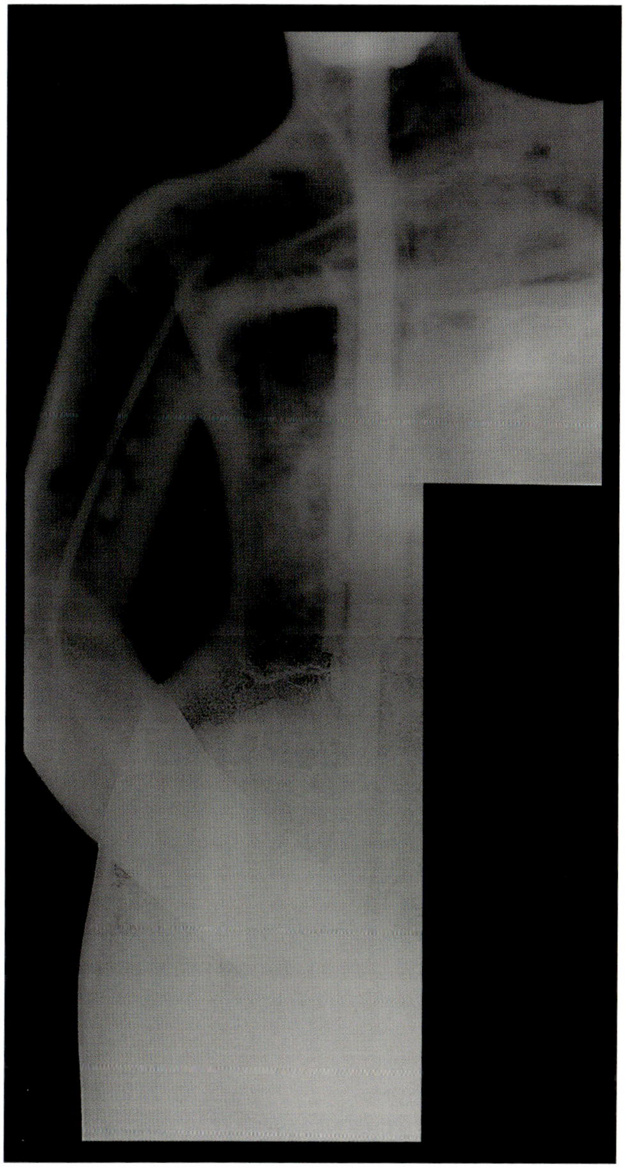

Figure 386 Assembled gamma radiographs. Francesco Primaticcio (Italian, 1504–1570), *Venus*, 1542, H. 192 cm (Musée national du château de Fontainebleau, France, inv. MR 3277). See Castelle 2016.

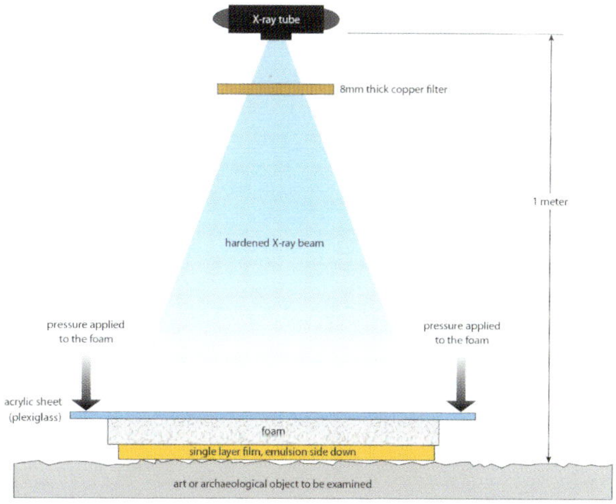

Figure 387 Schematic diagram of an electron emission radiography setup.

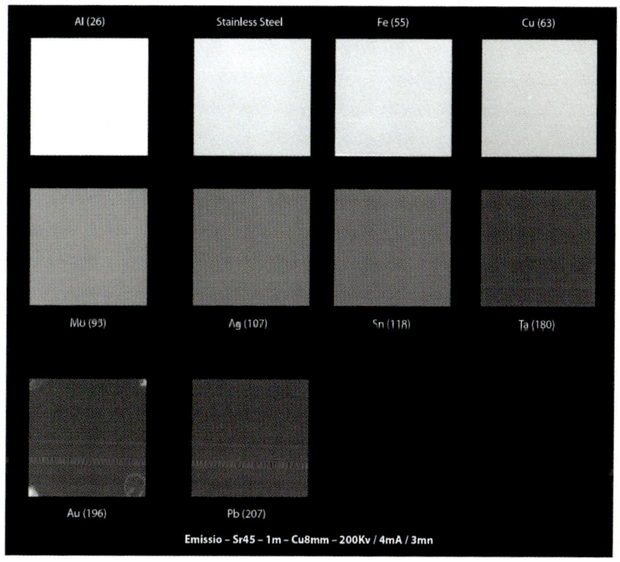

Figure 388 Electron emission radiographs from different metals showing that only the atomic number of the element controls the image density: the lightest sample is aluminum, the darkest is lead. The thickness of the sample has no impact. Operating conditions are indicated at the bottom of the image (source – distance object to detector – filter – voltage/current/exposure time).

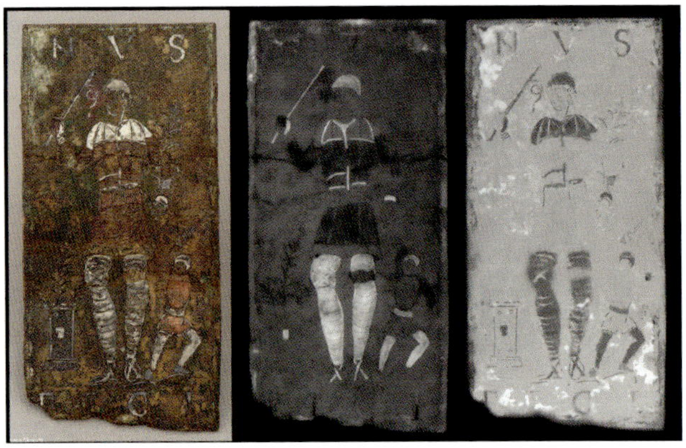

Figure 389 Comparison of electron emission radiography (right) versus X-ray radiography (middle) as seen on the Plaque à l'aurige, 4th century CE, H. 17 cm (Musée du Louvre, inv. Br3447). Daylight photography on the left. The electron emission image nicely reveals the inlays of gold and silver in the copper substrate. See Robcis et al. 2014.

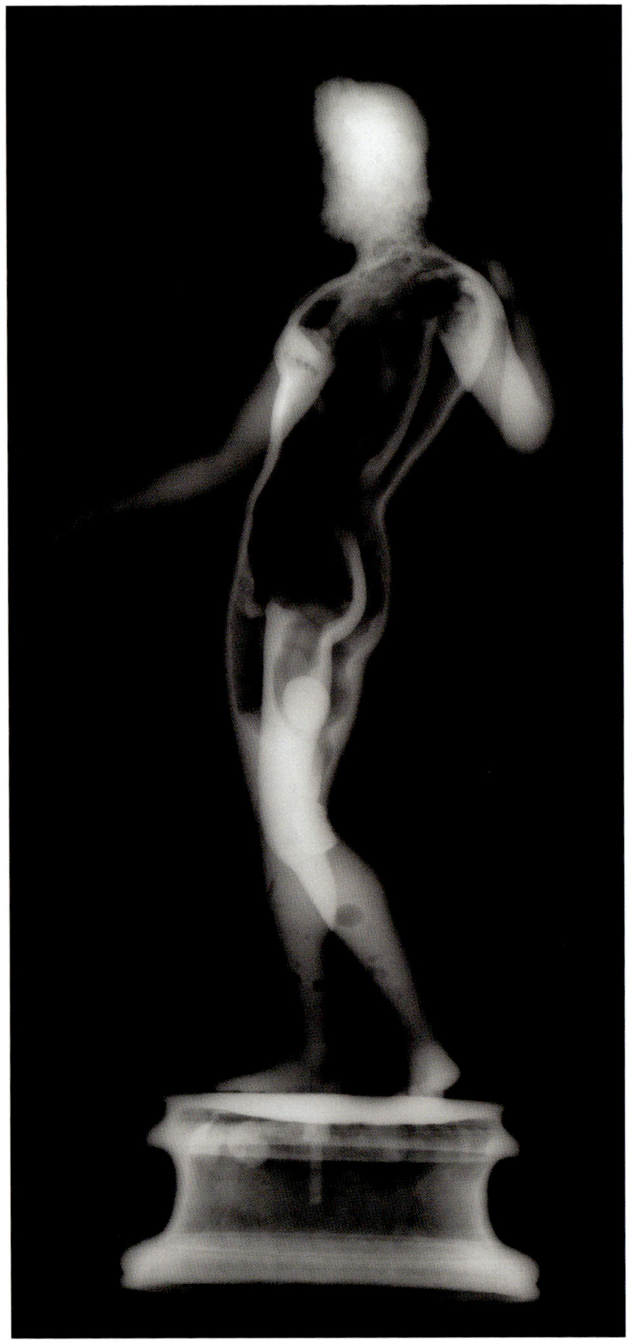

Figure 390 X-radiographic sequence (viewable in online edition). Jupiter (or Neptune?), Bavay (Nord, France), 1st century CE, H. 26.4 cm (Musée Archéologique de Bavay, France, inv. 1969 Br 15 / Biévelet 10). See Mille 2019a.

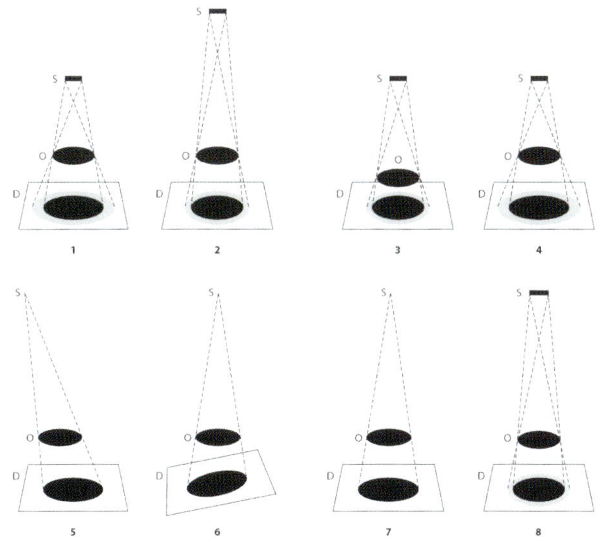

Figure 391 The influence of geometry on image quality for X-radiography: 1–2) the greater the distance between source and object, the sharper the image; 3–4) the closer the detector to the object, the sharper the image; 5–6) non-rectilinear arrangements can produce geometric distortion of the image; 7–8) the smaller the X-ray source (focal spot), the sharper the image.

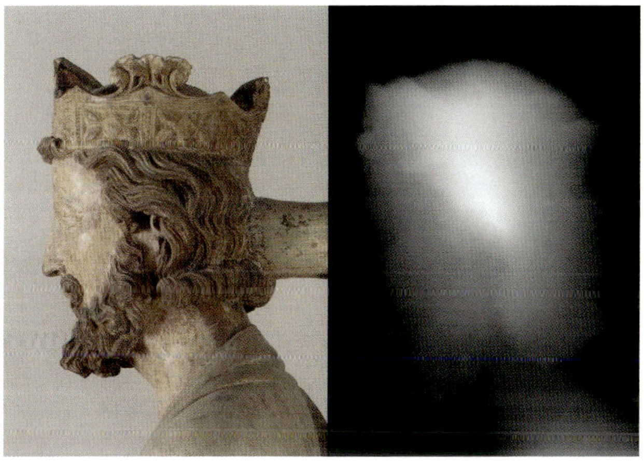

Figure 392 Head of a polychrome stone sculpture (calcareous) in visible-light photography and in X-radiography. The blurring of the radiograph is due to X-ray scattering generated by the object itself. *King Childebert*, Abbaye Saint Germain des Prés, Paris, 1239–44, total H. 1.9 m (Musée du Louvre, inv. ML 93).

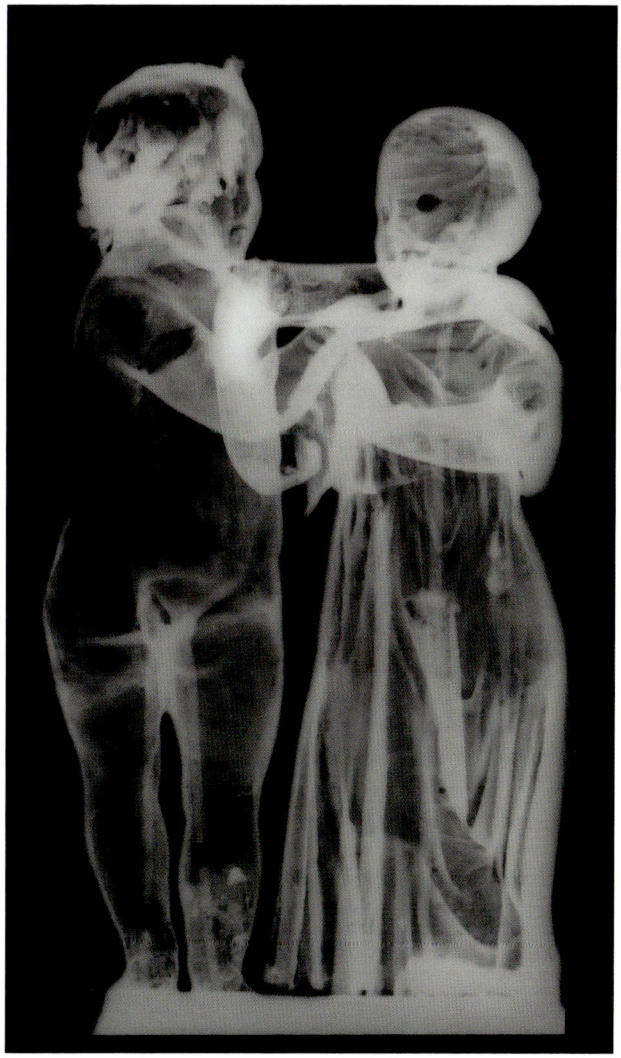

Figure 393 Because of the thickness of the metal wall and the relatively high density of the leaded copper alloy, high energy (400kV, 4mA) and a thick filter (10 mm copper) were necessary for X-radiography. The other operating conditions were: 10 min. exposure, 3 m source-to-object distance, MX125PB film. *Eros and Psyché*, Roman, 1st century BCE–1st century CE, H. 72 cm (Musée du Louvre, inv. Br 4105/MND 1035). See C2RMF Internal Report 2015A.

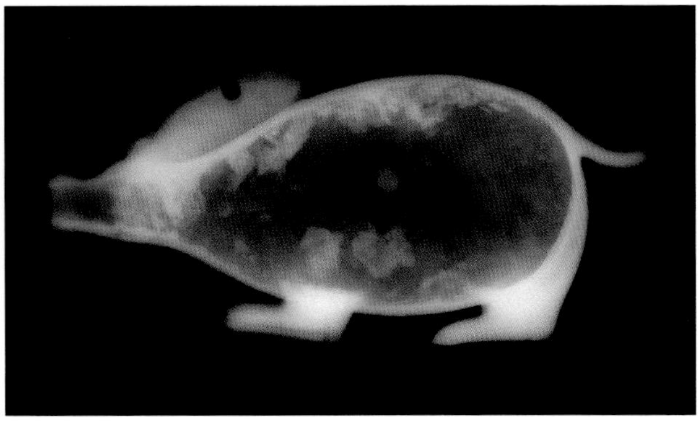

Figure 394 X-radiograph of a bronze hare or aardvark. Due to the relative thinness of the bronze walls (<2 mm), relatively low voltage (250 kV) proved sufficient for X-radiography. Other operating conditions (C2RMF): 2mA, 6 min. exposure, 2 m source-to-object distance, 4 mm copper filtering, M100PB film. *Bronze Hare or Aardvark*, Egypt, 11th century CE, L. 63 cm (Musée du Louvre, inv. OA 6675).

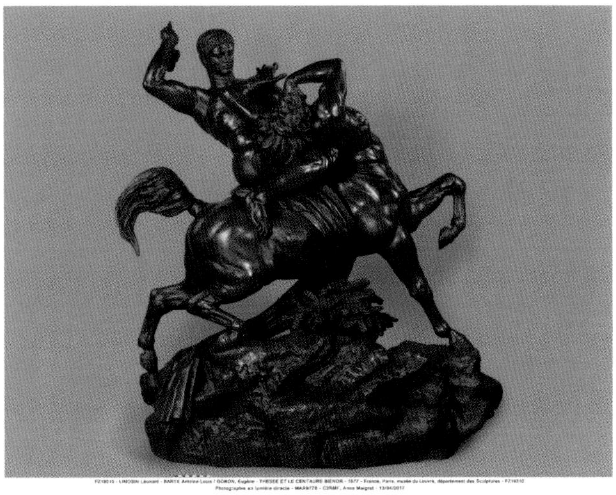

Figure 395 Antoine-Louis Barye (French, 1795–1875), *Theseus and the Centaur Bienor*, cast by Eugène Gonon (1814–1892) in 1877, H. 1.3 m (Musée du Louvre, inv. RF 3882). See C2RMF Internal Report 2016b. For a radiograph of this object, see fig. 32.

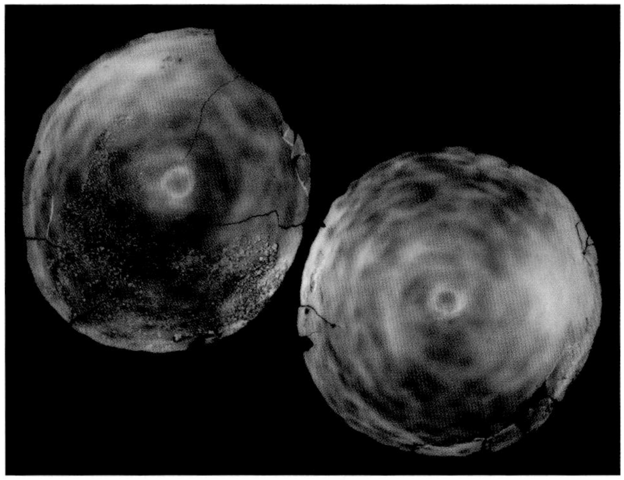

Figure 396 X-radiograph of a pair of breastplates. Contrast optimization in order to reveal hammering marks was obtained by discarding the X-ray beam filters. Other operating conditions were: 80kV, 4mA, 10 min. exposure, 1.5 m source-to-object distance, MX125PB film. Pair of *couvre seins*, Iran, early 2nd millennium BCE, 12 cm diameter (Musée du Louvre, inv. Sb 3055 a–b).

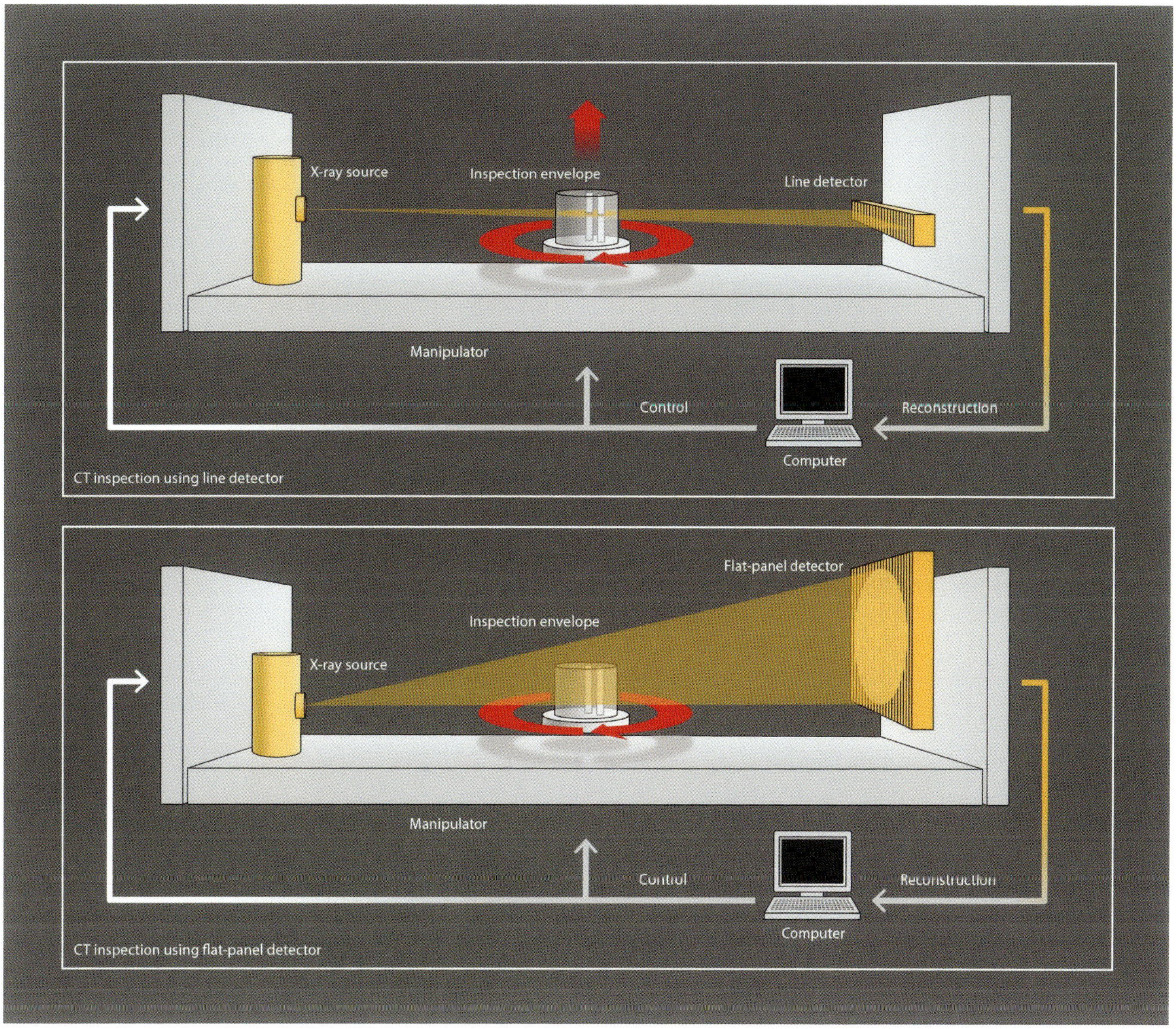

Figure 397 Typical setup of computerized tomography (CT), either with a line detector (top) or a flat panel detector (bottom).

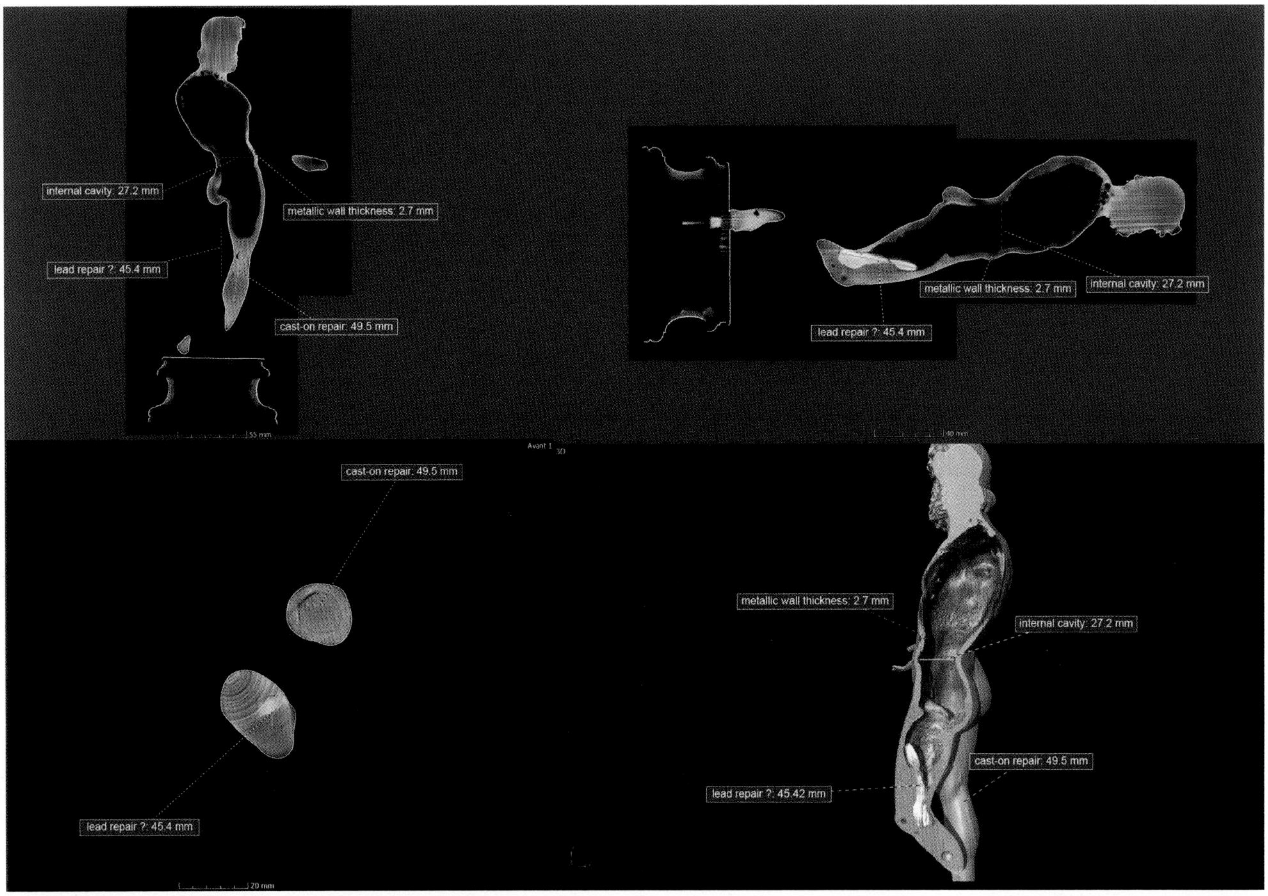

Figure 398 X-ray tomographic views of a bronze statuette. Software allows detailed measurements to be made of any feature. *Jupiter (or Neptune?)*, Bavay (Nord, France), 1st century CE, H. 26.4 cm (Musée Archéologique de Bavay, France, inv. 1969 Br 15 / Biévelet 10). See Mille 2019a.

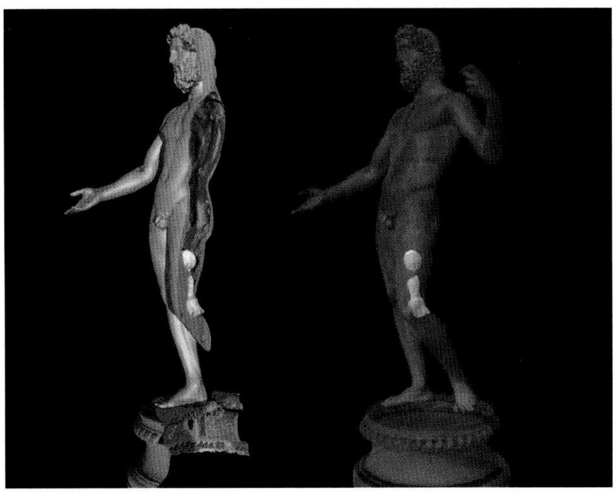

Figure 399 Segmentation of X-ray tomographic reconstructions can digitally isolate and highlight individual features, making them easier to visualize. *Jupiter (or Neptune?)*, Bavay (Nord, France), 1st century CE, H. 26.4 cm (Musée Archéologique de Bavay, France, inv. 1969 Br 15 / Biévelet 10). See Mille 2019a.

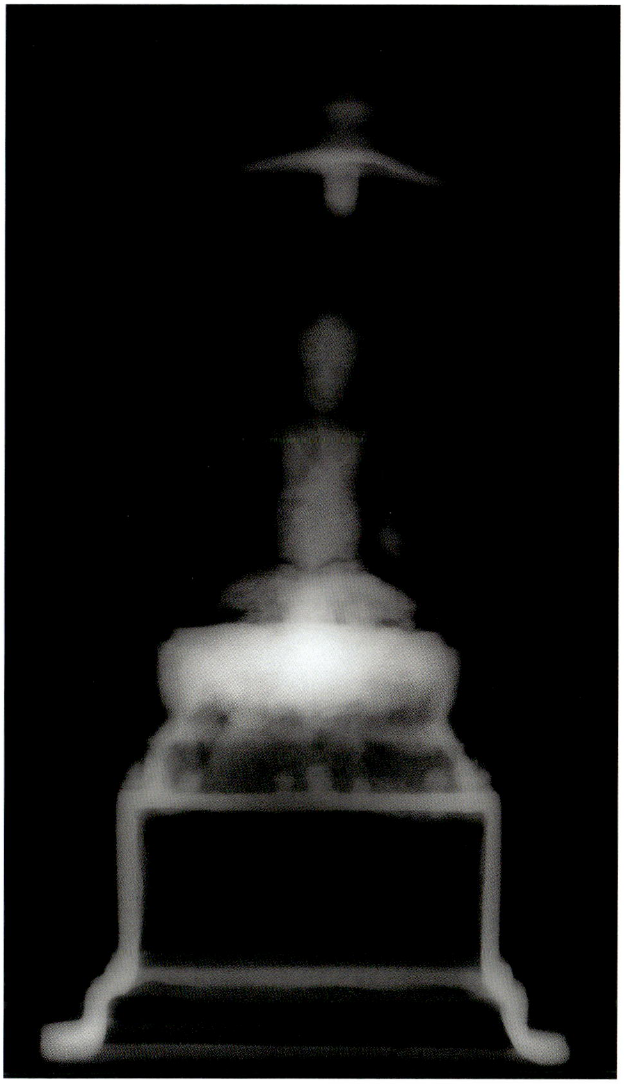

Figure 400 X-radiograph of frontal view. Kubera/Jambhala, Java, first half of the 10th century, H. 18 cm (Musée national des arts asiatiques – Guimet, Paris, inv. MG 3625). See Mechling et al. 2018.

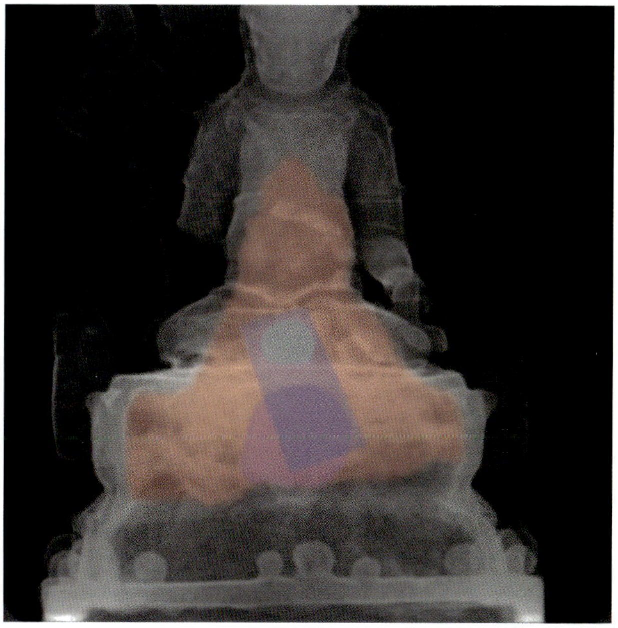

Figure 401 Annotated neutron radiograph of frontal view, showing more clearly than the X-radiograph (fig. 400) the contours of the sealing material (red overlay) and some of the consecration offerings (purple, red, and green overlays). For an even better visualization of the consecration offerings (in another statuette), refer to the tomographic images in figs. 403, 404. Kubera/Jambhala, Java, first half of the 10th century, H. 18 cm (Musée national des arts asiatiques – Guimet, Paris, inv. MG 3625). See Mechling et al. 2018.

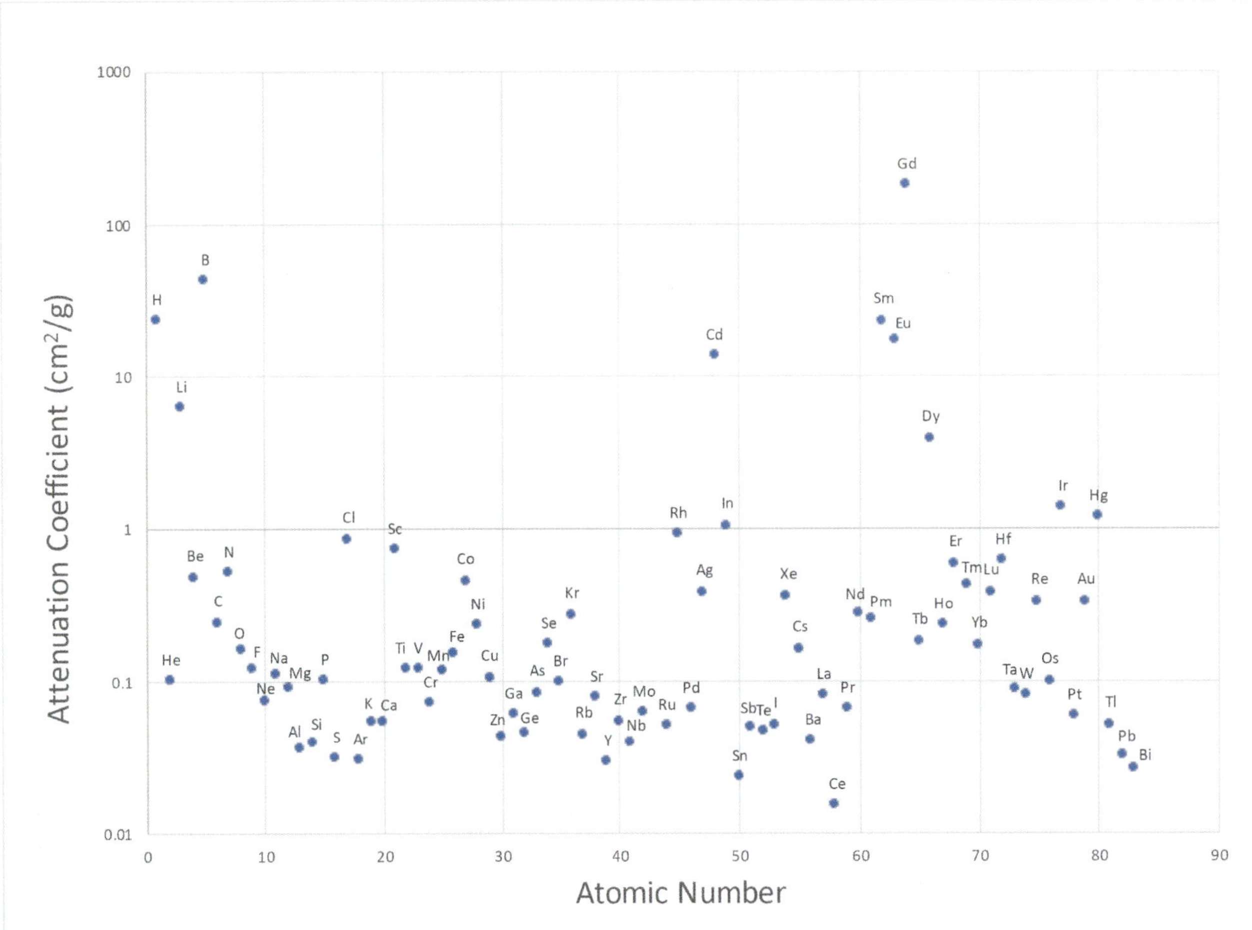

Figure 402 Mass attenuation coefficients for thermal neutrons as a function of atomic number of elements. A higher coefficient means the element is more opaque to neutrons.

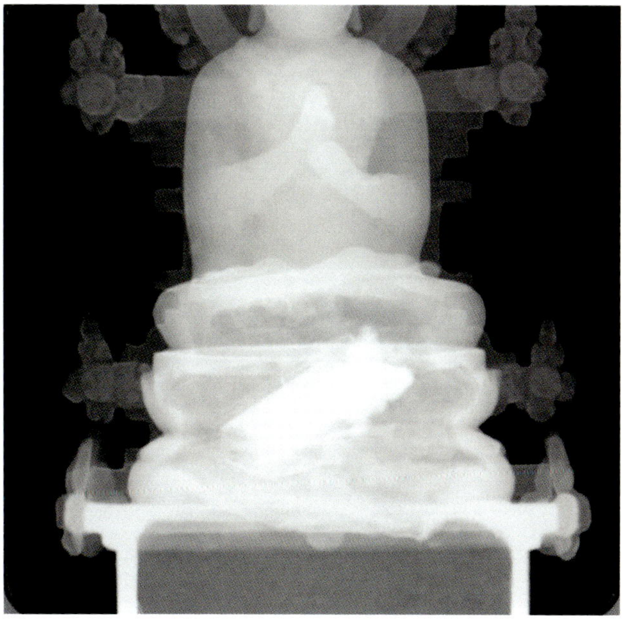

Figure 403 Neutron radiography. This series of thirty-seven neutron radiographs was made as the sculpture was rotated through 180° on a turntable. Most consecration offerings (coins, beads, folded metal sheet) can be observed (white shapes under the seat). *Vairocana*, Java, first half of the 10th century, H. 15.5 cm (Musée national des arts asiatiques – Guimet, Paris, inv. MG 18290). See Mechling et al. 2018.

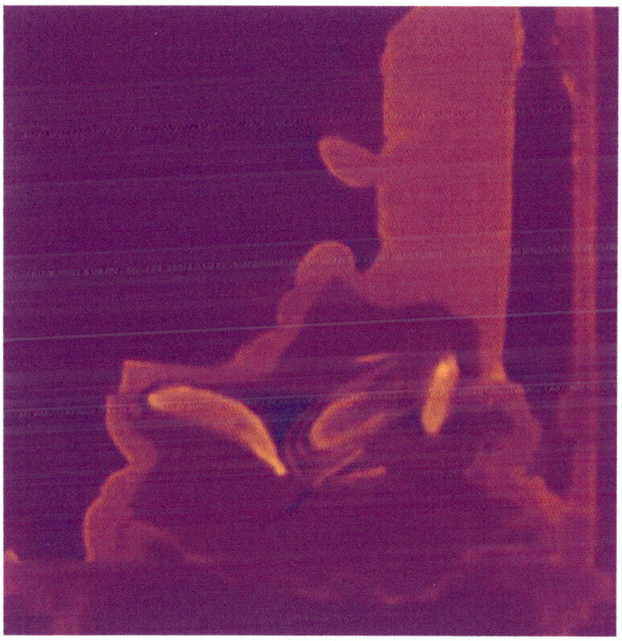

Figure 404 Neutron tomography (vertical slice) (viewable in online edition). The consecration offerings are revealed in the interior of the sculpture (in yellow). *Vairocana*, Java, first half of the 10th century, H. 15.5 cm (Musée national des arts asiatiques – Guimet, Paris, inv. MG 18290). See Mechling et al. 2018.

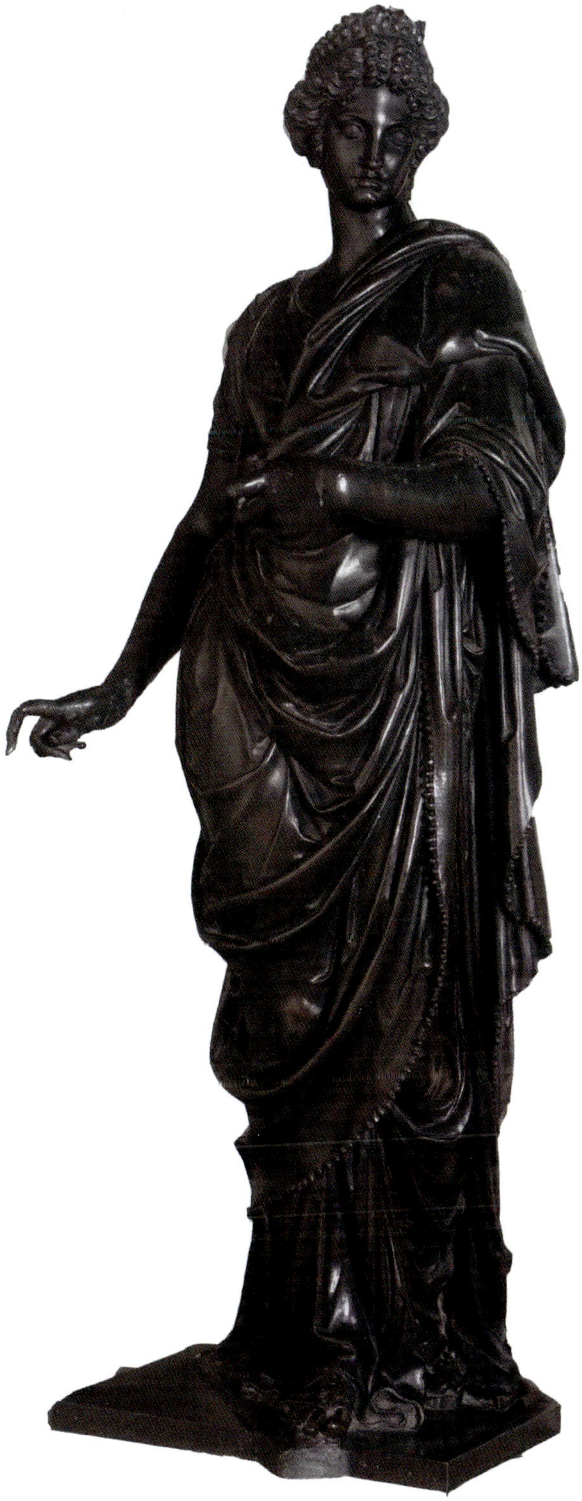

Figure 405 Ponce Jacquiot (French, 1515–1571), *Prudence*, Henri II and Catherine of Medici Funerary Monument, 1567, H. 190 cm (Saint-Denis Basilica, France). See Castelle 2016.

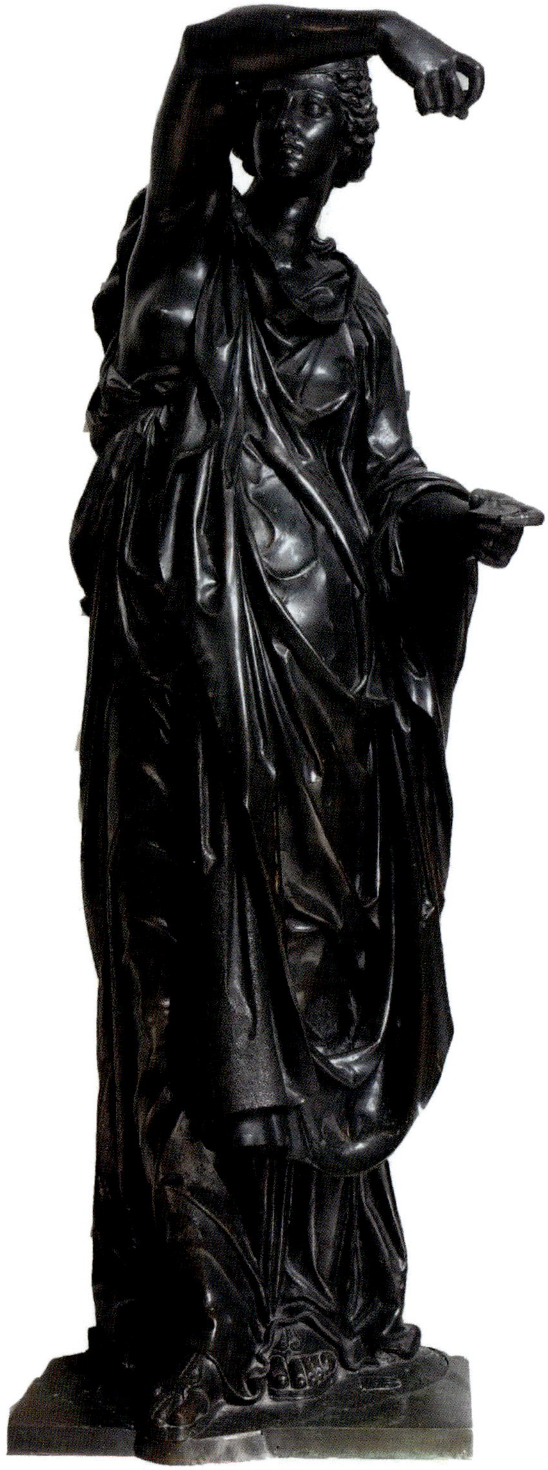

Figure 406 Ponce Jacquiot (French, 1515–1571), *Temperance*, Henri II and Catherine of Medici Funerary Monument, 1567, H. 187 cm (Saint-Denis Basilica, France). See Castelle 2016.

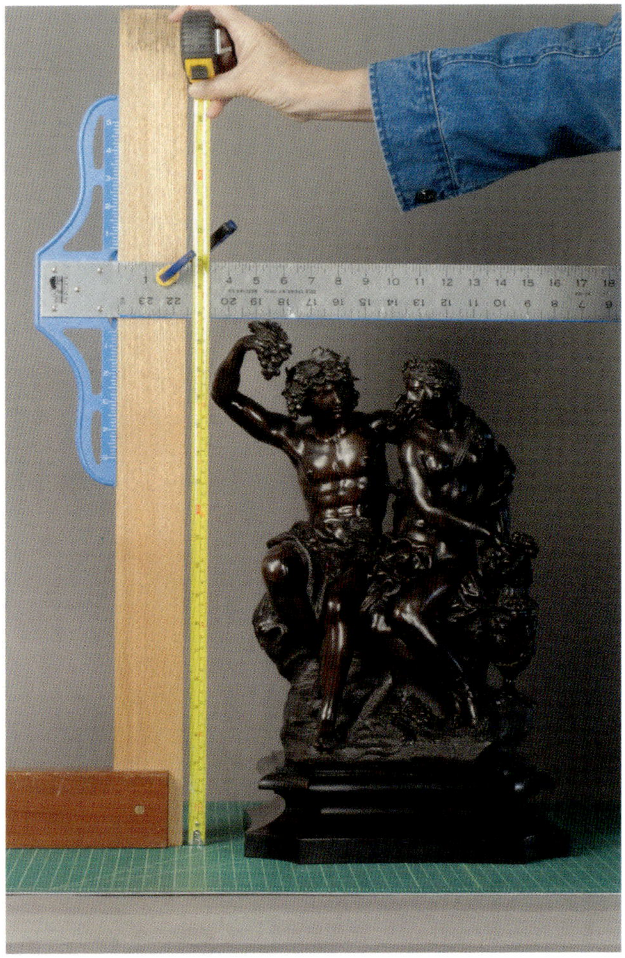

Figure 407 Measuring the height of a statuette using a combination of wooden framing squares, a T-square, and a measuring tape. Note that the metal edge of the T-square has been covered in tape to protect the surface of the bronze. Giuseppe Piamontini (Italian, 1663–1744), *Bacchus and Ariadne*, ca. 1700–1744, H. 40 cm, H. on base 47.9 cm (J. Paul Getty Museum, inv. 83.SB.333).

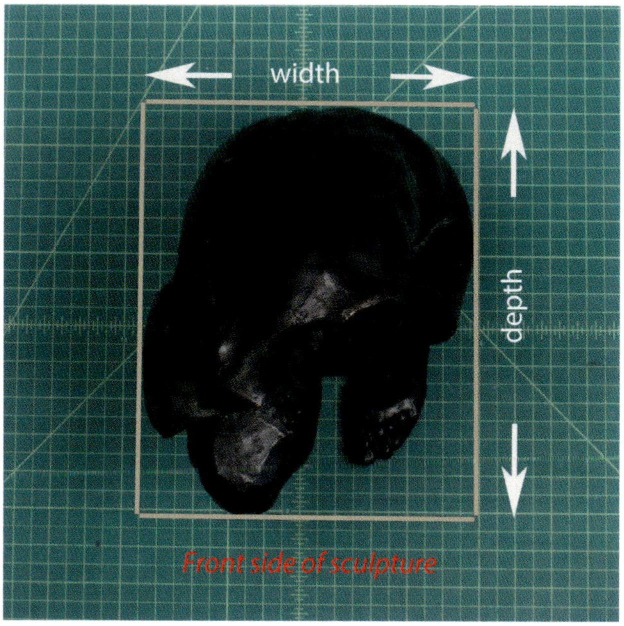

Figure 408 Measuring the width and depth of a statuette (viewed from above) by placing it in a virtual box. The gridded support on which the figure rests simplifies the determination of the box's perpendicular lines. Camille Claudel (French, 1864–1943), *Torso of a Crouching Woman*, model ca. 1884–85, cast by 1913, H. 35 cm (J. Paul Getty Museum, inv. 2018.32).

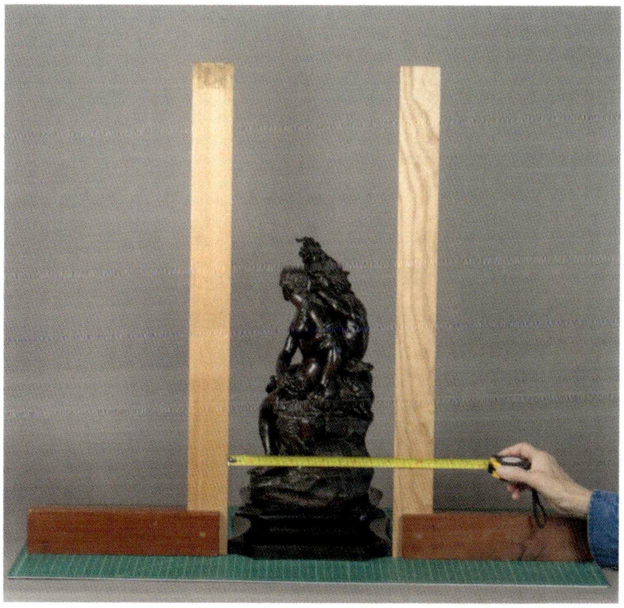

Figure 409 Measuring the depth of a statuette using a combination of wooden framing squares and a measuring tape. Giuseppe Piamontini (Italian, 1663–1744), *Bacchus and Ariadne*, ca. 1700–1744, H. 40 cm, H. on base 47.9 cm (J. Paul Getty Museum, inv. 83.SB.333).

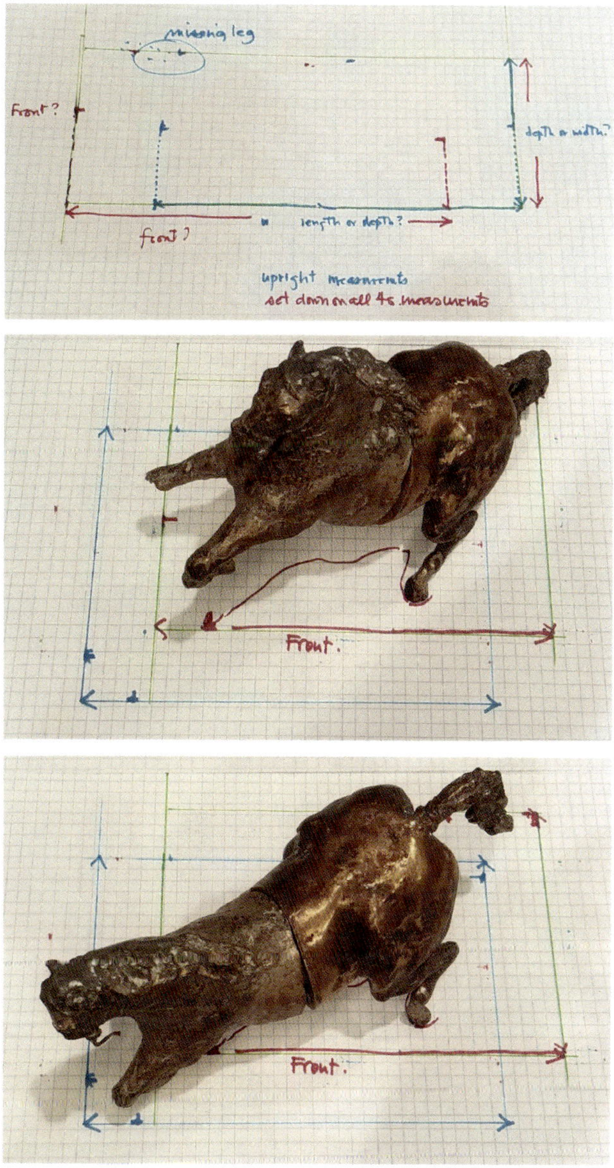

Figure 410 The measurements of a sculpture will vary significantly depending on the position one chooses to make them in—choices that are particularly challenging when there is no obvious primary view, or when an object is fragmentary. For the sake of argument, Francesca G. Bewer positioned the rearing horse, flawed with its front legs missing, in several ways and measured the width and length, projecting the farthest points onto gridded paper, to show the variations. The uppermost gridded sketch reflects measurements taken with the horse in alignment with the paper (red lines correspond to the horse upright on its hind legs, the blue ones with it posed on all fours). The bottom two images show measuring done with the horse at an angle, first upright (lines in blue), then lowered (lines in red). The height, of course, will vary in relation to the position. Andrew Lacey (British, b. 1969), equine study demonstration model, 2004, H. 20.3 cm horse upright and 15.5 cm with legs lowered (private collection), cast by the artist in 2004 for experimental purposes at AMTeC (Ancient materials and technologies) Renaissance bronze workshop, Chatham Dockyard, UK.

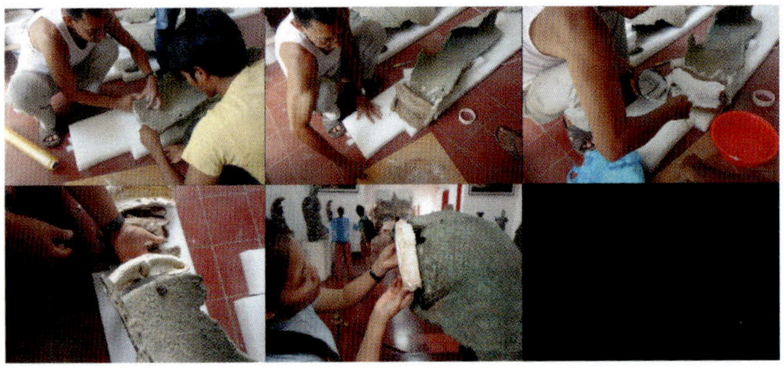

Figure 411 Plaster molding of the edge of a fragment from a monumental Reclining Vishnu to test whether it is part of the left arm of the bust. Left to right, top to bottom: a plastic film is laid on the bronze surface; an imprint of the edge is made using clay; the resulting clay mold is filled with plaster; view of the plaster imprint; test of the plaster imprint against the left shoulder. Reclining Vishnu, West Mebon (Khmer), 12th century, L. 220 cm (National Museum of Cambodia, inv. Ga 5387). See CAST:ING 2018.

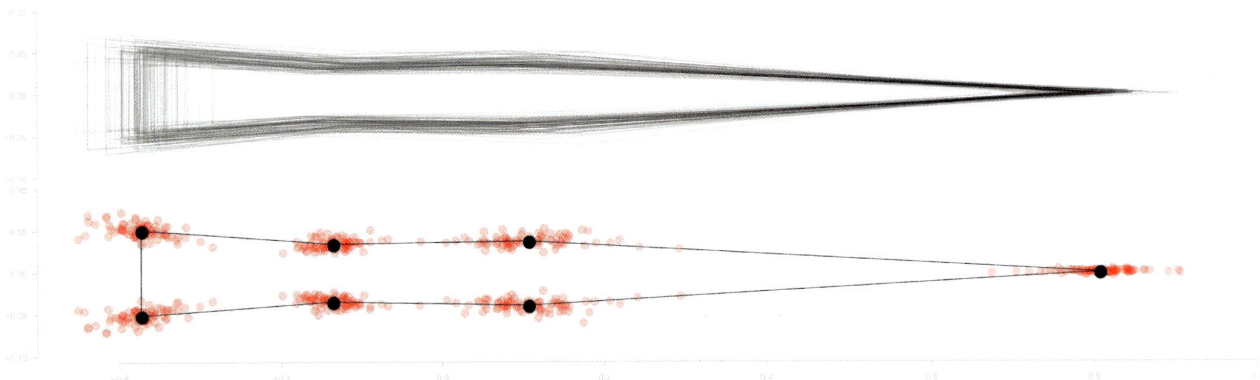

Figure 412 Example of geometric morphometric analysis on Iron Age iron lances. The various shapes of lances are superimposed using a specific protocol called generalized procrustes analysis (upper image). In the lower image, each point represents a normalized measure, for a given lance, at a predefined location or "landmark." See Birch and Martinón-Torres 2019, fig. 9.

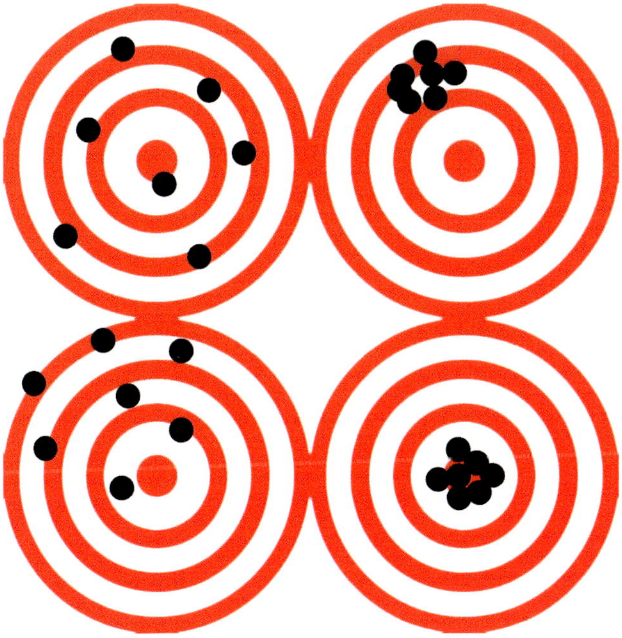

Figure 413 A representation of accuracy and precision. The top-left image shows a target hit with a high degree of accuracy but low precision. The top-right image shows the target hit with high precision but low accuracy. At bottom left, the target is hit with neither precision nor accuracy, and at bottom right the target is hit with both accuracy and precision.

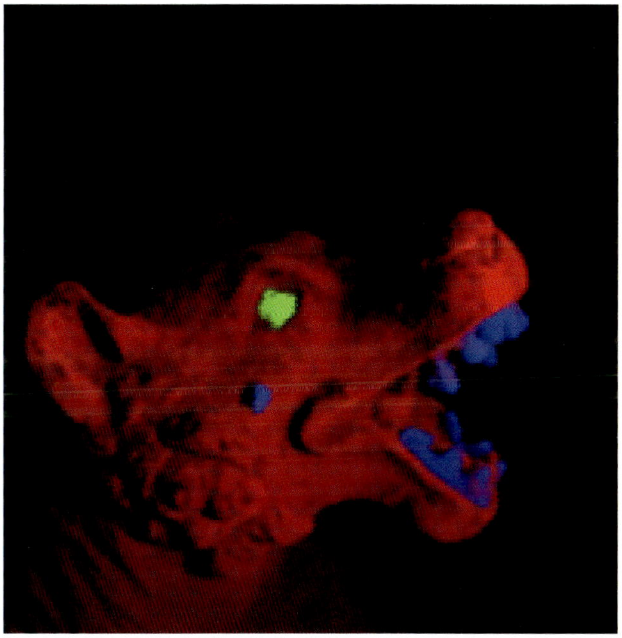

Figure 414 2D chemical mapping by XRF of the right profile of a bronze animal head showing gold inlay in the eye (green); silver teeth and silver inlay in the cheek (blue); and copper in the cast bronze substrate (red). *Statuette of a Dog or a Wolf*, Bavay (Nord, France), 1st century CE, H. 6.8 cm (Musée Archéologique de Bavay, France, inv. 1969 Br 23/ in. Biévelet 20). See Mille 2019a, 182–85; Laval, Calligaro, and Mille 2019.

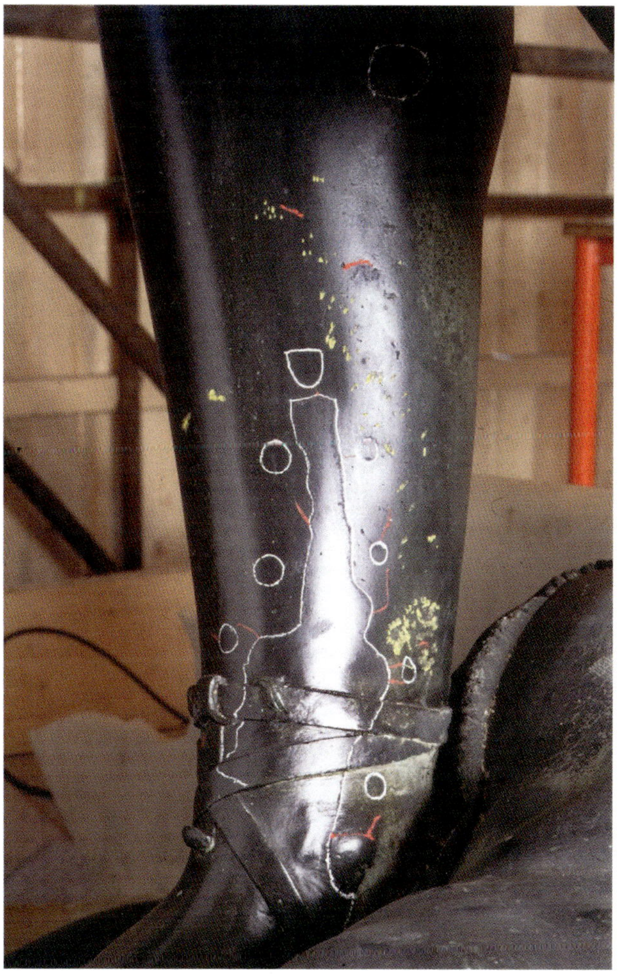

Figure 415 Mapping of cracks, defects, and cast-on metal on the heavily repaired right ankle of the Perseus monument was undertaken, before restoration, using an Eddy current probe (red and white lines) and a microwave probe (yellow lines and dots). Eddy currents detect all areas of discontinuity in a metallic surface. White lines are the outline of the cast-on repair, which was difficult to see and includes adjacent circular plugs that help lock the patch to the metallic wall. Microwaves detect air bubbles (yellow dots) or areas of discontinuity even under the surface. The round area at top right represents a sprue and was visible under raking light. Benvenuto Cellini (Italian, 1500–1571), *Perseus with the Head of Medusa*, 1545–54, H. 320 cm (bronze group); 199 cm (base, modern) (Loggia dei Lanzi, Piazza della Signoria, Florence). Morigi Restauratori, unpublished conservation report.

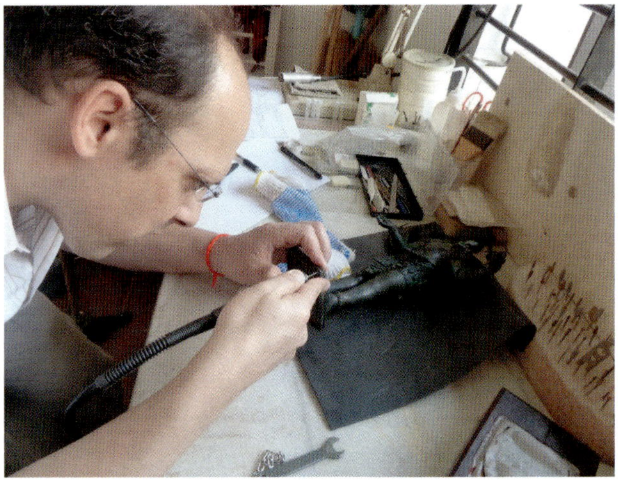

Figure 416 Drilling a sample from a statuette. In this case, a 1 mm diameter steel bit was used, drilling to a depth of 1 cm into the base. The metal drillings are collected on clean paper and carefully sorted under magnification to avoid any contamination by dust and/or corrosion products. *Five-Headed and Ten-Armed Śiva*, Angkor, 13th century, H. 38.5 cm (National Museum of Cambodia, inv. Ga 2778).

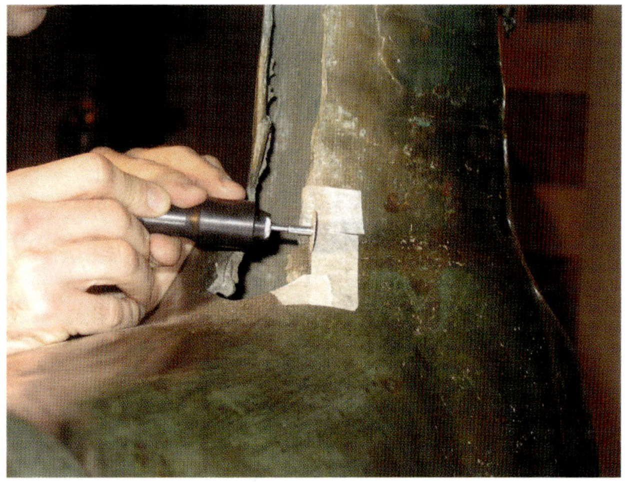

Figure 418 Sampling an antique bronze horse using a diamond wheel. The cross section was taken at the body-neck junction in order to study the assembly process (fusion welding). A mobile bronze mane (removed here) hides the sampling location. *Horse of Neuvy-en-Sullias*, France, 1st century BCE–1st century CE, H. 113 cm (Musée historique et archéologique de l'Orléanais, Orléans, France, inv. A6286). See Mille 2007.

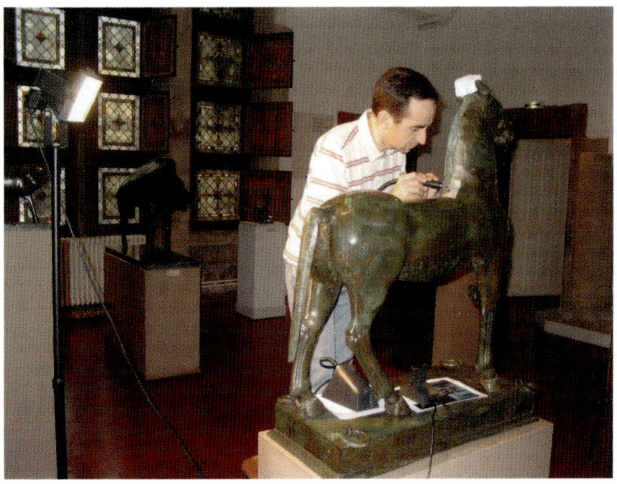

Figure 417 Sampling for metallographic study of welding. *Horse of Neuvy-en-Sullias*, France, 1st century BCE–1st century CE, H. 113 cm (Musée historique et archéologique de l'Orléanais, Orléans, France, inv. A6286). See Mille 2007.

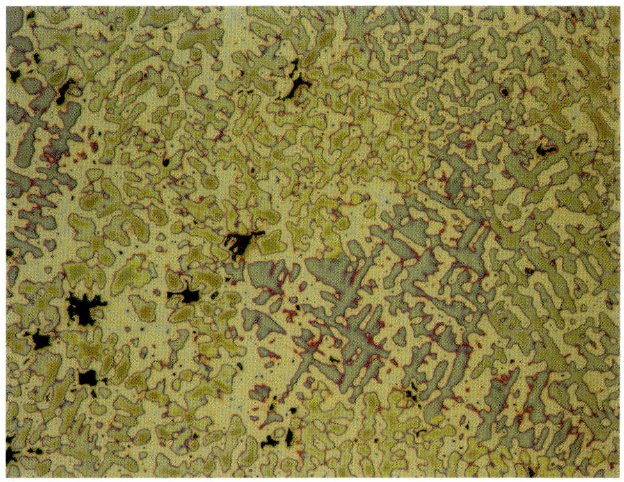

Figure 419 Bright field photomicrograph of an etched polished cross section of the welding zone of an experimental sample. See the dendritic microstructure of the primary cast. The outlines of dendrites appear colored; the interdendritic space is filled with eutectoïd phase alfa + delta (in blue). Sample E26, binary bronze CuSn10. See Azéma et al. 2012; Azéma and Mille 2013a.

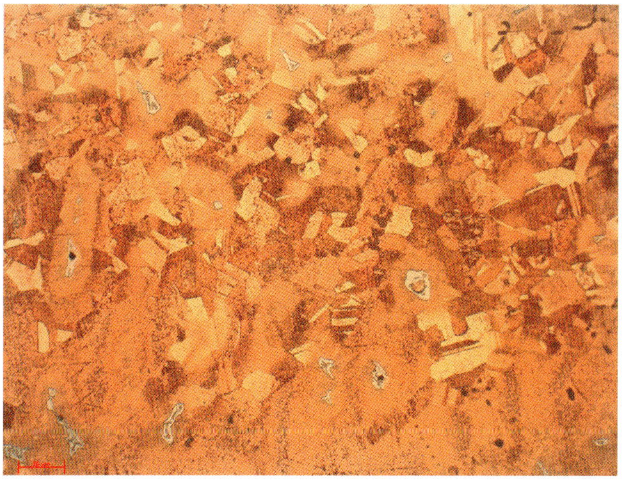

Figure 420 Bright field photomicrograph of an etched polished cross section of the welding zone of an experimental sample. See the dendritic microstructure and small grains due to recrystallization in the heat-affected zone (HAZ) generated by flow fusion welding. The outlines of dendrites appear brown; the interdendritic space is filled with eutectoïd phase alfa + delta (in blue). Sample E66, binary bronze CuSn10. See Azéma et al. 2012; Azéma and Mille 2013a.

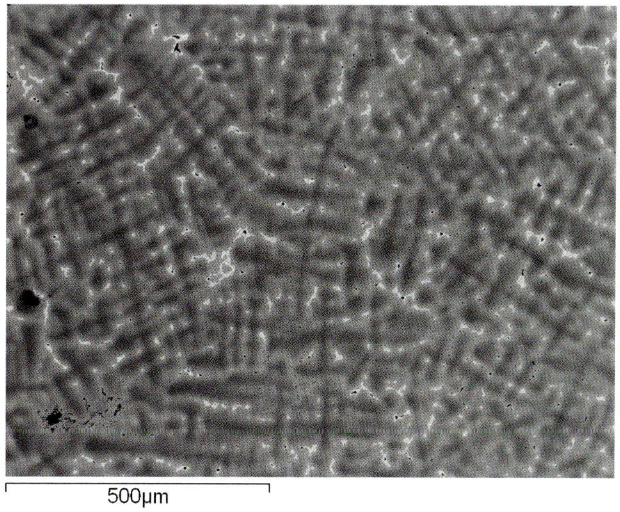

Figure 422 Backscattered electron micrograph of a polished cross section showing the dendritic microstructure of the primary cast of an experimental sample aimed to investigate fusion welding (E26, binary bronze CuSn10). The outlines of dendrites appear dark gray; the interdendritic space is filled with phase alfa + delta (eutectoïd composition) (light gray). See Azéma et al. 2017.

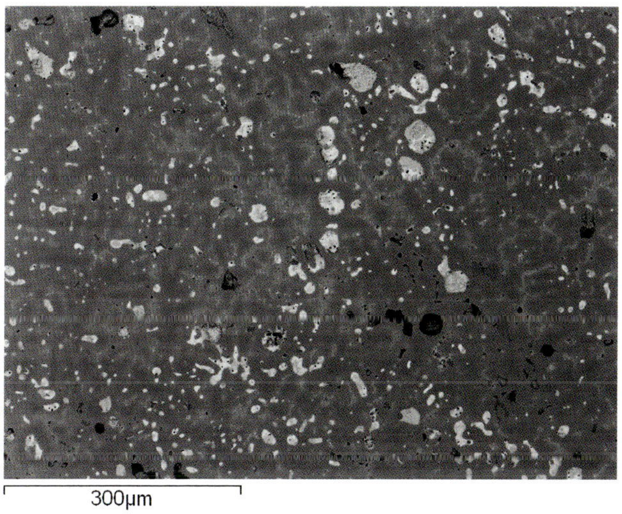

Figure 421 Backscattered electron micrograph of a polished cross section showing the dendritic microstructure of the primary cast. The outlines of dendrites appear dark gray; the interdendritic space is filled with phase alfa + delta (eutectoïd composition) and various inclusions (light gray) as well as nodules of lead (white oval shapes). Large bronze fragment from the hoard of Evreux, France, 1st century CE, L. 14 cm (Musée d'Evreux, France, inv. 4864). See Azéma et al. 2012; Azéma and Mille 2013a.

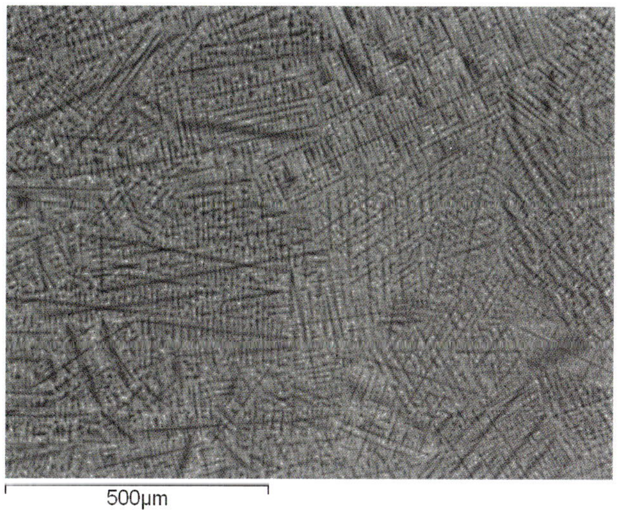

Figure 423 Backscattered electron micrograph of a polished cross section showing the dendritic microstructure and grains of the primary cast of an experimental sample aimed to investigate fusion welding (E54, binary bronze CuSn10). The outlines of dendrites appear dark gray; the interdendritic space is filled with phase alfa + delta (eutectoïd composition) (light gray). See Azéma et al. 2012; Azéma and Mille 2013a.

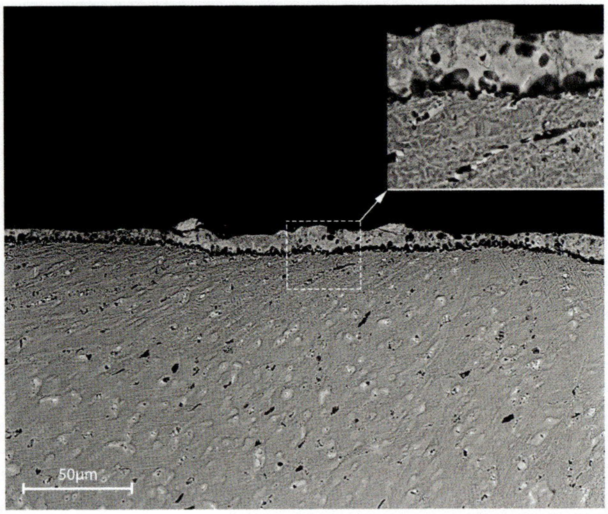

Figure 424 Backscattered electron micrograph showing a corroded tin layer on the surface of cast brass with a dendritic structure.

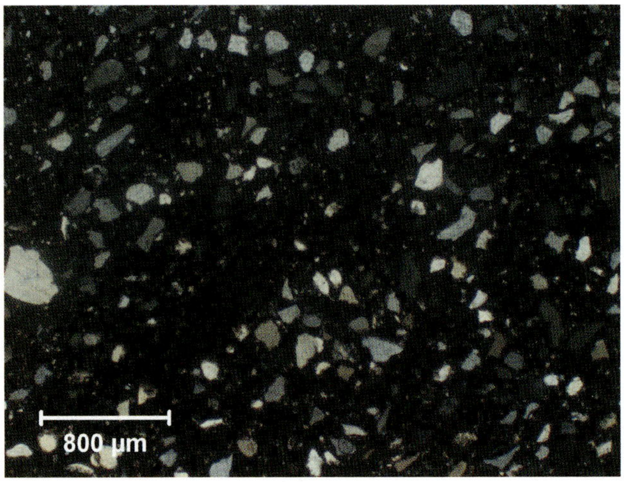

Figure 426 Photomicrograph under cross-polarized transmitted light of a thin section of a casting core. Francesco Primaticcio (Italian, 1504–1570), *Laocoön and His Sons*, 1542, H. 191 cm (Musée national du château de Fontainebleau, France, inv. MR 3290). See Castelle, Coquinot, and Bourgarit 2016.

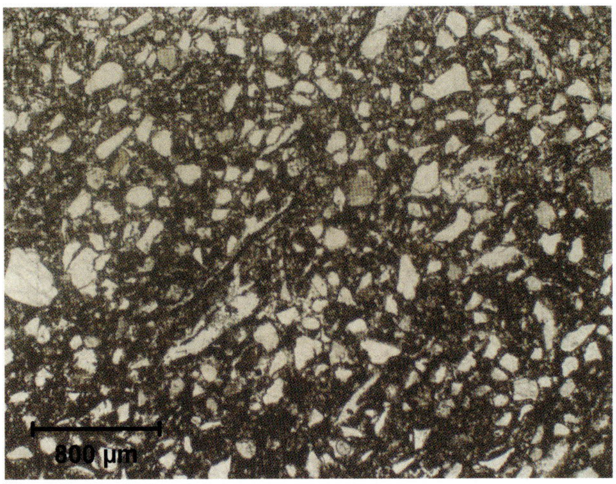

Figure 425 Photomicrograph under plane-polarized transmitted light of a thin section of a casting core. Francesco Primaticcio (Italian, 1504–1570), *Laocoön and His Sons*, 1542, H. 191 cm (Musée national du château de Fontainebleau, France, inv. MR 3290). See Castelle, Coquinot, and Bourgarit 2016.

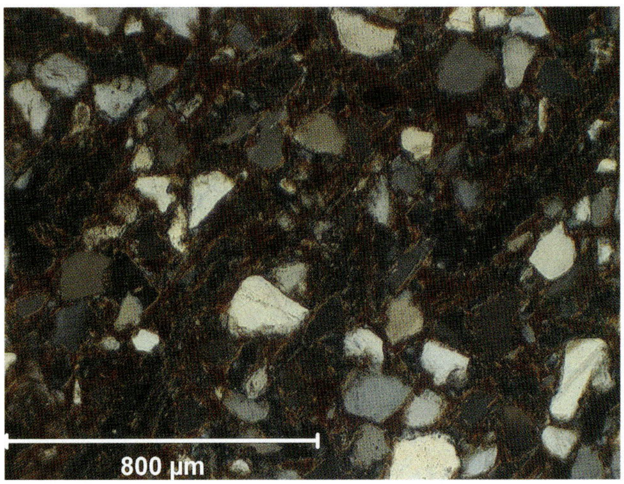

Figure 427 Photomicrograph under cross-polarized transmitted light of a thin section of the casting core showing its characteristic quartz distribution (see fig. 525, core recipe 1). Martin Lefort (French, dates unknown), *Justice*, 1571, H. 128 cm (Musée du Louvre, inv. MR1682). See Castelle, Coquinot, and Bourgarit 2016.

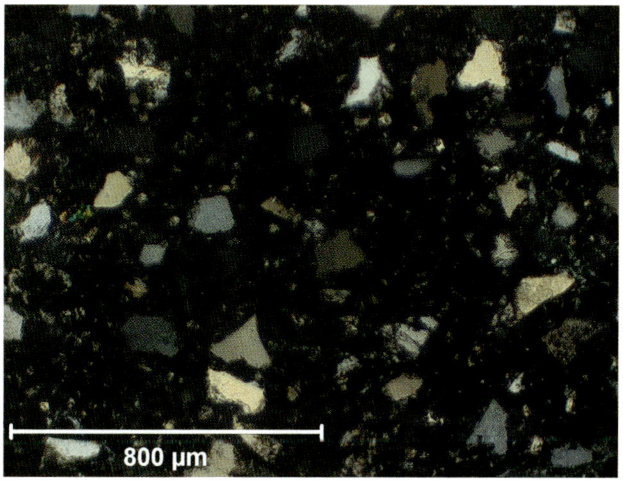

Figure 428 Photomicrograph under cross-polarized transmitted light of a thin section of the casting core showing its characteristic grain size distribution (see fig. 525, core recipe 2). Ponce Jacquiot (French, 1515–1571), *Temperance*, Henri II and Catherine of Medici Funerary Monument, 1567, H. 187 cm (Saint-Denis Basilica, France). See Castelle, Coquinot, and Bourgarit 2016.

Figure 429 Photomicrograph using cathodoluminescence of a thin section of a casting core. Francesco Primaticcio (Italian, 1504–1570), *Laocoön and His Sons*, 1542, H. 191 cm (Musée national du château de Fontainebleau, France, inv. MR 3290). See Castelle, Coquinot, and Bourgarit 2016.

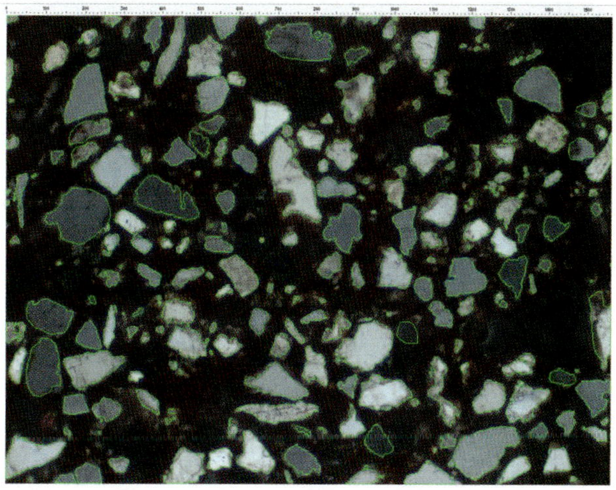

Figure 430 Photomicrograph under cross-polarized transmitted light of a thin section of a casting core. The different inclusions are sorted according to their gray level (green outlines) and characterized (shape, dimensions) using image analysis software. Francesco Primaticcio (Italian, 1504–1570), *Laocoön and His Sons*, 1542, H. 191 cm (Musée national du château de Fontainebleau, France, inv. MR 3290). See Castelle, Coquinot, and Bourgarit 2016.

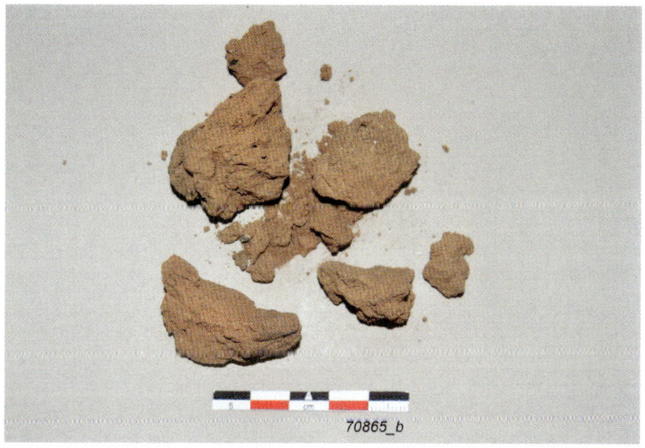

Figure 431 Example of a casting core sample taken from underneath. Francesco Primaticcio (Italian, 1504–1570), *Sleeping Ariadne*, 1542, L. 240 cm (Musée national du château de Fontainebleau, France, inv. MR 3284). See Castelle 2016.

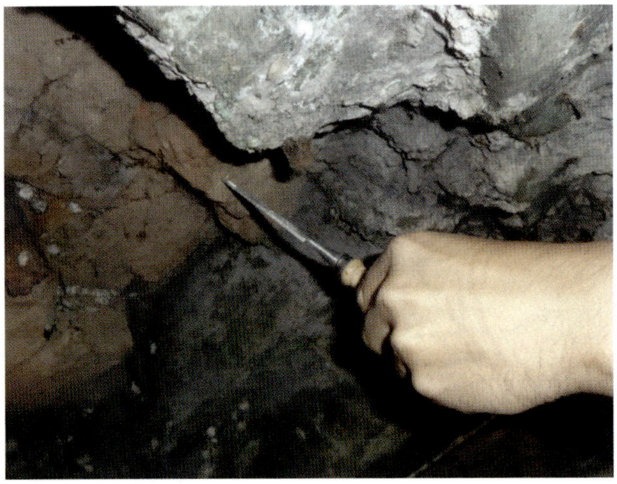

Figure 432 Sampling of casting core in process. Barthélemy Prieur (French, 1536–1611), *Diana the Huntress*, 1603, H. 200 cm (Musée du Louvre - Fontainebleau Castle, inv. RF261). See Castelle, Coquinot, and Bourgarit 2016.

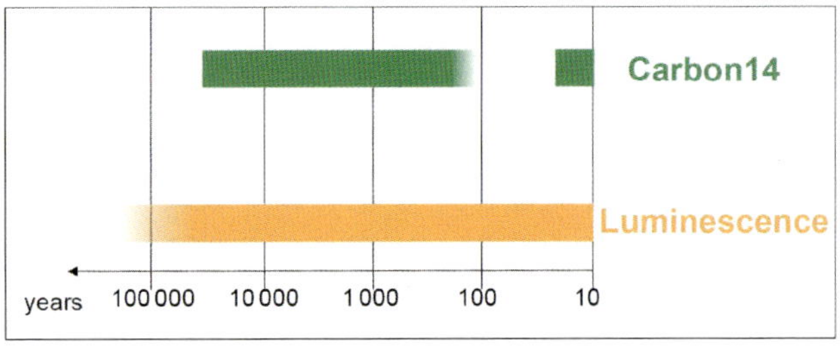

Figure 433 General age range of items, including bronze sculpture, that can be dated by radiocarbon and luminescence techniques.

VISUAL ATLAS

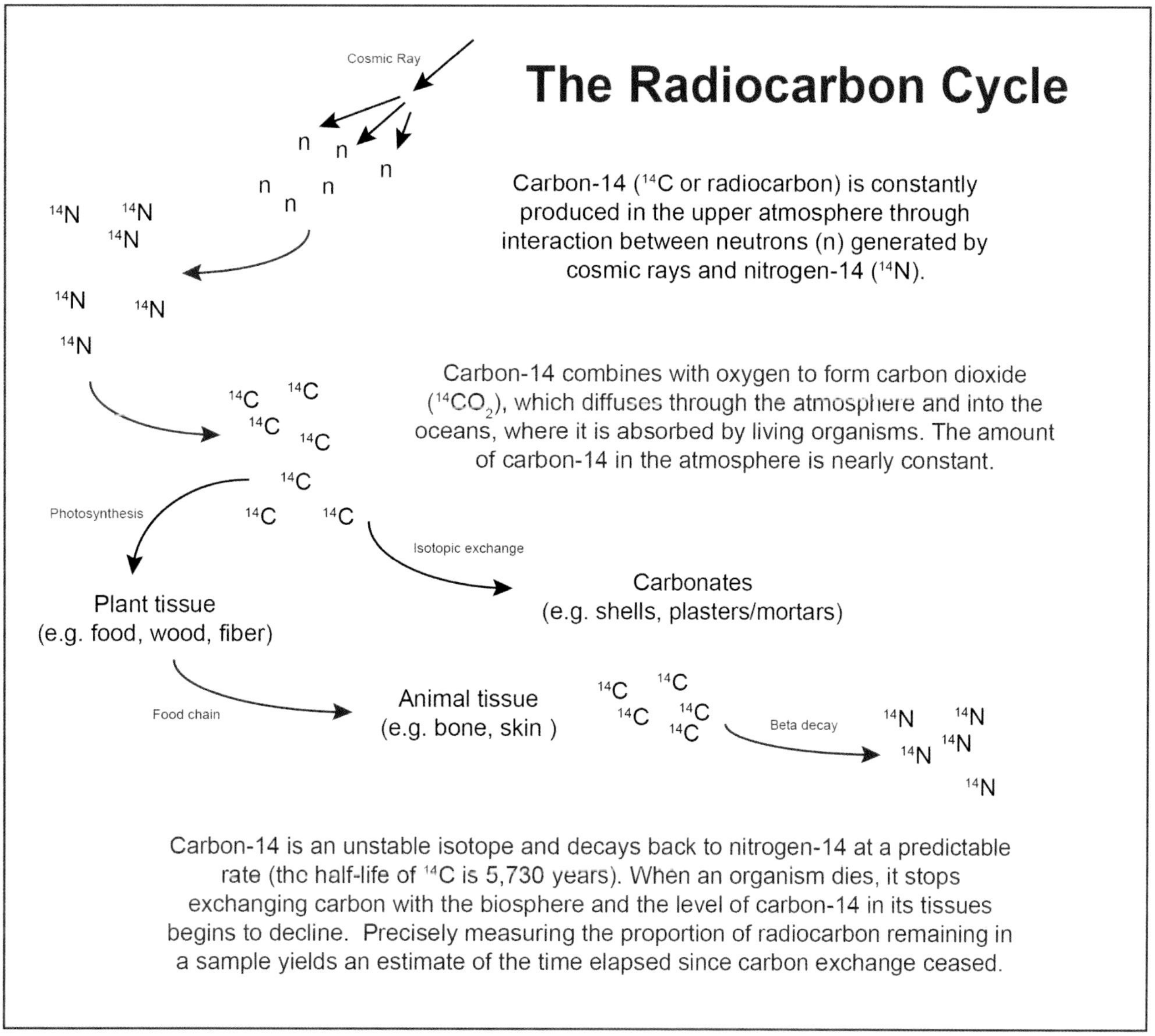

Figure 434 Diagram of the cycle of radiocarbon formation and decay in organic materials that underlies the principles of dating.

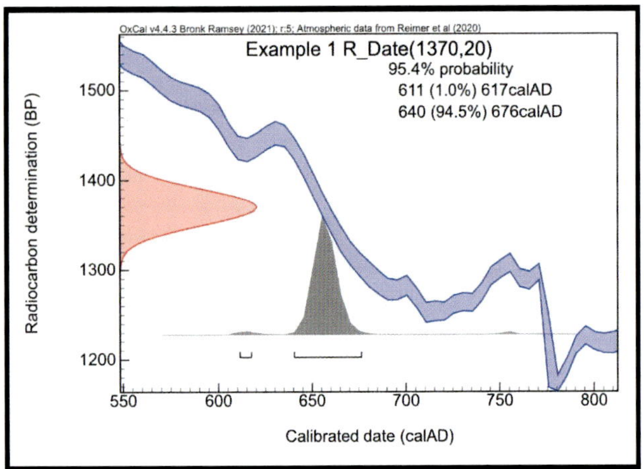

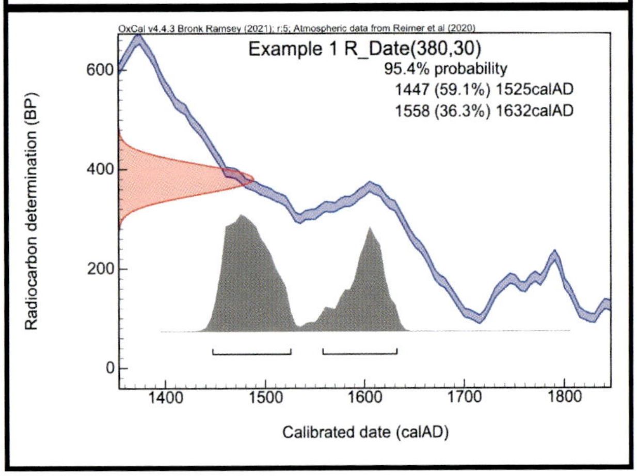

Figure 435 Two examples of radiocarbon dating results generated using OxCal 4.4 software (see Bronk Ramsey 2009). The radiocarbon determination is represented, along with its uncertainty, as the red curve along the vertical axis; the narrower the curve, the more precise was the instrumental measurement. The blue band represents the IntCal20 calibration curve, which accounts for minor variability in carbon-14 concentration in the atmosphere over time (see Reimer et al. 2020). The gray curves along the horizontal axis (calibrated date) represent the probability that the sample actually originated at any given time; the higher the curve, the greater the probability that the date below is correct. The brackets below the gray curves show the time period that is 95.4% certain to contain the true age. Depending on the precision of the measurement and the shape of the calibration curve, a radiocarbon analysis can yield results of varying precision, and can even yield results with more than one possible date range.

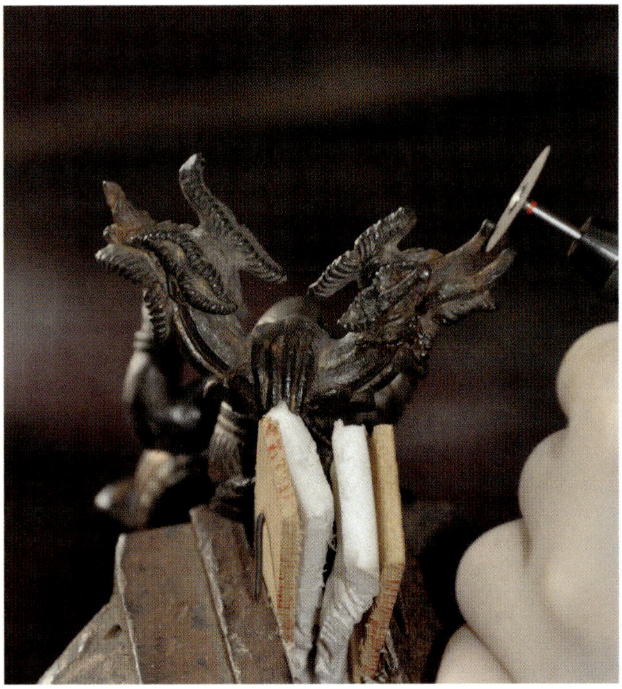

Figure 436 Sampling process on an iron armature from a bronze. Palanquin Hook in the Shape of a Monkey, late 12th–early 13th century, H. 17 cm (National Museum of Cambodia, inv. Ga 4745). See Leroy et al. 2021.

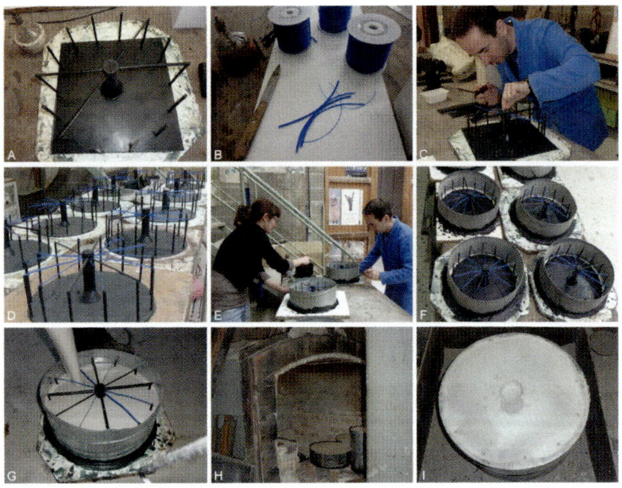

Figure 437 Design of wax models for castability tests: A) setup of the sprues and vents (4 mm and 6 mm in diameter); B) preparation of the wax wire (1 mm, 2 mm, and 3 mm in diameter); C) soldering of the wax wire; D) the wax models once completed; E) setup of the metal mold; F) molds once completed; G) molding using refractory plaster; G) molds in the stove for dewaxing and baking; H) molds ready for metal pouring. Experiments carried out at Coubertin foundry, France, 2015. See Mille 2017.

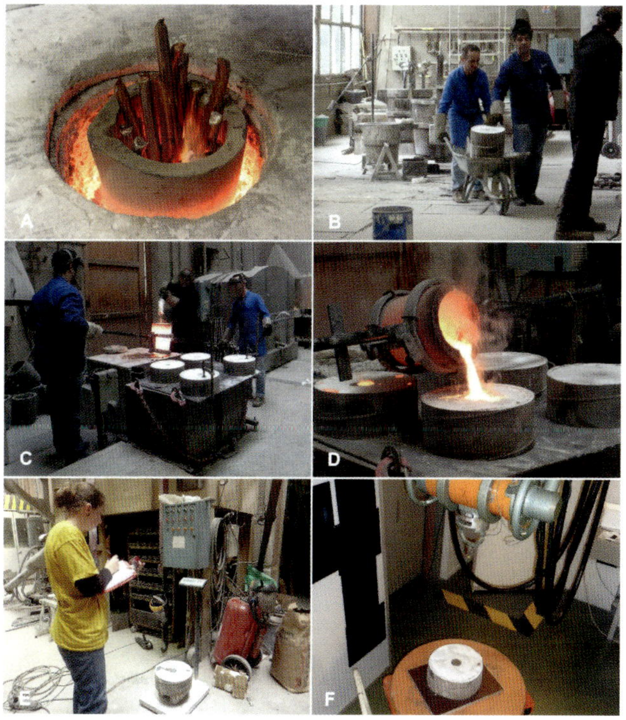

Figure 438 Castability tests devised by the C2RMF were carried out at the Coubertin foundry, France, in 2015. The distance that the molten metal travels in channels of different diameters is used as a measure of its castability. A) Copper and copper alloy rods of specific compositions were melted in preparation for the test; B) hollow channels of different diameters were made in refractory molds using wires of different gauges, and the baked molds were transferred to the furnace; C) the melt temperature was measured prior to pouring; D) the alloy was poured into the molds; E) each mold was weighed post-pour to gauge the quantity of metal; F) each mold was then radiographed at the C2RMF to assess how far the metal had traveled. See Mille 2017.

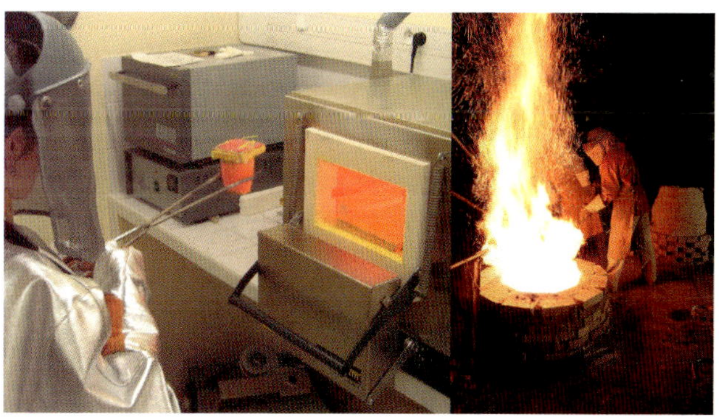

Figure 439 Brass-making experimental simulation. View of the furnace used for the laboratory experiments at C2RMF (left) and the large furnace used during field experiments at Barsy, Belgium (right). See Bourgarit and Thomas 2015; Thomas and Bourgarit 2018.

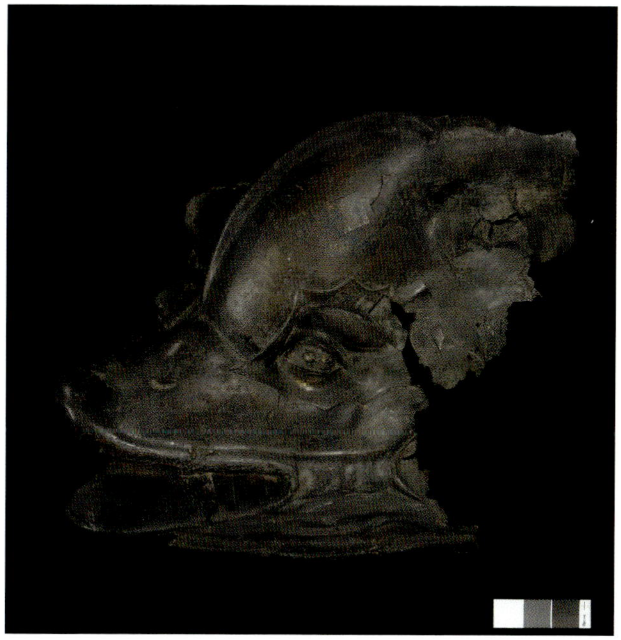

Figure 440 Front view of the first head. *Great Dolphins of Vienne*, Roman period, Vienne, France, 2nd century CE, max L. 260 cm (Musée des Beaux-Arts et d'Archéologie de Vienne, France, inv. R.1998.2.43). For a description of all visible technical evidence, refer to fig. 367.

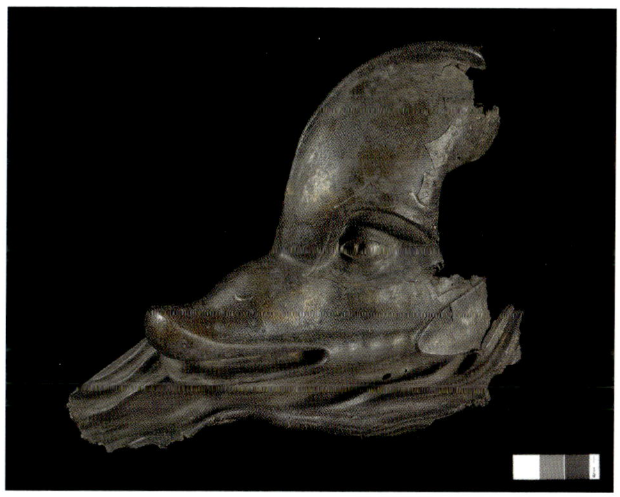

Figure 441 Front view of the second head. *Great Dolphins of Vienne*, Roman period, Vienne, France, 2nd century CE, max L. 260 cm (Musée des Beaux-Arts et d'Archéologie de Vienne, France, inv. R.1998.2.43). For a description of all visible technical evidence, refer to fig. 367.

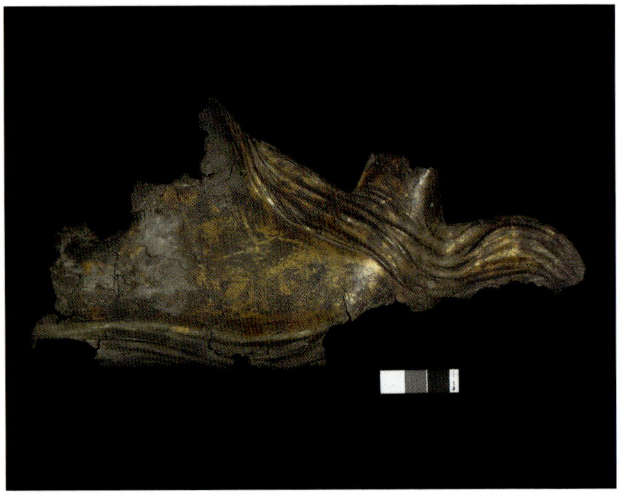

Figure 442 Front view of the body. *Great Dolphins of Vienne*, Roman period, Vienne, France, 2nd century CE, max L. 260 cm (Musée des Beaux-Arts et d'Archéologie de Vienne, France, inv. R.1998.2.43). For a description of all visible technical evidence, refer to fig. 367.

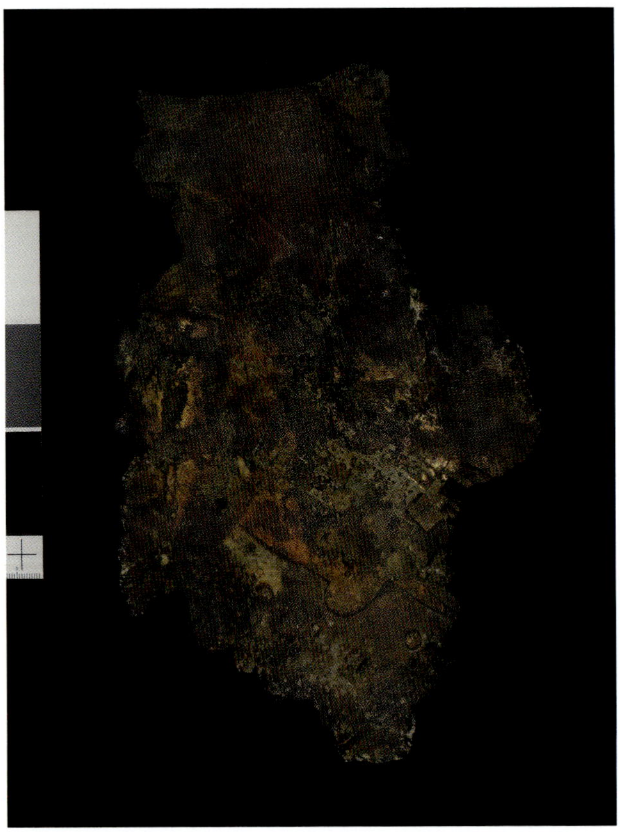

Figure 443 Front view of the upper fragment of the body. *Great Dolphins of Vienne*, Roman period, Vienne, France, 2nd century CE, max L. 260 cm (Musée des Beaux-Arts et d'Archéologie de Vienne, France, inv. R.1998.2.43). For a description of all visible technical evidence, refer to fig. 367.

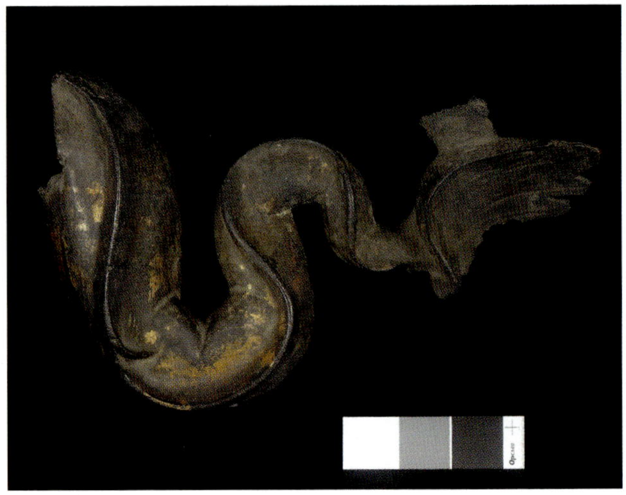

Figure 444 Front view of a fragment of the upper tail. *Great Dolphins of Vienne*, Roman period, Vienne, France, 2nd century CE, max L. 260 cm (Musée des Beaux-Arts et d'Archéologie de Vienne, France, inv. R.1998.2.43). For a description of all visible technical evidence, refer to fig. 367.

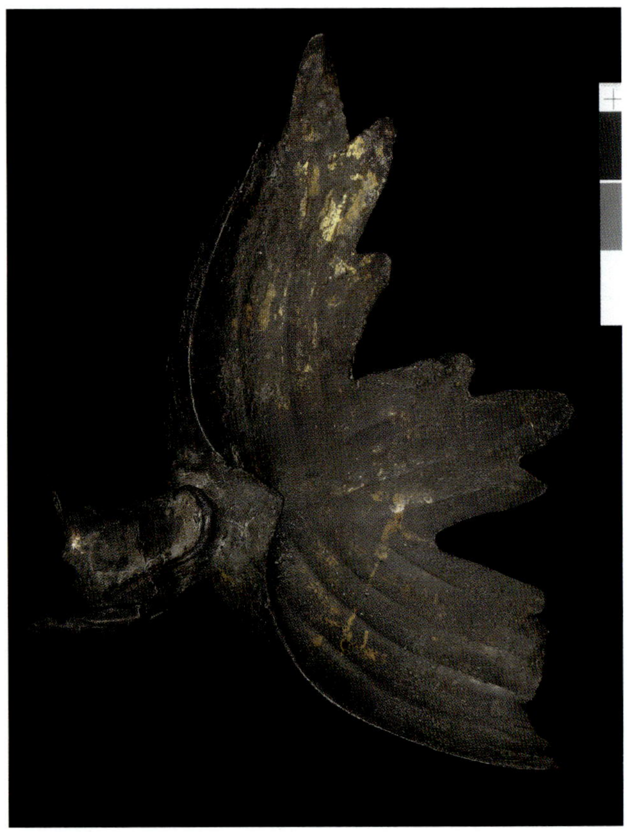

Figure 445 Front view of a fragment of the tip of the upper tail. *Great Dolphins of Vienne*, Roman period, Vienne, France, 2nd century CE, max L. 260 cm (Musée des Beaux-Arts et d'Archéologie de Vienne, France, inv. R.1998.2.43). For a description of all visible technical evidence, refer to fig. 367.

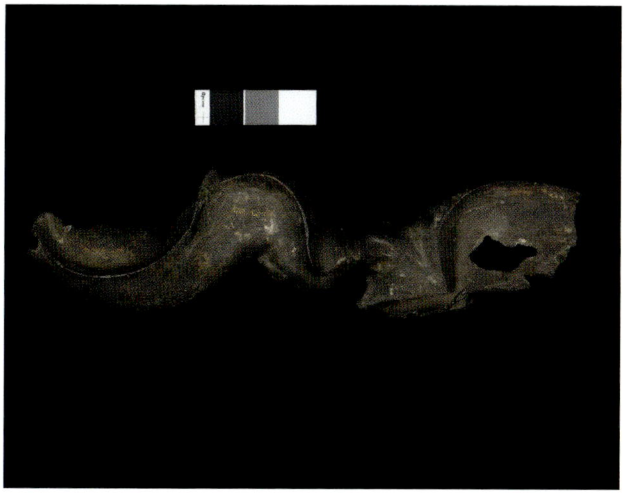

Figure 446 Front view of a fragment of the lower tail. *Great Dolphins of Vienne*, Roman period, Vienne, France, 2nd century CE, max L. 260 cm (Musée des Beaux-Arts et d'Archéologie de Vienne, France, inv. R.1998.2.43). For a description of all visible technical evidence, refer to fig. 367.

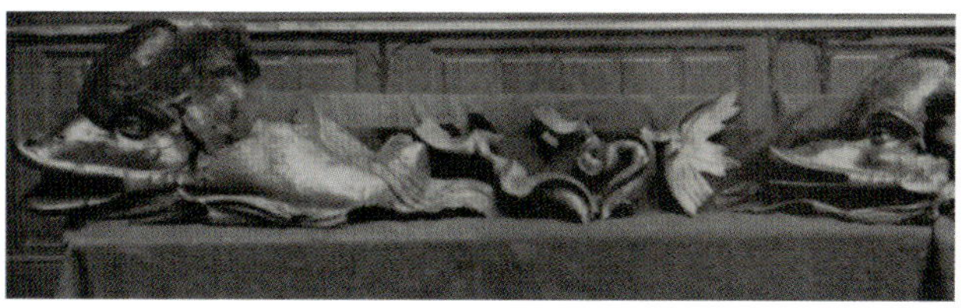

Figure 447 The Museum of Vienne's former presentation of the *Great Dolphins of Vienne*, Roman period, Vienne, France, 2nd century CE, max L. 260 cm (Musée des Beaux-Arts et d'Archéologie de Vienne, France, inv. R.1998.2.43). From Boucher 1964.

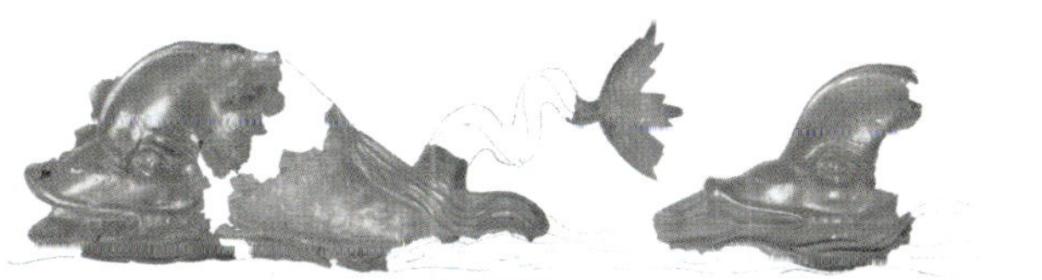

Figure 448 Former attempt at a reconstruction of the composition of the surviving fragments. *Great Dolphins of Vienne*, Roman period, Vienne, France, 2nd century CE, max L. 260 cm (Musée des Beaux-Arts et d'Archéologie de Vienne, France, inv. R.1998.2.43). For a description of all visible technical evidence, refer to fig. 367. From Boucher 1964.

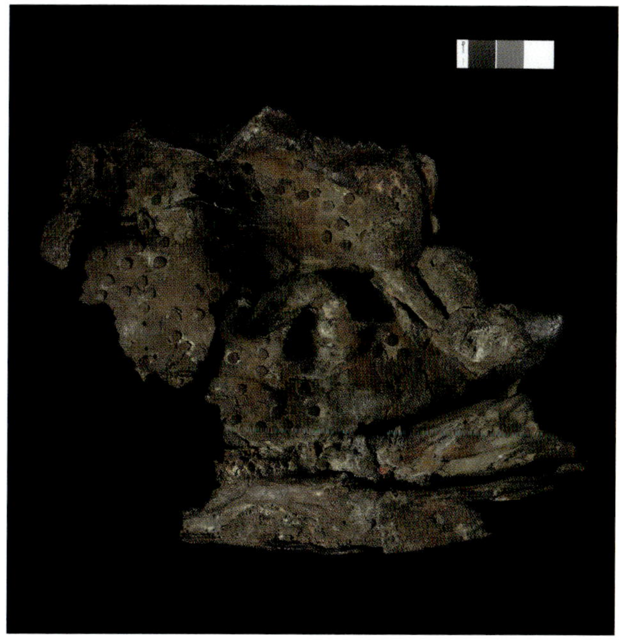

Figure 449 Rear view of the first head. *Great Dolphins of Vienne*, Roman period, Vienne, France, 2nd century CE, max L. 260 cm (Musée des Beaux-Arts et d'Archéologie de Vienne, France, inv. R.1998.2.43). For a description of all visible technical evidence, refer to fig. 367.

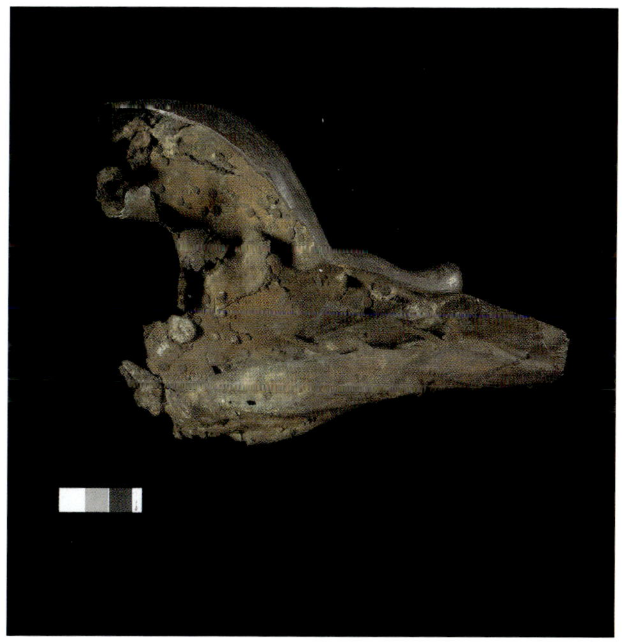

Figure 450 Rear view of the second head. *Great Dolphins of Vienne*, Roman period, Vienne, France, 2nd century CE, max L. 260 cm (Musée des Beaux-Arts et d'Archéologie de Vienne, France, inv. R.1998.2.43). For a description of all visible technical evidence, refer to fig. 367.

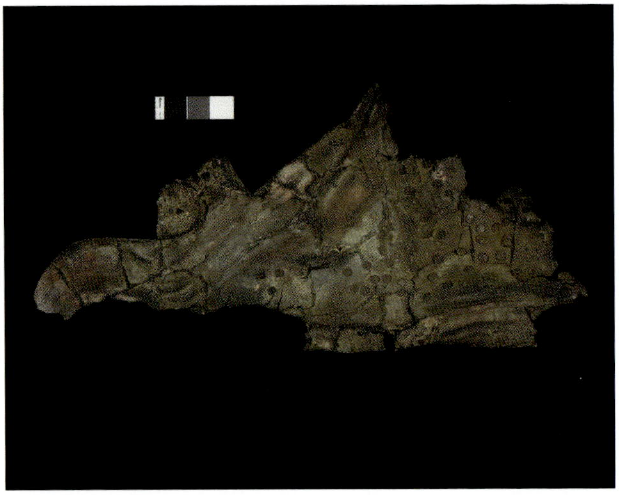

Figure 451 Rear view of the body. *Great Dolphins of Vienne*, Roman period, Vienne, France, 2nd century CE, max L. 260 cm (Musée des Beaux-Arts et d'Archéologie de Vienne, France, inv. R.1998.2.43). For a description of all visible technical evidence, refer to fig. 367.

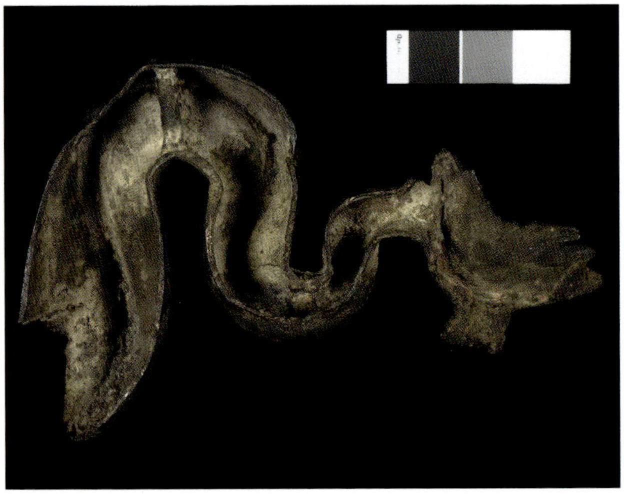

Figure 453 Rear view of a fragment of the upper tail. *Great Dolphins of Vienne*, Roman period, Vienne, France, 2nd century CE, max L. 260 cm (Musée des Beaux-Arts et d'Archéologie de Vienne, France, inv. R.1998.2.43). For a description of all visible technical evidence, refer to fig. 367.

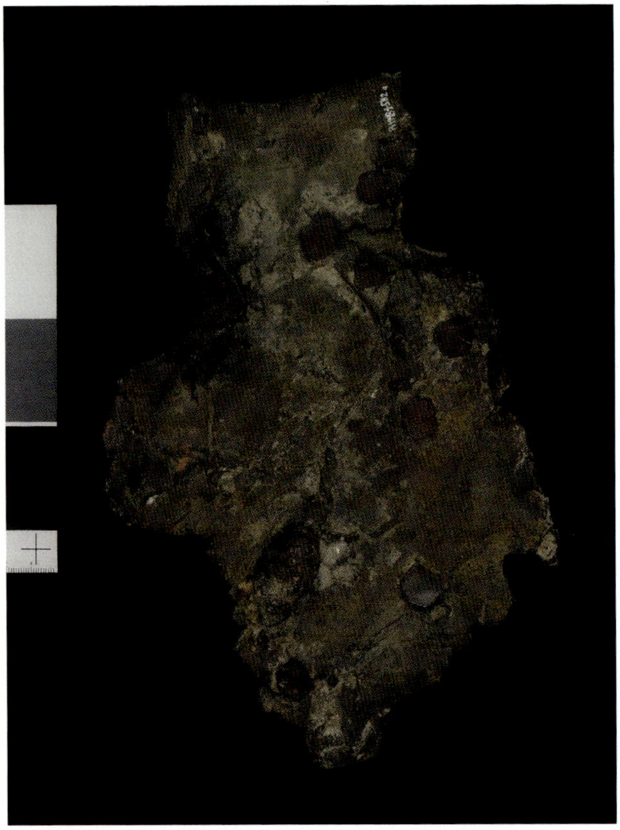

Figure 452 Rear view of the upper fragment of the body. *Great Dolphins of Vienne*, Roman period, Vienne, France, 2nd century CE, max L. 260 cm (Musée des Beaux-Arts et d'Archéologie de Vienne, France, inv. R.1998.2.43). For a description of all visible technical evidence, refer to fig. 367.

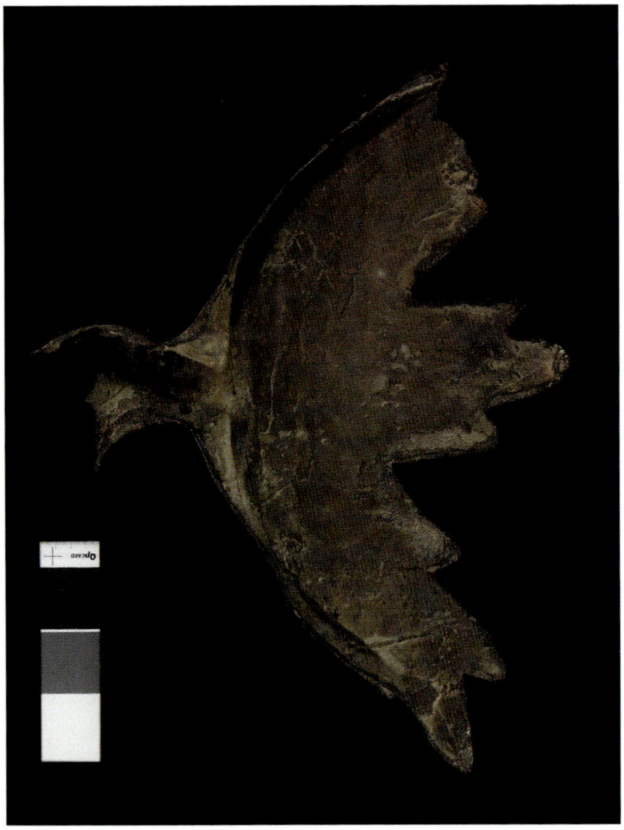

Figure 454 Rear view of a fragment of the tip of the upper tail. *Great Dolphins of Vienne*, Roman period, Vienne, France, 2nd century CE, max L. 260 cm (Musée des Beaux-Arts et d'Archéologie de Vienne, France, inv. R.1998.2.43). For a description of all visible technical evidence, refer to fig. 367.

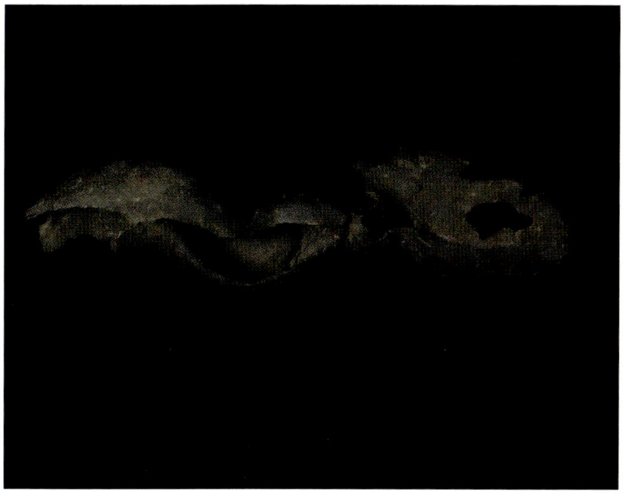

Figure 455 Rear view of a fragment of the lower tail. *Great Dolphins of Vienne*, Roman period, Vienne, France, 2nd century CE, max L. 260 cm (Musée des Beaux-Arts et d'Archéologie de Vienne, France, inv. R.1998.2.43). For a description of all visible technical evidence, refer to fig. 367.

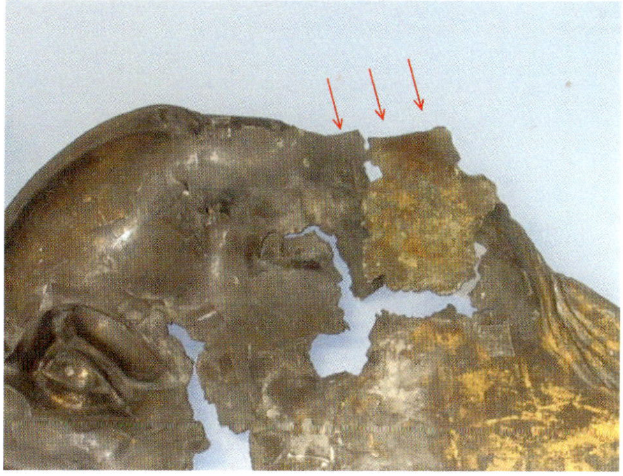

Figure 457 Detail of the connection of the first head to the two body fragments, also showing the neat cutout at the top, possibly the place for a rider (arrows). *Great Dolphins of Vienne*, Roman period, Vienne, France, 2nd century CE, max L. 260 cm (Musée des Beaux-Arts et d'Archéologie de Vienne, France, inv. R.1998.2.43).

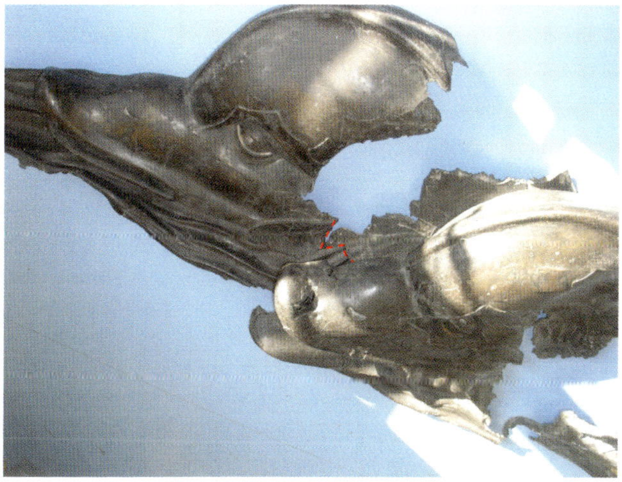

Figure 456 Assembly of the front views of the two heads showing that these two elements used to be connected (red line). *Great Dolphins of Vienne*, Roman period, Vienne, France, 2nd century CE, max L. 260 cm (Musée des Beaux-Arts et d'Archéologie de Vienne, France, inv. R.1998.2.43).

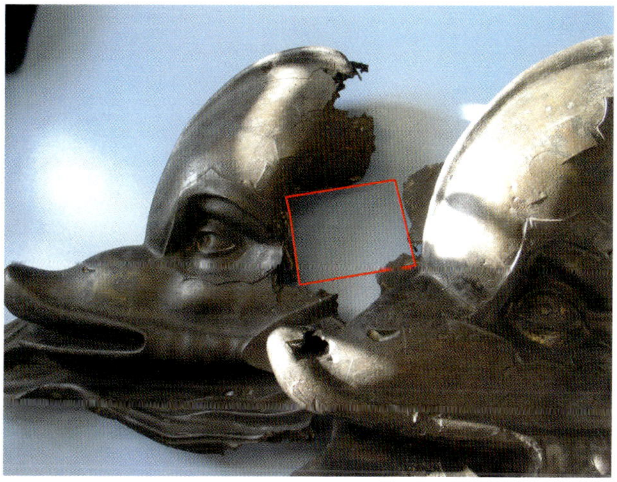

Figure 458 The front view of the two heads reveals a neat square hole between them (overlay), the result of the impact of a pile driver used in 1839 during the building of the Rhone riverbanks. *Great Dolphins of Vienne*, Roman period, Vienne, France, 2nd century CE, max L. 260 cm (Musée des Beaux-Arts et d'Archéologie de Vienne, France, inv. R.1998.2.43).

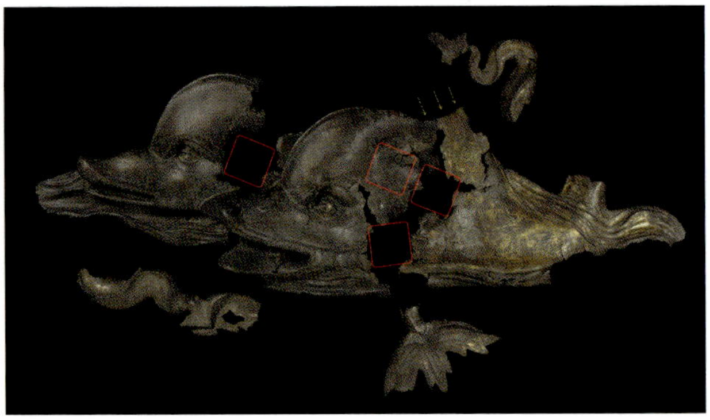

Figure 459 The new presentation of the frieze. *Great Dolphins of Vienne*, Roman period, Vienne, France, 2nd century CE, max L. 260 cm (Musée des Beaux-Arts et d'Archéologie de Vienne, France, inv. R.1998.2.43). The neat square hole (overlay) between the two heads is the result of the impact of a pile driver used in 1839 during the building of the Rhone riverbanks. Also visible is a neat cutout at the top, possibly the place for a rider (arrows).

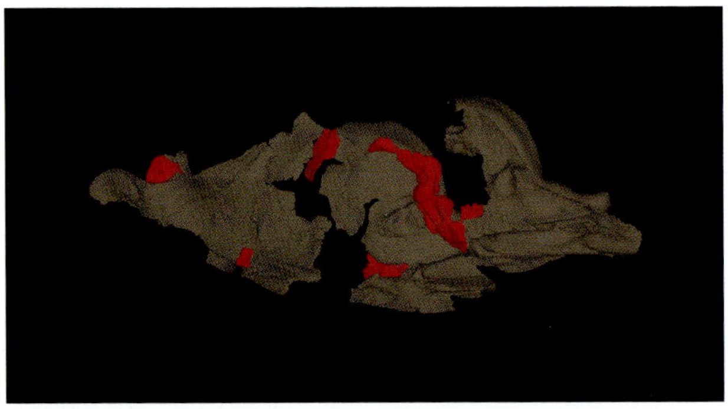

Figure 460 3D model of the rear surface of the surviving fragments showing the assembly of the different parts. The red bands represent flow-fusion welding zones. *Great Dolphins of Vienne*, Roman period, Vienne, France, 2nd century CE, max L. 260 cm (Musée des Beaux-Arts et d'Archéologie de Vienne, France, inv. R.1998.2.43).

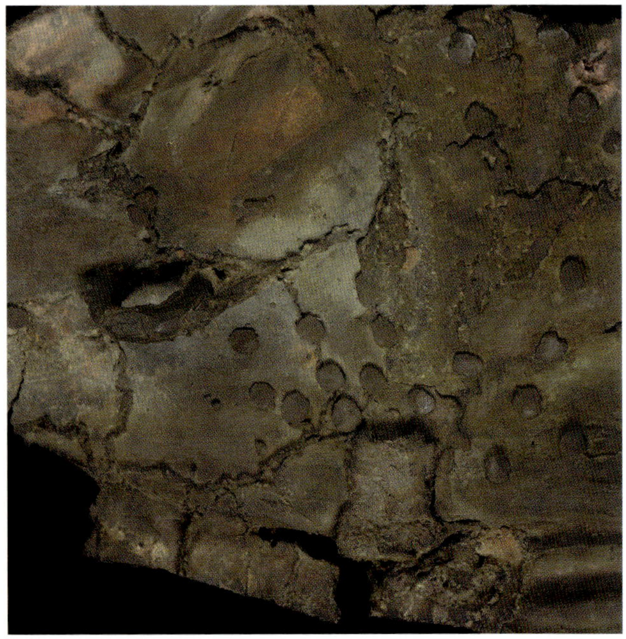

Figure 461 Rear surface of a body fragment of a dolphin showing a long weld joint and numerous rivet heads used to secure the largest polygonal repair patches on the front side. *Great Dolphins of Vienne*, Roman period, Vienne, France, 2nd century CE, max L. 260 cm (Musée des Beaux-Arts et d'Archéologie de Vienne, France, inv. R.1998.2.43). See technical sketch, fig. 367.

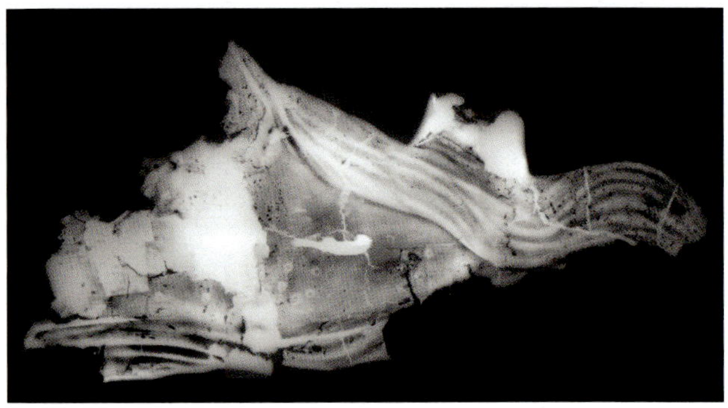

Figure 462 Detail of an X-radiograph of a dolphin's body in which the denser welding zone appears as a brighter white area in the center, the rivets show up as small white donuts with darker centers, and porosity appears as black flecks. *Great Dolphins of Vienne*, Roman period, Vienne, France, 2nd century CE, max L. 260 cm (Musée des Beaux-Arts et d'Archéologie de Vienne, France, inv. R.1998.2.43). For a description of all visible technological evidence, refer to figs. 367, 467.

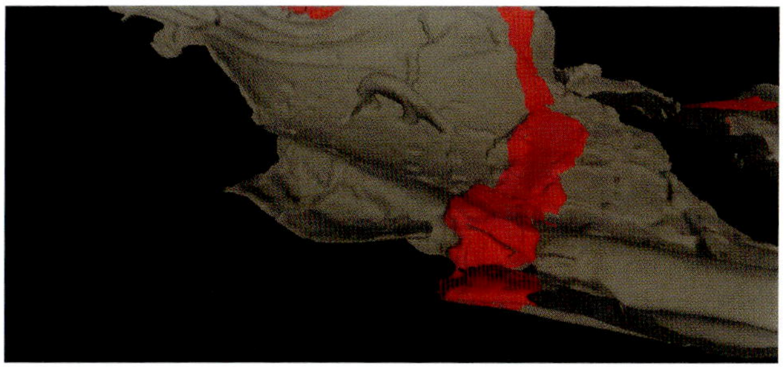

Figure 463 3D model of the rear surface of the surviving fragments showing the welding zones in red, and in the foreground, the metal reservoir in the form of a basin where the welding metal was poured in. *Great Dolphins of Vienne*, Roman period, Vienne, France, 2nd century CE, max L. 260 cm (Musée des Beaux-Arts et d'Archéologie de Vienne, France, inv. R.1998.2.43).

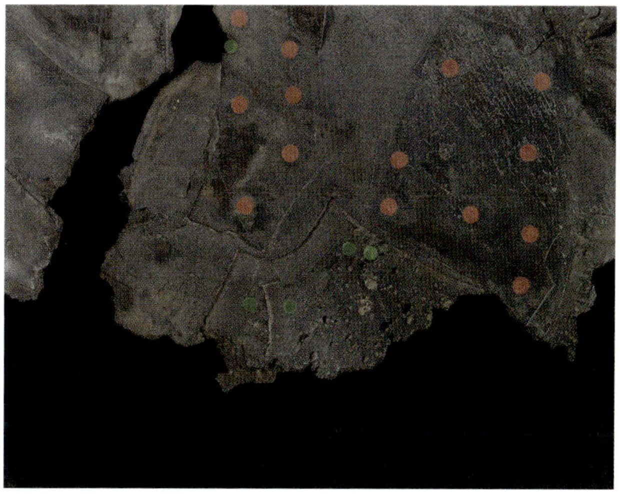

Figure 465 Detail of the front view of the body fragment showing the location of a lost repair polygonal patch that was fastened by rivets (now rivet holes, see green spots). Numerous patches and rivet heads (orange spots) are also visible at this location (see also fig. 466). *Great Dolphins of Vienne*, Roman period, Vienne, France, 2nd century CE, max L. 260 cm (Musée des Beaux-Arts et d'Archéologie de Vienne, France, inv. R.1998.2.43).

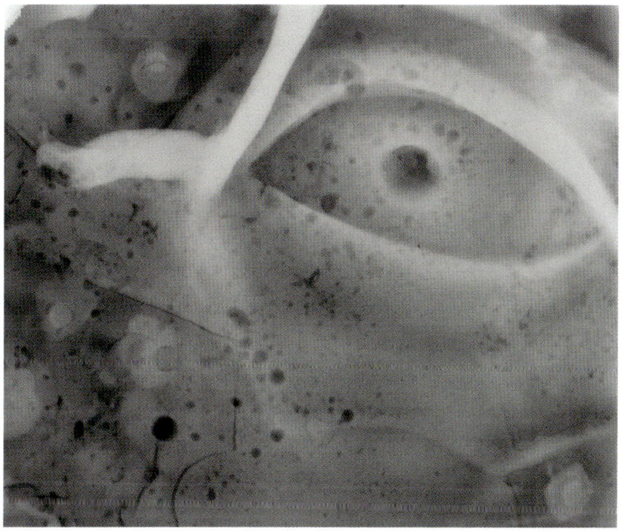

Figure 464 Detail of an X-radiograph of the first dolphin head, in which the extensive porosity appears as dark spots of various sizes and densities. *Great Dolphins of Vienne*, Roman period, Vienne, France, 2nd century CE, max L. 260 cm (Musée des Beaux-Arts et d'Archéologie de Vienne, France, inv. R.1998.2.43).

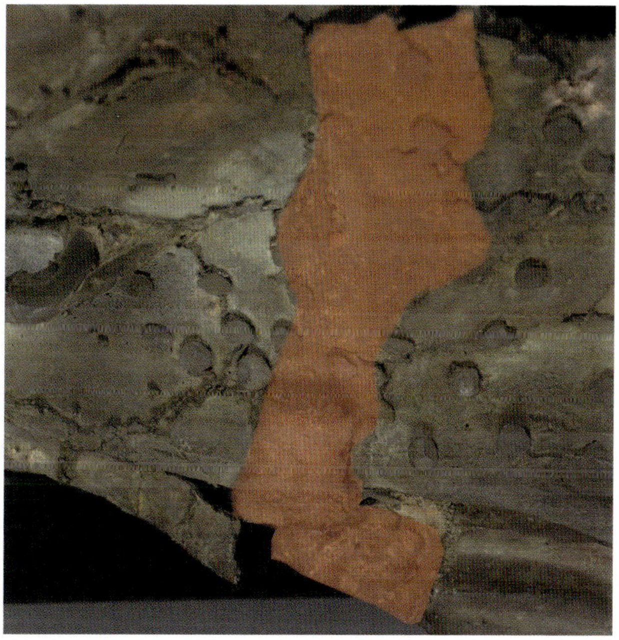

Figure 466 Detail of the back view of the body fragment showing the flow fusion weld (red overlay). *Great Dolphins of Vienne*, Roman period, Vienne, France, 2nd century CE, max L. 260 cm (Musée des Beaux-Arts et d'Archéologie de Vienne, France, inv. R.1998.2.43).

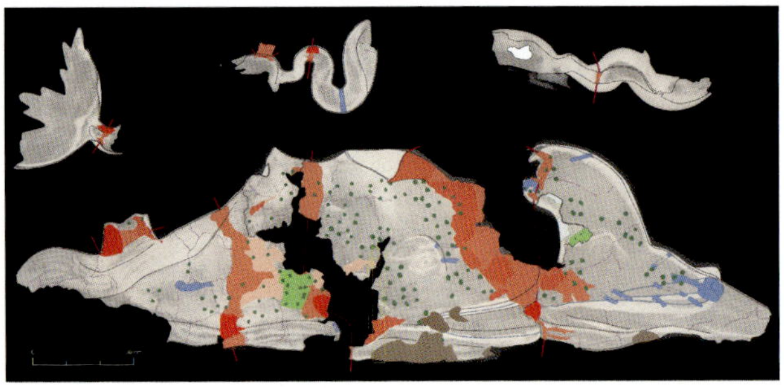

Figure 467 Annotated X-radiograph showing the technical features observed on the reverse (see fig. 367 for mapping of evidence visible on the front). *Great Dolphins of Vienne*, Roman period, Vienne, France, 2nd century CE, max L. 260 cm (Musée des Beaux-Arts et d'Archéologie de Vienne, France, inv. R.1998.2.43).

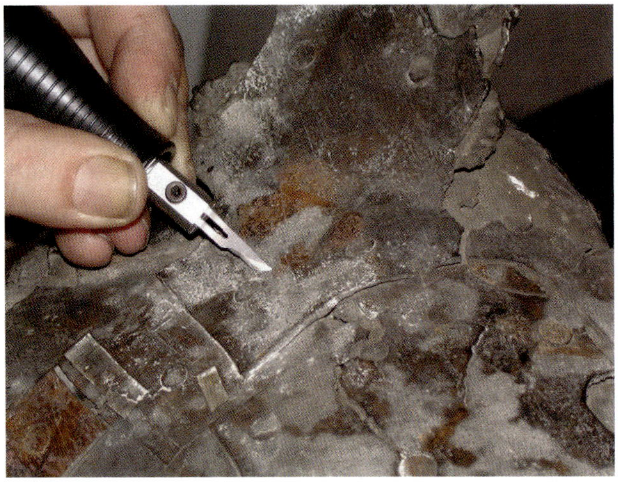

Figure 468 The body fragment was cleaned mechanically using an ultrasonic scalpel. Great *Dolphins of Vienne*, Roman period, Vienne, France, 2nd century CE, max L. 260 cm (Musée des Beaux-Arts et d'Archéologie de Vienne, France, inv. R.1998.2.43).

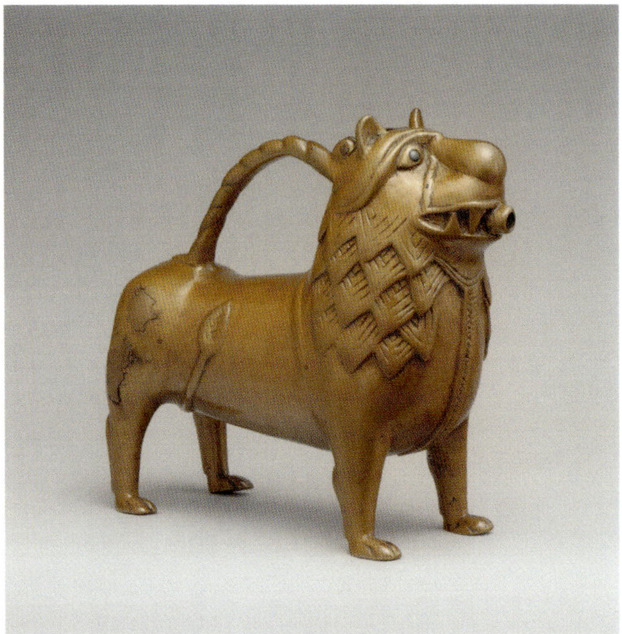

Figure 469 *Aquamanile in the Form of a Lion*, probably northern Germany, 12th century, H. 19.5 cm (Metropolitan Museum of Art, gift of Irwin Untermyer, 1964, inv. 64.101.1491).

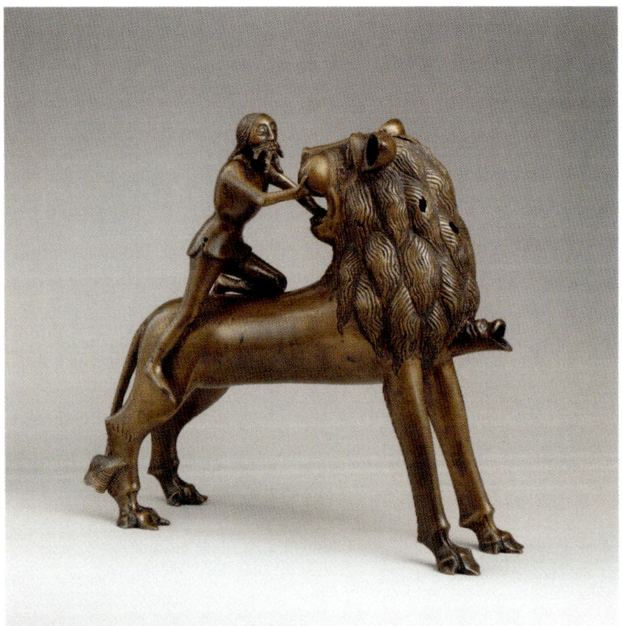

Figure 470 *Aquamanile in the Form of Samson and the Lion*, northern Europe (?), possibly Germany, ca. 1380–1400, H. 34.1 cm (Metropolitan Museum of Art, Robert Lehman Collection, 1975, inv. 1975.1.1412).

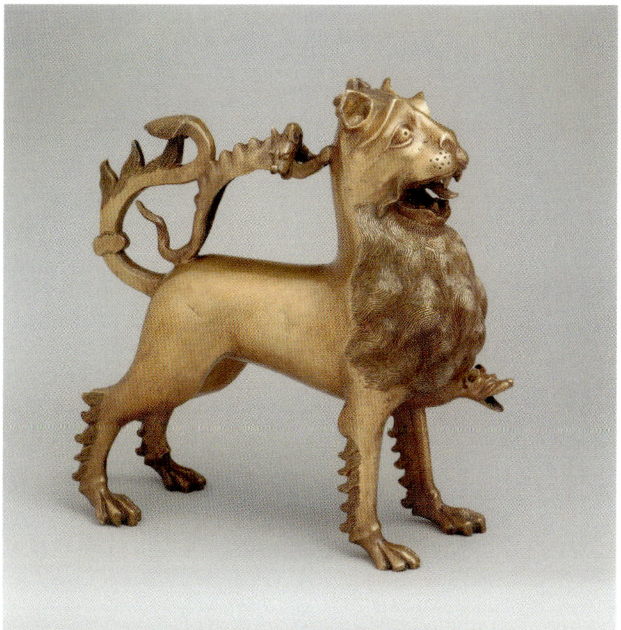

Figure 471 *Aquamanile in the Form of a Lion*, Nuremberg, Germany, ca. 1400, H. 31.9 cm (Metropolitan Museum of Art, The Cloisters Collection, 1994, inv. 1994.244).

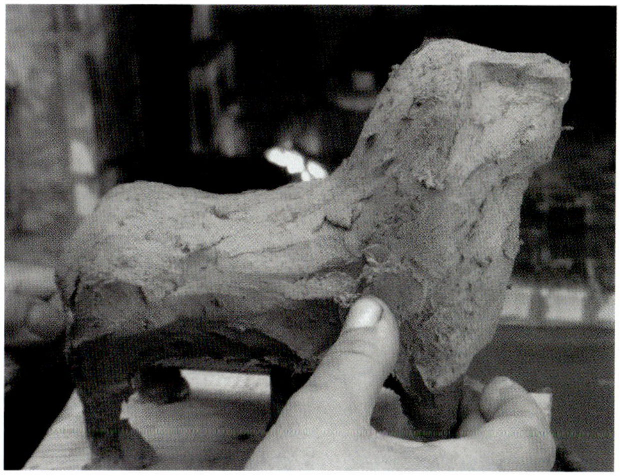

Figure 473 Modeling the clay core for an experimental casting. Ubaldo Vitali (American, b. 1944), Reproduction of a Lion Aquamanile, Maplewood, New Jersey, 2006, H. 19 cm, after *Aquamanile in the Form of a Lion*, probably northern Germany, 12th century, H. 19.5 cm (Metropolitan Museum of Art, gift of Irwin Untermyer, 1964, inv. 64.101.1491). See Barnet and Dandridge 2006.

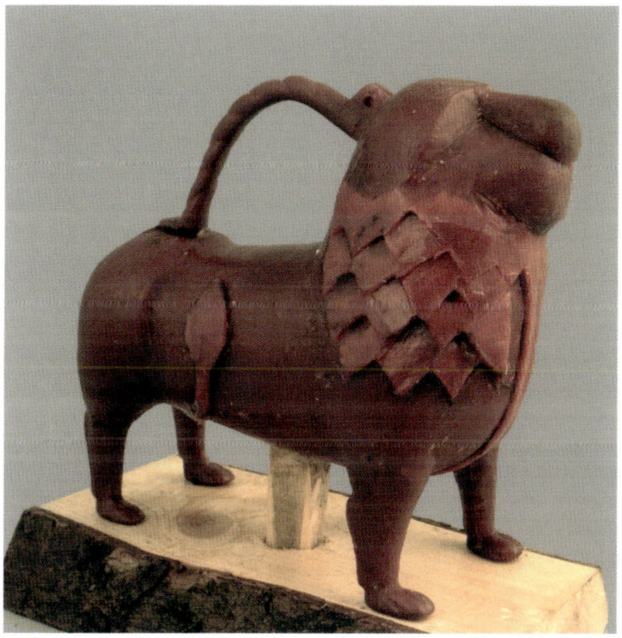

Figure 472 Partially fabricated wax model for an experimental casting. Additional wax is applied to the wax sheet in the reproduction for the modeling of the lion's mane. Ubaldo Vitali (American, b. 1944), Reproduction of a Lion Aquamanile, Maplewood, New Jersey, 2006, H. 19 cm, after *Aquamanile in the Form of a Lion*, probably northern Germany, 12th century, H. 19.5 cm (Metropolitan Museum of Art, gift of Irwin Untermyer, 1964, inv. 64.101.1491). See Dandridge 2006.

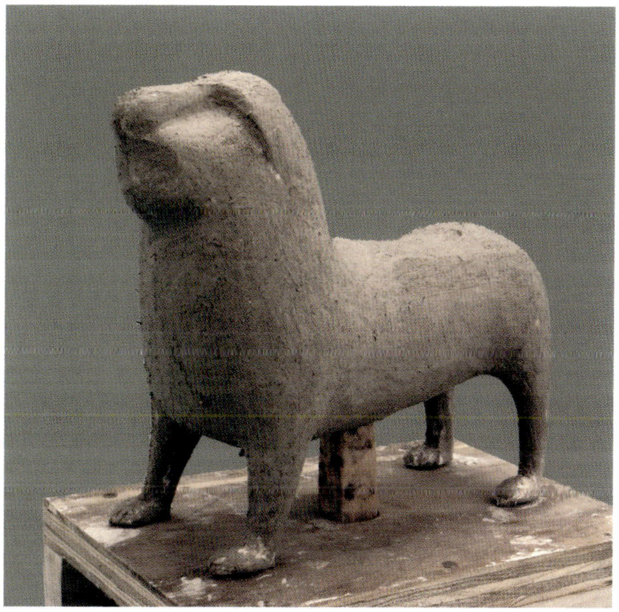

Figure 474 Completed core for an experimental casting. Ubaldo Vitali (American, b. 1944), Reproduction of a Lion Aquamanile, Maplewood, New Jersey, 2006, H. 19 cm, after *Aquamanile in the Form of a Lion*, probably northern Germany, 12th century, H. 19.5 cm (Metropolitan Museum of Art, gift of Irwin Untermyer, 1964, inv. 64.101.1491). See Dandridge 2006.

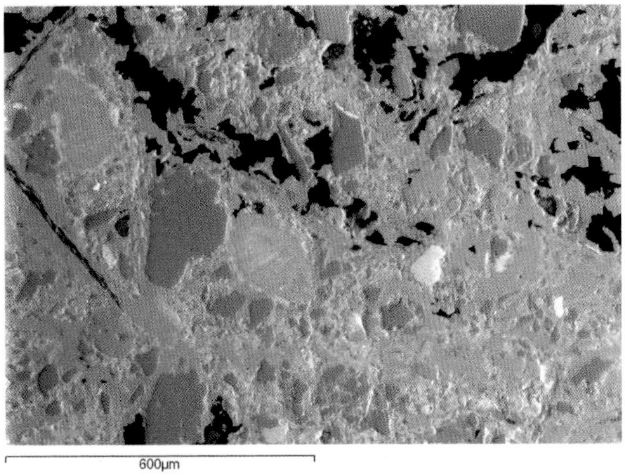

Figure 475 Backscattered electron micrograph of a polished cross section of a core sample. The large rounded grains are quartz and potassium feldspars, with the former being darker gray. The dark splintery shapes at the top left are pieces of organic material. *Aquamanile in the Form of a Lion*, probably northern Germany, ca. 1200, H. 21.2 cm (Metropolitan Museum of Art, The Cloisters Collection, 1947, inv. 47.101.52). See Dandridge 2006.

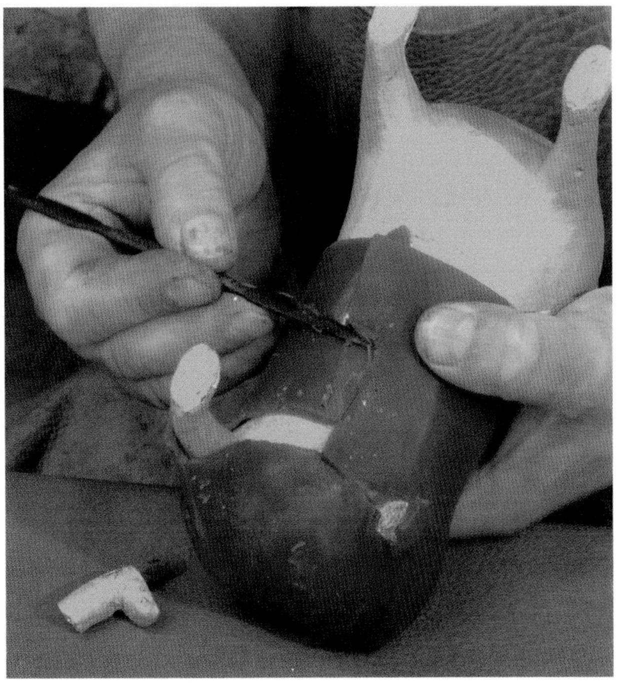

Figure 476 A wax sheet is cut and applied over the core with seams secured with a hot spatula. Ubaldo Vitali (American, b. 1944), Reproduction of a Lion Aquamanile, Maplewood, New Jersey, 2006, H. 19 cm, after *Aquamanile in the Form of a Lion*, probably northern Germany, 12th century, H. 19.5 cm (Metropolitan Museum of Art, gift of Irwin Untermyer, 1964, inv. 64.101.1491).

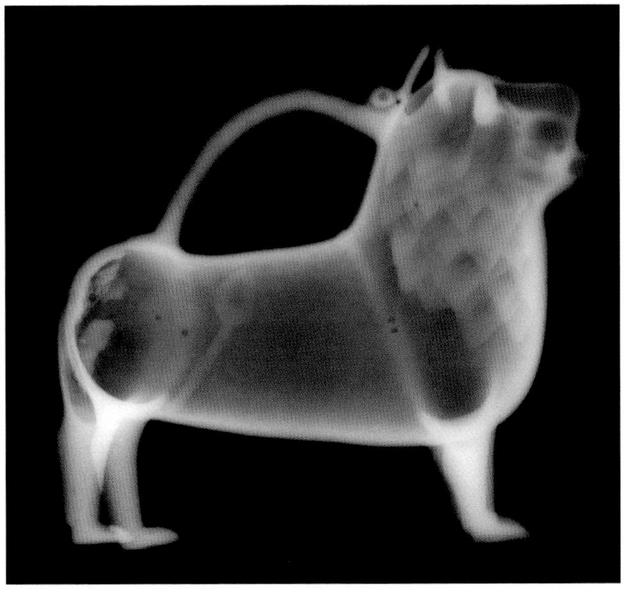

Figure 477 Radiograph. *Aquamanile in the Form of a Lion*, probably northern Germany, 12th century, H. 19.5 cm (Metropolitan Museum of Art, gift of Irwin Untermyer, 1964, inv. 64.101.1491). See Dandridge 2006.

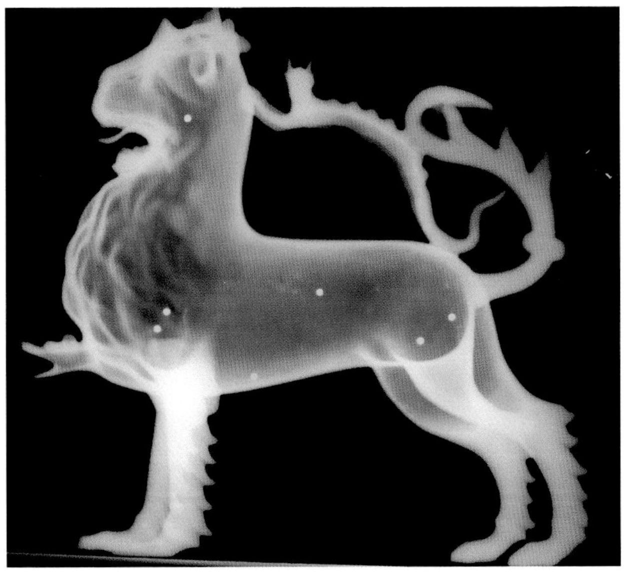

Figure 478 Radiograph confirming the use of an even layer of wax to form the lion's body and mane, as well as the location of the circular core plugs. *Aquamanile in the Form of a Lion*, Nuremberg, Germany, ca. 1400, H. 31.9 cm (Metropolitan Museum of Art, The Cloisters Collection, 1994, inv. 1994.244).

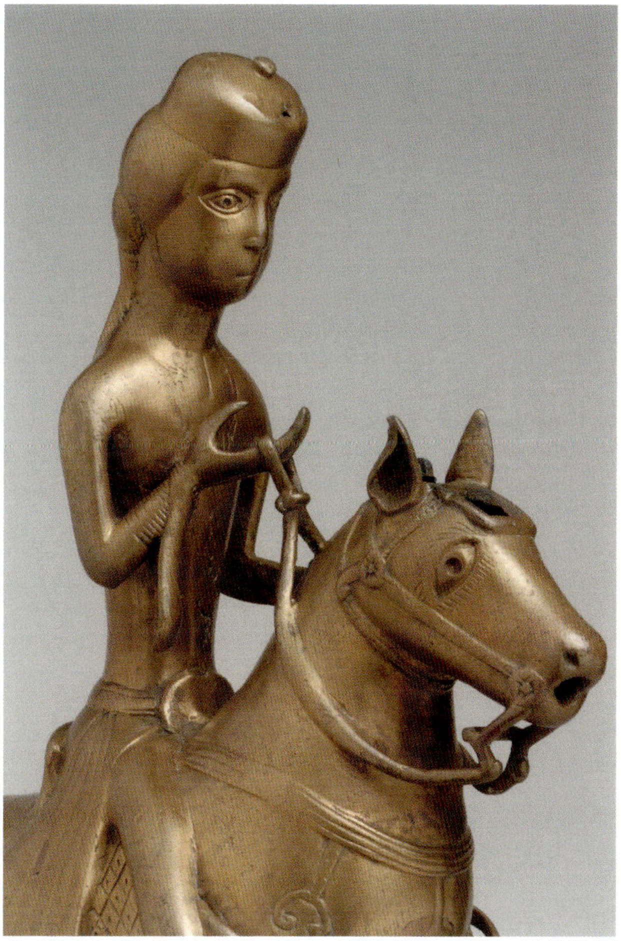

Figure 479 Detail. *Aquamanile in the Form of a Falconer on Horseback*, northern Germany, 13th century, H. 34.7 cm (Metropolitan Museum of Art, The Cloisters Collection, 1947, inv. 47.101.55). See Dandridge 2006.

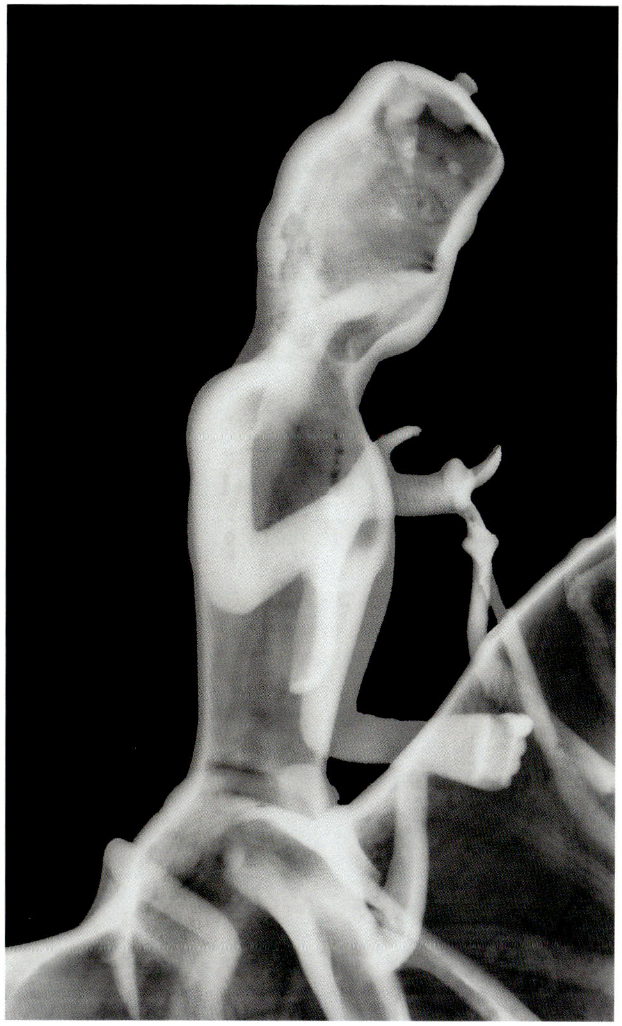

Figure 480 Detail from a radiograph of an aquamanile, showing the print for the original armature extending up through the center of the rider's torso and visible as a slightly less dense, darker channel within retained core material. *Aquamanile in the Form of a Falconer on Horseback*, northern Germany, 13th century, H. 34.7 cm (Metropolitan Museum of Art, The Cloisters Collection, 1947, inv. 47.101.55). See Dandridge 2006.

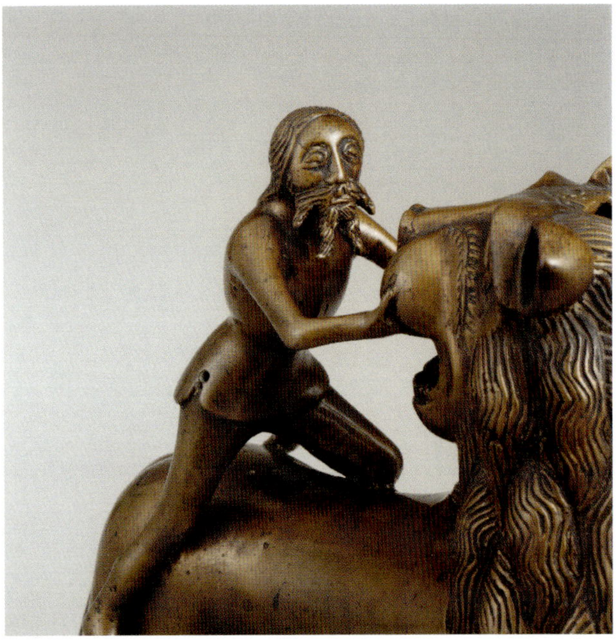

Figure 481 Detail. *Aquamanile in the Form of Samson and the Lion*, northern Europe (?), possibly Germany, ca. 1380–1400, H. 34.1 cm (Metropolitan Museum of Art, Robert Lehman Collection, 1975, inv. 1975.1.1412). See Dandridge 2006.

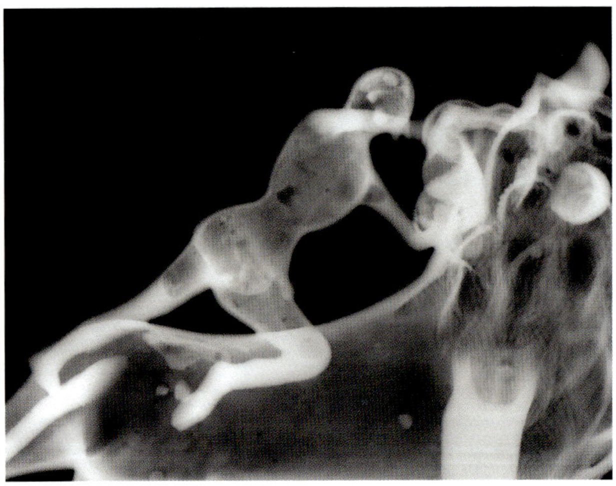

Figure 482 Detail from radiograph of an aquamanile showing that the form of Samson's core extends into his legs, arms, and head, and that the lion's mane was fully modeled in the core. *Aquamanile in the Form of Samson and the Lion*, northern Europe (?), possibly Germany, ca. 1380–1400, H. 34.1 cm (Metropolitan Museum of Art, Robert Lehman Collection, 1975, inv. 1975.1.1412). See Dandridge 2006.

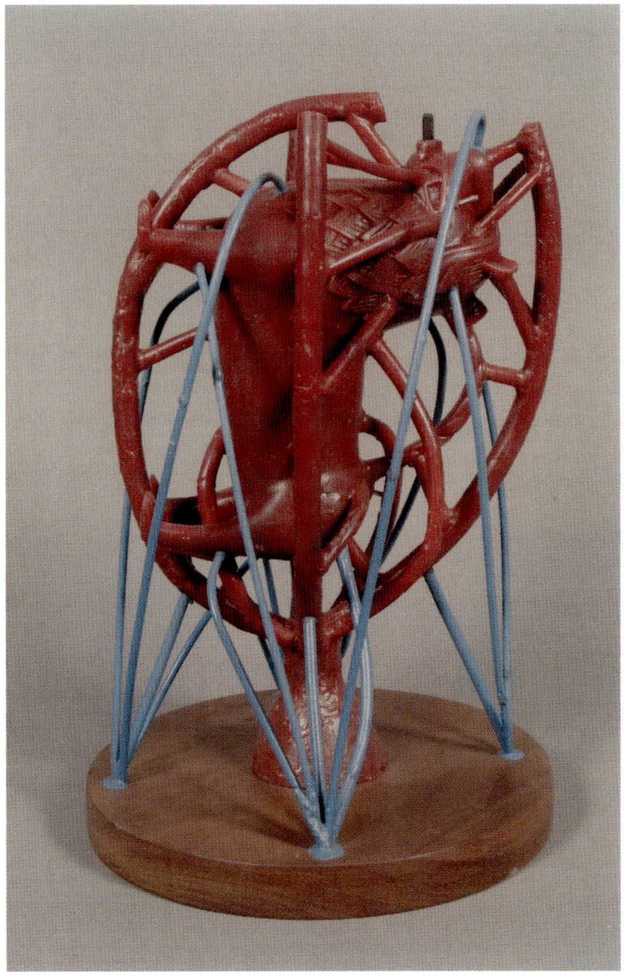

Figure 483 Fully sprued wax model for an experimental casting. Solid wax sprues and pouring cup that feed the metal through the mold are in red. Vents that allow gases to escape from the mold are in blue. Ubaldo Vitali (American, b. 1944), Reproduction of a Lion Aquamanile, Maplewood, New Jersey, 2006, H. 19 cm, after *Aquamanile in the Form of a Lion*, probably northern Germany, 12th century, H. 19.5 cm (Metropolitan Museum of Art, gift of Irwin Untermyer, 1964, inv. 64.101.1491).

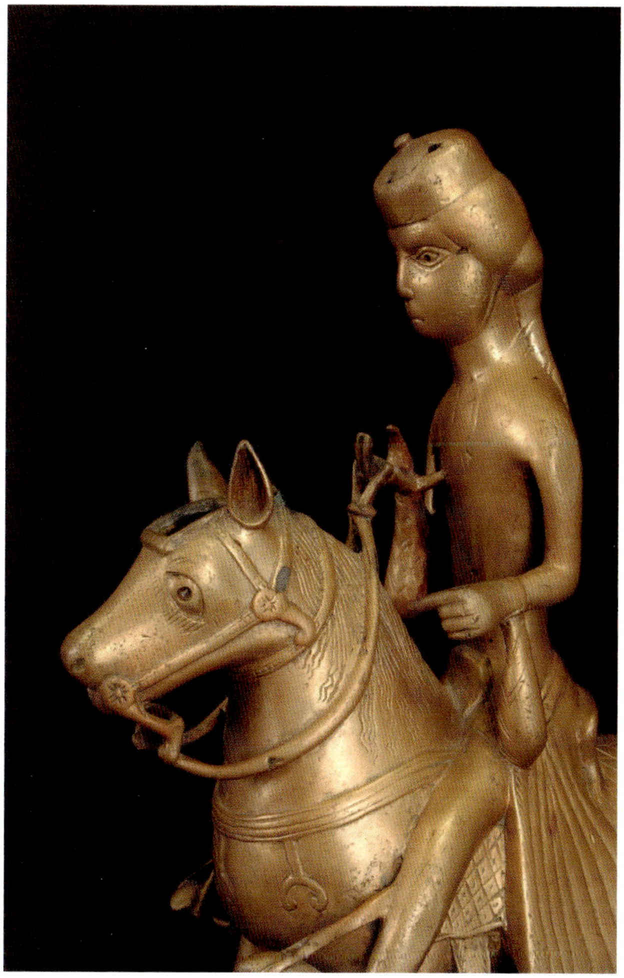

Figure 484 Detail from an aquamanile illustrating the rod-shaped strut extending out of the chest and supporting the falconer's right arm, as well as the exaggerated length of the glove flap on his opposite arm, enabling the circular flow of metal during the pour. *Aquamanile in the Form of a Falconer on Horseback*, northern Germany, 13th century, H. 34.7 cm (Metropolitan Museum of Art, The Cloisters Collection, 1947, inv. 47.101.55). See Dandridge 2006.

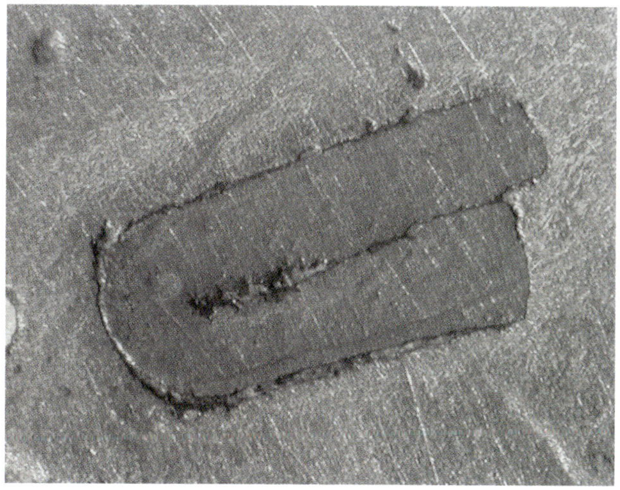

Figure 485 Detail showing a core pin fabricated from a folded piece of copper sheet. *Aquamanile in the Form of a Lion*, probably northern Germany, ca. 1200, H. 21.2 cm (Metropolitan Museum of Art, The Cloisters Collection, 1947, inv. 47.101.52). See Dandridge 2006.

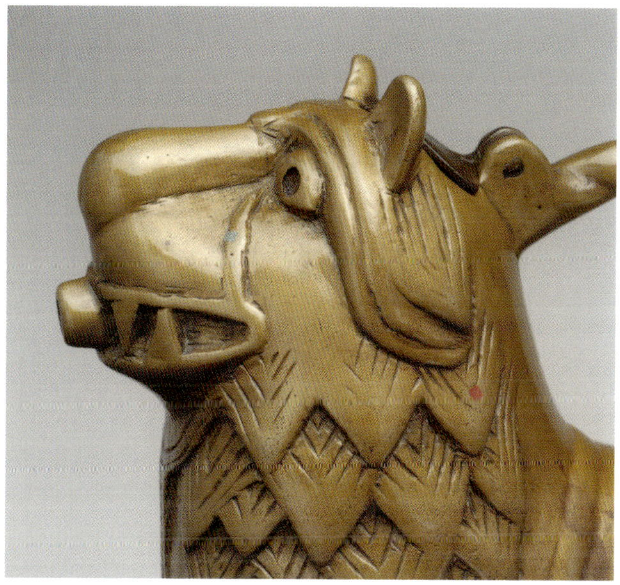

Figure 486 Detail showing both iron and copper core pins in the lion's head. *Aquamanile in the Form of a Lion*, probably northern Germany, 12th century, H. 19.5 cm (Metropolitan Museum of Art, gift of Irwin Untermyer, 1964, inv. 64.101.1491). See Dandridge 2006.

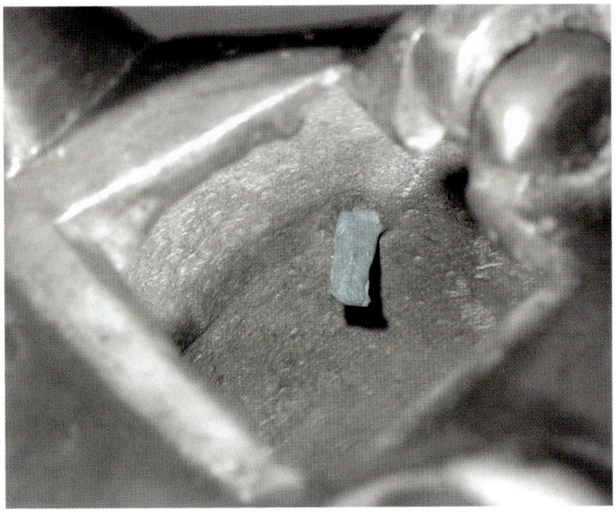

Figure 487 Interior detail showing the protruding tip of a surviving core pin that was fused to the surrounding metal. *Aquamanile in the Form of a Lion*, Germany (Lower Saxony), late 13th–early 14th century, H. 21.0 cm (Metropolitan Museum of Art, gift of Irwin Untermyer, 1964, inv. 64.101.1490). See Dandridge 2006.

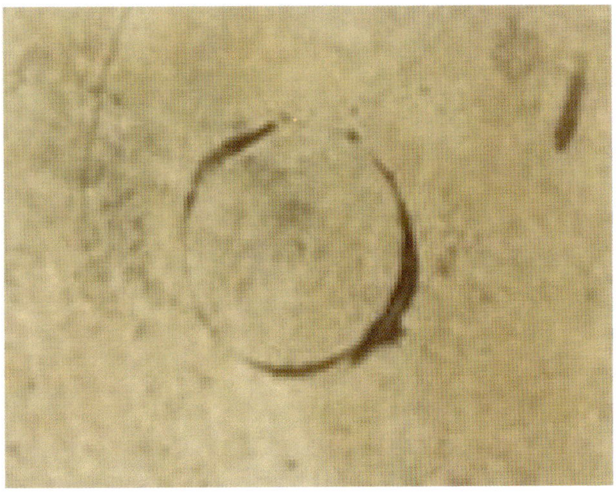

Figure 488 Detail of circular disc inserted as a core plug and secured with solder. *Aquamanile in the Form of a Lion*, Nuremberg, Germany, ca. 1400, H. 31.9 cm (Metropolitan Museum of Art, The Cloisters Collection, 1994, inv. 1994.244). See Dandridge 2006.

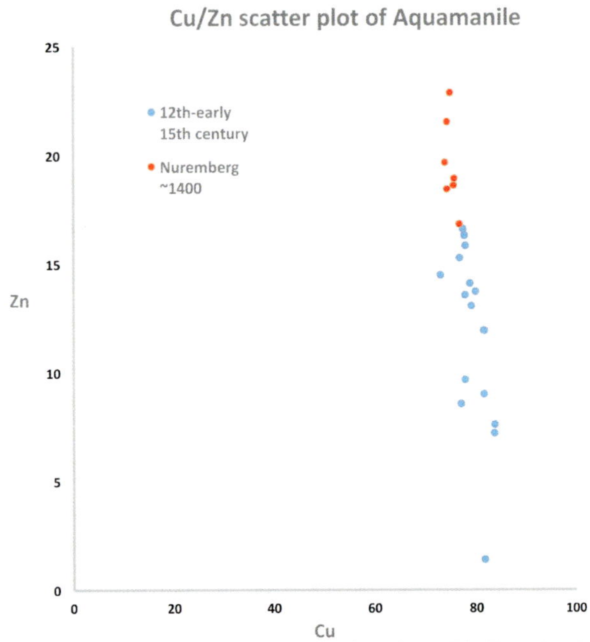

Figure 489 This scatter plot of alloys used in the study group analyzed by inductively coupled plasma-mass spectrometry (ICP-MS) illustrates an increased concentration of zinc in aquamanilia produced in Nuremberg at the beginning of the 15th century. See Dandridge 2006.

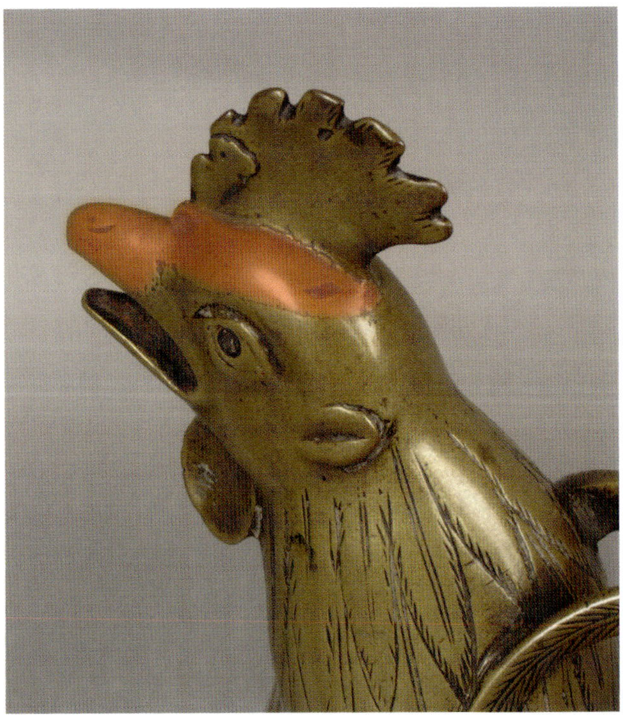

Figure 490 Detail showing the cast-in repair on the top of the animal's head (see red overlay). *Aquamanile in the Form of a Rooster*, Germany (Lower Saxony), 13th century, H. 25.2 cm (Metropolitan Museum of Art, The Cloisters Collection, 1989, inv. 1989.292). See Barnet and Dandridge 2006.

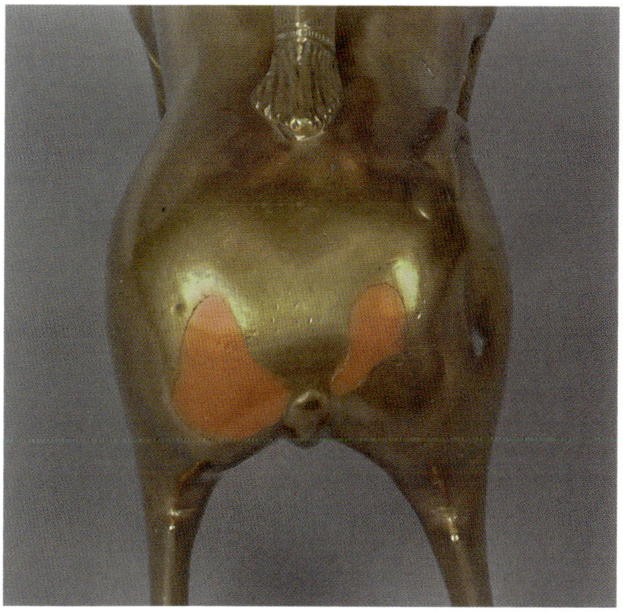

Figure 491 Detail showing the cast-in repairs across the animal's rump, also visible in the radiograph in fig. 180. *Aquamanile in the Form of a Lion*, probably northern Germany, ca. 1200, H. 21.2 cm (Metropolitan Museum of Art, The Cloisters Collection, 1947, inv. 47.101.52). See Dandridge 2006.

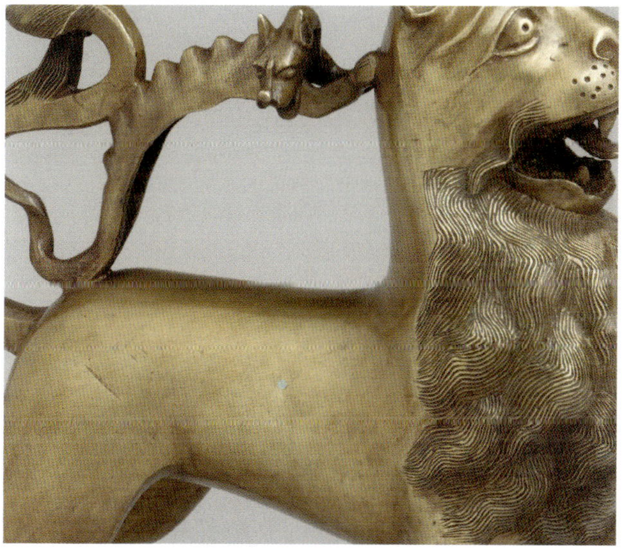

Figure 492 Detail of side showing the location of the core plug illustrated in fig. 488. *Aquamanile in the Form of a Lion*, Nuremberg, Germany, ca. 1400, H. 31.9 cm (Metropolitan Museum of Art, The Cloisters Collection, 1994, inv. 1994.244). See Barnet and Dandridge 2006.

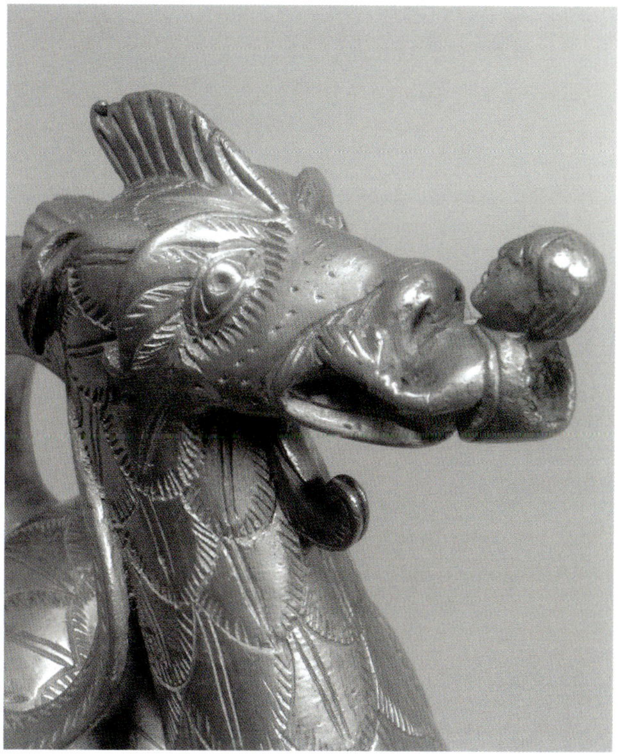

Figure 493 Detail illustrating the use of a spade drill to clarify the iris and a V-shaped graver to delineate the edges of the feathers, other details, and the punch work around the snout of a dragon. The faint parallel marks across the neck are indications of the use of a scraper to help smooth the surface. *Aquamanile in the Form of a Dragon*, northern Germany, ca. 1200, H. 21.2 cm (Metropolitan Museum of Art, The Cloisters Collection, 1947, inv. 47.101.51). See Barnet and Dandridge 2006.

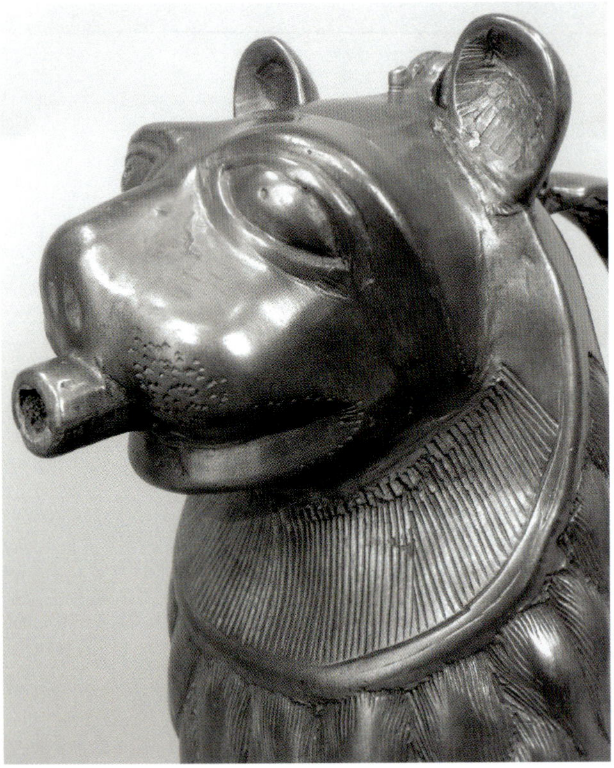

Figure 494 Detail of head illustrating where a V-shaped graver defined the lion's mane, a flat chisel refined the ear, and punch marks made whisker holes in the cheek. *Aquamanile in the Form of a Lion*, Germany (Lower Saxony), late 13th–early 14th century, H. 26.1 cm (Metropolitan Museum of Art, The Friedsam Collection, bequest of Michael Friedsam, 1931, inv. 32.100.198). See Barnet and Dandridge 2006.

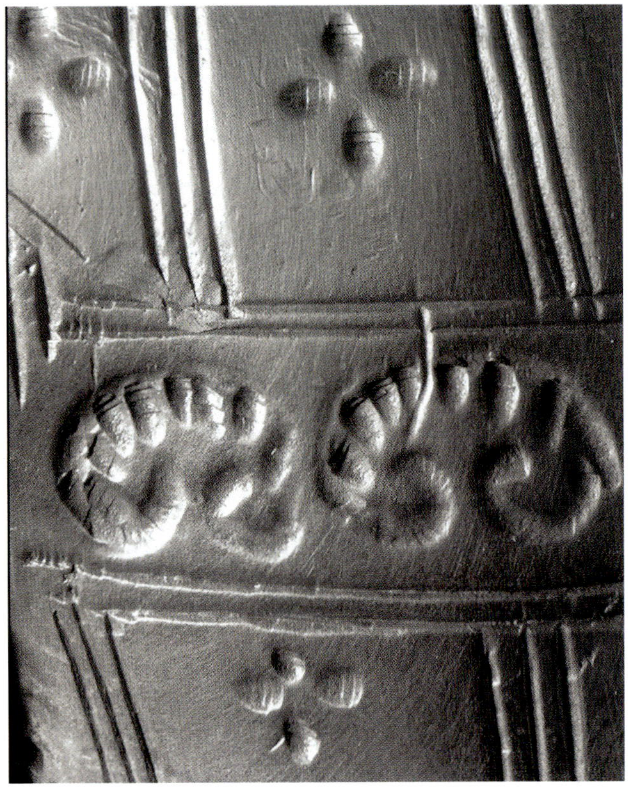

Figure 495 Detail from centaur's chest showing the use of a scorper to define U-shaped designs with the series of steps within each cut corresponding to the strike of the mallet or hammer. *Aquamanile in the Form of a Crowned Centaur Fighting a Dragon*, possibly Hildesheim, Lower Saxony, Germany, 1200–1225, H. 36.5 cm (Metropolitan Museum of Art, Rogers Fund, 1910, inv. 10.37.2). See Dandridge 2006.

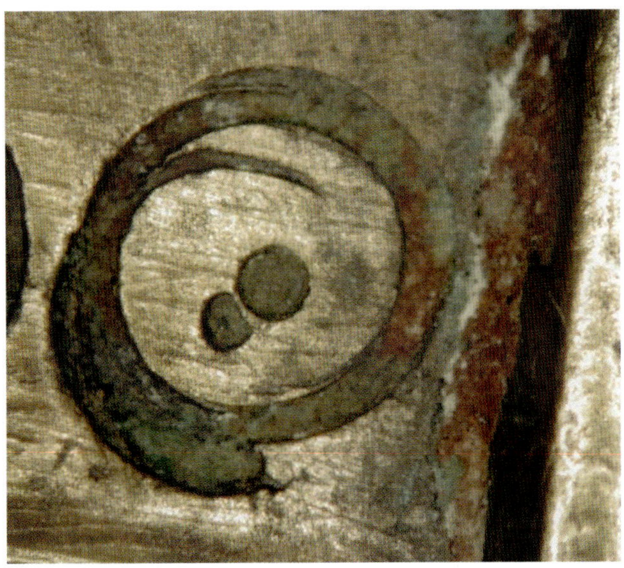

Figure 496 Detail showing double strike of a circle and dot punch used to ornament the belt. *Aquamanile in the Form of Phyllis and Aristotle*, southern Netherlands, late 14th or early 15th century, H. 32.5 cm (Metropolitan Museum of Art, Robert Lehman Collection, 1975, inv. 1975.1.1416). See Barnet and Dandridge 2006.

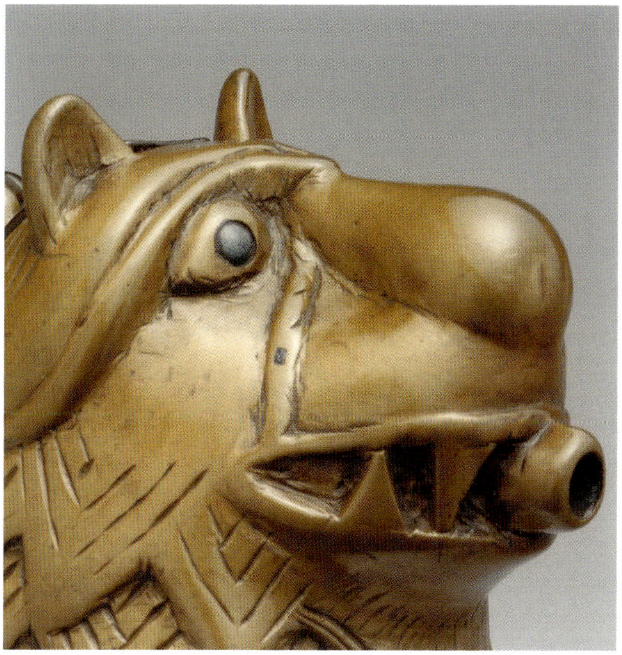

Figure 497 Detail showing inlaid eye made of colored glass. *Aquamanile in the Form of a Lion*, probably northern Germany, 12th century, H. 19.5 cm (Metropolitan Museum of Art, gift of Irwin Untermyer, 1964, inv. 64.101.1491). See Barnet and Dandridge 2006.

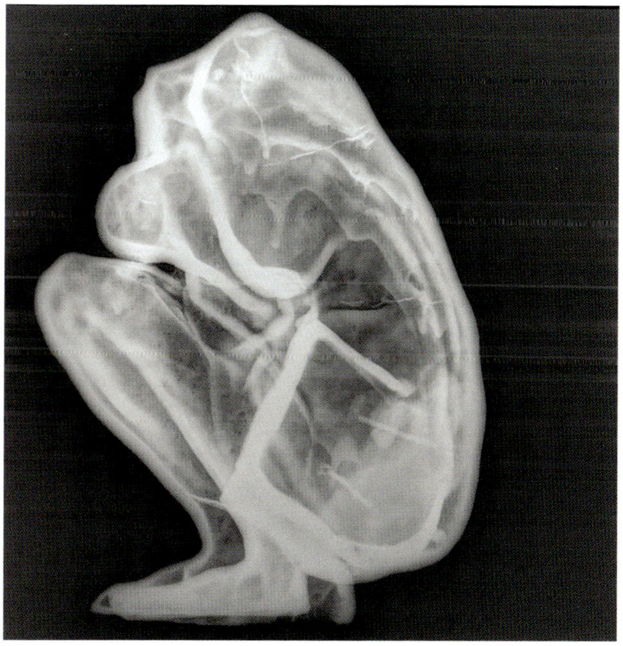

Figure 498 A digital (DR) 360° radiograph (viewable in online edition). Camille Claudel (French, 1864–1943), *Torso of a Crouching Woman*, model about 1884–85, cast by 1913, H. 35 cm (J. Paul Getty Museum, inv. 2018.32).

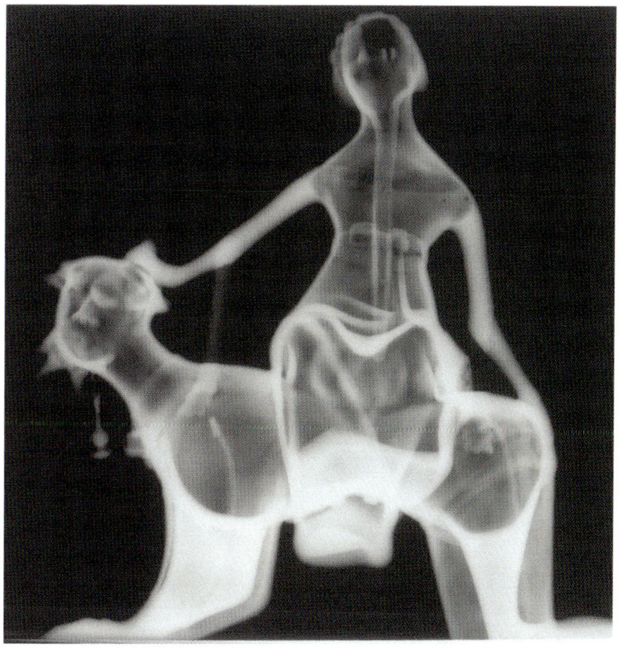

Figure 499 Radiograph showing the integration of square-sectioned iron armature elements extending up out of Aristotle's arm and leg. The sharp bend at their tops suggests a point of overlap with additional elements and a potentially different technical approach to the fabrication of armatures. *Aquamanile in the Form of Phyllis and Aristotle*, southern Netherlands, late 14th or early 15th century, H. 32.5 cm (Metropolitan Museum of Art, Robert Lehman Collection, 1975, inv. 1975.1.1416). See Barnet and Dandridge 2006.

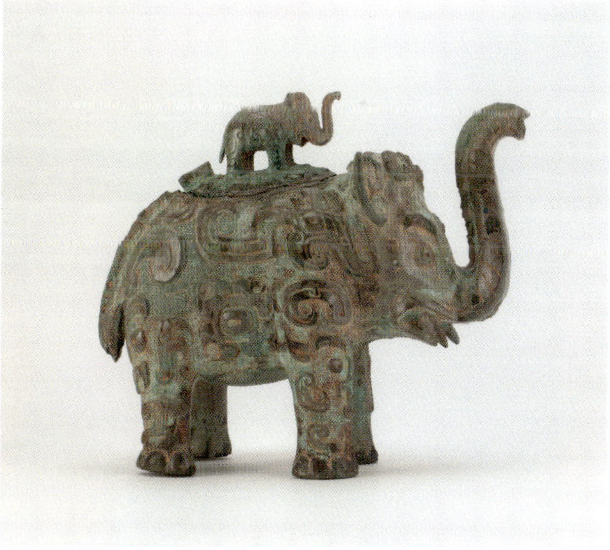

Figure 500 Side view. *Huo Ritual Vessel*, 11th century BCE, Middle Yangtze Valley, China, H. 17.2 cm (Freer Gallery of Art, Smithsonian Institution, Washington, DC, purchase—Charles Lang Freer Endowment, inv. F1936.6a–b / Department of Conservation and Scientific Research).

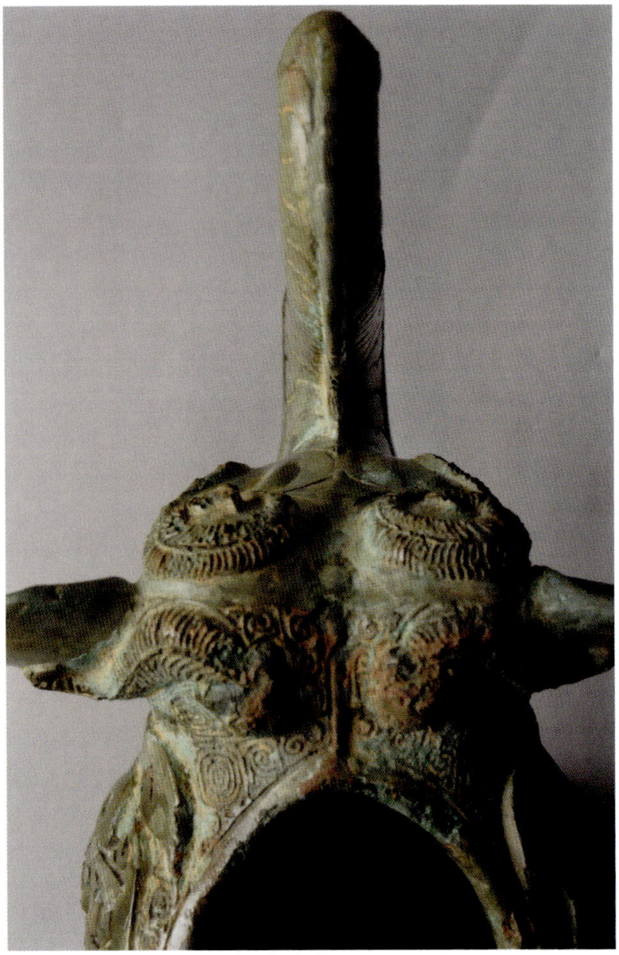

Figure 501 Detail of top, showing seam lines. *Huo Ritual Vessel*, 11th century BCE, Middle Yangtze Valley, China, H. 17.2 cm (Freer Gallery of Art, Smithsonian Institution, Washington, DC, purchase—Charles Lang Freer Endowment, inv. F1936.6a–b / Department of Conservation and Scientific Research).

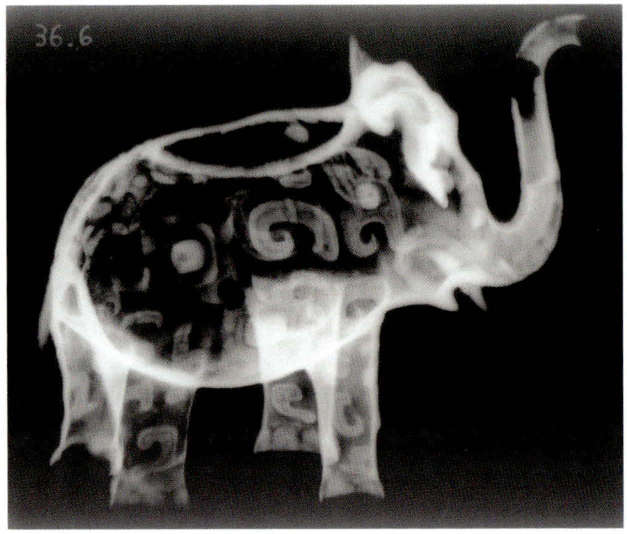

Figure 502 X-radiograph from the side without the lid. *Huo Ritual Vessel*, 11th century BCE, Middle Yangtze Valley, China, H. 17.2 cm (Freer Gallery of Art, Smithsonian Institution, Washington, DC, purchase—Charles Lang Freer Endowment, inv. F1936.6a–b / Department of Conservation and Scientific Research).

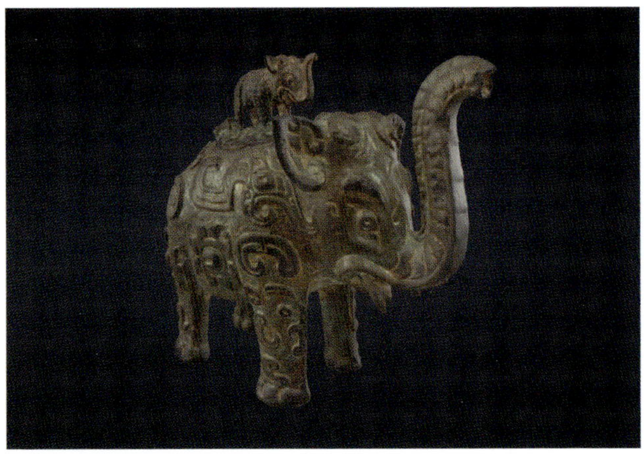

Figure 503 3D scan. *Huo Ritual Vessel*, 11th century BCE, Middle Yangtze Valley, China, H. 17.2 cm (Freer Gallery of Art, Smithsonian Institution, Washington, DC, purchase—Charles Lang Freer Endowment, inv. F1936.6a–b / Department of Conservation and Scientific Research).

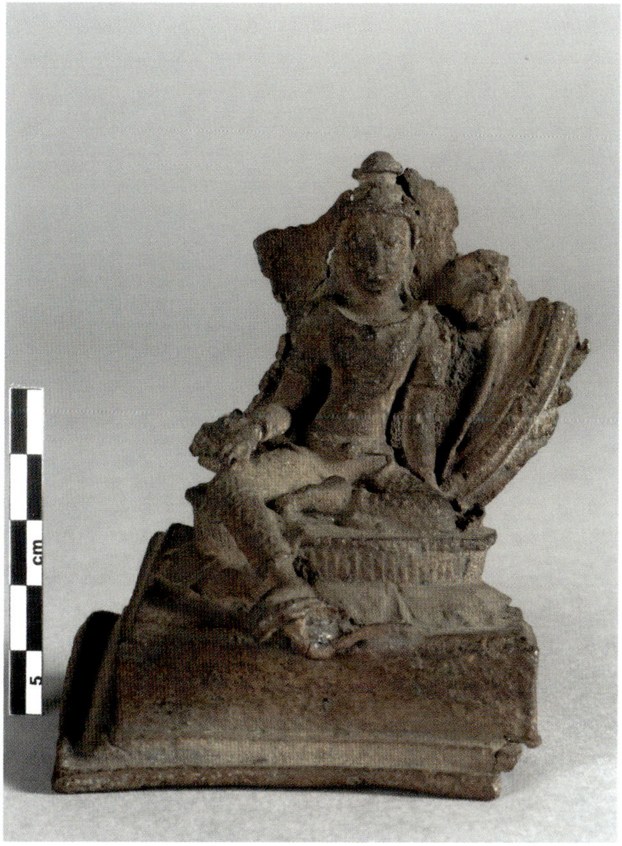

Figure 504 *Statuette of Bodhisattva*, style associated with Bangladesh but found in Java, 8th–early 9th century, H. 8.5 cm (Musée national des arts asiatiques – Guimet, Paris, inv. MA 507). See C2RMF Internal Report 2021.

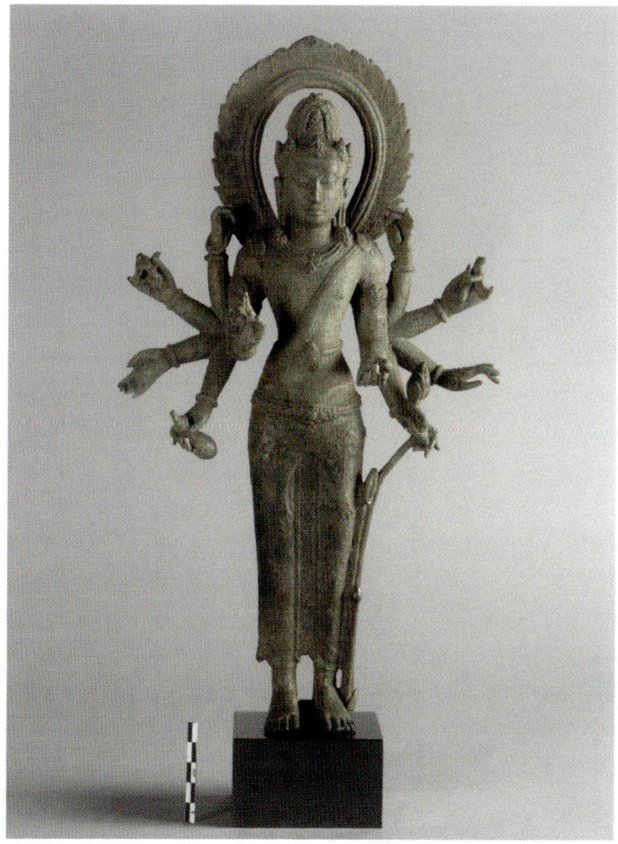

Figure 505 Front view. *Ten-Armed Avalokiteśvara*, Java, first half of the 9th century, H. 34 cm (Musée national des arts asiatiques – Guimet, Paris, inv. MG 3816). See Mechling et al. 2018.

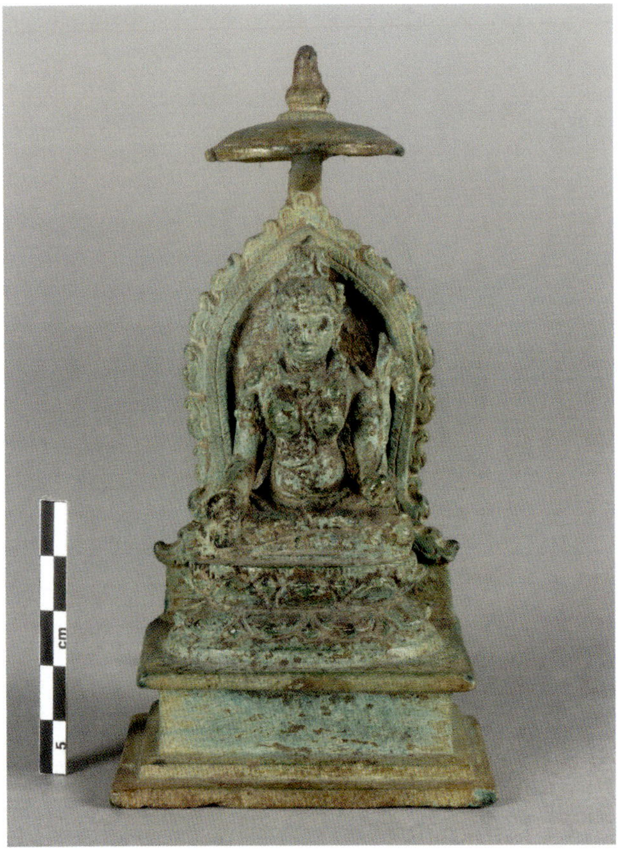

Figure 506 *Statuette of Vasudhārā*, style associated with Bangladesh but found in Java, second half of the 9th century, H. 13 cm (Musée national des arts asiatiques - Guimet, Paris, inv. MG 2255). See Mechling et al. 2018.

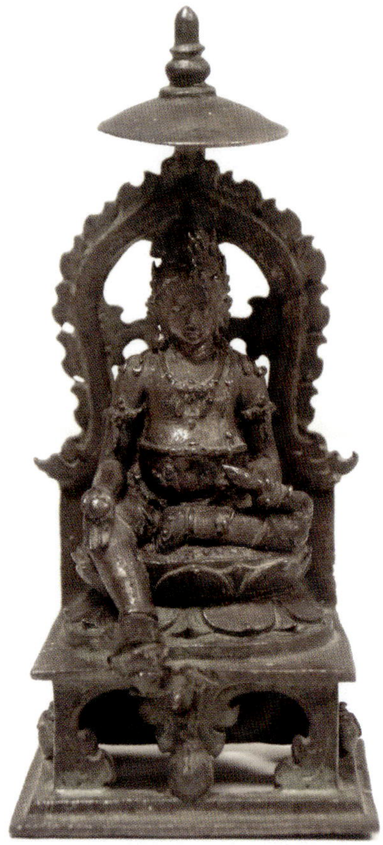

Figure 507 *Kubera/Jambhala*, Java, ca. late 9th–early 10th century, H. 17 cm (Musée national des arts asiatiques - Guimet, Paris, inv. MG 3619). See Mechling et al. 2018.

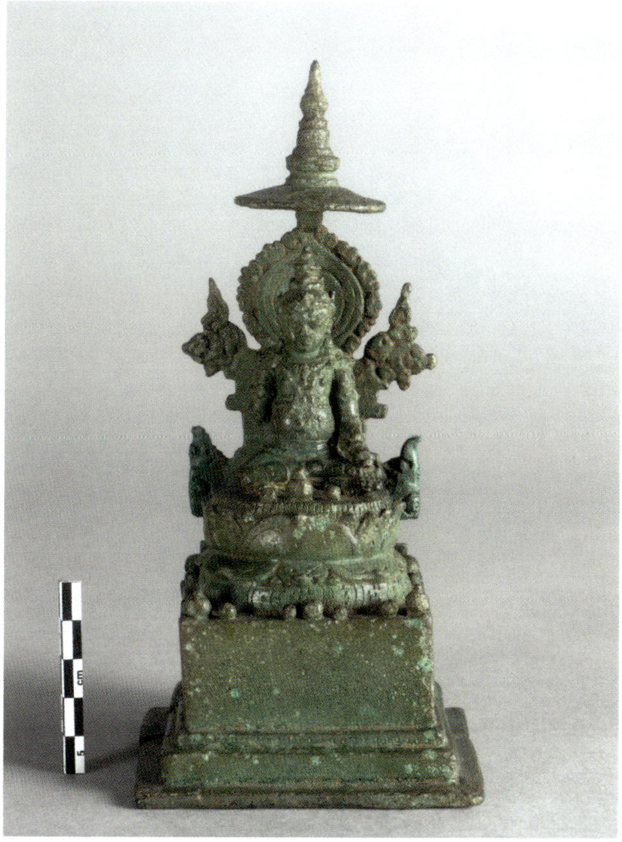

Figure 508 Front view. *Kubera/Jambhala*, Java, first half of the 10th century, H. 18 cm (Musée national des arts asiatiques - Guimet, Paris, inv. MG 3625). See Mechling et al. 2018.

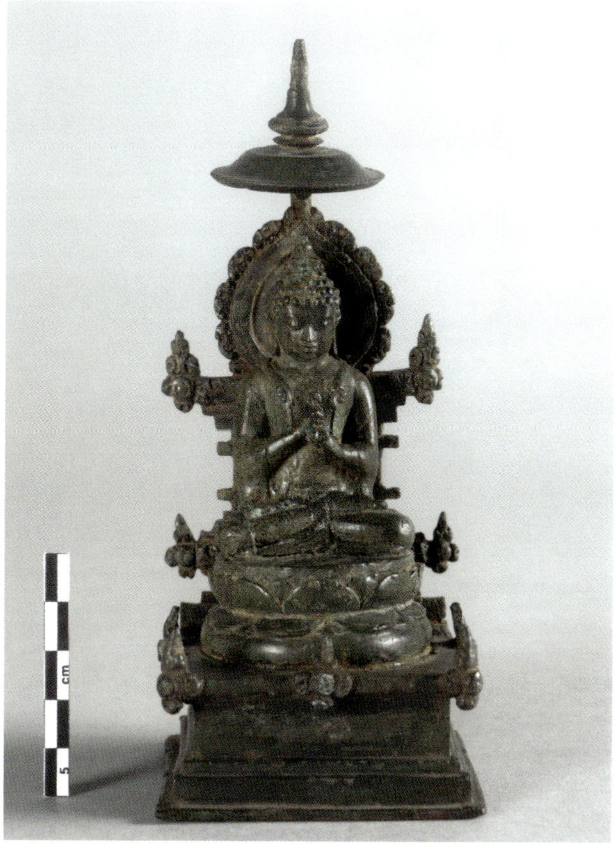

Figure 509 Front view. *Vairocana*, Java, first half of the 10th century, H. 15.5 cm (Musée national des arts asiatiques - Guimet, Paris, inv. MG 18290). See Mechling et al. 2018.

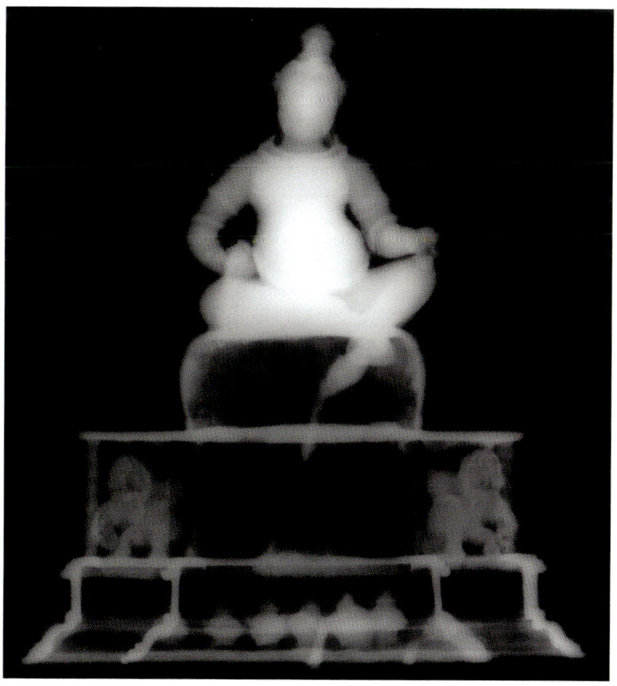

Figure 510 X-radiograph of frontal view. *Statuette of Jambhala*, central Java, first half of the 9th century, H. 28 cm (Musée national des arts asiatiques - Guimet, Paris, inv. MG 3814). See Mechling et al. 2018.

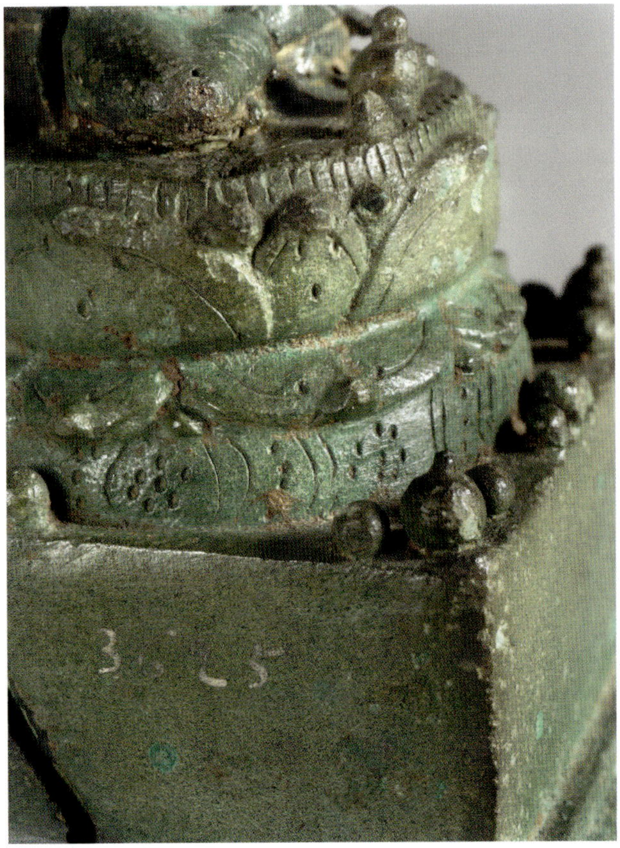

Figure 511 Detail showing engraving and punch marks. *Kubera/Jambhala*, Java, first half of the 10th century, H. 18 cm (Musée national des arts asiatiques - Guimet, Paris, inv. MG 3625). See Mechling et al. 2018.

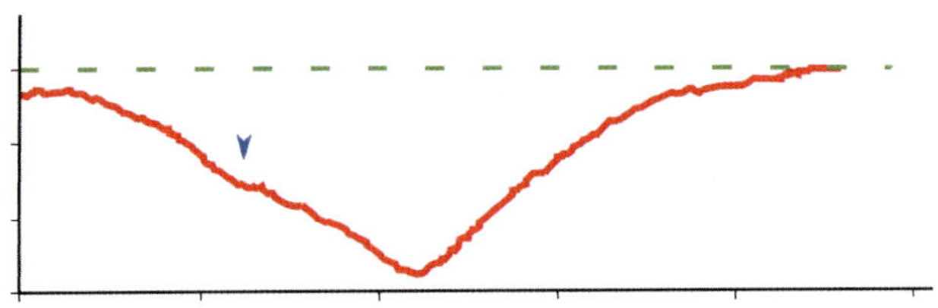

Figure 512 Profile of an engraving made in the metal on the front piece of the cloth of the *Statuette of Jambhala*, as measured by digital microscopy. The vertical axis reports the depth and the horizontal axis the width of the engraving (both in µm). The surface of the bronze is represented by the horizontal green dotted line. The cross section of the engraving and its dimensions are represented by the red curve. The maximum depth appears to be ~0.6 mm (600 µm), and the width at the surface is ~2 mm (2000 µm). See the altered V shape of the imprint (blue arrow), which demonstrates that the engraving in the metal was made by following a line already in the wax model. *Statuette of Jambhala*, central Java, first half of the 9th century, H. 28 cm (Musée national des arts asiatiques - Guimet, Paris, inv. MG 3814). See Mechling et al. 2018.

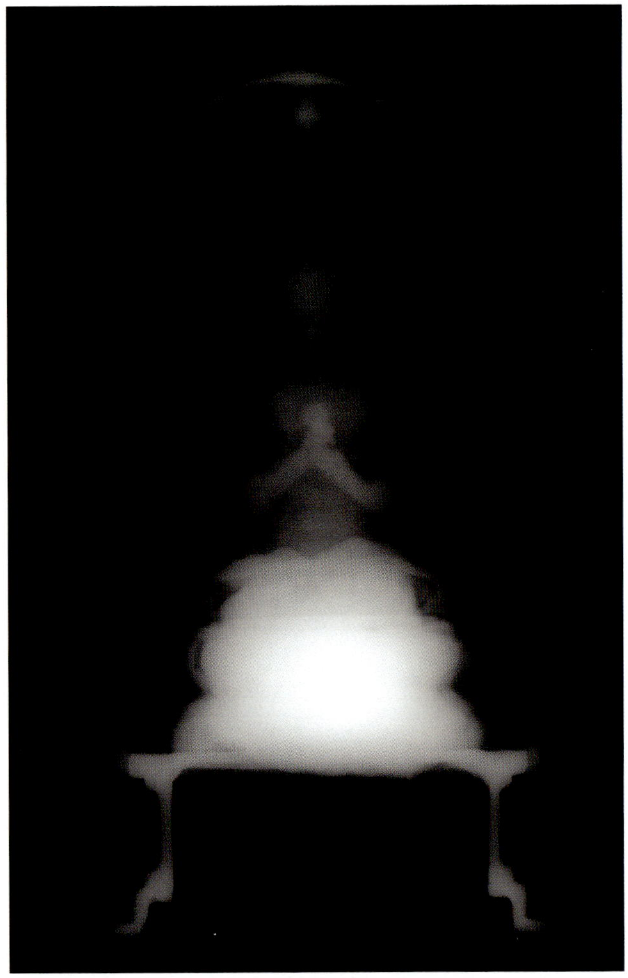

Figure 513 X-radiograph of frontal view. *Vairocana*, Java, first half of the 10th century, H. 15.5 cm (Musée national des arts asiatiques - Guimet, Paris, inv. MG 18290) See Mechling et al. 2018.

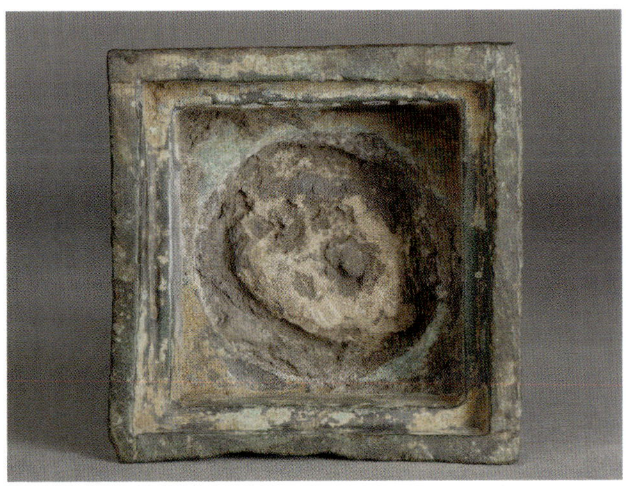

Figure 514 Bottom view showing areas of the preserved conservation deposits poking out of the metal fills. *Vairocana*, Java, first half of the 10th century, H. 15.5 cm (Musée national des arts asiatiques - Guimet, Paris, inv. MG 18290). See Mechling et al. 2018.

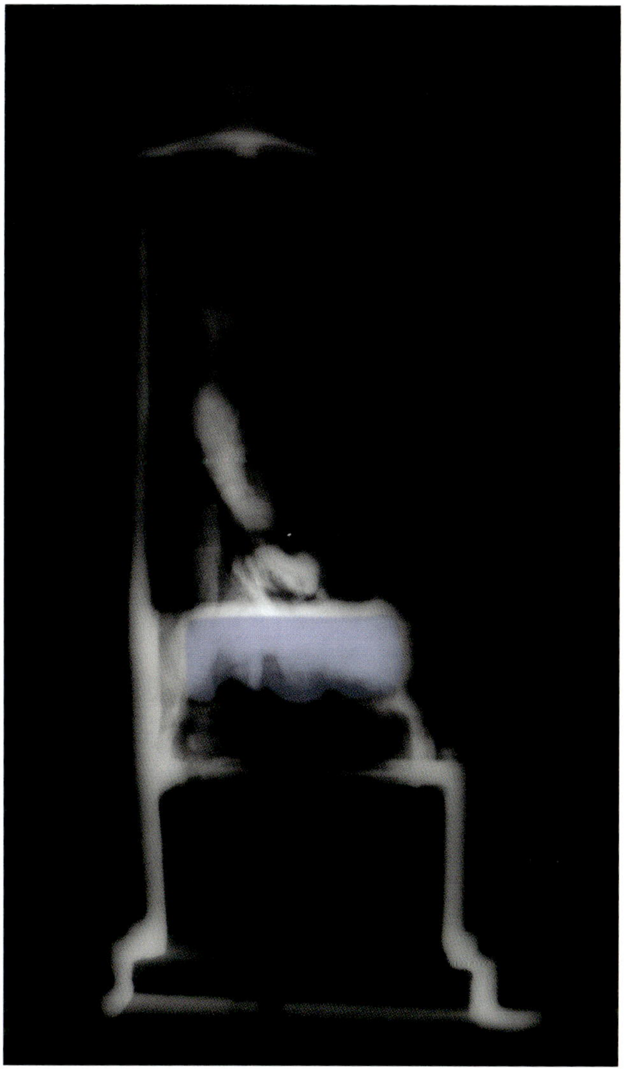

Figure 515 X-radiograph of side view. *Kubera/Jambhala*, Java, first half of the 10th century, H. 18 cm (Musée national des arts asiatiques – Guimet, Paris, inv. MG 3625). See Mechling et al. 2018.

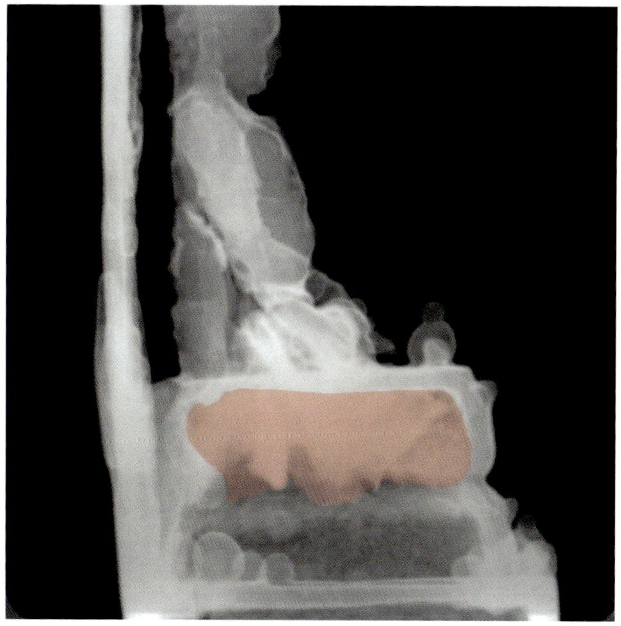

Figure 516 Annotated neutron radiograph of side view, showing more clearly than the X-radiograph (fig. 515) the contours of the sealing material (red overlay) and some of the consecration offerings. For an even better visualization of the consecration offerings, see tomographic images of a related figure (figs. 403, 404). *Kubera/Jambhala*, Java, first half of the 10th century, H. 18 cm (Musée national des arts asiatiques – Guimet, Paris, inv. MG 3625). See Mechling et al. 2018.

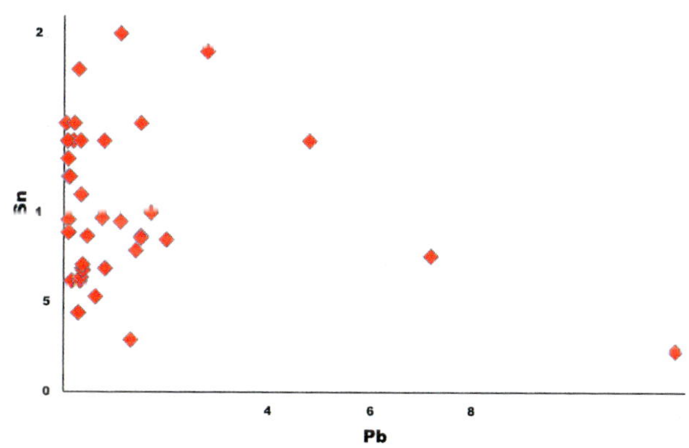

Figure 517 Tin and lead content of the alloys of the Javanese statuettes from the Musée national des arts asiatiques – Guimet, Paris. One dot represents one statuette (wt%, ICP-AES analysis on drillings). See Mechling et al. 2018.

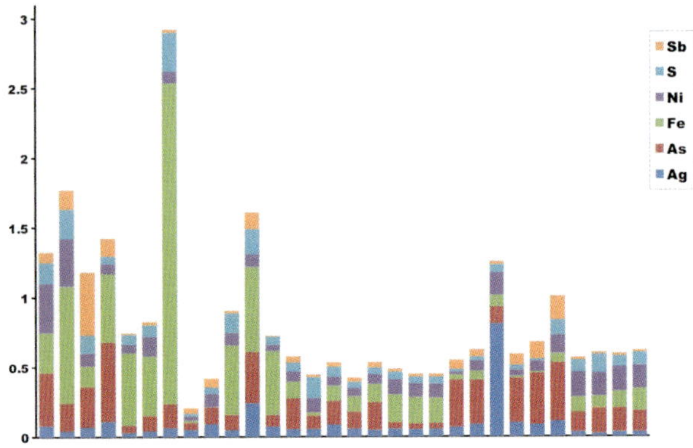

Figure 518 Main impurities in the metal of the Javanese statuettes from the Musée national des arts asiatiques - Guimet, Paris. One bar represents one statuette, sorted by chronological group (wt%, ICP-AES analysis on drillings). See Mechling et al. 2018.

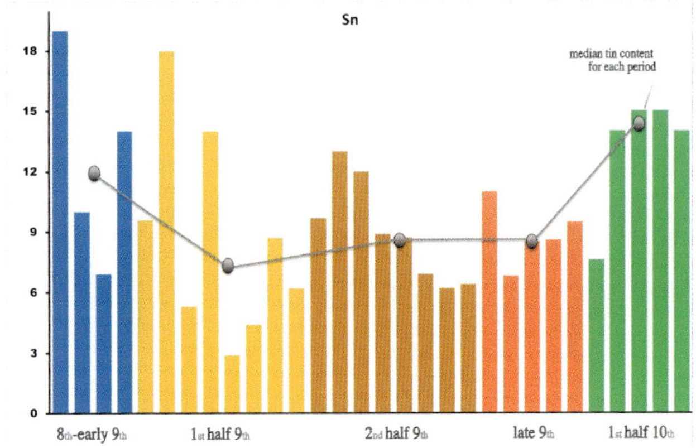

Figure 519 Tin content in the metal of Javanese statuettes from the Musée national des arts asiatiques - Guimet, Paris. One bar represents one statuette, sorted by chronological group (wt%, ICP-AES analysis on drillings). See Mechling et al. 2018.

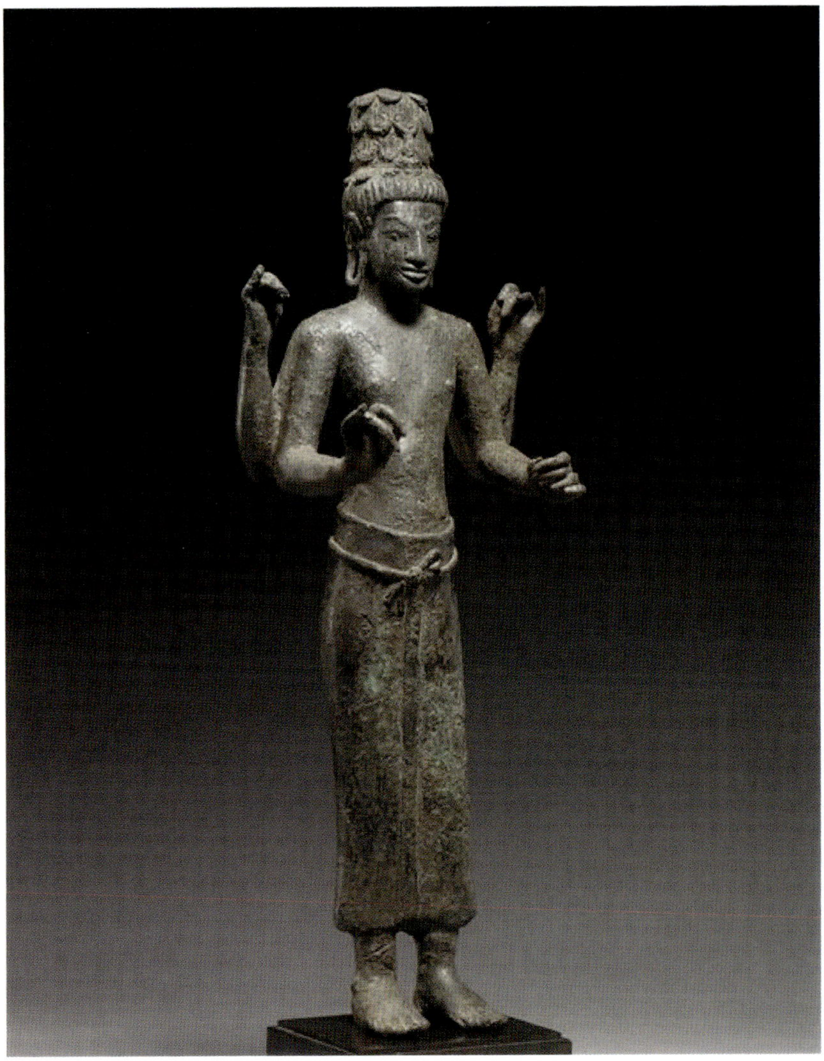

Figure 520 Example of Thai statuette cast in an unleaded tin bronze (11wt% Sn, 0.6wt% Pb, ICP-AES analyses) with a surface enriched in tin by depletion (~20wt% Sn, PIXE analysis). *Maitreya*, found in the Prasat Hin Khao Plai Bat II cache, northeast Thailand, late 7th–first half of the 9th century, H. 46 cm (Musée national des arts asiatiques - Guimet, Paris, inv. MA 3321). See Bourgarit et al. 2003; Mechling et al. 2018.

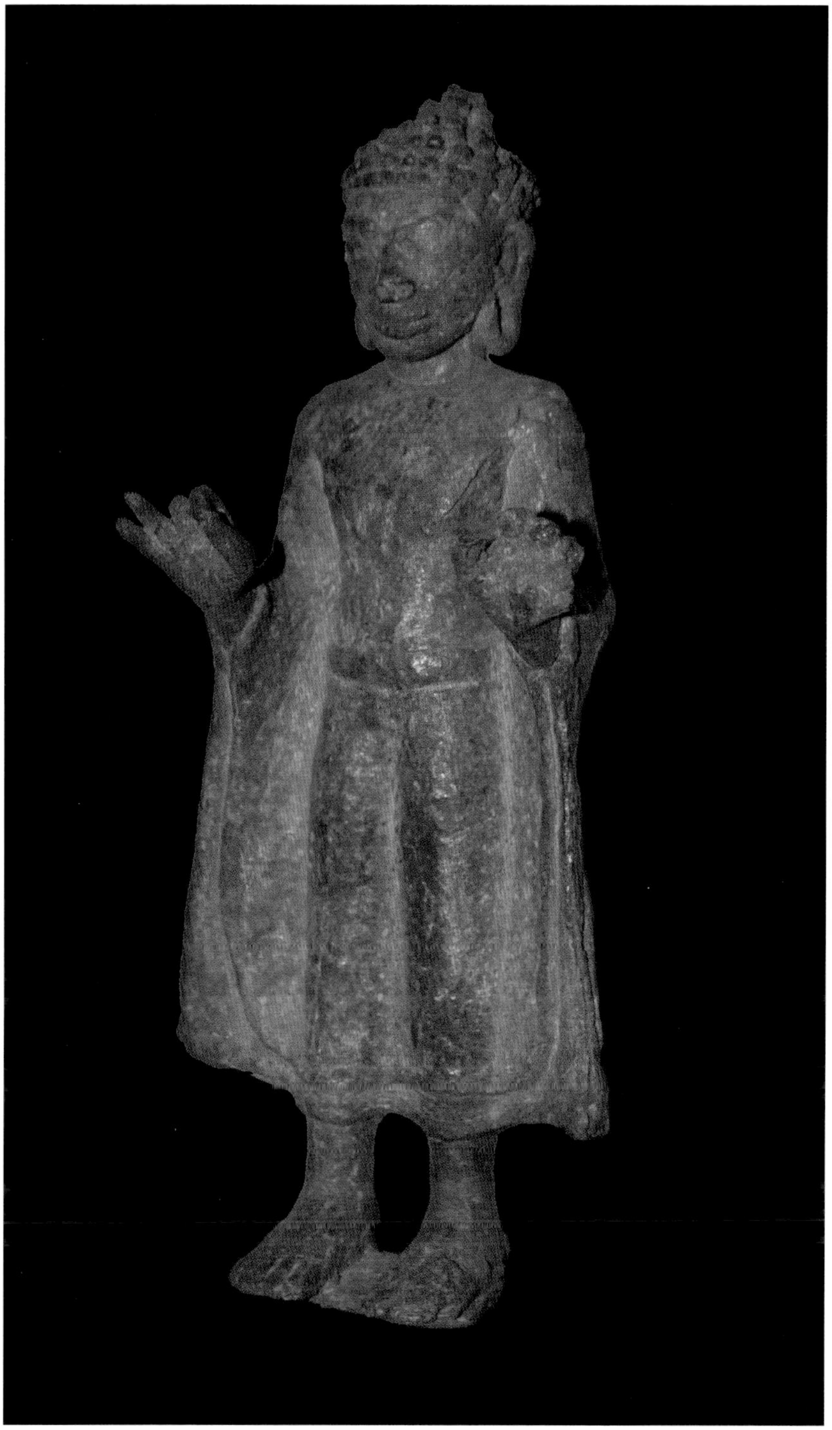

Figure 521 Example of a Thai statuette cast in an unleaded high-tin bronze (16wt% Sn, 0.4wt% Pb, ICP-AES analysis). *Buddha in Vitarkamudrā*, Mon culture of Dvāravatī, central Thailand, 7th–8th century, H. 19 cm (Musée national des arts asiatiques – Guimet, Paris, inv. MA 3785). See Bourgarit et al. 2003; Mechling et al. 2018.

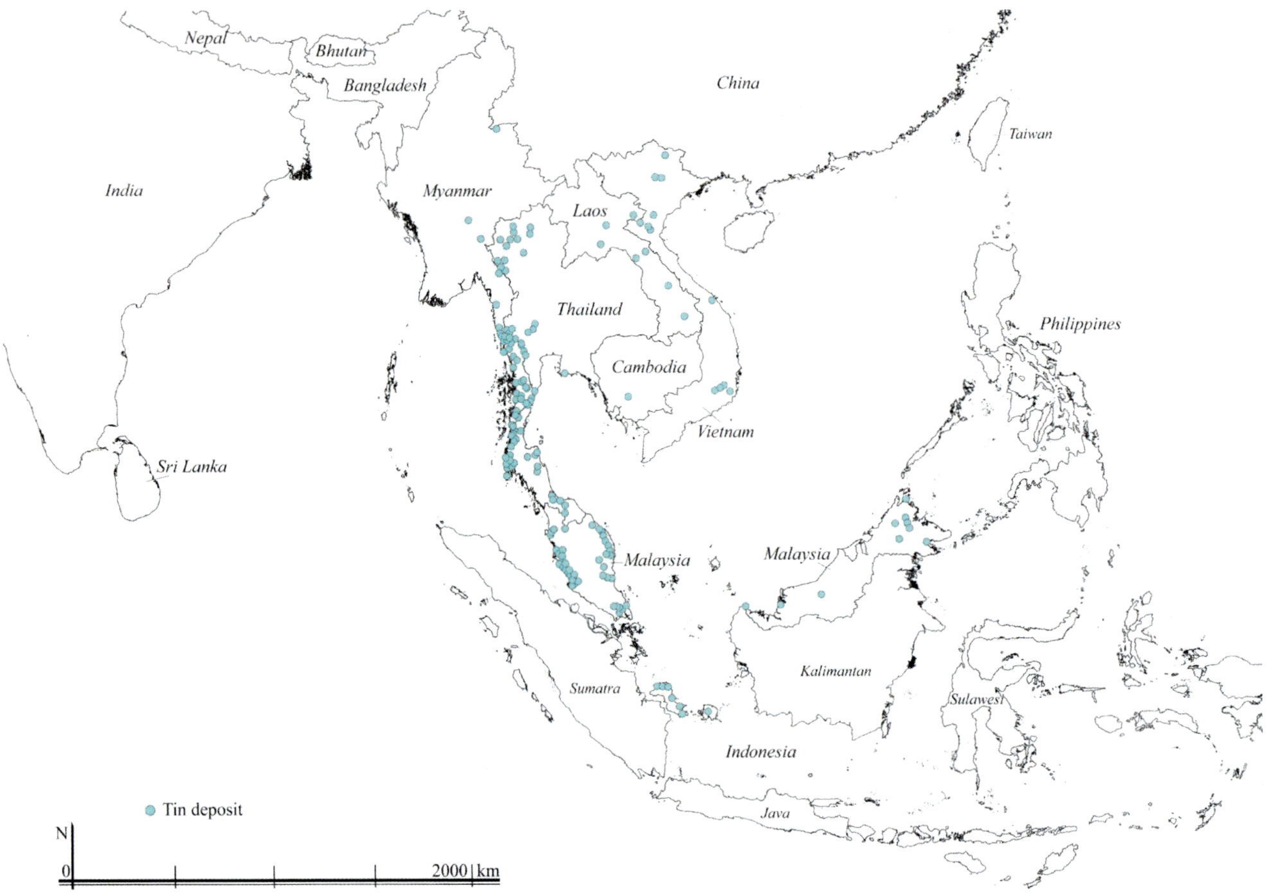

Figure 522 Geopolitical map showing tin deposit locations in Southeast Asia as revealed by geological surveys carried out since the 1960s (this does not include deposits in adjacent southern China and India). Courtesy of Sébastien Clouet, ongoing PhD work (2019–22): "Les mines d'Angkor, provenance et circulation des métaux non ferreux dans le Cambodge angkorien (IXe–XVe siècle)," Faculté des Lettres – Sorbonne Université, Paris.

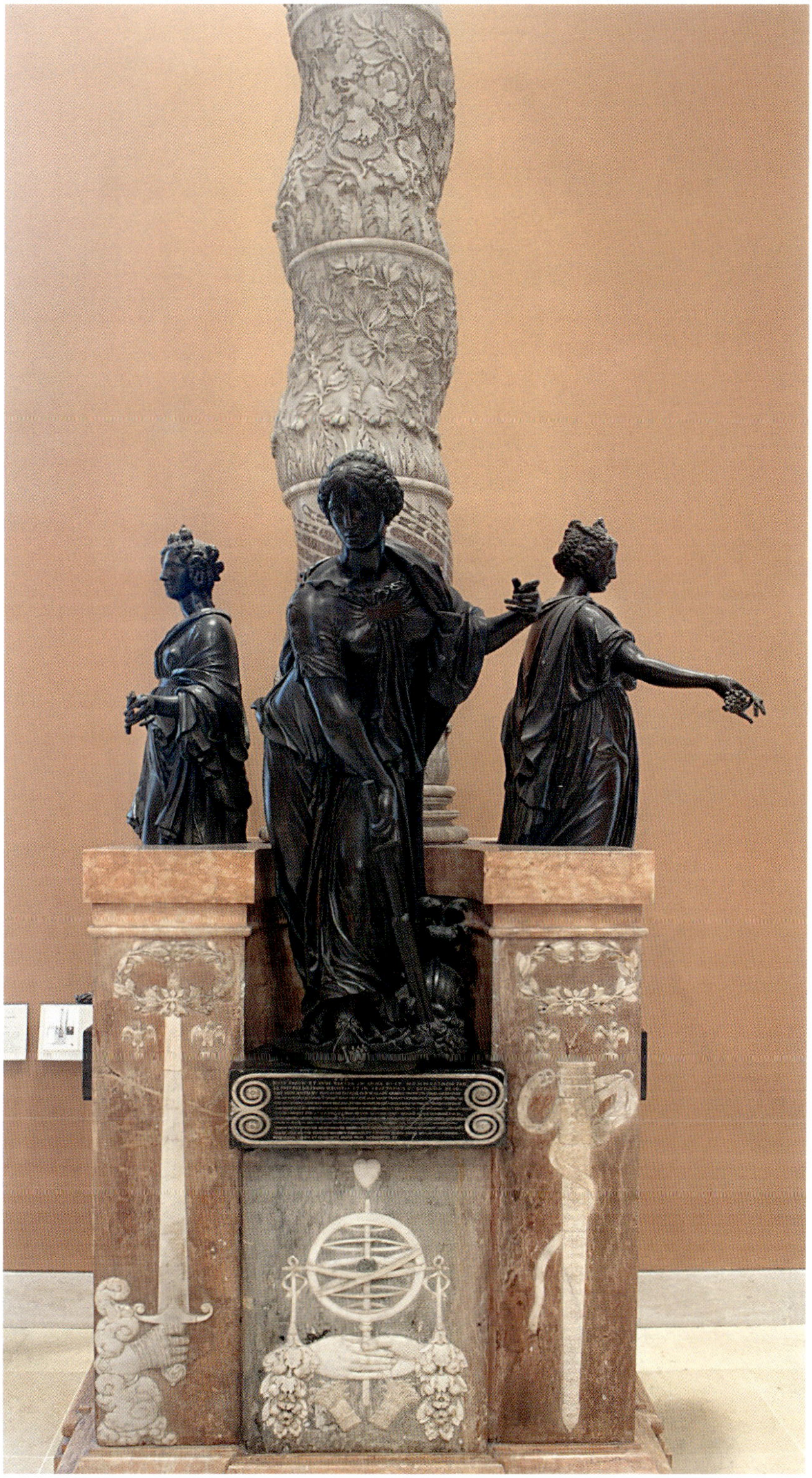

Figure 523 Jean Bullant (French, 1515–1578), *Anne de Montmorency Funerary Heart Monument*, 1571, H. of bronzes ca. 130 cm (Musée du Louvre, inv. MR1681–1683). See Seelig-Teuwen, Bourgarit, and Bewer 2014; Castelle et al. 2021.

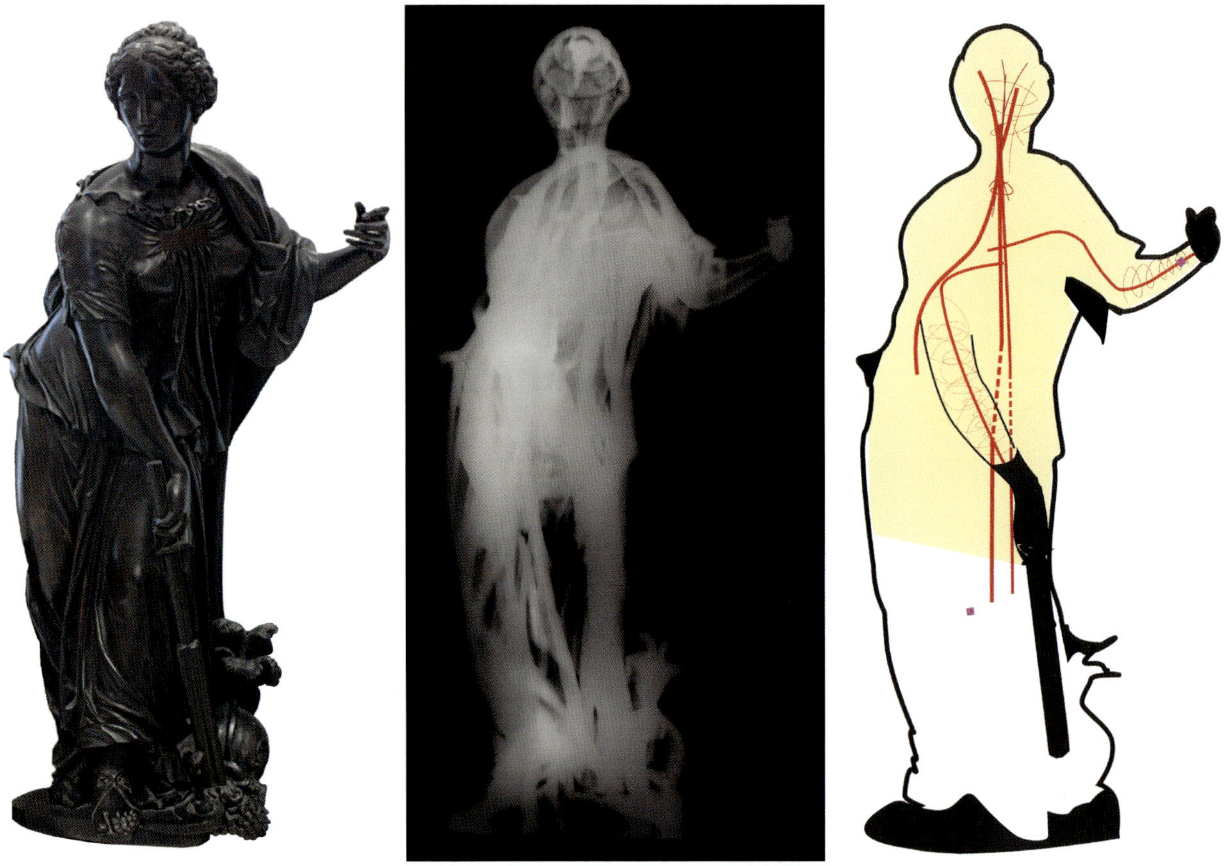

Figure 524 Overlay of interpretive technical diagram and both X-radiograph and daylight photography. Barthélemy Prieur (French, 1536–1611), *Peace*, cast in 1571 by Nicolas Péron, H. 128 cm (Musée du Louvre, inv. MR1683). See Case Study 5.

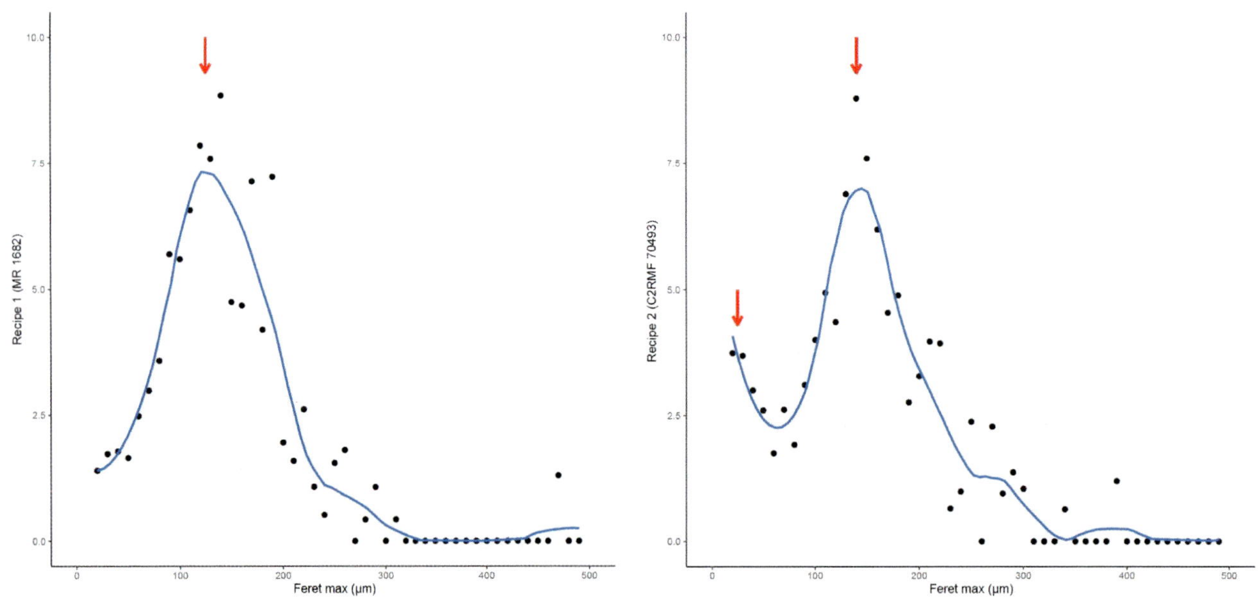

Figure 525 These two plots illustrate the different size distribution of quartz inclusions in the clay samples of two large early modern French bronzes. They represent the two distinct "recipes" for clay-based cores identified among a larger group of such bronzes. The size is represented on the horizontal axis by the maximum dimension of the grains (Feret). The number of quartz grains for each size is represented on the vertical axis by the surface ratio (the total surface of all the grains having the same size divided by the total surface of all grains). Recipe 1 contains only one size (red arrow on the left graph), while recipe 2 includes two sizes of quartz inclusions (see the two red arrows on the right graph). See Case Study 5, Castelle, Coquinot, and Bourgarit 2016.

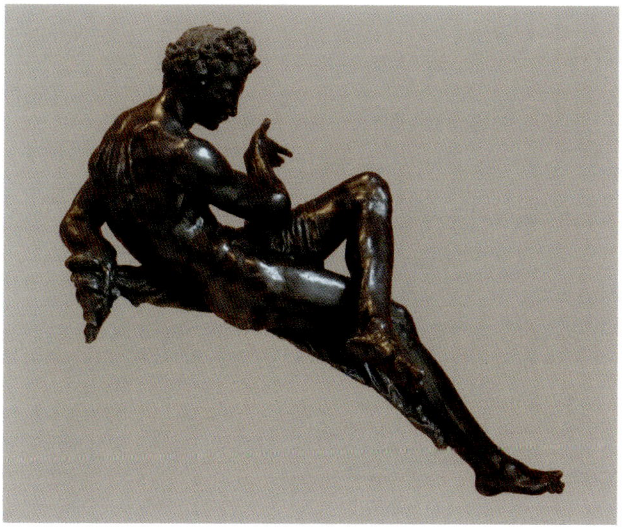

Figure 526 Barthélemy Prieur (French, 1536–1611), *Funerary Genius, Christophe de Thou Monument*, 1583–85, L. 109 cm (Musée du Louvre, inv. MR 1685). See Seelig-Teuwen, Bourgarit, and Bewer 2014.

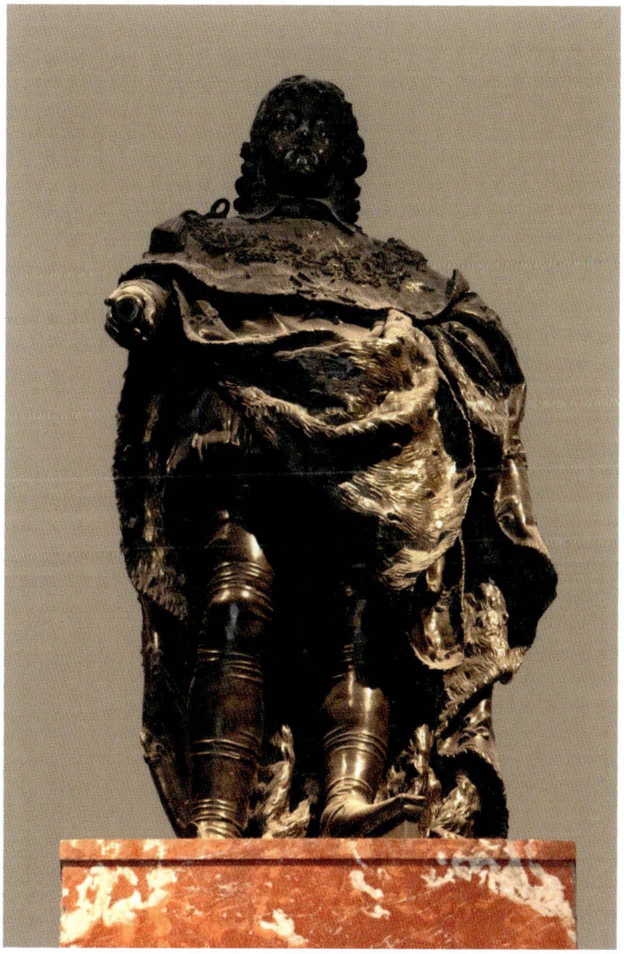

Figure 527 Simon Guillain (French 1581–1658), *Louis XIV as a Child*, ca. 1647, H. 153 cm (Musée du Louvre, inv. MR 3232). See C2RMF Internal Report 2021.

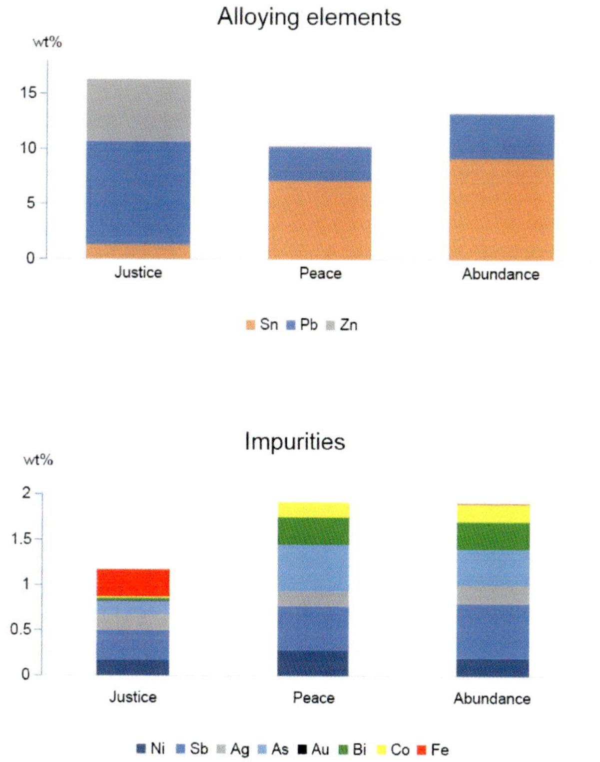

Figure 528 Histograms showing the alloy composition (ICP-AES analysis) of the three Virtues adorning the Anne de Montmorency Funerary Heart Monument, 1571, H. 128 cm (Musée du Louvre, inv. MR1681-1683). See Case Study 5, Castelle et al. 2021.

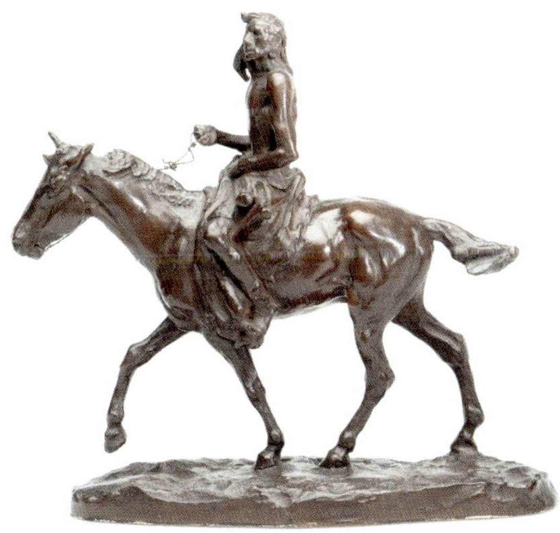

Figure 529 Charles Marion Russell (American, 1864–1926), *Medicine Whip*, modeled 1911, sand cast 1912–16, H. 14.8 cm (Gilcrease Museum, Tulsa, 0837.14).

VISUAL ATLAS

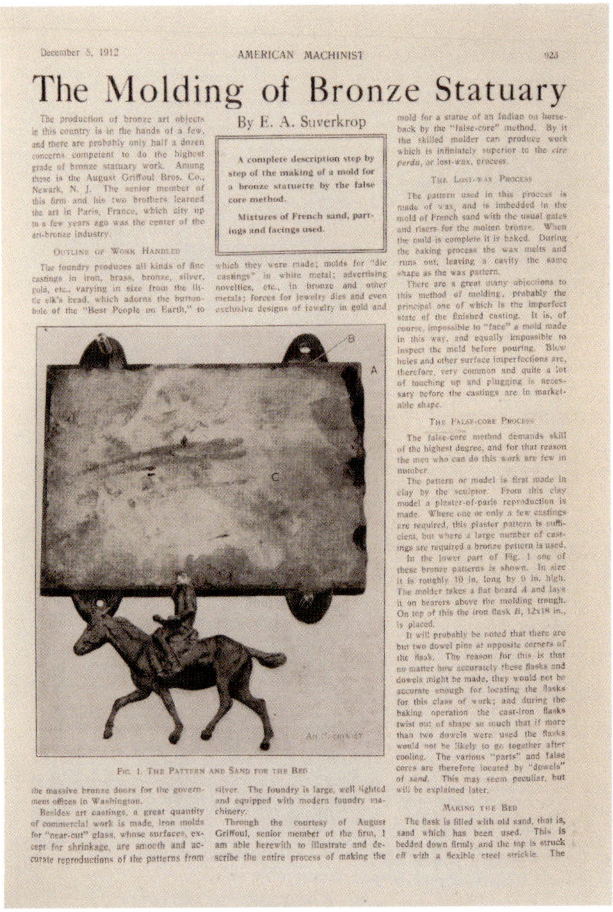

December 5, 1912 AMERICAN MACHINIST 923

The Molding of Bronze Statuary

By E. A. Suverkrop

> A complete description step by step of the making of a mold for a bronze statuette by the false core method.
>
> Mixtures of French sand, partings and facings used.

The production of bronze art objects in this country is in the hands of a few, and there are probably only half a dozen concerns competent to do the highest grade of bronze statuary work. Among these is the August Griffoul Bros. Co., Newark, N. J. The senior member of this firm and his two brothers learned the art in Paris, France, which city up to a few years ago was the center of the art-bronze industry.

Outline of Work Handled

The foundry produces all kinds of fine castings in iron, brass, bronze, silver, gold, etc., varying in size from the little elk's head, which adorns the buttonhole of the "Best People on Earth," to the massive bronze doors for the government offices in Washington.

Besides art castings, a great quantity of commercial work is made, iron molds for "near-cut" glass, whose surfaces, except for shrinkage, are smooth and accurate reproductions of the patterns from which they were made; molds for "die castings" in white metal; advertising novelties, etc., in bronze and other metals; forces for jewelry dies and even exclusive designs of jewelry in gold and silver. The foundry is large, well lighted and equipped with modern foundry machinery.

Through the courtesy of August Griffoul, senior member of the firm, I am able herewith to illustrate and describe the entire process of making the mold for a statue of an Indian on horseback by the "false-core" method. By it the skilled molder can produce work which is infinitely superior to the *cire perdu*, or lost-wax, process.

The Lost-wax Process

The pattern used in this process is made of wax, and is imbedded in the mold of French sand with the usual gates and risers for the molten bronze. When the mold is complete it is baked. During the baking process the wax melts and runs out, leaving a cavity the same shape as the wax pattern.

There are a great many objections to this method of molding, probably the principal one of which is the imperfect state of the finished casting. It is, of course, impossible to "face" a mold made in this way, and equally impossible to inspect the mold before pouring. Blow holes and other surface imperfections are, therefore, very common and quite a lot of touching up and plugging is necessary before the castings are in marketable shape.

The False-core Process

The false-core method demands skill of the highest degree, and for that reason the men who can do this work are few in number.

The pattern or model is first made in clay by the sculptor. From this clay model a plaster-of-paris reproduction is made. Where one or only a few castings are required, this plaster pattern is sufficient, but where a large number of castings are required a bronze pattern is used.

In the lower part of Fig. 1 one of these bronze patterns is shown. In size it is roughly 10 in. long by 9 in. high. The molder takes a flat board *A* and lays it on bearers above the molding trough. On top of this the iron flask *B*, 12x18 in., is placed.

It will probably be noted that there are but two dowel pins at opposite corners of the flask. The reason for this is that no matter how accurately these flasks and dowels might be made, they would not be accurate enough for locating the flasks for this class of work; and during the baking operation the cast-iron flasks twist out of shape so much that if more than two dowels were used the flasks would not be likely to go together after cooling. The various "parts" and false cores are therefore located by "dowels" of *sand*. This may seem peculiar, but will be explained later.

Making the Bed

The flask is filled with old sand, that is, sand which has been used. This is bedded down firmly and the top is struck off with a flexible steel strickle. The

Fig. 1. The Pattern and Sand for the Bed

Figure 530 The first page of the *American Machinist* story describing the molding of bronze statuary at the Griffoul foundry. The article shows, step by step, the production of Charles Russell's *Medicine Whip*. *American Machinist*, December 5, 1912, 923.

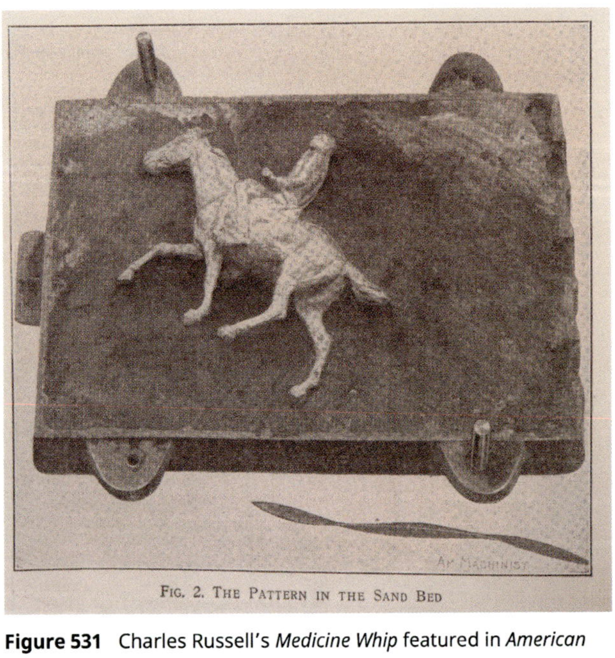

Fig. 2. The Pattern in the Sand Bed

Figure 531 Charles Russell's *Medicine Whip* featured in *American Machinist*, December 5, 1912, showing Griffoul foundry sand mold with chef-modèle on sand before embedding.

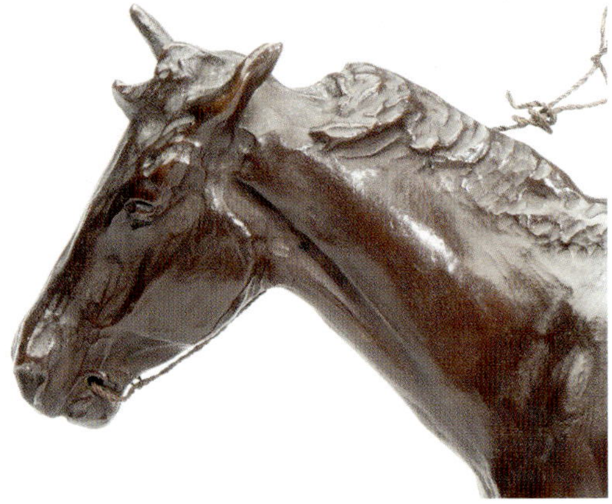

Figure 532 Detail side view of horse's head and neck showing artist's modeling. Charles Marion Russell (American, 1864–1926), *Medicine Whip*, modeled 1911, sand cast 1912–16, H. 14.8 cm (Gilcrease Museum, Tulsa, 0837.14).

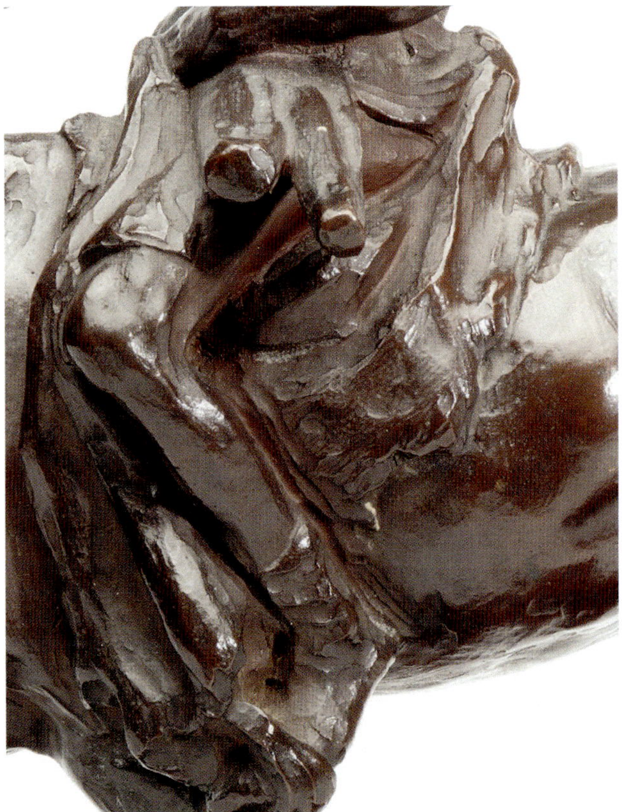

Figure 533 Detail side view of Indian's leg and horse's belly showing artist's modeling. Charles Marion Russell (American, 1864–1926), *Medicine Whip*, modeled 1911, sand cast 1912–16, H. 14.8 cm (Gilcrease Museum, Tulsa, 0837.14).

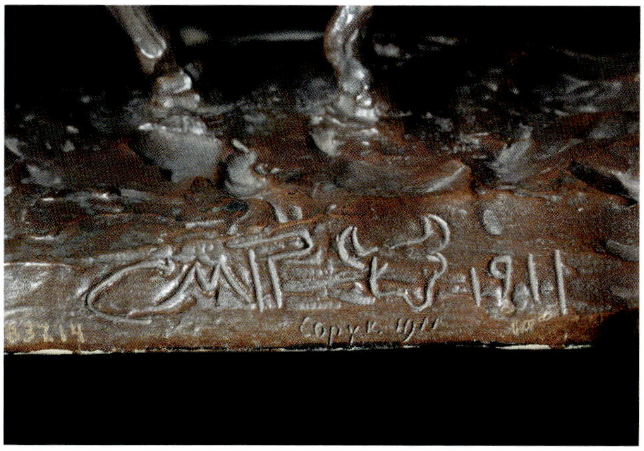

Figure 534 Detail of artist signature made in a soft material prior to casting with a partial fingerprint above and rows of possible punch marks made after casting. Charles Marion Russell (American, 1864–1926), *Medicine Whip*, modeled 1911, sand cast 1912–16, H. 14.8 cm (Gilcrease Museum, Tulsa, 0837.14).

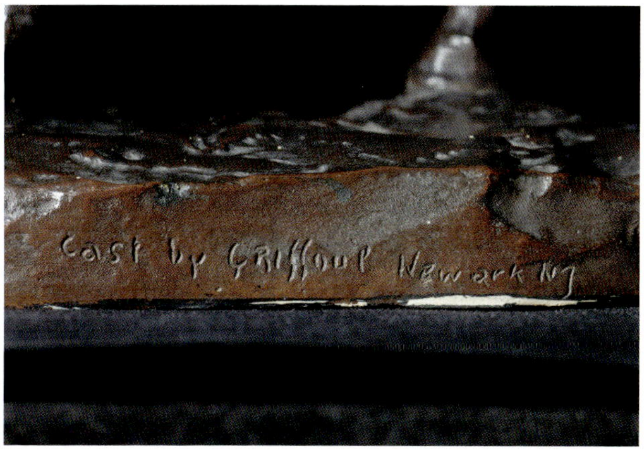

Figure 535 Detail showing the Griffoul foundry mark cast through from the model. Charles Marion Russell (American, 1864–1926), *Medicine Whip*, modeled 1911, sand cast 1912–16, H. 14.8 cm (Gilcrease Museum, Tulsa, 0837.14).

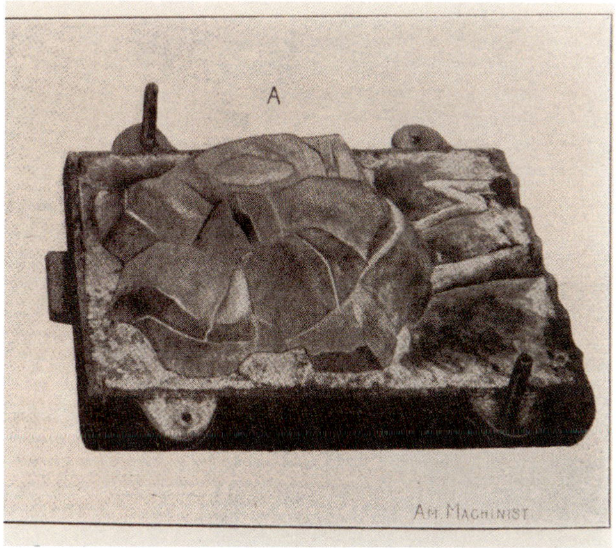

Figure 536 Sand mold showing sand piece mold pieces covering model prior to backfilling. Charles Russell's *Medicine Whip* featured in *American Machinist*, December 5, 1912.

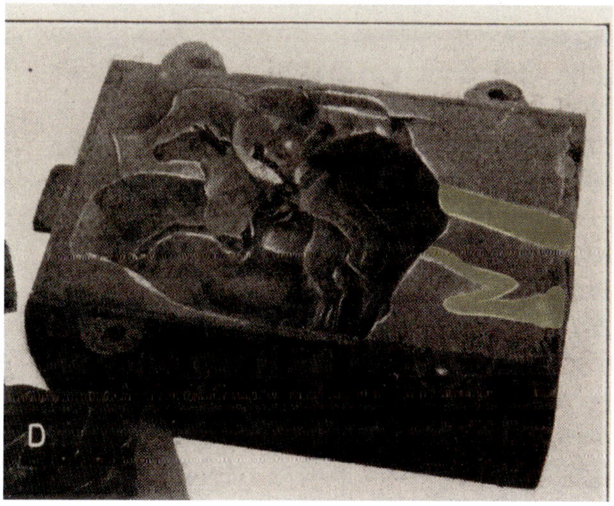

Figure 537 Sand mold disassembled and ready for baking. Channels on right side of image (yellow overlay) will direct molten metal into the mold and gases out. Charles Russell's *Medicine Whip* featured in *American Machinist*, December 5, 1912.

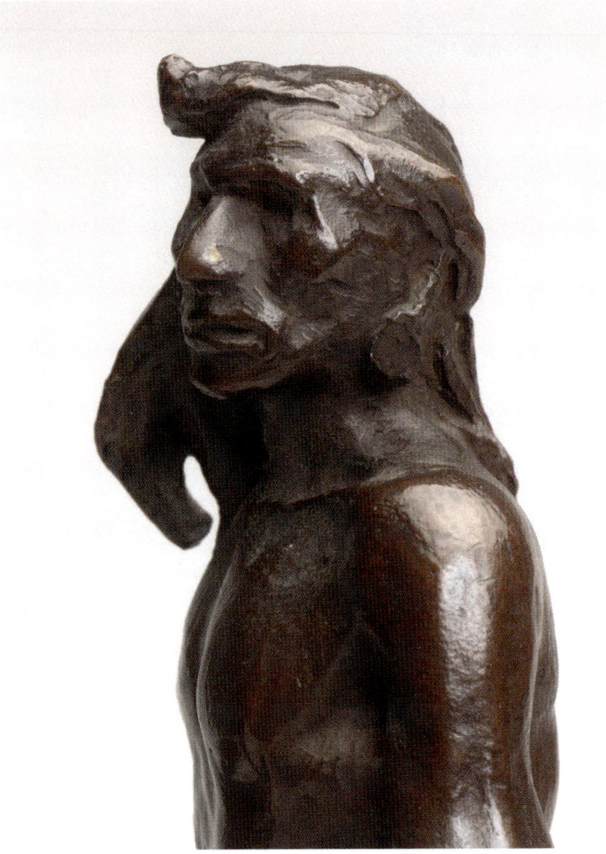

Figure 538 Detail side view of head and shoulders. Charles Marion Russell (American, 1864–1926), *Medicine Whip*, modeled 1911, sand cast 1912–16, H. 14.8 cm (Gilcrease Museum, Tulsa, 0837.14).

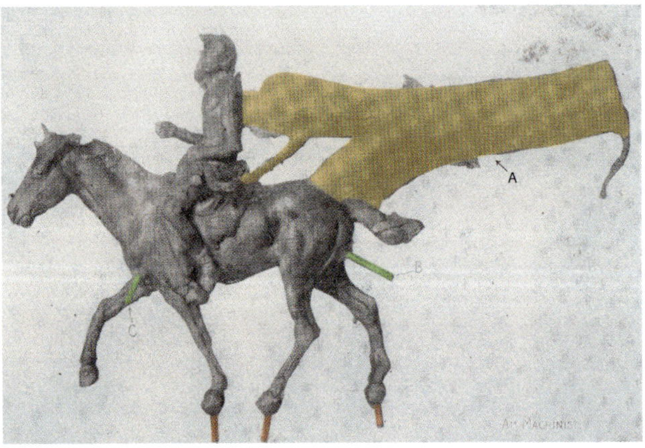

Figure 539 The bronze cast of *Medicine Whip* at the Griffoul foundry after removal from the sand mold, before fettling and finishing. The sprues (yellow overlay) have yet to be removed, and the ends of the core vents are protruding (green overlays). Threaded screws were cast into the hooves for mounting (orange overlay). *American Machinist*, December 5, 1912, fig. 14.

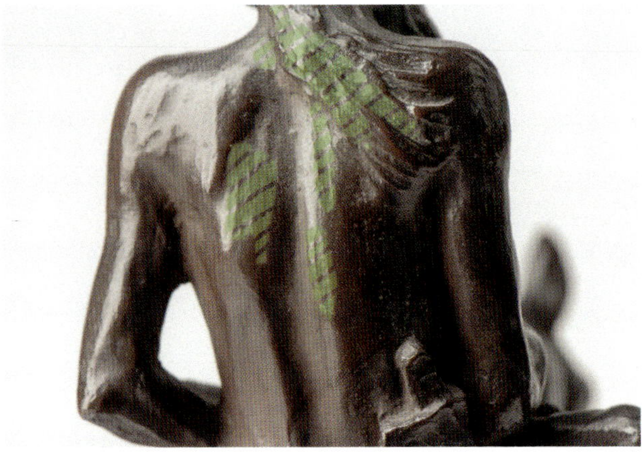

Figure 540 Detail of back showing tool marks where a sprue was removed. Charles Marion Russell (American, 1864–1926), *Medicine Whip*, modeled 1911, sand cast 1912–16, H. 14.8 cm (Gilcrease Museum, Tulsa, 0837.14).

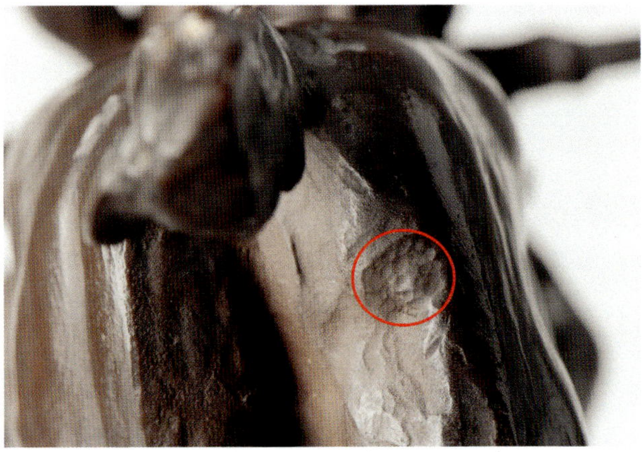

Figure 541 Detail of horse's rear showing tool marks (circled in red) where core vent hole was repaired after casting. Charles Marion Russell (American, 1864–1926), *Medicine Whip*, modeled 1911, sand cast 1912–16, H. 14.8 cm (Gilcrease Museum, Tulsa, 0837.14).

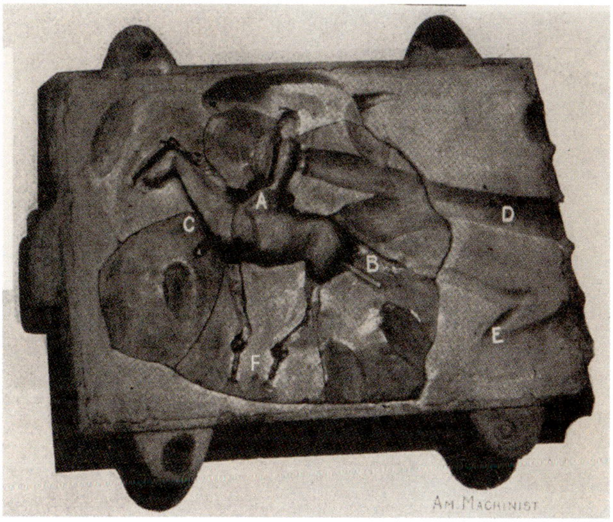

Figure 542 Annotated sand mold showing core (A); core vents protruding from core (B & C); channel cut to direct molten metal (D); vent (E); and screws inlaid to attach base after casting (F). Charles Russell's *Medicine Whip* featured in *American Machinist*, December 5, 1912.

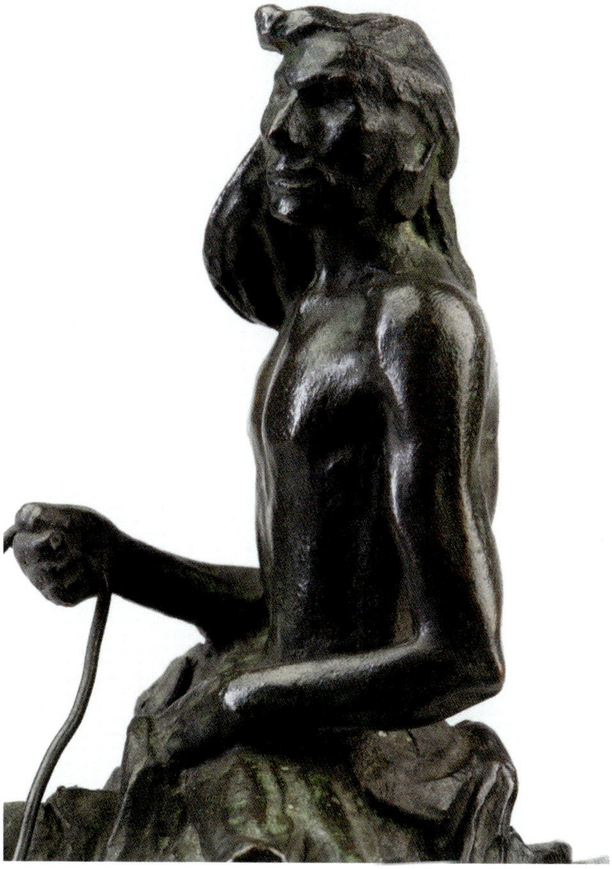

Figure 543 Detail of head and shoulders of the Indian. Charles Marion Russell (American, 1864–1926), *Medicine Whip*, modeled 1911, lost-wax cast by the California Art Bronze Foundry 1927–28, H. 27.6 cm (Amon Carter Museum of American Art, Fort Worth, inv. 1961.96).

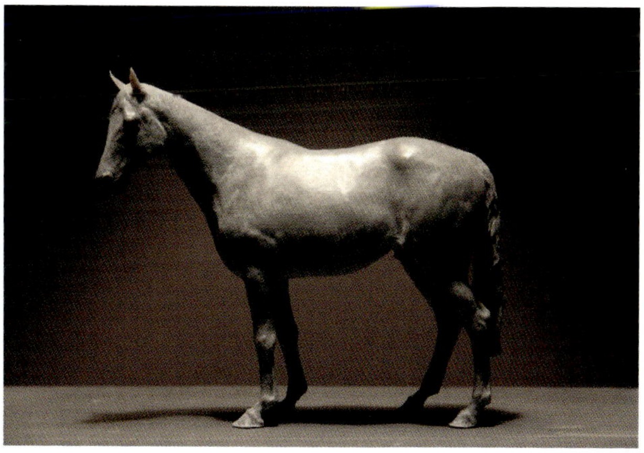

Figure 544 Reduced-scale 3D resin print of an artist-made model to be used as a foundry model for producing smaller bronze variants. Andrew Lacey (British, b. 1969), *The Anatomy of Bronze*, cast by the artist in 2019, Devon, UK, H. 45 cm (artist's collection).

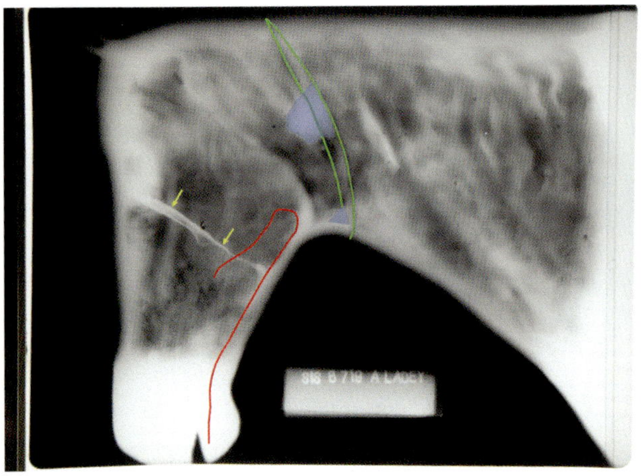

Figure 545 Annotated X-radiograph showing the flashing that formed along the juncture between different core materials—ceramic shell (in the muzzle) and plaster/clay mixture—as a white line of uneven density halfway up the horse's head (yellow arrows). An internal core support wire was inserted to connect the two core sections (red line). A line around the neck (green line) corresponds to TIG welding to join two separately cast parts. Remnants of sprues appear as denser patches (blue overlays). Note that the use of beam limiting shutters (used to reduce X-ray scatter, see II.3) have caused underexposure of the lower quarter and the right edge of the radiograph. Andrew Lacey (British, b. 1969), *The Anatomy of Bronze*, cast by the artist in 2019, Devon, UK, H. 45 cm (artist's collection).

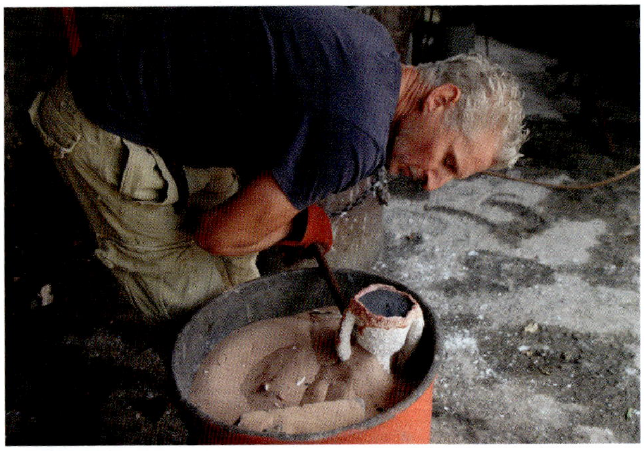

Figure 546 The artist-founder lifting the hot ceramic shell mold out of the casting pit after casting. All but the casting cup and sprue are buried in the sand. Andrew Lacey (British, b. 1969), *The Anatomy of Bronze*, cast by the artist in 2019, Devon, UK, H. 45 cm (artist's collection).

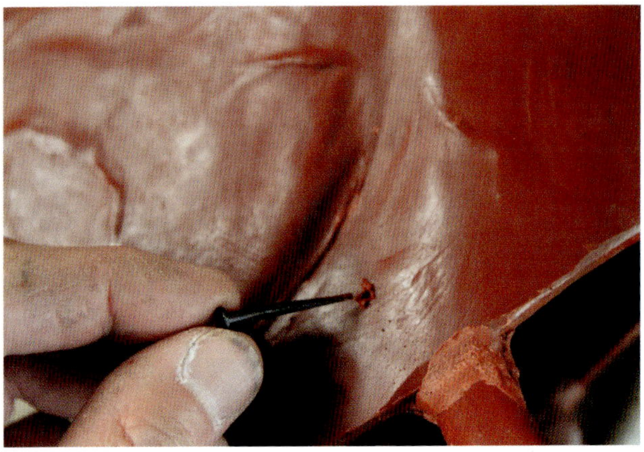

Figure 547 Core pin being pierced through the wax cast. Andrew Lacey (British, b. 1969), *The Anatomy of Bronze*, cast by the artist in 2019, Devon, UK, H. 45 cm (artist's collection).

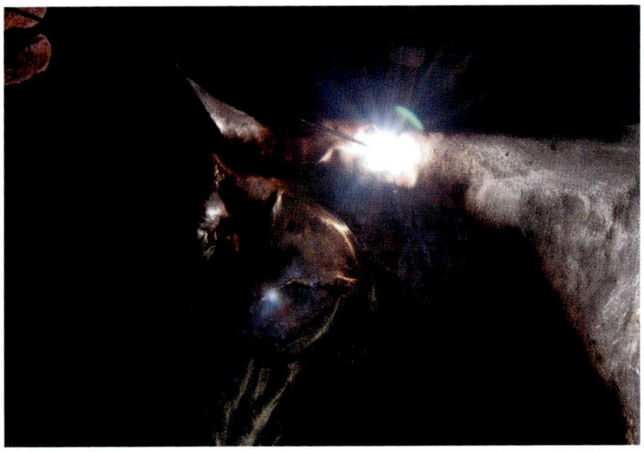

Figure 548 Bright arc of the TIG welder during the joining of two sections. Andrew Lacey (British, b. 1969), *The Anatomy of Bronze*, cast by the artist in 2019, Devon, UK, H. 45 cm (artist's collection).

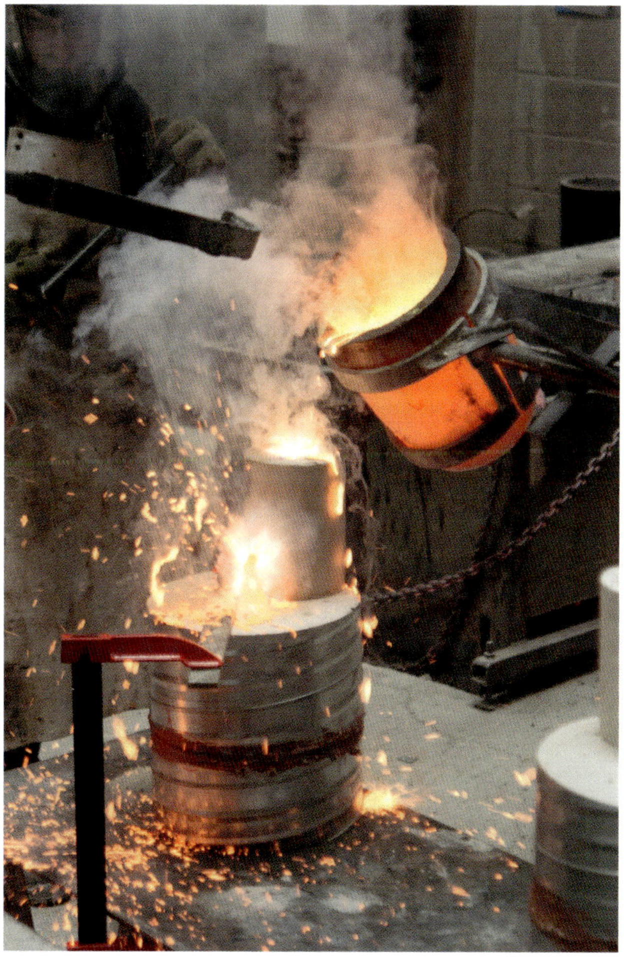

Figure 549 Molten bronze poured into a cylindrical plaster refractory mold with a separate pouring cup. Experimental reproduction cast of the head of the *Apollo of Lillebonne* during the 2016 CAST:ING meeting, Coubertin, France. Original: *Apollo of Lillebonne*, France, 1st century BCE–1st century CE, H. 193 cm (Musée du Louvre, inv. Br37 [NIII65]). See fig. 288.

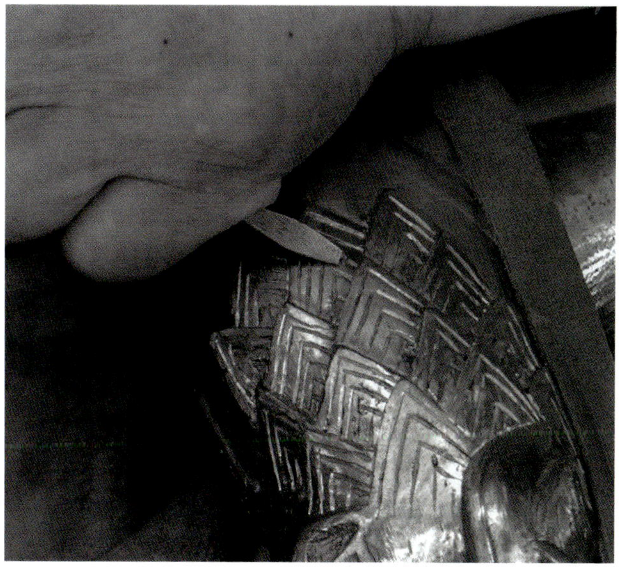

Figure 550 One aspect of chasing can involve enhancing detailed surface decoration. Here a V-shaped graver is used to remove metal in order to reinforce the lines of the lion's mane cast through from the model, where they had been incised into the wax. Ubaldo Vitali (American, b. 1944), Reproduction of a Lion Aquamanile, Maplewood, New Jersey, 2006, H. 19 cm, after *Aquamanile in the Form of a Lion*, probably northern Germany, 12th century, H. 19.5 cm (Metropolitan Museum of Art, gift of Irwin Untermyer, 1964, inv. 64.101.1491). See Dandridge 2006.

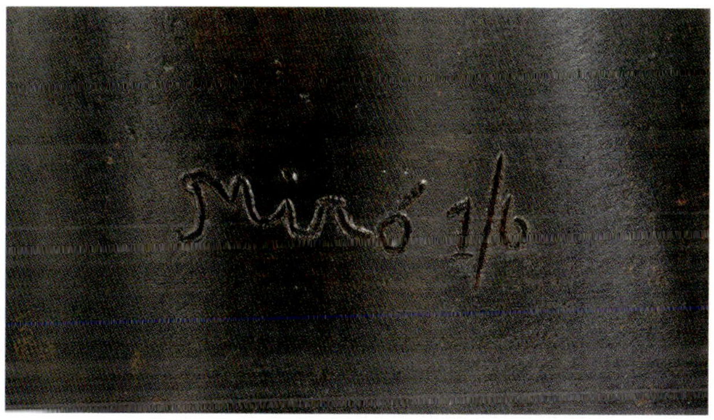

Figure 551 Detail showing the artist's signature and the edition number that were inscribed in the wax. Joan Miró (Spanish, 1893–1983), *Personnage*, designed 1976, cast 1985, H. 213.4 cm (J. Paul Getty Museum, inv. 2005.116).

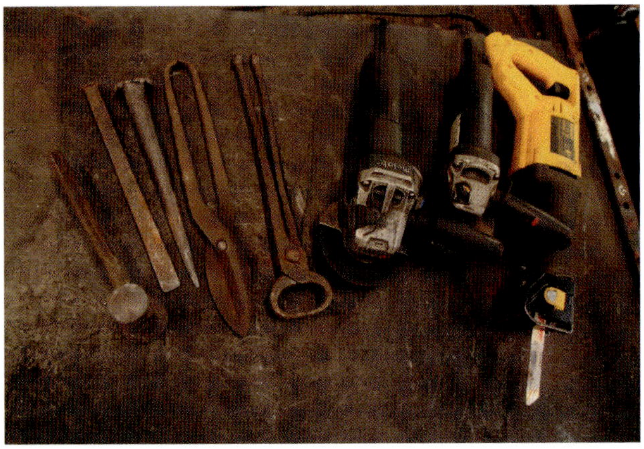

Figure 552 Examples of both hand and electric tools used in modern art foundries for fettling include (from the left): hammer and chisels, tin snips, grip pincher, angle grinder fitted with a cutting blade, rotary disk sander, reciprocating saw.

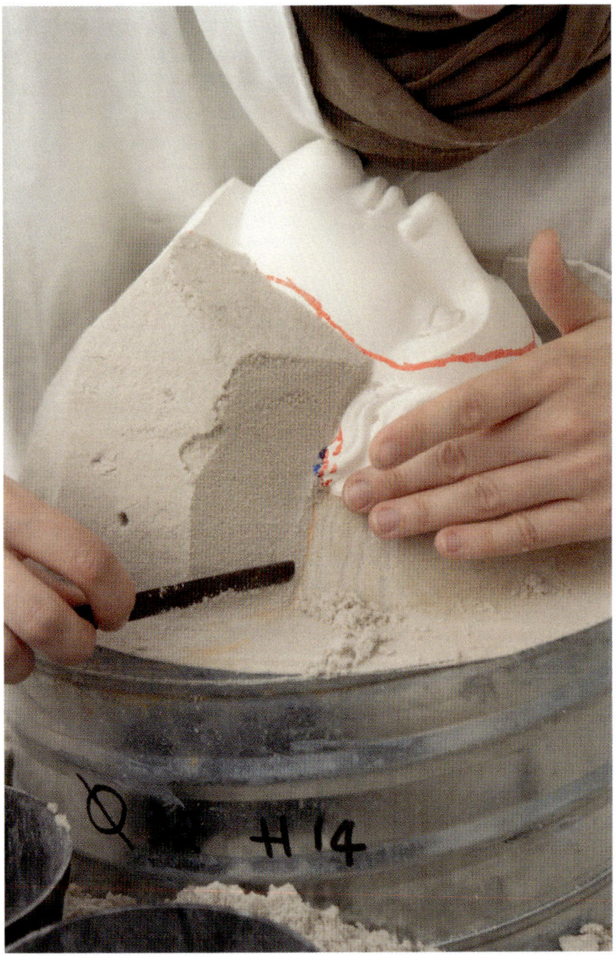

Figure 553 Plaster chef-modèle used to sand cast a two-thirds-scale bronze reproduction of the head of the *Apollo of Lillebonne* during the 2016 CAST:ING meeting, Coubertin, France. Original: *Apollo of Lillebonne*, France, 1st century BCE–1st century CE, H. 193 cm (Musée du Louvre, inv. Br37 [NIII65]). See fig. 288.

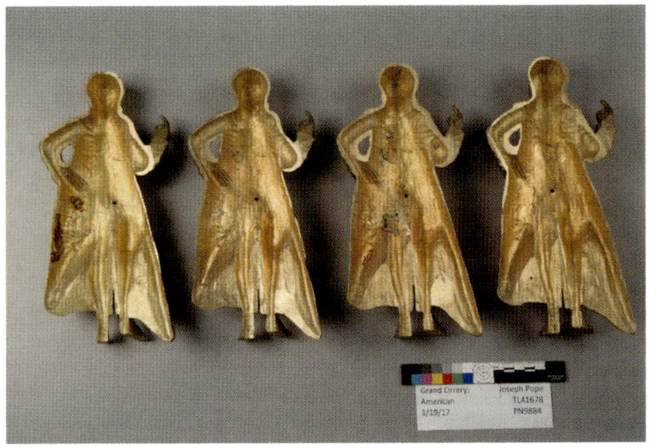

Figure 554 The backs of all four gilded brass reliefs show identical gouge marks—evidence that they were sand cast from the same carved wooden pattern. Relief figures of Benjamin Franklin, purportedly cast by Paul Revere (American, ca. 1735–1818) after a wood model by Simon Skillen, from Joseph Pope's (American, 1750–1826) *Grand Orrery*, 1776–87, H. 31 cm (Collection of Historical Scientific Instruments, Harvard University, inv. 0005).

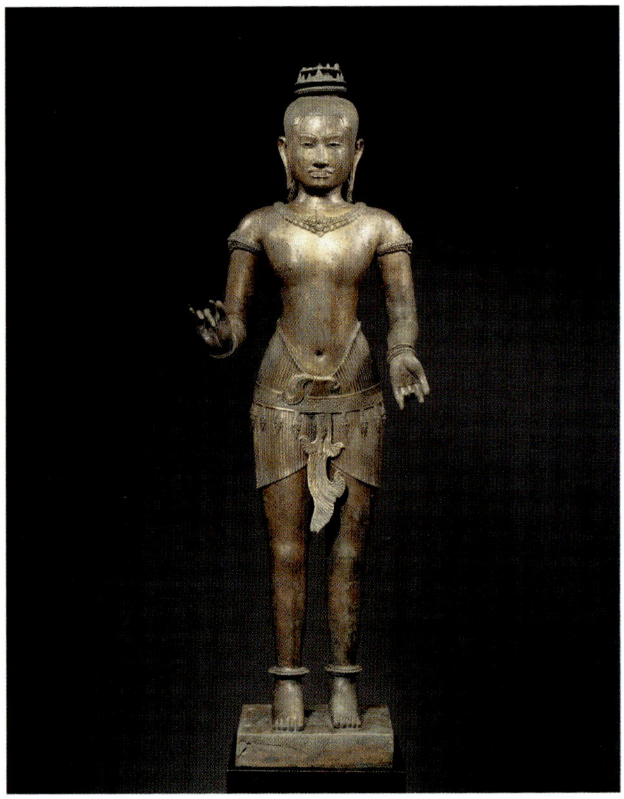

Figure 555 *Gilded Standing Divinity*, Cambodia, Angkor, Siem Reap province, 11th century, H. 130.8 cm (National Museum Bangkok).

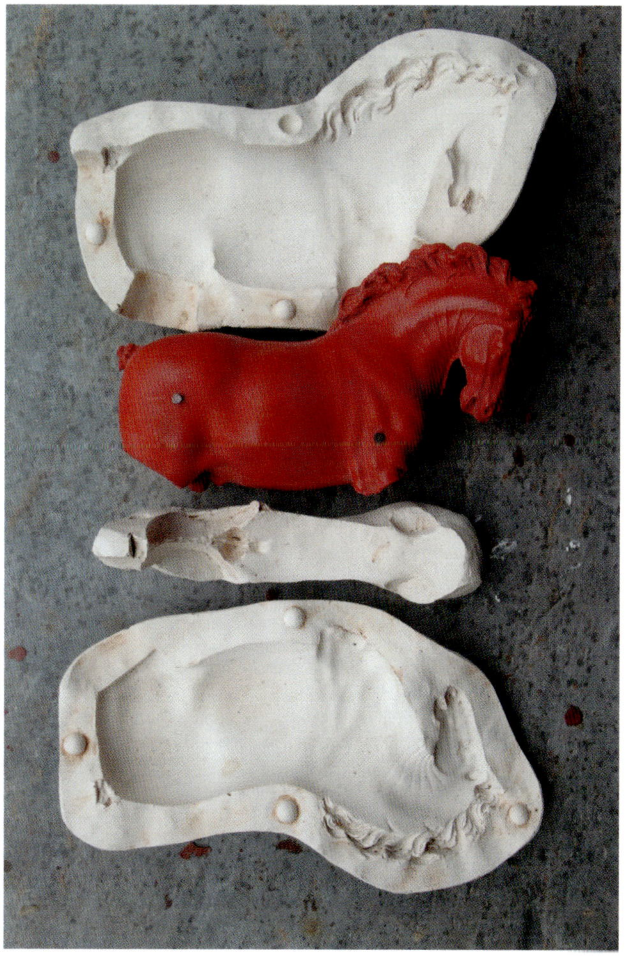

Figure 556 Wax inter-model formed using a plaster piece mold. Note the core pins inserted into the wax. Andrew Lacey, 2018.

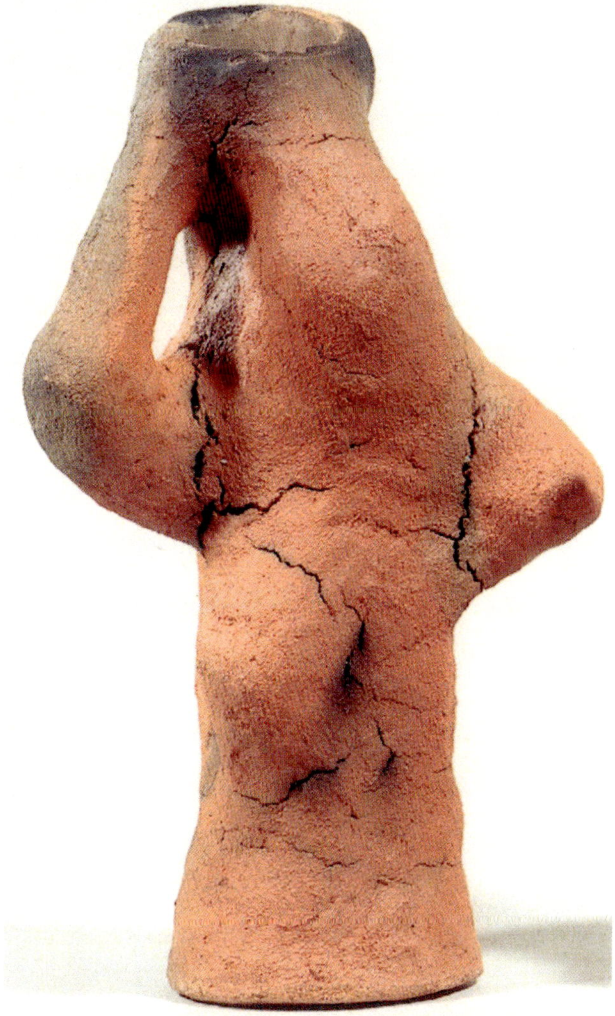

Figure 557 Investment mold used to cast a small bronze (mold height ~32 cm). Andrew Lacey, 1998.

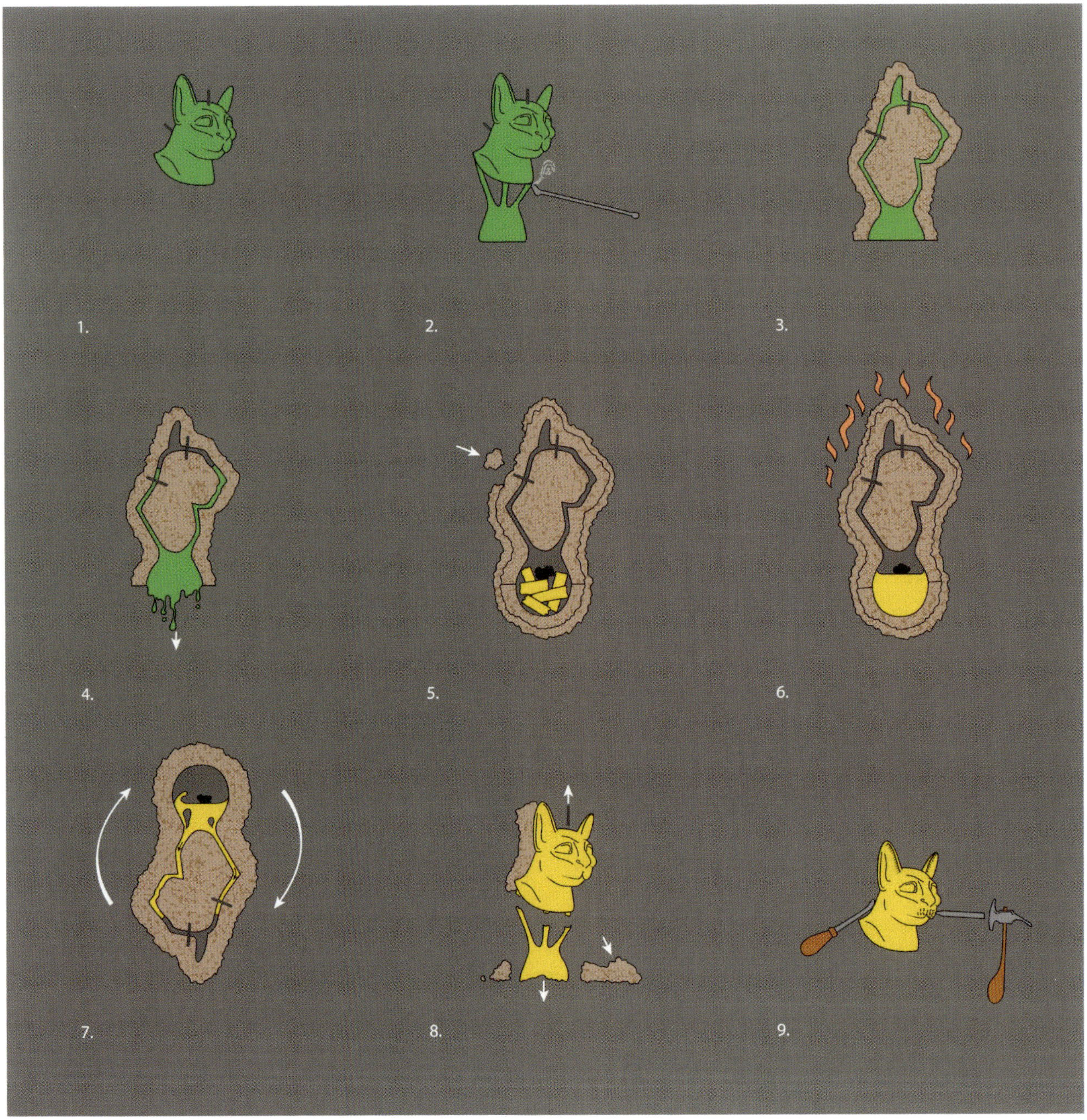

Figure 558 Diagram depicting a type of inversion casting: 1) core pins are inserted into a hollow wax model; 2) the sprue system is fused onto the wax model; 3) the wax model is covered in a refractory investment material; 4) the mold is then heated until dry and all traces of the wax melted and burned out; 5) meanwhile, a simple receptacle that will serve as a crucible is shaped of refractory material; small pieces of bronze and charcoal are placed in the crucible, which is luted to the mold to form an enclosed unit; 6) the mold-crucible is heated until the metal is molten; 7) the mold with its integral crucible is then inverted to allow the liquefied bronze to pour quickly into the mold; the charcoal helps to reduce oxidation; 8) fettling includes breaking off the refractory mold, removing the core pins, and cutting away the sprue system; 9) chasing may consist of polishing, burnishing, and the addition of details as needed; the surface color may also be enhanced with inlays, coating, plating, and/or patination.

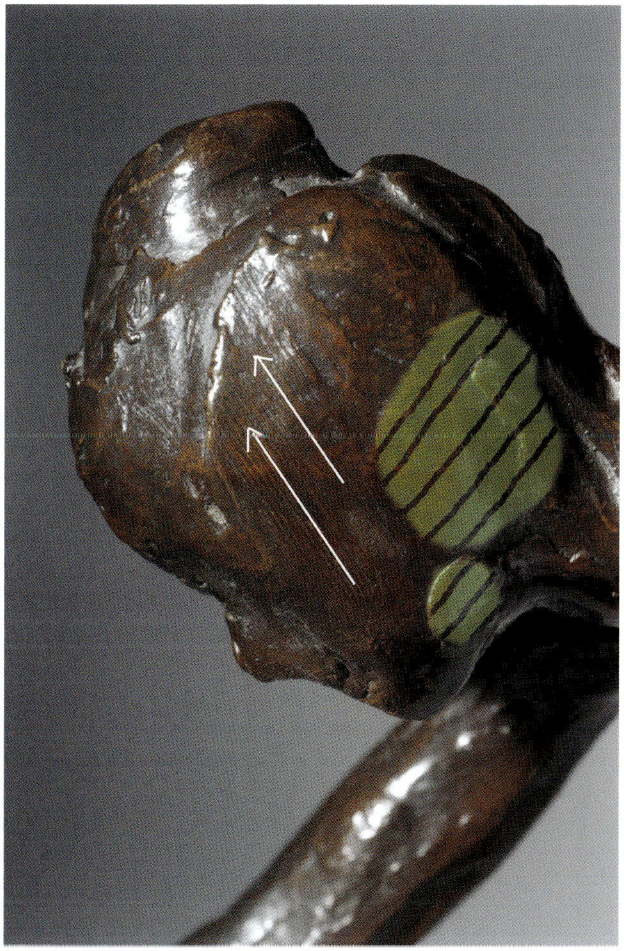

Figure 559 Detail of dancer's head showing how the bronze reproduces the blob of wax that Degas used in his original model to reinforce the jawline (green overlay). The cast also captures the quick gesture with which he pressed wax in to form the eye socket and pinched the displaced wax out into a fine ridge to form the nose, and the streak of his fingerprint up the side of the dancer's head (indicated by the white arrows). Edgar Degas (French, 1834–1917), *Grande Arabesque, Third Time,* ca. 1885–90, H. 40.6 cm (Harvard Art Museums / Fogg Museum, bequest from the Collection of Maurice Wertheim, Class of 1906, 1951.78).

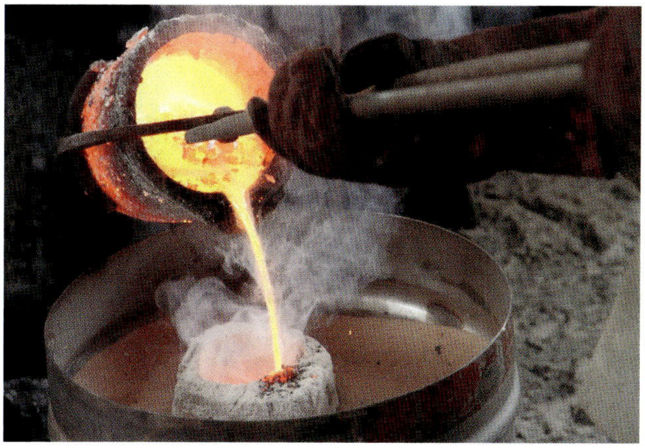

Figure 560 Bronze heated on a charcoal hearth being poured from a crucible into a plaster investment mold (Andrew Lacey studio, 2016).

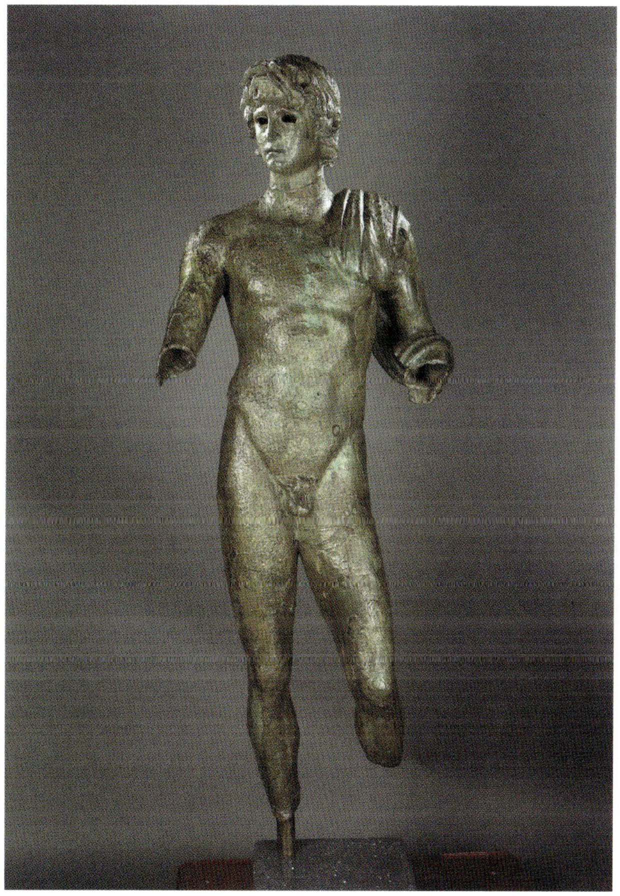

Figure 561 *The Youth of Agde*, France, 1st century CE, H. 140 cm (Musée de l'Ephèbe et d'Archéologie Sous-Marine de la Ville d'Agde, France, inv. 839). See Mille 2012.

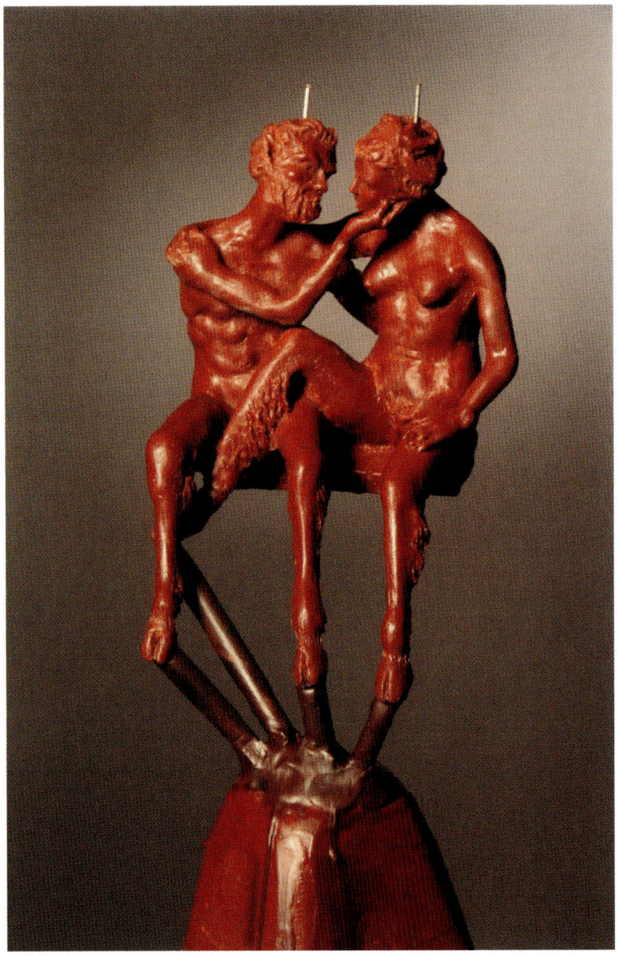

Figure 562 Andrew Lacey (British, b. 1969), sprued wax model reconstruction after Andrea Riccio (Italian, 1470–1532), *Satyr and Satyress*, 1510–20, H. 24 cm (Victoria and Albert Museum, presented by Art Fund, inv. A.8-1949).

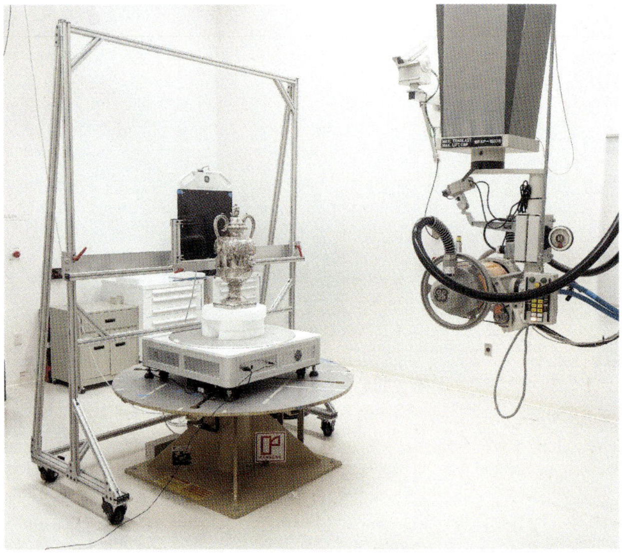

Figure 563 A digital turntable used in conjunction with a simple easel allows multiple 360° rotation series to be acquired in identical positions and composited in batches. The detector panel slides to a new position for each rotation. Because the detector stays in one plane, the image sets can be assembled by simple tiling, without the need for sophisticated alignment algorithms.

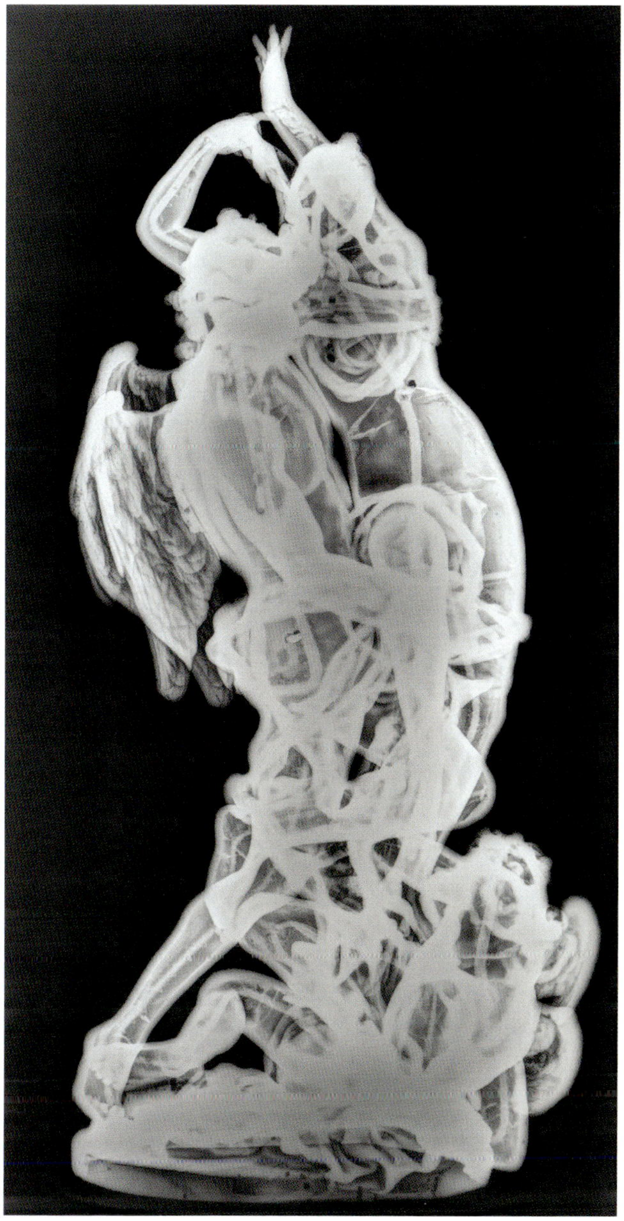

Figure 564 A digital (DR) 360° X-ray package (viewable in online edition) made using the set up shown in fig. 563. Each X-ray frame is made from 8 individual radiographs composited using Adobe After Effects software. After a model by Gaspard Marsy (French, 1624-1681), *Boreas Abducting Orithyia*, 1693–1710, H. 104.5 cm (J. Paul Getty Museum, inv. 88.SB.74).

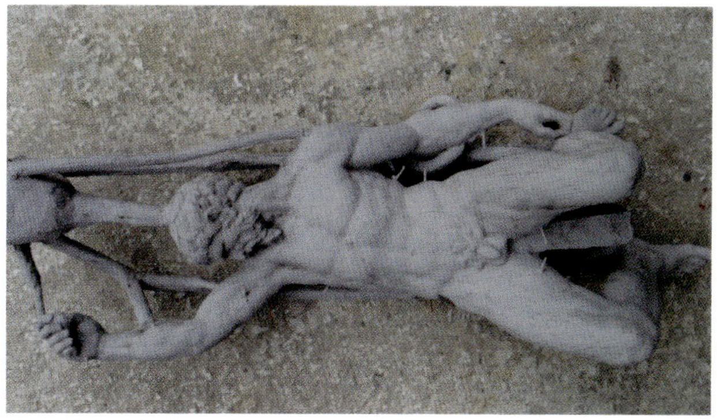

https://vimeo.com/672850991/6c63d94252

Video 1 Application of the first layer of investment material on a wax model by Andrew Lacey (British, b. 1969), filmed as part of "Experimental Reconstruction of the Rothschild Bronzes," 2015, Devon, UK. 2020 Siân Lewis and Andrew Lacey

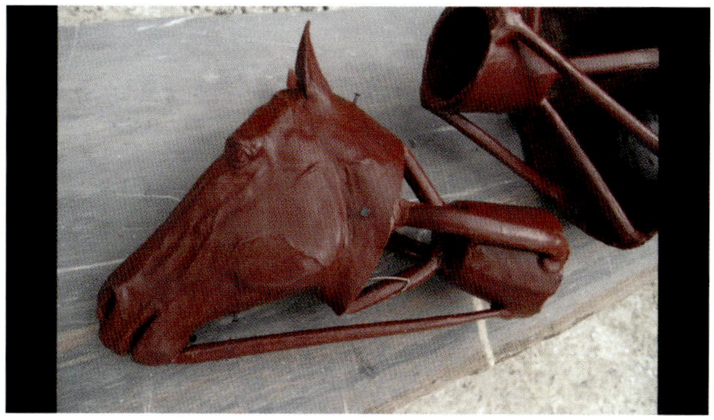

https://vimeo.com/672851118/418aee344b

Video 2 Contemporary artist-founder Andrew Lacey (British, b. 1969) offers insights on creating a bronze horse: material choices and process for the making of the wax model and the investment, 2019, Devon, UK. 2021 Siân Lewis and Andrew Lacey

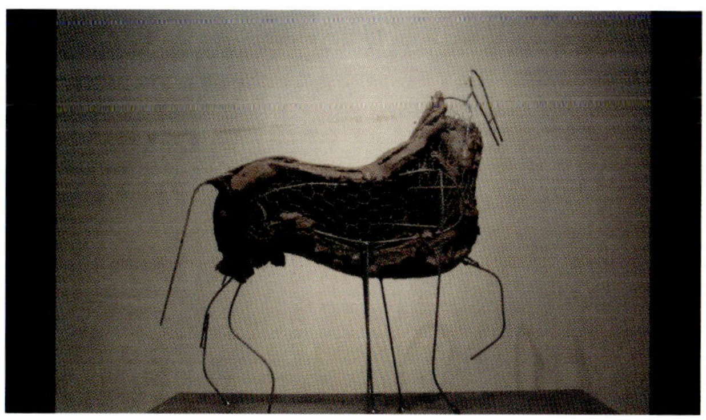

https://vimeo.com/672851307/f50cb47466

Video 3 Contemporary artist-founder Andrew Lacey (British, b. 1969) offers insights on creating a bronze horse: material choices and process for the making of the original model and the associated mold using 3D laser scanning, 2019, Devon, UK. 2021 Siân Lewis and Andrew Lacey

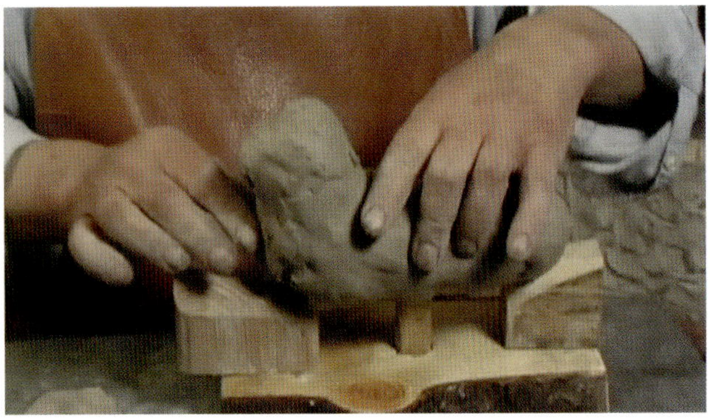

https://vimeo.com/672851374/1d4ca42c8b

Video 4 Fabrication of the clay-based core in preparation for direct lost-wax casting. Ubaldo Vitali (American, b. 1944), Reproduction of a Lion Aquamanile, Maplewood, New Jersey, 2006, H. 19 cm, after *Aquamanile in the Form of a Lion*, probably northern Germany, 12th century, H. 19.5 cm (Metropolitan Museum of Art, gift of Irwin Untermyer, 1964, inv. 64.101.1491).

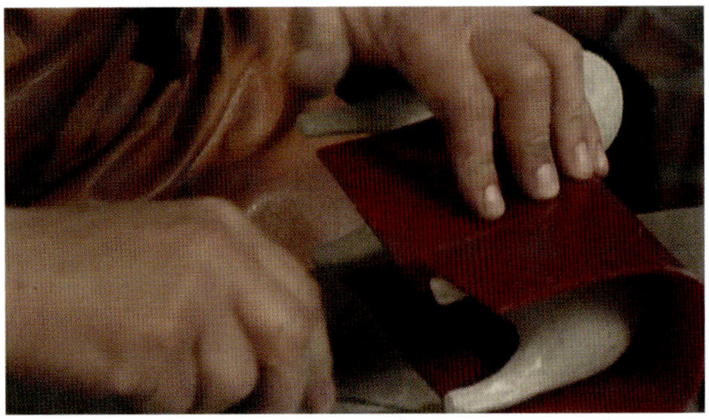

https://vimeo.com/672851443/c5ac6ba826

Video 5 Two methods of applying wax to a pre-made core include dipping the core in liquid wax and applying wax slabs. Parts such as scales, tail, and handle are added by welding the wax joints with heat. Ubaldo Vitali (American, b. 1944), Reproduction of a Lion Aquamanile, Maplewood, New Jersey, 2006, H. 19 cm, after *Aquamanile in the Form of a Lion*, probably northern Germany, 12th century, H. 19.5 cm (Metropolitan Museum of Art, gift of Irwin Untermyer, 1964, inv. 64.101.1491).

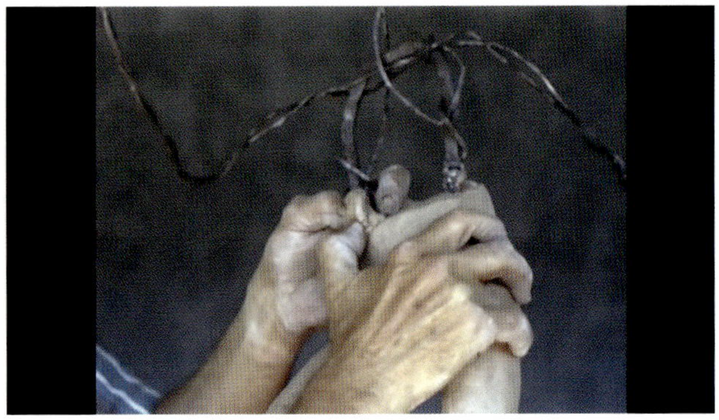

https://vimeo.com/816385794/dc744473bb

Video 6 Adriaen de Vries's (Netherlandish, 1556–1626) bronze casting technique: the direct lost-wax method. All parts of the process are described, from the setup of armatures to finishing.

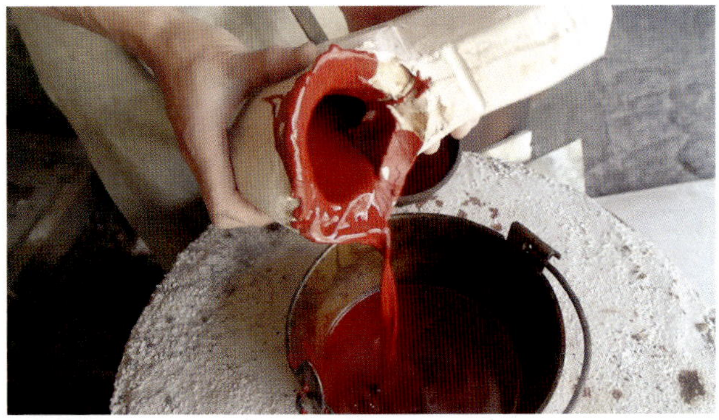

https://vimeo.com/672851502/f356f17315

Video 7 Short demonstration of slush molding wax into a plaster piece mold by Andrew Lacey (British, b. 1969), filmed as part of "Experimental Reconstruction of the Rothschild Bronzes," 2016, Devon, UK. 2020 Siân Lewis and Andrew Lacey

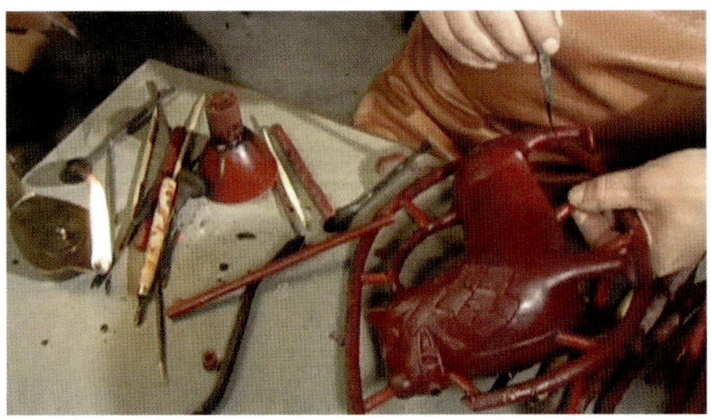

https://vimeo.com/672851620/a15f318198

Video 8 Applying solid wax sprues and a pouring cup to the completed wax model. Vents that allow gases to escape from the mold are in blue. Ubaldo Vitali (American, b. 1944), Reproduction of a Lion Aquamanile, Maplewood, New Jersey, 2006, H. 19 cm, after *Aquamanile in the Form of a Lion*, probably northern Germany, 12th century, H. 19.5 cm (Metropolitan Museum of Art, gift of Irwin Untermyer, 1964, inv. 64.101.1491.

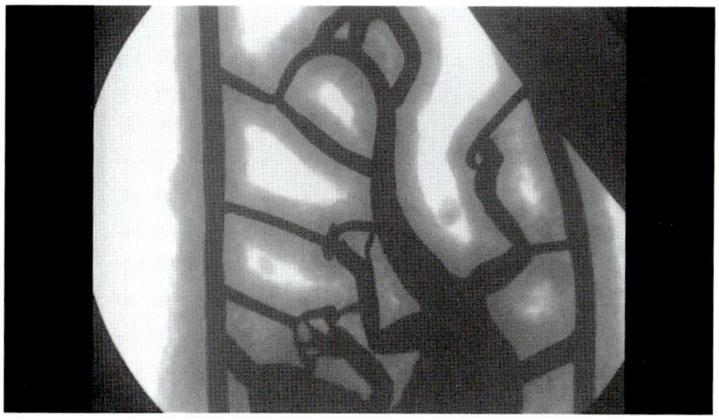

https://vimeo.com/672851728/ef8007af4c

Video 9 Real-time video of the X-radiography of the pouring of a bronze, showing the path of the molten 10 wt% tin bronze into a refractory mold of a life cast of a lizard. The pour results in shrinkage porosity on the surface of the bronze. It is unlikely to be gaseous porosity, as it was degassed prior to pouring and the holes are very large, with the last one being highly granular in texture, indicating slow cooling. Note also how quickly the mold is filled with bronze (see also fig. 27). Andrew Lacey (British, b. 1969), *Life Cast of a Lizard*, cast by the artist in 2015, Devon, UK, L. 15 cm (Victoria and Albert Museum, inv. NCOL.517-2015). 2020 Siân Lewis and Andrew Lacey

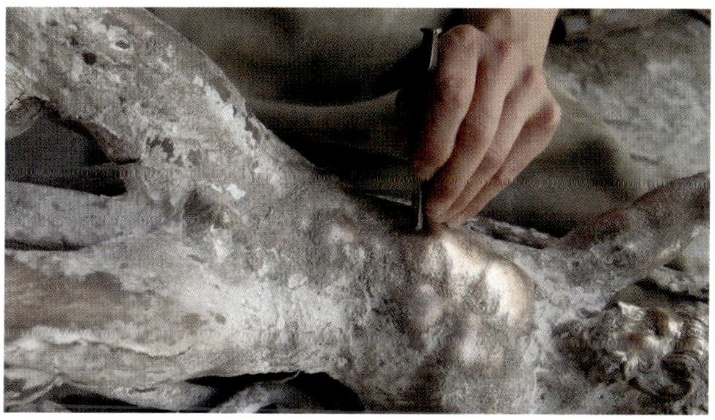

https://vimeo.com/672853119/c50b735e70

Video 10 Short demonstration of chasing using a punch to compress the flawed surface of a cast by Andrew Lacey (British, b. 1969), filmed as part of "Experimental Reconstruction of the Rothschild Bronzes," 2015, Devon, UK. 2020 Siân Lewis and Andrew Lacey

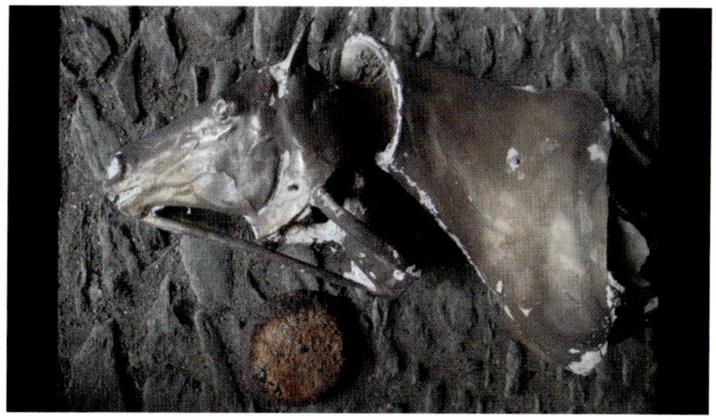

https://vimeo.com/672853278/b3f8d29d53

Video 11 Contemporary artist-founder Andrew Lacey (British, b. 1969) offers insights on creating a bronze horse: material choices and process for the metal pouring, fettling, and assembly steps, 2019, Devon, UK. 2021 Siân Lewis and Andrew Lacey

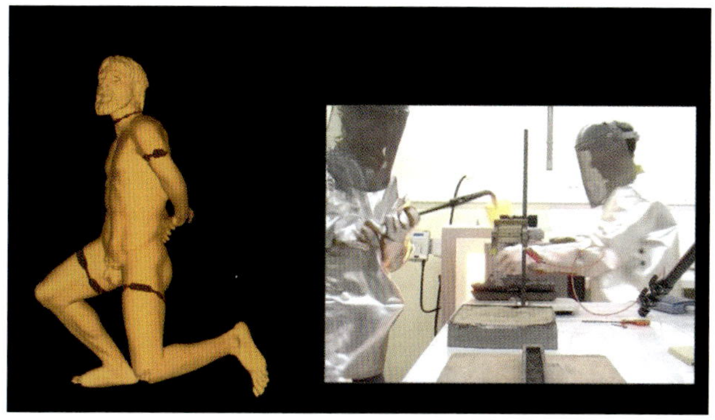

https://vimeo.com/800608712/1cd80c51ad

Video 12 Flow-fusion-welding laboratory experiment carried out at the C2RMF, Paris, showing the fabrication process for the *Captive Gaul*, Gaul, last quarter of the 1st century BCE, H. 63 cm (Musée départemental Arles Antique [MDAA], France, inv. Rho.2007.06.1962), excerpted from a didactic video presented at the exhibitions *Arles, les fouilles du Rhône*, Musée du Louvre (March 9–June 25, 2012), and *César, le Rhône pour mémoire, 20 ans de fouilles archéologiques*, Musée de l'Arles antique, France (November 2009–January 2011). See Azéma et al. 2013. For the complete original video visit https://youtu.be/vXPmorUKJuc.

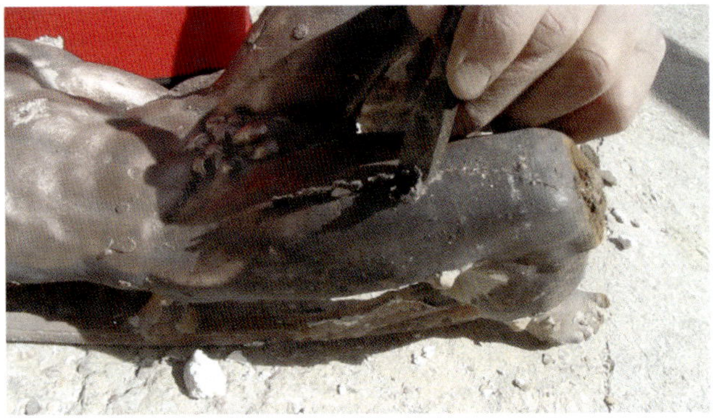

https://vimeo.com/672853423/0214a29d8c

Video 13 Short demonstration of fettling with a chisel to remove a mold flashing by Andrew Lacey (British, b. 1969), filmed as part of "Experimental Reconstruction of the Rothschild Bronzes," 2015, Devon, UK. For the complete original video visit https://vimeo.com/133884185. 2020 Siân Lewis and Andrew Lacey

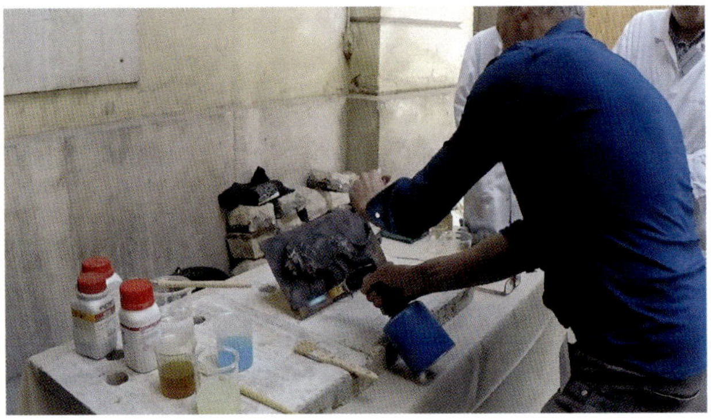

https://vimeo.com/672853511/33ca671c7c

Video 14 Chemical patination demonstrated by Andrew Lacey (British, b. 1969) on a replica casting of a lion relief after Antoine-Louis Barye (French, 1795–1875) at an October 2018 CAST:ING workshop on patination and gilding at the Laboratoire de Recherche des Monuments Historiques (LRMH), Champs sur Marne, France. To start, Lacey simultaneously heats the bronze relief while gradually applying a solution of potassium polysulfide to create a dark-black background color on the bronze. Next, a layer of ferric nitrate serves to build a brown undertone. This is followed by a green textured layer by stippling copper nitrate solution over the surface at a slightly lower temperature. The final image shows the relief with two different finishes—the left-hand side had hot wax applied to the surface, while the right-hand side was waxed cold—demonstrating how the type of wax application at the final stage can radically alter the appearance of a patina.

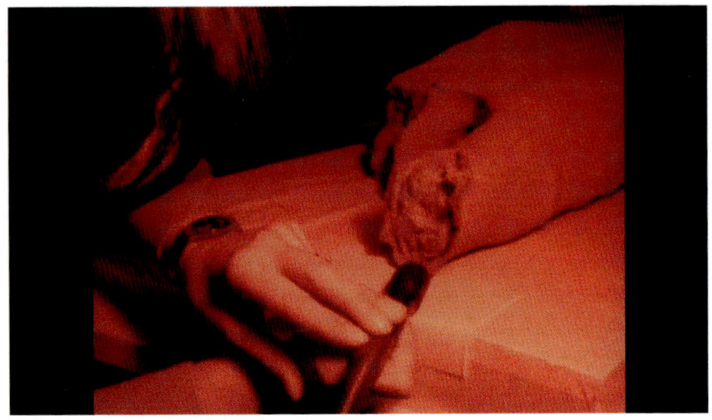

https://vimeo.com/672853890/fbf173a518

Video 15 Sampling the core of a large bronze under safelight for OSL dating at the C2RMF, Paris. The core has been sampled in two different locations: the feet and the arms. For each, several samples were taken at various distances from the bronze wall in order to check the possible effects of an airport X-ray scan during the bronze's transportation from Sana'a to Paris. *Bronze Statue of Hawtar'athat*, Yemen, 1st millennium BCE, H. 140 cm (National Museum of Sana'a, inv. YM 23206). See Mille et al. 2010.

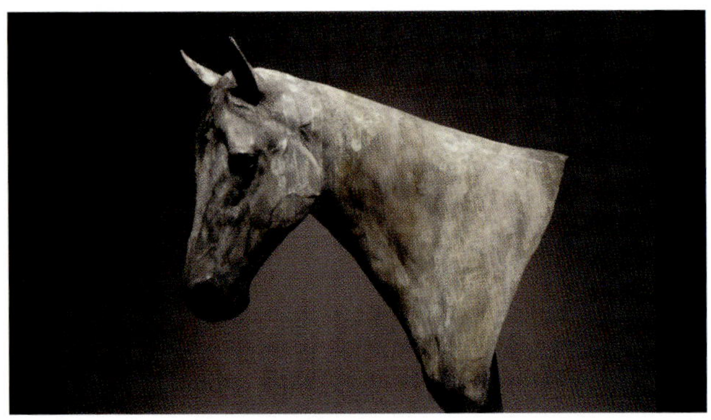

https://vimeo.com/672853935/66a501d3f2

Video 16 Contemporary artist-founder Andrew Lacey (British, b. 1969) offers insights on creating a bronze horse: material choices of and process patination, 2019, Devon, UK. 2021 Siân Lewis and Andrew Lacey

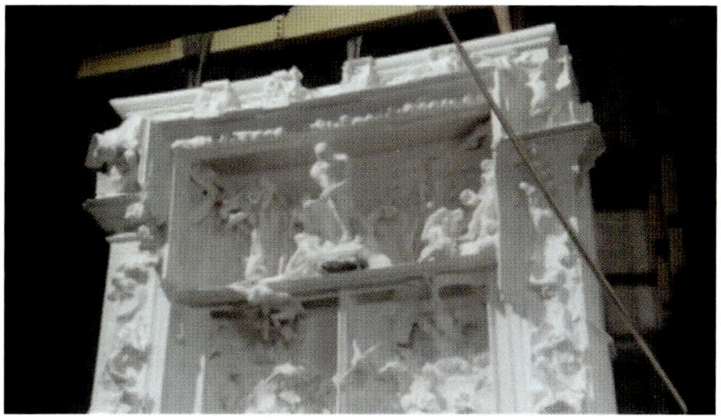

https://vimeo.com/801440105/7e9e74d298

Video 17 TV documentary (dir. Laurène L'Allinec for French National Television FR3, 1992, 20 min.) showing how the sixth edition of Auguste Rodin's (French, 1840–1917) *Doors of Hell* was cast at Fonderie de Coubertin in 1992, as requested by Musée Rodin for the Shizuoka Prefectural Museum of Art, Japan.

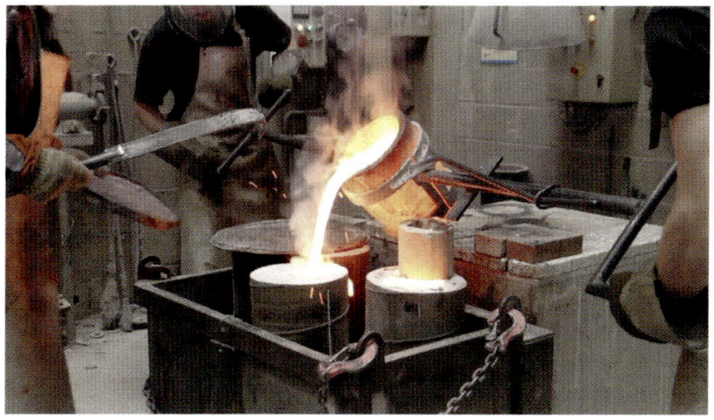

https://vimeo.com/672854085/be5c239063

Video 18 Arrangement and sequence of a bronze pour for small items in a modern foundry (dir. Christophe Béry), filmed during the July 2016 CAST:ING workshop at Fonderie de Coubertin, Saint-Rémy-lès-Chevreuse, France. Original: *Apollo of Lillebonne*, France, 1st century BCE–1st century CE, H. 193 cm (Musée du Louvre, inv. Br37 [NIII65]). See fig. 288.

Tables

This section includes fifteen synthetic tables created to provide an overview of the technical data.

Table 1
Duration of different steps in making a bronze sculpture

The time required for each step depends on many parameters, including the size of the sculpture, its desired finish quality, the casting process employed, the number of parts requiring assembly, the amount of repair and fettling necessary, the type of patina used, and more. Time also depends on the available technologies and the production environment (e.g., a specific commission in a small workshop versus industrial production of a replica in an industrial foundry). Some extreme examples are given here based on historical documents and direct knowledge of recent fabrications.

PROCESS STEP	GENERAL COMMENTS	SPECIFIC EXAMPLES	APPROX. DURATION	DETAILED PROCESS	ILLUSTRATIONS
lost-wax casting: making the wax model	Given the variety of methods used to prepare the wax, the varying complexity of models, and the range of sizes that may be encountered in lost-wax casting of bronze sculpture, time to form the wax model may vary greatly (see GI§2.2, GI§2.3).	Head of *Apollo of Lillebonne* (H. 18 cm), Fonderie Coubertin(1)	3–4 hours	Making a plaster piece mold of the 3D resin print took 2 hours, including setting of the plaster. Time to prepare the wax model depended on the process chosen: 1 hour was needed for slush molding; 1.5 hours for the indirect wax-slab process; 2 hours for lasagna.	figs. 15, 17, 24, 74, 377
		Horse head by Andrew Lacey (H. 45 cm)(5)	1 day	Making a a two-piece silicone rubber mold of the original clay model took 1 day; making the wax inter-model (slush molding) took 1.5 hours.	video 3
		Andrew Lacey's experimental replica of the *Satyr and Satyress* by Andrea Riccio (H. 18.5 cm) (after original in the V&A: A.8-1949, presented by Art Fund)	3 days	A direct process was used; the hand modeling took 3 days.	fig. 562
		Rodin's *Gates of Hell* (H. ~6 m), Fonderie Coubertin(2)	1.3 years	A total of almost 13,000 person-hours were recorded for molding the model, pouring wax in separate sections (two main sections and 136 small ones), assembly of waxes, insertion of armatures, and core. Considering 5 persons working simultaneously, this meant 64 weeks of work for the foundry.	video 17

PROCESS STEP	GENERAL COMMENTS	SPECIFIC EXAMPLES	APPROX. DURATION	DETAILED PROCESS	ILLUSTRATIONS
lost-wax casting: applying sprues and completing the refractory mold	Once sprueing is carried out (see GI§2.7) and the investment and core are set, the refractory mold is baked to dry it fully and melt out the wax (see GI§2.1.1).	Head of *Apollo of Lillebonne* (H. 18 cm), Fonderie Coubertin[(1)]	4 hours	Sprueing took 30 min.; pouring and setting of the plaster for both core and investment took 30 min.; 3 hours firing in a kiln were necessary to ensure that the plaster core and refractory mold were free of wax and moisture.	-
		Horse head by Andrew Lacey (H. 45 cm)	5 hours	Sprueing took 1 hour; making the ceramic shell investment of seven layers took 2.5 hours spread over 3 days of drying; dewaxing 30 min.; kiln firing 1.5 hours.	video 2
		Andrew Lacey's experimental replica of the *Satyr and Satyress* by Andrea Riccio (H. 18.5 cm) (after original in the V&A: A.8-1949, presented by Art Fund)	4 hours	Sprueing took 30 min.; making the investment 2 hours; dewaxing 10 min.; kiln firing 1.5 hours.	-
		Rodin's *Gates of Hell* (H. ~6 m), Fonderie Couberin[(2)]	3 months	A total of ~760 person-hours were recorded for making the ceramic-shell investment. Considering 5 persons working simultaneously, this meant 4 weeks of work for the foundry. 2 months were then needed to dry and bake the investment.	video 17
		Comparison of two monumental equestrian statues: François Girardon's monumental *Louis XIV* (cast by Balthasar Keller in 1699) and Edme Bouchardon's *Louis XV* (cast by Pierre Gor in 1758)[(3)]	1 month	Once the wax was withdrawn, it took 23 days to bake the refractory mold and the core for the Louis XIV versus more than 27 days for the Louis XV, including the cooling down. Note that the building of the huge brick refractory molds (whether buried [Louis XVI]; or not [Louis XV]) certainly required several days, although not specified by Boffrand or Mariette.	fig. 45
sand casting	The process here includes the formation of the chef-modèle, ramming the sand around the chef-modèle (including the formation of piece mold sections), formation of the core, and drying of the mold (see GI§2.4.1).	Head of *Apollo of Lillebonne* (H. 18 cm), Fonderie Coubertin[(1)]	1-2 days	Preparing the sand casting mold took <0.5 days; drying the mold took approximately 1.5 days.	fig. 62
		Two sculptures of two very different sizes by Santiago Calatra Valls, cast by Fonderie Coubertin: *Support de table homme assis* (H. 70 cm), and *Grand disque solaire fendu* (D. 300 cm)[(6)]	Small sculpture: 8 days; large one: 1.5 months	The formation of the plaster chef-modèle took 5 person-hours for the small sculpture, 35 person-hours for the large one (5/35). Ramming of the sand required 30/210 person-hours; forming of the core 25/175 person-hours. Drying of the sand was accelerated by using carbon dioxide. Considering one person alone for the small piece and two people working simultaneously for the big one, this meant ~8 days total work for the small sculpture and 1.5 months for the larger one.	-
melting the metal	Time needed to melt the metal can vary greatly and depends on the type of furnace and fuel used, the capacity of the crucible or furnace load, the quantity of metal, and the alloy composition.	Head of *Apollo of Lillebonne* (H. 18 cm), Fonderie Coubertin[(1)]	1 hour	Melting the batch of metal (~50 kg) in an induction furnace took 1 hour.	-
		Horse head by Andrew Lacey (H. 45 cm)	1.5 hours	1.5 hours needed to melt 45 kg of alloy in a gas furnace and degassed by poling with a hazel stick.	-
		Andrew Lacey's experimental replica of the *Satyr and Satyress* by Andrea Riccio (H. 18.5 cm) (after original in the V&A: A.8-1949, presented by Art Fund)	30 minutes	30 min. needed to melt 3 kg of alloy in a charcoal hearth using Cu90% Sn10%, degassed and cleaned by poling with a hazel stick (for a final sculpture weight of 1.98 kg).	-

PROCESS STEP	GENERAL COMMENTS	SPECIFIC EXAMPLES	APPROX. DURATION	DETAILED PROCESS	ILLUSTRATIONS
		Rodin's *Gates of Hell* (H. ~6 m), Fonderie Coubertin(2)	12–20 hours	8.5 tons of metal were needed for the two main pieces.Thanks to several improvements in the reverberatory furnace, the melting time dropped from 20 hours for the first edition in 1981, to 16 hours in 1992, to 12 hours in 1996. 500 kg of metal were needed for the small pieces, melted in a small gas furnace.	video 17
		Two sculptures of two very different sizes by Santiago Calatra Valls, cast by Fonderie Coubertin: *Support de table homme assis* (H. 70 cm), and *Grand disque solaire fendu* (D. 300 cm)(6)	Small sculpture: 6 hours; large one: 40 hours	150 kg of metal was melted for the small piece, 1,500 kg for the large one.	-
		Comparison of two monumental equestrian statues: François Girardon's monumental *Louis XIV* (cast by Balthasar Keller in 1699) and Edme Bouchardon's *Louis XV* (cast by Pierre Gor in 1758)	1–2 days	For the cast of *Louis XIV*, the furnace was heated for 40 hours, including 24 hours to melt the ~41 tons of metal (84,000 French pounds), versus 28 hours for the 29 tons of metal (60,000 French pounds) for the cast of Louis XV.	-
pouring of the metal	Pouring of the metal never exceeds several minutes, even for monumental bronzes (see GI§2.7).	Head of *Apollo of Lillebonne* (H. 18 cm), Fonderie Coubertin(1)	10 seconds	The pour lasted less than 10 sec. Two people were needed to take the crucible out of the induction furnace and to feed the mold. One more was needed to prevent slags from entering the mold.	video 18
		Horse head by Andrew Lacey (H. 45 cm)	30 seconds	The single-handed pour using an electric hoist lasted 30 sec.: ~10–12 sec. for the two casts with 8–10 sec. between to steady the crucible. The bronze was melted in a gas furnace and degassed by poling with a hazel stick.	video 11
		Andrew Lacey's experimental replica of the *Satyr and Satyress* by Andrea Riccio (H. 18.5 cm) (after original in the V&A: A.8-1949, presented by Art Fund)	6 seconds	This very small pour took no more than 6 sec. from a 3 kg crucible held in blacksmith tongs. The bronze was melted on a charcoal hearth using Cu90% Sn10%, degassed and cleaned by poling with a hazel stick.	video 9
		Rodin's *Gates of Hell* (H. ~6 m), Fonderie Coubertin(2)	55 seconds	To feed the main mold, 7 min. were required to transfer the metal from the reverberatory furnace to the quenouilles (holding) basin. Subsequent feeding of the mold took 55 sec.	video 17
		Two sculptures of two very different sizes by Santiago Calatra Valls, cast by Fonderie Coubertin: *Support de table homme assis* (H. 70 cm), and *Grand disque solaire fendu* (D. 300 cm)(6)	Small sculpture: 90 seconds; large one: 150 seconds	-	-
		Comparison of two monumental equestrian statues: François Girardon's monumental *Louis XIV* (cast by Balthasar Keller in 1699) and Edme Bouchardon's *Louis XV* (cast by Pierre Gor in 1758)	a few minutes	The duration of the pour is described as "en très peu de temps" (very little time) for the Louis XIV; the pour took 5:04 min. for the Louis XV.	-

PROCESS STEP	GENERAL COMMENTS	SPECIFIC EXAMPLES	APPROX. DURATION	DETAILED PROCESS	ILLUSTRATIONS
fettling, chasing, and repairs	The amount of fettling, chasing, and repairs undertaken depends on several parameters, including the success of the casting, the type and degree of chasing desired, as well as the number and type of repairs and assemblies undertaken (see I.4, I.5). Nevertheless, these are often the most time-consuming steps in the making of a bronze (see also fig. 3). For more on fettling and chasing see GI§2.9, I.6).	Head of *Apollo of Lillebonne* (H. 18 cm), Fonderie Coubertin[1]	6 hours	Fettling took 2 hours, chasing 4 hours. The hole in the face was relatively quickly repaired with a welding rod.	figs. 73, 121
		Horse head by Andrew Lacey (H. 45 cm)	6 hours	Fettling took 2 hours, chasing and repairs 4 hours.	video 11
		Andrew Lacey's experimental replica of the Satyr and Satyress by Andrea Riccio (H. 18.5 cm) (after original in the V&A: A.8-1949, presented by Art Fund)	2 hours	Fettling took 15 min., chasing and repairs 2 hours.	-
		Andrew Lacey's reconstruction of a figure (the old man) from one of a pair of sculptures formerly in the Rothschild Collection[4]	1–2 days	A more careful application of the first layers of the investment considerably reduced surface imperfections on the second replica, thus requiring half the time for chasing: 1 day instead of 2 days on the first replica.	video 13; figs. 4, 5, 105
		Rodin's *Gates of Hell* (H. ~6 m), Fonderie Coubertin[2]	1.4 years	A total of ~13,600 person-hours were recorded for fettling and chasing. Considering 5 people working simultaneously, this required 68 weeks. Note that 2 weeks were necessary for the bronze to cool down before fettling could start.	video 17
		Two sculptures of two very different sizes by Santiago Calatra Valls, cast by Fonderie Coubertin: *Support de table homme assis* (H. 70 cm), and *Grand disque solaire fendu* (D. 300 cm)[6]	Small sculpture: 1 week; large one: 3 weeks	5 person-hours were needed for the fettling of the small piece, versus 30 person-hours for the large one. Chasing lasted 32 versus 225 person-hours. Considering one person alone for the small piece and two working simultaneously for the big one, this meant ~1 week (respectively 3 weeks) of work for the foundry.	-
		Comparison of two monumental equestrian statues: François Girardon's monumental *Louis XIV* (cast by Balthasar Keller in 1699) and Edme Bouchardon's *Louis XV* (cast by Pierre Gor in 1758)	A few years	Due to major casting flaws, chasing and repairing the monumental cast of Louis XV took 5 years. No indications are given by Boffrand on the time needed to finish the cast of Louis XIV.	-
patination	For more on patination, see I.8. Other finishing processes such as metal plating and inlaying might be carried out instead of or in addition to patination.	Head of *Apollo of Lillebonne* (H. 18 cm), Fonderie Coubertin[1]	1 hour	Chemical patination took 1 hour.	-
		Horse head by Andrew Lacey (H. 45 cm)	1 hour	Chemical patination took 1 hour.	video 16
		Andrew Lacey's experimental replica of the Satyr and Satyress by Andrea Riccio (H. 18.5 cm) (after original in the V&A: A.8-1949, presented by Art Fund)	1 hour	Chemical patination took 1 hour.	-
		Rodin's *Gates of Hell* (H. ~6 m), Fonderie Coubertin[2]	4 weeks	1,560 person-hours were recorded for the chemical patination, representing 4 weeks of work.	video 17

PROCESS STEP	GENERAL COMMENTS	SPECIFIC EXAMPLES	APPROX. DURATION	DETAILED PROCESS	ILLUSTRATIONS
		Two sculptures of two very different sizes by Santiago Calatra Valls, cast by Fonderie Coubertin: *Support de table homme assis* (H. 70 cm), and *Grand disque solaire fendu* (D. 300 cm)[(6)]	Small sculpture: 2 days; large one: 2 weeks	Chemical patination took 2 days/2 weeks.	video 17

[1] Although the casting of reproductions of the Apollo heads illustrated here was carried out by inexperienced attendees of a training workshop in 2016, all of the reported durations have been adjusted for skilled craftspeople based on interviews with professional foundry workers.
[2] Three editions of the monumental *Gates of Hell* have been cast by the Fonderie Coubertin: 1981, 1992, and 1996. Details of the casting processes in use at the Fonderie Coubertin can be found in Dubos 2003; Dubos 2014.
[3] More details in Boffrand 1743, 54, 57; Mariette 1768,111; Desmas 2014.
[4] More details in Lacey 2018. See also in the present publication II.9§2.2.
[5] See Case Study 7 for a detailed description of the process.
[6] The plaster models provided by the artist were enlarged at the foundry.

Table 2
Common names of casting processes in English, French, German, and Italian

ENGLISH	FRENCH		GERMAN		ITALIAN	
	terms	**references**	**terms**	**references**	**terms**	**references**
lost-wax (casting or casting process)	procédé (ou fonte) à la cire perdue	Bewer, Bourgarit, and Bassett 2008; Lebon et al. 2016, 330; Mille 2012; Azéma 2013	Wachsausschmelzverfahren / Guss nach dem Wachsausschmelzverfahren	Weihrauch 1944	(fusione o processo) a cera persa	Giuffredi 2010, 112
direct lost-wax (casting or casting process)	procédé (ou fonte) direct(e) à la cire perdue	Bewer, Bourgarit, and Bassett 2008	Direktes Wachsausschmelzverfahren / Direktes Bronzegussverfahren	Lein 2000, 9	(fusione o processo) a cera persa con metodo diretto	Carruba 2006
indirect lost-wax (casting or casting process)	procédé (ou fonte) indirect(e) à la cire perdue	Bewer, Bourgarit, and Bassett 2008; Azéma 2013	Indirektes Wachsausschmelzverfahren / Indirektes Bronzegussverfahren	Lein 2000, 9	(fusione o processo) a cera persa con metodo indiretto	Formigli 1984
direct wax slabs (casting or casting process)	procédé (ou fonte) direct(e) avec plaques de cire	no reference	Direktes Wachsplattenverfahren	Janietz 2001, 9	fusione (o processo) diretta con lastre di cera (suggestion by F. Bewer)	no reference
indirect wax slabs (casting or casting process)	procédé (ou fonte) indirect(e) par estampage à la cire (dit à la galette ou à l'imprimé)	J. Dubos, pers. comm.; Rama 1988	Indirektes Wachsplattenverfahren	Janietz 2001, 9	fusione (o processo) indiretta con lastre di cera (suggestion by F. Bewer)	Formigli uses the term "sfoglia di cera" to describe the wax slab in this process in Formigli 1984, 109

ENGLISH	FRENCH		GERMAN		ITALIAN	
	terms	references	terms	references	terms	references
indirect wax painting (casting or casting process)	procédé (ou fonte) indirect(e) au pinceau (dit à la brosse)	J. Dubos, pers. comm.	Wachsauftragsverfahren	invented, no reference. In Eduard Uhenhuth's book on Vollständige Einleitung (1920), it is written: "Man streicht Wachs auf die innere Fläche"	fusione (o processo) indiretta a spennellatura (suggestion by F. Bewer)	no reference
slush molding (process)	procédé (ou fonte) au renversé (dit à l'italienne ou à la volée)	J. Dubos, pers. comm.; Rama 1988	Schwenkverfahren	Willer, Schwab, and Mirschenz 2017, 86	formatura a sciacquo	Giuffredi 2010, 113
lasagna (process)	procédé (ou fonte) en lasagne ou à l'épargne ou par estampage à la terre (dit à la plaque)	Bewer, Bourgarit, and Bassett 2008; J. Dubos, pers. comm. (pour "estampage à la terre")	Lasagne-Technik	Brepohl 2005, 182	(fusione o processo) alla "lasagna" (suggestion by F. Bewer)	no reference
cut-back core (process)	procédé (ou fonte) par tirage d'épaisseur (dit à la gauloise ou à la française)	Bewer, Bourgarit, and Bassett 2008; J. Dubos, pers. comm.; Rama 1988	Kernschälung	invented, no reference	(fusione o processo) con riduzione dello spessore dell'anima di fusione	no reference
sand casting	fonte au sable	Rolley 1994, 66; Baudry and Bozo 1978, 624; Bewer, Bourgarit, and Bassett 2008; Lebon et al. 2016, 330; Dubos 2003; Rama 1988	Sandformguss (casting); Sandguss / Sandgussverfahren (process)	Maaz 2010, 723	fusione a staffa OR/ALSO fusione a terra (suggestion by F. Bewer)	Giuffredi 2010, 129, Treccani
life casting	fonte sur le vif	Making and Knowing Project et al. 2020	Naturabguss	Lein 2004, 42–45	fusione dal vero	Giuffredi 2010, 176
piece-mold casting	fonte (en moule) à pièces	Rama 1988, 196	Teilformverfahren	Maaz 2010, 731	(fusione i processo) con stampi a tasselli (OR/ ALSO stampi composti) (suggestion by F. Bewer)	no reference

Table 3
Casting processes used in selected modern art foundries worldwide

Country		France		UK			Germany			Italy			USA		Cameroon	Burkina Faso	Mali	South Africa	India	Nepal	Thailand					Cambodia	China	
Foundry name (and/ or location)		Fonderie de Coubertin (St Rémy-lès-Chevreuse, Ile de France)	Fonderie Susse (Arcueil, Ile de France)	Andrew Lacey	Pangolin	Royal College of Art (London)	Hermann Noack (Berlin)	Strassacker (near Stuttgart)	Lauchhammer (near Dresden)	Fonderia Del Chiaro (Pietrasanta, Tuscany)	Fonderia Mariani (Pietrasanta, Tuscany)	Fonderia Battaglia (Milan)	Modern Art Foundry (Astoria, NY)	Shindoni (Tesuque, NM)	(Bamenda)	(Ouagadougou)	(Bamako)	Dionysus Sculpture Works (Pretoria)	(Swamimilai)		(Thonburi)	(Chiang Mai)	Brass worker village (Ubon)	Buranathai Buddha Image Factory (Pitsanulok)	Bangkok Noi Foundry (Bangkok)	It Sopheap foundry (Siem Reap)	Tongling Xinjiuding Bronze Culture Industry Co., Ltd. (铜陵新九鼎铜文化有限公司 | Tongling City, Anhui Province)	Various factories (Shanxi, Shandong, Jiangxi, Sichuan, and other provinces, e.g., Yuda Group, Shanxi province; Aerosun Corporation, Jiangsu province)
Source		Dubos, pers. comm. 2020; Dubos 2003; Dubos 2014	Ginisty 1987	Lacey, pers. comm. 2020	Lacey, pers. comm. 2020	Rome and Young 2003	Welter, pers. comm. 2020; Diehl 1927; Schulz and Baatz 1993	Kreutner, pers. comm. 2020	Welter, pers. comm. 2020; Schmidt 2011	Morigi, pers. comm. 2020	Morigi, pers. comm. 2020	Morigi, pers. comm. 2020	Bassett, pers. comm. 2020 w/ Jeffrey Spring	Lacey, pers. comm. 2020	Coutant 2005	Armbruster 1993		Nolene Gerber, pers. comm. w/ Bewer 2021	Levy et al. 2008	Craddock 2015; Furger 2017	Strahan 1997					It Sopheap, pers. comm.	Strahan, pers. comm. w/ Bewer 2021	Haiping Lian, pers. comm. w/ Bewer 2021
lost-wax processes	direct lost wax	×	×	×	×	×	×	×		×	×	×	×	×	×	×	×				×	×	×	×	×	×		
	direct wax slab			×	×			×		×	×	×							×	×								
	slush molding	×	×	×	×	×		×	×	×	×	×	×	×				×	×		×				×	×	×	×
	indirect wax slab	×	×			×		×		×	×	×																
	indirect wax painting	×	×	×	×	×	×	×	×	×	×	×	×	×														
	lasagna																											
	cut-back core	×																										
other processes	sand	×	×		×	×	×	×	×															×	×			×
	piece-mold			×																								

Table 4

Physical characteristics relating to different casting processes observable on a bronze sculpture

For more details on casting processes please refer to the General Introduction (§2). Warning: It is essential to read the full details in the text (I.1§2) to avoid any confusion. Whereas the presence of a particular feature can rule out a particular process, the reverse is not true. For instance, the absence of a fingerprint on a cast does not discount the lost-wax process as the method of facture. Further, the features mentioned here are the most common ones encountered for a given process carried out under "normal" conditions, but any accident may imply diversion from the usual process and generate unexpected features. If, for example, in a direct lost-wax process some of the core has to be reworked in order to adjust the inner thickness of the wax before pouring the bronze, small spherical globules and/or fingerprints may appear on the internal surface of the finished bronze.

LOCATION OF EVIDENCE	**DISTINCTIVE FEATURES**	**PRESSING MOLD PARTS onto a REMOVABLE, HARD PATTERN**			**EMBEDDING of COMBUSTIBLE MODEL in MOLD and BURNING IT OUT**						
		sand casting	**clay piece mold**	**cuttlebone**	**lost wax**						**life cast (solid)**
					direct		**indirect**				
					traditional	**wax slab**	**lasagna**	**cut-back core**	**slush molding**	**wax slab**	
external surface	fingerprints	X			X	X	X	X	X	X	
	seam line (due to wax piece mold, refractory mold, etc.)	X	X	X			X	X	X	X	X
internal surface	soft/round contours	depends on the model and mold material	X			X			X		
	geometric contours	possibly	X		possibly		possibly	X			
	small spherical globules					if using a poured core		if using a poured core	possibly	possibly	
	fingerprints	rare	X		X	X	possibly		X	X	
	brush or other tool marks	X	possibly		X	X	X		X	X	
	drip marks					X			X	X	
	core flashing	relatively uncommon and straight	X		if hollow cast or relief	X	X	X	X	X	
in the metal wall	wax-to-wax joints					X			X	X	
	inclusion of ash										X

Table 5

Examination and analytical techniques used in the study of bronze sculpture, sorted by process steps

Refer to tables 10 and 13 and the relevant chapters in volume II for the signification of acronyms, and advantages and limitations of each examination and analytical technique. For each process step, there are numerous risks of misidentification/misinterpretation, so refer to the relevant chapters for more details.

PROCESS STEPS	METHODS OF EXAMINATION(1)		METHODS OF ANALYSIS(2)		PROVIDE INFORMATION ON:	SEE
	Usual techniques and tools	**Rarer techniques(3)**	**Usual techniques and tools**	**Rarer techniques(4)**		
Casting processes	Naked-eye examination (good lighting at different angles, including raking light and RTI), portable loupe, stereo microscope, magnet, borescope or endoscope, X-radiography	3D scanning/ photogrammetry, high energy X-rays (1 MeV and beyond) or gamma or neutron radiography, computerized tomography (CT), ultrasonic testing, thermography	-	Carbon-14 dating (of iron armatures)	Method of fabrication of the models, degree of modeling, type of cast (hollow, solid, direct, indirect, other), casting method, possible absolute and relative dating	I.1 II.8
Core	Naked-eye examination, borescope or endoscope, stereo microscope, X-radiography	High-energy X-rays (1 MeV and beyond) or gamma or neutron radiography, computerized tomography (CT)	• Petrographic analysis (PLM) • Microchemical tests, XRD, XRF, SEM	• Cathodoluminescence microscopy • Elemental analysis: LIBS, PIXE, PIGE, EPMA, AAS, ICP-AES, (LA-)ICP-MS, NAA • Luminescence and carbon-14 dating	Identification of core materials, method of fabrication, provenance of core materials, possible absolute and relative dating	II.7 II.8
Metals(5)	Naked-eye examination, olfactory testing	-	• Elemental analysis without sampling (XRF) • Elemental analysis with sampling (microchemical tests, XRF, SEM-EDS)	• Elemental analysis without sampling (LIBS, PIXE, PIGE, LA-ICP-MS) • Elemental analysis with sampling (EPMA, NAA, AAS, ICP-AES, ICP-MS) • Metallography • Testing of physical and chemical properties	Identification of the alloy, characterization of its properties, provenance of raw materials, possible relative dating	I.2 II.5
Casting defects	Naked-eye examination (good lighting at different angles, including raking light and RTI), portable loupe, stereo microscope, borescope or endoscope, X-radiography	3D scanning/ photogrammetry, RTI, high-energy X-rays (1 MeV and beyond) or gamma or neutron radiography, computerized tomography (CT), ultrasonic testing, thermography, eddy currents, image analysis	-	-	Quantity, form, distribution and location of defects, causes of flaws	I.3 fig. 150 (defects diagram)

PROCESS STEPS	METHODS OF EXAMINATION[1]		METHODS OF ANALYSIS[2]		PROVIDE INFORMATION ON:	SEE
	Usual techniques and tools	**Rarer techniques[3]**	**Usual techniques and tools**	**Rarer techniques[4]**		
Repairs	Naked-eye examination (good lighting at different angles, including raking light and RTI, and UV light for nonmetallic fills), portable loupe, stereo microscope, borescope or endoscope, X-radiography	3D scanning/ photogrammetry, RTI, high-energy X-rays (1 MeV and beyond) or gamma or neutron radiography, computerized tomography (CT), ultrasonic testing, thermography, eddy currents, image analysis	• Elemental analysis without sampling (microchemical tests, XRF) • Elemental analysis with sampling (XRF, SEM-EDS)	• Elemental analysis without sampling (LIBS, PIXE, PIGE, LA-ICP-MS) • Elemental analysis with sampling (EPMA, AAS, ICP-AES, ICP-MS) • Structural analysis without sampling (for nonmetallic fills) (XRD, FTIR) • Structural analysis with sampling (for nonmetallic fills) (GC-MS, FTIR, Raman) • Metallography	Techniques and materials used for repairing, skills of the founder, history of the piece (later repairs), possible relative attribution/dating	I.4 fig. 167 (repairs diagram)
Assemblies	Naked-eye examination (good lighting at different angles, including raking light and RTI), portable loupe, stereo microscope, borescope or endoscope, X-radiography	High-energy X-rays (1 MeV and beyond) or gamma or neutron radiography, computerized tomography (CT), ultrasonic testing, thermography, eddy currents	• Elemental analysis without sampling (microchemical tests, XRF) • Elemental analysis with sampling (XRF, SEM-EDS)	• Elemental analysis without sampling (LIBS, PIXE, PIGE, LA-ICP-MS) • Elemental analysis with sampling (EPMA, AAS, ICP-AES, ICP-MS) • Metallography	Casting plan, assembly techniques and materials, possible relative dating	I.5 fig. 201 (assemblies diagram)
Tool marks	Naked-eye examination (good lighting at different angles, including raking light and RTI), portable loupe, stereo microscope, borescope or endoscope (for inside marks), X-radiography	Silicone rubber impression, 3D scanning/ photogrammetry, RTI, high-energy X-rays (1 MeV and beyond) or gamma or neutron radiography, computerized tomography (CT), digital microscopy, rugisometry/tribology techniques (interferometric microscopy, profilometry), microhardness, neutron diffraction	-	-	Degree and type of cold working/working on the model, recurrence of tool marks, artist's involvement in the process, history of piece (original appearance, later interventions, present condition)	I.6 fig. 233 (flow chart)
Gilding and plating	Naked-eye examination (good lighting at different angles, including raking light), borescope or endoscope (for drips of gilding on interior), portable loupe, stereo microscope, X-radiography	High-energy X-rays (1 MeV and beyond) or gamma or neutron radiography, computerized tomography (CT)	• Elemental analysis without sampling (microchemical tests, XRF) • Elemental analysis with sampling (XRF, SEM-EDS)	• Elemental analysis without sampling (LIBS, PIXE, PIGE, RBS, LA-ICP-MS) • Elemental analysis with sampling (EPMA, AAS, ICP-AES, ICP-MS) • Depth chemical profiling (LA-ICP-MS, RBS)[6] • Metallography	Techniques and materials, including thickness of plating, history of piece (original appearance, later surface treatment, present condition)	I.7 II.6 fig. 304 (flow chart)

PROCESS STEPS	METHODS OF EXAMINATION[1]		METHODS OF ANALYSIS[2]		PROVIDE INFORMATION ON:	SEE
	Usual techniques and tools	Rarer techniques[3]	Usual techniques and tools	Rarer techniques[4]		
Patina	Naked-eye examination (good lighting at different angles, including raking light and RTI, and UV light), portable loupe, borescope or endoscope (for corrosion on interior), stereo microscope	-	• UV light • Elemental analysis without sampling (microchemical tests, XRF) • Elemental analysis with sampling (XRF, SEM-EDS)	• Elemental analysis without sampling (LIBS, PIXE, PIGE, RBS, LA-ICP-MS) • Elemental analysis with sampling (EPMA, AAS, ICP-AES, ICP-MS) • Depth chemical profiling (LA-ICP-MS, RBS, NRA, SIMS, Auger spectroscopy, GDL) • Structural analysis without sampling (XRD, FTIR) • Structural analysis with sampling (GC-MS, FTIR, Raman) • Metallography • Color measurement	Techniques and materials, including thickness and layering of patina, history of piece (original appearance, later surface treatment, present condition), possible relative dating	I.8 II.6 figs. 305, 307, 338 (patina diagram and flow charts)
Inlays and overlays	Naked-eye examination (good lighting at different angles, including raking light and RTI and UV light), portable loupe, stereo microscope, X-radiography	3D scanning, RTI, high-energy X-rays (1 MeV and beyond) or gamma or neutron radiography, computerized tomography (CT)	• UV light • Elemental analysis without sampling (microchemical tests, XRF) • Elemental analysis with sampling (XRF, SEM-EDS)	• Elemental analysis without sampling (LIBS, PIXE, PIGE, RBS, LA-ICP-MS) • Elemental analysis with sampling (EPMA, AAS, ICP-AES, ICP-MS) • Structural analysis without sampling (XRD, FTIR) • Structural analysis with sampling (GC-MS, FTIR, Raman) • Metallography	History of piece (original appearance, later surface treatment, present condition)	I.9 fig. 340 (inlays diagram)

1 For more details on examination techniques, see II.1, II.2, II.3.
2 For more details on analytical techniques, see II.5, II.6.
3 More details in Boffrand 1743, 54, 57; Mariette 1768,111; Desmas 2014.
4 By "rarer" techniques, we mean techniques less available in most laboratories devoted to cultural heritage. Note that this may vary from country to country.
5 This includes repairs.
6 Techniques such as Secondary Ions Mass Spectrometry (SIMS), Auger spectroscopy, and glow discharge ICP are not mentioned here since they are devoted exclusively to modern materials science (thin films in electronics, notably).

Table 6
Main properties of different copper alloys used for bronze sculpture

Castability ratings are based on Schmidt and Schmidt 1992, 1155; weldability ratings are based on Rogers 1993, 1874; color names are based on Tyler and Black 1992, 815. All contents in weight %.

	melting temperature	castability	machinability	weldability (fusion welding)	ability for fire gilding	ability for chemical patina	corrosion resistance in air	color
unalloyed copper	high	poor	poor	poor	good	good	poor	soft pink
arsenical alloy (e.g., 3% As)	medium	medium	?	?	?	?	?	soft pink

	melting temperature	castability	machinability	weldability (fusion welding)	ability for fire gilding	ability for chemical patina	corrosion resistance in air	color
tin-bronze (e.g., 10% Sn)	medium	medium	poor	good	medium	poor	good	bronze-gold
zinc-containing alloy (e.g., 15% Zn)	medium	medium	poor	poor	medium	medium	medium	gold
leaded alloy (e.g., 10% Sn & Pb)	medium	good	good	depends on other additions	poor	depends on other additions	depends on other additions	bronze-gold
phosphor bronze (e.g., 0.02% P)	high	good	poor	excellent	good	good	poor	soft pink
silicon bronze (e.g., 4% Si)	medium	good	medium	excellent	?	poor	excellent	gold

Table 7

Chemical symbols of main copper alloying elements and impurities encountered in bronze sculpture

Each element may impact the physical and chemical properties of the metal. The effects are various and depend on how the different elements are combined; see St. John 1931.

	Symbol	English	French	German	Italian
	Cu	copper	cuivre	Kupfer	rame
main alloying elements known for historical and contemporary bronzes	Sn	tin	étain	Zinn	stagno
	Zn	zinc	zinc	Zink	zinco
	Pb	lead	plomb	Blei	piombo
alloying elements specific to contemporary bronzes	Si	silicon	silicium	Silizium	silicio
	P	phosphorous	phosphore	Phosphor	fosforo
main impurities in historical bronzes	Ag	silver	cuivre	Silber	argento
	As	arsenic	arsenic	Arsen	arsenico
	Au	gold	or	Gold	oro
	Bi	bismuth	bismuth	Bismut	bismuto
	Co	cobalt	cobalt	Kobalt	cobalto
	Fe	iron	fer	Eisen	ferro
	Ni	nickel	nickel	Nickel	nichel
	S	sulphur	soufre	Schwefel	zolfo
	Sb	antimony	antimoine	Antimon	antimonio

Table 8

Common names used to describe copper alloys for bronze sculpture

As determined by eighteen CAST:ING group members, who include specialists in a wide range of disciplines and fields. Their agreement or disagreement on nomenclature is reported through the percentage of people using a given name. The preferred names of the author are reported in the gray cells. Arsenical bronzes, aluminum bronzes, and more exotic alloys such as copper-antimony, cupro-nickel, and tumbaga are not reported since they are rarely if ever used in bronze sculpture. For exotic prehistoric and historic copper alloys, see Northover 1998. The author wishes to recognize the contributors to this table, namely J. Bassett, A. Boulton, M. Castelle, P. Dandridge, L. Garenne-Marot, J. Kreutner, A. Lacey, S. La Niece, E. Lebon, B. Mille, L. Morigi, P. Motture, D. Reid, D. Robcis, D. Strahan, S. Sturman, and the late J.-M. Welter. All contents in weight %.

Type	Name	Description	Most common compositions (weight %)	Comment	Agreement among the 17 CAST:ING specialists (%)
copper	copper	copper with no intentionally added elements	various (see comment)	Various definitions depending on the CAST:ING member: <1% Sn, <1% Pb, >95% Cu, or an alloy with ≤3% of Sn, Zn, and/or Pb, may be called copper. The 1% threshold is definitely arbitrary.	71
			<1% Sn, <1% Pb		6
	impure copper		>96% Cu, <1% Sn, <1% Zn, <2% Pb		6
	low alloyed copper	copper with minor addition of other elements	>95% Cu		6
	mixed copper	copper with minor addition of other elements	>92% Cu, 1–2% Sn, 1–2% Zn	rarely used	6
bronze	bronze	copper alloyed with tin	5–13%	The minimum amount of tin for an alloy to be called bronze is difficult to state (see "bronze"). In the literature devoted to bronzes (sculpture and other implements), it is rare to encounter a bronze with <5% Sn.	76
	low-tin bronze	copper alloyed with small amounts of tin	5–8% Sn	This designation has been suggested by 20% of the members, yet they do not agree on the definition. For some, it should hold <5% Sn, for others, 5–8%.	6
			<5% Sn		6
			2–7% Sn		6
	medium-tin bronze	copper alloyed with moderate amounts of tin	8–15% Sn		6
			7–14% Sn		6
	high-tin bronze	bronze with high tin	from 12–15% Sn (depending on the author) to ≥25% Sn	Typical alloys for bells, but encountered in sculptures as well (India, Southeast Asia, etc.).	76
	bell metal	copper-tin alloy with tin >15%	>15% Sn		12
brass	brass	copper alloyed with zinc	5–15% Zn	The minimum amount of zinc for an alloy to be called brass is difficult to state. In the literature devoted to bronzes (sculpture and other implements), it is rare to encounter a brass with <6–7% Zn.	65
	low-zinc brass	copper alloyed with small amounts of zinc	5–10% Zn	This designation has been suggested by one-third of the members. Yet whereas there is consensus on the minimum content (5% Zn), the maximum content varies from 15% to 30% according to the member.	6
			5–30% Zn		18
			2–7% Zn		6
	medium-zinc brass	copper alloyed with moderate amounts of zinc	10–20% Zn		6

Type	Name	Description	Most common compositions (weight %)	Comment	Agreement among the 17 CAST:ING specialists (%)
	high-zinc brass	brass with large amounts of zinc	>30% Zn		24
			15–30% Zn		53
			>20% Zn		6
			>14% Zn		6
	latten		Zn as the primary alloying element, with <1% Sn and <1% Pb		12
ternary alloys	gunmetal	copper alloyed with zinc and tin	4–8% Sn, 4–8% Zn	This term is widely used in the literature, yet it can be anachronistic, as guns did not appear before the Late Middle Ages. Some authors prefer "red brass." See also comment in "bronze."	24
	red brass	copper alloyed with zinc and tin	4–8% Sn, 4–8% Zn	See comments for "gunmetal" and "bronze."	24
leaded alloys	leaded alloy	any copper alloy with lead added	3–10% Pb	Same comment as for brass regarding the minimum amount of lead.	65
			Bronze and brass alloys with a Pb concentration of ≥4%, and Cu alloys with a Pb content of ≥2%, are described as "leaded."		6
	low-lead alloys	any copper alloy with small amounts of lead added	<1% Pb		6
	high-lead alloy	any copper alloy with large amounts of lead	15–30% Pb		41
			>10% Pb		6
quarternary alloy	quarternary alloy	any copper alloyed with tin, zinc, and lead		Easy way to describe an alloy containing tin, zinc, and lead as intentionally added elements without having to decide on a specific name.	24
arsenic	arsenical copper or arsenical bronze	any copper alloy containing high amounts of arsenic	1–3% As		18
contemporary alloys	silicon bronze	copper alloyed with silicon	≤5% Zn	Welds easily, but difficult to chase and patinate.	12
			3.5–4.5% Si, 0.5% Mg, 0.5% Fe	Developed around the 1880s–90s, silicon bronze was first used for the underwater telegraph cables connecting the UK to the continent. Silicon adds ductility to the alloy and lowers the viscosity of the melt, while magnesium and iron remove oxygen from the melt.	6
	silicon brass	copper alloyed with silicon and zinc	>5% Zn		6
	phosphor bronze	bronze with phosporous added intentionally	0.5–11% Sn	Developed around the 1880s–90s, phosphorus increases wear resistance and helps remove oxygen from the melt.	6

Table 9
Hardness and machinability of various modern as-cast copper alloys

After Tyler and Black 1992, table 3. Machinability is scaled against the most machinable alloy, namely "free-cutting brass C3600" (35.5% Zn, 3% Pb). Names of alloys are those most commonly used for bronze sculpture.

Alloy name (author's nomenclature)	Alloy number (Amercian Society of Metals [ASM] nomenclature)	Composition (wt%)	Hardness (Brinell 500kg)	Machinability (%)
pure copper	C80100	99.95% Cu	44	10
tin bronze	C90200	7% Sn	70	20
tin bronze	C90700	11% Sn	80	20
leaded tin bronze	C92700	10% Sn, 2% Pb	77	45
leaded tin bronze	C93700	10% Sn, 10% Pb	60	80
ternary alloy	C90500	10% Sn, 2% Zn	75	30
quaternary alloy	C92600	10% Sn, 1% Pb, 2% Zn	70	40
quaternary alloy	C83300	1.5% Sn, 1.5% Pb, 4% Zn	35	35
quaternary alloy	C83600	5% Sn, 5% Pb, 5% Zn	60	84
quaternary alloy	C83800	4% Sn, 6% Pb, 7% Zn	60	90
brass	C83400	10% Zn	?	60
high-zinc brass	C85400	1% Sn, 3% Pb, 29% Zn	50	80
silicon bronze	C87200	4% Si (89% min Cu)	85	40

Table 10
Scientific analytical techniques used in the study of bronze sculpture

		Method of analysis	Materials	Sampling requirements and preparation	Advantages	Drawbacks and limitations	See
Microchemical testing	**Minimally invasive techniques**	Microchemical/ solvent tests	• Core: elemental composition, testing for the presence of carbonates and gypsum • Alloy • Repairs • Patina • Inlays and overlays • Gilding and other metal plating	Tests can often be performed under the microscope using very small samples	• Economical, rapid, very precise way to qualitatively determine the presence of specific elements or substances • Quantitative analysis is possible in specialized laboratories	• The range of compounds that can be identified is limited • In many cases, tests can only confirm the presence or absence of a given material or element • Only those elements are found that are specifically looked for • Can be misleading, as it may indicate the presence of a substance whose quantity may be relatively minor, or even an incidental or accidental inclusion • Detection limits are rarely known • Best used in conjunction with other methods	II.6§3.4

		Method of analysis	Materials	Sampling requirements and preparation	Advantages	Drawbacks and limitations	See
Elemental anlaysis	**Non- or minimally invasive techniques**	X-ray fluorescence analysis (XRF)	• Core • Alloy • Repairs • Patina • Inlays and overlays • Gilding and other metal plating	• Useful information can be obtained without sampling • Quantitative analysis of alloys may require polishing a spot (2–10 mm diameter) to remove corrosion and expose base metal	• Rapid, multi-element characterization • Inexpensive • Many analyses can be made over large areas • Quantitative alloy analysis is possible • Elemental mapping possible	• Rigorous quantitative analysis is difficult • Geometry can affect readings (flat surfaces are ideal; curves can create issues) • Not effective for light elements such as carbon, oxygen, and sodium • Only a thin surface layer is analyzed for most materials	II.5§2.1, II.6§2.2, II.7§2.1.4
		Particle induced X-ray emission (PIXE) and particle-induced gamma-ray emission (PIGE)	• Core • Alloy • Repairs • Patina • Inlays and overlays • Gilding and other metal plating	• Useful information can be obtained without sampling • Quantitative analysis of alloys may require polishing a spot (0.5–5 mm diameter) to remove corrosion and expose base metal	Relatively high spatial resolution (50 μm) enables mapping of complex patina or gilding layers as well as specific surface features (repairs, assemblies, inlays, overlays)	• Expensive • Rarely available as a service analysis; limited to a few laboratories • Most laboratories cannot analyze objects larger than a few cm high • Only a very thin surface layer is analyzed for most materials	II.5§2.2, II.6§2.2, II.7§2.1.5
		Rutherford backscattering spectrometry (RBS) and nuclear reaction analysis (NRA)	• Patina • Gilding and other metal plating	Useful information can be obtained without sampling	Depth profiling possible (analysis of different strata for patina and plating)	• Expensive, not widely available (few laboratories can do such analysis) • Not suitable for complex layered stuctures • Needs to be combined with other techniques Note: for NRA there is no issue of decontamination, as there is with neutron analysis	II.6§2.3
		Laser-induced breakdown spectroscopy (LIBS)	• Core • Alloy • Patina • Inlays and overlays • Gilding and other metal plating	Useful information can be obtained without sampling	• Rapid, multi-element characterization • Inexpensive • Able to analyze light elements • Depth profiling is possible (analysis of different strata)	• Minimally destructive; creates many small ablation craters (<1 mm diameter) • Small analysis spot may yield unrepresentative results • Not widely accepted as a method for rigorous quantitative analysis	II.5, II.6§2.3
		Laser ablation inductively coupled plasma – mass spectrometry (LA-ICP-MS)	• Core • Alloy • Patina • Inlays and overlays • Gilding and other metal plating	Sampling may be necessary if object is not small (see "atomic spectroscopies" below)	• Rapid, multi-element characterization • Able to analyze light elements • Depth profiling is possible (analysis of different strata)	• Minimally destructive; creates many small ablation craters (<<1 mm diameter) • Small analysis spot may yield unrepresentative results • Quantitative analysis of copper alloys is difficult due to matrix effects • Only for small sculptures if sampling is not possible (although specific extensions are under development that allow analysis of large sculptures)	II.6§2.3

		Method of analysis	Materials	Sampling requirements and preparation	Advantages	Drawbacks and limitations	See
		Neutron diffraction	• Core • Alloy	No sampling required	• Able to measure the composition of the metal throughout the thickness of the metal wall • Possible to detect evidence of cold working	• Highly specialized technique requiring large-scale research facilities, typically nuclear reactors • Only certain elements can be detected and quantified using neutron diffraction • The error of measurement is relatively high compared to other techniques • Some objects may need to be quarantined for decontamination (de-activation) at the neutron facility for anywhere from several minutes to several days before returning to the owner depending on a number of parameters, including neutron flux and energy, alloy composition, etc., which can create temporary radioactivity of the material	II.5§2.3.3
	Invasive techniques	Scanning electron microscopy (SEM-EDS or SEM-WDS)	• Core • Alloy • Repairs • Patina • Inlays and overlays • Gilding and other metal plating	Small samples, <1 mm^2, usually embedded in cross section and carefully polished	• Best spatial resolution of all techniques (beam spot size of a few nm) • Enables precise mapping of complex patina or gilding layers	Expensive	II.5§3.4, II.6§3.3, II.7§2.1.3
		Atomic spectroscopies: atomic absorption spectrometry (AAS) and inductively coupled plasma – atomic emission spectroscopy (ICP-AES)	• Core • Alloy	• 50–100mg of drillings (or pieces) for cores • approx. 20 mg of clean metal drillings (or pieces) for alloys	• High sensitivity for trace element analysis • Much more sensitive than XRF or SEM-EDS	More laborious, more expensive, and less easily accessible than EDXRF	II.5§3.1
		Inductively coupled plasma – mass spectrometry (ICP-MS)	• Core • Alloy	• 50–100mg of drillings (or pieces) for cores • approx. 20 mg of clean metal drillings (or pieces) for alloys	The most sensitive technique for elemental analysis (approx. 100 times more sensitive than ICP-AES)	• The most laborious of the invasive spectroscopic techniques • More expensive and less accessible than atomic spectroscopies (including ICP-AES), although in the 2010s it has greatly evolved and is slowly replacing atomic spectroscopies	II.5§3.2, II.7§2.1.1
		Neutron activation analysis (NAA)	• Core • Alloy	50–100 mg of sample material (alloy or core)	Relatively good sensitivity and accuracy on a large range of elements	• Expensive, not available as a service analysis for bronzes • Some objects may need to be quarantined for decontamination (see "neutron diffraction" above)	II.5§3.3, II.7§2.1.2
Isotope analysis (mainly Pb isotopes)	**Invasive technique**	Multi-collector – inductively coupled plasma – mass spectrometry (MC-ICP-MS)	• Alloy • Metallic repairs • Metallic inlays and overlays	Small sample of material (approx. 20 mg for alloys)	Useful for provenancing of metals	• Relatively expensive • Requires a database and a good knowledge of the prehistoric/historic context to be useful • Lead isotopes have limited utility for intentionally leaded alloys	II.5§5

		Method of analysis	Materials	Sampling requirements and preparation	Advantages	Drawbacks and limitations	See
Structural analysis*	**Invasive techniques**	Optical microscopy (metallography)	Metal	Polished samples of a few mm^2 to 1 cm^2 are usually required for the study of microstructure	• Microstructure can be used to distinguish as-cast metal, cold-worked metal, and annealed metal • Can help to identify joinery techniques such as welding and brazing • Can be used to characterize corrosion and patina	Large sample size	II.5§6
		Fourier transform infrared spectroscopy (FTIR) and Raman spectroscopy	• Organic and inorganic fills for repair • Organic and inorganic patina • Organic and inorganic inlays and overlays	• Samples <<1 mm^2 required • Microscopic samples suffice if FTIR microscopy is available	Rapid and easily accessible; most laboratories have this technique (at least FTIR)	• Spectra require interpretation by a scientist experienced in artworks • Reference spectra are required for comparison; some materials may not have adequate reference spectra • Complex mixtures generate complex spectra that are not easy to interpret; best for pure materials or simple mixtures • Often complemented by other techniques	II.6§2.4, II.6§3.4
		Gas chromatography with mass spectrometry (GC/MS or pyrolysis (Py)-GC/MS)	• Organic fills for repair • Organic patina • Organic inlays and overlays	Microscopic samples required	• Qualitative and quantitative analysis (sensitivity typically in ppb) of organic materials • Gas chromatography has the advantage over infrared spectrometry that it permits the identification of complex mixtures	• Chromatograms and spectra require interpretation by an experienced scientist • Reference spectra are required for comparison; some materials may not have adequate reference spectra	II.6§3.4
		X-ray diffraction (XRD)	• Core • Inorganic fills for repairs • Inorganic patina • Inorganic inlays and overlays (other than metals)	Microscopic samples required, although portable instruments are becoming available that may enable in situ measurements	• Quite rapid • Very informative, often a necessary step in the identification of compounds • Widely available	• Some mineral compounds are difficult to detect (SnO_2, sulfur compounds, etc.) • Complex mixtures may be hard to decipher, though software to aid interpretation is improving	II.6§2.4, II.6§3.4
Petrography**	**Invasive techniques**	Optical microscopy (petrography)	Core	Samples of a few mm^2 to 1 cm^2 are required for thin section preparation	• Rapid • Widely available • Used in conjunction with digital image analysis, many distinctive characteristics of a core can be quantified (such as porosity and particle size, shape and distribution)	May not be sufficient to fully identify the material	II.7§2.1

		Method of analysis	Materials	Sampling requirements and preparation	Advantages	Drawbacks and limitations	See
		Cathodoluminescence (CL) microscopy	Core	Polished samples of a few mm^2 to 1 cm^2 are usually required	• Most scanning electron microscopes can be fitted with a CL detector • Rapid imaging • Can be useful for classifying mineral species and core types, usually in conjunction with other techniques such as SEM-EDS	• Rarely applied to the study of core material in bronzes to date • Relies on comparative analysis; currently few documented samples are available for comparison • Not widely available in cultural heritage laboratories	II.7§2 .2.2
Dating	**Invasive techniques**	Trapped charge dating; thermoluminescence dating (TL) and optically stimulated luminescence (OSL)	Core: mineral constituents	100–200 mg of core material uncontaminated by radiation (including UV radiation); samples must be taken in the dark	• An absolute dating method (if the annual dose can be measured) • Applies to all cores containing silicates (quartz, feldspars, etc.) • Although not routinely used for bronzes, it is becoming more widespread • New, single-crystal OSL techniques offer the potential for more precise and robust dating	• Calculations require estimation of many parameters, leading to imprecision (results of analyses of the same material by different laboratories do not always coincide) • Analysis of core that was not sufficiently heated during casting can yield an overestimate of age • Exposure to subsequent sources of radiation (such as X-rays during radiographic exams) can yield an overestimate of age • Few labs currently have the equipment and experience necessary	II.8§1
		Radiocarbon dating (carbon-14)	Core: organic constituents	2–5 mg	• Simple collection procedure if the core is accessible and has substantial organic content • Readily available as a service analysis in many radiocarbon laboratories	Errors are possible if old organic matter was naturally present in the core's raw materials, particularly in clays	II.8§2
		Radiocarbon dating (carbon-14)	Armature (carbon in iron)	100–1500 mg depending on the carbon content (0.8 %wt of C => 125 mg, 0.3 %wt of C=> 333 mg , etc.)	Currently the only accepted method to date iron armatures	• Large sample required • Very recent technique, not available as a service analysis, few laboratories are able to undertake such analysis	II.8§2

***** See II.6§2.1 for a definition of "structural analysis."

****** Although petrography enables structural analysis, it is presented separately, since for bronze sculpture it is exclusively devoted to core analysis.

Table 11

The four main types of defects on a bronze sculpture

Each defect type may in turn be divided into several subcategories (see [fig. 150](/visual-atlas/150/)).

TYPE of DEFECT	DESCRIPTION
Avoidable defects occurring BEFORE the casting	model mishandling (sagging of a wax model placed in an overheated environment, wear on the surface, etc.)
	misfortunes (core displacement, mold cracks, etc.)
Defects inherent to the process	shrinkage
	protrusions (sprues, etc.)
	seam lines
	holes (armatures, core pins, etc.)
Avoidable defects occurring AFTER the casting	downstream processing (incompetent or incomplete finishing)
Later damage	mechanical damage (reuses, lost parts, etc.)
	corrosion
	surface disfigurement (vandalism, harsh cleaning)

Table 12

Trilingual summary of the seven most common casting defect categories

As adopted by the metallurgical industry (after Mascré, Thomas, and Hénon 1952 and Hénon, Mascré, and Blanc 1971). In industry, each category is hierarchically subdivided into three levels (not shown here), and thus any defect can be precisely characterized using a three-character code.

Category (after Hénon, Mascré, and Blanc 1971)	English	French	German	Example
A	metallic projections	excroissances métalliques	metallische Auswüchse	flashing
B	cavities	cavités	Hohlräume	pores
C	discontinuities	défauts linéaires	Unterbrechung des Zusammenhangs	cold shut
D	defective surface	défauts de surface	fehlerhafte Oberfläche	orange peel
E	incomplete casting	pièce incomplète	unvollständiges Stück	hole
F	incorrect dimensions or shape	dimension ou forme incorrecte	ungenaue Maβe oder Form	shrinkage
G	inclusions or structural anomalies	inclusions ou anomalies internes	Einschlüsse und Gefügeanomalien	inclusions

Table 13

Technical examination techniques used in the study of bronze sculpture

Method of examination	Surfaces	Structure	Materials	Importance and advantages	Drawbacks and limitations*	See
Naked-eye examination	examination, description, and measurement of technical features on the inner and outer surfaces, including: • tool marks • color • coating • casting artifacts • ondition issues, etc.	identification, description, and measurement of visible joints, repairs, internal supports, inlays, etc.	examination and description of materials, identification of good sampling locations	• important first step of any examination/ analysis • essential for comparison with other images (X-ray, UV, etc.); universally accessible, allows good overview of an artifact	limited to visually accessible areas	II.1§2 II.1§3 II.2§1 II.4
Stereo microscopy and digital microscopy	provides information on the surface of bronzes: • tool marks • color • coatings • restorations	provides information on the structure of the metal, joints, repairs, surface coatings, core material	useful for the removal, preparation, and examination of samples of core material, surface coatings, corrosion products, and metal, and for observation of microchemical tests	• useful supplement to naked-eye examination and generally essential for sampling • can reveal many significant details not visible to the naked eye	• if not already available, acquisition of a stereo microscope can be expensive • many microscopes are set up for viewing samples on a horizontal stage that cannot be adapted to sculpture • binocular microscopes on stands that allow free movement of the head can be very expensive	II.2§2.1
Reflectance transformation imaging (RTI)	• can greatly enhance subtle surface details on bronzes such as tool marks, defects, repairs, and inscriptions • . augments naked-eye examination and photography	can greatly enhance subtle surface characteristics associated with joints in the sculpture or the model	may show up microstructure of metal and/or pseudomorphs	can be done with a minimum investment in apparatus and software beyond standard photography equipment	investment required in time and training	II.2§2.3

Method of examination	Surfaces	Structure	Materials	Importance and advantages	Drawbacks and limitations*	See
Photogrammetry and 3D laser or structured light scans	• both methods can provide a detailed 3D model of the surface of a sculpture • photogrammetry and most 3D scanners will capture surface color and texture	3D scanning allows detailed and sophisticated measurement and comparison between related pieces; with proper setup photogrammetry will also allow precise measurements; useful in deciphering fabrication methods	-	• creation of 3D models allows direct and sophisticated comparison of dimensions between related sculptures (often in disparate locations) without having to bring them together physically • photogrammetry can be done with a minimum investment in apparatus beyond standard photography equipment	• for both methods, considerable investment is required in time and training • 3D scanning apparatus can be expensive to acquire • obviously, accurate comparison between related pieces requires that data be acquired with sufficient accuracy	II.2§5
Ultraviolet examination	helps to discern certain types of original coatings, adhesives, fillers, and overpainting	may aid in the detection of structural repairs and restorations	• fluorescence under UV is used to detect the presence of organic materials such as resins, oils, proteins, and aliphatic glues • certain pigments and minerals may also be detectable	low-cost and accessible method that allows rapid examination of objects with minimal equipment and training	requires a darkened examination space and eye protection; documentation with UV photography requires greater investment in equipment and training; does not allow identification of exact composition of materials	II.2§3.1
Endoscopy	examination and description of inner surfaces	identification and description of internal evidence of casting and assembly processes (joints, repairs, internal supports, armature, core pins, core material remains)	can help identify the type of material used for core material or armature	useful aid for visual examination of areas that are difficult to access	may be hard to maneuver, and to gauge the exact location and proportions of features observed; access may be limited	II.2§2.2

Method of examination	Surfaces	Structure	Materials	Importance and advantages	Drawbacks and limitations*	See
X- and gamma radiography	examination, description, and rough measurement of technical features on the inner surfaces, including casting artifacts, inlays, condition issues, etc.	• reveals internal structure and technical features (e.g., joints, repairs, internal supports) • reveals internal structure and technical features (e.g., joints, repairs, internal supports)	records differences in physical and chemical density; as a result it may: • discriminate between different inlays • reveal metal plating such as gilding • reveal the presence or absence of core	radiography is a key tool in technical studies of bronzes: • can provide a great deal of detailed information related to fabrication method (especially of hollow casts) that other methods cannot provide, particularly if no access to inside of hollow cast and endoscopy is not possible • radiographs are often very aesthetic and didactic • real-time radiography is helpful in selecting angles	• equipment required is expensive and not standard in museums, including dedicated space • requires skilled technicians experienced with industrial methods and bronze sculpture • serious health and safety issues, especially with portable equipment and gamma radiography (both higher kV than medical) • radiographs can be difficult to interpret, requiring training and experience; multiple views are helpful; multiple exposures of each view may also be helpful, particularly for sculptures with large variations in metal thickness • requires comparison with object • will render future luminescence dating of the core material difficult	II.3§1
Neutron radiography	although some surface features might be revealed, X-ray is much preferable	reveals artifacts made of a lower-density material than their surrounding: typically organic fibers, paper, etc. trapped in the bronze, or glass, copper, silver, and gold implements embedded in a lead sealing, core material	records differences in atomic number; as a result it may discriminate between different materials (inlays, metal plating, core materials, other materials trapped in the internal cavity)	very informative, particularly for visualization of low-density (organic or core) material in the interior of a sculpture	• requires a specialized facility, typically a nuclear reactor, with highly trained staff • very few examples of this technique for bronze sculpture for comparison • insurance and transportation costs can be significant • spatial resolution and contrast usually inferior to X-radiography • interpretation of radiographs requires specialized training • will make future luminescence dating of the core material difficult • some objects may need to be quarantined for decontamination (de-activation), that is, kept at the neutron facility for anywhere between several minutes to several days before returning to their owner. The quarantine depends on a number of parameters, including neutron flux and energy, alloy composition, etc., which can create temporary radioactivity of the material. Copper alloys are actually less impacted (unless they contain large amounts of gold or arsenic). For example, no quarantine was needed for the Javanese statuettes in Case Study 4.	II.3§3

Method of examination	Surfaces	Structure	Materials	Importance and advantages	Drawbacks and limitations*	See
Computed tomography (CT scanning using X-ray or neutrons)	tomography can provide, if properly calibrated, a dimensionally accurate model of the inner and outer surfaces of a sculpture	can provide detailed and dimensionally accurate models of internal structural elements	can discriminate between different inlays	• tomographic 3D models and associated renderings may be easier to interpret than conventional radiographs • accurate measurement of dimensions, including wall thickness, is easier than with conventional radiographs	• requires expensive equipment rarely found in museums • lengthy exposure to radiation • postprocessing requires significant investment in time by highly skilled technicians • resolution is generally lower than with conventional radiography • will make future luminescence dating of the core material difficult	II.3§2
Other nondestructive testing techniques (ultrasonic [US] testing, thermography, eddy currents)		reveal defects, repairs, assemblies; US enables to measure the wall thickness of hollow bronzes	reveal metal plating such as gilding	• access to most of the sculpture thus good representativity of thickness variation • low to middle cost of device (standard camera for termographyn, circa 2,000€ in 2018 for US testing) • private companies may carry out US testing for relatively low cost	• the measurement of wall thickness by US may be altered by any kind of perturbation in the metal wall (porosity, crack, large inclusion, etc.). The latter may echo the signal and may be confused with the internal surface. Calibration is mandatory. A skilled operator is needed • US is slightly invasive: need for a binding medium between the bronze surface and the detector of the instrument, typically a gel (water)	II.2§3.2 II.4§2.3.3 II.5§2.3.2

* Time and training are mentioned below. This may involve hiring an expert if the necessary skills are not available within the institution performing the examination.

Table 14
Common abrasives used on modern bronze sculpture

stone	pumice stone
	sandstone
papers and cloth	cabinet paper
	sandpaper
	emery paper
	emery cloth
blasting media	sand
	iron grit
	glass bead

	copper oxide
	aluminum oxide
	zirconium oxide
	crushed walnut shell
	crushed olive seed
flexible pads	steel wool (coarse to fine)
	Scotch-Brite™ (coarse to fine)
powders	ground pumice
	tripoli (gesso de tripoli)
	aluminum oxide
	trent sand
	carborundum
	jeweler's rouge
	pumice powder

Table 15
Appropriate usage for various detectors used in radiography of bronze objects

	RADIATION SOURCE			
	low-kV X-ray (<200 kV)	**high-kV X-ray (200–450 kV)**	**gamma source (^{192}Ir = ^{60}Co)**	**accelerator**
film without screen	X			
film with lead screens		X	X	X
photostimulable phosphor imaging plates	X	X		
pixelated semiconductor detector with Gd2O2S converter	X	X	X	
pixelated semiconductor detector with CsI converter		X	X	X

Bibliography

One overall bibliography serves as reference for all sections of the *Guidelines* and is unique in its diversity, as it draws on the expertise of an interdisciplinary cast of contributing members. Yet while many of the key publications on bronze casting and related disciplines have been included, the bibliography does not pretend to cover every publication about this expansive field. The bulk of the publications cited are in English and French, reflecting the makeup of the primary organizers of the publication, but every effort has been made to incorporate important works in other languages.

Adamski, Pernot, and Bouet 2020
Adamski, F., M. Pernot, and A. Bouet. 2020. "Étude de la pratique de réparation des grands bronzes à partir d'un stock de métal cuivreux." In *Barzan IV, L'entrepôt de La Palisse à Barzan (Charente-Maritime), port des Santons, et les entrepôts urbains et périurbains dans les provinces Gauloises*, edited by Alain Bouet, 419–25. Bordeaux, France: Ausonius.

Adil and DePhillips 1991
Adil, Carol P., and Henry A. DePhillips. 1991. *Paul Wayland Bartlett and the Art of Patination*. Wethersfield, CT: Paul Wayland Bartlett Society.

Agricola [1556] 1950
Agricola, Georgius. (1556) 1950. *De re metallica libri XII*. Translated and edited by Herbert Clark Hoover and Lou Henry Hoover, 1912. New York: Dover.

Aitken 1985
Aitken, M. J. 1985. *Thermoluminescence Dating*. Orlando: Academic Press.

Aitken 1998
Aitken, M. J. 1998. *An Introduction to Optical Dating: The Dating of Quaternary Sediments by the Use of Photon-Stimulated Luminescence*. Oxford: Oxford University Press.

Alberghina et al. 2011
Alberghina, M. F., R. Barraco, M. Brai, T. Schillaci, and L. Tranchina. 2011. "Comparison of LIBS and μ-XRF Measurements on Bronze Alloys for Monitoring Plasma Effects." *Journal of Physics: Conference Series* 275 (January): 012017. https://doi.org/10.1088/1742-6596/275/1/012017.

Alessandri et al. 2013
Alessandri, O. Colacicchi, M. Ferretti, and Edilberto Formigli. 2013. "L'occhio pesto del pugile. Indagini archaeometriche sulla statua in bronzo del Museo Nazionale Romano." In *Colore e Luce Nella Statuaria Antica in Bronzo*, edited by Edilberto Formigli, 25–31. Rome: L'Erma di Bretschneider.

Allen et al. 2022
Allen, Denise, Linda Borsch, James David Draper, Jeffrey Fraiman, and Richard E. Stone, eds. 2022. *Italian Renaissance and Baroque Bronzes in the Metropolitan Museum of Art*. New York: The Metropolitan Museum of Art. https://www.metmuseum.org/art/metpublications/Italian_Renaissance_and_Baroque_Bronzes_in_the_Metropolitan_Museum_of_Art.

Allison 1993
Allison, Ann Hersey. 1993. "The Bronzes of Pier Jacopo Alari-Bonacolsi, Called Antico." *Jahrbuch der kunsthistorischen Sammlungen in Wien* 89/90:35–310.

Alscher 1987
Alscher, Ludger, ed. 1987. *Lexikon der Kunst: Architektur, Bildende Kunst, Angewandte Kunst, Industrieformgestaltung, Kunsttheorie*. Vol. 1. Leipzig, Germany: E. A. Seemann.

American Foundrymen's Society 1984
American Foundrymen's Society (AFS), ed. 1984. *Casting Copper-Base Alloys*. 3rd ed. Des Plaines, IL: AFS.

American Foundry Society (AFS) n.d.
American Foundry Society (AFS). n.d. "Glossary of Metalcasting Terms." American Foundry Society. Accessed April 12, 2021. https://www.afsinc.org/metalcasting-terms.

Ammen 1980
Ammen, Charles William. 1980. *The Metalcaster's Bible*. New York: McGraw-Hill and TAB Books.

Andrieu et al. 2000
Andrieu, Sandra, Bernard Bayle, Michel Pernot, and Jean-Marie Welter. 2000. "Influence de diverses inclusions sur le comportement d'un bronze CuSn9 sous différentes sollicitations mécaniques." *Revue de Métallurgie* 97 (5): 599–612. https://doi.org/10.1051/metal/200097050599.

Anguisola 2018
Anguisola, Anna. 2018. *Supports in Roman Marble Sculpture: Workshop Practice and Modes of Viewing*. Cambridge, UK: Cambridge University Press.

Anheuser 1997
Anheuser, Kilian. 1997. "The Practice and Characterization of Historic Fire Gilding Techniques." *JOM: Journal of the Minerals, Metals & Materials Society* 49 (11): 58–62.

Anheuser 1999
Anheuser, Kilian. 1999. *Im Feuer vergoldet: Geschichte und Technik der Feuervergoldung und der Amalgamversilberung*. AdR-Schriftenreihe zur Restaurierung und Grabungstechnik 4. Stuttgart, Germany: Theiss.

Arias 2012
Arias, Elena. 2012. "Esculturas de Leone y Pompeo Leoni: Técnicas escultóricas sobre metal." In *Leone & Pompeo Leoni: actas del Congreso Internacional = Proceedings of the International Symposium: Madrid, Museo Nacional del Prado, octubre de 2011*, edited by Stephan F. Schröder, 160–72. Madrid: Museo Nacional del Prado.

Armbruster 1993
Armbruster, Barbara. 1993. "A Study on Lost Wax Casting Processes in Mali and Burkina-Faso in Western Africa." In *Antiche officine del bronzo: Materiali, strumenti, tecniche: atti del seminario di studi ed esperimenti, Murlo, 26–31 luglio 1991*, edited by Edilberto Formigli, 153–64. Siena, Italy: Nuova Immagine.

Arminjon and Bilimoff 1998
Arminjon, Catherine, and Michèle Bilimoff. 1998. *L'art du métal. Vocabulaire technique*. Paris: Editions du patrimoine.

Arteaga et al. 2021
Arteaga, Yoko, Clotilde Boust, Angèle Dequier, and Jon Yngve Hardeberg. 2021. "Image-Based Goniometric Appearance Characterisation of Bronze Patinas." *Color and Imaging Conference* 29 (1): 294–99. https://doi.org/10.2352/issn.2169-2629.2021.29.294.

Arteaga et al. 2022
Arteaga, Yoko, Diane Marchioni, Stephanie Courtier, Clotilde Boust, and Jon Y. Hardeberg. 2022. "Appearance-Based Evaluation of Varnish Removal Methods in Gilded Surfaces." Preprint. In Review. https://doi.org/10.21203/rs.3.rs-1933665/v1.

Artioli 2010
Artioli, Gilberto. 2010. *Scientific Methods and Cultural Heritage: An Introduction to the Application of Materials Science to Archaeometry and Conservation Science*. Oxford and New York: Oxford University Press.

Artioli et al. 2003
Artioli, G., M. Dugnani, T. Hansen, L. Lutterotti, A. Pedrotti, and G. Sperl. 2003. "Crystallographic Texture Analysis of the Iceman and Coeval Copper Axes by Non-Invasive Neutron Powder Diffraction." In *La mummia dell'età del rame. 2. Nuove ricerche sull'uomo venuto dal ghiaccio*, edited by A. Fleckinger, 9–22. Bolzano, Italy: Folio Verlag.

Artioli et al. 2020
Artioli, G., C. Canovaro, P. Nimis, and I. Angelini. 2020. "LIA of Prehistoric Metals in the Central Mediterranean Area: A Review." *Archaeometry* 62 (S1): 53–85. https://doi.org/10.1111/arcm.12542.

Ashley-Smith 1977
Ashley-Smith, Jonathan. 1977. "The Scientific Examination of Small Bronzes." *The Conservator* 1:57.

Association Technique de Fonderie, Commission Ingénieurs et Techniciens 1979
Association Technique de Fonderie, Commission Ingénieurs et Techniciens. 1979. *Dictionnaire idéologique de fonderie / Indexed Foundry Dictionary / Systematisches Giessereiwörterbuch*. Meudon: Techniques des industries de la Fonderie.

Aucouturier 2007
Aucouturier, Marc. 2007. "Patinage et patines." *Techniques de l'ingénieur*.

Aucouturier and Darque-Ceretti 2015
Aucouturier, Marc, and Evelyne Darque-Ceretti. 2015. "Materials Surface Science Applied to the Investigation of Cultural Heritage Artefacts." *Procedia Materials Science*, International Congress of Science and Technology of Metallurgy and Materials, SAM – CONAMET 2014, 9 (Supplement C): 31–47. https://doi.org/10.1016/j.mspro.2015.04.005.

Aucouturier et al. 2003
Aucouturier, Marc, Michel Keddam, Luc Robbiola, and Hisasi Takenouti. 2003. "Les patines des alliages de cuivre: Processus naturel ou œuvre de l'homme?." *Techné* 18:86–94.

Aucouturier, Mathis, and Robcis 2017
Aucouturier, Marc, François Mathis, and Dominique Robcis. 2017. "Les bronzes noirs antiques – nouvelles observations et mécanismes de création." *Techné* 45:114–23.

Avery 2011
Avery, Victoria. *Vulcan's Forge in Venus' City: The Story of Bronze in Venice 1350–1650*. Oxford: Oxford University Press, 2011.

Avery 2018
Avery, Victoria, ed. 2018. *Michelangelo: Sculptor in Bronze: The Rothschild Bronzes*. London: Philip Wilson Publishers.

Azéma 2013
Azéma, Aurélia. 2013. "Les techniques de soudage de la grande statuaire antique en bronze: etude des paramètres thermiques et chimiques contrôlant le soudage par fusion au bronze liquide." PhD diss., Pierre et Marie-Curie, Paris VI. https://tel.archives-ouvertes.fr/tel-00918829.

Azéma and Mille 2013a
Azéma, Aurélia, and Benoît Mille. 2013a. "Etude archéométallurgique du dépôt de grands bronzes du sanctuaire gallo-romain du Vieil-Évreux (Eure)." In *Grands dieux! Chefs-d'œuvre de la statuaire antique*, edited by François Bridey, 11–17. Évreux, France: Musée d'Art, Histoire et Archéologie.

Azéma and Mille 2013b
Azéma, Aurélia, and Benoît Mille. 2013b. "Un point sur la technique de fabrication des grands bronzes antiques." *L'actualité chimique* 377:43–44.

Azéma, Descamps-Lequime, and Mille 2018
Azéma, Aurélia, Sophie Descamps-Lequime, and Benoît Mille. 2018. "Fragment de doigt colossal et Main colossale." In *Un rêve d'Italie: la collection du marquis Campana*, edited by Françoise Gaultier, Laurent Haumesser, and Anna Trofimova, 43–44. Paris: LienArt.

Azéma et al. 2011
Azéma, A., Benoît Mille, Patrick Echegut, and Domingos de Sousa Meneses. 2011. "An Experimental Study of the Welding Techniques Used on Large Greek and Roman Bronze Statues." *Historical Metallurgy* 45:71–80.

Azéma et al. 2012
Azéma, Aurélia, Benoît Mille, Fabien Pilon, Jean-Claude Birolleau, and Laurent Guyard. 2012. "Étude archéométallurgique du dépôt de grands bronzes du sanctuaire gallo-romain du Vieil-Evreux (Eure)." *ArcheoSciences. Revue d'archéométrie*, no. 36, 153–73. https://doi.org/10.4000/archeosciences.3670.

Azéma et al. 2013
Azéma, Aurélia, Florent Angelini, Benoît Mille, Gaëlle Framezelle, and Daniel Chauveau. 2013. "Ultrasonic Phased Array Contribution to the Knowledge of the Flow Fusion Welding Process Used to Make the Roman Large Bronze Statues." *Welding in the World* 57 (4): 477–86.

Azéma et al. 2017
Azéma, Aurélia, Daniel Chauveau, Gaëlle Porot, Florent Angelini, and Benoît Mille. 2017. "Pour une meilleure compréhension du procédé de soudage de la grande statuaire antique en bronze: Analyses et modélisation expérimentale." *Techné* 45:73–83.

Baca 2008
Baca, Murtha, ed. 2008. *Introduction to Metadata*. Los Angeles: Getty Research Institute.

Baca 2016

Baca, Murtha. 2016. "Introduction to Metadata." InteractiveResource. Los Angeles: Getty Research Institute. http://www.getty.edu/publications/intrometadata.

Bader and Théret 1961

Bader, O., and M. Théret. 1961. *Dictionnaire de métallurgie*. Paris: Eyrolles.

Bagley 1987

Bagley, R. W. 1987. *Shang Ritual Bronzes in the Arthur M. Sackler Collections*. Cambridge, MA: Harvard University Press.

Baglione 1642

Baglione, Giovanni. 1642. *Le vite de' pittori scultori et architetti. Dal pontificato di Gregorio XIII del 1572 in fino a'tempi di Papa Urbano Ottavo nel 1642*. Rome: Nella stamperia d'Andrea Fei. https://archive.org/details/gri_vitedepittor00bagl.

Baldwin et al. 2013

Baldwin, David, John Birkett, Owen Facey, and Gilleon Rabey. 2013. *The Forensic Examination and Interpretation of Tool Marks*. Chichester, West Sussex: Wiley-Blackwell.

Barbour and Glinsman 2015

Barbour Daphne and Lisha Deming Glinsman. 2015. "Auguste Rodin's Lifetime Bronze Sculptures in the Simpson Collection." In *Facture: Conservation, Science, Art History*, edited by Daphne Barbour and E. Melanie Gifford, 2, Art in Context: 54–81. Washington, D.C and New Haven: National Gallery of Art distributed by Yale University Press.

Barbour and Sturman 2017

Barbour, Daphne, and Shelley Sturman. 2017. "Casting Degas's Sculptures into Bronze: A Closer Look." *Facture: Conservation, Science, Art History* 3:78–111.

Barnet and Dandridge 2006

Barnet, Peter, and Pete Dandridge, eds. 2006. *Lions, Dragons, and Other Beasts: Aquamanilia of the Middle Ages, Vessels for Church and Table*. New Haven, CT: Yale University Press.

Baruffetti, De Marchi, and Ciatti 2021

Baruffetti, Maria, Andrea De Marchi, and Marco Ciatti. 2021. "Per un glossario interdisciplinare rivolto alla conservazione della scultura in bronzo." PhD diss., Università degli Studi di Firenze.

Bassett 2008

Bassett, Jane. 2008. *The Craftsman Revealed: Adriaen de Vries, Sculptor in Bronze*. Los Angeles: Getty Conservation Institute. https://www.getty.edu/publications/virtuallibrary/9780892369195.html.

Bassett and Bewer 2014

Bassett, Jane, and Francesca G. Bewer. 2014. "The Cut-Back Core Process in Late 17th- and 18th-Century French Bronzes." In *French Bronze Sculpture: Materials and Techniques 16th–18th Century*, edited by David Bourgarit, Jane Bassett, Francesca G. Bewer, Geneviève Bresc-Bautier, Philippe Malgouyres, and Guilhem Scherf, 205–14. Paris: Archetype.

Bassett and Fogelman 1997

Bassett, Jane, and Peggy Fogelman. 1997. *Looking at European Sculpture: A Guide to Technical Terms*. Los Angeles: J. Paul Getty Museum in collaboration with the Victoria and Albert Museum.

Bassett and Scherf 2014

Bassett, Jane, and Guilhem Scherf. 2014. "Jean-Antoine Houdon: Sculptor and Founder." In *French Bronze Sculpture: Materials and Techniques 16th–18th Century*, edited by David Bourgarit, Jane Bassett, Francesca G. Bewer, Geneviève Bresc-Bautier, Philippe Malgouyres, and Guilhem Scherf, 107–24. Paris: Archetype.

Battaglia 1961

Battaglia, S. 1961. *Grande dizionario della lingua italiana*. Edited by Giorgio Bárberi Squarotti. Turin, Italy: Unione tipografico-editrice torinese. http://www.gdli.it/.

Baudin 2011

Baudin, Thierry. 2011. "Analyse EBSD Déformation et recristallisation des matériaux métalliques." *Techniques de l'ingénieur Essais métallographiques des métaux et alliages base documentaire*: TIB343DUO (ref. article: m4139). https://doi.org/10.51257/a-v1-m4139.

Baudry and Bozo 1978

Baudry, Marie Thérèse, and Dominique Bozo. 1978. *La sculpture: Méthode et vocabulaire*. Principes d'analyse scientifique. Paris: Impr. nationale.

Baulez 1991

Baulez, Christian. 1991. "Le luminaire de la princesse Kinsky", *L'Objet d'art* 247: 84–99.

Baxandall 1965

Baxandall, Michael. 1965. "A Masterpiece by Hubert Gerhard." *Victoria and Albert Museum Bulletin* 1 (2): 1–17.

Baxandall 1966

Baxandall, Michael. 1966. "Hubert Gerhard and the Altar of Christoph Fugger: The Sculpture and its Making." *Münchner Jahrbuch der bildenden Kunst* 3 (17): 127–44.

Baxter 2001

Baxter, Michael J. 2001. "Statistical Modelling of Artefact Compositional Data." *Archaeometry* 43 (1): 131–47.

Baxter 2003

Baxter, Michael J. 2003. *Statistics in Archaeology*. London and New York: Oxford University Press.

Baxter 2006

Baxter, Michael J. 2006. "A Review of Supervised and Unsupervised Pattern Recognition in Archaeometry." *Archaeometry* 48 (4): 671–94.

Bayley 1991

Bayley, Justine. 1991. "Alloy Nomenclature." In *Medieval Finds from Excavations in London*. Vol. 3, *Dress Accessories c. 1150–c. 1450*, edited by Geoff Egan and Frances Pritchard, 13–17. London: HMSO.

Beale 1975

Beale, Arthur. 1975. "A Technical View of Nineteenth-Century Sculpture." In *Metamorphoses in Nineteenth-Century Sculpture*, 28–55. Cambridge, MA: Harvard University Press.

Beale 1996

Beale, Arthur. 1996. "Understanding, Restoring, and Conserving Ancient Bronzes with the Aid of Science." In *The Fire of Hephaistos: Large Classical Bronzes from North American Collections*, edited by Carol C. Mattusch, 65–80. Cambridge, MA: Harvard University Art Museums.

Bearzi 1950

Bearzi, Bruno. 1950. "Considerazioni di tecnica sui San Ludovico e la Giuditta di Donatello." *Bollettino d'Arte* 16:119–23.

Bearzi 1968

Bearzi, Bruno. 1968. "Tecnica fusoria di Donatello." In *Donatello e il suo tempo. Atti del VIII Convegno Internazionale di Studi sul Rinascimento, Firenze, Padova, 25 settembre–1 ottobre, 1966*, 97–105. Florence: Istituto Nazionale di Studi sul Rinascimento, Palazzo Strozzi.

Beeley 2001

Beeley, Peter. *Foundry Technology*. Oxford: Butterworth-Heinemann, 2001.

Beentjes 2019

Beentjes, Theodorus Petrus Cornelis. 2019. "Casting Rodin's Thinker – Sand Mould Casting, the Case of the Laren Thinker and Conservation Treatment Innovation." PhD diss., Universiteit van Amsterdam. https://dare.uva.nl/search?identifier=69db04ce-74ba-4e9c-95e1-a05f69e9d694.

Beguin et al. 1982

Beguin, G., J. Liszak-Hours, L. Hurtel, J. M. Malfoy, M. Menu, and F. Drilhon. 1982. "Objets himâlayens en métal du musée guimet: étude en laboratoire." *Annales du laboratoire de recherche des musées de France* 1982:28–83.

Bennett and Sargentson 2008
Bennett, Shelley M., and Carolyn Sargentson. 2008. *French Art of the Eighteenth Century at the Huntington*. San Marino, CA: Huntington Library; New Haven, CT: Yale University Press.

Benzonelli, Freestone, and Martinón-Torres 2017
Benzonelli, A., I. C. Freestone, and M. Martinón-Torres. 2017. "A Better Shade of Black: Effects of Manufacturing Parameters on the Development of Ancient Black Bronzes." *Archaeometry* 59 (6): 1034–49.

Berger et al. 2018
Berger, Daniel, Elin Figueiredo, Gerhard Brügmann, and Ernst Pernicka. 2018. "Tin Isotope Fractionation during Experimental Cassiterite Smelting and Its Implication for Tracing the Tin Sources of Prehistoric Metal Artefacts." *Journal of Archaeological Science* 92 (April): 73–86.

Berman 1974
Berman Harold. 1974. *Bronzes: Sculptors and Founders, 1800–1930*. Chicago: Abage.

Bernardoni 2009
Bernardoni, Andrea. 2009. "Leonardo and the Equestrian Monument for Francesco Sforza: The Story of an Unrealized Monument Sculpture." In *Leonardo da Vinci and the Art of Sculpture*, edited by Gary M. Radke, 94–135. Los Angeles: J. Paul Getty Museum; New Haven, CT, and London: Yale University Press.

Berns 2016
Berns, Roy S. 2016. *Color Science and the Visual Arts: A Guide for Conservators, Curators, and the Curious*. Los Angeles: Getty Trust Publications.

Bertelli et al. 2013
Bertelli, Piero, Edilberto Formigli, Ferdinando Marinelli, Bernardino Morsani, and Alessandro Pacini. 2013. "Ricostruzioni di archeologia sperimentale." In *Colore e luce nella statuaria antica in bronzo*, edited by Edilberto Formigli, 91–95. Rome: L'Erma di Bretschneider.

Bertholon 2004
Bertholon, Régis. 2004. "Le concept de surface d'origine des objets métalliques archéologiques: Une rencontre délicate entre sciences humaines et sciences de la matière." *Alliage* 53–54:155–63.

Berthoud 1977
Berthoud, Thierry. 1977. "Application de la spectrométrie de masse à étincelles à l'analyse d'objets de cuivre anciens." PhD diss., Université Paris 6.

Bevington and Robinson 2003
Bevington, Philip R., and D. Keith Robinson. 2003. *Data Reduction and Error Analysis for the Physical Sciences*. Boston: McGraw-Hill.

Bewer 1995
Bewer, Francesca G. 1995. "Studying the Technology of Renaissance Bronzes." In *Materials Issues in Art and Archaeology IV: Symposium Held May 16–21, 1994, Cancun, Mexico*, edited by James R. Druzik, José Luis Galván Madrid, I. C. Freestone, and George Segan Wheeler, 701–9. Materials Research Society Symposium Proceedings 352. Warrendale, PA: Materials Research Society.

Bewer 1996a
Bewer, Francesca G. 1996a. "Del formare e del getto: Vom Modellieren und vom Giessen die Herstellung von Bronzestatuetten im 16. Jahrhundert." In *Von allen Seiten schön*, edited by Volker Krahn, 82–91. Cologne: Letter-Stiftung.

Bewer 1996b
Bewer, Francesca G. 1996b. "A Study of the Technology of Renaissance Bronze Statuettes." PhD diss., University of London.

Bewer 1998
Bewer, Francesca G. 1998. "Kunststücke von gegossenem Metall – Adriaen de Vries's Bronze Technique." In *Adriaen de Vries (1556–1626): Imperial Sculptor*, edited by Frits Scholten, 64–77. Los Angeles: Getty Trust Publications and J. Paul Getty Museum.

Bewer 1999
Bewer, Francesca G. 1999. "Bronze Casts after Bozzetti and Modelli by Bernini." *Harvard University Art Museums Bulletin* 6 (3): 162–67.

Bewer 2001
Bewer, Francesca G. 2001. "The Sculpture of Adriaen de Vries: A Technical Study." In *Small Bronzes in the Renaissance*, edited by Debra Pincus, 158–93. Studies in the History of Art 62. Washington, DC: National Gallery of Art; New Haven, CT: Yale University Press.

Bewer 2010
Bewer, Francesca G. 2010. *A Laboratory for Art: Harvard's Fogg Museum and the Emergence of Conservation in America, 1900–1950*. Cambridge, MA, and New Haven, CT: Harvard Art Museums.

Bewer 2012
Bewer, Francesca G. 2012. "A Chemist under a Spell: Rutherford John Gettens's Early Encounters with Chinese Bronzes." In *Scientific Research on Ancient Asian Metallurgy (Proceedings of Studies of Ancient Asian Metallurgy, Fifth Forbes Symposium of the Freer Gallery of Art, October 2010, Washington, D.C.)*, edited by Paul Jett, Blythe McCarthy, and Janet G. Douglas, 3–11. London: Archetype.

Bewer and MacNamara 2012
Bewer, Francesca G., and Molly MacNamara. 2012. "The Portrait Busts of Cosimo I & Bindo Altoviti from the Inside Out: Cellini's Practice of the Art of Bronze Casting." In *Marks of Identity: New Perspectives on Sixteenth-Century Italian Sculpture*, edited by Dimitrios Zikos, 62–81. Boston: Isabella Stewart Gardner Museum.

Bewer and Scott 1995
Bewer, Francesca G., and David A. Scott. 1995. "A Bronze Sculpture Attributed to Louis-Simon Boizot and Platinum Coating Methods." *Archaeometry* 37 (2): 351–61.

Bewer, Bourgarit, and Bassett 2008
Bewer, Francesca G., David Bourgarit, and Jane Bassett. 2008. "Les bronzes français (XVIe–XVIIIe siècle): Notes techniques." In *Bronzes français de la Renaissance au Siècle des lumières*, edited by Geneviève Bresc-Bautier and Guilhem Scherf, 28–41. Paris: Musée du Louvre, Somogy.

Bewer, Bourgarit, and Bassett 2009
Bewer, Francesca G., David Bourgarit, and Jane Bassett, "French Bronzes (16th–18th centuries): Notes on Technique." In *Cast in Bronze: French Sculpture from Renaissance to Revolution*, ed. Geneviève Bresc-Bautier and Guilhem Scherf with James David Draper. New York: Metropolitan Museum of Art; Los Angeles: J. Paul Getty Museum.

Bewer et al. 2003
Bewer, Francesca G., E. van Binnebeke, R. Van Langh, and R. E. Stone. 2003. "Towards a History of Tetrode's Casting Technique." In *Willem van Tetrode, Sculptor (c. 1525–1580) Guglielmo Fiammingo Scultore*, edited by Frits Scholten, 93–112. Zwolle, the Netherlands: Waanders.

Bewer, Stone, and Sturman 2007
Bewer, Francesca G., Richard. E. Stone, and Shelley G. Sturman. 2007. "Reconstructing the Casting Technique of Lorenzo Ghiberti's Gates of Paradise." In *The Gates of Paradise: Lorenzo Ghiberti's Renaissance Masterpiece*, edited by Gary M. Radke, 156–82. Atlanta: High Museum of Art; New Haven, CT: Yale University Press.

Bian et al. 2010
Bian, Junguo, Jeffrey H. Siewerdsen, Xiao Han, Emil Y. Sidky, Jerry L. Prince, Charles A. Pelizzari, and Xiaochuan Pan. 2010. "Evaluation of Sparse-View Reconstruction from Flat-Panel-Detector Cone-Beam CT." *Physics in Medicine and Biology* 55 (22): 6575–99.

Bilak et al. 2016
Bilak, Donna, Jenny Boulboullé, Joel Klein, and Pamela H. Smith. 2016. "The Making and Knowing Project: Reflections, Methods, and New Directions." *West 86th: A Journal of Decorative Arts, Design History, and Material Culture* 23 (1): 35–55.

Birch and Martinón-Torres 2019
Birch, Thomas, and Marcos Martinón-Torres. 2019. "Shape as a Measure of Weapon Standardisation: From Metric to Geometric Morphometric Analysis of the Iron Age 'Havor' Lance from Southern Scandinavia." *Journal of Archaeological Science* 101:34–51.

Biringuccio [1540] 1942
Biringuccio, Vannoccio. (1540) 1942. *The Pirotechnia of Vannoccio Biringuccio*. Translated by Cyril Stanley Smith and Martha Teach Gnudi. New York: American Institute of Mining and Metallurgical Engineers.

Biringuccio [1540] 1977
Biringuccio, Vannoccio. (1540) 1977. *De la pirotechnia*. Edited by Adriano Carugo. Milano: Edizioni Il Polifilo.

Biringuccio [1540] 1990
Biringuccio, Vannoccio. (1540) 1990. *The Pirotechnia of Vannoccio Biringuccio*. Translated by Cyril Stanley Smith and Martha Teach Gnudi. New York: American Institute of Mining and Metallurgical Engineers.

Bish and Post 1989
Bish, David L., and Jeffrey E. Post, eds. 1989. "Modern Powder Diffraction." Special issue, *Reviews in Mineralogy* 20.

Boas and Fleischmann 2012
Boas, F. Edward, and Dominik Fleischmann. 2012. "CT Artifacts: Causes and Reduction Techniques." *Imaging in Medicine* 4 (2): 229–40.

Boffrand 1743
Boffrand, Germain. 1743. *Description de ce qui a été pratiqué pour fondre en bronze d'un seul jet la figure équestre de Louis XIV, élevée par la ville de Paris dans la place de Louis le Grand, en 1699*. Paris: Guillaume Cavelier. https://archive.org/details/gri_33125010863229/page/n3.

Bol 1985
Bol, Peter C. 1985. *Antike Bronzetechnik: Kunst und Handwerk antiker Erzbildner*. München: Beck.

Bolland 1894
Bolland, Simpson. 1894. *The Encyclopedia of Founding and Dictionary of Foundry Terms Used in the Practice of Moulding*. New York: Wiley.

Bordenave and Mirebeau 2018
Bordenave, Aurélie, and Isabelle Mirebeau. 2018. *Diffusons les neutrons. Orphée, réacteur nucléaire de recherche à Saclay = Let's Scatter Neutrons; Orphée, Nuclear Research Reactor in Saclay*. Saclay, France: Laboratoire Léon Brillouin (CEA-CNRS). https://fr.calameo.com/read/0056806668657b7f21394.

Born 1985
Born, Hermann, ed. 1985. *Archeologische Bronzen, alte Kunst, moderne Technik*. Berlin: Dietrich Reimer Verlag.

Bossema et al. 2024
Bossema, Francien G., Willem Jan Palenstijn, Arlen Heginbotham, Madeline Corona, Tristan van Leeuwen, Robert van Liere, Jan Dorscheid, et al. 2024. "Enabling 3D CT-Scanning of Cultural Heritage Objects Using Only in-House 2D X-Ray Equipment in Museums." *Nature Communications* 15 (1): 3939. https://doi.org/10.1038/s41467-024-48102-w.

Bott and Willer 2014
Bott, Kati, and Frank Willer. 2014. "Gold – Ein unvergängliches Material: Vergoldungstechniken an Bronzestatuen." In *Gebrochener Glanz: römische Großbronzen am UNESCO-Welterbe Limes*, edited by Stephanie Müller, 207–9. Mainz, Germany: Nünnerich-Asmus.

Boucher 1964
Boucher, Stéphanie. 1964. "Les grands dauphins de bronze doré du Musée Municipal de Vienne (Isère)." *Gallia* 22 (1): 23–35.

Boulton 2006
Boulton, Ann. 2006. "The Art Bronze Foundry of Antoine-Louis Barye." In *Untamed: The Art of Antoine-Louis Barye*, edited by William R. Johnston and Simon Kelly, 66–73. Baltimore: Walters Art Museum.

Boulton 2007
Boulton, Ann. 2007. "The Making of Matisse's Bronzes." In Dorothy Kosinski, *Matisse: Painter as Sculptor*, 73–95. Baltimore: Baltimore Museum of Art; Dallas: Dallas Museum of Art and Nasher Sculpture Center; New Haven, CT: Yale University Press.

Boulton 2018
Boulton, Ann. 2018. "Made in New York: Italian and French Art Bronze Casting." *Gilcrease Museum Journal* 25 (1): 38–53.

Bouquillon et al. 2006
Bouquillon, Anne, Sophie Descamps, Antoine Hermary, and Benoît Mille. 2006. "Une nouvelle étude de l'Apollon Chatsworth." *Revue Archéologique* 42 (2): 227–61.

Bourgarit 2019
Bourgarit, David. 2019. "Mineralogy of Slags: A Key Approach for Our Understanding of Ancient Copper Smelting Processes." In *The Contribution of Mineralogy to Cultural Heritage*, edited by Gilberto Artioli and Roberti Oberti. European Mineralogical Union Notes in Mineralogy 20. Mineralogical Society of Great Britain and Ireland. https://www.minersoc.org/emu-notes-20-5.html.

Bourgarit and Bauchau 2010
Bourgarit, David, and Fanny Bauchau. 2010. "The Ancient Brass Cementation Processes Revisited by Extensive Experimental Simulation." *Journal of Metals* 62 (3): 27–33.

Bourgarit and Mille 2003
Bourgarit, David, and Benoît Mille. 2003. "The Elemental Analysis of Ancient Copper-Based Artefacts by Inductively-Coupled-Plasma Atomic-Emission Spectrometry: An Optimized Methodology Reveals Some Secrets of the Vix Crater." *Measurement Science and Technology* 14 (9): 1538-1555. https://doi.org/10.1088/0957-0233/14/9/306.

Bourgarit and Mille 2014
Bourgarit, David, and Benoît Mille. 2014. "Provenance du cuivre et alliages." In *Circulation et provenance des matériaux dans les sociétés anciennes*, edited by Philippe Dillmann and Ludovic Bellot-Gurlet, 103–30. Sciences Archéologiques. Paris: Editions des archives contemporaines.

Bourgarit and Thomas 2011
Bourgarit, David, and Nicolas Thomas. 2011. "The Ancient Brass Cementation Processes: From Laboratory to Field Experiments." *Historical Metallurgy* 45 (1): 8–16.

Bourgarit and Thomas 2012
Bourgarit, David, and Nicolas Thomas. 2012. "Late Medieval Copper Alloying Practices: A View from a Parisian Workshop of the 14th Century AD." *Journal of Archaeological Science* 39 (10): 3052–70.

Bourgarit and Thomas 2015
Bourgarit, David, and Nicolas Thomas. 2015. "Ancient Brasses: Misconceptions and New Insights." In *Archaeometallurgy in Europe III, Proceedings of the 3rd International Conference, Deutsches Bergbau-Museum Bochum, June 29–July 1*, edited by Andreas Hauptmann and Diana Modarressi-Tehrani, 255–62. Bochum, Germany: Deutsches Bergbau Museum.

Bourgarit, Bewer, and Bresc-Bautier 2014
Bourgarit, David, Francesca G. Bewer, and Geneviève Bresc-Bautier. 2014. "Francesco Bordoni: Spécificités techniques chez un sculpteur-fondeur du 17e siècle." In *French Bronze Sculpture: Materials and Techniques 16th–18th Century*, edited by David Bourgarit, Jane Bassett, Francesca G. Bewer, Geneviève Bresc-Bautier, Philippe Malgouyres, and Guilhem Scherf, 1–17. Paris: Archetype.

Bourgarit et al. 2003
Bourgarit, David, Benoît Mille, Thierry Borel, Pierre Baptiste, and Thierry Zéphir. 2003. "A Millennium of Khmer Bronze Metallurgy." In *Scientific Research in the Field of Asian Art: Proceedings of the First Forbes Symposium at the Freer Gallery of Art*, edited by Paul Jett, 103–26. London: Archetype.

Bourgarit et al. 2014
Bourgarit, David, Jane Bassett, Francesca G. Bewer, Geneviève Bresc-Bautier, Philippe Malgouyres, and Guilhem Scherf, eds. 2014. *French Bronze Sculpture: Materials and Techniques 16th–18th Century*. Paris: Archetype.

Bovenkerk, Braun, and Dörflinger 1987
Bovenkerk, Konrad, Herwig Braun, and Rolf Dörflinger. 1987. Fachkunde Metall. 48. Europa-Fachbuchreihe für Metallberufe. Wuppertal: Europa Lehrmittel.

Bower, Lundstrom, and Hendin 2019
Bower, Nathan. W., Craig. C. Lundstrom, and David B. Hendin. 2019. "MC-ICP-MS Analyses of Tin Isotopes in Roman-Era Bronze Coins Reveal Temporal and Spatial Variation." *Archaeometry* 61 (4): 891–905.

Bray et al. 2015
Bray, Peter J., Aurélie Cuénod, Chris Gosden, Peter Hommel, Ruiliang Liu, and A. Mark Pollard. 2015. "Form and Flow: The 'Karmic Cycle' of Copper." *Journal of Archaeological Science* 56 (supplement C): 202–9.

Brepohl 2005
Brepohl, Erhard, éd. 2005. *Benvenuto Cellini: Traktate über die Goldschmiedekunst und die Bildhauerei = I trattati dell'oreficeria e della scultura*. Cologne: Böhlau.

Bresc-Bautier 1989
Bresc-Bautier, Geneviève. 1989. "Problèmes du bronze français, fondeurs et sculpteurs à Paris (1600–1660)." *Archives de l'art français* 30:11–50.

Bresc-Bautier 2003
Bresc-Bautier, Geneviève. 2003. "Parisian Casters in the Sixteenth Century." In *Large Bronzes in the Renaissance*, edited by Peta Motture, 95–113. New Haven, CT, and London: Yale University Press.

Bresc-Bautier and Scherf 2008
Bresc-Bautier, Geneviève, and Guilhem Scherf, eds. 2008. *Bronzes français de la Renaissance au siècle des lumières*. Paris: Musée du Louvre, Somogy.

Bresc-Bautier and Scherf 2009
Bresc-Bautier, Geneviève, and Guilhem Scherf with James David Draper, eds. 2009. *Cast in Bronze: French Sculpture from Renaissance to Revolution*. New York: Metropolitan Museum of Art; Los Angeles: J. Paul Getty Museum.

Bronk Ramsey 2009
Bronk Ramsey, Christopher. 2009. "Bayesian Analysis of Radiocarbon Dates." *Radiocarbon* 51 (1): 337–60.

Brooks 1991
Brooks, Randall C. 1991. "Origins, Usage, and Production of Screws: An Historical Perspective." *History and Technology* 8 (1): 51–76.

Brunhuber 1986
Brunhuber, Ernst. 1986. *Giesserei Lexikon*. Edited by Stephan Hasse. Berlin: Fachverlag Schiele & Schön.

Brunhuber 1988
Brunhuber, Ernst. 1988. *Gießerei-Fachwörterbuch = Foundry Dictionary*. 4th ed. Berlin: Schiele & Schön.

Brunhuber 2001
Brunhuber, Ernst. 2001. *Giesserei Lexikon*. Edited by Stephan Hasse. Berlin: Schiele & Schön.

Bruni 1994
Bruni, Francesco. 1994. *La fusione artistica a cera persa*. Venezia-Marghera: Edizioni Arte.

Buchanan 1903
Buchanan, John F. 1903. *Foundry Nomenclature: The Moulder's Pocket Dictionary*. London: E. & F. N. Spon.

Bunk, Sahm, and Ratka 1999
Bunk, Wolfgang G. J., Peter R. Sahm, and Markus Ratka. 1999. "Numerical Simulation of Historical Casting Processes for Bronze." In *The Beginnings of Metallurgy: Proceedings of the International Conference, Bochum 1995*, edited by Andreas Hauptmann, Ernst Pernicka, Thilo Rehren, and Ünsal Yalçin, 301–3. Bochum, Germany: Deutsches Bergbau-Museum.

Buratti et al. 2016
Buratti, Andrea, Soufian Ben Achour, Christopher Isenberg, and Robert Schmitt. 2016. "Frequency-Based Method to Optimize the Number of Projections for Industrial Computed Tomography." In *6th Conference on Industrial Computed Tomography (ICT) 2016, 9–12 February 2016*, 1–8. Wels, Austria: NDT.net. https://www.ndt.net/article/ctc2016/papers/ICT2016_paper_id12.pdf.

Burd and Greene 1948
Burd, David Q., and Roger S. Greene. 1948. "Tool Mark Comparisons in Criminal Investigations." *Journal of Criminal Law and Criminology* 39 (3): 379–91.

Burger, Bourgarit, Frotté et al. 2010
Burger, Emilien, David Bourgarit, Vincent Frotté, and Fabien Pilon. 2010. "Kinetics of Iron-Copper Sulphides Oxidation in Relation to Protohistoric Copper Smelting." *Journal of Thermal Analysis and Calorimetry* 103 (1): 249–56.

Burger, Bourgarit, Wattiaux et al. 2010
Burger, E., D. Bourgarit, A. Wattiaux, and M. Fialin. 2010. "The Reconstruction of the First Copper Smelting Processes in Europe during the 4th to the 3rd Millennium BC: Where Does the Oxygen Come From?" *Applied Physics A* 100 (3): 713–24.

Burmester and Koller 1987
Burmester, A., and J. Koller. 1987. "Known and New Corrosion Products on Bronzes: Their Identification and Assessment, Particularly in Relation to Organic Protective Coatings." In *Recent Advances in the Conservation and Analysis of Artifacts*, edited by James Black, 97–103. London: Summer Schools Press [for] University of London Institute of Archaeology.

C2RMF Internal Report 1997
"C2RMF Internal Report." 1997. #2410.

C2RMF Internal Report 1998
"C2RMF Internal Report." 1998. #2424.

C2RMF Internal Report 2002
"C2RMF Internal Report." 2002. #3086.

C2RMF Internal Report 2004
"C2RMF Internal Report." 2004. #3239.

C2RMF Internal Report 2008a
"C2RMF Internal Report." 2008a. #12629.

C2RMF Internal Report 2008b
"C2RMF Internal Report." 2008b. #8098.

C2RMF Internal Report 2009
"C2RMF Internal Report." 2009. #18072.

C2RMF Internal Report 2012
"C2RMF Internal Report." 2012. #24058.

C2RMF Internal Report 2015a
"C2RMF Internal Report." 2015a. #26159.

C2RMF Internal Report 2015b
"C2RMF Internal Report." 2015b. #29770.

C2RMF Internal Report 2016a
"C2RMF Internal Report." 2016a. #30190.

C2RMF Internal Report 2016b
"C2RMF Internal Report." 2016b. #35153.

C2RMF Internal Report 2017
"C2RMF Internal Report." 2017. #33054.

C2RMF Internal Report 2018a
"C2RMF Internal Report." 2018a. #37090.

C2RMF Internal Report 2018b
"C2RMF Internal Report." 2018b. #37541.

C2RMF Internal Report 2021
"C2RMF Internal Report." 2021. #47260.

Calligaro et al. 2000
Calligaro, Thomas, Jean-Claude Dran, Evanthia Ioannidou, Brice Moignard, Laurent Pichon, and Joseph Salomon. 2000. "Development of an External Beam Nuclear Microprobe on the AGLAE Facility of the Louvre Museum." *Nuclear Instruments and Methods in Physics Research Section B* 161/162/163:328–33.

Calligaro et al. 2011
Calligaro, Thomas, Yvan Coquinot, Laurent Pichon, and B. Moignard. 2011. "Advances in Elemental Imaging of Rocks Using the AGLAE External

Microbeam." *Nuclear Instruments and Methods in Physics Research Section B-Beam Interactions with Materials and Atoms* 269 (20): 2364–72.

Camm 1942
Camm, Frederick James. 1942. *Screw-Thread Manual*. Brooklyn: Chemical Publishing. https://catalog.hathitrust.org/Record/001616194.

Campbell 1991
Campbell, John. 1991. *Castings*. Oxford and Boston: Butterworth-Heinemann.

Campbell 2003
Campbell, John. 2003. *Castings: The New Metallurgy of Cast Metals*. 2nd ed. Oxford: Butterworth-Heinemann.

Campbell 2004
Campbell, John. 2004. *Castings Practice: The 10 Rules of Castings*. Oxford: Butterworth-Heinemann.

Campbell 2011
Campbell, John. 2011. *Complete Casting Handbook*. Oxford and Waltham, MA: Butterworth-Heinemann Elsevier.

Campbell 2015
Campbell, John. 2015. *Complete Casting Handbook: Metal Casting Processes, Metallurgy, Techniques, and Design*. 2nd ed. Amsterdam: Butterworth-Heinemann.

Cannell and Sabau 2007
Cannell, Nick, and Adrian S. Sabau. 2007. "Predicting Pattern Tooling and Casting Dimensions for Investment Casting, Phase III." In *Final Technical Report ORNL/TM-2007/204*. Oak Ridge, TN: US Department of Energy, Oak Ridge National Laboratory. https://www.osti.gov/bridge.

Carlisle 2005
Carlisle, Rodney. 2005. *Scientific American Inventions and Discoveries: All the Milestones in Ingenuity, from the Discovery of Fire to the Invention of the Microwave Oven*. Hoboken, NJ: Wiley.

Carò 2010
Carò, Federico. 2010. "Petrographic and Mineralogical Analysis of the Casting Core of a Chinese Bronze Buddha Maitreya." *Metropolitan Museum Journal* 1:155–61.

Carré 2005
Carré, Dominique, ed. *Alexandre Dumas, de Villers-Cotterêts au Panthéon, du Panthéon à Villers-Cotterêts*. Paris : Ministère de la Culture et de La Communication, Service National de Travaux. Un chantier.

Carruba 2006
Carruba, Anna Maria. 2006. *La Lupa capitolina: Un bronzo medievale*. Rome: De Luca.

Castaing et al. 2004
Castaing, Jacques, and Antoine Zink with Thierry Borel and Elisa Porto. 2004. "Influence de l'irradiation aux rayons X sur la luminescence et la datation des terres cuites." *Techné* 19:130–33.

Castelle 2016
Castelle, Manon. 2016. "Les techniques de la fabrication de la grande statuaire en bronze entre 1540 et 1660 en France." PhD diss., Université Paris-Saclay / Versailles Saint-Quentin-en-Yvelines.

Castelle, Bourgarit, and Bewer 2018
Castelle, Manon, David Bourgarit, and Francesca G. Bewer. 2018. "The Lasagna Method for Lost Wax Casting of Large 16th-Century Bronzes: Searching for the Sources." In *Cuivre, bronzes et laitons médiévaux: Histoire, archéologie et archéométrie des productions en laiton, bronze et autres alliages à base de cuivre dans l'Europe médiévale (12e–16e Siècles) = Medieval Copper, Bronze and Brass: History, Archaeology and Archaeometry of the Production of Brass, Bronze and Other Copper Alloy Objects in Medieval Europe (12th–16th Centuries)*, edited by Nicolas Thomas and Pete Dandridge, 297–308. Études et Documents, Archéologie 39. Namur, Belgium: Agence wallonne du Patrimoine.

Castelle, Coquinot, and Bourgarit 2016
Castelle, Manon, Yvan Coquinot, and David Bourgarit. 2016. "Casting Cores of French Bronze Statues of the 16th and 17th Centuries: Identification of Regional Practices and Artistic Choices." *Microchemical Journal* 126:121–31.

Castelle et al. 2019
Castelle, Manon, Marc Bormand, Yannick Vandenberghe, and David Bourgarit. 2019. "Two of a Kind: Shining New Light on Bronze Spiritelli Attributed to Donatello." *Studies in Conservation* 65 (4): 200–211.

Castelle et al. 2021
Castelle, Manon, David Bourgarit, Geneviève Bresc-Bautier, and Lise Cadot-Leroux. 2021. "Three Is the Magic Number: Techniques and Materials of the Prestigious Anne de Montmorency Bronze Virtues." *Studies in Conservation* 28 (April): 1–14.

CAST:ING 2018
CAST:ING. 2018. "Angkorian Founders and Bronze Casting Skills: First Technical Investigation of the West Mebon Viṣṇu." *Bulletin de l'École française d'Extrême-Orient* 104 (1): 303–41.

"Casting in Bronze" 1891
"Casting in Bronze." 1891. *Iron Age* 47:864–69.

Caumont et al. 2006
Caumont, Olivier, Xavier Margarit, Benoît Mille, Paolo Piccardo, and Claude Rolley. 2006. "Un bras d'empereur romain en bronze à Essegney (Vosges)." *Revue archéologique de l'est* 55 (177): 173–95. https://journals.openedition.org/rae/898.

Caussé, Mille, and Tansu 2020
Caussé, René, Benoît Mille, and Margaux Tansu. 2020. "La facture des trompes: étude acoustique et métallurgique, essai de modélisation." In *Cornua de Pompéi, trompettes romaines de la gladiature*, edited by Christophe Vendries, 79–98. Rennes, France: Presses Universitaires de Rennes.

Cellini [1568] 1967
Cellini, Benvenuto. 1568 (1967). *The Treatises of Benvenuto Cellini on Goldsmithing and Sculpture*. Translated by Charles Robert Ashbee. New York: Dover Publications.

Cellini [1568] 1983
Cellini, Benvenuto. (1568) 1983. *Due Trattati: Uno Intorno Alle Otto Principali Arti Dell'oreficeria: L'altro in Materia Dell'arte Della Scultura, Dove Si Veggono Infiniti Segreti Nel Lavorar Le Figure Di Marmo, & Nel Gettarle Di Bronzo*. Edited by Antonio Altomonte. Modena: Edizioni Aldine.

Cellini 1956
Cellini, Benvenuto. 1956. *The Autobiography of Benvenuto Cellini*. Translated by George Bull. Harmondsworth, UK: Penguin.

Cellini 1996
Cellini, Benvenuto. 1996. *Benvenuto Cellini: La vita*. Edited by Lorenzo Bellotto. Parma, Italy: Fondazione Pietro Bembo / Ugo Guanda Editore.

Cesareo et al. 2009
Cesareo, Roberto, Marcia A. Rizzutto, Antonio Brunetti, and Donepudi V. Rao. 2009. "Metal Location and Thickness in a Multilayered Sheet by Measuring Kα/Kβ, Lα/Lβ and Lα/Lγ X-Ray Ratios." *Nuclear Instruments and Methods in Physics Research Section B: Beam Interactions with Materials and Atoms* 267 (17): 2890–96.

CETIM 2017
CETIM. 2017. Upublished booklet given to attendees of the 2017 tomography training. France: Centre Techniques des Industries Mécaniques (CETIM).

Chankow 2012
Chankow, Nares. 2012. "Neutron Radiography." In *Nondestructive Testing Methods and New Applications*, edited by Mohammad A. Omar, 73–100. Rijeka, Croatia: InTech. https://www.intechopen.com/books/nondestructive-testing-methods-and-new-applications/neutron-radiography.

Chapoulie, Robert, and Casenave 2016
Chapoulie, Rémy, Béatrice Robert, and Sandrine Casenave. 2016. "The Cathodoluminescence Phenomenon Used for the Study of Ancient Ceramics

and Stones." *Cities of Memory: International Journal on Culture and Heritage at Risk* 1 (1): 53–72.

Chase 1991
Chase, W. T. 1991. *Ancient Chinese Bronze Art: Casting the Precious Sacral Vessel*. New York: China Institute in America.

Chase 1994
Chase, W. Thomas 1994. "Chinese Bronzes: Casting, Finishing, Patination, and Corrosion." In *Ancient and Historic Metals: Conservation and Scientific Research*, edited by David A. Scott, Jerry Podany, and Brian B. Considine, 85–117. Los Angeles: Getty Conservation Institute. https://www.getty.edu/publications/virtuallibrary/0892362316.html.

Chesnel de la Charbouclais 1857
Chesnel de la Charbouclais, Louis Pierre François Adolphe. 1857. *Dictionnaire de technologie, etymologie et définition des termes employés dans les arts et métiers*. Paris: J.-P. Migne.

Ciasca 1922
Ciasca, Raffaele, ed. 1922. *Statuti dell'Arte dei Medici e speziali, editi a spese della Camera di Commercio e Industria di Firenze*. Florence: A. Vallecchi. http://archive.org/details/statutidellarted00arteuoft.

Clarke et al. 2005
Clarke, Mark, Joyce Townsend, Ad Stijnman, and Instituut Collectie Nederland, eds. 2005. *Art of the Past: Sources and Reconstruction: Proceedings of the First Symposium of the Art Technological Source Research Study Group*. London: Archetype.

Collinet 2021
Collinet, Annabelle. 2021. *Précieuses matières: L'art du métal dans le monde iranien médiéval*. Dijon: Éditions Faton.

Comité international du dictionnaire technique de fonderie 1955
Comité international du dictionnaire technique de fonderie. 1955. *Dictionnaire de fonderie* [German, Spanish, English, Dutch, Italian, Norwegian, Swedish]. Milan: Edition Provisoire.

Considine et al. 2010
Considine, Brian B., Julie Wolfe, Katrina Posner, and Michel Bouchard. 2010. *Conserving Outdoor Sculpture: The Stark Collection at the Getty Center*. Los Angeles: J. Paul Getty Museum.

Cooper and Hecker 2003
Cooper, Harry, and Sharon Hecker. 2003. *Medardo Rosso: Second Impressions*. New Haven, CT: Yale University Press.

Cottrell 1967
Cottrell, A. H. 1967. "The Nature of Metals." *Scientific American* 217 (3): 90–100.

Coules, Orrock, and Seow 2019
Coules, H. E., P. J. Orrock, and C. E. Seow. 2019. "Reflectance Transformation Imaging as a Tool for Engineering Failure Analysis." *Engineering Failure Analysis* 105 (November): 1006–17. https://doi.org/10.1016/j.engfailanal.2019.07.037.

Coutant 2005
Coutant, F. 2005. "La Fonderie de Mile 2, Bamenda, Cameroun." Unpublished PDF.

Couture-Rigert, Sirois, and Moffatt 2013
Couture-Rigert, Doris E., P. Jane Sirois, and Elizabeth A. Moffatt. 2013. "An Investigation into the Cause of Corrosion on Indoor Bronze Sculpture." *Studies in Conservation* 57 (3): 142–63.

Cowell, La Niece, and Rawson 2003
Cowell, Michael, Susan La Niece, and Jessica Rawson. 2003. "A Study of Later Chinese Metalwork." In *Scientific Research in the Field of Asian Art: Proceedings of the First Forbes Symposium at the Freer Gallery of Art*, edited by Paul Jett, 80–89. London: Archetype.

Craddock 1981
Craddock, Paul T. 1981. "The Copper Alloys of Tibet and Their Background." In *Aspects of Tibetan Metallurgy*, edited by W. Andrew Oddy and Wladimir Zwalf, 1–33. British Museum Occasional Paper 15. London: British Museum.

Craddock 1985
Craddock, Paul T. 1985. "Medieval Copper Alloy Production and West African Bronze Analyses: Part I." *Archaeometry* 27 (1): 17–41.

Craddock 1990
Craddock, Paul T. 1990. "The Art and Craft of Faking: Copying, Embellishing and Transforming." In *Fake? The Art of Deception*, edited by Mark Jones, Nicolas Barker, and the British Museum, 101–27. London: British Museum Publications.

Craddock 1992
Craddock, Paul T. 1992. "A Short History of the Patination of Bronze." In *Why Fakes Matter: Essays on Problems of Authenticity*, edited by Mark Jones, 63–70. London: Trustees of the British Museum, British Museum Press.

Craddock 2009
Craddock, Paul T. 2009. *Scientific Examination of Copies, Fakes and Forgeries*. Oxford: Elsevier.

Craddock 2015
Craddock, Paul T. 2015. "The Metal Casting Traditions of South Asia: Continuity and Innovation." *Indian Journal of History of Science* 50 (1): 55–82.

Craddock 2017
Craddock, Paul T. 2017. "From Egypt to Greece via India: New Insights into Bronze Casting Technology in Antiquity." In *The RITaK Conferences 2013–2014*, edited by Petra Eisenach, T. Stöllner, and Arne Windler, 229–42. Bochum, Germany: Verlag Marie Leidorf GmbH.

Craddock and Giumlia-Mair 1993
Craddock, Paul T., and Alessandra Giumlia-Mair. 1993. "*Hsmn-Km*, Corinthian Bronze, *Shakudo*: Black Patinated Bronze in the Ancient World." In *Metal Plating and Patination: Cultural, Technical and Historical Developments*, edited by Susan La Niece and Paul T. Craddock, 101–27. Oxford, London, Boston [etc.]: Butterworth-Heinemann.

Craddock and Picton 1986
Craddock, P. T., and J. Picton. "Medieval Copper Alloy Production and West African Bronze Analyses, Part II." *Archaeometry* 28 (1): 3–32.

Craddock, Wayman, and Jull 2002
Craddock, P. T., M. L. Wayman, and A. J. T. Jull. 2002. "The Radiocarbon Dating and Authentication of Iron Artifacts." *Radiocarbon* 44 (3): 717–32.

CTIF 2022
Centre Technique des Industries de la fonderie (CTIF). 2022. "Glossaire de La Fonderie." Glossaire de La Fonderie. Accessed November 12, 2022. http://www.ctif.com/connaissances/base-de-connaissances/glossaire-fonderie/.

Cuénin 1994
Cuénin, Pierre. 1994. "Moulage - noyautage." *Techniques de l'ingénieur fonderies et moulages: aspects généraux* TIB352DUO (m3512). https://www.techniques-ingenieur.fr/base-documentaire/materiaux-th11/fonderies-et-moulages-aspects-generaux-42352210/moulage-noyautage-m3512/.

Cuénin 1997a
Cuénin, Pierre. 1997a. "Contrôle. Qualité." *Techniques de l'ingénieur fonderies et moulages: aspects généraux* TIB352DUO (m3530). https://www.techniques-ingenieur.fr/base-documentaire/materiaux-th11/fonderies-et-moulages-aspects-generaux-42352210/controle-qualite-m3530/.

Cuénin 1997b
Cuénin, Pierre. 1997b. "Parachèvement. Finition." *Techniques de l'ingénieur fonderies et moulages: aspects généraux* TIB352DUO (m3525). https://www.techniques-ingenieur.fr/base-documentaire/materiaux-th11/fonderies-et-moulages-aspects-generaux-42352210/parachevement-finition-m3525/.

Cuomo di Caprio 2007
Cuomo di Caprio, Ninina. 2007. *La ceramica in archeologia 2: antiche tecniche di lavorazione e moderni metodi d'indagine*. Rome: L'Erma di Bretschneider.

Cüppers 1994
Cüppers, Barbara. 1994. "Metalleinlagen als farbige Ziertechnik." In *Das Wrack. Der antike Schiffsfund von Mahdia*, edited by Gisela Hellenkemper Salies with Hans-Hoyer von Prittwitz und Gaffron and Gerhard Bauchhenß, 1:1013–16. Kataloge des Rheinischen Landesmuseums Bonn. Cologne: Rheinland Verlag.

Daehner and Lapatin 2015
Daehner, Jens M., and Kenneth Lapatin, eds. 2015. *Power and Pathos: Bronze Sculpture of the Hellenistic World*. Los Angeles: J. Paul Getty Museum.

Dal Sasso et al. 2014
Dal Sasso, Gregorio, Lara Maritan, Sandro Salvatori, Claudio Mazzoli, and Gilberto Artioli. 2014. "Discriminating Pottery Production by Image Analysis: A Case Study of Mesolithic and Neolithic Pottery from Al Khiday (Khartoum, Sudan)." *Journal of Archaeological Science* 46 (June): 125–43.

Dandridge 2006
Dandridge, Pete. 2006. "Exquisite Objects, Prodigious Technique. Aquamanilia, Vessels of the Middle Ages." In *Lions, Dragons, and Other Beasts: Aquamanilia of the Middle Ages, Vessels for Church and Table*, edited by Peter Barnet and Pete Dandridge, 35–56. New Haven, CT: Yale University Press.

Dandridge 2018
Dandridge, Pete. 2018. "The Hildesheim Baptismal Font. A Window into Medieval Workshop Practices." In *Cuivre, bronzes et laitons médiévaux: histoire, archéologie et archéométrie des productions en laiton, bronze et autres alliages à base de cuivre dans l'europe médiévale (12e–16e siècles) = Medieval Copper, Bronze and Brass: History, Archaeology and Archaeometry of the Production of Brass, Bronze and Other Copper Alloy Objects in Medieval Europe (12th–16th Centuries)*, edited by Nicolas Thomas and Pete Dandridge, 205–18. Études et Documents, Archéologie 39. Namur, Belgium: Agence wallonne du Patrimoine.

Darblade-Audoin and Tavoso 2008
Darblade-Audoin, Maria Pia, and Olivier Tavoso with the collaboration of Guy Alfonso and Benoît Mille. 2008. "Le pied de bronze colossal de Clermont-Ferrand." *Monuments et mémoires de la Fondation Eugène Piot*, no. 87, 31–68.

d'Arcet 1818
Arcet, Jean-Pierre-Joseph d'. 1818. *Mémoire sur l'art de dorer le bronze au moyen de l'amalgame d'or et de mercure*. Paris: De l'imprimerie de Mme. Veuve Agasse.

Darque-Ceretti and Aucouturier 2012
Darque-Ceretti, Évelyne, and Marc Aucouturier. 2012. *Dorure: décor et sublimation de la matière*. Paris: Mines ParisTech – Presses des Mines.

Darque-Ceretti, Felder, and Aucouturier 2011
Darque-Ceretti, Évelyne, Eric Felder, and Marc Aucouturier. 2011. "Foil and Leaf Gilding on Cultural Artifacts: Forming and Adhesion." *Matéria (Rio de Janeiro)* 16 (1): 540–59. https://doi.org/10.1590/S1517-70762011000100002.

Davis 1998
Davis, Joseph R., ed. 1998 *Metals Handbook*. 2nd ed. Materials Park, OH: ASM International.

Day and Allen 2014
Day, J., and D. Allen. 2014. "Casting Sculpture and Cannons in Bronze: Jehan Barbet's Angel of 1475 in The Frick Collection." *French Bronze Sculpture: Materials and Techniques 16th–18th Century*, edited by David Bourgarit, Jane Bassett, Francesca G. Bewer, Geneviève Bresc-Bautier, Philippe Malgouyres, and Guilhem Scherf, 187–204. Paris: Archetype.

Day et al. 2010
Day, Julia, Jens Stenger, Katherine Eremin, Narayan Khandekar, and Virginia Budny. 2010. "Gaston Lachaise's Bronze Sculptures in the Fogg Museum." *Journal of the American Institute for Conservation* 49 (1): 1–26.

De Bois 1999
De Bois, Guy. 1999. *La ciselure et ses techniques*. Paris: H. Vial.

Debonliez and Malepeyre 1979
Debonliez, G., and François Malepeyre. 1979. *Nouveau manuel complet du bronzage des métaux et du plâtre*. Paris: L. Laget.

Dehejia et al. 2002
Dehejia, Vidya, et al. 2002. *The Sensuous and the Sacred: Chola Bronzes from South India*. New York; Seattle: American Federation of Arts; University of Washington Press.

Delange, Meyohas, and Aucouturier 2005
Delange, Elisabeth, Marie-Emmanuelle Meyohas, and Marc Aucouturier. 2005. "The Statue of Karomama, a Testimony of the Skill of Egyptian Metallurgists in Polychrome Bronze Statuary." *CULHER Journal of Cultural Heritage* 6 (2): 99–113.

Dellepiane et al. 2006
Dellepiane, Matteo, Massimiliano Corsini, Marco Callieri, and Roberto Scopigno. 2006. "High Quality PTM Acquisition: Reflection Transformation Imaging for Large Objects." In *CIPA / VAST / EG / EuroMed 2006: 37th CIPA International Workshop dedicated on e-Documentation and Standardisation in Cultural Heritage: incorporating: the 7th International Symposium on Virtual Reality, Archaeology and Intelligent Cultural Heritage, the 4th EUROGRAPHICS Workshop on Graphics and Cultural Heritage, 1st Euro-Med Conference on IT in Cultural Heritage, Nicosia, Cyprus, October 30-November 4, 2006*, edited by Marinos Ioannides, David B. Arnold, Franco Niccolucci, and Katerina Mania, 179–86. Aire-la-Ville, Switzerland: Eurographics Association. http://dx.doi.org/10.2312/VAST/VAST06/179-186

Delon 1877
Delon, C. 1877. *Le cuivre et le bronze*. Paris: Hachette. https://gallica.bnf.fr/ark:/12148/bpt6k65387071.r=delon le cuivre et le bronze?rk=21459;2.

Delqué-Količ et al. 2017
Delqué-Količ, Emmanuelle, Stéphanie Leroy, Gaspard Pagès, and Julie Leboyer. 2017. "Iron Bar Trade between the Mediterranean and Gaul in the Roman Period: 14C Dating of Products from Shipwrecks Discovered off the Coast of Saintes-Maries-de-La-Mer (Bouches-Du-Rhône, France)." *Radiocarbon* 59 (2): 531–44.

Descamps-Lequime 2005
Descamps-Lequime, Sophie. 2005. "L'encrier de Vaison-La-Romaine et la patine volontaire des bronzes antiques." *Monuments et mémoires de la Fondation Eugène Piot* 84 (1): 5–30.

Descamps-Lequime 2015
Descamps-Lequime, Sophie. 2015. "The Color of Bronze: Polychromy and the Aesthetics of Bronze Surfaces." In *Power and Pathos: Bronze Sculpture of the Hellenistic World*, edited by Jens M. Daehner and Kenneth Lapatin, 150–66. Los Angeles: J. Paul Getty Museum.

Descamps-Lequime 2019
Descamps-Lequime, Sophie. 2019. "La couleur des bronzes antiques: recherches récentes." *Technè: La science au service de l'histoire de l'art et de la préservation des biens culturels*, no. 48 (December): 72–82. https://doi.org/10.4000/techne.2323.

Descamps-Lequime and Mille 2017
Descamps-Lequime, Sophie, and Benoît Mille. 2017. "Progrès de la recherche sur la statuaire antique en bronze." *Technè* 45:4–13.

Descamps-Lequime, Biron, and Langlois 2017
Descamps-Lequime, Sophie, Isabelle Biron, and Juliette Langlois. 2017. "Les yeux d'Auguste et de Livie au musée du Louvre: Une nouvelle technique antique identifiée." *Techné* 45:84–99. https://doi.org/10.4000/techne.1304.

Descamps-Lequime, Mille, and Psalti 2019
Descamps-Lequime, Sophie, Benoît Mille, et Nancy Psalti. "Projet d'une nouvelle étude technologique de l'Aurige de Delphes: les premiers résultats." In *Proceedings of the XXth International Congress on Ancient Bronzes: Resource, reconstruction, representation, role*, edited by Philipp Baas, 45–55. BAR International series 2958. Oxford: BAR Publishing, 2019.

Descamps-Lequime, Mille, and Robcis 2008
Descamps-Lequime, Sophie, Benoît Mille, and Dominique Robcis. 2008. "L'enfant à la bulle: Histoire moderne d'un bronze antique." *Techné*, nos. 27/28: 11–20.

Desmas 2014
Desmas, Anne-Lise. 2014. "Boffrand's and Mariette's Descriptions of the Casting of Louis XIV and Louis XV on Horseback." In *French Bronze Sculpture: Materials and Techniques 16th–18th Century*, edited by David Bourgarit, Jane Bassett, Francesca G. Bewer, Geneviève Bresc-Bautier, Philippe Malgouyres, and Guilhem Scherf, 234–340. Paris: Archetype.

Deutsches Kupferinstitut 2005
Deutsches Kupferinstitut. 2005. *Trennen und Verbinden von Kupfer und Kupferlegierungen*. Informationsdruck i.16. Düsseldorf: Deutsches Kupferinstitut. https://www.kupferinstitut.de/fileadmin/user_upload/kupferinstitut.de/de/Documents/Shop/Verlag/Downloads/Verarbeitung/i016.pdf.

Devogelaere 2017
Devogelaere, Jonathan. 2017. "The Colour Palette of Antique Bronzes: An Experimental Archaeology Project." *Experimental Archaeology* 2017 (2). https://exarc.net/ark:/88735/10289.

Diderot et al. 1751
Diderot, Denis, Jean le Rond d'Alembert et al. 1751. *Encyclopédie; ou Dictionnaire raisonné des sciences, des arts et des métiers*. 17 vols. Paris: Briasson. http://enccre.academie-sciences.fr/encyclopedie/.

Diehl 1927
Diehl, Gustav Eugen. 1927. *Die Bildgiesserei Noack*. Berlin: G. E. Diehl.

Diemer 1996
Diemer, Dorothea. 1996. "Handwerkgeheimnisse der Vischer-Werkstatt." *Münchner Jahrbuch der bildenden Kunst* 3 (47): 24–54.

Dillmann and L'Héritier 2007
Dillmann, Philippe, and Maxime L'Héritier. 2007. "Slag Inclusion Analyses for Studying Ferrous Alloys Employed in French Medieval Buildings: Supply of Materials and Diffusion of Smelting Processes." *Journal of Archaeological Science* 34 (11): 1810–23.

Dillon 2002
Dillon, Jo. 2002. "Technical Reports" and "Glossary of Technical Terms." In Avery, Victoria, *Renaissance and Baroque Bronzes from the Fitzwilliam Museum, Cambridge*, 276–88, 299–303. London: Daniel Katz Ltd.; Gli Ori.

Doeringer, Mitten, and Steinberg 1970
Doeringer, Suzannah F., David Gordon Mitten, and Arthur Richard Steinberg, eds. 1970. *Art and Technology: A Symposium on Classical Bronzes*, Cambridge, MA: MIT Press.

Dolfini and Crellin 2016
Dolfini, Andrea, and Rachel J. Crellin. 2016. "Metalwork Wear Analysis: The Loss of Innocence." *Journal of Archaeological Science* 66 (February): 78–87.

Dran, Calligaro, and Salomon 2000
Dran, Jean-Claude, Thomas Calligaro, and Joseph Salomon. 2000. "Particle-Induced X-Ray Emission." In *Modern Analytical Methods in Art and Archaeology*, edited by Enrico Ciliberto and Giuseppe Spoto, 135–66. New York: Wiley-Interscience.

Drayman-Weisser 1994
Drayman-Weisser, Terry. 1994. "A Perspective on the History of the Conservation of Archaeological Copper Alloys in the United States." *Journal of the American Institute for Conservation* 33 (2): 141–52.

Drayman-Weisser 2000
Drayman-Weisser, Terry, ed. 2000. *Gilded Metals: History, Technology and Conservation*. London: Archetype in association with the American Institute of Conservation of Historic and Artistic Works.

Drescher 1958
Drescher, Hans. 1958. *Der Überfangguss. Ein Beitrag zur vorgeschichtlichen Metalltechnik*. Vol. 39. Mainz, Germany: Verlag des Romisch-Germanischen Zentralmuseums.

Droth et al. 2005
Droth, Martina, Frits Scholten, Michael Wayne Cole, Penelope Curtis, and Henry Moore Institute. 2005. *Bronze: The Power of Life and Death*. Leeds, UK: Henry Moore Institute.

Dubos 2003
Dubos, Jean. 2003. "Fondeur d'art Aujourd'hui." *Techné* 18:49–56.

Dubos 2014
Dubos, Jean. 2014. "Fonte à la cire perdue en moule carapace. A travers les recherches et les réalisations de la fonderie de Coubertin." In *French Bronze Sculpture: Materials and Techniques 16th–18th Century*, edited by David Bourgarit, Jane Bassett, Francesca G. Bewer, Geneviève Bresc-Bautier, Philippe Malgouyres, and Guilhem Scherf, 241–49. Paris: Archetype.

Dumoulin et al. 2017
Dumoulin, J.-P., C. Comby-Zerbino, E. Delqué-Količ, C. Moreau, I. Caffy, S. Hain, M. Perron et al. 2017. "Status Report on Sample Preparation Protocols Developed at the LMC14 Laboratory, Saclay, France: From Sample Collection to 14C AMS Measurement." *Radiocarbon* 59 (3): 713–26.

Dungworth and Doonan 2013
Dungworth, David, and Roger Doonan, eds. 2013. *Accidental and Experimental Archaeometallurgy*. London: Historical Metallurgy Society.

Dupré and Hendrikson 2023
Hendriksen, Marieke M. A. and Dupré, Sven, "Introduction: Mapping Technique in the Arts," in *The Making of Technique in the Arts. Theories and Practice from the Sixteenth to the Twentieth Century*, eds. Sven Dupré and Marieke Hendriksen, Turnhout, 2023, pp. 9-21.

Dupré et al. 2020
Dupré, Sven, Anna Harris, Julia Kursell, Patricia Lulof, and Maartje Stols-Witlox. 2020. *Reconstruction, Replication and Re-Enactment in the Humanities and Social Sciences*. Amsterdam: Amsterdam University Press.

Dussubieux, Golitko, and Gratuze 2016
Dussubieux, Laure, Mark Golitko, and Bernard Gratuze, eds. 2016. *Recent Advances in Laser Ablation ICP-MS for Archaeology*. Berlin: Springer-Verlag.

Eggert 1994
Eggert, Gerhard. 1994. "Schwarzfärbung oder Korrosion. Das Rätsel der schwarzen Bronzen aus chemischer Sicht." In *Das Wrack. Der antike Schiffsfund von Mahdia*, edited by Gisela Hellenkemper Salies with Hans-Hoyer von Prittwitz und Gaffron and Gerhard Bauchhenß, 2:1033–40. Kataloge des Rheinischen Landesmuseums Bonn. Cologne: Rheinland Verlag.

Ekins and Edwards 1997
Ekins, Roger, and Philip Edwards. 1997. "Point on the Meaning of 'Sensitivity.'" *Clinical Chemistry* 43 (10): 1824–31.

Ekserdjian 2012
Ekserdjian, David. 2012. *Bronze*. London: Royal Academy of Arts; New York: Harry N. Abrams.

Espi and Aucouturier 2001
Espi, L., and M. Aucouturier. 2001. "Surface Modification Issues in Art." *Revue de Métallurgie* 98 (9): 751–66.

Evelyn 1995
Evelyn, Peta. 1995. "Hubert Le Sueur's Equestrian Bronzes at the Victoria and Albert Museum." *Burlington Magazine* 137 (1103): 85–92.

Evelyn 2000
Evelyn, Peta. 2000. "The Equestrian Bronzes of Hubert le Sueur." In *Giambologna tra Firenze e l'Europa. Atti del Convegno Internazionale, Firenze, Istituto universitario olandese di storia dell'arte, 1995*, edited by Sabine Eiche, 141–56. Florence: Centro Di.

Ezrati 2015
Ezrati, Jean-Jacques. 2015. *Éclairage d'exposition: Musées et autres espaces*. Paris: Eyrolles.

Fagg and Underwood 1949
Fagg, William, and Leon Underwood. 1949. "An Examination of the So-Called 'Olokun' Head of Ife, Nigeria." *Man* 49:1–7.

Fairchild 2013
Fairchild, Mark D. 2013. *Color Appearance Models*. Chichester, UK: Wiley.

Falaschi 2017
Falaschi, Eva. 2017. "Looking at the Bronze of Lost Sculptures: The Reception of the Delphic Monument of the Admirals in the Imperial Age." In *Artistry in Bronze: The Greeks and Their Legacy: XIXth International Congress on Ancient Bronzes*, edited by Jens M. Daehner, Kenneth Lapatin, and Ambra Spinelli, 94–97. Los Angeles: J. Paul Getty Museum and Getty Conservation Institute. http://www.getty.edu/publications/artistryinbronze/the-artist/11-falaschi/.

Falchetti 2003
Falchetti, Ana Maria. 2003. "The Seed of Life: The Symbolic Power of Gold-Copper Alloys and Metallurgical Transformations." In *Gold and Power in Ancient Costa Rica, Panama, and Colombia*, edited by Jeffrey Quilter and John W. Hoopes, 345–81. Washington, DC: Dumbarton Oaks Research Library and Collection. http://www.humanas.unal.edu.co/colantropos/files/5414/7422/2423/The_seed_of_life__Falchetti.pdf.

Félibien 1690
Félibien, André. 1690. *Des principes de l'architecture, de la sculpture, de la peinture et des autres arts qui en dépendent: Avec un dictionnaire des termes propres à chacun de ces arts*. 2nd ed. Paris: Chez la veuve de Jean Baptiste Coignard et Chez Jean Baptiste Coignard fils. https://archive.org/details/principesdelarch00fl.

Felicísimo and Polo 2022
Felicísimo, Ángel M., and María-Eugenia Polo. 2022. "Measurement and Control of Colour Fidelity in Scanned 3D Models for Heritage Conservation." *Journal of Cultural Heritage* 56 (July): 159–66. https://doi.org/10.1016/j.culher.2022.06.010.

Fishlock 1962
Fishlock, David. 1962. *Metal Colouring*. Teddington, UK: Robert Draper.

Fleming 1971
Fleming, S. J. 1971. "New Techniques of Thermoluminescence Dating of Ancient Pottery: II: The Predose Method." In *Proceedings of the Third International Conference on Luminescence Dosimetry, held at the Danish AEC Research Establishment Risö 11–14 October 1971*, edited by V. Mejdahl, 3:895–929. Risø Report 249. Roskilde, Denmark: Risø National Laboratory. https://orbit.dtu.dk/en/publications/proceedings-of-the-third-international-conference-on-luminescence-3.

Fleming and Tilch 1993
Fleming, Eckart, and Werner Tilch. 1993. *Formstoffe und Formverfahren*. Leipzig, Germany: Deutscher Verlag für grundstoffindustrie.

Fogelman and Fusco 2002
Fogelman, Peggy, and Peter Fusco. 2002. *Italian and Spanish Sculpture: Catalogue of the J. Paul Getty Museum Collection*. Los Angeles: J. Paul Getty Museum. http://www.getty.edu/publications/virtuallibrary/0892366893.html.

Fontein, Soekmono, and Sedyawati 1990
Fontein, Jan, R. Soekmono, and Edi Sedyawati. 1990. *The Sculpture of Indonesia*. Washington, DC: National Gallery of Art.

Fontein, Soekmono, and Suleiman 1971
Fontein, Jan, R. Soekmono, and Sayawati Suleiman. 1971. *Ancient Indonesian Art of the Central and Eastern Javanese Periods*. New York: Asia Society.

Formigli 1984
Formigli, Edilberto. 1984. "Due bronzi da Riace." *Bolletino d'Arte* 2, special series 3. Rome: Libreria dello Stato. http://www.bollettinodarte.beniculturali.it/opencms/multimedia/BollettinoArteIt/documents/1475238151057_Due_Bronzi_di_Riace_I_(Serie_speciale_1984).pdf.

Formigli 1993
Formigli, Edilberto. 1993. "Antiche terre di fusione." In *Antiche officine del bronzo: materiali, strumenti, tecniche: atti del Seminario di studi ed esperimenti, Murlo, 26–31 luglio 1991*, edited by Edilberto Formigli, 69–98. Siena, Italy: Nuova Immagine.

Formigli 1999a
Formigli, Edilberto. ed. 1999a. *I grandi bronzi antichi: Le fonderie e le tecniche di lavorazione dall'età arcaica al rinascimento*. Siena, Italy: Nuova Immagine.

Formigli 1999b
Formigli, Edilberto. 1999b. "Le antiche terre di fusione, i problem di formatura dei grandi bronzi e la tecnica di fusione dei bronzi di Riace." In *I grandi bronzi antichi: Le fonderie e le tecniche di lavorazione dall'età arcaica al rinascimento: Atti dei Seminari di Studi ed Esperimenti, Murlo 24–30 luglio 1993 e 1–7 luglio 1995*, edited by Edilberto Formigli, 67–74. Siena, Italy: Nuova Immagine.

Formigli 1999c
Formigli, Edilberto. 1999c. "Resoconto degli esperimenti di saldatura per colate e di rifinitura a freddo sui grandi bronzi antichi." In *I grandi bronzi antichi: le fonderie e le tecniche di lavorazione dall'età arcaica al rinascimento: atti dei Seminari di Studi ed Esperimenti, Murlo 24–30 luglio 1993 e 1–7 luglio 1995*, edited by Edilberto Formigli, 317–34. Siena, Italy: Nuova Immagine.

Formigli 1999d
Formigli, Edilberto. 1999d. "Due millenni di storia della metallotecnica nei restauri alla statua equestre del Marco Aurelio in Roma." In *I grandi bronzi antichi: le fonderie e le tecniche di lavorazione dall'età arcaica al Rinascimento: atti dei seminari di studi ed esperimenti, Murlo 24-30 luglio 1993 e 1-7 luglio 1995*, edited by Edilberto Formigli, 147–90. Siena, Italy: Nuova immagine.

Formigli 2007
Formigli, Edilberto. 2007. "Ghiberti and the Art of Chasing." In *The Gates of Paradise: Lorenzo Ghiberti's Renaissance Masterpiece*, edited by Gary M. Radke, 118–33. Atlanta: High Museum of Art.

Formigli 2010
Formigli, Edilberto. 2010. "La storia della tecnologia dei grandi bronzi." In *La lupa capitolina: Nuove prospettive di studio: Incontro-dibattito in occasione della pubblicazione del volume di Anna Maria Carruba, La lupa capitolina: Un bronzo medievale: Sapienza, Università di Roma, Roma 28 febbraio 2008*, edited by Gilda Bartoloni, 15–24. Supplementi e monografie della rivista *Archeologia classica* 5. Roma: L'Erma di Bretschneider.

Formigli 2012
Formigli, Edilberto. 2012. "Ghiberti's Saint Matthew and Roman Bronze Statuary: Technical Investigations during Restoration." *Studies in the History of Art* 76:245–56.

Formigli 2013a
Formigli, Edilberto. 2013a. *Colore e luce nella statuaria antica in bronzo*. Rome: L'Erma di Bretschneider.

Formigli 2013b
Formigli, Edilberto. 2013b. "Die 'Natürlichen' und künstlichen Patinae." In *Zurück zur Klassik. Ein neuer Blick auf das alte Griechenland*, edited by Vinzenz Brinkman, 289–94. Munich: Hirmer.

Formigli 2013c
Formigli, Edilberto. 2013c. "Le tecniche del colore nella statuaria antica in bronzo." In Colore e Luce Nella Statuaria Antica in Bronzo, edited by Edilberto Formigli, 1–23. Rome: L'Erma di Bretschneider.

Formigli and Hackländer 1999
Formigli, Edilberto, and Nele Hackländer. 1999. "Resocento della ricostruzione sperimentale di una fossa di fusione ellenistica e della cottura du un modello dell'adorante di Berlino." In *I grandi bronzi antichi: le fonderie e le tecniche di lavorazione dall'età arcaica al rinascimento: atti dei Seminari di Studi ed Esperimenti, Murlo 24–30 luglio 1993 e 1–7 luglio 1995*, edited by Edilberto Formigli, 299–316. Siena, Italy: Nuova Immagine.

François 2004
François, Dominique. 2004. "Essais mécaniques des métaux: introduction." In *Techniques de l'ingénieur. Essais mécaniques sur les métaux et alliages* TIB531DUO (m4150). https://www.techniques-ingenieur.fr/base-documentaire/materiaux-th11/essais-mecaniques-sur-les-metaux-et-alliages-42531210/essais-mecaniques-des-metaux-m4150/.

Frapiccini 2017
Frapiccini, Nicoletta. 2017. "Roman Bronze Figurines of Deities in the National Archaeological Museum of the Marche (Ancona)." In *Artistry in Bronze: The Greeks and Their Legacy: XIXth International Congress on Ancient Bronzes*, edited by Jens M. Daehner, Kenneth Lapatin, and Ambra Spinelli. Los Angeles: J. Paul Getty Museum and Getty Conservation Institute. http://www.getty.edu/publications/artistryinbronze/statuettes/17-frapiccini/.

Frel 1982
Frel, Jiří. 1982. *The Getty Bronze*. Malibu, CA: J. Paul Getty Museum.

Fucito 2013
Fucito, Luisa. 2013. "Methods and Materials Used for Patination at the Fonderia Chiurazzi." In *The Restoration of Ancient Bronzes: Naples and Beyond*, edited by Erik Rissser and David Saunders, 137–42. Los Angeles: J. Paul Getty Museum. http://www.getty.edu/publications/virtuallibrary/1606061542.html.

Furger 2017
Furger, Alex R. 2017. *The Gilded Buddha: The Traditional Art of the Newar Metal Casters in Nepal*. Basel, Switzerland, and Frankfurt, Germany: Librum.

Gale 2009
Gale, Noël H. 2009. "A Response to the Paper of A. M. Pollard: What a Long, Strange Trip It's Been: Lead Isotopes and Archaeology." In *From Mine to Microscope*, edited by Andrew J. Shortland, Ian C. Freestone, and Thilo Rehren, 191–95. Oxford: Oxbow Books.

Gänsicke 2016
Gänsicke, Susanne. 2016. "Bodies in Bronze: Four South Indian Temple Bronzes in the Museum of Fine Arts, Boston." In *Proceedings of the Interim Meeting of the ICOM-CC Metals Working Group, September 26–30, 2016 New Delhi, India*, edited by Raghu Menon, Claudia Chemello, and Achal Pandya, 14–21. Paris; New Delhi: ICOM-CC & Indira Gandhi National Centre of the Arts.

Gänsicke and Sharma 2021
Gänsicke, Susanne, and Nutandhar Sharma. 2021. "The Preservation of Metalwork in the Kathmandu Valley: Reflections on Past and Current Practices." In *Transcending Boundaries: Integrated Approaches to Conservation. ICOM-CC 19th Triennial Conference Preprints, Beijing, 17–21 May 2021*, edited by Janet Bridgeland, 1–9. Paris: International Council of Museums. https://www.icom-cc-publications-online.org/4274/The-preservation-of-metalwork-in-the-Kathmandu-Valley--Reflections-on-past-and-current-practices.

Garenne-Marot, Robion, and Mille 2003
Garenne-Marot, Laurence, Caroline Robion, and Benoît Mille. 2003. "Cuivre, alliages de cuivre et histoire de l'empire du Mali. À propos de trois figurines animales d'un tumulus du delta intérieur du Niger (Mali)." *Techné* 18:74–85.

Garnier 1903
Garnier, Jean. 1903. *Nouveau manuel complet du ciseleur, contenant la description des procédés de l'art de ciseler et repousser les métaux ductiles*. Paris: Encyclopédie-Roret; L. Mulo. http://gallica.bnf.fr/ark:/12148/bpt6k6576990q/f17.image.r=roret ciseleur#.

Gaudiuso et al. 2010
Gaudiuso, Rosalba, Marcella dell'Aglio, Olga de Pascale, Giorgio S. Senesi, and Alessandro de Giacomo. 2010. "Laser Induced Breakdown Spectroscopy for Elemental Analysis in Environmental, Cultural Heritage and Space Applications: A Review of Methods and Results." *Sensors* 10 (8): 7434–68. https://doi.org/10.3390/s100807434.

Gaurico [1504] 1999
Gaurico, Pomponius. 1504 (1999). *De Sculptura*. Translated by Paolo Cutolo. Florence, Italy: Filippo Giunta.

Gauricus [1504] 1886
Gauricus, Pomponius. 1504 (1886). *De Sculptura*. Translated by Heinrich Brockhaus. Leipzig, Germany: F. A. Brockhaus.

Gauricus [1504] 1969
Gauricus, Pomponius. 1504 (1969). *De Sculptura*. Translated by André Chastel and Robert Klein. Geneva and Paris: Droz.

Gavrilov, Maev, and Almond 2014
Gavrilov, D., R. Gr. Maev, and D. P. Almond. 2014. "A Review of Imaging Methods in Analysis of Works of Art: Thermographic Imaging Method in Art Analysis." *Canadian Journal of Physics* 92 (4): 341–64.

Gehres and Querré 2018
Gehres, B., and G. Querré. 2018. "New Applications of LA-ICP-MS for Sourcing Archaeological Ceramics: Microanalysis of Inclusions as Fingerprints of Their Origin." *Archaeometry* 60 (4): 750–63.

Gesamtverband Deutscher Metallgießereien 1982
Gesamtverband Deutscher Metallgießereien, Verein Deutscher Giessereifachleute. 1982. *Guss aus Kupfer und Kupferlegierungen. Technische richtlinien*. Düsseldorf: Deutsches Kupfer-Institut.

Gettens 1933
Gettens, Rutherford J. 1933. "Mineralization, Electrolytic Treatment, and Radiographic Examination of Copper and Bronze Objects from Nuzi." *Technical Studies in the Field of the Fine Arts* 1 (3): 118–42.

Gettens 1967
Gettens, Rutherford J. 1967. "Joining Methods in the Fabrication of Ancient Chinese Bronze Ceremonial Vessels." In *Application of Science in Examination of Works of Art; Proceedings of the Seminar: September 7-16, 1965*, 205–17. Boston: Museum of Fine Arts, Boston.

Gettens 1969
Gettens, Rutherford J. 1969. *The Freer Chinese Bronzes*. Vol. II Technical Studies. Freer Gallery of Art, Oriental Studies, No. 7. Washington: Smithsonian Institution.

Gettens 1970
Gettens, Rutherford J. 1970. "Patina: Noble and Vile." In *Art and Technology: A Symposium on Classical Bronzes*, edited by Suzannah Doeringer, David Gordon Mitten, and Arthur Steinberg, 57–68. Cambridge, MA: MIT Press.

Ginisty 1987
Ginisty, Monique. 1987. *Susse fondeur et les maîtres du bronze. 150 ans au service de la sculpture*. Arcueil, France: Susse Fondeur.

Giuffredi 2006
Giuffredi, Augusto. 2006. *Manuale delle techniche di formatura e fonderia*. Florence: Alinea.

Giuffredi 2010
Giuffredi, Augusto. 2010. *Formatura e fonderia: guida ai processi di lavorazione*. Florence: Alinea.

Giumlia-Mair 1995
Giumlia-Mair, Alessandra. 1995. "The Appearance of Black Patinated Copper-Gold-Alloys in the Mediterranean Area in the Second Millennium B.C." In *Prehistoric Gold in Europe: Mines, Metallurgy and Manufacture*, edited by Giulio Morteani and Jeremy P. Northover, 425–34. NATO ASI Series. Dordrecht, the Netherlands: Springer Netherlands.

Giumlia-Mair 1999
Giumlia-Mair, Alessandra R. 1999. "Studi tecnici sulla testa del cavallo romano da Augusta Vindelicum." In In In *I grandi bronzi antichi: Le fonderie e le tecniche di lavorazione dall'età arcaica al rinascimento: Atti dei Seminari di Studi ed Esperimenti, Murlo 24–30 luglio 1993 e 1–7 luglio 1995*, edited by Edilberto Formigli, 255–70. Siena, Italy: Nuova Immagine.

Giumlia-Mair 2002
Giumlia-Mair, Alessandra. 2002. "Zosimos the Alchemist: Manuscript 6.29, Cambridge, Metallurgical Interpretation." In *I bronzi antichi: Produzione e tecnologia. Atti del XV Congresso Internazionale sui Bronzi Antichi*, edited by Alessandra Giumlia-Mair, 317–23. Montagnac, France: Editions Monique Mergoil.

Glinsman and Hayek 1993
Glinsman, Lisha A., and Lee-Ann C. Hayek. 1993. "A Multivariate Analysis of Renaissance Portrait Medals: An Expanded Nomenclature for Defining Alloy Composition." *Archaeometry* 35 (1): 49–67.

"Glossaire de la fonderie et des domaines connexes | CTIF" n.d.
"Glossaire de la fonderie et des domaines connexes | CTIF." n.d. Accessed April 14, 2021. https://www.ctif.com/connaissances/base-de-connaissances/glossaire-fonderie/.

Goemaere et al. 2014
Goemaere, Éric, Denis Henrotay, Olivier Collette, Mark Golitko, Thomas Delbey, and Thierry Leduc. 2014. "Caractérisation de la céramique médiévale d'Autelbas (Arlon, Belgique) et identification de la source de la matière première." *ArchéoSciences*, no. 38 (December): 31–47.

Gonon 1876
Gonon, E. 1876. "L'art de fondre en bronze à cire perdue." Paris. https://inha.revues.org/3522.

Goodison 1974
Goodison, Nicholas. 1974. *Ormolu, the Work of Matthew Boulton*. London: Phaidon.

Graedel, Nassau, and Franey 1987
Graedel, T. E., K. Nassau, and J. P. Franey. 1987. "Copper Patinas Formed in the Atmosphere: Introduction." *Corrosion Science* 27 (7): 639–58.

Grandpierre and Avale 1867
Grandpierre and Avale. 1867. *Rapport des délégations ouvrières - Fondeurs en cuivre - Exposition Universelle de 1867 à Paris.* Paris: Librairie A. Morel.

Grandpierre and Dargent 1862
Grandpierre and Dargent. 1862. "Fondeurs en cuivre." In *Rapport des délégués des ouvriers parisiens à l'Exposition de Londres en 1862 publiés par la Commission ouvrière*, 269–98. Paris: Chez M. Chabaud, président de la Commission ouvrière, 34 rue Dauphine et chez MM. Wanschooten, Grandpierre et Dargent, membres de la Commision. http://cnum.cnam.fr/CGI/fpage.cgi?8XAE135/283/100/902/0/0.

Grant and Patterson 2018
Grant, Alistair, and Angus Patterson. 2018. *The Museum and the Factory: The V&A, Elkington and the Electrical Revolution*. V&A 19th-Century Series. London: Lund Humphries.

Gratuze, Blet-Lemarquand, and Barrandon 2001
Gratuze, B., M. Blet-Lemarquand, and J.-N. Barrandon. 2001. "Mass Spectrometry with Laser Sampling: A New Tool to Characterize Archaeological Materials." *Journal of Radioanalytical and Nuclear Chemistry* 247 (3): 645–56.

Grimwade 1999
Grimwade, Mark. 1999. "The Surface Enrichment of Carat Gold Alloys: Depletion Gilding." *Gold Technology* 26:16–23.

Grissom and Harvey 2003
Grissom, Carol A., and Ronald S. Harvey. 2003. "The Conservation of American War Memorials Made of Zinc." *Journal of the American Institute for Conservation* 42 (1): 21–38.

Grodecki 1986
Grodecki, Catherine. 1986. *Documents du Minutier central des notaires de Paris. Histoire de l'art au XVIème siècle (1540–1600)*, Vol. 2, *Sculpture, peinture et gravure, broderie, faïence et émaillerie, orfèvrerie et joaillerie, armes et armures*. Paris: Archives nationales.

Grünauer 2005
Grünauer, F. 2005. "Design, Optimization, and Implementation of the New Neutron Radiography Facility at FRM-II." PhD diss., Technische Universität München.

Grzywacz 2006
Grzywacz, Cecily. 2006. *Monitoring for Gaseous Pollutants in Museum Environments: Tools for Conservation.* Los Angeles: Getty Conservation Institute.

Guettier 1858
Guettier, André. 1858. *De la fonderie telle qu'elle existe aujourd'hui en France et de ses nombreuses applications à l'industrie*. Paris: Chez Carilian-Cœury et Dalmont; Angers, France: Chez Cosnier et Lachèse; Metz, France: Chez Mme Thiel. https://books.google.fr/books?id=FmEUAAAAQAAJ.

Guettier 1867
Guettier, André. 1867. "Bronzes et fontes d'art - ouvrages d'art en métaux." In *Études sur l'Exposition de 1867 ou les archives de l'industrie au XIXe siècle - Description générale, encyclopédique, méthodique et raisonnée de l'état actuel des arts, des sciences, de l'industrie et de l'agriculture, chez toutes les nations - Recueil de travaux techniques, théoriques, pratiques et historiques par MM. les rédacteurs des Annales du Génie civil avec la collaboration de savants, d'ingénieurs et de professeurs français et étrangers*, edited by Eugène Lacroix, 6th fascicule, 2:211–26. Paris: Librairie scientifique, industrielle et agricole.

Gwinnett and Gorelick 1998
Gwinnett, John, and Leonard Gorelick. 1998. "A Brief History of Drills and Drilling." *Beads: Journal of the Society of Bead Researchers* 10:49–56.

Halleux 1973
Halleux, Robert. 1973. "L'orichalque et le laiton." *L'antiquité classique* 42 (1): 64–81.

Halmshaw 1995
Halmshaw, R. 1995. *Industrial Radiology: Theory and Practice.* 2nd ed. London: Chapmann & Hall.

Hamilton 2018
Hamilton, Emily. 2018. "All That Glitters Is Not Gold: Repatinating the Sculpture of Jean Arp." SFMOMA.org, March 2018. https://www.sfmoma.org/read/all-glitters-not-gold-repatinating-sculptures-jean-arp/.

Hamm 1924
Hamm, Georges. 1924. *Manuel du bronze d'art, ciselure et gravure*. Paris: Librairie J.-B. Baillière et fils.

Hancock et al. 2019
Hancock, R. G. V., K. Michelaki, W. C. Mahaney, and S. Aufreiter. 2019. "Justification for Reassessing Elemental Analysis Data of Ceramics, Sediments and Lithics Using Rare Earth Element Concentrations and Ratios." *Archaeometry* 61 (6): 1430–45.

Hanson and Pell-Walpole 1951
Hanson, D., and W. T. Pell-Walpole. 1951. *Chill-Cast Tin Bronze*. London: Edward Arnold.

Hardt and Röttger 1981
Hardt, P., and H. Röttger. 1981. *Neutron Radiography Handbook: Nuclear Science and Technology*. Dordrecht, the Netherlands: Springer Netherlands.

Haynes 1992
Haynes, Denys Eyre Lankester. 1992. *The Technique of Greek Bronze Statuary*. Mainz, Germany: Philipp von Zabern.

Hecker 2017
Hecker, Sharon. 2017. "The Afterlife of Sculptures: Posthumous Casts and the Case of Medardo Rosso (1858–1928)." *Journal of Art Historiography* 16 (June): 1–18.

Hecker 2020
Hecker, Sharon, ed. 2020. *Finding Lost Wax: The Disappearance and Recovery of an Ancient Casting Technique and the Experiments of Medardo Rosso*. Studies in Art and Materiality 3. Leiden, the Netherlands, and Boston: Brill.

Hecker 2021
Hecker, Sharon, ed. 2021. *Finding Lost Wax - The Disappearance and Recovery of an Ancient Casting Technique and the Experiments of Medardo Rosso*. Studies in Art and Materiality 3. Leiden; Boston: Brill.

Heeb and Ottaway 2014
Heeb, Julia, and Barbara S. Ottaway. 2014. "Experimental Archaeometallurgy." In *Archaeometallurgy in Global Perspective*, edited by Benjamin W. Roberts and Christopher P. Thornton, 161–92. New York: Springer.

Heginbotham 2014
Heginbotham, Arlen. 2014. "*Bronzes Dorés*: A Technical Approach to Examination and Authentication of French Gilt Bronze." In *French Bronze Sculpture: Materials and Techniques 16th–18th Century*, edited by David Bourgarit, Jane Bassett, Francesca G. Bewer, Geneviève Bresc-Bautier, Philippe Malgouyres, and Guilhem Scherf, 150–65. Paris: Archetype.

Heginbotham and Solé 2017
Heginbotham, Arlen, and V. Armando Solé. 2017. "CHARMed PyMca, Part I: A Protocol for Improved Inter-Laboratory Reproducibility in the Quantitative ED-XRF Analysis of Copper Alloys." *Archaeometry* 59 (4): 714–30. https://doi.org/10.1111/arcm.12282.

Heginbotham, Bassett et al. 2014
Heginbotham, Arlen, Jane Bassett, David Bourgarit, Chris Eveleigh, Lisha Glinsman et al. 2014. "The Copper CHARM Set: A New Set of Certified

Reference Materials for the Standardization of Quantitative X-Ray Fluorescence Analysis of Heritage Copper Alloys." *Archaeometry* 57 (5): 856–68.

Heginbotham, Beltran et al. 2014
Heginbotham, Arlen, Vincent Beltran, Briana Feston, Ruven Pillay, and Christina Simms. 2014. "An Evaluation of Protective Coatings for Brass in Indoor Environments, with an Emphasis on Boulle Marquetry." In *Furniture Finishes: Past, Present and Future of Transparent Wood Coatings*, edited by Miko Vasques Dias, 84–103. Amsterdam: Stichting Ebenist.

Heginbotham, Erdmann, and Hayek 2018
Heginbotham, Arlen, Robert Erdmann, and Lee-Ann C. Hayek. 2018. "The Dating of French Gilt Bronzes with ED-XRF Analysis and Machine Learning." *Journal of the American Institute for Conservation* 57 (4): 149–68.

Heginbotham et al. 2011
Heginbotham, Arlen, Aniko Bezur, Michel Bouchard, Jeffrey M. Davis, Katherine Eremin, James H. Frantz, Lisha Glinsman et al. 2011. "An Evaluation of Inter-Laboratory Reproducibility for Quantitative XRF of Historic Copper Alloys." In *Metal 2010: Proceedings of the Interim Meeting of the ICOM-CC Metal Working Group, October 11–15, 2010, Charleston, South Carolina, USA*, edited by Paul Mardikian, Claudia Chemello, Christopher Watters, and Peter Hull, 244–55. Clemson, SC: Clemson University Press. https://repository.si.edu/handle/10088/90692

Heginbotham et al. 2019
Heginbotham, Arlen, David Bourgarit, Julia Day, Jan Dorscheid, Joe Godla, Lynn Lee, Arie Pappot, and Dominique Robcis. 2019. "CHARMed PyMca, Part II: An Evaluation of Interlaboratory Reproducibility for ED-XRF Analysis of Copper Alloys." *Archaeometry* 61 (6): 1333–52. https://doi.org/10.1111/arcm.12488.

Heginbotham, Millay, and Quick 2006
Heginbotham, Arlen, Victoria Millay, and Michael Quick. 2006. "The Use of Immunofluorescence Microscopy and Enzyme-Linked Immunosorbent Assay as Complementary Techniques for Protein Identification in Artists' Materials." *Journal of the American Institute for Conservation* 45 (2): 89–105.

Heikamp and Paolozzi Strozzi 2014
Heikamp, Detlef, and Beatrice Paolozzi Strozzi, eds. 2014. *Baccio Bandinelli. Scultore e Maestro (1493–1560)*. Florence: Museo Nazionale del Bargello.

Heilmeyer 1994
Heilmeyer, W. D. 1994. "Zur Oberfläche antiker Großbronzen." In *Das Wrack. Der antike Schiffsfund von Mahdia*, edited by Gisela Hellenkemper Salies with Hans-Hoyer von Prittwitz und Gaffron and Gerhard Bauchhenß, 2:801–7. Kataloge des Rheinischen Landesmuseums Bonn. Cologne: Rheinland Verlag.

Hemingway and Abramitis 2017
Hemingway, Seán A., and Dorothy Abramitis. 2017. "The Use of Inlays in Early Greek Bronzes." In *Artistry in Bronze: The Greeks and Their Legacy: XIXth International Congress on Ancient Bronzes*, edited by Jens M. Daehner, Kenneth Lapatin, and Ambra Spinelli, 114–22. Los Angeles: J. Paul Getty Museum and Getty Conservation Institute. http://www.getty.edu/publications/artistryinbronze/statuettes/14-hemingway-abramitis/.

Hénon, Mascré, and Blanc 1971
Hénon, Guy, Claude Mascré, and Georges Blanc. 1971. *Recherche de la qualité des pièces de fonderie*. Paris: Editions techniques des industries de la fonderie.

Henss and Lehmann 2016
Henss, M., and E. Lehmann. 2016. "The Canned Buddha: Neutron Radiography and Tomography of Tibetan Buddhist Metal Images." *Orientations* 47 (5): 77–81.

Herbert 1984
Herbert, Eugenia W. 1984. *Red Gold of Africa: Copper in Precolonial History and Culture*. Madison: University of Wisconsin Press.

Hickel 1963
Hickel, Erika. 1963. "Chemikalien im Arzneischatz deutscher Apotheken des 16. Jahrhunderts, unter besonderer Berücksichtigung der Metalle." PhD diss., Technische Hochschule Carolo-Wilhelmina zu Braunschweig.

Hill 1982
Hill, Dorothy K. 1982. "Note on the Piecing of Bronze Statuettes." *Hesperia* 51 (3): 277–83.

Hiorns 1907
Hiorns, Arthur H. 1907. *Metal-Colouring and Bronzing*. London and New York: Macmillan.

Hollinshead 2002
Hollinshead, Mary Berg. 2002. "Extending the Reach of Marble: Struts in Greek and Roman Sculpture." In *The Ancient Art of Emulation*, edited by Elaine K. Gazda, 117–52. Ann Arbor: University of Michigan Press.

Holmes and Harbotte 1991
Holmes, Lore L., and Garman Harbotte. 1991. "Provenance Study of Cores from Chinese Bronze Vessels." *Archeomaterials* 5 (2): 165–84.

Hua 2013
Hua Jueming | 华觉明. 2013. "技术史的研究方法——以失蜡法学案为例 = Research Methodology for History of Technology: Using Lost-Wax Casting as an Example." 自然科学史研究 = *Studies in the History of Natural Sciences* 32 (3): 347–59. https://www.ixueshu.com/document/214c9856c461100728794d8529613e51318947a18e7f9386.html.

Huang et al. 2015
Huang, Xing, Wei Qian, Wei Wei, Jingning Guo, and Naitao Liu. 2015. "3D Numerical Simulation on the Flow Field of Single Tuyere Blast Furnaces: A Case Study of the Shuiquangou Iron Smelting Site Dated from the 9th to 13th Century in China." *Journal of Archaeological Science* 63:44–58.

Hughes 1993
Hughes, Richard. 1993. "Artificial Patination." In *Metal Plating and Patination: Cultural, Technical and Historical Developments*, edited by Susan La Niece and Paul Craddock, 1–18. Oxford, London, Boston [etc.]: Butterworth-Heinemann.

Hughes and Rowe 1989
Hughes, Richard, and Michael Rowe. 1989. *The Colouring, Bronzing, and Patination of Metals: A Manual for the Fine Metalworker and Sculptor; Cast Bronze, Cast Brass, Copper and Copper-Plate, Gilding Metal, Sheet Yellow Brass, Silver and Silver-Plate*. 4th ed. London: Crafts Council.

Hüls, Petri, and Föll 2019
Hüls, C. Matthias, Ingo Petri, and Helmut Föll. 2019. "Absolute Dating of Early Iron Objects from the Ancient Orient: Radiocarbon Dating of Luristan Iron Mask Swords." *Radiocarbon* 61 (5): 1229–38. https://doi.org/10.1017/RDC.2019.13.

Hunt 1973
Hunt, L. B. 1973. "The Early History of Gold Plating." *Gold Bulletin* 6 (1): 16–27. https://doi.org/10.1007/BF03215178.

Hunt and Pointer 2011
Hunt, R. W. G, and Michael R. Pointer. 2011. *Measuring Colour*. 4th ed. Hoboken, NJ: Wiley.

Hunter 2002
Hunter, Erica C. D. 2002. "Beautiful Black Bronzes: Zosimos' Treatises in Cam. Mm.6.29." In *I bronzi antichi: Produzione e tecnologia. Atti del XV Congresso Internazionale sui Bronzi Antichi*, edited by Alessandra Giumlia-Mair, 655–60. Montagnac, France: Editions Monique Mergoil.

Ioannidou et al. 2000
Ioannidou, E., D. Bourgarit, T. Calligaro, J. C. Dran, M. Dubus, J. Salomon, and P. Walter. 2000. "RBS and NRA with External Beams for Archaeometric Applications." *Nuclear Instrument and Methods B* 161/162/163:730–36.

Ishikawa 1990
Ishikawa, Kaoru. 1990. *Introduction to Quality Control*. Tokyo: 3A Corp.

ISO 2007
ISO. 2007. "International Vocabulary of Metrology — Basic and General Concepts and Associated Terms (VIM)." *ISO/IEC Guide* 99:2007. Geneva: International Organization for Standardization. https://www.iso.org/standard/45324.html.

ISO 2008
ISO. 2008. "Uncertainty of Measurement — Part 3: Guide to the Expression of Uncertainty in Measurement (GUM:1995)." *ISO/IEC Guide* 98-3:2008. Geneva: International Organization for Standardization. https://www.iso.org/standard/50461.html.

Jacobs et al. 2011
Jacobs, Z., M. C. Meyer, R. G. Roberts, V. Aldeias, H. Dibble, and M. A. El Hajraoui. 2011. "Single-Grain OSL Dating at La Grotte Des Contrebandiers ('Smugglers' Cave'), Morocco: Improved Age Constraints for the Middle Paleolithic Levels." *Journal of Archaeological Science* 38 (12): 3631–43.

Janietz 2000
Janietz, B. 2000. *Ein Depot zerschlagener Grossbronzen aus Augusta Raurica. Die Rekonstruktion der Gewandfiguren*. Forschungen in Augst 30. Augst, Switzerland: Römerstadt Augusta Raurica.

Janietz 2001
Janietz, B. 2001. "Untersuchungen an geometrischen Bronzen." Phd diss., Albert-Ludwigs-Universität. 2001. https://freidok.uni-freiburg.de/fedora/objects/freidok:199/datastreams/FILE1/content.

JCGM/WG 1 2008
JCGM/WG 1, Working Group 1 of the Joint Committee for Guides in 2008. "Evaluation of Measurement Data — Guide to the Expression of Uncertainty in Measurement." *JCGM* 100:2008. Sèvres, France: Bureau International des Poids et Mesures. https://www.bipm.org/documents/20126/2071204/JCGM_100_2008_E.pdf/cb0ef43f-baa5-11cf-3f85-4dcd86f77bd6.

Jett 1993
Jett, Paul. 1993. "A Study of the Gilding of Chinese Buddhist Bronzes." In *Metal Plating and Patination: Cultural, Technical and Historical Developments*, edited by Susan La Niece and Paul Craddock, 193–200. Oxford, London, Boston [etc.]: Butterworth-Heinemann.

Jett 2010
Jett, Paul. 2010. "Buddhist Bronzes in Cambodia: A Newly Discovered Cache." *Orientations* 41 (5): 48–52.

Johnston-Feller 2001
Johnston-Feller, Ruth. 2001. *Color Science in the Examination of Museum Objects: Nondestructive Procedures*. Tools for Conservation. Los Angeles: Getty Conservation Institute. http://www.getty.edu/conservation/publications_resources/pdf_publications/pdf/color_science.pdf.

Jones 1990
Jones, Mark, ed. 1990. *Fake? The Art of Deception*. London: British Museum Publications.

Junge 2007
Junge, Peter. 2007. "Age Determination of Commemorative Heads: The Example of the Berlin Collection." In *Benin. Kings and Rituals: Court Arts from Nigeria*, edited by Barbara Plankensteiner, 185–98. Ghent, Belgium: Snoeck.

Karydas 2007
Karydas, Andreas G. 2007. "Application of a Portable XRF Spectrometer for the Noninvasive Analysis of Museum Metal Artefacts." *Annali di chimica* 97 (7): 419–32.

Kienlin, Bischoff, and Opielka 2006
Kienlin, T. L., E. Bischoff, and H. Opielka. 2006. "Copper and Bronze during the Eneolithic and Early Bronze Age: A Metallographic Examination of Axes from the Northalpine Region." *Archaeometry* 48 (3): 453–68.

Killick 2015
Killick, David. 2015. "Archaeometallurgy as Archaeology." In *Archaeometallurgy in Europe III, Proceedings of the 3rd International Conference, Deutsches Bergbau-Museum Bochum, June 29–July 1*, edited by Andreas Hauptmann and Diana Modarressi-Tehrani, 295–300. Bochum, Germany: Deutsches Bergbau Museum.

Kipper 1995
Kipper, Patrick V. 1995. *Patinas for Silicon Bronze*. Loveland, CO: Path.

Kirby et al. 2011
Kirby, Daniel. P., Narayan Khandekar, Julie Arslanoglu, and Ken Sutherland. 2011. "Protein Identification in Artworks by Peptide Mass Fingerprinting." In *ICOM Committee for Conservation 16th Triennial Conference, Lisbon, Portugal, 19–23 September 2011*, edited by Janet Bridgland. Almada, Portugal: Critério.

Kleinbauer 1976
Kleinbauer, W. Eugene. 1976. *A Byzantine Revival: The Inlaid Bronze Doors of Constantinople*. Cambridge, MA: Archaeology.

Knitel 1987
Knitel, Otto. 1987. *Die Gießer zum Maximiliangrab: Handwerk und Technik*. Innsbruck, Austria: self-published.

Koch and Newell 1963
Koch, Zbigniew J., and W. C. Newell, eds. 1963. *Vocabulary of Foundry Practice, In Six Languages: English, Czech, German, French, Polish, Russian*. Oxford; New York: Pergamon Press.

Koller and Baumer 2000
Koller, Johann, and Ursula Baumer. 2000. "Organische Überzüge auf Metallen: Teil 1, Lacke." *Arbeitsblätter für Restauratoren. Gruppe 19. Naturwissenschaftliche Untersuchungen* 33 (1): 201–25.

Kölschbach et al. 2000
Kölschbach, S., G. Woelk, A. Hauptmann, and W. G. I. Bunk. 2000. "Experimente zur Simulation prähistorischer Kupfergewinnung: Zur Verfahrenstechnik von Windöfen." *Metalla* 7 (7): 5–22.

Konstam and Hoffmann 2002
Konstam, N., and H. Hoffmann 2002. "Casting the Riace Bronzes: Modern Assumptions and Ancient Facts." *Oxford Journal of Archaeology* 21 (2): 153–65.

Konstam and Hoffmann 2004
Konstam, N., and H. Hoffmann 2004. "Casting the Riace Bronzes (2): A Sculptor's Discovery." *Oxford Journal of Archaeology* 23 (4): 397–402.

Koutsoudis et al. 2014
Koutsoudis, Anestis, Blaž Vidmar, George Ioannakis, Fotis Arnaoutoglou, George Pavlidis, and Christodoulos Chamzas. 2014. "Multi-Image 3D Reconstruction Data Evaluation." *Journal of Cultural Heritage* 15 (1): 73–79.

Kozachuk et al. 2016
Kozachuk, Madalena, Alexandra Suda, Lisa Ellis, Mary Walzak, Mark Biesinger, Sheila Macfie, Robert Hudson, Andrew Nelson, Ronald Martin, and Arlen Heginbotham. 2016. "Possible Radiation-Induced Damage to the Molecular Structure of Wooden Artifacts Due to Micro-Computed Tomography, Handheld X-Ray Fluorescence, and X-Ray Photoelectron Spectroscopic Techniques." *Journal of Conservation and Museum Studies* 14 (1): n.p. https://doi.org/10.5334/jcms.126.

Kroustallis et al. 2008
Kroustallis, Stefanos, Joyce Townsend, Elena Cenalmor Bruquetas, and Ad Stijnman, eds. 2008. *Art Technology: Sources and Methods: Proceedings of the Second Symposium of the Art Technological Source Research Study Group*. London: Archetype.

Kucera 2004
Kucera, Matthias. 2004. "Das Experiment in der Archäologie." In *Archäologie in Europa, Bilanz 2004*, edited by Corinna Endlich, 7–13. Oldenburg, Germany: Isensee.

Laboulaye 1861
Laboulaye, Charles. 1861. *Complément du dictionnaire des arts et manufactures*. Paris: Librairie scientifique, industrielle et agricole, de E. Lacroix.

Lacey 2018
Lacey, Andrew. 2018. "The Sculptor at Work: Recreating the Rothschild Bronzes." In *Michelangelo: Sculptor in Bronze: The Rothschild Bronzes*, edited by Victoria Avery, 174–99. London: Philip Wilson.

Lacey and Lewis 2020
Lacey, Andrew, and Siân Lewis. 2020. "In Pursuit of Magic." In *Secrets of Craft and Nature in Renaissance France: A Digital Critical Edition and English Translation of BnF Ms. Fr. 640*, edited by Pamela H. Smith, Naomi Rosenkranz, Tianna

Helena Uchacz, Tillmann Taape, Clément Godbarge, Sophie Pitman, Jenny Boulboullé, Joel Klein, Donna Bilak, Marc Smith, and Terry Catapano. New York: Making and Knowing Project. https://edition640.makingandknowing.org/#/essays/ann_501_ad_20.

Lahusen and Formigli 2006
Lahusen, Götz, and Edilberto Formigli. 2006. "Römische Herrscher und Bürger: Die Bronzestatuen aus Herculaneum und ihre Restaurierung im 18. Jahrhundert." *Antike Welt* 37 (2): 8–14.

Lambert 2002
Lambert, Daniel. 2002. *Moulage et fonderie d'art: Du modèle au bronze final*. Dourdan, France: Vial.

Lamouche 2021
Lamouche, Emmanuel. 2021. *Les fondeurs de bronze dans la Rome des papes (1585–1630)*. Rome: Editions de l'Ecole Française de Rome.

Lang and Middleton 2005
Lang, Janet, and Andrew Middleton, eds. 2005. *Radiography of Cultural Material*. 2nd ed. Oxford: Butterworth-Heinemann; Burlington, MA: Elsevier.

La Niece 1989
La Niece, Susan. 1989. "White Copper." *Journal of the Historical Metallurgy Society* 23 (1): 9–15.

La Niece 1990
La Niece, Susan. 1990. "Silver Plating on Copper, Bronze and Brass." *Antiquaries Journal* 70 (1): 102–14.

La Niece 1993
La Niece, Susan. 1993. "Silvering." In *Metal Plating and Patination: Cultural, Technical and Historical Developments*, edited by Susan La Niece and Paul Craddock, 201–10. Oxford, London, Boston [etc.]: Butterworth-Heinemann.

La Niece 1995
La Niece, Susan. 1995. "Depletion Gilding from 3rd Millennium BC Ur." *Iraq* 57:41–47.

La Niece and Carradice 1989
La Niece, Susan, and Ian Carradice. 1989. "White Copper: The Arsenical Coinage of the Libyan Revolt 241–238 BC." *Journal of the Historical Metallurgy Society* 23 (1): 9–15.

La Niece et al. 2002
La Niece, Susan, Fleur Shearman, John Taylor, and Anthony Simps. 2002. "Polychromy and Egyptian Bronze: New Evidence for Artificial Coloration." *Studies in Conservation* 47 (2): 95–108.

Lankheit 1962
Lankheit, Klaus. 1962. *Florentinische Barockplastik. Die Kunst am Hofe der letzten Medici, 1670–1743*. Italienische Forschungen, III, No. 2. Munich: Bruckmann.

Lanteri, Agresti, and Pelosi 2019
Lanteri, Luca, Giorgia Agresti, and Claudia Pelosi. 2019. "A New Practical Approach for 3D Documentation in Ultraviolet Fluorescence and Infrared Reflectography of Polychromatic Sculptures as Fundamental Step in Restoration." *Heritage* 2 (1): 207–15.

La Rivière 1721
La Rivière, Jean de. 1721. *Secrets concernans les arts et metiers. Nouvelle édition, revûë, corrigée & considérablement augmentée*. Nancy, France: Chez Jean de la Rivière. https://www.archive.org/details/secretsconcernan01lari.

Larsen 1987
Larsen, E. B. 1987. "SEM-Identification and Documentation of Tool Marks and Surface Textures on the Gundestrup Cauldron." In *Recent Advances in the Conservation and Analysis of Artifacts*, edited by James Black, 393–409. London: Summer Schools Press.

Larsson 1979
Larsson, Lars. 1979. "Zur Giambologna-Ausstellung in Edinburgh, London und Wien." *Kunstchronik* 32 (4): 129–37.

Lauffenburger 2006
Lauffenburger, J. 2006. "The Language of Bronze Casting." In William R. Johnston, Antoine-Louis Barye, and Simon Kelly, *Untamed: The Art of Antoine-Louis Barye*, 74–77. Baltimore: Walters Art Museum; Munich: Prestel.

Lauffenburger, Strahan, and Gates 2016
Lauffenburger, J., D. Strahan, and G. Gates. 2016. "Two Bàyon-Period Khmer Bronzes in the Walters Art Museum: Technical Study Revisited." In *Metal 2016, Proceedings of the Interim Meeting of the ICOM-CC Metals Working Group, September 26–30, 2016, New Delhi, India*, edited by R. Menon, C. Chemello, and A. Pandya, 30–39. Paris: ICOM-CC & Indira Gandhi National Centre of the Arts.

Launay 1827
Launay d'Avranches, J.-B. 1827. *Manuel du fondeur sur tous les métaux ou traité des opérations de fonderie*. Paris: Roret. http://gallica.bnf.fr/ark:/12148/bpt6k62000p.r=.langFR.

Laval, Calligaro, and Mille 2019
Laval, Eric, Thomas Calligaro, and Benoît Mille. 2019. "Mise en évidence de la polychromie métallique des statuettes de Bavay par cartographie de fluorescence X." In *Nouveaux regards sur le trésor des bronzes de Bavay*, edited by Laure de Chavagnac and Benoît Mille, 196–97. Milan: Silvana Editoriale.

Lebon 2003
Lebon, Elisabeth. 2003. *Dictionnaire des fondeurs de bronze d'art. France 1890–1950*. Perth, Australia: Marjon éditions.

Lebon 2012
Lebon, Elisabeth. 2012. *Le Fondeur et le sculpteur: Technique du bronze et histoire de l'art*. Paris: Publications de l'Institut national d'histoire de l'art.

Lebon 2016
Lebon, Elisabeth. 2016. "Le mystère de la Porte des Lions... (Paris, porte des Lions, grande galerie du Louvre – 1847/1867)." Fontesdart.org, May 31, 2016, http://www.fontesdart.org/fontes-n100-elisabeth-lebon/.

Lebon 2019
Lebon, Elisabeth. 2019. "Delhommeau." In *Sculpture of "Les Animaliers" 1900–1950*, edited by Edward Horswell, 110–13. London: Sladmore Gallery.

Lebon 2020
Lebon, Elisabeth. 2020. "Sculptor-Founders in Late Nineteenth-Century France: The Role of Lost Wax Casting." In *Finding Lost Wax: The Disappearance and Recovery of an Ancient Casting Technique and the Experiments of Medardo Rosso*, edited by Sharon Hecker, 18–65. Studies in Art and Materiality 3. Leiden, the Netherlands: Brill.

Lebon et al. 2016
Lebon, Elisabeth, Cecile Godefroy, Virginie Perdrisiot, and Diana Widmaier Picasso. 2016. "Glossaire." In *Picasso: Sculptures*. Paris: Musée national Picasso, Somogy; Brussels: Palais des beaux-arts Bruxelles.

Le Bonheur 1971
Le Bonheur, Albert. 1971. *La sculpture indonésienne au Musée Guimet*. Paris: Presses universitaires de France

Lechtman 1979
Lechtman, Heather. 1979. "A Pre-Columbian Technique for Electrochemical Replacement Plating of Gold and Silver on Copper Objects." *JOM: Journal of the Minerals, Metals & Materials Society* 31 (12): 154–60.

Lechtman 1984
Lechtman, Heather. 1984. "Pre-Columbian Surface Metallurgy." *Scientific American* 250 (6): 56–63.

Lechtman 1996
Lechtman, Heather. 1996. "Arsenic Bronze: Dirty Copper or Chosen Alloy? A View from the Americas." *Journal of Field Archaeology* 23 (4): 477–514.

Lechtman and Steinberg 1970
Lechtman, Heather, and Arthur Steinberg. 1970. "Bronze Joining: A Study in Ancient Technology." In *Art and Technology: A Symposium on Classical Bronzes*, edited by Suzannah F. Doeringer, David Gordon Mitten, and Arthur Richard Steinberg, 5–35. Cambridge, MA: MIT Press.

Lein 2000
Lein, Edgar. 2000. "Die Kunst des Bronzegießens, ihre Darstellung in Traktaten und die Bedeutung von Bronze." In *Bronze- und Galvanoplastik. Geschichte – Materialanalyse – Restaurierung*, edited by Birgit Meissner, Anke Doktor, and Martin Mach, 9–24. Arbeitsheft 5. Dresden: Landesamt für Denkmalpflege Sachsen. http://www.lfd.sachsen.de/download/AH5_Bronze_Galvanoplastik.pdf.

Lein 2004
Lein, Edgar. 2004. *Ars Aeraria. Die Kunst des Bronzegießens und die Bedeutung von Bronze in der florentinischen Renaissance*. Mainz, Germany: von Zabern.

Leithe-Jasper 1986
Leithe-Jasper, Manfred. 1986. *Renaissance Master Bronzes from the Collection of the Kunsthistorisches Museum, Vienna*. Washington, DC: Scala Books.

Lemonnier 1992
Lemonnier, Pierre. 1992. *Elements for an Anthropology of Technology*. Anthropological Papers 88. Ann Arbor : Museum of Anthropology, University of Michigan.

Le Normand-Romain 2007
Le Normand-Romain, Antoinette. 2007. *The Bronzes of Rodin: Catalogue of Works in the Musée Rodin*. Paris: Musée Rodin; Éditions de la Réunion des Musées Nationaux.

Léon 2014
Léon, Yoanna. 2014. "La circulation des savoir-faire techniques des céramiques." In *Circulation et provenance des matériaux dans les sociétés anciennes*, edited by Philippe Dillmann and Ludovic Bellot-Gurlet, 225–36. Sciences Archéologiques. Paris: Editions des archives contemporaines.

Leonardo 1490
Leonardo da Vinci. 1490. *The Madrid Codices*. Translated by L. Reti. New York: McGraw-Hill.

Leoni 1979
Leoni, Massimo. 1979. "Observations on Ancient Bronze Casting." In *The Horses of San Marco, Venice*, edited by Guido Perocco, translated by John Wilton-Ely and Valerie Wilton-Ely, 180–181. London: Thames and Hudson.

Leroy et al. 2013
Leroy, Stéphanie, Emmanuelle Delque-Kolic, Jean-Pascal Dumoulin, Christophe Moreau, and Philippe Dillmann. 2013. "Datation radiocarbone des alliages ferreux anciens." In *Sciences des matériaux du patrimoine culturel: actes du colloque, Paris, 20 et 21 Novembre 2012*, 57–63. Paris: Ministère de la Culture et de la Communication. https://www.culture.gouv.fr/Media/Thematiques/Enseignement-superieur-et-recherche/Archives-anciens-programmes/PNRCC/Actes-colloque-SMPC2/Actes-du-colloque-SMPC-Paris-nov.2012-144p.-5-6Mo.

Leroy et al. 2018
Leroy, Stéphanie, Mitch Hendrickson, Sylvain Bauvais, Enrique Vega, Théo Blanchet, Alexandre Disser, and Emmanuelle Delque-Kolic. 2018. "The Ties That Bind: Archaeometallurgical Typology of Architectural Crampons as a Method for Reconstructing the Iron Economy of Angkor, Cambodia (Tenth to Thirteenth C.)." *Archaeological and Anthropological Sciences* 10:2137–57.

Leroy et al. 2021
Leroy, Stéphanie, E. Delqué-Količ, Brice Vincent, Pierre Baptiste, Forest McGill, Enrique Vega, Jean-Pascal Dumoulin, Marion Perron, Philippe Dillmann, and Lucille Beck. 2021. "Le fer comme moyen de datation des bronzes khmers : première approche de prélèvement in situ." *Techné* 52: in preparation.

Leroy, Hendrickson et al. 2015
Leroy, Stéphanie, Mitch Hendrickson, Emmanuelle Delqué-Kolic, Enrique Vega, and Philippe Dillmann. 2015. "First Direct Dating for the Construction and Modification of the Baphuon Temple Mountain in Angkor, Cambodia." *PLOS ONE* 10 (11): e0141052. https://doi.org/10.1371/journal.pone.0141052.

Leroy, L'Héritier et al. 2015
Leroy, Stéphanie, Maxime L'Héritier, Emmanuelle Delqué-Kolic, Jean-Pascal Dumoulin, Christophe Moreau, and Philippe Dillmann. 2015. "Consolidation or Initial Design? Radiocarbon Dating of Ancient Iron Alloys Sheds Light on the Reinforcements of French Gothic Cathedrals." *Journal of Archaeological Science* 53 (January): 190–201.

Lesoult 1986
Lesoult, G. 1986. "Solidification – cristallisation et microstructure." *Techniques de l'ingénieur. Matériaux métalliques*, M58.

Lespinasse and Bonnardot 1879
Lespinasse, René de, and François Bonnardot. 1879. *Les métiers et corporations de la ville de Paris: XIIIe siècle. Le livre des métiers d'Etienne Boileau*. Paris: Imprimerie nationale. https://gallica.bnf.fr/ark:/12148/bpt6k110190t.image.

Letardi et al. 2016
Letardi, Paola, Barbara Salvadori, Monica Galeotti, Andrea Cagnini, Simone Porcinai, Alessandra Santagostino Barbone, and Antonio Sansonetti. 2016. "An In Situ Multi-Analytical Approach in the Restoration of Bronze Artefacts." *Microchemical Journal* 125:151–58.

Levy and Saurat 2005
Levy, Robert, and Marcel Saurat. 2005. "Revêtements en production industrielle: contrôles essentiels." *Techniques de l'ingénieur Traitements de surface des métaux: contrôle et préparation* TIB362DUO (m1680).

Levy et al. 2008
Levy, Thomas Evan, Alina M. Levy, D. Radhakrishna Sthapathy, D. Srikanda Sthapathy, and D. Swaminatha Sthapathy. 2008. *Masters of Fire: Hereditary Bronze Casters of South India*. Bochum, Germany: DBM.

Lie and Bewer 2014
Lie, Henry, and Francesca G. Bewer. 2014. "Ex Aere Factum: Technical Notes on Ancient Bronzes." In *Ancient Bronzes through a Modern Lens: Introductory Essays on the Study of Ancient Mediterranean and Near Eastern Bronzes*, edited by Susanne Ebbinghaus, 38–63. Cambridge, MA: Harvard Art Museums.

Li et al. 2011
Li, Xiuzhen Janice, Marcos Martinón-Torres, Nigel D. Meeks, Yin Xia, and Kun Zhao. 2011. "Inscriptions, Filing, Grinding and Polishing Marks on the Bronze Weapons from the Qin Terracotta Army in China." *Journal of Archaeological Science* 38 (3): 492–501.

Li et al. 2012
Li, Xiuzhen Janice, Marcos Martinón-Torres, Nigel Meeks, and Yin Xia. 2012. "Scanning Electron Microscopy Imaging of Tool Marks on Qin Bronze Weapons Using Silicone Rubber Impressions." In *Historical Technology, Materials and Conservation: SEM and Microanalysis*, edited by Nigel Meeks, Caroline Cartwright, Andrew Meek, and Aude Mongiatti, 62–68. London: Archetype and the British Museum.

Lindsay, Barbour, and Sturman 2010
Lindsay, Suzanne G., Daphne Barbour, and Shelley Sturman. 2010. *Edgar Degas Sculpture*. Washington, DC: National Gallery of Art; Princeton, NJ: Princeton University Press.

Lins 2000
Lins, Andrew. 2000. "Gilding Techniques of the Renaissance and After." In *Gilded Metals: History, Technology and Conservation*, edited by Terry Draymann-Weisser, 241–66. London: Archetype in association with the American Institute of Conservation of Historic and Artistic Works.

Lins and Malenka 2000
Lins, Andrew, and Sally Malenka. 2000. "The Use of Mercury Salts in Gold Electroplating." In *Gilded Metals: History, Technology and Conservation*, edited by Terry Draymann-Weisser, 267–82. London: Archetype in association with the American Institute of Conservation of Historic and Artistic Works.

Lins and Oddy 1975
Lins, Andrew, and W. A. Oddy. 1975. "The Origins of Mercury Gilding." *Journal of Archaeological Science* 2:365–73.

Littre 1873
Littre, M. P. Emile. 1873. *Dictionnaire de la langue Francaise.* n.p.: n.p.

Liu 2015
Liu Yu et al. | 刘煜. 2015) "殷墟青銅器金屬芯撐技術研究 = A Technical Study of Metal Chaplets Used on Bronzes from Yinxu." *南方文物 = Cultural Relics in Southern China* 4:97–111.

Liu et al. 2013
Liu, Siran, Kai Wang, Quanfa Cai, and Jianli Chen. 2013. "Microscopic Study of Chinese Bronze Casting Moulds from the Eastern Zhou Period." *Journal of Archaeological Science* 40 (5): 2402–14. https://doi.org/10.1016/j.jas.2012.11.010.

Livingood and Cordell 2009
Livingood, Patrick C., and Ann S. Cordell. 2009. "Point/Counter Point: The Accuracy and Feasibility of Digital Image Techniques in the Analysis of Ceramic Thin Sections." *Journal of Archaeological Science* 36 (3): 867–72.

Lohuizen-de Leeuw 1984
Lohuizen-de Leeuw, J. E. van. 1984. *Indo-Javanese Metalwork.* Stuttgart, Germany: Linden-Museum.

Lombardi 2002
Lombardi, Gianni. 2002. "A Petrographic Study of the Casting Core of the Lupa Capitolina Bronze Sculpture (Rome, Italy) and Identification of Its Provenance." *Archaeometry* 44 (4): 601–12.

Lombardi 2009
Lombardi, Gianni. 2009. "The Casting Core Composition and Provenance of the Goljamata Kosmatka (Bulgaria) Bronze Head." *Journal of Archaeological Science* 36 (2): 520–27.

Lombardi and Vidale 1998
Lombardi, Gianni, and Massimo Vidale. 1998. "From the Shell to Its Content: The Casting Cores of the Two Bronze Statues from Riace (Calabria, Italy)." *Journal of Archaeological Science* 25:1055–66.

Lopes et al. 2016
Lopes, Fabio, Fabio Luiz Melquiades, Carlos Roberto Appoloni, Roberto Cesareo, Marcia Rizzutto, and Tiago F. Silva. 2016. "Thickness Determination of Gold Layer on Pre-Columbian Objects and a Gilding Frame, Combining PXRF and PLS Regression." *X-Ray Spectrometry* 45 (6): 344–51.

Lunsingh Scheurleer and Klokke 1988
Lunsingh Scheurleer, Pauline, and Marijke J. Klokke, eds. 1988. *Divine Bronze: Ancient Indonesian Bronzes from AD 600 to 1600.* Leiden, the Netherlands: Brill.

Maaz 2010
Maaz, Bernhard. 2010. *Skulptur in Deutschland zwischen französischer Revolution und erstem Weltkrieg.* Berlin and Munich: Deutscher Kunstverlag.

Magnus 1967
Magnus, Albertus. 1967. *Book of Minerals.* Translated by Dorothy Wyckoff. Oxford: Clarendon.

Mahan and Kay 2012
Mahan, Shannon, and John Kay. 2012. "Building on Previous OSL Dating Techniques for Gypsum: A Case Study from Salt Basin Playa, New Mexico and Texas." *Quaternary Geochronology* 10 (July): 345–52.

Maish 2017
Maish, Jeffrey. 2017. "The Getty Herm of Dionysos: Technical Observations, Review, and Interpretation." In *Artistry in Bronze: The Greeks and Their Legacy: XIXth International Congress on Ancient Bronzes*, edited by Jens M. Daehner, Kenneth Lapatin, and Ambra Spinelli. Los Angeles: J. Paul Getty Museum and Getty Conservation Institute. http://www.getty.edu/publications/artistryinbronze/conservation-and-analysis/41-maish/.

Making and Knowing Project et al. 2020
Making and Knowing Project, Pamela H. Smith, Naomi Rosenkranz, Tianna Helena Uchacz, Tillmann Taape, Clément Godbarge, Sophie Pitman, Jenny Boulboullé, Joel Klein, Donna Bilak, Marc Smith, and Terry Catapano, eds. 2020. *Secrets of Craft and Nature in Renaissance France: A Digital Critical Edition and English Translation of BnF Ms. Fr. 640*. New York: Making and Knowing Project. http://edition640.makingandknowing.org.

Mannes et al. 2015
Mannes, D., Florian Schmid, J. Frey, K. Schmidt-Ott, and E. Lehmann. 2015. "Combined Neutron and X-Ray Imaging for Non-Invasive Investigations of Cultural Heritage Objects." *Physics Procedia* 69 (January): 653–60. https://doi.org/10.1016/j.phpro.2015.07.092.

Marabelli 1994
Marabelli, Maurizio. 1994. "The Monument of Marcus Aurelius: Research and Conservation." In *Ancient and Historic Metals: Conservation and Scientific Research*, edited by David A. Scott, Jerry Podany, and Brian B. Considine, 1–19. Los Angeles: Getty Conservation Institute. https://www.getty.edu/publications/virtuallibrary/0892362316.html.

Marabelli 1995
Marabelli, Maurizio. 1995. *Conservazione e restauro dei metalli d'arte*. Rome: Accademia Nazionale dei Lincei.

Maral, Bourgarit, and Amarger 2014
Maral, Alexandre, David Bourgarit, and Antoine Amarger. 2014. "Keller et les autres: Les fondeurs des Jardins de Versailles." In *French Bronze Sculpture: Materials and Techniques 16th–18th Century*, edited by David Bourgarit, Jane Bassett, Francesca G. Bewer, Geneviève Bresc-Bautier, Philippe Malgouyres, and Guilhem Scherf, 76–94. Paris: Archetype.

Marconi 2004
Marconi, Nicoletta. 2004. *Edificando Roma Barocca. Macchine, Apparati, Maestranze e Cantieri Tra XVI e XVIII Secolo*. Città di Castello, Italy: Edimond.

Margreiter et al. 2022
Margreiter, Raphael, Jonas Baumann, Ioanna Mantouvalou, Martin Radtke, Uwe Reinholz, and Erik Strub. 2022. "Investigations on Fire-Gilding." *Archaeometry* 64 (6): 1465–78. https://doi.org/10.1111/arcm.12797.

Mariette 1768
Mariette, Pierre-Jean. 1768. *Description des travaux qui ont précédé, accompagné et suivi la fonte en bronze d'un seul jet de la statue équestre de Louis XV, le Bien-Aimé, dressée sur les mémoires de M. Lempereur*. Paris: Imprimerie de P. G. Le Mercier. https://archive.org/details/gri_33125010908289/mode/2up.

Martin 2001
Martin, Graham. 2001. "Alloy Analysis" and "Appendix II." In Motture, Peta, *Bells and Mortars and Related Utensils: Catalogue of Italian Bronzes in the Victoria and Albert Museum*, 51–52 and 208–13. London: V&A Publications.

Martin, Incerti, and Mercier 2015
Martin, Loic, Sebastien Incerti, and Norbert Mercier. 2015. "DosiVox: Implementing Geant 4-Based Software for Dosimetry Simulations Relevant to Luminescence and ESR Dating Techniques." *Ancient TL* 33 (1): 10.

Martinón-Torres et al. 2007
Martinón-Torres, Marcos, Roberto Valcárcel Rojas, Jago Cooper, and Thilo Rehren. 2007. "Metals, Microanalysis, and Meaning: A Study of Metal Objects Excavated from the Indigenous Cemetery of El Chorro de Maíta, Cuba." *Journal of Archaeological Science* 34 (2): 194–204.

Mascré, Thomas, and Hénon 1952
Mascré, Claude, Auguste le Thomas, and Guy Hénon. 1952. *Album de défauts de fonderie*. Paris: Éditions techniques des industries de la fonderie.

Maskinaktiebolaget Karlebo 1982
Maskinaktiebolaget Karlebo. 1982. *Technical Dictionary: Swedish, German, English, French, Spanish, Portuguese, Italian, Finnish, Serbo-Croatian, Turkish, Greek, Russian: Covering Technical Terms in the Field of Machine Tools, Small Tools, and Foundry*. Stockholm: Maskinaktiebolaget Karlebo.

Mason et al. 2016
Mason, A. H., W. G. Powell, H. A. Bankoff, R. Mathur, A. Bulatović, V. Filipović, and J. Ruiz. 2016. "Tin Isotope Characterization of Bronze Artifacts of the Central Balkans." *Journal of Archaeological Science* 69 (May): 110–17.

Mathis et al. 2005
Mathis, F., S. Descamps, D. Robcis, and M. Aucouturier. 2005. "Original Surface Treatment of Copper Alloy in Ancient Roman Empire: Chemical Patination on a Roman Strigil." *Surface Engineering* 21:346–51.

Mathis et al. 2007
Mathis, F., S. Descamps, E. Delange, J. Salomon, S. Pagès-Camagna, M. Dubus, D. Robcis, and M. Aucouturier. 2007. "Corrosion Patina or Voluntary Patina? Contribution of Non-Destructive Analyses to the Surface Study of Copper Based Archaeological Objects." In *Corrosion of Metallic Heritage Artefacts: Investigation, Conservation and Prediction of Long Term Behaviour*, edited by P. Dillmann, G. Beranger, P. Piccardo, and H. Matthiessen, 219–38. European Federation of Corrosion Series 48. Cambridge, UK: Woodhead.

Mattusch 1975
Mattusch, Carol C. 1975. "Casting Techniques of Greek Bronze Sculpture: Foundries and Foundry Remains from the Athenian Agora with Reference to Other Ancient Sources." PhD diss., University of North Carolina.

Mattusch 1988
Mattusch, Carol C. 1988. *Greek Bronze Statuary: From the Beginnings through the Fifth Century B.C.* Ithaca, NY: Cornell University Press.

Mattusch 1990
Mattusch, Carol C. 1990. "A Trio of Griffins from Olympia." *Hesperia* 59 (3): 549–60.

Mattusch 1996
Mattusch, Carol C. 1996. *The Fire of Hephaistos: Large Classical Bronzes from North American Collections*. Cambridge, MA: Harvard University Art Museums.

Mattusch 2005
Mattusch, Carol C. 2005. *The Villa dei Papiri at Herculaneum. Life and Afterlife of a Sculpture Collection*. Los Angeles: J. Paul Getty Museum.

Mattusch 2006
Mattusch, Carol C. 2006. "The Seated Hermes from the Villa of the Papyri at Herculaneum: Ancient and Modern Lives." In *Mittel und Wege. Zur Bedeutung von Material und Technik in der Archäologie*, edited by Astrid Dostert and Franziska Lang, 167–79. Berlin: Bibliopolis.

Mattusch 2009
Mattusch, Carol C. 2009. "Guests, Hosts, and Politics at Herculaneum." In *New Perspectives on Etruria and Early Rome*, edited by Sinclair Bell and Helen Nagy, 264–73. Madison: University of Wisconsin Press.

Mattusch 2013
Mattusch, Carol C. 2013. "Appearances Can Be Deceiving: The Presentation of Bronzes from Herculaneum and Pompeii." In *The Restoration of Ancient Bronzes: Naples and Beyond*, edited by Erik Risser and David Saunders, 30–42. Los Angeles: J. Paul Getty Museum. http://www.getty.edu/publications /virtuallibrary/1606061542.html.

Mattusch 2014
Mattusch, Carol C. 2014. "Artists and Workshops: The Craft and the Product." In *Ancient Bronzes through a Modern Lens: Introductory Essays on the Study of Ancient Mediterranean and Near Eastern Bronzes*, edited by David Gordon Mitten, Susanne Ebbinghaus, and Lisa M. Anderson, 114–31. Cambridge, MA: Harvard Art Museums.

McCamy 1996
McCamy, C. S. 1996. "Observation and Measurement of the Appearance of Metallic Materials. Part I. Macro Appearance." *Colour Research and Application* 21 (4): 292–304.

McCreight 1996
McCreight, Tim. 1996. *Practical Casting: A Studio Reference*. Boylston, MA: Brynmorgen Press.

McGrath 2005
McGrath, Jinks. 2005. *The Jeweler's Directory of Decorative Finishes: From Enameling and Engraving to Inlay and Granulation*. London: KP Books.

Measday, Walker, and Pemberton 2017
Measday, Danielle, Charlotte Walker, and Briony Pemberton. 2017. "A Summary of Ultra-Violet Fluorescent Materials Relevant to Conservation." *AICCM National Newsletter*, no. 137 (March): https://aiccm.org.au/network -news/summary-ultra-violet-fluorescent-materials-relevant-conservation/.

Mechling et al. 2018
Mechling, Mathilde, Brice Vincent, Pierre Baptiste, and David Bourgarit. 2018. "The Indonesian Bronze-Casting Tradition: Technical Investigations on Thirty-Nine Indonesian Bronze Statues (7th–11th C.) from the Musée national des arts asiatiques – Guimet, Paris." *Bulletin de l'École française d'Extrême-Orient* 104:63–139.

Meeks 1986
Meeks, N. D. 1986. "Tin-Rich Surfaces on Bronze: Some Experimental and Archaeological Considerations." *Archaeometry* 28 (2): 133–62.

Meeks 1993
Meeks, Nigel. 1993. "Surface Characterization of Tinned Bronze, High-Tin Bronze, Tinned Iron and Arsenical Bronze." In *Metal Plating and Patination: Cultural, Technical and Historical Developments*, edited by Susan La Niece and Paul Craddock, 247–75. Oxford, London, Boston [etc.]: Butterworth-Heinemann.

Meissner, Haber, and Mach 2000
Meissner, Birgit, Georg J. Haber, and Martin Mach. 2000. "Der Leipziger Mendebrunnen – Historie und Restaurierung." In *Bronze- und Galvanoplastik. Geschichte – Materialanalyse – Restaurierung*, edited by Birgit Meissner, Anke Doktor, and Martin Mach, 92–108. Arbeitsheft 5. Dresden: Landesamt für Denkmalpflege Sachsen. http://www.lfd.sachsen.de/download/AH5_Bronze _Galvanoplastik.pdf.

Mélard et al. 2016
Mélard, Nicolas, Clotilde Boust, Gabrielle Cogné, and Anne Maigret. 2016. "Comparison of Imaging Techniques Used in the Microanalysis of Paleolithic Mobiliary Art." *Journal of Archaeological Science: Reports* 10 (December): 903–9.

Mende 2013
Mende, Ursula. 2013. *Die mittelalterlichen Bronzen im Germanischen Nationalmuseum: Bestandskatalog*. Nuremberg, Germany: Germanisches Nationalmuseum.

Mercuri et al. 2017
Mercuri, Fulvio, Noemi Orazi, Ugo Zammit, Augusto Giuffredi, Carlo Stefano Salerno, Cristina Cicero, and Stefano Paoloni. 2017. "The Manufacturing Process of the Capitoline She Wolf: A Thermographic Method for the Investigation of Repairs and Casting Faults." *Journal of Archaeological Science: Reports* 14:199–207.

Mercuri et al. 2018
Mercuri, F., G. Caruso, N. Orazi, U. Zammit, C. Cicero, O. Colacicchi Alessandri, M. Ferretti, and S. Paoloni. 2018. "Interface Thermal Conductance Characterization by Infrared Thermography: A Tool for the Study of Insertions in Bronze Ancient Statuary." *Infrared Physics and Technology* 90 (May): 31–39.

Merkel 1990
Merkel, J. F. 1990. "Experimental Reconstruction of Bronze Age Copper Smelting Based on Archaeological Evidence from Timna." In *The Ancient Metallurgy of Copper: Archaeology-Experiment-Theory*, edited by Beno Rothenberg, 78–122. London: Institute for Archaeo-Metallurgical Studies.

Merwe and Stuiver 1968
Merwe, Nicolas J. van der, and Minze Stuiver. 1968. "Dating Iron by the Carbon-14 Method." *Current Anthropology* 9 (1): 48–53.

Meyer, Thomas, and Wyss 2014
Meyer, Olivier, Nicolas Thomas, and Michaël Wyss. 2014. "Témoins archéologiques d'un atelier de bronzier fonctionnant à Saint-Denis à la fin du XVIe siècle." In *French Bronze Sculpture: Materials and Techniques 16th–18th Century*, edited by David Bourgarit, Jane Bassett, Francesca G. Bewer, Geneviève Bresc-Bautier, Philippe Malgouyres, and Guilhem Scherf, 215–24. Paris: Archetype.

Michel 1922
Michel, Jacques. 1922. *Coloration des métaux: Bronzage, patinage, oxydation, marbrage, irisation, nielle*. Paris: Librarie Général Scientifique et Industrielle, Desforges.

Michelucci 2006
Michelucci, Maurizio, ed. 2006. *Apoxyomenos: The Athlete of Croatia*. Florence and Milan: Giunti.

Mietzsch 2009
Mietzsch, Andreas. 2009. *Bronzeguss – Handwerk für die Kunst*. Berlin: Zentralverlag.

Milam, Reedy, and Sussman 1988
Milam, Billie, Chandra L. Reedy, and Carol Sussman. 1988. *Technical Analysis of Renaissance Bronzes for Provenance Studies: Pilot Project*. Malibu, CA: Getty Conservation Institute. http://hdl.handle.net/10020/gci_pubs/tech_analysis_renaissance_bronzes.

Mille 2007
Mille, Benoît. 2007. "Etude technique du cheval de bronze de Neuvy-en-Sullias." In *Le Cheval et la danseuse, à la redécouverte du trésor de Neuvy-En-Sullias, catalogue de l'exposition présentée au Musée Des Beaux-Arts d'Orléans du 13 Mars au 26 août 2007*, edited by C. Gorget and J. P. Guillaumet, 88–99, 264–65. Paris: Somogy.

Mille 2012
Mille, Benoît. 2012. "The Casting Techniques of Antique South Arabian Large Bronze Statues." In *Scientific Research on Ancient Asian Metallurgy: Proceedings of Fifth Forbes Symposium at the Freer Gallery of Art, Washington, Smithsonian Institution, 28–29/10/2010*, edited by P. Jett, B. McCarthy, and J. G. Douglas, 225–47. London: Archetype.

Mille 2017
Mille, Benoît. 2017. "D'une amulette en cuivre aux grandes statues de bronze, évolution des techniques de fonte à la cire perdue, de l'Indus à la Méditerranée, du 5e millénaire au 5e siècle av. J.-C." PhD diss., Université de Paris-Nanterre et Université de Fribourg. http://www.theses.fr/2017PA100057.

Mille 2019a
Mille, Benoît. 2019a. "Forme, matière et couleurs de la petite statuaire du Trésor de Bavay, mise en évidence d'innovations techniques au service de la création artistique romaine." In *Nouveaux regards sur le Trésor des bronzes de Bavay*, edited by Laure de Chavagnac and Benoît Mille, 166–89. Milan: Silvana Editoriale.

Mille 2019b
Mille, Benoît. 2019b. "Les fragments de statues équestres découverts à Augst / Augusta Raurica (Suisse)." In *Nouveaux regards sur le Trésor des bronzes de Bavay*, edited by Laure de Chavagnac and Benoît Mille, 248–59. Milan: Silvana Editoriale.

Mille and Bourgarit 2000
Mille, Benoît, and David Bourgarit. 2000. "L'analyse des alliages anciens à base de cuivre: État des connaissances et développement d'un protocole d'analyse par ICP-AES." *Revue d'archéométrie* 24:13–26.

Mille and Bourgarit 2003
Mille, Benoît, and David Bourgarit. 2003. "Composition élémentaire des bronzes de la tombe de Vix." *La tombe princière de Vix*, edited by Claude Rolley, 253–54. Paris: Picard.

Mille and Descamps-Lequime 2017
Mille, Benoît, and Sophie Descamps-Lequime. 2017. "A Technological Reexamination of the Piombino Apollo." In *Artistry in Bronze: The Greeks and Their Legacy: XIXth International Congress on Ancient Bronzes*, edited by Jens M. Daehner, Kenneth Lapatin, and Ambra Spinelli, 349–60. Los Angeles: J. Paul Getty Museum and Getty Conservation Institute. http://www.getty.edu/publications/artistryinbronze/conservation-and-analysis/42-mille-descamps/.

Mille and Robcis 2009
Mille, Benoît, and Dominique Robcis. 2009. "Note sur la technique de fabrication du captif." In *César, le Rhône pour mémoire, vingt ans de fouilles dans le fleuve à Arles, catalogue de l'exposition du 24 octobre 2009 au 19 septembre 2010 au Musée Départemental Arles Antique*, edited by Luc Long and Pascale Picard, 162–65. Arles: Actes Sud.

Mille and Robcis 2012
Mille, Benoît, and Dominique Robcis. 2012. "Le cas des grands bronzes antiques. Etudier pour restaurer ou restaurer pour étudier?" In *La restauration des peintures et des sculptures. Connaissance et reconnaissance de l'œuvre*, edited by Pierre-Yves Kairis, Béatrice Sarrazin, and François Trémolières, 101–15. Paris: Armand Colin.

Mille and Serneels 2012
Mille, Benoît, and V. Serneels. 2012. "Pas de bras, pas de chocolat! L'envers du décor à Arconciel." In *Archeoquiz, Enquêtes archéologiques fribourgeoises*, edited by Anne-Francine Auberson and Dominique Bugnon, 104–7. Fribourg, Switzerland: Service Archéologique de l'Etat de Fribourg.

Mille, Besenval, and Bourgarit 2004
Mille, Benoît, Roland Besenval, and David Bourgarit. 2004. "Early 'Lost-Wax Casting' in Balochistan (Pakistan): The 'Leopards Weight' from Shahi-Tump." In *Persiens Antike Pracht, Bergbau – Handwerk – Archäologie*, edited by T. Stöllner, R. Slotta, and A. Vatandoust, 1:274–80. Bochum, Germany: Deutsches Bergbau Museum.

Mille, Bourgarit, and Besenval 2005
Mille, Benoît, David Bourgarit, and Roland Besenval. 2005. "Metallurgical Study of the 'Leopards Weight' from Shahi-Tump (Pakistan)." In *South Asian Archaeology 2001*, edited by Catherine Jarrige and Vincent Lefèvre, 1:237–44. Paris: Editions Recherches sur les Civilisations.

Mille et al. 2010
Mille, Benoît, I. Gajda, F. Demange, C. Pariselle, Y. Coquinot, E. Porto, O. Tavoso, and Antoine Zink. 2010. "Hawtar'athat, Fils de Radaw'il Du Lignage de Shalalum. Une grande statue de bronze du Royaume de Saba' (Yémen)." *Monuments et mémoires de la Fondation Eugène Piot* 89: 5–68. https://doi.org/10.3406/piot.2010.1726.

Mille et al. 2012
Mille, Benoît, Loretta Rossetti, Claude Rolley (†), Edilberto Formigli, and Michel Pernot. 2012. "Les deux statues d'enfant en bronze (Cap d'Adge): Étude Iconographique et Technique." In *"Bronzes grecs et romains, recherches récentes" – Hommage à Claude Rolley*, edited by Sophie Descamps-Lequime, Stéphane Verger, and Martine Denoyelle. Actes de Colloques. Paris: Publications de l'Institut national d'histoire de l'art. http://books.openedition.org/inha/3949.

Ming 2001
Ming Juxin | 明举新. 2001. *汉英冶金工业词典* = Han ying yejin gongye cidian = *A Comprehensive Chinese-English Dictionary of Metallurgical Industry*. 长沙市: 中南大学出版社 = Changsha shi: Zhongnan daxue chuban she [Zhong Nan University Press].

Ming 2010
Ming Juxin | 明举新. 2001. *汉英冶金工业词典* = Han ying yejin gongye cidian = *A Comprehensive Chinese-English Dictionary of Metallurgical Industry*. 3rd ed. 长沙市: 中南大学出版社 = Changsha shi: Zhongnan daxue chuban she [Zhong Nan University Press].

Mitten and Doeringer 1967
Mitten, David G., and Suzannah F. Doeringer. 1967. *Master Bronzes from the Classical World*. Mainz, Germany: Von Zabern.

Mödlinger and Sabatini 2016
Mödlinger, M., and B. Sabatini. 2016. "A Re-Evaluation of Inverse Segregation in Prehistoric As-Cu Objects." *Journal of Archaeological Science* 74:60–74.

Mödlinger, Asmus, and Ghiara 2024
Mödlinger, Marianne, Bastian Asmus, et Giorgia Ghiara. "The 'Schwarze Mander' of the Court Church in Innsbruck, Austria: Manufacture and Production of Monumental Brass Statues in the Renaissance." *International Journal of Metalcasting*, 22 March 2024. https://doi.org/10.1007/s40962-024-01299-4.

Mödlinger et al. 2017
Mödlinger, Marinne, Maikel H. G. Kuijpers, Dennis Braekmans, and Daniel Berger. 2017. "Quantitative Comparisons of the Color of CuAs, CuSn, CuNi, and CuSb Alloys." *Journal of Archaeological Science* 88:14–23.

Morigi 1990
Morigi, Giovanni. 1990. *Fusione a cera persa con tecnica parzialmente indiretta. Tecnica impiegata nei bronzi a tutto tondo della Fontana del Nettuno di Giambologna, 1566.* Bologna, Italy: Commune di Bologna, Assessorato alla cultura.

Morigi 1999
Morigi, Giovanni. 1999. "Indagini sulla tecnica di fusione, sulle tecniche di 'saldatura' e di lavorazione delle superficie." In *La porta di Bonanno nel duomo di Pisa*, edited by Ottavio Banti, 155–68. Pontedera, Italy: Bandecchi and Vivaldi.

Morigi 2018
Morigi, Lorenzo. 2018. "Relazione di restauro del busto in bronzo rappresentante l'imperatore Antonino Pio." Florence: Muzeo Nazionale del Bargello.

Morigi and Morigi 2004
Morigi, Giovanni, and Lorenzo Morigi. 2004. "Il bronzo di Benvenuto Cellini a Boston. Il restauro del busto di Bindo Altoviti." *Kermes* 53:20–24.

Morigi and Morigi 2008
Morigi, Giovanni, and Lorenzo Morigi. 2008. "Note tecniche sul restauro del monumento a Bartolomeo Colleoni di Andrea del Verrocchio." In *L'Industria artistica del bronzo del rinascimento a Venezia e nell'Italia settentrionale: Atti del Convegno Internazionale di Studi; Venezia, Fondazione Giorgio Cini, 23 e 24 Ottobre 2007*, edited by Matteo Ceriana, 459–79. Verona, Italy: Scripta.

Motture 2001
Motture, Peta. 2001. *Bells and Mortars and Related Utensils: Catalogue of Italian Bronzes in the Victoria and Albert Museum*. London: V&A Publishing.

Motture 2003a
Motture, Peta, ed. 2003a. *Large Bronzes in the Renaissance*. New Haven, CT, and London: Yale University Press.

Motture 2003b
Motture, Peta, 2003b. "The Production of Firedogs in Renaissance Venice." In *Large Bronzes in the Renaissance*, edited by Peta Motture, 277–308. New Haven, CT, and London: Yale University Press.

Motture 2019
Motture, Peta. 2019. *The Culture of Bronze: Making and Meaning in Italian Renaissance Sculpture*. London: V&A Publishing.

Motture, Jones, and Zikos 2013
Motture, Peta, Emma Jones, and Dimitrios Zikos, eds. 2013. *Carvings, Casts and Collectors: The Art of Renaissance Sculpture*. London: V&A Publications.

Moulherat et al. 2002
Moulherat, C., M. Tengberg, J.-F. Haquet, and B. Mille. 2002. "First Evidence of Cotton at Neolithic Mehrgarh, Pakistan: Analysis of Mineralized Fibres from a Copper Bead." *Journal of Archaeological Sciences* 29 (12): 1393–401.

Mugnaini et al. 2014
Mugnaini, Sonia, Marco Giamello, Anastasia Pisani, and Salvatore Siano. 2014. "Casting Cores Used to Craft Large Bronze Masterpieces of the Florentine Renaissance and Mannerism." *Journal of Archaeological Science* 47 (July): 85–98.

Müller 1902
Müller, Ludwig. 1902. *Die Bronzewaaren-Fabrikation*. 2nd ed. Vienna, Pest, and Leipzig: A. Hartleben's Verlag.

Müller 2014
Müller, Stephanie, ed. 2014. *Gebrochener Glanz: römische Großbronzen am UNESCO-Welterbe Limes*. Mainz, Germany: Nünnerich-Asmus

Munsell Color 2000
Munsell Color. 2000. *Munsell Soil Color Charts.* New York: Munsell Color.

Murry 1994
Murry, Guy. 1994. "Soudage et soudabilité métallurgique des métaux." *Techniques de l'ingénieur* MC2, no. M715: 1–35.

Museum of Fine Arts (Boston), Research Laboratory 1959
Museum of Fine Arts (Boston), Research Laboratory. 1959. *Application of Science in Examination of Works of Art: Proceedings of the Seminar: September 15–18, 1958.* Boston: Museum of Fine Arts.

Museum of Fine Arts (Boston) Research Laboratory 1967
Museum of Fine Arts (Boston) Research Laboratory. 1967. *Application of Science in Examination of Works of Art: Proceedings of the Seminar September 7–16, 1965.* Boston: Museum of Fine Arts.

Museum of Fine Arts (Boston) Research Laboratory 1970
Museum of Fine Arts (Boston) Research Laboratory. 1970. *Application of Science in Examination of Works of Art: Proceedings of the Seminar at the Museum of Fine Arts, Boston, September 15–18, 1958.* New York: Arno.

Nassau 1987
Nassau, Kurt. 1987. "The Fifteen Causes of Color: The Physics and Chemistry of Color." *Color Research & Application* 12 (1): 4–26.

Neff 2011
Neff, David. 2011. *Casting Defects Handbook: Copper and Copper-Base Alloys.* Schaumburg, IL: American Foundry Society.

Neri and Giannichedda 2018
Neri, Elisabetta, and Enrico Giannichedda. 2018. "La place des cloches dans les productions en alliages cuivreux: Spécificités techniques à travers les vestiges archéologiques d'ateliers italiens." In *Cuivre, Bronzes et Laitons Médiévaux: Histoire, Archéologie et Archéométrie Des Productions en Laiton, Bronze et Autres Alliages à Base de Cuivre Dans l'Europe Médiévale (12e–16e Siècles) = Medieval Copper, Bronze and Brass: History, Archaeology and Archaeometry of the Production of Brass, Bronze and Other Copper Alloy Objects in Medieval Europe (12th–16th Centuries)*, edited by Nicolas Thomas and Pete Dandridge, 227–38. Études et Documents, Archéologie 39. Namur, Belgium: Agence wallonne du Patrimoine.

Newbury, Notis, and Newbury 2005
Newbury, Brian D., Michael R. Notis, and Dale E. Newbury. 2005. "Revisiting the Zinc Composition Limit of Cementation Brass." *Historical Metallurgy* 39 (2): 75–81.

Newman 2006
Newman, Richard. 2006. "Analysis of Core and Investment Samples from Some Aquamanilia." In *Lions, Dragons, and Other Beasts: Aquamanilia of the Middle Ages, Vessels for Church and Table*, edited by Peter Barnet and Pete Dandridge, 57–63. New Haven, CT: Yale University Press.

Newman 2011
Newman, Richard. 2011. *Conservation and Care of Museum Collections.* MFA Highlights. Boston: Museum of Fine Arts, Boston.

Nicholson 1979
Nicholson, E. D. 1979. "The Ancient Craft of Gold Beating." *Gold Bulletin* 12 (4) (December): 161–66.

Nord and Billström 2018
Nord, Anders G., and Kjell Billström. 2018. "Isotopes in Cultural Heritage: Present and Future Possibilities." *Heritage Science* 6 (25): https://heritagesciencejournal.springeropen.com/articles/10.1186/s40494-018-0192-3. https://doi.org/10.1186/s40494-018-0192-3.

Northover 1998
Northover, J. Peter. 1998. "Exotic Alloys in Antiquity." In *Metallurgica Antiqua*, edited by Thilo Rehren, Andreas Hauptamnn, and James D. Muhly, 113–21. Bochum, Germany: Deutsches Bergbau-Museum.

Northover and Northover 2012
Northover, Shirley. M., and J. Peter Northover. 2012. "Applications of Electron Backscatter Diffraction (EBSD) in Archaeology." In *Historical Technology, Materials and Conservation: SEM and Microanalysis*, edited by Nigel Meeks, Caroline Cartwright, Andrew Meek, and Aude Mongiatti, Archetype Publications, 76–85. London. http://oro.open.ac.uk/27052/5/Applns EBSD in archae revised with figs.pdf.

Northover and Salter 1990
Northover, J. Peter, and Chris J. Salter. 1990. "Decorative Metallurgy of the Celts." *Materials Characterization* 25:109–23.

Notis 1988
Notis, Michael R. 1988. "The Japanese Alloy *Shakudo*: Its History and Its Patination." In *The Beginning of the Use of Metals and Alloys: Papers from the*

Second International Conference on the Beginning of the Use of Metals and Alloys, Zhengzhou, China, 21–26 October 1986, edited by Robert Maddin, 315–27. Cambridge, MA: MIT Press.

Notis and Wang 2017
Notis, Michael R., and DongNing Wang. 2017. "Ancient Chinese Bronze Casting Methods: The Dilemma of Choice." *Materials Research Society. MRS Advances* 2 (33/34): 1743–68.

Nouet 2017
Nouet, Rachel. 2017. "Archéologie de l'empreinte: Techniques de fixation des statues en Grèce égéenne, de l'époque archaïque à la fin de l'époque hellénistique (VIIè - Ier siècle av. J.-C.)." PhD diss., École doctorale Archéologie, Paris.

Nouet forthcoming
Nouet, Rachel. Forthcoming. *Archéologie de l'empreinte – Techniques de fixation des statues dans la Grèce égéenne (VIIe-Ier av. J.-C.)*. Athens: Ecole française d'Athènes.

Oddy 1993
Oddy, W. Andrew. 1993. "Gilding of Metals in the Old World." In *Metal Plating and Patination: Cultural, Technical and Historical Developments*, edited by Susan La Niece and Paul Craddock, 171–81. Oxford, London, Boston [etc.]: Butterworth-Heinemann.

Oddy 1995
Oddy, W. Andrew. 1995. "A History of Gilding with Particular Reference to Statuary." In *Proceedings of the Objects Specialty Group Session 23rd Annual Meeting in St. Paul, Minnesota, June 10, 1995*. Objects Specialty Group Postprints. http://resources.conservation-us.org/osg-postprints/postprints/v03/oddy/.

Oddy 2000
Oddy, W. Andrew. 2000. "A History of Gilding with Particular Reference to Statuary." In *Gilded Metals: History, Technology and Conservation*, edited by Terry Draymann-Weisser, 1–19. London: Archetype in association with the American Institute of Conservation of Historic and Artistic Works.

Oddy, Bimson, and La Niece 1981
Oddy, W. Andrew, Mavis Bimson, and Susan La Niece. 1981. "Gilding Himalayan Images: History Tradition and Modern Techniques." In *Aspects of Tibetan Metallurgy*, edited by W. A. Oddy and W. Zwalf, 87–101. British Museum Occasional Paper 15. London: British Museum.

Oddy, La Niece, and Meeks 1981
Oddy, W. Andrew, Susan La Niece, and Nigel D. Meeks. 1981. "Diffusion-Bonding as a Method of Gilding in Antiquity." *MASCA Journal* 1 (8): 239–41.

Oddy, Pearce, and Green 1988
Oddy, W. Andrew, Philippa Pearce, and Lorna Green. 1988. "An Unusual Gilding Technique on Some Egyptian Bronzes." In *Conservation of Ancient Egyptian Materials*, edited by Sarah C. Watkins and Carol E. Brown, 35–39. London: UK Institute for Conservation, Archaeology Section.

Odegaard, Carroll, and Zimmt 2005
Odegaard, Nancy, Scott Carroll, and Werner S. Zimmt. 2005. *Material Characterization Tests for Objects of Art and Archaeology*. London: Archetype.

Office québécois de la langue française n.d.
Office québécois de la langue française. n.d. *Le grand dictionnaire terminologique (GDT)*. http://www.granddictionnaire.com/.

Ogden 2000
Ogden, Jack. 2000. "Metals." In *Ancient Egyptian Materials and Technology*, edited by Paul T. Nicholson and Ian Shaw. Cambridge, UK, and New York: Cambridge University Press.

Okumura and Araujo 2019
Okumura, Mercedes, and Astolfo G. M. Araujo. 2019. "Archaeology, Biology, and Borrowing: A Critical Examination of Geometric Morphometrics in Archaeology." *Journal of Archaeological Science* 101 (January): 149–58.

Orazi et al. 2016
Orazi, Noemi, Fulvio Mercuri, Ugo Zammit, Stefano Paoloni, Massimo Marinelli, Augusto Giuffredi, and Carlo Stefano Salerno. 2016. "Thermographic Analysis of Bronze Sculptures." *Studies in Conservation* 61 (4): 236–44.

Ottaway and Seibel 1998
Ottaway, B. S., and S. Seibel. 1998. "Dust in the Wind: Experimental Casting of Bronze in Sand Moulds." In *Paléométallurgie des cuivres*, edited by M. C. Frère-Sautot and J. P. Millotte, 59–63. Monographies Instrumentum 5. Montagnac, France: M. Mergoil.

Outram 2005
Outram, Alan K. 2005. "Publishing Archaeological Experiments: A Quick Guide for the Uninitiated." *(Re)Construction and Experiment in Archaeology* 2:107–10.

Pacheco and Mille 2019
Pacheco, Claire, and Benoît Mille. 2019. "Le Nouvel AGLAE pour scruter l'épiderme du chien (ou loup ?) de Bavay." In *Nouveaux regards sur le Trésor des bronzes de Bavay*, edited by Laure de Chavagnac and Benoît Mille, 192–93. Milan: Silvana Editoriale.

Page et al. 2016
Page, M. J., C. Boust, N. Mélard, D. Robcis, and G. Obein. 2016. "3D Surface Acquisition: Comparison of Two Microtopographic Equipments When Measuring Materials of Cultural Heritage." In *Proceedings of the 4th CIE Expert Symposium on Colour and Visual Appearance: 6–7 September 2016, Prague, Czech Republic*, edited by Commission internationale de l'éclairage, 111–20. Vienna: CIE.

Pal 2007
Pal, Pratapaditya. 2007. "Faith and Form: Religious Sculpture in Ancient Kashmir." In *The Arts of Kashimir*, edited by Asia Society, 60–99. New York: Asia Society and Cincinnati Art Museum.

Palmer 1929
Palmer, R. H. 1929. *Foundry Practice: A Text Book for Executives, Molders, Students and Apprentices*. 4th ed. New York: John Wiley & Sons. http://archive.org/details/foundrypracticet00palm.

Panazza 2011
Panazza, Pierfabio. 2011. "La Vittoria alata di Brescia: Repliche, calchi e fortuna iconografica nel XIX secolo." In *Grandi bronzi romani dall'Italia settentrionale: Brescia, Cividate Camuno e Verona*, edited by A. Salcuni, M. Bolla, and E. Formigli, 25–51. Bonn, Germany: Habelt.

Paolozzi Strozzi 2008
Paolozzi Strozzi, Beatrice, ed. 2008. *Donatello: il David restaurato*. Florence: Giunti.

Papet 2001
Papet, Edouard, ed. 2001. *A fleur de peau: le moulage sur nature au XIXe siècle*. Paris: Réunion des Musées Nationaux.

Papet 2018
Papet, Édouard. 2018. *En couleurs: La sculpture polychrome en France, 1850–1910*. Vanves; Paris: Hazan; Musée d'Orsay.

Pearsall 2007
Pearsall, Judy. 2007. *新牛津英汉双解大词典 = The New Oxford English-Chinese Dictionary*. Shanghai: Shanghai Foreign Language Education Press.

Pecchioli 1999
Pecchioli, Roberto. 1999. "Indagine radiografica sulla statua in bronzo da Punta del Serrone (Brindisi)." In *I grandi bronzi antichi: Le fonderie e le tecniche di lavorazione dall'età arcaica al rinascimento: Atti dei Seminari di Studi ed Esperimenti, Murlo 24–30 luglio 1993 e 1–7 luglio 1995*, edited by Edilberto Formigli, 191–202. Siena, Italy: Nuova Immagine.

Peek 2021
Peek, Philip M. 2021. *The Lower Niger Bronzes: Beyond Igbo-Ukwu, Ife and Benin*. London: Routledge.

Peetermans et al. 2012
Peetermans, Steven, Robert van Langh, Eberhard Lehmann, and Arie Pappot. 2012. "Quantification of the Material Composition of Historical Copper Alloys

by Means of Neutron Transmission Measurements." *Journal of Analytical Atomic Spectrometry* 27 (10): 1674–79.

Peltz 2011
Peltz, Uwe. 2011. "Technischer Vergleich." In *Der Xantener Knabe. Technologie, Ikonographie, Funktion und Datierung*, edited by Uwe Peltz and Hans-Joachim Schalles, 111–39. Darmstadt, Germany: Von Zabern.

Peltz 2015
Peltz, Uwe. 2015. "Von der Faszination zum Forschungsgegenstand." In *New Research on Ancient Bronzes: Acta of the XVIIIth International Congress on Ancient Bronzes*, edited by Eckhard Deschler-Erb and Philippe Della Casa, 257–66. Zurich Studies in Archaeology 10. Zurich: Chronos.

Penny 1993
Penny, Nicholas. 1993. *The Materials of Sculpture*. New Haven, CT: Yale University Press.

Pernicka 2014
Pernicka, Ernst. 2014. "Provenance Determination of Archaeological Metal Objects." In *Archaeometallurgy in Global Perspective*, edited by Benjamin W. Roberts and Christopher P. Thornton, 239–68. New York: Springer.

Pernicka and Berswordt-Wallrabe 2008
Pernicka, Ernst, and Silke von Berswordt-Wallrabe, eds. 2008. *Original – Copy – Fake? Examining the Authenticity of Ancient Works of Art, Focusing on African and Asian Bronzes and Terracottas; International Symposium Stiftung Situation Kunst/ Ruhr-University Bochum, February 17th and 18th, 2007*. Mainz, Germany: Von Zabern.

Pernot 2000
Pernot, M. 2000. "Forming Bronze by Plastic Deformation around the 1st Millennium BC in Western Europe." In *J. J. Jonas Symposium on Thermomechanical Processing of Steel*, edited by S. Yue and E. Essadiqi, 615–26. Montreal, Quebec: Metallurgical Society of the Canadian Institute of Mining, Metallurgy and Petroleum.

Perrault 2006
Perrault, Gilles. 2006. "Les contrefaçons de l'œuvre de Diego Giacometti." *Revue Experts* 71 (May): https://www.gillesperrault.com/les-contrefacons-de-lœuvre-de-diego-giacometti/.

Philibert 2014
Philibert, Jean. 2014. "Cent cinquante ans de metallographie." *SF2M Info*, March 2014, 1–2.

Pichon et al. 2014
Pichon, L., B. Moignard, Q. Lemasson, C. Pacheco, and P. Walter. 2014. "Development of a Multi-Detector and a Systematic Imaging System on the AGLAE External Beam." "The 13th International Conference on Particle Induced X-ray Emission (PIXE 2013)." Special issue, *Nuclear Instruments and Methods in Physics Research Section B: Beam Interactions with Materials and Atoms* 318 (January): 27–31.

Pilon et al. 2017
Pilon, Fabien, Alexandre Labet, Karine Vielle, Olivier Vigneau, Guy Granier, Philippe Bienvenu, and Jean-Michel Mermet. 2017. *La validation de méthode en spectrométrie d'émission optique à source plasma: De l'échantillon au résultat. La validation de méthode en spectrométrie d'émission optique à source plasma*. Les Ulis: EDP Sciences. https://doi.org/10.1051/978-2-7598-2083-2.

Pingeot 2002
Pingeot, Anne. 2002. "Degas and His Castings." In *Degas Sculptures: Catalogue Raisonné of the Bronzes*, edited by Joseph S. Czestochowski and Anne Pingeot, 27–37. Memphis: Torch Press and International Arts.

Pitthard et al. 2011
Pitthard, Václav, Richard Stone, Sabine Stanek, Martina Griesser, Claudia-Kryza-Gersch, and Helene Hanzer. 2011. "Organic Patinas on Renaissance and Baroque Bronzes: Interpretation of Compositions of the Original Patination by Using a Set of Simulated Varnished Bronze Coupons." *Journal of Cultural Heritage* 12 (1): 44–53.

Plankensteiner 2010
Plankensteiner, Barbara. 2010. *Benin. Visions d'Afrique*. Milan: 5 Continents.

Plessis, le Roux, and Guelpa 2016
Plessis, Anton du, Stephan Gerhard le Roux, and Anina Guelpa. 2016. "Comparison of Medical and Industrial X-Ray Computed Tomography for Non-Destructive Testing." *Case Studies in Nondestructive Testing and Evaluation* 6 (November): 17–25. https://doi.org/10.1016/j.csndt.2016.07.001.

Pliny the Elder [77 CE] 1857
Pliny the Elder. 1857. *Natural History*, translated by John Bostock and Henry Thomas Riley. London: H.G. Bohn. https://www.google.com/books/edition/The_Natural_History_of_Pliny/9zwZAAAAYAAJ?hl=en&gbpv=0.

Polkinghorne et al. 2014
Polkinghorne, Martin, Brice Vincent, Nicolas Thomas, and David Bourgarit. 2014. "Casting for the King: The Royal Palace Bronze Workshop of Angkor Thom." *Bulletin de l'École française d'Extrême-Orient* 100:327–58.

Pollard 2009
Pollard, A. Mark. 2009. "What a Long, Strange Trip It's Been: Lead Isotopes and Archaeology." In *From Mine to Microscope*, edited by Andrew J. Shortland, Ian C. Freestone, and Thilo Rehren, 181–89. Oxford: Oxbow Books.

Pollard et al. 2007
Pollard, A. M., B. Stern, C. M. Batt, and S. M. M. Young. 2007. "Inductively Coupled Plasma–Mass Spectrometry (ICP–MS)." In *Analytical Chemistry in Archaeology*, 195–214. Cambridge Manuals in Archaeology. Cambridge, UK: Cambridge University Press.

Pollard et al. 2018
Pollard, A. M., P. Bray, A. Cuénod, P. Hommel, Y.-K. Hsu, R. Liu, L. Perucchetti, J. Pouncett, and M. Saunders. 2018. "The Flame GIS-Database." In *Beyond Provenance: New Approaches to Interpreting the Chemistry of Archaeological Copper Alloys*, edited by A. M. Pollard, 171–86. Studies in Archaeological Sciences 6. Leuven, Belgium: Leuven University Press. https://doi.org/10.2307/j.ctv7xbs5r.10.

Pope et al. 1969
Pope, John A., Rutherford J. Gettens, James Cahill, and Noel Barnard. *The Freer Chinese Bronzes*, Vol. 1. Washington, DC: Smithsonian Institution.

Pope-Hennessy 1985
Pope-Hennessy, John. 1985. *Cellini*. Segrate, Italy: Mondadori.

Pozeilov 2015
Pozeilov, Yosi A. 2015. *Digital Imaging and Documentation for Art Conservation*. 4th ed. Edited by Laura Cherry. Digital Photography for Art Conservation. Self-published, Lulu.com.

Radcliffe and Penny 2004
Radcliffe, Anthony, and Nicholas Penny. 2004. *Art of the Renaissance Bronze, 1500–1650*. London: Philip Wilson.

Radivojević et al. 2018
Radivojević, M., J. Pendić, A. Srejić, M. Korać, C. Davey, A. Benzonelli, M. Martinón-Torres, N. Jovanović, and Z. Kamberović. 2018. "Experimental Design of the Cu-As-Sn Ternary Colour Diagram." *Journal of Archaeological Science* 90:106–19.

Rajkolhe and Khan 2014
Rajkolhe, Rajesh, and J. G. Khan. 2014. "Defects, Causes and Their Remedies in Casting Process: A Review." *International Journal of Research in Advent Technology* 2 (3): 375–83.

Rama 1988
Rama, Jean-Pierre. 1988. *Le bronze d'art et ses techniques*. Dourdan, France: Editions H. Vial.

Raub 1993
Raub, Christoph. 1993. "The History of Electroplating." In *Metal Plating and Patination: Cultural, Technical and Historical Developments*, edited by Susan La Niece and Paul Craddock, 284–290. Oxford, London, Boston [etc.]: Butterworth-Heinemann.

Reedy 1987
Reedy, Chandra L. 1987. "Modern Statues and Traditional Methods: A Casting Workshop in Chamba, Himachal Pradesh." *Expedition* 29 (3): 4754.

Reedy 1991a
Reedy, Chandra L. 1991a. "The Opening of Consecrated Tibetan Bronzes with Interior Contents: Scholarly, Conservation and Ethical Considerations." *Journal of the American Institute for Conservation* 30 (1): 13–34.

Reedy 1991b
Reedy, Chandra L. 1991b. "Petrographic Analysis of Casting Core Materials for Provenancing Studies of Copper Alloy Sculptures." *Archaeomaterials* 5:121–63.

Reedy 1997
Reedy, Chandra L. 1997. *Himalayan Bronzes: Technology, Style and Choices*. Newark, DE: University of Delaware; London and Cranbury, NJ: Associated University Presses.

Reedy 2006
Reedy, Chandra L. 2006. "Review of Digital Image Analysis of Petrographic Thin Sections in Conservation Research." *Journal of the American Institute for Conservation* 45 (2): 127–46.

Reedy and Meyers 1987
Reedy, Chandra, and Pieter Meyers. 1987. "An Interdisciplinary Method for Employing Technical Data to Determine Regional Provenance of Copper Alloy Statues," 173–78. In *Recent Advances in the Conservation and Analysis of Artifacts*. University of London, Institute of Archaeology Jubilee Conservation Conference Papers. London: Summer Schools Press.

Reedy and Meyers 2007
Reedy, Chandra L., and Pieter Meyers. 2007. "New Methods for Analyzing Thin Sections of Casting Core Materials: A Case Study with Southeast Asian Bronzes." In *Scientific Research on the Sculptural Arts of Asia: Proceedings of the Third Forbes Symposium at the Freer Gallery of Art*, edited by Janet G. Douglas, Paul Jett, and John Winter, 103–14. London: Archetype.

Reedy et al. 2014
Reedy, Chandra L., Jenifer Anderson, Terry J. Reedy, and Yimeng Liu. 2014. "Image Analysis in Quantitative Particle Studies of Archaeological Ceramic Thin Sections." *Advances in Archaeological Practice* 2 (4): 252–68.

Rehren 2014
Rehren, Thilo. 2014. "Crossing Boundaries." In *Sharing Archaeology: Academe, Practice and the Public*, edited by Peter Stone and Zhao Hui, 36–46. New York: Routledge.

Reimer et al. 2020
Reimer, Paula J., William E. N. Austin, Edouard Bard, Alex Bayliss, Paul G. Blackwell, Christopher Bronk Ramsey, Martin Butzin et al. 2020. "The IntCal20 Northern Hemisphere Radiocarbon Age Calibration Curve (0–55 Cal KBP)." *Radiocarbon* 62 (4): 725–57. https://doi.org/10.1017/RDC.2020.41.

Reindell and Tommasi 1999
Reindell, Ingrid, and Gabriele Tommasi. 1999. "Brevi cenni sulla fusione della porta della cattedrale di Trani." In *I grandi bronzi antichi: le fonderie e le tecniche di lavorazione dall'età arcaica al rinascimento: atti dei Seminari di Studi ed Esperimenti, Murlo 24–30 luglio 1993 e 1–7 luglio 1995*, edited by Edilberto Formigli, 351–64. Siena, Italy: Nuova Immagine.

Reliance Foundry n.d.
Reliance Foundry. n.d. "Foundry Glossary." https://www.reliance-foundry.com/castings/glossary-casting.

Reuter and Schneider 1971
Reuter, Horst, and Philipp Schneider. 1971. *Gussfehler-Atlas*. 2nd ed. Düsseldorf: Giesserei-Verlag.

Revankar 2000
Revankar, G. 2000. "Introduction to Hardness Testing." In *ASM Metal Handbook*. Vol. 8, *Mechanicial Testing and Evaluation*, 416–28. Materials Park, OH: ASM International.

Rich 1988
Rich, Jack. 1988. *The Materials and Methods of Sculpture*. New York: Dover.

Rickard 1932
Rickard, T. A. 1932. "The Nomenclature of Copper and Its Alloys." *Journal of the Royal Anthropological Institute of Great Britain and Ireland* 62:281–90.

Riederer 1982
Riederer, Josef. 1982. "Die naturwissenschaftiche Untersuchung der Bronzen der Staatlichen Sammlung Ägyptischer Kunst in München." *Berliner Beiträge zur Archäometrie* 7:2–23.

Riederer 2002
Riederer, Josef. 2002. "Die Datenbank zu Metallanalysen des Rathgen-Forschungslabors. Ein Beitrag zur Erschließung kulturgeschichtlicher Objekte." *Jahrbuch Preussischer Kulturbesitz* 39:329–44.

Riederer 2004
Riederer, Josef. 2004. "Thin Section Microscopy Applied to the Study of Archaeological Ceramics." *Hyperfine Interactions* 154:143–58.

Risser and Saunders 2017
Risser, Erik, and David Saunders. 2017. "Tiberius from Herculaneum: Methods of Assembling a Monumental Bronze Portrait." In *Artistry in Bronze: The Greeks and Their Legacy: XIXth International Congress on Ancient Bronzes*, edited by Jens M. Daehner, Kenneth Lapatin, and Ambra Spinelli, 61–68. Los Angeles: J. Paul Getty Museum and Getty Conservation Institute. http://www.getty.edu/publications/artistryinbronze/large-scale-bronzes/7-risser/.

Rivero 2016
Rivero, Olivia. 2016. "Master and Apprentice: Evidence for Learning in Palaeolithic Portable Art." *Journal of Archaeological Science* 75:89–100.

Robbiola 2015
Robbiola, Luc. 2015. "Les patines naturelles des bronzes - quelques développements majeurs." In *La conservation-restauration des métaux archéologiques: Des premiers soins à la conservation durable*, edited by S. Clerbois, C. Cappucci, J. Moulin, L. Nonne, S. Piermarini, and M.-H. Schumacher, 45–50. Les Dossiers de l'Institut Du Patrimoine Wallon (IPW) 15. Namur, Belgium: IPW Institut du Patrimoine wallon.

Robbiola and Hurtel 1997
Robbiola, Luc, and L. P. Hurtel. 1997. "Standard Nature of the Passive Layers of Buried Archaeological Bronze: The Example of Two Roman Half-Length Portraits." In *METAL 95: International Conference on Metals Conservation*, edited by Ian D. MacLeod, Stéphane L. Pennec, and Luc Robbiola, 109–17. London: Ames & James Science.

Robbiola and Portier 2006
Robbiola, Luc, and Richard Portier. 2006. "A Global Approach to the Authentication of Ancient Bronzes Based on the Characterization of the Alloy-Patina-Environment System." *Journal of Cultural Heritage* 7 (1): 1–12.

Robbiola, Blengino, and Fiaud 1998
Robbiola, L., J. M. Blengino, and C. Fiaud. 1998. "Morphology and Mechanisms of Formation of Natural Patinas on Archaeological Cu-Sn Alloys." *Corrosion Science* 40 (12): 2083–111.

Robcis, Bourgarit, and Mille 2000
Robcis, Dominique, David Bourgarit, and Benoît Mille. 2000. "Tsuba, de l'art à la matière." *Techné* 13/14:84–88.

Robcis et al. 2014
Robcis, D., M. Aucouturier, S. Descamps, J. Marsac, E. Lambert, J. Langlois, Q. Lemasson, B. Moignard, C. Pacheco, and L. Pichon. 2014. "La prise en compte des surfaces et patines antiques dans la restauration des métaux archéologiques: Étude de la Plaque à l'Aurige." *Les dossiers de l'institut du patrimoine Wallon* 15:67–75.

Robcis et al. 2017
Robcis, Dominique, Sophie Descamps-Lequime, Nathalie Pingaud, and Benoît Mille. 2017. "Dorure et redorure de la statuaire antique en bronze: Une nouvelle lecture de l'Apollon de Lillebonne." *Techné* 45:100–113. https://doi.org/10.4000/techne.1324.

Rogers 1993
Rogers, M. Ned. 1993. "Welding of Copper Alloys." In *Welding, Brazing and Soldering*, edited by David LeRoy Olson, Thomas A. Siewert, Stephen Liu, and Glen R. Edwards, 6:0. Materials Park, OH: ASM International.

Rohnstock 1999
Rohnstock, U. 1999. "Adorante da Rodi: le tecniche applicate sul bronzo antico dal XVI al XX secolo." In *I grandi bronzi antichi: le fonderie e le tecniche di lavorazione dall'età arcaica al rinascimento: atti dei Seminari di Studi ed Esperimenti, Murlo 24–30 luglio 1993 e 1–7 luglio 1995*, edited by Edilberto Formigli, 402–3. Siena, Italy: Nuova Immagine.

Rolley 1994
Rolley, Claude. 1994. *La sculpture grecque*. Vol. 1, *Des origines au milieu du Ve siècle*. Les manuels d'art et d'archéologie antiques. Paris: Picard.

Rome and Young 2003
Rome, Richard, and Hamish Young. 2003. *Fine Art Metal Casting: An Illustrated Guide to Mould Making and Lost Wax Processes*. London: Robert Hale.

Rostoker and Dvorak 1991
Rostoker, W., and J. R. Dvorak. 1991. "Some Experiments with Co-Smelting to Copper Alloys." *Archaeomaterials* 5 (1): 5–20.

Ruault 1991
Ruault, Pierre André. 1991. *Radiologie Industrielle*. 2 vols. Paris: Publications de la soudure Autogène.

Runfola 2014
Runfola, Matthew. 2014. *Patina: 300+ Coloration Effects for Jewelers and Metalsmiths*. Loveland, CO: Interveave.

Rybczynski 2005
Rybczynski, Witold. 2005. *One Good Turn: A Natural History of the Screwdriver and the Screw*. New York: Scribner.

Salter and Gilmour n.d.
Salter, C. J., and B. J. J. Gilmour. n.d. "Glossary of Terms Used in the Study of Ancient Metal-Working and Associated Processes." Materials Science-Based Archaeology Group, Department of Materials, University of Oxford.

Saussus and Thomas 2019
Saussus, Lise, and N. Thomas. 2019. *Un atelier d'orfèvre autour de l'an mil. Travail du cuivre, de l'argent et du fer à Oostvleteren (Flandre-Occidentale, Belgique)*. Collection d'archéologie Joseph Mertens 18. Louvain-la-Neuve, Belgium: Presses universitaires de Louvain.

Schmidt 2011
Schmidt, Martin H., ed. 2011. *Kunstguss in Lauchhammer*. Regardeur Series 6. Norderstedt, Germany: BoD.

Schmidt and Schmidt 1992
Schmidt, Robert F., and Donald G. Schmidt. 1992. "Selection and Application of Copper Alloy Castings." In *ASM Metal Handbook*. Vol. 2, *Properties of Non-Ferrous Alloys*, 346–55. Materials Park, OH: ASM International.

Schmidtling 2008
Schmidtling II, Ronald C. 2008. "Core Analysis." In Jane Bassett, *The Craftsman Revealed: Adriaen de Vries, Sculptor in Bronze*, 35–44. Los Angeles: Getty Conservation Institute. http://www.getty.edu/publications/virtuallibrary/9780892369195.html.

Schmutzler, Eggert, and Kuhn-Wawezinek 2018
Schmutzler, Britta, Gerhard Eggert, and Charlotte F. Kuhn-Wawezinek. 2018. "Copper (II) Hydroxide on Artefacts: Corrosion, Conservation, Colourants." *Studies in Conservation* 62 (2): 61–67.

Scholten 1998
Scholten, Frits, ed. 1998. *Adriaen de Vries (1556–1626): Imperial Sculptor*. Los Angeles: Getty Trust Publications and J. Paul Getty Museum.

Schorsch, Becker, and Caro 2019
Schorsch, Deborah, Lawrence Becker, and Federico Caro. 2019. "Enlightened Technology: Casting Divinity in the Gupta Age." *Arts of Asia* 49 (2): 131–43.

Schrenk 1994
Schrenk, Janet L. 1994. "The Royal Art of Benin: Surfaces, Past and Present." In *Ancient and Historic Metals: Conservation and Scientific Research*, edited by David A. Scott, Jerry Podany, and Brian B. Considine, 51–62. Los Angeles: Getty Conservation Institute. https://www.getty.edu/publications/virtuallibrary/0892362316.html.

Schulz and Baatz 1993
Schulz, Paul Otto, and Ulrich Baatz. 1993. *Bronzegiesserei Noack*. Ravensburg, Germany: Otto Maier.

Schwartz and Aircraft 1951
Schwartz, M. M., and S. Aircraft. 1951. *Metals Handbook: Welding, Brazing and Soldering*. ASTM 6. Materials Park, OH: ASM International.

Schweizer 1994
Schweizer, François. 1994. "Bronze Objects from Lake Sites: From Patina to 'Biography.'" In *Ancient and Historic Metals: Conservation and Scientific Research*, edited by David A. Scott, Jerry Podany, and Brian B. Considine, 33–50. Los Angeles: Getty Conservation Institute. https://www.getty.edu/publications/virtuallibrary/0892362316.html.

Scott 1983
Scott, David A. 1983. "Depletion Gilding and Surface Treatment of Gold Alloys from the Narino Area of Ancient Colombia." *Journal of the Historical Metallurgy Society* 17 (2): 99–115.

Scott 1986
Scott, David A. 1986. "Fusion Gilding and Foil Gilding in Pre-Hispanic Colombia and Ecuador." In *Metalurgia de America precolombina*, edited by Clemencia Plazas de Nieto, 283–325. Bogotá, Colombia: Banco de la Republica.

Scott 1991
Scott, David A. 1991. *Metallography and Microstructure of Ancient and Historic Metals*. Marina del Rey, CA: Getty Conservation Institute in association with Archetype Books.

Scott 2000
Scott, David A. 2000. "A Review of Gilding Techniques in Ancient South America." In *Gilded Metals: History, Technology and Conservation*, edited by Terry Drayman-Weisser, 203–22. London: Archetype in association with the American Institute of Conservation of Historic and Artistic Works.

Scott 2002
Scott, David A. 2002. *Copper and Bronze in Art: Corrosion, Colorants, Conservation*. Los Angeles: Getty Publications.

Scott 2014
Scott, David A. 2014. "Metallography and Microstructure of Metallic Artifacts." In *Archaeometallurgy in Global Perspective: Methods and Syntheses*, edited by Benjamin W. Roberts and Christopher P. Thornton, 67–89. New York: Springer.

Scott and Swartz Dodd 2002
Scott, David A., and Lynn Swartz Dodd. 2002. "Examination, Conservation, and Analysis of a Gilded Bronze Osiris." *Journal of Cultural Heritage* 3 (4): 333–45.

***Secrets concernant les arts et métiers* 1810**
Secrets concernant les arts et métiers (Nouvelle édition, revue, corrigée et considérablement augmentée). 1810. Avignon, France: E. T. Chaillot. https://books.google.fr/books?id=J3U7AAAAcAAJ&printsec=frontcover&dq=.

Seelig-Teuwen, Bourgarit, and Bewer 2014
Seelig-Teuwen, Regina, David Bourgarit, and Francesca G. Bewer. 2014. "Barthélemy Prieur Fondeur, son atelier, ses méthodes de travail." In *French Bronze Sculpture: Materials and Techniques 16th - 18th Century*, edited by David Bourgarit, Jane Bassett, Francesca G. Bewer, Geneviève Bresc-Bautier, Phillippe Malgouyres, and Guilhem Scherf, 18–38. London: Archetype.

Selwyn 1950
Selwyn, A. 1950. *The Retail Jeweller's Handbook*. London: Heywood.

Selwyn and Canadian Conservation Institute 2004
Selwyn, Lyndsie, and Canadian Conservation Institute. 2004. *Metals and Corrosion: A Handbook for the Conservation Professional*. Ottawa: Canadian Conservation Institute.

Sequenzia, Fatuzzo, and Oliveri 2021
Sequenzia, Gaetano, Gabriele Fatuzzo, and Salvatore Massimo Oliveri. 2021. "A Computer-Based Method to Reproduce and Analyse Ancient Series-Produced Moulded Artefacts." *Digital Applications in Archaeology and Cultural Heritage* 20 (March): e00174. https://doi.org/10.1016/j.daach.2021.e00174.

Shanxi Provincial Institute of Archaeology 1993
Shanxi Provincial Institute of Archaeology | 山西省考古研究所编. 1993. 侯马铸铜遗址 = *The Houma Bronze Casting Site*. 北京 = Beijing: 文物出版社 = Cultural Relics Press. https://book.douban.com/subject/3828980/.

Shearman 2010
Shearman, Fleur. 2010. "Patina and Polychromy: Reconstructing Some Technical Histories for Small Egyptian Bronze Sculpture in the British Museum." In *Decorated Surfaces on Ancient Egyptian Objects: Technology, Deterioration and Conservation; Proceedings of a Conference Held in Cambridge, UK, on 7–8 September 2007*, edited by J. Dawson, C. Rozeik, and M. M. Wright, 46–57. London: Archetype.

Siddalingswami and Dulange 2015
Siddalingswami, S. Hiremath, and S. R. Dulange. 2015. "Advanced Techniques in Casting Defects and Rejection Analysis: A Study in an Industry." *International Journal of Innovations in Engineering, Research and Technology* 2 (9): 1-9. https://doi.org/10.5281/zenodo.1467541.

Sidky, Kao, and Pan 2006
Sidky, E. Y., C.-M. Kao, and X. Pan. 2006. "Accurate Image Reconstruction from Few-Views and Limited-Angle Data in Divergent-Beam CT." *Journal of X-Ray Science and Technology* 14 (2): 119–39.

Singer and Fletcher 2004
Singer, Martha C., and Valerie Fletcher. 2004. "The Conservation of Arp's Bronzes: Preserving the Sculpture's History or the Artist's Intent?" In *Modern Art, New Museums: Contributions to the Bilbao Congress, 13–17 September 2004*, edited by Ashok Roy and Perry Smith, 73–76. London: International Institute for Conservation of Historic and Artistic Works.

Smith 1967
Smith, Cyril Stanley. 1967. "The Interpretation of Microstructures of Metallic Artifacts." In W. J. Young, *Application of Science in Examination of Works of Art*, 20–52. Boston: Research Laboratory of the Museum of Fine Arts.

Smith 1970
Smith, Cyril Stanley. 1970. "An Examination of the Arsenic-Rich Coating on a Bronze Bull from Horoztepe." In *Application of Science in Examination of Works of Art*, edited by William J. Young, 96–102. Boston: Museum of Fine Arts, Boston.

Smith 1981
Smith, Cyril Stanley. 1981. "On Art, Invention and Technology." In *A Search for Structure: Selected Essays on Science, Art, and History*, 325–31. Cambridge, MA: MIT Press.

Smith 2008a
Smith, Dylan. 2008. *Bronze and Boxwood: Renaissance Masterpieces from the Robert H. Smith Collection*. Washington, DC: Board of Trustees, National Gallery of Art.

Smith 2008b
Smith, Dylan. 2008b. "I bronzi di Severo da Ravenna: Un approcio tecnologico per la cronologia." In *L'Industria artistica del bronzo del rinascimento a Venezia e nell'Italia settentrionale: Atti del Convegno Internazionale di Studi; Venezia, Fondazione Giorgio Cini, 23 e 24 Ottobre 2007*, edited by Matteo Ceriana, 49–80. Verona, Italy: Scripta.

Smith 2012
Smith, Dylan. 2012. "Handheld X-Ray Fluorescence Analysis of Renaissance Bronzes: Practical Approaches to Quantification and Acquisition." In *Handheld XRF for Art and Archaeology*, edited by Aaron N. Shugar and Jennifer L. Mass, 37–74. Leuven, Belgium: Leuven University Press.

Smith 2013
Smith, Dylan. 2013. "Reconstructing the Casting Technique of Severo da Ravenna's *Neptune*." *Facture* 1:167–81.

Smith 2015
Smith, Dylan. 2015. "Riccio in Relief: Documented Sculpture as a Technical Context for *The Entombment*." *Facture: Conservation, Science, Art History* 2: pages unknown.

Smith and Beentjes 2010
Smith, Pamela H., and Tonny Beentjes. 2010. "Nature and Art, Making and Knowing: Reconstructing Sixteenth-Century Life-Casting Techniques." *Renaissance Quarterly* 63:128–79.

Smith and Hawthorne 1974
Smith, Cyril Stanley, and John G. Hawthorne. 1974. *Mappae Clavicula: A Little Key to the World of Medieval Techniques*. Philadelphia: American Philosophical Society.

Smith and Sepponen 2019
Smith, Dylan, and Wendy Sepponen. 2019. "'Comparable to the Very Tips of Their Spades': Technical and Political Connections among Serial Busts of Charles V." *Facture: Conservation, Science, Art History* 4:34–65.

Smith and Sturman 2011
Smith, Dylan, and Shelley G. Sturman. 2011. "The Art and Innovation of Antico's Bronzes: A Technical Investigation." In *Antico: The Golden Age of Renaissance Bronzes*, edited by Eleonora Luciano, 157–85. Washington, DC: National Gallery of Art; London: Paul Holberton.

Smith et al. 2011
Smith, A., H. Botha, F. C. de Beer, and E. Ferg. 2011. "The Examination, Analysis and Conservation of a Bronze Egyptian Horus Statuette." *Nuclear Instruments and Methods in Physics Research Section A: Accelerators, Spectrometers, Detectors and Associated Equipment* 651 (1): 221–28.

Snodgrass 2000
Snodgrass, Amy. 2000. "Analysis of Gilding and Other Decorative Metals on Chosen Bronzes." In *From the Parts to the Whole: Acta of the 13th International Bronze Congress, Held at Cambridge, Massachusetts, May 28–June 1, 1996*, edited by Carol C. Mattusch, Amy Brauer, and Sandra E. Knudsen, 277–81. Cambridge, MA: Journal of Roman Archaeology.

Somigli 1958
Guglielmo Somigli. 1958. *Notizie storiche sulla fusione del Perseo con alcuni documenti inediti di Benvenuto Cellini*. Milan: Associazione Italiana Metallurgia.

Stalker and Parker 1960
Stalker, John, and George Parker. 1960. *A Treatise of Japanning and Varnishing*. Chicago: Quadrangle Books.

Stephens et al. 2021
Stephens, J. A., M. N. Ducea, D. J. Killick, and J. Ruiz. 2021. "Use of Non-Traditional Heavy Stable Isotopes in Archaeological Research." *Journal of Archaeological Science* 127 (March): 105334.

Stephens et al. 2022
Stephens, Catherine H., Eric M. Breitung, Corina Geiculescu, and Thomas Strange. 2022. "Determination of Nitrile Gloves Appropriate for Use When Dry Handling Art." *Journal of the American Institute for Conservation* 61 (3): 162–71. https://doi.org/10.1080/01971360.2021.1913543.

Stewart 1994
Stewart, Rick. 1994. *Charles M. Russell, Sculptor*. Fort Worth: Amon Carter Museum; New York: H. N. Abrams.

Stewart 2014
Stewart, Andrew. 2014. "Why Bronze?" In *Power and Pathos: Bronze Sculpture of the Hellenistic World*, edited by Jens M. Daehner and Kenneth Lapatin, 34–47. Los Angeles: J. Paul Getty Museum.

Stijnman 2005
Stijnman, Ad. 2005. "Style and Technique Are Inseparable: Art Technological Sources and Reconstructions." In *Art of the Past: Sources and Reconstruction: Proceedings of the First Symposium of the Art Technological Source Research Study Group*, edited by Mark Clarke, Joyce H. Townsend, and Ad Stijnman, 1–8. London: Archetype.

Stijnman 2012
Stijnman, Ad. 2012. "Some Thoughts on Realia: Material Sources for Art Technological Source Research." In *The Artist's Process: Technology and Interpretation: Proceedings of the Fourth Symposium of the Art Technological Source Research Working Group*, edited by Sigrid Eyb-Green, 18–20. London: Archetype.

St. John 1931
St. John, H. M. 1931. "The Influence of Impurities in Foundry Brasses and Bronzes." *Materials Engineering* 2 (October): 242–45.

Stoltman et al. 2018
Stoltman, James B., Zhanwei Yue, Zhichun Jing, Jigen Tang, James H. Burton, and Mati Raudsepp. 2018. "New Insights into the Composition and Microstructure of Ceramic Artifacts Associated with the Production of Chinese Bronzes at Yinxu, the Last Capital of the Shang Dynasty." *Archaeological Research in Asia* 15 (September): 88–100.

Stone 1981
Stone, Richard E. 1981. "Antico and the Development of Bronze Casting in Italy at the End of the Quattrocento." *Metropolitan Museum Journal* 16:87–116.

Stone 2001
Stone, Richard E. 2001. "A New Interpretation of the Casting of Donatello's Judith and Holofernes." In *Small Bronzes in the Renaissance*, edited by Debra Pincus, 55–67. Studies in the History of Art 62. Washington, DC: National Gallery of Art; New Haven, CT: Yale University Press.

Stone 2006
Stone, Richard E. 2006. "Severo Calzetta da Ravenna and the Indirectly Cast Bronze." *Burlington Magazine* 148:810–19.

Stone 2008
Stone, Richard E. 2008. "Riccio: Technology and Connoisseurship." In *Andrea Riccio: Renaissance Master of Bronze*, edited by Denise Allen, 81–96. London: Philip Wilson.

Stone 2010
Stone, Richard E. 2010. "Organic Patinas on Small Bronzes of the Italian Renaissance." *Metropolitan Museum Journal* 45 (January): 107–24.

Stone 2011
Stone, Richard E. 2011. "Understanding Antico's Patinas." In *Antico: The Golden Age of Renaissance Bronzes*, edited by Eleonora Luciano, 178–82. Washington, DC: National Gallery of Art; London: Paul Holberton.

Stone, White, and Indictor 1990
Stone, R., R. White, and N. Indictor. 1990. "The Surface Composition of Some Italian Renaissance Bronzes." In *ICOM Committee for Conservation 9th Triennial Meeting, Dresden, German Democratic Republic, 26–31 August 1990*, 568–73. Paris: ICOM Committee for Conservation.

Strahan 1997
Strahan, Donna K. 1997. "Bronze Casting in Thailand." In *The Sacred Sculpture of Thailand: The Alexander B. Griswold Collection*. Baltimore: Walters Art Gallery, 27–41.

Strahan 1998
Strahan, Donna. 1998. "A Technical Study of 11–12th c. Khmer Bronze Sculpture." In *The Fourth International Conference on the Beginning of the Use of Metals and Alloys (BUMA-IV), May 25–27, 1998, Matsue, Shimane, Japan, The Japan Institute of Metals, Sendai City, Japan*, edited by Nihon Kinzoku Gakkai, 97–102. Sendai, Japan: BUMA-IV Organizing Committee / Japan Institute of Metals.

Strahan 2010
Strahan, Donna. 2010. "Piece Mold Casting: A Chinese Tradition for Fourth- and Fifth-Century Bronze Buddha Images." *Metropolitan Museum Studies in Art, Science, and Technology* 1:133–53.

Strahan 2019
Strahan, Donna. 2019. "Debating the Use of Lost-Wax Casting in Ancient China." Forbes Lecture on Scientific Research in the Field of Asian Art. Washington, DC: Freer Gallery of Art and Arthur M. Sackler Gallery, Smithsonian. https://www.freersackler.si.edu/wp-content/uploads/2017/10/Forbes-Lecture-lost-wax-ancient-china.pdf.

Strahan and Maines 2000
Strahan, Donna K., and Christopher A. Maines. 2000. "Lacquer as an Adhesive for Gilding on Copper Alloy Sculpture in Southeast Asia." In *Gilded Metals: History, Technology and Conservation*, edited by Terry Draymann-Weisser, 185–202. London: Archetype in association with the American Institute of Conservation of Historic and Artistic Works.

Sturman 2001
Sturman, Shelley. 2001. "A Group of Giambologna Female Nudes: Analysis and Manufacture." In *Small Bronzes in the Renaissance*, edited by Debra Pincus, 120–41. Studies in the History of Art 62. Washington, DC: National Gallery of Art; New Haven, CT: Yale University Press.

Sturman 2004
Sturman, Shelley. 2004. "Selected Bronzes from the Quentin Collection: Introduction to Technical Reports." In *European Bronzes from the Quentin Collection*, edited by Manfred Leithe-Jasper and Patricia Wengraf, 303. London: Scala.

Sturman et al. 2009
Sturman, Shelley G., Simona Cristanetti, Debra Pincus, Karen Serres, and Dylan Smith. 2009. "'Beautiful in Form and Execution': The Design and Construction of Andrea Riccio's Paschal Candlestick." *Burlington Magazine* 151 (October): 666–72.

Suverkrop 1912
Suverkrop, E. A. 1912. "The Molding of Bronze Statuary." *American Machinist* 37 (23): 923–28.

Sylvester 2008
Sylvester, Paul. 2008. "Matrix Effects in Laser Ablation-ICP-MS." In *Laser Ablation-ICP-MS in the Earth Sciences: Current Practices and Outstanding Issues*, edited by Paul Sylvester, Pap/Cdr edition, 67–78. Quebec: Mineralogical Association of Canada.

Tabor, Molinarib, and Juleff 2005
Tabor, G. R., D. Molinarib, and G. Juleff. 2005. "Computational Simulation of Air Flows through a Sri Lankan Wind-Driven Furnace." *Journal of Archaeological Science* 32 (5): 753–66.

Taylor 1987
Taylor, R. E. 1987. "Radiocarbon Dating in Historical Perspective." In *Radiocarbon Dating: An Archaeological Perspective*, edited by R. E. Taylor, 147–70. Orlando: Academic Press.

Texier et al. 2016
Texier, A., A. Azema, M. Bouichoi, J. Legendre, and C. Bottineau. 2016. "The Vendôme Column: Restoration of a Monumental Bronze in Paris." In *Proceeedings of the Interim Meeting of the ICOM-CC Metals Working Group, Sept. 26–30, 2016*, edited by R. Menon, C. Chemello and A. Pandya, 298–305. New Delhi: ICOM-CC and Indira Gandhi National Centre of the Arts.

Theophilus [ca. 1122] 1979
Theophilus. (ca. 1122) 1979. *On Divers Arts: The Foremost Medieval Treatise on Painting, Glassmaking, and Metalwork*. Translated by John G. Hawthorne and Cyril Stanley Smith. New York: Dover.

Thomas 2009
Thomas, Nicolas. 2009. "Les ateliers urbains de travail du cuivre et de ses alliages au bas moyen âge: Archéologie et histoire d'un site parisien du XIVe siècle dans la Villeneuve du Temple (1325–1350)." PhD diss., Université Paris 1.

Thomas and Bourgarit 2014
Thomas, N., and D. Bourgarit. 2014. "Les techniques de production des batteurs et fondeurs mosans au Moyen Age (XIIe-XVIe siècles)." In *L'or des dinandiers. Fondeurs et batteurs mosans au Moyen Age*, edited by N. Thomas, I. Leroy, and J. (dir.) Plumier, 43–63. Cahiers de la MPPM 7. Bouvignes-Dinant, Belgium: Maison du Patrimoine Médiéval Mosan.

Thomas and Bourgarit 2018
Thomas, Nicolas, and David Bourgarit. 2018. "La peine emporte-t-elle le profit? Économie de la production du laiton par cémentation au Moyen Âge." In *Cuivre, Bronzes et Laitons Médiévaux: Histoire, Archéologie et Archéométrie Des Productions En Laiton, Bronze et Autres Alliages à Base de Cuivre Dans l'Europe Médiévale (12e–16e Siècles) = Medieval Copper, Bronze and Brass: History, Archaeology and Archaeometry of the Production of Brass, Bronze and Other Copper Alloy Objects in Medieval Europe (12th–16th Centuries)*, edited by Nicolas Thomas and Pete Dandridge, 71–88. Études et Documents, Archéologie 39. Namur, Belgium: Agence wallonne du Patrimoine.

Thomas and Dandridge 2018
Thomas, Nicolas, and Pete Dandridge, eds. 2018. *Cuivre, Bronzes et Laitons Médiévaux: Histoire, Archéologie et Archéométrie Des Productions En Laiton, Bronze et Autres Alliages à Base de Cuivre Dans l'Europe Médiévale (12e–16e Siècles) = Medieval Copper, Bronze and Brass: History, Archaeology and Archaeometry of the Production of Brass, Bronze and Other Copper Alloy Objects in Medieval Europe (12th–16th Centuries)*, Études et Documents, Archéologie 39. Namur, Belgium: Agence wallonne du Patrimoine.

Thoury et al. 2016
Thoury, M., B. Mille, T. Séverin-Fabiani, L. Robbiola, M. Réfrégiers, J-F Jarrige, and L. Bertrand. 2016. "High Spatial Dynamics-Photoluminescence Imaging Reveals the Metallurgy of the Earliest Lost-Wax Cast Object." *Nature Communications* 7 (1): 13356. https://doi.org/10.1038/ncomms13356.

Treccani n.d.
"Treccani - La Cultura Italiana - Enciclopedia." n.d. Accessed April 4, 2023. https://www.treccani.it/enciclopedia/.

Turner 1982
Turner, Eric. 1982. *An Introduction to Brass*. London: Her Majesty's Stationery Office and Victoria & Albert Museum.

Tylecote, Ghaznavi, and Boydell 1977
Tylecote, R. F., H. A. Ghaznavi, and P. J. Boydell. 1977. "Partitioning of Trace Elements between the Ores, Fluxes, Slags and Metal during the Smelting of Copper." *Journal of Archaeological Sciences* 4:305–33.

Tyler and Black 1992
Tyler, Derek E., and William T. Black. 1992. "Introduction to Copper and Copper Alloys." In *ASM Handbook*. Vol. 2, *Properties and Selection: Non-Ferrous Alloys and Special-Purpose Materials*, 759–839. Materials Park, OH: ASM International.

Tzachou-Alexandri 2000
Tzachou-Alexandri, O. 2000. "Some Remarks on the Bronze God of Artemision." In *From the Parts to the Whole: Acta of the 13th International Bronze Congress, Held at Cambridge, Massachusetts, May 28–June 1, 1996*, edited by Carol C. Mattusch, Amy Brauer, and Sandra E. Knudsen, 86–95. Cambridge, MA: Journal of Roman Archaeology.

Uhlenhuth 1920
Uhlenhuth, Eduard. 1920. *Vollständige Anleitung zum Formen und Giessen*. 8th ed. Vienna and Leipzig, Germany: A. Hartleben's Verlag.

Untracht 1968
Untracht, Oppi. 1968. *Metal Techniques for Craftsmen: A Basic Manual on the Methods of Forming and Decorating Metals*. New York: Doubleday.

Untracht 1982
Untracht, Oppi. 1982. *Jewelry: Concepts and Technology*. London: Robert Hale.

Van Langh 2012
Van Langh, Robert. 2012. "Technical Studies of Renaissance Bronzes: The Use of Neutron Imaging and Time-of-Flight Neutron Diffraction in the Studies of the Manufacture and Determination of Historical Copper and Alloys." PhD diss., Delft University of Technology. http://resolver.tudelft.nl/uuid:3898ae1c-7934-4d85-a0db-628760739b50.

Van Langh et al. 2009
Van Langh, Robert, E. Lehmann, S. Hartmann, A. Kaestner, and Fritz Scholten. 2009. "The Study of Bronze Statuettes with the Help of Neutron-Imaging Techniques." *Analytical and Bioanalytical Chemistry* 395 (7): 1949–59. https://doi.org/10.1007/s00216-009-3058-4.

Van Langh et al. 2011
Van Langh, R., L. Bartoli, J. Santisteban, and D. Visser. 2011. "Casting Technology of Renaissance Bronze Statuettes: The Use of TOF-Neutron Diffraction for Studying Afterwork of Renaissance Casting Techniques." *Journal of Analytical Atomic Spectrometry* 26 (5): 892–98.

Vasari [1550] 1960
Vasari, Giorgio. (1550) 1960. *Vasari on Technique; Being the Introduction to the Three Arts of Design, Architecture, Sculpture and Painting, Prefixed to the Lives of the Most Excellent Painters, Sculptors and Architects.* Translated by Louisa S. Maclehose. New York: Dover.

Vasari [1550] 1996
Vasari, Giorgio. (1550) 1996. *Lives of the Most Eminent Painters, Sculptors and Architects*. Translated by Gaston du C. de Vere. New York: Everyman's Library.

Vauxcelles 1905
Vauxcelles, Louis. 1905. "La fonte à cire perdue." *Art et décoration* 8:195.

Velde 2005
Velde, Bruce. 2005. "Use of Image Analysis in Determining Multi-Source Ceramic Materials." In *Pottery Manufacturing Processes: Reconstitution and Interpretation: Acts of the XIVth UISPP Congress, University of Liège, Belgium, 2–8 September 2001*, edited by Alexandre Livingstone Smith, Dominique Bosquet, and Rémi Martineau, 95–99. BAR International Series 1349. Oxford: Archaeopress.

Verma 2019
Verma, A. K. 2019. "A Method Based on Structure-from-Motion Photogrammetry to Generate Sub-Millimetre-Resolution Digital Elevation Models for Investigating Rock Breakdown Features." *Earth Surface Dynamics* 7:45–66.

Vincent 2012
Vincent, Brice. 2012. "Samrit. Étude de la métallurgie du bronze dans le Cambodge angkorien (fin du XIe-début du XIIIe siècle)." PhD diss., Université Paris 3. http://www.theses.fr/2012PA030032.

Vincent 2014
Vincent, B. 2014. "Searching for the Bronze Workshops of Angkorian Cambodia: Petrographic Study Applied to Casting Cores of 11th–12th C. Khmer Bronzes" (unpublished report). Washington, DC: Freer Gallery of Art and Arthur M Sackler Gallery.

Vincent, Bourgarit, and Jett 2012
Vincent, Brice, David Bourgarit, and Paul Jett. 2012. "Khmer Bronze Metallurgy during the Angkorian Period (12th–13th Centuries): Technical Investigation of a New Selected Corpus of Artifacts from the National Museum of Cambodia, Phnom Penh." In *Scientific Research on Ancient Asian Metallurgy*, edited by Paul Jett, Blythe McCarthy, and Janet G. Douglas, 124–53. London: Archetype.

von Schroeder 1981
von Schroeder, U. 1981. *Indo-Tibetan Bronzes*. Hong Kong: Visual Dharma.

Waksman 2014
Waksman, Yona. 2014. "Etudes de Provenance de Céramiques." In *Circulation et provenance es matériaux dans les sociétés anciennes*, edited by Philippe Dillmann and Ludovic Bellot-Gurlet, 195–216. Sciences Archéologiques. Paris: Editions des archives contemporaines.

Wallack 1840
Wallack, August. 1840. *Vollständiges Handbuch des Gürtlers und Bronzearbeiters*. Vol. 108. Neuer Schauplatz der Künste und Handwerke. Weimar, Germany: Bernh. Fr. Voigt.

Ward 2008
Ward, Gerald W. R. 2008. *The Grove Encyclopedia of Materials and Techniques in Art*. Oxford and New York: Oxford University Press.

Warda 2011
Warda, Jeffrey, ed. 2011. *The AIC Guide to Digital Photography and Conservation Documentation*. 2nd ed. Washington, DC: American Institute for Conservation of Historic and Artistic Works.

Warren 2014
Warren, Jeremy, 2014. *Medieval and Renaissance Sculpture: A Catalogue of the Collection in the Ashmolean Museum*, Oxford, Vol. I. Sculptures in Metal. Oxford: Ashmolean Museum, University of Oxford.

Warren 2016
Warren, Jeremy. 2016. *The Wallace Collection: Catalogue of Italian Sculpture*. 2 vols. London: Holberton.

Washizuka, Tomii, and Friello 1997
Washizuka, Hiromitsu, Reiko Tomii, and Kathleen M. Friello. 1997. *Enlightenment Embodied: The Art of the Japanese Buddhist Sculptor (7th–14th Centuries)*. New York: Japan Society.

Wasserman 1969
Wasserman, Jeanne L. 1969. *Daumier Sculpture: A Critical and Comparative Study*. Greenwich, CT: New York Graphic Society.

Wasserman 1975
Wasserman, Jeanne L., ed. 1975. *Metamorphoses in Nineteenth-Century Sculpture*. Cambridge, MA: Fogg Art Museum and Harvard University Press.

Watson 1786
Watson, R. 1786. *Chemical Essays*. Vol. 4. Cambridge, UK: J. Archdeacon.

Weihrauch 1944
Weihrauch, Hans Robert. 1944. "Bronze, Bronzeguß, Bronzeplastik." In *Reallexikon zur Deutschen Kunstgeschichte*, 2:1182–216. http://www.rdklabor.de/w/?oldid=92409.

Weihrauch 1967
Weihrauch, Hans Robert. 1967. *Europäische Bronzestatuetten*. Braunschweig, Germany: Klinkhardt & Biermann.

Weil 1977
Weil, Phoebe Dent. 1977. "A Review of the History and Practice of Patination." In *Corrosion and Metal Artifacts: A Dialogue between Conservators and Archaeologists and Corrosion Scientists*, edited by B. Floyd Brown, Harry C. Burnett, W. Thomas Chase, Martha Goodway, Jerome Kruger, and Marcel Pourbaix, 77–92. Washington, DC: National Bureau of Standards. https://archive.org/details/corrosionmetalar479brow.

Weil 1996
Weil, Phoebe Dent. 1996. "A Review of the History and Practice of Patination." In *Historical and Philosophical Issues in the Conservation of Cultural Heritage*, edited by Nicholas Stanley-Price, Mansfield Kirby Talley, and Alessandra Melucco Vaccaro, 394–414. Readings in Conservation. Los Angeles: Getty Conservation Institute.

Weisman and Reedy 2002
Weisman, Billie Milam, and Chandra L. Reedy. 2002. "Technical Studies on Renaissance Bronzes." In *Materials Issues in Art and Archaeology VI: Symposium Held November 26–30, 2001, Boston, Massachusetts, USA*, edited by Pamela B. Vandiver, Martha Goodway, and Jennifer L. Mass, 483–95. Materials Research Society Symposium Proceedings 712. Warrendale, PA: Materials Research Society.

Welter 2007
Welter, Jean Marie. 2007. "Negative and Positive Impact of Minor Elements up to the 0/00 Range on Downward Properties." In *First Meeting on Minor Element Contaminants in Copper Metallurgy*, edited by Jaime Alvarez, 95–102. Concepción, Chile: University of Concepción.

Welter 2008
Welter, Jean-Marie. 2008. "Les bronzes français du xvie au xviiie siècle: Est ce du bronze?." In *Bronzes français de la Renaissance au Siècle des lumières*, edited by Geneviève Bresc-Bautier and Guilhem Scherf, 42–45. Paris: Musée du Louvre, Somogy.

Welter 2014
Welter, Jean-Marie. 2014. "Keller and His Alloy: Copper, Some Zinc, and a Little Bit of Tin." In *French Bronze Sculpture: Materials and Techniques 16th–18th Century*, edited by David Bourgarit, Jane Bassett, Francesca G. Bewer, Geneviève Bresc-Bautier, Philippe Malgouyres, and Guilhem Scherf, 95–106. London: Archetype.

Welter 2018
Welter, Jean-Marie. 2018. "Medieval and Renaissance Italian Statuary Copper Alloys." In *Cuivre, Bronzes et Laitons Médiévaux: Histoire, Archéologie et Archéométrie Des Productions En Laiton, Bronze et Autres Alliages à Base de Cuivre Dans l'Europe Médiévale (12e–16e Siècles) = Medieval Copper, Bronze and Brass: History, Archaeology and Archaeometry of the Production of Brass, Bronze and Other Copper Alloy Objects in Medieval Europe (12th–16th Centuries)*, edited by Nicolas Thomas and Pete Dandridge, 271–83. Études et Documents, Archéologie 39. Namur, Belgium: Agence wallonne du Patrimoine.

Welter 2019
Welter, Jean-Marie. 2019. "Contact Tinning: A Millennia-Old Plating Technology." *Archaeometry* 61 (10): 906–920.

Werner 1972
Werner, O. 1972. *Spektralanalytische und metallurgische Untersuchungen anindischen Bronzen*. Leiden, the Netherlands: E. J. Brill.

Wikipedia n.d.
Wikipedia. n.d. "Glossary of Sculpting." https://en.wikipedia.org/wiki/Glossary_of_sculpting.

Wilkinson 2000
Wilkinson, Alan G. 2000. *The Sculptures of Jacques Lipchitz*. London: Thames and Hudson.

Willer 1994
Willer, Frank. 1994. "Fragen zur intentionellen Schwarzpatina an den Mahdiabronzen." In *Das Wrack. Der antike Schiffsfund von Mahdia*, edited by Gisela Hellenkemper Salies with Hans-Hoyer von Prittwitz und Gaffron and Gerhard Bauchhenß, 2:1023–31. Kataloge des Rheinischen Landesmuseums Bonn. Cologne: Rheinland Verlag.

Willer 2016
Willer, Frank. 2016. "Exkurs zur Antiken Überfang- bzw. Angusstechnik." In *Materialien einer Gusswerkstatt Egon Schiele Schön-Verlag von der Qubbet el-Hawa*, edited by Martin Fitzenreiter, Frank Willer, and Johannes Auenmüller, 139–40. Berlin: EB-Verlag.

Willer and Meijers 2014
Willer, Frank, and Ronny Meijers. 2014. "Hightech trifft Antike: Römischen Bronzegießern auf der Spur." In *Gebrochener Glanz: römische Großbronzen am UNESCO-Welterbe Limes*, edited by Stephanie Müller, 166–82. Mainz, Germany: Nünnerich-Asmus.

Willer, Schwab, and Mirschenz 2016a
Willer, Frank, Roland Schwab, and Manuela Mirschenz. 2016a. "Archäometrische Untersuchungen zur Herstellungstechnik." In *Römische Bronzestatuen am Limes*, edited by Frank Willer, Roland Schwab, and Manuela Mirschenz and Gerwulf Schneider, 81–117. Darmstadt, Germany: Philipp Von Zabern.

Willer, Schwab, and Mirschenz 2016b
Willer, Frank, Roland Schwab, and Manuela Mirschenz. 2016b. Römische Bronzestatuen am Limes: Archäometrische Untersuchungen Zur Herstellungstechnik." *Bonner Jahrbücher* 216:57–187. https://journals.ub.uni-heidelberg.de/index.php/bjb/article/view/70701/64059.

Willer, Schwab, and Mirschenz 2017
Willer, Frank, Roland Schwab, and Manuela Mirschenz. 2017. "Ergebnisse der archäometrischen und herstellungstechnischen Forschungen and den 'Limesbronzen.'" In *Römische Grossbronzen am UNESCO-Welterbe Limes*, edited by M. Kemkes, 79–105. Darmstadt, Germany: Konrad. https://landesmuseum-bonn.lvr.de/media/lmb/forschung/bilder_4/restaurierung/Mirschenz_M.Schwab_R._Willer_F._Ergebnisse_der_archaeometrischen_und_herstellungstechnischen_Forschungen_an_den_Limesbronzen.pdf.

Woodward 1997
Woodward Jr., Hiram W. 1997. *The Sacred Sculpture of Thailand: The Alexander B. Griswold Collection*. Baltimore: Walters Art Gallery.

Yamazaki et al. 2014
Yamazaki, E., S. Nakai, Y. Sahoo, T. Yokoyama, H. Mifune, T. Saito, J. Chen, N. Takagi, N. Hokanishi, and A. Yasuda. 2014. "Feasibility Studies of Sn Isotope Composition for Provenancing Ancient Bronzes." *Journal of Archaeological Science* 52 (December): 458–67.

***Ying Han Chi Shu Tz'u Tien* 1978**
Ying Han Chi Shu Tz'u Tien = An English-Chinese Dictionary of Technology. 1978. Pei-ching, China: Kuo fang kung yeh ch'u pan she.

Young 1973
Young, Williams, ed. 1973. *Application of Science in Examination of Works of Art: Proceedings of the Seminar: June 15–19, 1970*. Boston: Museum of Fine Arts.

Young and Pernicka 1999
Young, Suzanne M. M., and Ernst Pernicka, eds. 1999. “Trace Element Fingerprinting of Ancient Copper: A Guide to Technology or Provenance?” In *Metals in Antiquity*, edited by Suzanne M. M. Young, A. Mark Pollard, Paul Budd and Robert A. Ixer, 163–71. BAR International Series 792. Oxford: Archaeopress.

Yue and Yinxu Xiaomintun Archaeological Team 2008
Yue Zhanwei and Yinxu Xiaomintun Archaeological Team. 2008. “The Shang Bronze Foundry-Site at Xiaomintun in Anyang City.” Translated by Mo Runxian. *Chinese Archaeology* 8:16–21. http://www.kaogu.cn/uploads/soft/Chinese Archaeology/8/The Shang Bronze Foundry-Site at Xiaomintun in Anyang City .pdf.

Zawisza et al. 2011
Zawisza, Beata, Katarzyna Pytlakowska, Barbara Feist, Marzena Polowniak, Andrzej Kita, and Rafal Sitko. 2011. “Determination of Rare Earth Elements by Spectroscopic Techniques: A Review.” *Journal of Analytical Atomic Spectrometry* 26 (12): 2373–90.

Zenghelis 1929
Zenghelis, C. 1929. “Contribution à l’étude des bronzes antiques.” *Mouseion: Bulletin de l’Office International des Musées, Institut de Coopération Intellectuelle de la Société des Nations* 3 (September): 113.

Zhang 2010
Zhang Changping et al. | 张昌平. 2010. “二里冈文化至殷墟文化时期青铜器范型技术的发展 = Development of Clay Mold-Making Technology for Bronze Casting from Erligang Culture to Yinxu Culture.” *考古* = *Kaogu* 8:79–86. Republished in English in *Chinese Archaeology* 11 (2011): 148–54.

Zikos 2002
Zikos, Dimitrios. 2002. “Giambologna’s Land, House, and Workshops in Florence.” *Mitteilungen des Kunsthistorischen Institutes in Florenz* 46:357–408. http://www.jstor.org/stable/27654576.

Zikos 2012
Zikos, Dimitrios, ed. 2012. *Marks of Identity: New Perspectives on Sixteenth-Century Italian Sculpture*. Boston: Isabella Stewart Gardner Museum.

Zimmer et al. 2011
Zimmer, Gerhard, Monika Wirth, Rainer Ellerbrok, and Andreas Bührig-Polaczek. 2011. “Stückungstechnik griechischer Bronzestatuen-Kunst und technologischer Fortschritt.” *Giesserei* 98 (6): 264–73.

Zimmerman, Yuhas, and Meyers 1974
Zimmerman, D. W., M. P. Yuhas, and P. Meyers. 1974. “Thermoluminescence Authenticity Measurements on Core Material from the Bronze Horse of the New York Metropolitan Museum of Art.” *Archaeometry* 16 (1): 19–30.

Zwicker 1993
Zwicker, Ulrich. 1993. “Investigation on Antique Techniques for Joining and Repairing Bronzecastings.” In *Antiche officine del bronzo: materiali, strumenti, tecniche: atti del seminario di studi ed esperimenti, Murlo, 26–31 luglio 1991*, edited by Edilberto Formigli, 137–52. Siena, Italy: Nuova Immagine.

About the Contributors

Jane Bassett

Jane Bassett (Head of Decorative Arts and Sculpture Conservation, J. Paul Getty Museum, Los Angeles) completed her graduate studies in conservation at New York State College at Buffalo. A conservator at the Getty Museum since 1991, she has co-curated many exhibitions focused on sculptural processes, including *La Roldana's St. Ginés: The Making of a Polychrome Sculpture*, *Bernini and Marble*, and *Foundry to Finish* on the direct lost-wax casting process. She authored *The Artist Revealed: Adriaen de Vries, Sculptor in Bronze* (2008) and coedited *French Bronze Sculpture: Materials and Techniques 16th–18th Century* (2014).

Francesca G. Bewer

Francesca G. Bewer (Research Curator for Conservation and Technical Studies Programs and Director, Summer Institute for Technical Studies in Art, Harvard Art Museums, Cambridge, Massachusetts) undertook her graduate theses at the Warburg Institute, University of London (MPhil 1986) and the Institute of Archaeology, University College London (PhD 1996), focusing on Renaissance bronze technology. She has published widely on art technology and conservation history. She authored *A Laboratory for Art: Harvard's Fogg Museum and the Emergence of Conservation in America (ca. 1900–1950)* (2010) and coedited *French Bronze Sculpture: Materials and Techniques 16th–18th Century* (2014) and *The Explicit Material: Inquiries on the Intersection of Curatorial and Conservation Cultures* (2019).

David Bourgarit

David Bourgarit (Archaeometallurgist, Centre de Recherche et de Restauration des Musées de France [C2RMF], Paris, and Laboratory TEMPS-CNRS-Nanterre University) has a background in physics, with a PhD on the physical metallurgy of a specific titanium alloy. Since 1996 he has been a researcher at the C2RMF, where he has been investigating metallic artifacts from almost all periods and regions. His primary research interests are in the technological approach to copper metallurgy, with a focus on the provenance of copper and fabrication techniques. He coedited *French Bronze Sculpture: Materials and Techniques 16th–18th Century* (2014).

Arlen Heginbotham

Arlen Heginbotham (Conservator of Decorative Arts and Sculpture, J. Paul Getty Museum, Los Angeles) received his AB in East Asian studies from Stanford University, his MA in art conservation from Buffalo State College, and his PhD in earth sciences from Vrije Universiteit Amsterdam. His research interests include the use of X-ray fluorescence spectroscopy as a tool for studying copper alloy artifacts, microscopic and chemical wood identification, X-radiographic dendrochronology, and the technical study of Asian export lacquer. He coauthored, with Gillian Wilson, the collection catalogue *French Rococo Ébénisterie in the J. Paul Getty Museum* (2021).

Andrew Lacey

Andrew Lacey (Artist, Founder, and Independent Scholar) studied archaeometallurgy at the Institute of

Archaeology, University College London (MSc 1998), focusing on interview techniques for gaining insight into the working practices of London fine art foundries. His work as a contemporary artist is highly experimental; casting all his own work in bronze enables creative interception in all stages of the process, further developing visual and material symbologies that are embodied within the sculpture. His work as an independent scholar is a creative collaboration among conservators, historians, makers, and curators, and explores possibilities of hypothetical facture via processes of experimental reconstruction.

Peta Motture

Peta Motture (formerly Senior Curator, Sculpture, Victoria and Albert Museum, London) joined the V&A in 1985 and retired in 2022. Since 1990 she has specialized primarily in Renaissance sculpture, with a focus on Italy. She has published widely on medieval and later sculpture, notably bronzes, often collaborating with colleagues in the technical study of sculpture. Her books include the *V&A Catalogue of Italian Bells and Mortars* (2001), *Large Bronzes in the Renaissance* (2003), and *The Culture of Bronze: Making and Meaning in Italian Renaissance Sculpture* (2019).

◆ ◆ ◆

Ann Boulton

Ann Boulton (Associate Conservator, Gilcrease Museum, Tulsa, Oklahoma) has been a practicing objects conservator for thirty-five years. She received her MA in art conservation from Buffalo State College in 1985. While objects conservator at the Baltimore Museum of Art, she researched nineteenth- and twentieth-century French art bronze technology in her study of the sculpture of Antoine-Louis Barye and Henri Matisse. In 2009 she shifted her research focus to American bronze sculpture at the Gilcrease and curated the exhibition *Frontier to Foundry* (2014). She contributed the essay "A Tale of Two Foundries" to Sharon Hecker's *Finding Lost Wax: The Disappearance and Recovery of an Ancient Casting Technique and the Experiments of Medardo Rosso* (2021).

Clotilde Boust

Clotilde Boust (Research Engineer, Centre de Recherche et de Restauration des Musées de France [C2RMF], Paris, and PSL University UMR 8247 PCMTH CNRS) has a background in physics and photography with a PhD on optronics. Since 2013 she has been heading the C2RMF imaging group in producing images for art conservation (UV/VIS/IR imaging, X-rays, and surface 3D). She is responsible for online publications on scientific imaging for Cultural Heritage (copa.hypotheses.org), contributes to conferences as color or electronic imaging, and has published in books centered on the Villa of Diomede (2020), Genoese School drawings at the Louvre (2017), and the Bayeux Tapestry (2016).

Manon Castelle

Manon Castelle is a conservation scientist specializing in copper alloys in ancient and historical metallurgy. She received a PhD from the Université de Versailles Saint-Quentin-en-Yvelines in 2016 for her work on French bronze statuary techniques, 1540 to 1660. She then worked on bronze artifacts excavated from the Celtic prince of Lavau tomb (fifth century BCE, France) and, as part of a postdoctoral project (UVSQ), on copper-based seal matrices from the French Archives Nationales and the Musée des beaux-arts de Lyon. In 2018 she obtained the two-year Migelien Gerritzen Fellowship at the Rijksmuseum and continued her research on European bronze objects from the fifteenth to the eighteenth centuries.

Yi Chen

Yi Chen received her PhD in Chinese archaeology from the University of Oxford. She is a former curator of early Chinese collections and now a visiting researcher at the British Museum. In addition, she is an academic advisor of the Dresden Porcelain Project of the Staatliche Kunstsammlungen Dresden and a J. S. Lee Memorial Fellow (2022/2023), hosted by The Art Museum, Chinese University of Hong Kong. Before she joined the British Museum in 2015, she was the Christensen Fellow in Chinese Painting at the Ashmolean Museum of Art and Archaeology in Oxford.

Annabelle Collinet

Annabelle Collinet (Curator, Medieval Iranian World Collections, Musée du Louvre, and Associate Researcher of the CéSor – Centre d'études en sciences sociales du religieux, EHESS/CNRS) completed her PhD at Paris 1 Panthéon-Sorbonne and is a specialist in the arts of fire of the Iranian and Indian worlds (Islamic period). She is the coauthor of *Nishapur Revisited* (2013) and coeditor of *Bodies and Artefacts: Relics and Other Devotional Supports in Shia Societies in the Indic and Iranian Worlds* (2020), and just completed *Précieuses matières. Les arts du metal dans le monde iranien mediéval*, vol. 1, X^{e}–XIIIe siècles (2021).

Pete Dandridge

Pete Dandridge (Conservator Emeritus, Metropolitan Museum of Art, New York) came to the Metropolitan Museum in 1979 from the Cooperstown Graduate Program. He was principal conservator of the ivories, enamels, and metalwork in the Department of Medieval Art and The Cloisters with a focus on the technical history of those materials and the capabilities of the associated artists. He contributed to *Small Wonders: Late-Gothic Boxwood Micro-Carvings from the Low Countries* (2016) and was an author and coeditor for both *Medieval Copper, Bronze and Brass* (2013) and the exhibition *Lions, Dragons, and Other Beasts: Aquamanilia of the Middle Ages, Vessels for Church and Table* (2006); he also co-curated the latter.

Susanne Gänsicke

Susanne Gänsicke (Head of Antiquities Conservation, J. Paul Getty Museum, Los Angeles) holds a degree in archaeological conservation from Römisch Germanisches Zentralmuseum, Mainz (1987). She was objects conservator at the Museum of Fine Arts, Boston, from 1990 to 2016, where she remains a project member of Auloi from Meroë, funded by the European Research Council. In 2016 she received an individual grant for fieldwork in Nepal from the Asian Cultural Council of New York. Her research interests include materials and manufacturing techniques of ancient and historic metalwork, and the intersection of traditional craftsmanship and conservation. She coauthored, with Yvonne Markowitz, *Looking at Jewelry* (2019).

Joachim Kreutner

Joachim Kreutner (Metals Conservator, Bavarian Nationalmuseum, Munich) received his degree in conservation, restoration, and art technology at the Technische Universität, Munich. Since 2016 he has been supervisor to the metals conservation team at the Bavarian Nationalmuseum. His research interests are focused on preventative conservational conditions of museum silver collections and the technique of bronze casting. He is deputy spokesperson for the conservation working group at Deutscher Museumsbund.

Susan La Niece

Susan La Niece (Archaeometallurgist, British Museum Department of Scientific Research, 1977–2017) graduated in archaeology and ancient history, Manchester University, with further studies in archaeological science and metallurgy. She is a fellow of the Society of Antiquaries of London. Her research interest is the history of metal technology and workshop techniques. She has published on metalwork of many periods and cultures, including medieval Europe and the Islamic world, bronze casting, gold, niello, gilding, silvering, and decorative patination. She was coeditor of *Metal Plating and Patination* (1993) and *Metals and Mines* (2007) and the author of *Gold* (2009). She is the editor of the journal *Jewellery Studies*.

Elsa Lambert

Elsa Lambert (Radiologist of Cultural Heritage, Centre de Recherche et de Restauration des Musées de France [C2RMF], Paris) joined the C2RMF in 2001. For eleven years she worked as a scientific photographer for the study of paintings. Her work as a photographer has been published in *Inside Mona Lisa* (2004) and *Grünewald* (2012), among others; the latter received the art book prize in 2013. In 2012 she took over objects radiography and tomography activities of the C2RMF, thus bringing her knowledge to the service of studies, restoration, and expertise of works of art and archaeological artifacts.

Elisabeth Lebon

Elisabeth Lebon holds a doctorate in art history from Paris I Panthéon-Sorbonne. She is an independent researcher in nineteenth- and twentieth-century French sculpture and the author of the catalogues raisonnés of Antoine Pevsner (in collaboration with Pierre Brullé), Charles Despiau, and Jean Joire. She specializes in the history of art foundries and casting processes in France, and is the author of *Dictionnaire des fondeurs de bronze d'art. France 1890–1950* (2003), *Le Fondeur et le sculpteur: Technique du bronze et histoire de l'art* (2012), and *Fonte au sable – fonte à cire perdue: histoire d'une rivalité* (2012). She has been brought to work on a wide range of artists through this specialty.

Stéphanie Leroy

Stéphanie Leroy (Archaeometallurgist, Senior Researcher at CNRS, Laboratoire Archéomatériaux et Prévision de l'Altération-LAPA/ Institut de Recherche sur les

ArchéoMATériaux-IRAMAT) has a PhD in archaeometry, focusing on the physico-chemical characteristics of iron for the provenance investigation of ancient ferrous artifacts. As a researcher at LAPA since 2013, she develops transdisciplinary methods, such as on the origin identification and dating of iron to generate knowledge on the manufacturing processes, and on production and distribution networks of ancient ferrous metals. Since 2014, she has expanded her regional and disciplinary range with comparative work in Southeast Asia.

Linda Ying-Chun Lin

Linda Ying-Chun Lin (Objects Conservator, Fine Arts Museum of San Francisco) received her MA in conservation of cultural heritage materials from UCLA / Getty Interdepartmental Program in 2010. She was formerly the conservator for arts of Asia at the Newark Museum, New Jersey. She has translated articles in the areas of archaeology, conservation, and technical research published in both Chinese and English journals. Her most recent translation projects include ancient Chinese bronze-casting technology and Chinese lacquer, in collaboration with the Smithsonian Institution's National Museum of Asian Art.

Jeffrey Maish

Jeffrey Maish (formerly Conservator of Antiquities, J. Paul Getty Museum, Los Angeles) received his MA in conservation from Buffalo State College in 1987, joined the Getty Museum in 1988, and retired from there in 2021. He has published a range of technical studies on ancient bronzes from the Getty collections and on loan. He was a resident of the American Academy Rome in 2012, focusing on bronze technology and radiography, and continues his research on ancient bronze production and the interpretation of casting features. Additional research interests include the techniques of Attic pottery and relief line production.

Mathilde Mechling

Mathilde Mechling received her PhD in languages, civilizations, and Asian societies in 2020 from University Sorbonne Nouvelle and Leiden University. Her thesis focused on Hindu and Buddhist bronze statuary from the Indonesian archipelago, developing an interdisciplinary methodology combining stylistic and iconographic analysis, archaeometallurgy, archaeology, and religious aspects. She published "The Indonesian Bronze-Casting Tradition: Technical Investigations on Thirty-Nine Indonesian Bronze Statues (7th–11th c.) from the Musée national des arts asiatiques – Guimet, Paris" (2018) with David Bourgarit (C2RMF), Brice Vincent (EFEO), and Pierre Baptiste (Musée Guimet). She is adapting her PhD thesis into a book publication.

Benoît Mille

Benoît Mille (Archaeometallurgist, Centre de Recherche et de Restauration des Musées de France [C2RMF], Paris, and Laboratory TEMPS-CNRS-Nanterre University) received his PhD in 2017 on the earliest history of lost-wax casting from the Universities of Nanterre and Fribourg. Since 1993 he has been a researcher at the C2RMF, mostly investigating the beginnings of metallurgy (south of France, Pakistan, Chile) and the evolution of large bronze casting techniques (Egypt, South Arabia, Greece, Rome). He coedited *"Bronzes grecs et romains, recherches récentes" – Hommage à Claude Rolley* (2012), *Bronzes grecs et romains: études récentes sur la statuaire antique* (2017), *Launac et le launacien* (2017), and *Nouveaux regards sur le trésor des bronzes de Bavay* (2019).

Lorenzo Morigi

Lorenzo Morigi (Independent Restorer) joined the company owned by his father in 1990, specializing in the restoration of outdoor bronze sculptures and metal objects. He studied chemistry for conservators at ICCROM in Rome. He has been responsible for planning, direct restoration, and technological study of many Italian bronze sculptures from ancient Rome to the modern era, and specializes in the conservation of contemporary sculptures, also with thorough experience in consolidation and restoration of waterlogged wood. He has reproduced medieval bronze doors and sculptures in art foundries. He collaborates with art handling companies in conceiving, designing, and building special art crating with reverse engineering. He works with 3D scanners and digital fabrication.

Ruven Pillay

Ruven Pillay (Research Scientist, Centre de Recherche et de Restauration des Musées de France [C2RMF], Paris) holds an MPhys in physics from the University of Manchester, an MSc in computer science from the University of Edinburgh, and a PhD in hyperspectral imaging from NTNU, Norway. His research interests include the application of advanced imaging, data processing and visualization, and other techniques to the study of art. In addition to his work at the C2RMF he has more than twenty-five years of experience working in major art galleries, and has also worked at the National Gallery in London, the National Museum in Stockholm, and as an invited scholar at the J. Paul Getty Museum in Los Angeles.

Chandra L. Reedy

Chandra L. Reedy (Professor, University of Delaware) received a PhD in archaeology from UCLA in 1986, specializing in archaeological science and material culture of Asia. She has special interest in the copper alloy casting technologies of northwest India and of Tibetan regions, and published *Himalayan Bronzes: Technology, Style, and Choices* in 1997. She has also participated in projects studying bronzes of Sri Lanka, Southeast Asia, and Renaissance Europe. She has done extensive research on thin-section petrography of clay core materials from bronzes, incorporating image analysis methods, and published *Thin-Section Petrography of Stone and Ceramic Cultural Materials* in 2008.

David J. Reid

David J. Reid (Founder and Enquirer, Semi-Retired) studied chemistry, physics, mathematics, and astronomy at the Universities of Auckland and Canterbury, and has owned and operated a number of bronze foundries around the world. He received the Queen Elizabeth II Research Grant (1979) to travel the world to study art casting techniques, and he holds a UK patent (1990) for an innovative method of metal casting. He was a research fellow at Central Saint Martin's College of Art (1992–96) and has developed a method for melting and casting silver or bronze using a domestic microwave oven. He continues to teach his accessible foundry techniques in many countries.

Dominique Robcis

Dominique Robcis (Head Conservator, Metal Team of the Research Department of the Centre de Recherche et de Restauration des Musées de France [C2RMF], Paris) received his MA in medieval history and metal conservation before joining the C2RMF. He specializes in metal techniques and conservation, and his technological studies focus on jewelry and surface treatments such as gilding, patina, and tool marks, mainly in European antiquity but also for Japanese productions. He has worked on collection objects from a number of French museums, including the Louvre and the Musée des Arts Décoratifs, Paris. He is in charge of digital microscopy equipment and a p-XRF.

Donna Strahan

Donna Strahan (Head of the Department of Conservation and Scientific Research, Freer and Sackler Galleries of Art, Smithsonian Institution, Washington, DC) holds a BA in Chinese language and an MA in conservation of ethnographic and archaeological objects from George Washington University. From 2006 to 2014 she was the conservator of Asian objects at the Metropolitan Museum of Art, New York. Her major interests are the technology and preservation of ancient metals and Asian lacquer. She has lectured and published widely, including *Wisdom Embodied: Chinese Buddhist and Daoist Sculpture in the Metropolitan Museum of Art*, coauthored with Denise Patry Leidy (2010).

Shelley Sturman

Shelley Sturman (formerly Head of Object Conservation, National Gallery of Art, Washington, DC) received her BA and MA from Brandeis University and her MS in conservation from the University of Delaware. She has a special interest is interdisciplinary investigation of Renaissance bronze casting with curatorial colleagues. She is honored to have performed technical examinations and published on Lorenzo Ghiberti's Gates of Paradise, Andrea Riccio's Paschal Candlestick, Jacopo Sansovino's Sacristy Door, Giambologna's female nudes and Mercury, bronzes by Antico and Niccolò Roccatagliata, the Budapest Horse, and masterpieces in the Claudia Quentin and Robert H. Smith collections. Other research and publication interests include original Edgar Degas waxes and contemporary sculpture.

Jeremy Warren

Jeremy Warren (Honorary Curator of Sculpture, Ashmolean Museum, Oxford, and Sculpture Research Curator, the National Trust) is a specialist in Renaissance and later European sculpture. His numerous publications include the catalogues *Medieval and Renaissance Sculpture in the Ashmolean Museum* (2014) and *Italian Sculpture in the Wallace Collection* (2016), as well as articles on the sculptors Antico, Giovanni Bandini, Giambologna, Vincenzo and Gian Gerolamo Grandi, Leone Leoni, and Severo da Ravenna. Exhibitions include *Beauty and Power: Renaissance and Baroque Bronzes from the Peter Marino Collection* (2010). He has also written extensively on the history of collecting.

Jean-Marie Welter

Jean-Marie Welter (Independent Scholar, Luxembourg) held an engineering diploma from École Polytechnique, Paris (1966), and graduated as a Dr. rer. nat. from the Technische Hochschule, Munich (1969). He started his professional career in academia, producing and studying metallic materials. In 1985 he joined the copper industry and became corporate director of research and development of a major European group (KME) fabricating semi-products of copper and copper alloys. At that time, he started investigating historical copper artifacts, which he continued to do within the context of the history of the copper industry, its products, and their applications including exploring its historic use in sculpture until his death in 2022.

Antoine Zink

Antoine Zink (Archaeometer, Centre de Recherche et de Restauration des Musées de France [C2RMF], Paris) specializes in luminescence dating. After a PhD in physics applied to archaeology with Professor R. Visocekas and a two-year AvH fellowship at the TU Bergakademie Freiberg, he founded the first luminescence dating laboratory in the Iberian Peninsula, Instituto tecnologico e nuclear in Sacavem, Portugal. Since 2002 he has been in charge of luminescence dating at the C2RMF, and his research centers on the dating of collection objects and Bayesian analysis.

Illustration Credits

Figs. 1, 48, 51–52, 56, 69, 99: F. G. Bewer

Figs. 2, 4–5, 105, 154, 237, 270, 560, 562: A. Lacey and S. Lewis

Figs. 3, 24, 36, 60, 233–34, 240, 262–63, 266, 411, 519: © C2RMF / David Bourgarit

Figs. 7, 21, 27, 33, 77, 85, 88–89, 98, 118, 161–62, 165, 251, 544–48: © Andrew Lacey. Photo: A. Lacey and S. Lewis

Figs. 8, 34, 38, 97, 425–32, 524–25: © C2RMF / Manon Castelle

Figs. 9, 217–18, 220, 332, 531–34, 536–38: Gilcrease Museum, Tulsa, Oklahoma

Figs. 10, 271, 374–75: Lorenzo Morigi

Figs. 12, 307: © C2RMF / Benoît Mille and Thierry Borel

Figs. 13–14: KHM-Museumsverband, Weltmuseum Vienna

Fig. 15: © C2RMF / D. Bourgait and V. Fournier

Figs. 17, 19, 22, 74, 549, 553: © C2RMF / Vanessa Fournier

Figs. 20, 81–82, 257, 561: © C2RMF / Anne Chauvet

Fig. 29: Photo: Paul Craddock

Fig. 30: © President and Fellows of Harvard College / F. G. Bewer

Figs. 31–32, 37, 63–65, 80, 91, 120–21, 126, 190, 199, 382–84, 388, 390–94, 396, 398, 400, 510, 513, 515: © C2RMF / Elsa Lambert

Fig. 39: Palmer, 1929

Fig. 43: Courtesy of the Rijksmuseum

Figs. 44, 104, 180, 247, 250, 252–53, 256, 277, 472–74, 476–78, 480, 482–83, 485, 487–91, 493–94, 499, 550: © The Metropolitan Museum of Art / Photo: Pete Dandridge, The Sherman Fairchild Center for Objects Conservation

Fig. 45: J. Paul Getty Trust / A.-L. Desmas

Fig. 46: Statens Museum for Kunst, Copenhagen

Figs. 47, 223, 308, 310, 351–52, 359, 362, 469–71, 479, 481, 484, 486, 492, 497, 555; Videos 4–5, 8: © The Metropolitan Museum of Art

Figs. 49, 68, 83, 172, 175, 191, 254, 265, 274–75, 297, 309, 344–45, 347, 350, 526: © C2RMF / Dominique Bagault

Figs. 50, 294: Image © The Metropolitan Museum of Art / photo: Marco Leona / Department of Scientific Research

Fig. 53: Harvard Art Museums / Fogg Museum, Bequest of Grenville L. Winthrop, Photo © President and Fellows of Harvard College, 1943.1154 / F. G. Bewer

Figs. 54, 62, 67, 72–73, 87, 95, 129, 135, 156, 168, 235–36, 238, 241, 260, 276, 287–88, 339, 395, 405–6, 504–6, 508–9, 511, 514, 523, 527: © C2RMF / Anne Maigret

Fig. 55: © Andrew Lacey

Fig. 61: Photo: Andrew Silla

Fig. 66: The Metropolitan Museum of Art, Gift of Irwin Untermyer, 1964 (64.101.1492) / Image © The Metropolitan Museum of Art

Fig. 71: Harvard Art Museums / Arthur M. Sackler Museum, Gift of Benjamin and Lilian

Hertzberg, Photo © President and Fellows of Harvard College, 2004.205

Fig. 76: F. G. Bewer, with permission from the Collection of Mr. and Mrs. J. Tomilson Hill

Fig. 78: Royal Collection Trust / © Her Majesty Queen Elizabeth II 2021. Photo: F. G. Brewer

Figs. 84, 86, 440–46, 449–55, 459, 461, 465–66: © C2RMF / Daniel Vigears

Figs. 90, 92, 127, 194, 211, 215, 387, 462, 464; Video 15: © C2RMF / Thierry Borel

Figs. 94, 140, 312, 314: Laboratoire de Recherche des Monuments Historiques (LRMH) / A. Azéma

Fig. 96: Su concessione del Ministero della Cultura – Museo Nazionale del Bargello. Photo: F. G. Bewer

Fig. 100: Giovanni and Lorenzo Morigi

Figs. 101, 106, 554: Collection of Historical Scientific Instruments, Harvard University, inv. no. 0005; Photo: Tony Sigel

Figs. 102–3, 232, 326–27: The Walters Art Museum, Baltimore

Figs. 107–8: Harvard Art Museums / Fogg Museum, The Henry Dexter Sharpe Collection, Photo © President and Fellows of Harvard College, 1956.150

Fig. 109: © Comité Rodin

Figs. 110–17, 500–503: Freer Gallery of Art, Smithsonian Institution, Washington, DC: Purchase — Charles Lang Freer Endowment, F1936.6a–b / Department of Conservation and Scientific Research, Photo: Donna Strahan

Figs. 119, 184: © C2RMF / Manon Castelle and Anne Maigret

Figs. 123, 320, 323, 325, 336: Courtesy National Gallery of Art, Washington, DC

Fig. 124: Photo: Frits Scholten

Figs. 125, 163, 173: Photo: J. Paul Getty Trust / Jane Bassett

Figs. 130, 258: Harvard Art Museums / Fogg Museum, Partial gift of Max Falk and partial purchase through the Director's Acquisition Fund, Photo © President and Fellows of Harvard College, 1998.1

Fig. 133: Samples courtesy of Jean-Marie Welter

Fig. 134: Jürgen Vogel / LandesMuseum Bonn

Fig. 136: Photo: J. Paul Getty Trust / Kenneth Lapatin

Figs. 137, 145, 148, 205, 227–28, 298, 341, 456–58: © C2RMF / Benoît Mille

Fig. 141: © Paul and Didier Landowski / ARS

Figs. 142–43, 146–47, 364, 419–23: © C2RMF / Aurélia Azéma

Figs. 149, 259: Antoine Amarger

Fig. 150: J. M. Welter and C2RMF / David Bourgarit

Figs. 151, 373, 520–21: Courtesy Musée National des Arts Asiatiques - Guimet

Figs. 152–53, 171, 178, 188: The Huntington Art Museum. The Arabella D. Huntington Memorial Art Collection. Image: Courtesy of the Huntington Art Museum, San Marino, California

Figs. 155, 159: Photo: © Nationalmuseum, Stockholm

Fig. 157: Harvard Art Museums / Arthur M. Sackler Museum, Gift of Mrs. Edward Jackson Holmes, Photo © President and Fellows of Harvard College, 1952.20

Fig. 158: © C2RMF / Alexis Komenda

Fig. 164: Harvard Art Museums / Fogg Museum, Bequest from the Collection of Maurice Wertheim, Class of 1906. Photo © President and Fellows of Harvard College, 1951.79

Fig. 166: © C2RMF / B. Hirard

Fig. 167: CAST:ING, 2018, Jane Bassett.

Fig. 169: CReA-Patrimoine, Université libre de Bruxelles. Collection comte Simon du Chastel, donated by Nathalie du Chastel / Photo: L.Garenne-Marot

Figs. 170, 342, 356: Musée du Louvre / C2RMF / Islametal / Annabelle Collinet

Fig. 177: © C2RMF / David Bourgarit and Francesca G. Bewer

Fig. 179: Su concessione del Museo Archeologico Nazionale di Firenze (Direzione regionale Musei della Toscana)

Fig. 181: Illustration based on concept sketch by Jane Bassett; final illustration by Gary Hespenheide

Fig. 185: Toledo Museum of Art (Toledo, Ohio), Purchased with funds from the Libbey Endowment, Gift of Edward Drummond Libbey, 1966.126. Image Credit: Chris Ridgway

Fig. 187: © Grünes Gewölbe, Staatliche Kunstsammlungen Dresden

Fig. 189: © Successió Miró / Artists Rights Society (ARS), New York / ADAGP, Paris. The J. Paul Getty Museum, Los Angeles, Gift of Fran and Ray Stark

Fig. 200: The Huntington Art Museum. Adele S. Browning Memorial Collection, gift of Mildred Browning Green and Honorable Lucius Peyton Green

Fig. 201: D. Strahan and C2RMF / Dominique Robcis and David Bourgarit

Fig. 202: © C2RMF / Benoît Mille and Aurélia Azéma

Fig. 203: © 2012 The Henry Moore Foundation. All Rights Reserved. / ARS, New York / DACS, London. The J. Paul Getty Museum, Los Angeles, Gift of Fran and Ray Stark

Fig. 204: © C2RMF / D. Kolin, G. Blaise, Dominique Bagault

Fig. 206: © The Metropolitan Museum of Art / Photo: Donna Strahan, The Sherman Fairchild Center for Objects Conservation

Fig. 207: Photo and photo annotations (left): F. G. Bewer. Diagram (right): Lorenzo Morigi

Figs. 208–9: Freer Gallery of Art Study Collection, Smithsonian Institution, Washington, DC: Purchase — Charles Lang Freer Endowment, FSC-B-56

Fig. 210: Image courtesy Clark Art Institute. clarkart.edu

Fig. 214: Freer Gallery of Art, Smithsonian Institution, Washington, D.C.: Gift of Charles Lang Freer, F1914.21a–h

Fig. 222: Stefanie Rocknak. Photo: Marjee Levine

Figs. 225–26: Arthur M. Sackler Gallery, Smithsonian Institution, Washington, DC, The Alice S. Kandell Collection, S2018.32

Fig. 230: © C2RMF et Institut de Soudure

Fig. 231: © C2RMF

Fig. 239: Jean Dubos

Fig. 242: Isabella Stewart Gardner Museum, Boston (S26e21)

Figs. 246, 293: © Victoria and Albert Museum, London

Figs. 248, 255: © Barbara Plankensteiner, 2006

Fig. 249: Collection of Historical Scientific Instruments, Harvard University. Photo: Tony Sigel

Figs. 261, 559: Harvard Art Museums / Fogg Museum, Bequest from the Collection of Maurice Wertheim, Class of 1906, Photo © President and Fellows of Harvard College, 1951.78

Figs. 264, 268–69, 416: Photo: Brice Vincent

Fig. 272: Experimental work at AMTeC; micro-photo by Dana Goodburn-Brown

Figs. 273, 279, 285–86, 512: © C2RMF / Mathilde Mechling and David Bourgarit

Fig. 278: The Metropolitan Museum of Art, Gift of Irwin Untermyer, 1964 (64.101.1492) / Image © The Metropolitan Museum of Art / Photo: Pete Dandridge, The Sherman Fairchild Center for Objects Conservation

Fig. 280: The Metropolitan Museum of Art, Robert Lehman Collection, 1975 (1975.1.1413) / Image © The Metropolitan Museum of Art / Photo: Pete Dandridge, The Sherman Fairchild Center for Objects Conservation

Fig. 281–83, 290, 301, 417–18, 468: © C2RMF / Dominique Robcis

Fig. 289: Joachim Kreutner, BNM Munich

Fig. 291: Arthur M. Sackler Gallery, Smithsonian Institution, Washington, DC: Purchase - Friends of the Freer and Sackler Galleries, S1997.28

Figs. 292, 296, 299, 343, 349, 360: © The Trustees of the British Museum. All rights reserved.

Fig. 295: Freer Gallery of Art, Smithsonian Institution, Washington, DC / Department of Conservation and Scientific Research, metallographic reference collection ME90069D

Fig. 300: Copyright © Hellenic Ministry of Culture and Sports – Archaeological Receipts Fund) (N. 3028/2002). Ephorate of Antiquities of Athens City, Ancient Agora Museum.

Fig. 302: © C2RMF / Nathalie Pingaud

Fig. 305: CAST:ING, 2018, Jane Bassett

Fig. 311: J-G. Aubert - Laboratoire Arc'Antique - Grand Patrimoine de Loire-Atlantique

Fig. 316: © 2022 Artist Rights Society (ARS), New York

Fig. 318: Photo: Richard Valencia

Figs. 319, 322: Photograph © 2024 Museum of Fine Arts, Boston.

Fig. 321: © The Field Museum, Cat. No. 337093, Photographer Ann Boulton

Fig. 324: Harvard Art Museums / Fogg Museum, The Henry Dexter Sharpe Collection, Photo

© President and Fellows of Harvard College, 1956.169

Fig. 328: Photo: © Alexis Daflos / Nationalmuseum

Fig. 329: Photo: Ann Boulton

Fig. 346: Su concessione del Ministero della Cultura – Museo Nazionale Romano

Figs. 348, 522: Courtesy Sébastien Clouet

Fig. 353: Patrimonio Nacional, Real Monasterio de San Lorenzo de El Escorial, 10034776

Figs. 354, 361: © C2RMF / Thomas Calligaro

Fig. 355: © Skulpturensammlung, Staatliche Kunstsammlungen Dresden / Photo by Elke Estel / Hans-Peter Klut

Fig. 357: Los Angeles County Museum of Art, Gift of Mr. and Mrs. Eric Lidow (M.76.109a–b), Photo © Museum Associates / LACMA

Fig. 358: © C2RMF / Dominique Bagault and Benoît Mille, courtesy of CEA-LETI / Vincent Moulin

Fig. 365: photo by Nicola Dell'Aquila and Federico Taverni / Museo Egizio

Fig. 368: Sophie Descamps and Benoît Mille

Figs. 369, 402, 413, 424, 498: Photo: J. Paul Getty Museum / Arlen Heginbotham

Fig. 371: CC-BY license

Fig. 372: Juan Enrique Agudo, Pedro J. Pardo, Héctor Sánchez, Ángel Luis Pérez and María Isabel Suero via CC BY 3.0 license

Fig. 376: Image courtesy Falmouth University

Fig. 377: © C2RMF (image by Creaform)

Figs. 380–81: Archeovision / P. Mora and L. Espinasse

Figs. 385–86: © C2RMF (images by CEA)

Fig. 389: © C2RMF / Anne Chauvet and Elsa Lambert

Fig. 397: Drawing by Gary Hespenheide

Figs. 401, 403–4, 516: images by CEA / François Ott, processing by © C2RMF / Elsa Lambert

Fig. 412: Image by Thomas Birch, found originally in his PhD thesis (2014): "The provenance and technology of Iron Age war booty from southern Scandinavia," University of Aberdeen

Fig. 414: © C2RMF / Eric Laval and Benoît Mille

Fig. 433: © C2RMF / Pascale Richardin

Fig. 434: Centre de Datation par le radiocarbone / C. Oberlin

Fig. 435: LSCE-LMC14 / E. Delqué-Količ / Graphs generated using OxCal4.3; https://c14.arch.ox.ac.uk/oxcal/OxCal.html

Fig. 436: IRAMAT-LAPA / S. Leroy

Figs. 437–38: Benoît and Florence Mille

Fig. 439: C2RMF / David Bourgarit (left image) and Expérimentation archéologique Nicolas Thomas Barsy 2012, photo Nicolas Méreau @ Inrap/ Awap (right image)

Figs. 447–48: Collection Musées de Vienne. © DR

Figs. 460, 463: © C2RMF / David Kolin, Guillaume Blaise

Fig. 475: © The Metropolitan Museum of Art / Photo: Richard Newman, Scientific Research Lab, Museum of Fine Arts, Boston

Figs. 498, 563, 564: J. Paul Getty Trust / Arlen Heginbotham

Figs. 517–18: David Bourgarit and Mathilde Mechling

Fig. 528: © C2RMF / Manon Castelle and David Bourgarit

Fig. 542: from the collection of Gilcrease Museum, Tulsa, Oklahoma, Replicated by Permission of American Machinist and Endeavor Business Media

Fig. 543: Amon Carter Museum of American Art, Fort Worth, inv. 1961.96

Videos 1, 7, 10, 13: © 2020 Andrew Lacey and Siân Lewis

Videos 2–3, 9, 11, 16,18: © 2021 Andrew Lacey and Siân Lewis

Video 6: © J. Paul Getty Trust and Sandpail Productions

Video 12: © Centre de Recherche et de Restauration des Musées de France (C2RMF) and Musées Departemental Arles Antique

Video 14: © 2018 D. Bouchardon / Laboratoire de Recherche des Monuments Historiques (LRMH)

Video 17: © 1992 Laurène L'Allinec. Permission courtesy of Laurène L'Allinec